SEMIARID LANDS AND DESERTS

BOOKS IN SOILS, PLANTS, AND THE ENVIRONMENT

Series Editor
G. STOTZKY
Department of Biology
New York University
New York, New York

Soil Biochemistry, Volume 1, edited by A. D. McLaren and G. H. Peterson

Soil Biochemistry, Volume 2, edited by A. D. McLaren and J. Skujiņš

Soil Biochemistry, Volume 3, edited by E. A. Paul and A. D. McLaren

Soil Biochemistry, Volume 4, edited by E. A. Paul and A. D. McLaren

Soil Biochemistry, Volume 5, edited by E. A. Paul and J. N. Ladd

Soil Biochemistry, Volume 6, edited by Jean-Marc Bollag and G. Stotzky

Organic Chemicals in the Soil Environment, Volume 1, edited by C. A. I. Goring and J. W. Hamaker

Organic Chemicals in the Soil Environment, Volume 2, edited by C. A. I. Goring and J. W. Hamaker

Humic Substances in the Environment, by M. Schnitzer and S. U. Khan

Microbial Life in the Soil: An Introduction, by T. Hattori

Principles of Soil Chemistry, by Kim H. Tan

Soil Analysis: Instrumental Techniques and Related Procedures, edited by Keith A. Smith

Soil Reclamation Processes: Microbiological Analyses and Applications, edited by Robert L. Tate III and Donald A. Klein

Symbiotic Nitrogen Fixation Technology, edited by Gerald H. Elkan

Soil-Water Interactions: Mechanisms and Applications, edited by Shingo Iwata, Toshio Tabuchi, and Benno P. Warkentin

Soil Analysis: Modern Instrumental Techniques, Second Edition, edited by Keith A. Smith

Soil Analysis: Physical Methods, edited by Keith A. Smith and Chris E. Mullins

Growth and Mineral Nutrition of Field Crops, edited by N. K. Fageria, V. C. Baligar, and Charles Allan Jones

Semiarid Lands and Deserts: Soil Resource and Reclamation, edited by J. Skujiņš

Plant Roots: The Hidden Half, edited by Yaov Waisel, Amram Eshel, and Uzi Kafkafi

Additional Volumes in Preparation

SEMIARID LANDS AND DESERTS

SOIL RESOURCE AND RECLAMATION

edited by

J. Skujiņš

Department of Biology
Utah State University
Logan, Utah

Marcel Dekker, Inc. New York • Basel • Hong Kong

Library of Congress Cataloging-in-Publication Data

Semiarid lands and deserts: soil resource and reclamation / edited by J. Skujiņš.
p. cm. -- (Books in soils, plants and the environment)
Includes bibliographical references and index.
ISBN 0-8247-8388-3
1. Arid soils. 2. Arid regions agriculture. 3. Desert reclamation. I. Skujiņš. J. II. Series
S592.17.A73S46 1991
631.4'715'4--dc20 90-28512
CIP

This book is printed on acid-free paper.

MARCEL DEKKER, INC.
270 Madison Avenue, New York, New York 10016

Current printing (last digit):
10 9 8 7 6 5 4 3 2 1

PRINTED IN THE UNITED STATES OF AMERICA

Foreword

Arid lands have been the site of notable advances in the social and economic development of the human race despite the rigor of the climate. Following the invention of agriculture somewhere in the Middle East, three great hydraulic civilizations arose in the arid and hyperarid Indus Valley, Mesopotamia, and Egypt and brought advancements in the arts and humanities as well as in mathematics and astronomy. From Xian in the semiarid loessial plateau region, the Han people developed a prosperous agriculture and began the unification of China. The Aztecs and Incas left an indelible imprint on history as they established their far-flung empires in the drylands of the Americas. Romans learned how to manage runoff waters in semiarid and arid North Africa and made the area into the granary of Rome. Modern-day visitors to Sbeitla in Tunisia and the ancient forts on the edge of the Sahara marvel at the ability of the Romans to utilize scarce water resources in what is today a barren and hostile environment. By comparison, the rise of advanced civilizations in the humid regions came a millennium or more after it had occurred in the arid regions.

The mystique associated with arid regions led UNESCO (United Nations Educational, Scientific, and Cultural Organization) in the 1950s to create the Major Project on Scientific Research on Arid Lands. The project had for its dominant theme the conduct of research that would allow the wise utilization of arid lands—one-third of the Earth's surface, regions that are vast and largely unoccupied, where extensive mineral deposits, picturesque landscapes, unpolluted air, a unique plant and animal life, abundant solar energy, and endless open spaces still exist.

Research in the drylands had been carried out long before UNESCO organized its global project. What UNESCO did was to put a global context on the research, publicize the gaps in information, and expand research funding. The United Nations Conference on Desertification in Nairobi, Kenya, in 1977 focused further attention on many of the serious problems that dwellers in the arid regions faced and the need for research to assist policymakers in making development decisions.

This book brings together in one volume a wealth of information. Some of it had been published previously but was widely scattered in journals, proceedings volumes, and obscure reports; some of it is new; all of it deals with arid region soils.

Soils are the outermost layer of the Earth's exposed crusts. They are the biochemically altered product of natural processes. Soils are alive in that they are host to an almost infinite variety of microorganisms, macroorganisms, and plant roots. Their chemical, physical, and biological properties are complex in their interactions. Understanding those interactions provides a fascinating challenge to researchers seeking insights to a world in which the principal reactions are carried out at the microscopic and submicroscopic scale.

Some characteristics are common to arid region soils: most have small amounts of organic matter (humus); surface soils range in acidity from slightly acid to alkaline; accumulations of calcium carbonate (lime) can be found in the upper two meters in temperature zone soils; biological activity between rains is low; and gravelly soils have a gravel surface called desert pavement or desert armor. Nutrient contents are low, but the soils can be highly productive when irrigated and fertilized. Salinity is generally low in all except gypsiferous and playa (closed depression) soils.

The principal exception to the typical condition is the strong acidity of coarse-textured soils in areas, such as the semiarid regions south of the Sahara in Africa, where rainfall is concentrated in a few months during the hot season. These soils usually suffer pronounced nutrient deficiencies.

Human-induced soil degradation has been a problem for at least two thousand years. Agricultural activities, mining, industrial operations, municipal waste disposal, and even recreational pursuits have damaged soils, sometimes reversibly, sometimes irreversibly. With the increase in public awareness of environmental pollution and its health and economic consequences, concern is growing over the capacity of soils to absorb and immobilize contaminants and still remain productive. There are no more quick and easy answers in the face of a rapidly growing global population.

Field studies on arid region soils tend to take longer than similar studies conducted in the humid regions. The primary reason is the wide annual variation in weather conditions in the drylands. It commonly requires several years to determine a statistically significant effect of an imposed treatment under nonirrigated conditions because of that climatic variability. In addition, annual and seasonal fluctuations in rainfall, especially, are such that vegetative rehabilitation of dryland soils seldom is simple, even if a good technique is available.

We believe the information in this book will lead to a better understanding of the properties of arid region soils, the threats to their environmental stability, and the means to reclaim those that have been degraded. These are the only soils we will have in our lifetimes. We must use them judiciously and sustainably.

H. E. Dregne
Horn Professor
ICASALS, Texas Tech University
Lubbock, Texas

Preface

For a long time arid lands have been considered agriculturally and economically unimportant. A large segment of the human population, nevertheless, depends on the meager resources arid lands provide. Exploitation of these lands by the ever-increasing population continuously degrades the soil, resulting in markedly decreased productivity. The degradation is enhanced by natural disasters and, as a result, large areas are becoming desertified, thus further decreasing the overall available resources for sustenance and economic development.

The gravity of desertification became evident during the tragic Sahelian draught from 1969 to 1974. Since then considerable advances have been made through basic and applied sciences, to help us understand the processes involved and find applicable methodologies, not only to reverse arid land degradation but also to improve agricultural productivity of the existing primary resource—the soil. These efforts have resulted in vastly expanded knowledge and information about these problems.

This volume is a compendium of current information by recognized experts on arid soils and has a two-pronged purpose. First, it provides a comprehensive background of the various soils and biota of arid regions; second, it provides a detailed account of the current understanding of degradation processes and includes methodologies for arid land maintenance and rehabilitation and for increasing its agricultural value.

The chapters in Part I of this volume provide descriptions of arid soil distribution, soil genesis and classification, and of its chemical and physical properties. Also in Part I are chapters on arid soil flora, fauna, and microbiology and the effects that these components have on nutrient cycling. Emphasis is placed on their relationships with climatic conditions.

Part II begins with a treatise on the causes of desertification, followed by discussions of current aspects of arid soil management: the use of controlled grazing and reduced tillage, the application of succession theory in revegetation, the utilization of nitrogen-fixing plant associations (rhizobial and actinorhizal plants), and the importance of mycorrhizae and cyanobacteria and of algal cultivation for food and fodder. Also included are chapters on the importance of desert crust maintenance and development, application of organic matter, and inorganic fertilization. Lastly, a discussion on water management is included.

Because both the successful rehabilitation and the efficient use of arid lands are of worldwide importance, the authors describe their subject matter as it pertains to the global arid environment, rather than limit themselves to sectional situations.

During the last decade a number of detailed and encyclopedic volumes have appeared about arid lands and soils. Until now, however, no publication has described the arid soils, the processes involved, and the methodologies for their utilization and rehabilitation in an easily manageable concise manner. This volume now fills that gap by providing the most recent information on arid soils and more than 2000 useful references.

With this in mind, this volume is directed to all those who are actively involved in working with arid lands, their preservation, improvement, and agricultural utilization. It will be of decided value to soil scientists, ecologists and environmental scientists, agronomists, planners and decision makers, and field workers. The editor and contributors to this volume trust that the information presented herein will be used by all those concerned with improving the arid land productivity, thus providing improved food production for the increasing population that depends on these meager soil resources.

J. Skujiņš

Contents

PART II. RECLAMATION AND IMPROVEMENT OF FERTILITY

Contributors

Edith B. Allen Department of Biology and Systems Ecology Research Group, San Diego State University, San Diego, California

William A. Dick-Peddie Department of Biology, New Mexico State University, Las Cruces, New Mexico

H. G. Diem Laboratoire de Biotechnologie des Systèmes Symbiotiques, Forestiers Tropicaux (ORSTOM/CTFT), Nogent-sur-Marne, France

Y. R. Dommergues Laboratoire de Biotechnologie des Systèmes Symbiotiques, Forestiers Tropicaux (ORSTOM/CTFT), Nogent-sur-Marne, France

Harold E. Dregne Texas Tech University, Lubbock, Texas

Wallace H. Fuller Department of Soil and Water Science, University of Arizona, Tucson, Arizona

George H. Hargreaves Department of Agricultural and Irrigation Engineering, Utah State University, Logan, Utah

David M. Hendricks Department of Soil and Water Science, University of Arizona, Tucson, Arizona

Carlton H. Herbel U.S. Department of Agriculture and Department of Animal and Range Sciences, New Mexico State University, Las Cruces, New Mexico

Thomas L. Kieft Department of Biology, New Mexico Institute of Mining and Technology, Socorro, New Mexico

William G. Knight The Earth Sciences Group and Department of Agronomy, Colorado State University, Fort Collins, Colorado

Gjert Knutsen Department of Microbiology and Plant Physiology, University of Bergen, Bergen, Norway

Blaine Metting Cyanotech Corporation, Woodinville, Washington

L. E. Moser Department of Agronomy, University of Nebraska, Lincoln, Nebraska

I. L. Pepper Department of Soil and Water Science, University of Arizona, Tucson, Arizona

G. A. Peterson Department of Agronomy, Colorado State University, Fort Collins, Colorado

Rex D. Pieper U.S. Department of Agriculture and Department of Animal and Range Sciences, New Mexico State University, Las Cruces, New Mexico

P. Reddell Division of Soils, CSIRO, Townsville, Queensland, Australia

E. F. Redente Range Science Department, Colorado State University, Fort Collins, Colorado

F. B. Reeves Department of Biology, Colorado State University, Fort Collins, Colorado

Zohrab A. Samani Civil Engineering Department, New Mexico State University, Las Cruces, New Mexico

Michael J. Singer Department of Land, Air and Water Resources, University of California, Davis, California

Gaylord V. Skogerboe International Irrigation Center, Department of Agricultural and Irrigation Engineering, Utah State University, Logan, Utah

Yosef Steinberger Department of Life Sciences, Bar Ilan University, Ramat Gan, Israel

Paul W. Unger Conservation and Production Research Laboratory, Agricultural Research Service, U.S. Department of Agriculture, Bushland, Texas

R. P. Upchurch Arboretum Affairs Department, College of Agriculture, University of Arizona, Tucson, Arizona

Neil E. West Department of Range Science and the Ecology Center, Utah State University, Logan, Utah

S. E. Williams Department of Plant, Soil and Insect Sciences, University of Wyoming, Laramie, Wyoming

Part I

SOILS AND BIOTA OF ARID LANDS

1

Semiarid and Arid Lands: A Worldwide Scope

WILLIAM A. DICK-PEDDIE *New Mexico State University, Las Cruces, New Mexico*

I. INTRODUCTION

From ancient times to the present, humans have tried to utilize the arid lands of the world with only marginal success. We have been more successful in our attempts to utilize semiarid lands. However, heavy and/or prolonged use of semiarid lands has often led to a reduction in available moisture for vegetation and has initiated succession from semiarid types of vegetation toward arid types of vegetation. Such human-caused succession of vegetation from semiarid to arid or from arid to extremely arid is currently referred to as *desertification*. Le Houerou (1986) used the term "desertization" for this process in northern Africa and emphasized that the process is caused by human beings, not climate.

The transition from semiarid to arid lands is often subtle. These zones of subtle transition between semiarid and arid and the utilization or attempted utilization by humans have resulted in a plethora of classifications and terms for arid regions of the world and their vegetation. Consequently, before attempting a worldwide overview of lands possessing the soil resources covered in this volume, it is necessary to review terms used for vegetation of the world's semiarid and arid lands and to consider what constitutes desert, semiarid, and arid environments.

II. TERMS AND DEFINITIONS USED AROUND THE WORLD FOR XERIC ENVIRONMENTS AND THEIR VEGETATION

There are differences around the world in the concept of desert as well as differences in terms and concepts used for describing and classifying the vegetation of arid and semiarid lands. Even the criteria used to determine arid and semiarid environments differ from hemisphere to hemisphere. These differences have often been addressed in the literature during the last 15-20 years. Some of this same literature contains detailed descriptions of major arid and semiarid regions of the world, including their vegetation. Material from the publications edited by Specht (1979), di Castri et al. (1981), Bourliere (1983), West (1983a), and Evenari et al. (1985, 1986) constitutes the bulk of information in this chapter.

A. Xeric Environments

The term "desert" implies aridity. There is no worldwide agreement as to what constitutes arid and as to what gradations occur within the concept of arid. This is one of the reasons that desert is a less than precise term. The volumes edited by West (1983a) and Evenari et al. (1985, 1986) are part of a series of volumes on world ecosystems and are considered mutually exclusive in their coverage. Maps in these volumes, depicting areas covered by the volumes, exhibit considerable overlap in portions of Asia, North America, and South America. These overlaps are likely due, in part, to a lack of consensus among scientists as to implications of the terms "desert" and "arid," as well as such modifiers as temperate, hot, and semi. There is a figure by Shmida (1985) that depicts areas of cold, warm, and foggy desert regions of the world. Shmida's cold desert boundaries correspond fairly well with those of the arid-temperate and extremely arid-temperate regions depicted on the inside cover map of the West (1983a) volume. West's (1983a) temperate zones cover more area than Shmida's (1985) cold deserts because West included "semiarid" regions on his map. The coverage by the contributors to all three volumes actually has little overlap, but the overlap of map areas is mentioned here to demonstrate the semantic and/or conceptual difficulties around the world with the terms *desert, arid,* and *semiarid.*

Shmida (1985) equated extremely arid environments with extreme deserts, arid environments with deserts or true deserts, and semiarid environments with semideserts. Mares et al. (1985) equated arid and semiarid environments with semideserts. This results in equating extremely arid environments with deserts (true deserts). Walter and Box (1983b) equated desert steppe to semidesert environment and considered semidesert to be a transition zone (ecotone) between steppe and desert.

Some authors use isohyets of annual precipitation to place limits on the various xeric zones. There is great variation when annual precipitation is used as the single criterion. Table 1 lists the desert types and annual precipitation boundaries of four authors. From this table it is evident that isohyets of precipitation are not a consistent criterion to use for limits of various types of xeric environments.

Following are three sets of desert types commonly used in different parts of the world and the xeric environments in which they are considered to occur:

Semiarid	Arid	Extremely arid
	Semidesert	Desert
Semidesert	Desert	True desert-extreme desert
	Desert	Desert

TABLE 1 Terms and Limits for Xeric Environments Used by Several Authors[a]

	Annual precipitation (mm): 0 50 100 150 200 250 300 350 400
All deserts[a]	Extensive desert---X True desert----X------Semidesert----------------X
Thar desert[b] (India)	X----Desertic zone----------X---- Subdesertic zone---------X
North Africa[c]	Desertic zone-X-----------------Arid zone-------------X
Middle East[d]	Desert (arid)-X-----------Semidesert-------------X

[a]Shmida (1985).

[b]Gupta (1986).

[c]Le Houerou (1986).

[d]Orshen (1986)

Both Petrov (1976) and Walter (1973) categorized the world's arid environments. However, it appears to the author that the most appropriate classification of the world's xeric environments for use in this volume is that of Meigs (1953). Meigs divided xeric environments into extremely arid, arid, and semiarid. These "homoclimates" were based upon the moisture index system of Thornthwaite (1948). Figure 1 is after Meigs and modified from Shmida (1985). It illustrates the xeric areas of the continents that contain soil resources discussed in this volume. In some instances, the zones of semiarid homoclimates in Fig. 1 could be expanded on their mesic boundaries to accommodate areas of human-caused aridity, such as the overgrazing of xeric grasslands and savannas and the clearing of forests and woodlands.

The symbols Meigs used for his homoclimates were E = extremely arid, A = arid, and S = semiarid. Meigs's classification code included symbols that further subdivided these basic zone: a = no marked season of precipitation; b = summer precipitation; and c = winter precipitation. The letter code was followed by two digits. The first digit indicated mean temperature of the coldest month and the second digit indicated mean temperature of the warmest month. Digits used were 0 = <0°C; 1 = 0-10°C; 2 = 10-20°C; 3 = 20-30°C; and 4 = >30°C. For example, Meigs's code for the Gobi Desert was Ab04, which translates to arid; summer precipitation; mean temperature of coldest month, <0°C; and mean temperature of warmest month, >30°C. Meigs's symbols have not been included on the maps of Fig. 1 because climatic features are included in descriptions of the world's major xeric regions. Meigs's code has been explained to re-emphasize the imprecision of the term "desert" and its modifiers and the use of annual precipitation as the criterion for identification of the various xeric regions of the world.

In summary, the most universally accepted classification of xeric environments is that of Meigs, and in most instances, the world's deserts tend to be located in his arid and extremely arid homoclimates.

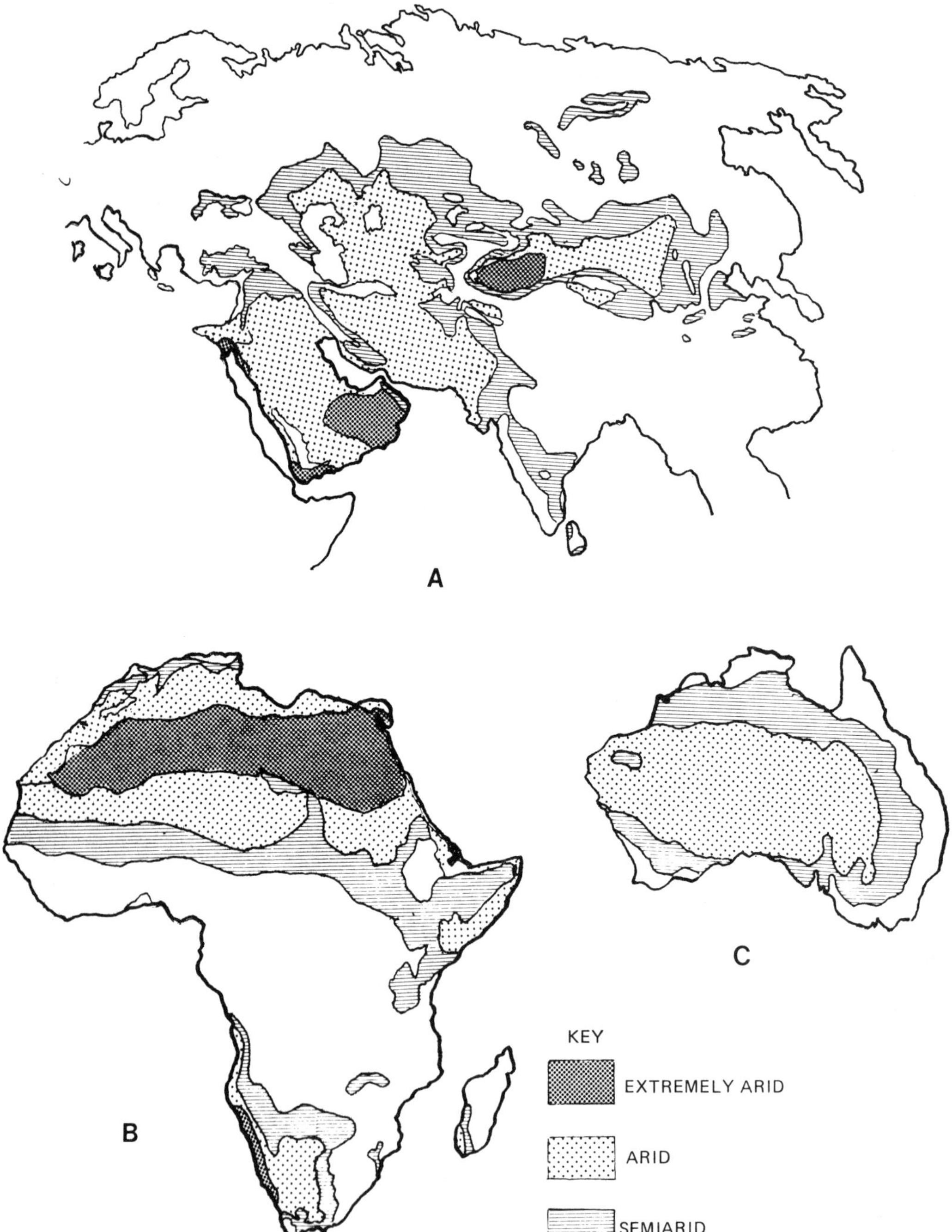

FIGURE 1 Semiarid, arid, and extremely arid regions of the continents: A, Asia; B, Africa; C, Australia; D, North America; E, South America. (After Shmide, 1985.)

D

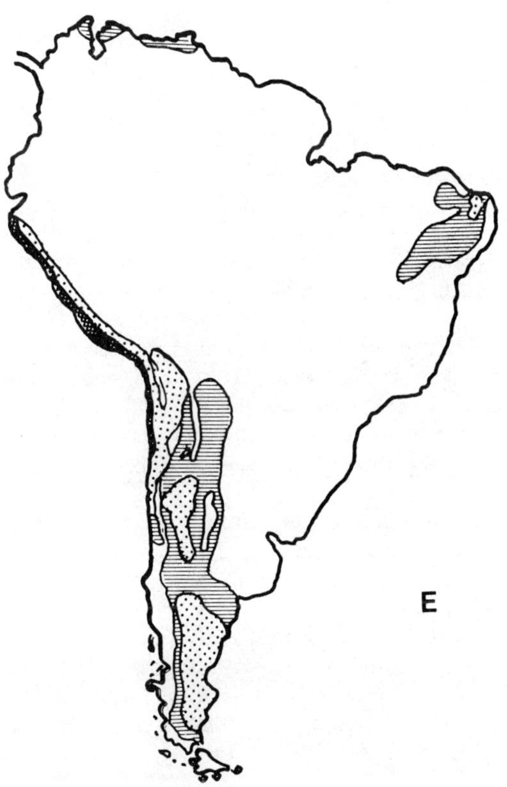

E

B. Xeric Vegetation

Part of the problem of differences in definitions and concepts of xeric vegetation stems from the different criteria used as the basis for descriptive terms for environments and units of vegetation. Many terms used in classifying vegetation are based upon the vegetation itself, such as the life form of the predominant plants. Examples are forest (trees), grassland (grasses), and scrubland (shrubs). These are usually further refined with climatic and/or physiognomic modifiers, such as tropical, rain, coniferous, deciduous, tall, low, open, and closed. Other terms reflect the general appearance of the vegetation on the landscape, such as savanna. Some terms originally applied to extensive amounts of a particular type of vegetation on a continent. For example, large expanses of grassland have been called steppe in Asia, pampa in South America, and prairie in North America. Other examples are dense scrublands called mallee in Australia, fynbos in Africa, maquis in Eurasia, and chaparral in North America.

Sometimes a vegetation name is based solely upon climate (macroenvironment), such as tundra and desert. Vegetation type names based upon climate tend to be general and less precise than names based upon other features or combinations of features. Scientists usually assume that desert must carry a modifier if used as a term for vegetation, and so there are many and varied modifiers used, such as cold, temperate, hot, tropical, salt, succulent, ephemeral, and foggy. It is apparent that even with modifiers the term "desert" lacks precision. Even when the same terms are used by various authors, the vegetation encompassed by the terms may vary greatly. For example, "steppe," unmodified, traditionally has meant grassland (West, 1983h). However, it is used by Shmida (1985) for habitats of scattered dwarf shrubs with ground cover of 10-30%. Shmida equated steppe in the Middle East to chamaephyte semidesert, which is more xeric than grassland but less xeric than true desert. Other authors consider steppe a term for low open or scattered vegetation of any kind and use modifiers to indicate life form or physiognomy. Le Houerou (1986) modified steppe in North Africa as follows: gramineous steppe, chamaephytic steppe, crassulescent steppe, nonhalophilous steppe, succulent steppe, and pseudosteppe. West (1983d) used shrub steppe in western North America for transitional (ecotonal) zones between grassland and xeric scrubland. Savanna also has inconsistent uses around the world. When forests open up with upper canopies not touching and the trees are not tall, the vegetation type may be referred to as a woodland. When a woodland opens up and has only a scattering of trees in a matrix of grass or shrubs, the vegetation may be called a savanna (Africa and California), steppe forest (Asia), or thorn scrub (Mexico).

Major vegetation types tend to occur along a gradient of available moisture. Available moisture is used here as the amount of moisture available to the plants that is necessary for their establishment and maintenance. The gradient can be steep or gradual. Consequently there may be a narrow zone of transition from one vegetation type to another in one area, and in another area the zone (ecotone) may be so broad and extensive that it is not considered a transition zone at all but a vegetation type in itself. One such vast transition zone is the zone between the Saharan Desert and the Soudanian Savanna in North Africa, which is called the Sahel (Monod,

1986). Mares et al. (1985) considered the Chaco of northern Argentina and northwestern Paraguay an ecotone (transition zone) between the forest type (semiarid) and the shrub type (arid) of that region.

There are areas or pockets found in most arid regions supporting vegetation that is very different from surrounding vegetation. This vegetation is associated with drainage systems, both surface and subsurface. Such vegetation can be thought of as *riparian* vegetation. Riparian habitats may be intermittent (ephemeral) drainages referred to as *wadis*, *oueds*, or *arroyo secos* (Evenari, 1985). It has been the custom in the United States to drop "seco" and refer only to "arroyo." These drainages are bordered by vegetation comprised of species that require more available moisture than those of surrounding nonriparian habitats. Many of the world's arid landscapes have internal drainage and do not flow into systems that eventually reach an ocean. These "closed basins" may contain ephemeral lakes or ponds referred to as *playas*, *chotts*, *salars*, or *salinas* (Evenari, 1985). In some regions, such as northern Africa, underground hydrology may yield upwellings or water tables only slightly below the surface, resulting in ephemeral or permanent ponds referred to as *oases*. The characteristic high evaporation rates of arid regions result in the high salinity or alkalinity of many riparian habitats. In turn, these habitats are vegetated by species having a tolerance for high salinity or alkalinity in the soil solution. These salt-tolerant plants are often called *halophytes*.

When looking closely at the vegetation of semiarid and arid lands, moisture availability is not always directly correlated with average annual precipitation. In addition to quantity of precipitation, season of precipitation, form of precipitation, evapotranspiration rates, length of growing season, topography, and soil surface type are a few of the other factors that may influence the amount of available moisture for plants.

In this volume we are interested in soils associated with vegetation found on the dry (xeric) end of the "available moisture" gradient. Table 2 presents a comparison of terms used around the world for vegetation types found along a gradient of available moisture, from highest (open woodland) to lowest (forbland). The positions of grassland and scrubland on the gradient in Table 2 could be reversed or shown as similar in their moisture requirements. The vegetation found at the upper end of the available moisture gradient is that of semiarid or more mesic environments. It is the xeric portions of these vegetation types in various arid regions of the world that have been degraded by humans to an even more xeric form (desertification).

TABLE 2 Terms and Descriptions Used for Vegetation Associated with Semiarid and Arid Regions of the World

Vegetation type[a]	Names used for type
Open woodland: low trees or tall shrubs, canopies not touching, 20-80% cover	Savanna (Africa and California); steppe forest (Asia); thorn forest (Mexico); Chaco (Argentina); semiarid shrub woodland (Australia).

TABLE 2 (continued)

Vegetation type[a]	Names used for type
Scrubland: dense shrubs, canopies close or touching, shrubs 0.5-7 m tall, more than 50% cover	Maquis (France and Israel), machie (Italy), garrigue (northern Africa); mattoral (Spain and Chile); fynbos (southern Africa); chaparral (California); shibeliek (central Asia); mallee, acacia shrubland, and heathland (Australia)
Open scrubland: scattered shrubs 0.5-2 m tall, 3-50% cover	Marginal maquis (around Mediterranean); semidesert (Near East); espinal (Argentina); arid and semiarid low woodland (Australia); desert grassland, desert chaparral, shrub-steppe, interior chaparral, montane scrub (western United States)
Grassland: shrub cover less than 10%	Steppe (central Asia); veld (southern Africa); pampa (central South America); prairie (central North America); semidesert (Near East)
Low scrubland: scattered or no grass, low shrubs, 10-30% cover	Steppe (Near East); chamaephyte semidesert (Near East); true desert (Near East); xeric heathland (Australia); jaral (Chile); coastal sage (California); desert (Africa and North and South America)
Forbland: scattered dwarf shrubs, less than 10% cover, ephemeral annuals	Extreme desert (Asia); ephemeral desert (Near East); annual desert (Near East); desert (Africa, North and South America)

[a]Based upon physiognomy of life form.

III. CHARACTERISTICS OF ARID AND SEMIARID ENVIRONMENTS

A. Moisture Availability

As pointed out earlier, arid and semiarid environments fall at the xeric end of the available moisture gradient. Cloudsley-Thompson and Chadwick (1964) listed four primary causes for arid and semiarid regions as follows: (1) latitudinal position, (2) orographic effects, (3) position in continent, and (4) cold ocean currents. Figure 2 (after Shmida, 1985) illustrates the world's arid environments resulting from the four primary causes of Cloudsley-Thompson and Chadwick. The majority of arid and semiarid regions of the world are located between latitudes 25 and 35° (Fig. 1), where high pressures cause warm air to descend, resulting in dry stable air masses. Orographic aridity is common in North and South America, where high mountain ranges extend perpendicularly to prevailing air-mass movements. Air masses are cooled as they are forced up mountains. This reduces their water-holding capacity, and most of the moisture is precipitated at the high elevations of the windward slopes. The now already relatively dry air masses warm as they descend on the leeward side of the mountain ranges, increasing

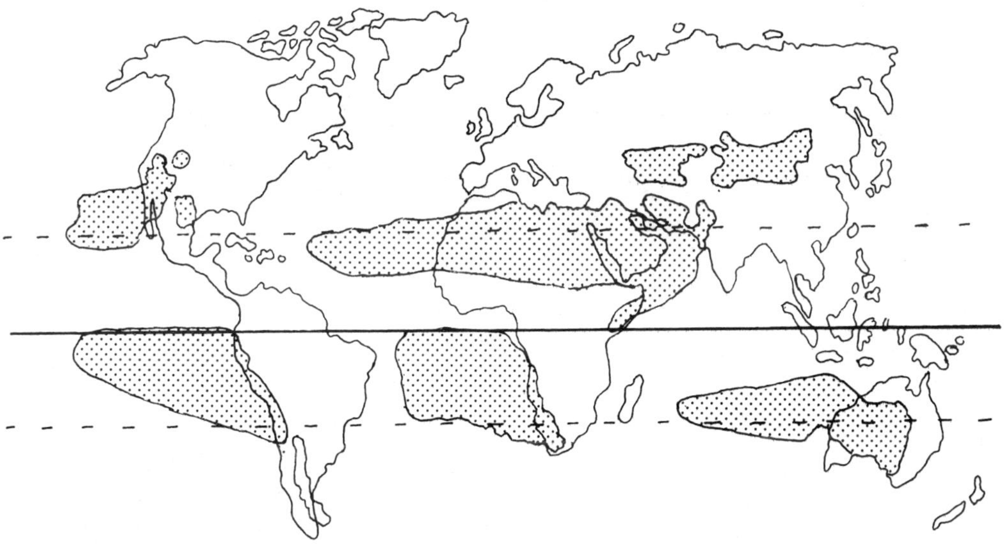

FIGURE 2 Arid climates of the world over continents and oceans. (After Shmida, 1985.)

their water-holding capacity and reducing the chance of any precipitation. This orographic aridity is referred to as the *rain shadow effect*. The third cause, position in a continent, is responsible for much of the arid and semiarid environment of central Asia, where sheer distance from oceans lessens the chance of encountering moisture-laden air masses. The last of Cloudsley-Thompson and Chadwick's causes, cold ocean currents, results in the coastal arid regions of Chile and Peru and those of interior northern Argentina. Here cold ocean currents in close proximity to the coast supply dry air, which comes on shore, but as the mass is forced up the mountain slopes there is no moisture to be lost as the air mass cools. Not only is there no "approach effect" precipitation on the coastal slopes, but the lee, rain shadow effect is even more intensified.

Cloudsley-Thompson and Chadwick's causes of aridity are limited to the quantity of precipitation. On the borders of these low-precipitation zones, arid conditions may extend into areas of slightly higher precipitation due to combinations of other factors. If there is a limited growing season (temperate zones) and some of the annual precipitation falls outside that growing season, much of this off-season moisture does not remain in the soil profile long enough to be available for growth during the following growing season. If rains during the growing season are torrential, there is likely to be extensive runoff and comparatively less available moisture for plant growth. If a significant portion of the annual precipitation is in the form of snow, much potential available moisture may be lost to the atmosphere via evaporation and/or sublimation. Consistent fog, such as the *garua* of coastal Peru and northern Chile, may compensate for extremely low precipitation and result in more available moisture than might be expected from the quantity of annual precipitation (Rauh, 1985). Certain soil chemicals may render less available moisture for plant growth than might be expected from the average annual rainfall, as in the dry heathlands of Australia (Specht, 1981). These features are some of the reasons that the arid and semiarid zones of the world do not conform directly to isohyets of average annual precipitation.

The activities of humans and their animals in areas bordering semiarid or arid zones have often reduced available moisture for plant growth. This reduction may have been caused by increased runoff, decreased infiltration, or decreased water-holding capacity of the soil (loss of topsoil via erosion), for example, but the resulting "desertification" has added and is adding semiarid and arid hectares to the world's arid lands. Instances of this phenomenon are detailed in the following descriptions of the various arid and semiarid regions of the world.

B. Physical Features

The physical features of semiarid regions are highly varied because much of the semiarid vegetation is found on the margins of less arid types, such as grasslands and scrublands, for the reasons just explained. On the other hand, arid landscapes, land forms, and habitats have a great deal in common. Evenari (1985) gave a good overview of physical features of the world's "hot deserts," including the following almost poetic summary: "The desert landscape is dominated by sombre-hued plains stretching into infinity, from horizon to horizon, by mountains, escarpments and rock plateaux incised steeply by 'rivers' flowing nowhere, by isolated rock domes rising like forlorn giants from unending plains, by vast masses of sand in sand fields, and varied dunes of all sizes and configurations."

Most arid lands contain or adjoin irregular (mountainous) topography. The erosional history of these lands has usually been extreme and often dramatic. Many times past erosion has left isolated islands supporting remnant segments of past surfaces. These remnants may be referred to as *inselbergs* (Asia), *buttes* (North America), or *bornhardts* (Africa).

Another result of long periods of extensive erosion is the formation downslope of plains, which may be small, local, and relatively young, called *alluvial fans*, "toe slopes" of whole mountains called *bajadas* in South and North America, or erosion slopes of mountain ranges, often referred to as *piedmont slopes* or *piedmont plains*.

As indicated earlier, ephemeral (intermittent) drainages called wadis, arroyos, or oueds may drain into a system that eventually reaches an ocean or into a closed basin with no outlet to the sea. In either case, on the lower (down) side of an alluvial fan, a bajada, or a piedmont slope, where the rate of flow slows from a reduction in slope or from spreading out, small particles carried in the flow finally drop out, forming accumulations of sand. These accumulations are susceptible to wind transport and may result in dune systems. The term "erg" is used for a sand desert area. The common crescent-shaped sand dunes are called *barkhans*. Low, often partially stabilized sand hills are referred to as *nebkas*. Sand-tolerant vegetation is said to be *psammophilic*.

Some surfaces left from geologic erosion are composed of gravel and small stones. These particles may be covered with black-brown desert varnish (desert patina). Evenari (1985) considers such a desert pavement a *hammada* if the surface composed of rocks, gravel, and small stones was formed in situ rather than having been transported. If a similar surface has been transported, he calls it a *reg*. Some scientists, including Evenari, consider that 40-50% of the surface of the world's hot deserts is comprised of regs and hammadas.

Finally, there are the low spots of arid landscapes, where water collects and usually evaporates between inundations or where the water table is close to the surface of a poorly drained area. These low spots may be small and localized or extensive, such as portions of the Caspian lowland biome (Walter and Box, 1983a), the *mansar* of Pakistan (Gupta, 1986), and the salt desert shrublands of the western United States (West, 1983e).

IV. DESCRIPTIONS OF MAJOR ARID AND SEMIARID REGIONS OF THE WORLD

Table 3 lists, by continents, the major desert regions of the world with their main geomorphology and vegetation types (after Shmida, 1985). The coundaries of these regions conform closely to the arid and extremely arid homoclimates of Meigs (1953); see Fig. 1. The following descriptions follow the classification and order of Table 3. The summary descriptions discuss, by continents, the extent and vegetation of the semiarid, arid, extremely arid, and riparian habitats within each region.

A. Asia

Western portions of the Asiatic semiarid and arid environments are in regions commonly referred to as Eurasia, the Near East, or the Middle East. The lower latitude or "tropical-semitropical" arid areas of Asia are found in the southwestern portion of the continent (Fig. 1). They are included in the Arab, Iran, Syrian, and Thar regions of Table 3.

Arab, Iran, and Syrian Regions

These regions were covered by Orshan (1986) and Abd el Rahman (1985) as the deserts of the Middle East and the deserts of the Arabian Peninsula, respectively. Orshan's Middle Eastern deserts overlap with Abd el Rahman's Arabian Peninsula arid regions in northern Saudi Arabia and Kuwait. Orshan mapped five major habitats and vegetation types for the semiarid and arid regions of the Middle East. These were steppes (semideserts) in the north from southwestern Turkey through Syria, northern Israel, and Jordan, east through Iraq down to north of Baghdad. The steppe vegetation covers basically the same areas in this portion of the continent as the semiarid regions shown in Fig. 1. Orshan's desert type lies south of the steppe and extends from the Red Sea to the Persian Gulf, across northern, central, and south-central Saudi Arabia. The third type, sand desert, occupies a small area in northern Sinai and an extensive area in north-central to southeastern Saudi Arabia. The sand desert is surrounded by a band of Orshan's savanoid desert. The savanoid desert also forms a broad band on the western shore of the Persian Gulf. The fifth type is the salt marshes, and it is scattered in Iraq from the northwest to east and south of Baghdad.

TABLE 3 Major Arid Regions of the Continents and Their Main Geomorphology and Vegetation Types[a]

Regions	Geomorphology types	Vegetation types
North America		
Great Basin	Basins and mountains	*Artemisia-Atriplex* steppes and pygmy open woodland
Sonora	Plateau, crystalline mountains	Thorny succulent savanna
Chihuahua	Limestone plateau	Open shrubland and thickets
Baja California	Diverse, coastal, sands	Thorny succulent savanna and dwarf shrubs
South America		
Patagonia	Plains, dissected low plateau	Perennial grassland
Peru-Chile	Coastal, badlands, salinas	Succulent desert dwarf shrubs
Monte	Basins, mountains, salinas	Thorny succulent savanna
Asia		
Arab	Huge sands, limestone plateau	Chenopod-*Zygophyllum* desert, dwarf shrubs, *Artemisia* and Poaceae on sands
Syrian	Limestone plateau, ergs	*Artemisia*-chenopod steppes and desert dwarf shrubs
Thar	Sand, salinas, gravels	Thorny-rattanoid savanna
Iran	Salinas, ergs, interplains	Chenopod-*Artemisia* steppes and pygmy open woodland
Turkestan	Loess, sandy plains	*Artemisia-Stipa* steppes and pygmy open woodland
Takla Makan	Interplains, ergs	Chenopod desert dwarf shrubs
Gobi	Basins, mountains, sands	Chenopod-*Tamarix* desert shrubland
Africa		
Sahara	Huge sands, ergs, diverse	Chenopod-*Zygophyllum* desert dwarf shrubs, *Artemisia* and Poaceae on sands
Sahel	Diverse plains, plateau	Thorny-rattanoid savanna
Somali	Coastal sands, old relief	Thorny savanna and desert dwarf shrubs
Namib	Coastal sands, diverse	Succulent desert dwarf shrubs
Kalahari	Old diverse plateaux, mountains	Thorny-succulent savanna

TABLE 3 (continued)

Regions	Geomorphology types	Vegetation types
Australia	Huge plains, salines, old relief, sands	Chenopod shrublands and sclerophyl evergreen low woodland

Source: Modified from Shmida (1985).

The boundaries of steppes in the Middle East have changed because of the activities of humans and their animals. The steppes have increased on their mesic borders at the expense of the Mediterranean steppe forest vegetation type. This type of vegetation is dominated by scattered trees in the genera *Pistacia*, *Amygdalus*, *Rhamnus*, and *Pyrus*, with grasses in between. According to Orshan, grazing has removed all grasses and has prevented the regeneration of trees. On the xeric borders of the steppe, the human presence has promoted the advance of desert, that is, desertification. Steppe vegetation is dominated by shrub species in the genera *Artemisia*, *Pyrethrum*, *Thymelaea*, *Noaea*, *Salvia*, and *Gymnocarpos*. Scattered grasses, often tussock types of *Poa*, *Aristida*, *Lasiurus*, and *Panicum*, are usually present. The shrub *Hammada scoparia* often dominates loessial and loesslike plains and plateaus as far as east-central Saudi Arabia. The sparse desert vegetation of rocky hills is usually dominated by species of *Zygophyllum*, *Reaumuria*, and *Gymnocarpos*. Regs may be dominated by species of *Anabasis*. Steppe vegetation in central Saudi Arabis is often dominated by *Rhanterium* shrubs and *Stipa* grasses.

Savanoid deserts is the term used by Orshan (1986) to apply to a riparian vegetation along wadis or where the water table is relatively close to the surface. Deep-rooted trees and shrubs grow with a dwarf shrub understory in these situations, giving a savannalike appearance, hence "savanoid." The major tree-shrub genera are *Acacia*, *Hammada*, *Anabasis*, *Gymnocarpos*, *Lycium*, and *Ziziphus*. Saline riparian vegetation of sinks, playas, and closed basins is dominated by plants in the genera *Halocnemium*, *Suaeda*, *Frankenia*, *Seidlitzia*, *Phoenix*, *Nitraria*, *Tamarix*, *Althrocnemium*, *Atriplex*, and *Salsola*.

Sand deserts are areas covered with deep sand. Here water infiltrates and percolates deep enough to avoid evaporation and as a result there is more available moisture than on nonsandy sites. Dominant species of this vegetation are in the genera *Calligonum*, *Retama*, *Haloxylon*, *Cornulaca*, *Panicum*, *Hammada*, *Anabasis*, *Aristida*, *Rhanterium*, and *Ephedra*. The central sandy area of the Arabian Peninsula is dominated by shrubs of *Artemisia* and *Calligonum*. Perennial herbs in the genera *Monsonia* and *Scrophularia* are common on most sands of the areas of the Middle East and Arabian Peninsula.

There is a savanna composed of *Juniperus* and *Acacia* trees with *Panicum* and *Aristida* tussock grasses at elevations of 1000-1500 m along the Red Sea in Saudi Arabia.

Thar Region

Gupta (1986) subdivided the semiarid and arid environments of the Thar region of northwest India and south-central Pakistan into a desertic zone and a subdesertic zone. In the desertic zone, the sand dunes and interdune sites support a sparse, scattered, low shrub vegetation dominated by shrubs in the genus *Calligonum*. *Calligonum* codominates with *Prosopis* and *Capparis* on interdunal sites. Sandstone hills of the desertic zones are dominated by *Acacia*, *Salvadora*, and *Commiphora* shrubs. As the name implies, the subdesertic zone is more mesic than the desertic zone. Gupta further subdivided subdesertic into areas with annual rainfall of 200-400 mm and areas with 400-500 mm. As in the desertic zone, the dominant shrubs belong to the genera *Prosopis*, *Capparis*, *Acacia*, *Calligonium*, and *Salvadora*. In addition, some sites are codominated by species in *Ziziphus*, *Lycium*, *Maytenus*, *Euphorbia*, and *Anogeissus*. The plants dominating in the more mesic (400-500 mm precipitation areas) tend to be taller and often treelike. Gupta indicated that the vegetation of these sites was originally thorn forest or woodland but that humans and their animals have greatly reduced the treelike forms, resulting in a low shrub vegetation. Because of its cactuslike growth form and spines, *Euphorbia caducifolia* is often an "increaser" on these disturbed sites.

Typical of arid regions worldwide, many riparian habitats of the Thar are highly saline and are vegetated by saline-tolerant plants. Treelike species of *Tamarix* are common halophytes found on these sites. Another halophyte, *Haloxylon*, is common on floodplain deposits.

Iran, Turkistan, Takla Makan, and Gobi Regions

These arid regions are mostly in the temperate zone of Asia, from the western Eurasian portion to the far eastern Mongolian portion of the continent. Walter and Box (1983c) stated that there is a natural division of these temperate arid areas of Asia. The western (middle Asia) portion, including western Russia and northern Iran and Afghanistan, receives moisture mostly in the winter from storms moving eastward from the Atlantic Ocean. The eastern (central Asia) portion receives only summer moisture, from frontal and cyclonic storms from eastern Asia.

Walter and Box (1983a) treated the Caspian lowland region as an ecotone between the eastern European steppe and middle Asian desert. The region can be thought of as a massive riparian closed basin even though it slowly drains into the Caspian Sea. The region includes salt lakes, playas, salt marshes, and salt swamps. Elevated portions forming broad plains account for 50-60% of the region. The dominant shrubs are in the genera *Artemisia*, *Kochia*, and *Camphorosma*. Perennials in the genera *Tulipa*, *Poa*, and *Ornithogalum* are common. The slopes from the elevated plains, termed "microslopes" by Walter and Box, have more available moisture than the uplands and are vegetated by plants mostly from the genera *Tanacetum*, *Poa*, and *Agropyron*. In the microdepressions the available moisture is enough to produce steppe vegetation dominated by *Festuca*, *Stipa*, *Koeleria*, and *Agropyron* grasses. The cover reaches 75-95% in these "microdepressions." The common halophytes of saline areas in decreasing order of salinity are the genera *Salicornia*, *Halocnennum*, *Petrosimonia*, *Atriplex*, *Anabasis*, *Nanophytum*, *Artemisia*, and *Kochia*.

Walter and Box (1983b) termed the arid region east of the Caspian Sea in Russia as central Kazakhstan in the north down to the Aral Sea. The area south of the Aral into northern Iran and Afghanistan they called middle Asian desert.

The northern portion of central Kazakhstan has the most available moisture, and the region becomes successively more xeric to the south. The northern portions are vegetated by desert-steppe (semidesert) plants. The dominant grasses are in the genera *Festuca*, *Stipa*, and *Poa*. The shrubs on rocky shallow soil sites are mostly species belonging to *Caragana*, *Spiraea*, and *Artemisia*. The desert zone vegetation can be distinguished from the semidesert by the absence of grass and an increase in dwarf shrubs such as species in *Salsola*, *Ferula*, *Caragana*, and *Artemisia*. On calcareous and saline sites, most of the plants are in the genera *Anabasis*, *Artemisia*, *Atraphaxis*, *Atriplex*, *Halocnemum*, *Kalidium*, *Nanophyton*, *Nitraria*, and *Tamarix*.

The Middle Asian deserts were subdivided by Walter and Box into six vegetation types based upon soil types. These can be grouped into semidesert, desert, and riparian vegetation. The semidesert includes gypsophilic dwarf shrub vegetation and psammophytic (sand) vegetation.

Many of the gypsophilic plants are thorny or spiny. These include *Salsola*, *Gaillonia*, *Sisymbrium*, *Noaea*, *Cousinia*, *Acanthophyllum*, and *Atraphaxis*. Other predominant nonspiny species are mostly in the genera *Artemisia*, *Reaumuria*, *Caragana*, *Anabasis*, *Convolvulus*, *Sophora*, *Ephedra*, and *Calligonum*. There are spectacular nonwoody plants, often 2 m tall, mostly in the family Apiaceae.

Sand (psammophilic) communities are dominated by shrubs in *Calligonum* 30 species), *Astragalus* (20 species), *Ammodendron* (5 species), *Artemisia*, and *Ephedra*. Sand areas of the southern Karakum region have such thick vegetation that it is difficult to think of them as desert areas. There are trees from 3 to 8 m tall in the genera *Ammodendron*, *Haloxylon*, *Salsola*, and *Calligonum*. There are many spring ephemerals, including species of *Allium*, *Tulipa*, *Iris*, *Rheum*, and *Eremurus*.

The desert areas have ephemeral and "takyr" vegetation. The ephemeral deserts are on loess. Characteristic dominants are *Poa bulbosa* and *Carex pachystylis*. Soils are shallow and very wet in the spring but quickly dry out after the rainy season. Ephemerals include species of *Geranium*, *Tulipa*, *Ranunculus*, *Gentiana*, *Draba*, *Alyssum*, *Trigonella*, *Delphinium*, *Papaver*, *Veronica*, *Erodium*, *Bromus*, *Agropyron*, and *Phleum*. Takyr sites are closed basin riparian sites. They are clayey and, once swollen with water, are water impermeable. Then after the water evaporates, the soils form large cracks and only lichens can be found on these sites.

Halophytic desert vegetation is comprised of many genera, some having many species, such as *Salsola* (20 species), *Zygophyllum* (14 species), *Suaeda* (9 species), and *Atriplex* (4 species). Important species are also in the genera *Haloxylon*, *Halostachys*, *Halocnemum*, *Salicornia*, and *Kalidium*.

Floodplain riparian vegetation is extensive and varied. Walter and Box (1983c) described five major floodplain vegetation types. Floodplain forest vegetation is populated by species of *Populus*, *Salix*, *Elaeagnus*, *Tamarix*, and *Halimodendron*. Species of *Clematis*, *Asparagus*, and *Cynanchum* become lianalike and climb to the tops of the trees. Floodplain shrub thicket vegetation is dominanted by *Tamarix*, with subdominant species in the genera

Lycium and *Halimodendron* being the most common. Floodplain tall grass and forb vegetation is actually marshlike and contains species of *Typha*, *Phragmites*, *Carex*, and *Schoenoplectus*. Floodplain meadow vegetation is swardlike and composed of grass species mostly of *Aeluropus*, *Cynodon*, *Calamagrostis*, and *Erianthus*. Common meadow forb genera are *Alhagi*, *Peganum*, *Glycyrrhiza*, and *Karelinia*. Floodplain ka'ir vegetation is actually early stage successional vegetation and, depending upon stage, has species of the floodplain types already described.

Breckle (1983) described the arid regions of southern middle Asia (western Asia) as temperate deserts and semideserts of Afghanistan and Iran. These areas are arid links with the semitropical-tropical deserts of that continent. Breckles stated the "Vast areas of inner Iran are almost bare of vegetation." He divided the arid vegetation based upon substrate conditions as follows: halophytic; gypsophilous; psammophytic (on deep sand); vegetation on loess; vegetation on scree, gravel, and rocks; mountain vegetation; and vegetation of oases and river margins (riparian). Halophytic and gypsophilous could also be considered riparian vegetation of closed basins (salines or alkali sinks).

Semidesert vegetation on gravel, rocks, and scree (hammadas and regs) is dominated in both Iran and Afghanistan by the shrubs *Artemisia herba-alba* and *Zygophyllum eurypterum*. Breckle (1983) indicated that much of the extensive "artemisian semidesert" of both countries is a recent extension (desertification) at the expense of steppes. Erg (psammophytic) vegetation is dominated by shrubs in the genera *Haloxylon*, *Aellenia*, and *Calligonum*, along with perennials in the genera *Stipagrostis*, *Cyperus*, and *Pennisetum*.

Vegetation on a substrate of loess is low and dominated by the perennials *P. bulbosa*, *Psoralea drpacea*, and many species of *Astragalus*. As in Eurasian loessial arid lands, the vegetation is rich in spring "geophytes" whose cultivars are well known in the Northern Hemisphere. The most common genera are *Crocus*, *Tulipa*, *Allium*, *Anemona*, *Corydalis*, and *Gagea*.

Semidesert open woodlands can be found at altitudes of around 1700 m and higher through most of Afghanistan. The trees are primarily species of *Pistacia*, *Amygdalus*, *Cercis*, and *Quercus*. Understory shrubs and subshrubs are species of *Amygdalus*, *Acantholium*, *Artemisia*, *Cerasus*, *Colutes*, *Ephedra*, and *Rosa*. Breckle indicated that much of the area that previously supported semidesert woodland is now covered with sagebrush from long-term disturbance by humans.

Vegetation above 3000 m often consists of dwarf shrubs, cushion plants, sedges, and grasses. At elevations of 4000 m or more, drier sites may be dominated by thorny subshrubs of "tragacanthoid character" in *Astragalus*, *Acantholium*, *Euphorbia*, *Onobrychis*, and *Platytaenia* genera. Where there is an increase in summer rains, grass-sedge steppes of *Kobresia*, *Carex*, and *Poa* may dominate at these elevations.

Riparian vegetation may be found on relatively nonsaline habitats of oasis and wadi margins. There are also many alkaline-saline habitats supporting a halophytic vegetation. The centers of *sebkhas* (clay pans in shallow depressions) are usually very saline with species of *Halocnemum* and *Seidlitzie* dominating. Moving out to progressively less saline conditions, species of *Haloxylon*, *Anabasis*, *Aeluropus*, *Atriplex*, *Limonium*,

Lycium, *Alhagi*, and *Salsola* tend to occur, in that order. In Iran, wadi margins are often dominated by one or more species of *Tamarix*, along with *Prosopis*, *Calligonum*, and *Haloxylon*.

Walter and Box (1983d) described the vegetation of the complicated basin and range systems of middle to central Asia as "orobiomes." Arid and semiarid vegetation is found at lower elevations on xeric south- to west-facing slopes and tundralike xeric vegetation at higher elevations. The basins (valleys) either support riparian vegetation or are agricultural lands.

Moving northeast into central Asia, the arid regions are usually isolated by mountain ranges. There are both arid and extremely arid regions with hot and cool summers. All winters are cold. The cool summer region is found mostly on the high Tibetan plateau.

The regions with hot summers include the Gobi area of Mongolia to the north and the extremely arid area of the Tarim Basin to the south and west (Fig. 1A). The Tarim Basin includes the Takla Makan desert. The northern Mongolian region supports steppes and semideserts. Xeric conditions increase from the higher areas of the north to the lower southern areas. In a map of the vegetation zones of Outer Mongolia, Hilbig (1983) illustrated a montane coniferous zone bordered to the south by a zone of montane forest steppes and steppes. This belt graded to the south into a grass steppe zone, which in turn gave way south and west to a semidesert zone. A desert zone lay south of the semidesert belt.

The Gobi desert-steppe supports shrub and semishrub species of *Aragana*, *Atraphaxis*, *Amygdalus*, *Oxytropis*, and *Artemisia*. The steppe and grass steppe are dominated by *Stipa* grasses and species of *Allium*. The semidesert and desert zones are covered by low shrub and semishrub species of *Anabasis*, *Haloxylon*, *Salsola*, *Kochia*, *Nanophyton*, *Reaumuria*, *Zygophyllum*, *Nitraria*, *Oxytropis*, *Caragana*, *Atraphaxis*, *Calligonum*, and *Artemisia*. Riparian vegetation of oases may have trees of *Populus*, *Elaeagnus*, *Tamarix*, and an occasional *Ulmus* (Hilbig, 1983).

The Tarim Basin is the most arid part of central Asia. Walter and Box (1983f) stated, "The sand desert Takla Makan is generally completely devoid of vegetation." The water table is close to the surface, and occasionally shrubs can subsist as "phreatophytes." These shrubs are mostly in the genera *Ephedra*, *Haloxylon*, *Nitraria*, and *Tamarix*. These plants often cause the formation of nebka dunes. Floodplain (riparian) vegetation is dominated by the same species as in the Gobi region.

Cool summer arid regions of the high Tibetan plateau support steppe to semidesert vegetation. Dominant shrub genera are *Ceretoides*, *Kochia*, *Reaumuria*, *Oxytropis*, *Astragalus*, *Acantholium*, *Artemisia*, and the endemic genera *Myricaria* and *Microula*. Grasses of *Agropyron*, *Avenastrum*, *Festuca*, and *Elymus* are common. Upper elevations of southern central Asia often support meadow and cushion plant communities of a tundra type of vegetation.

B. Africa

Africa has more area of extremely arid environment than any other continent. The extremely arid, arid, and semiarid environments are located across the northern one-third and in the southwestern and southern one-fourth of the continent (Fig. 1B).

Northern Africa, Sahara, and Sahel Regions

The abiotic and biotic features of northern Africa, including the Sahara, have been well described by Le Houerou (1986). Le Houerou considered the desert zone to be those areas with less than 100 mm annual precipitation and the arid zone to be areas of 100-400 mm annual precipitation. Le Houerou's arid zone includes most of the semiarid areas of other authors, as alluded to earlier.

Le Houerou subdivided the xeric environments of northern Africa into seven subzones based upon precipitation: upper arid subzone (400-300 mm), middle arid subzone (300-200 mm), lower arid subzone (200-100 mm), upper desert subzone (100-50 mm), middle desert subzone (50-25 mm), lower desert subzone (25-10 mm), and true desert (less than 10 mm). He described forest (trees over 5 m high and 100+ stems ha^{-1}) in the upper arid subzone of Morocco, Algeria, Tunisia, and Libya. The predominant tree genera are *Pinus*, *Juniperus*, *Tetraclinis*, and *Argania*. There is little forest left in northern Africa. The middle arid subzone is occupied by matorral (garrigue) vegetation of dense tall shrubs or small trees.

Le Houerou subdivided steppe vegetation into gramineous steppe, chamaephytic steppe, crassulescent steppe, nonhalophilous succulent steppe, and pseudosteppe. Steppe vegetation can be found in all arid subzones. The most mesic steppe vegetation is the gramineous steppe dominated by *Stipa* species and subdominant grass species of *Lygeum*, *Panicum*, and *Stipagrostis*. The next more xeric vegetation is the chamaephytic steppe composed of low shrub and half-shrub species of *Anabasis*, *Acacia*, *Anthyllis*, *Artemisia*, *Gymnocarpos*, *Hammada*, *Noaea*, *Rhanterium*, and *Thymelaea*.

The upper desert subzone has a contracted pattern of vegetation, limited to sand and riparian habitats. The dominants are similar to those of the chamaephytic steppe. Species of *Calligonum* and *Tamarix* can also be found. Heavy sand has species of *Stipa*, *Pennisetum*, and *Panicum* grasses along with shrub species of *Calligonum* and *Ephedra*.

Riparian vegetation is found in most zones and is called desert savanna with tree species of *Acacia*, *Pistacia*, *Ziziphus*, *Maerua*, and *Balanites*. Vegetation of the middle and lower desert subzones is limited to riparian sites (such as wadis and sinks). Dominants are mostly in the genera *Acacia*, *Artemisia*, *Atriplex*, *Fagonia*, *Reaumuria*, *Ziziphus*, *Alhagi*, *Anabasis*, *Hammada*, *Salsola*, *Tamarix*, *Capparis*, *Salvadora*, *Psoralea*, and *Zygophyllum*. The only vegetation of the true desert subzone is the occasional occurrence of a cover of annuals.

Le Houerou indicated that the depletion of vegetation from the impact of humans and their animals has been going on in northern Africa since Roman times. Forests were destroyed for crop cultivation, and upon the abandonment of cultivation, subsequent use of the land for grazing prevented re-establishment of forests.

Le Houerou gave a well-documented view of succession following disturbance in northern Africa. His desertization (desertification) sqeuence for northern Africa was as follows:

1. Primeval open forest or woodland of *Pinus halepensis* or *Tetraclinis articulata* with *Juniperus phoenicea* and *Stipa tenacissima*: deforestation

2. Shrubland of *J. phoenicea*, *Rosmarinus officinalis*, and *S. tenacissima*: deforestation
3. Gramineous steppe of *S. tenacissima*: overexploitation
4. Chamaephytic steppes of *Artemisia herba-alba* (silty to loamy soils) or *Artemisia campestris* (sandy soils): clearing, overstocking
5. Weeds of cultivation and fallows of *Cynodon dactylon*; depleted steppes of *Stipa capensis* and *Peganum harmala*

The xeric lands of northeastern Africa (Egypt and the Sudan) were described by Ayyad and Ghabbour (1986). They classified the climatic and geomorphological variations of the Sudano-Egyptian deserts into hyperarid, arid, and semiarid. These in turn were variously subdivided, resulting in a total of 10 regions. Each of the three major categories included a riparian portion. Wetlands were included in the classification.

In the hyperarid regions of Egypt, surfaces of mixed gravel and sand are dominated by species of *Indigofera*. Psammophilous (heavy sand) sites have species of *Panicum*, *Calotropis*, *Alhagi*, *Hyoscyamus*, and *Desmostachya*. The species dominating the slopes of low hills change with the amount of available moisture, which increases with elevation. The genera containing these species in order from lower to higher sites are *Euphorbia*, *Carallum*, *Delonix*, *Moringa*, *Acacia*, *Dodonaea*, *Dracaena*, *Euclea*, *Pistacia*, and *Rhus*. Vegetation of the arid areas is similar to the upper hill vegetation of hyperarid areas but with the addition of species of *Centaurea*, *Diospyros*, *Maytenus*, *Indigofera*, and *Maerua*.

Vegetation of the semiarid regions is mostly a savannalike type of trees and grasses. Dominant tree species are in *Acacia*, *Juniperus*, *Olea*, *Combretum*, *Anogeissus*, *Maba*, *Boscia*, *Cadaba*, *Albizia*, *Dalbergia*, *Commiphore*, and *Terminalia*. Major grasses are in *Cymbopogon*, *Hyparrhenia*, *Sehima*, *Sorghum*, *Andropogon*, *Sefaria*, *Cenchrus*, *Aristida*, *Eragrostis*, *Pennisetum*, and *Elionurus*.

Riparian vegetation is found in hyperarid, arid, and semiarid regions of Egypt and the Sudan. The vegetation of highly saline riparian habitats is dominated by species of *Limonium*, *Zygophyllum*, *Nitraria*, *Suaeda*, *Tamarix*, and *Salicornia*. Wadi vegetation is dominated by tree and shrub species of *Moringa*, *Acacia*, *Lycium*, *Aerva*, *Tamarix*, *Artemisia*, *Leptadenia*, *Cassia*, *Indigofera*, *Salvadora*, *Calotropis*, *Ephedra*, *Haloxylon*, *Retama*, *Anabasis*, *Salix*, *Ficus*, *Cordia*, and *Tamarindus*.

In northern Africa there is a transcontinental ecotone belt between the Saharan Desert to the north and the Soudanian savanna region to the south. This belt is referred to as the Sahel. Monod (1986) pointed out that the width (north-south) of this ecotone has expanded and is expanding at the expense of the Soudanian savanna. The Sahel is the most extensive example of desertification in the world.

The Sahel vegetation is largely an open grassland type with scattered thorn shrubs and trees. Dominant tree and shrub genera are *Combretum*, *Balanites*, *Acacia*, *Commiphora*, *Guiera*, and *Diospyros*. Predominant grass genera are *Aristida*, *Stipagrostis*, *Panicum*, *Cenchrus*, *Schoenfeldia*, *Andropogon*, *Cymbopogon*, and *Hyparrhenia*.

Riparian vegetation of "soaks," springs, pools, and wadis is dominated by woody species of *Acacia*, *Celtis*, *Combretum*, *Crateva*, *Ficus*, *Maerua*,

Malacantha, *Balanites*, *Cordia*, *Commiphora*, *Tamarindus*, *Ziziphus*, *Albizia*, *Grewia*, *Capparis*, and *Salvadora*.

Namib and Kalahari Regions

The other extremely arid region of Africa is on the southwest coast and is called the Namib Desert (Fig. 1B). Walter described the vegetation based upon predominant geomorphological features.

According to Walter (1986), the fog common over the "outer" Namib results in a reduction of evapotranspiration rates for plants. The major feature of the outer Namib is an extensive peneplain constituting a platform that slopes inland from the coast up to 1000 m. The surface is a hammada overlying a gypsum hardpan. The only permanent vegetation is composed of lichens. There are over 20 species of lichens, including crustose, foliose, and fruticose types. Occasionally perennial plants can become established and form nebkas. The nebka species are in *Arthraeria*, *Salsola*, *Psilocaulon*, and *Zygophyllum*. Niches and catchments in rock outcrops support an interesting assortment of succulent plants, including species of *Hoodia*, *Lithops*, *Trichocaulon*, *Cotyledon*, *Aloe*, and *Euphorbia*. There are dune areas of chains and fields of barkhans. They are without vegetation. On the edges of the dune fields only three species can be found. They are *Acanthosicyos horrida*, *Eragrostis spinosa*, and *Stipagrostis subulicola*.

Farther west the "inner" Namib is out of the fog belt but moisture from irregular summer rains is caught in scattered pockets. *Stipagrostis* grasses abound, forming an annual grassland.

Riparian vegetation can be found when erosion runnels and rivulets meet to form deeper and wider "riviere" (arroyos). Tree and shrub species of *Acacia*, *Azima*, *Euclea*, *Sesbania*, *Boscia*, *Asclepias*, *Commiphora*, *Salvadora*, *Lycium*, *Ziziphus*, *Parkinsonia*, and *Adenolobus* are common. The interesting *Welwitschia* plant is found on a large ephemeral floodplain area in the transition zone from the outer to the inner Namib.

Moving inland (east) from the Namib region, the environment is arid or semiarid. This is the region of the Karoo and Kalahari. A succulent dwarf shrub vegetation is found in the western Karoo region south of the Namib. South of this region on the southwest tip and cape is a region with a semiarid Mediterranean type of climate. This region supports fynbos or matorral vegetation. An open dwarf shrub vegetation is found to the east of the Namib and western Karoo on loamy sand to sandy loam of flats and gentle slopes. Taller shrubs and grasses co-dominate with dwarf shrubs on steeper slopes of broken rocky sites. In areas near the coast with more moderate winter temperatures, the dwarf shrub vegetation is often composed of succulents.

Riparian vegetation of "drainage lines" in the Karoo and southern Kalahari is composed primarily of species of *Acacia*, *Boscia*, and *Euclea*.

The arid lands of Madagascar were described by Rauh (1986). There is a narrow arid belt along the southwestern coastal region, banded to the east and north by a region of semiarid climate. The vegetation of the arid areas is dominated by large succulents. The species are from *Alluaudia*, *Adansonia*, *Moringa*, *Euphorbia*, *Pachypodium*, *Didierea*, and *Aloe*.

C. Australia

Most of Australia has a semiarid or arid environment but has no extremely arid environment. The semiarid region tends to be an outside band surrounding, except in the west, the central arid region (Fig. 1C). The semiarid region contains semiarid shrub woodland, arid and semiarid low woodland, eucalypt shrubland (mallee), and arid tussock grassland vegetation. The vegetation of the arid region is primarily arid hummock grassland, acacia shrubland, and shrub steppe (Williams and Calaby, 1985).

The shrub steppe is composed of chenopods primarily in the genera *Atriplex*, *Maireana*, *Sclerolaena*, *Chenopodium*, and *Rhagodia*. Grasses of *Aristida*, *Chloris*, *Eragrostis*, and *Sporobolus* are scattered with the shrubs of the shrub steppe. Acacia shrublands cover over 1.5 million km^2 from the east-central portion of the arid region to the west-central coast. There are 10 species of *Acacia* that codominate in various parts of the acacia scrublands. Common grasses of the acacia scrublands are species of *Aristida*, *Chloris*, *Dichanthium*, *Digitaria*, *Eragrostis*, and *Eriachne*. The arid hummock grasslands, also referred to as "spinifex" grasslands, cover over 1.5 million km^2 in Australia. The *Triodia* and *Plectrachne* hummocks are widespread, and the ground between is bare except for ephemerals after rain. In various portions of this vegetation type 11 species of *Triodia* share dominance. Shrubs are mixed in the hummock grasses in the dune fields of southwestern Queensland. Shrub species are mostly in the genera *Cassia*, *Dodonaea*, and *Eremophila*.

Arid tussock grasslands are found in the northeastern portions of the semiarid region. This type of vegetation occupies 500,000 km^2, and the tussocks are formed by grass species of *Astrebia*. Williams and Calaby (1985) stated, "This grassland [is] one of the few true grasslands in Australia."

The eastern portion of the semiarid region is vegetated by semiarid shrub woodlands. This type has various species of low (5-10 m tall) *Eucalyptus* trees and a shrub understory dominated by species of *Apophyllum*, *Atalaya*, *Casuarina*, *Eremophila*, and *Geijera*. Arid and semiarid low woodlands of dwarf *Eucalyptus* communities are found in the northern semiarid belt. The shrub layer contains species of *Acacia*, *Atalaya*, *Grevillea*, *Terminalia*, and *Petalostigma*.

Eucalypt shrubland (mallee) occupies the southwestern portion of the semiarid region. Here the various species of *Eucalyptus* have many stems arising from lignotubers that support an umbrella of olive-green leaves. Undershrubs may be species in the genera *Cassia*, *Maireana*, and *Heterodendrum*. Ground flora is dominated by species of *Triodia* hummock grass.

D. North America

The xeric regions of North America are mostly semiarid and arid environments (Fig. 1D) and would be considered semideserts by most authors outside North America. West (1983h) used the term "semidesert" for the North American temperate (cold) arid regions.

Great Basin Region

West (1983b) subdivided this region into semidesert (arid) and steppe (semiarid). The arid areas were divided into Great Basin-Colorado Plateau sagebrush semidesert and Colorado Plateau-Mohavian blackbrush semidesert (West, 1983c, f). Because of the mountainous terrain, the sagebrush desert is disjunct (not contiguous). Most occurs in Nevada, with lesser amounts in Utah, northern Arizona, southwestern Colorado, and north-central New Mexico. The blackbrush desert is largely restricted to southeastern Utah. Both these semidesert areas are covered by a low open scrubland type of vegetation, one dominated by various species of sagebrush (*Artemisia*) shrubs and the other by the shrub, blackbrush (*Coleogyne ramosissima*). Of the 10-40% cover on the sagebrush semidesert, sagebrush comprises 70%. Both semideserts have perennial bunchgrasses and perennial forbs scattered among the shrubs.

West (1983d, g) subdivided the semiarid portions into western intermountain sagebrush steppe and southwestern Utah galleta-threeawn shrub steppe. His sagebrush steppe region was extensive and included much of eastern Washington and Oregon, northwestern Nevada, and southern Idaho, southwest and north-south through central Wyoming, with minor amounts in northeastern Utah and northwestern Colorado. As the name implies, the galleta-threeawn shrub steppe is restricted to Utah. These semiarid shrub steppes can be considered transition zones (ecotones) between grasslands and scrublands. West indicated that it is difficult to determine how much is typical of a pristine (climax) type and how much is the result of livestock-induced succession from prior grassland. In some areas the grassland is gone and the sagebrush steppe integrades with woodland (usually pinyon-juniper) or yellow pine forest. The species composition is similar to that of the arid regions, but shrub dominance is shared by grasses.

An extensive portion of the Great Basin region is covered by riparian habitats in the form of arroyo systems and closed basins. West (1983e) called this type of habitat intermountain salt-desert shrubland. The closed basins often contain playas and/or alkali sinks. Vegetation is very sparse on this riparian scrubland. Where water spreads and evaporates on what West calls upland sites, shrubs in the genera *Atriplex* and *Ceretoides* are the dominant plants. Salines (alkali sinks) have halophytes in the genera *Sarcobatus*, *Allenrolfea*, *Suaeda*, and *Distichlis*.

Sonora, Chihuahua, and Baja California Regions

Authors in North America usually separate the North American arid zones differently than those of Table 3. MacMahon and Wagner (1985) gave a detailed view of these arid lands. They used the hot desert-cold desert approach, relegating the Great Basin to cold desert and the rest to hot desert. Their hot deserts are the Chihuahuan, Sonoran, and Mohave. The Mohave Desert links the Sonoran Desert to the south and the Great Basin Desert to the north and northeast. Most of the Chihuahuan Desert is in Mexico, and it is separated from the southern Sonoran to the west by the Sierra Madre Occidental. The three North American hot deserts all have the creosote brush (*Larrea tridentata*) as a dominant. The three hot deserts are separated primarily on the basis of the small groups of dominant plants unique to each desert.

There is considerable variation within the Sonoran Desert. The Sonoran extends from east-central Arizona southwest into Mexico, including most of the states of Sonora and Baja California. The northeastern and southern (mainland) portions of the Sonoran are the most mesic and tend to have a savanna or thorn woodland type of vegetation. MacMahon and Wagner (1985) stated that well-developed sites usually are composed of three layers (synusia) of perennial plants. The upper layer is dominated by tall or treelike shrubs (>150 cm). The middle layer is dominated by medium-sized shrubs, and the lower layer is dominated by low or semi-shrubs. Tall treelike plants of the upper layer are mostly in the genera *Eysenhardtia*, *Acacia*, *Bursera*, *Cercidium*, *Olneya*, and *Canotia*. Tall cacti are usually in the genera *Cereus*, *Lamaireocereus*, and *Pachycereus*. Dominant shrub species of the middle layer are in the genera *Larrea*, *Fouquieria*, *Yucca*, *Guaiacum*, *Jatropha*, and *Parthenium*. There are also cacti in this layer, many in the genus *Opuntia*. The lower layer is dominated by low shrubs or semishrubs of *Ambrosia*, *Encelia*, *Calliandra*, *Xanthocephalum*, and many cacti.

The Mohave and Chihuahuan deserts lack the large number of tall treelike species of the Sonoran but share most of the tall and low shrub genera. Tall and low scrubland vegetation covers the bajades and mesas of the northwestern (Sonoran-Mohave) and southeastern (Chihuahuan) portions of the arid regions of North America. Annual and perennial forbs (nongrass herbs) are also common summer components of the vegetation. Where the arid environments merge with semiarid environments, transitional (ecotone) vegetation is found. Where the ecotone is between grassland and xeric scrubland, the vegetation is sometimes referred to as "desert grassland" or "shrub steppe" (West, 1983h; Dick-Peddie, 1989). These areas contain scattered grasses, most commonly in the genera *Aristida*, *Sporobolus*, and *Bouteloua*. The shrubs are a mix of those found in arid zones, such as *Larrea*, *Yucca*, and *Ziziphus*, and those of lower montane scrublands, such as *Agave*, *Dasylerion*, *Berberis*, and *Nolina*. It is difficult to determine how much of the vegetation of the semiarid regions of North America is "natural" or "pristine" or successional disturbance types resulting from grazing by domestic livestock (MacMahon and Wagner, 1985; Dick-Peddie, 1989). The desert grassland type has been and apparently still is increasing in area in North America.

MacMahon and Wagner (1985) included riparian vegetation in an azonal vegetation category. Riparian vegetation of the arid regions of North America is dominated by shrubs in many of the same genera as in riparian areas of other continents. Examples are *Prosopis*, *Atriplex*, *Lycium*, *Acacia*, *Prunus*, and *Celtis*. Other riparian dominants are in the genera *Cercidium*, *Bursera*, *Hymenoclea*, *Fallugia*, and *Chrysothamnus*.

There are only two relatively small areas of extremely arid habitats in North America. One is Death Valley, including the Funeral and Panamint Mountains. The other is in southwestern Arizona and southeastern California and down the east coast of Baja California. This extremely arid area was called the Lower Colorado subdivision of the Sonoran Desert by Shreve and Wiggins (1964). These two extremely arid areas support a sparse low scrubland vegetation with only 10-30% cover. Creosote bush and the low shrub *Ambrosia dumosa* dominate the nonriparian areas. Riparian areas are arroyos and closed basins that are usually highly saline. Arroyos have scattered smoke trees (*Dalea spinosa*) and the alkali sinks (closed basins) are dominated by plants in the genera *Pluchea*, *Allenrolfea*, *Suaeda*, *Prosopis*, and *Atriplex*.

E. South America

Patagonia Region

The Patagonian xeric region occupies most of southern Argentina and has no extremely arid areas, only arid and semiarid. Its climate is temperate (Fig. 1E). In the chapter on Patagonia, Soriano (1983) used the title Deserts and Semi-Deserts of Patagonia. He equated arid environments with semideserts, and consequently, in spite of the title, there are no deserts in Patagonia. According to Soriano (1983), virtually all of Patagonia has been subjected to heavy grazing by sheep, and as in North America, it is difficult to ascertain whether the vegetation is pristine (climax).

The semidesert region of Patagonia was divided by Soriano (1983) into three major districts: occidental, central, and San Jorge Gulf. A fourth district, Payunia, is in a northwestern arm of Patagonia and considered by the authors as transitional with semitropical "monte" habitats. As with the hot deserts of North America, the districts of the semideserts of Patagonia are based largely on different assemblages of dominant shrubs.

The occidental district is a narrow central-west to northwest area. The vegetation is mostly steppe, with a cover of 20-60% perennial tussock grasses. The tussock grasses are taller than the scattered shrubs and are mostly *Stipa* and *Poa*. Dominant shrub genera are *Senecio*, *Adesmia*, *Acaena*, and *Mulinum*. Genera containing common subdominant shrubs are *Verbena*, *Berberis*, and *Prosopis*. These shrubs are often in riparianlike sites. Several species of cacti are found in this district.

The central district is by far the largest of the three and occupies the center of Patagonia from the north to the south. The diagnostic dominants of the central district are shrubs in the genus *Nassauvia*. Upland sites (plains, plateaus, and hills) are dominated by shrubs in the genera *Chuquiraga*, *Nassauvia*, *Stipa*, and *Poa*. Cover in this district is less than 35%. Again, as in the occidental district, plants in riparian sites are mostly from *Verbena*, *Berberis*, *Lycium*, and *Prosopis*.

The third district, San Jorge Gulf, is a narrow crescent-shaped zone on the east paralleling the gulf. The geomorphology of this district is plateaus (mesas), with numerous narrow valleys (arroyos) dissecting them. The mesas support "steppe" with grasses from *Festuca*, *Poa*, and *Stipa* and shrubs from the genera *Adesmia*, *Senecio*, and *Nardophyllum*. Slopes are covered with a dense low scrub with a cover of over 60%. The dominant low shrubs are in the genera *Trevoa* and *Colliguaya*. Valley bottoms (arroyos), closed basins, and coastal plains contain riparian plants, many of which are halophytic. Common genera are *Prosopis*, *Atriplex*, *Distichlis*, *Suaeda*, *Lycium*, and *Lepidophyllum*.

Peru-Chile Region

Xeric regions of Peru and Chile contain both arid and extremely arid areas (Fig. 1E). The region forms a narrow north-south band between the ocean and the higher elevations of the western slopes of the Andes. When air over the cold Humbolt current comes ashore, it carries little moisture. Eastern trade winds lose their moisture on the east side of the Andes, so the Peruvian-Chilean xeric region is subjected to both cold air and rain shadow effects (Rauh, 1985). The average annual precipitation is <50 mm.

Even though this region is one of the driest in the world, a frequent phenomenon in winter is a heavy fog from the coast up to an altitude of 500-700 m. It moves inland through valleys for distances of up to 40 km from the coast. As sea winds rise up coastal hills, the fog even condenses to form rain. This consistent winter fog is called *garua*, and the vegetation that develops under it is called *lomas* (Rauh, 1985). Lomas vegetation is found in both extremely arid and arid zones of the Peruvian-Chilean coastal band. In many of the extremely arid portions the genus *Tillandsia* dominates and the vegetation is referred to as *Tillandsia* lomas. Stands of cacti (cacti lomas) are mixed in on heavy sand, many of which are prostrate and definitely fog adapted. The lomas vegetation continues upslope to the upper limit of the garua (fog). This upper limit is found at an elevation of about 400 m on the continental side of the coastal cordillera in central Peru and at an elevation of approximately 1200 m in southern Peru and northern Chile. The regions just above these upper limits of the garus vegetation are extremely arid, and according to Ruah (1985), these "rock and debris deserts" have no vegetation at all, with the exception of some riparian plants along streams and arroyos. This barren zone may reach up to 2000 m in elevation on the Andean slopes and in some places has a width of 1000 m in elevation. The upper boundaries of the rock and debris desert give way to scattered cacti and dwarf shrubs, but the zone is still extremely dry.

The arid regions of coastal Peru and Chile support varied grassless steppe and savanna vegetation. The western slopes of the coastal cordillera in central and southern Peru support what Rauh terms a "wood lomas" vegetation. Wood lomas areas are also dissected by arroyos, and these riparian habitats contain treelike plants in the genera *Acacia, Caesalpinia,* and *Carica.* On slopes, treelike plants are found in the genera *Eugenia, Trichocereus,* and *Capparis.* Wild species of tomatoes and potatoes can be found on stony ground. The top of the coastal ranges, where the garua actually forms rain, may have open woodlands, called lomas forests by Rauh. These trees have hard wood good for fuel. The dry season in the mountains coincides with the garua season on the coastal cordillera, and livestock is driven down to use the lomas vegetation. Rauh (1985) stated that the lomas forests have been greatly reduced from fuel harvest and livestock grazing. Copious rains fall at irregular intervals about every 5-12 years, north of the garua fog zone along the Peruvian coast. In this region "tree deserts" can be found. The barrel-trunked *Ceiba* is found here along with tall candelabra and columinar cacti. A shrub-tree species of *Capparis* is also found here. There are dense stands of mesquite (*Prosopis juliflora*) on sandy-loamy soil.

There is great variation in the riparian vegetation of the arid regions of coastal Peru and Chile. Dominant riparian genera are *Salix, Cordia,* and *Valesia.* Halophytic genera, such as *Salicornia, Sesuvium,* and *Batis,* along with the grasses *Distichlis* and *Sporobolus* are found in brackish habitats of both extremely arid and arid regions.

Monte, Puna, and Chaco Regions

There are a number of arid and semiarid regions in Argentina north of Patagonia. All were described by Mares et al. (1985). They were called the chaci, the puna, and the monte. These regions contain no areas of extremely arid environment (see Fig. 1E).

The puna region is located in northwestern Argentina and southern Bolivia at elevations of 3400-4500 m. Even though the puna is located at 25° south, the altitude results in a temperate (cold) arid vegetation similar to that of Patagonia. Such dominant shrub genera as *Adesma*, *Chuquiraga*, *Fabiana*, *Lepidophyllum*, *Mulinum*, and *Nardophyllum* are shared with the low shrub areas of Patagonian arid areas.

Mares et al. (1985) indicated that the monte region of Argentina is similar in the structural and species diversities of its vegetation to that of the Sonoran arid region of North America. The dominant forms of the monte appear to have originated in the semiarid chaco in the same fashion as in North America, where the Sinaloan region supplied major elements to the Sonoran region. The steep slopes above upper bajades are dominated by large cacti of the *Cereus* and *Trichocereus* genera; species of *Acacia*; and ground bromeliads (*Deuterocochina* and *Dyckia*). Downslope this vegetation grades into vegetation dominated by shrubs in the genera *Larrea* and *Bulnesia*. Riparian habitats common in the monte region are dominated by plants in the genera *Salix*, *Celtis*, *Prosopis*, *Cercidium*, *Montiera*, and *Larrea* (the species *L. divaricata* is riparian).

The vegetation of the semiarid regions surrounding the monte often exhibit a subtle gradation with monte vegetation. These gradual transitions are due in part to heavy use by humans and their animals (Mares et al., 1985). The monte area appears to have been expanded at the expense of the semiarid chaco's xerophyllous forest (thorn forest) and espinal (xerophyllous woodland). This is another example of desertification.

Xerophyllous forests and espinal areas are found on the Argentina semiarid plains, east of the monte. The xerophyllous forest area can be considered an ecotone between the monte and the chaco to the north, even though the monte and chaco have direct contact in La Rioja Province. *Acacia*, *Prosopis*, and *Cassia* genera contribute most of the dominant species of the xerophyllous forest. The espinal area lies east of the southern portion of the monte. It is dominated by species of *Acacia* and *Prosopis*. In some places there are pure stands of *Prosopis caldenia*. These are more savanna-like than forests with various species of *Stipa* as dominant grass cover.

The semiarid chaco area extends from northeastern Argentina into northwestern Paraguay and southern Bolivia. The open xerophyllous forest is dominated by trees in the genera *Schinopsis* and *Aspidosperma*, often with *Valesia* shrubs in the understory. Xeric sites are dominated by cacti in the genera *Cereus*, *Cleistocactus*, and *Opuntia*. Areas impacted by domestic animals support a diverse forb flora similar to the desert grassland areas in the southwestern United States. Actually, many dominant forbs of the two regions are from the same genera, such as *Verbesina*, *Zinnia*, *Solanum*, *Physalis*, *Gomphrena*, and *Beloperone*.

V. SUMMARY

Xeric lands of the world and their vegetation have been differentiated and classified very differently over the years. This has resulted in a plethora of terms for vegetation types (Table 2). It has also resulted in varied definitions and limits for terms used for xeric lands. Because of these varied uses and because this volume deals with world soils of xeric lands, I considered Meigs's classification of xeric holoclimates (1953) as the most appropriate to use for the organization of this overview. Meigs divided the world's xeric holoclimates into semiarid, arid, and extremely arid regions.

Most workers in Europe, Asia, and South America consider true deserts to be located in extremely arid regions and semideserts to be found in arid regions. With this in view, there are no true deserts in Australia. Death Valley and the southwestern portion of the Sonoran Desert from southwestern Arizona and southeastern California into northern Baja California, Mexico, are the only true desert areas in North America (Fig. 1A-B). However, in this overview I have attempted to use "desert" on a continent as it has usually been used by scientists on that continent.

Major vegetation types, along with their dominant species, have been described by continents for the semiarid, arid, and extremely arid regions of the world. Vegetation along drainages, around oases, and over shallow water tables is similar around the world because available moisture is greater than on surrounding areas, and, in most cases, species have evolved a tolerance for relatively high alkaline and/or saline conditions. This "riparian" vegetation has been described separately for each area.

Human activities and those of domesticated animals have been responsible for shifting the vegetation of many areas from that found on semiarid regions to that found on arid regions. Examples are the Sahel region of northern Africa, desert grasslands and some of the shrub steppes of North America, much of the monte of Argentina, the Thar region of Pakistan and India, and much of the native grassland areas in Australia. In most areas of the world, these transition zones of increasing aridity have shifted or are shifting from slightly arid to semiarid or from semiarid to arid. The vegetation of these areas has gone or is going from woodlands, grasslands, or grass savannas to shrub steppes, shrub savannas, or desert scrublands. These human-caused rather than climate-caused shifts (desertification) are currently of great concern and interest around the world.

REFERENCES

Abd el Rahman, A. A. (1986). The deserts of the Arabian Peninsula. In *Ecosystems of the World*, vol. 12B, *Hot Deserts and Arid Shrublands*, M. Evenari, I. Noy-Meir, and D. W. Goodall (Eds.). Elsevier Science Publishers, Amsterdam, pp. 29-54.

Ayyad, M. A., and Ghabbour, S. I. (1986). Hot deserts of Egypt and the Sudan. In *Ecosystems of the World*, vol. 12B, *Hot Deserts and Arid Shrublands*, M. Evenari, I. Noy-Meir, and D. W. Goodall (Eds.). Elsevier Science Publishers, Amsterdam, pp. 149-202.

Bourliere, F. (Ed.) (1983). *Ecosystems of the World*, vol. 13, *Tropical Savannas*. Elsevier Science Publishers, Amsterdam.

Breckle, S.-W. (1983). The temperate deserts and semi-deserts of Afghanistan and Iran. In *Ecosystems of the World*, vol. 5, *Temperate Deserts and Semi-Deserts*, N. E. West (Ed.). Elsevier Science Publishers, Amsterdam, pp. 271-319.

Cloudsley-Thompson, J. L., and Chadwick, M. J. (1964). *Life in Deserts.* G. T. Foulis and Co., London.

Di Castri, F., Goodall, D. W., and Specht, R. L. (Eds.) (1981). *Ecosystems of the World*, vol. 11, *Mediterranean-Type Shrublands.* Elsevier Science Publishers, Amsterdam.

Dick-Peddie, W. A. (1989). *Vegetation of New Mexico: Past, Present, and Future.* University of New Mexico Press, Alburquerque.

Evenari, M. (1985). The desert environment. In *Ecosystems of the World*, vol. 12A, *Hot Deserts and Arid Shrublands*, M. Evenari and D. W. Goodall (Eds.). Elsevier Science Publishers, Amsterdam, pp. 1-22.

Evenari, M., Noy-Meir, I., and Goodall, D. W. (Eds.) (1985). *Ecosystems of the World*, vol. 12B, *Hot Deserts and Arid Shrublands, A.* Elsevier Science Publishers, Amsterdam.

Evenari, M., Noy-Meir, I., and Goodall, D. W. (Eds.) (1986). *Ecosystems of the World*, vol. 12B, *Hot Deserts and Arid Shrublands, B.* Elsevier Science Publishers, Amsterdam.

Gupta, R. K. (1986). The Thar Desert. In *Ecosystems of the World*, vol. 12B, *Hot Deserts and Arid Shrublands*, M. Evenari, I. Noy-Meir, and D. W. Goodall (Eds.). Elsevier Science Publishers, Amsterdam, pp. 55-99.

Hilbig, W. (1983). The deserts of central Asia. In *Ecosystems of the World*, vol. 5, *Temperate Deserts and Semi-Deserts*, N. E. West (Ed.). Elsevier Science Publishers, Amsterdam, pp. 197-207.

Le Houerou, H. N. (1986). The desert and arid zones of northern Africa. In *Ecosystems of the World*, vol. 12B, *Hot Deserts and Arid Shrublands*, M. Evenari, I. Noy-Meir, and D. W. Goodall (Eds.). Elsevier Science Publishers, Amsterdam, pp. 101-147.

MacMahon, J. A., and Wagner, F. H. (1985). The Mohave, Sonoran, and Chihuahuan deserts of North America. In *Ecosystems of the World*, vol. 12A, *Hot Deserts and Arid Shrublands*, M. Evenari, I. Noy-Meir, and D. W. Goodall (Eds.). Elsevier Science Publishers, Amsterdam, pp. 105-202.

Mares, M. A., Morello, J., and Goldstein, G. (1985). The Monte desert and other subtropical semi-arid biomes of Argentina, with comments on their relations to North America. In *Ecosystems of the World, Hot Deserts and Arid Shrublands*, M. Evenari, I. Noy-Meir, and D. W. Goodall (Eds.). Elsevier Science Publishers, Amsterdam, pp. 203-237.

Meigs, P. (1953). World distribution of arid and semi-arid homoclimates. In *Review of Research of Arid Zone Hydrology*, Arid Zone Programme, 1. UNESCO, Paris, pp. 203-209.

Monod, T. (1986). The Sahel zone north of the equator. In *Ecosystems of the World*, vol. 12B, *Hot Deserts and Arid Shrublands*, M. Evenari, I. Noy-Meir, and D. W. Goodall (Eds.). Elsevier Science Publishers, Amsterdam, pp. 203-243.

Orshan, G. (1986). The deserts of the Middle East. In *Ecosystems of the World*, vol. 12B, *Hot Deserts and Arid Shrublands*, M. Evenari, I. Noy-Meir, and D. W. Goodall (Eds.). Elsevier Science Publishers, Amsterdam, pp. 1-28.

Petrov, M. P. (1976). *Deserts of the World.* Wiley and Sons, Toronto, Ontario.

Rauh, W. (1985). The Peruvian-Chilean desert. In *Ecosystems of the World*, vol. 12A, *Hot Deserts and Arid Shrublands*, M. Evenari, I. Noy-Meir, and D. W. Goodall (Eds.). Elsevier Science Publishers, Amsterdam, pp. 239-267.

Rauh, W. (1986). The arid region of Madagascar. In *Ecosystems of the World*, vol. 12B, *Hot Deserts and Arid Shrublands*, M. Evenari, I. Noy-Meir, and D. W. Goodall (Eds.). Elsevier Science Publishers, Amsterdam, pp. 361-377.

Shmida, A. (1985). Biogeography of the desert flora. In *Ecosystems of the World*, vol. 12A, *Hot Deserts and Arid Shrublands*, M. Evenari, I. Noy-Meir, and D. W. Goodall (Eds.). Elsevier Science Publishers, Amsterdam, pp. 23-77.

Shreve, F., and Wiggens, I. L. (1964). *Vegetation and Flora of the Sonoran Desert*, 2 vols. Stanford University Press, Stanford, California.

Soriano, A. (1983). Deserts and semi-deserts of Patagonia. In *Ecosystems of the World*, vol. 5, *Temperate Deserts and Semi-Deserts*, N. E. West (Ed.). Elsevier Science Publishers, Amsterdam, pp. 423-460.

Specht, R. L. (Ed.) (1979). *Ecosystems of the World*, vol. 9A, *Heathlands and Related Shrublands: Descriptive Studies*. Elsevier Science Publishers, Amsterdam.

Specht, R. L. (1981). Heathlands. In *Australian Vegetation*, R. H. Groves (Ed.). Cambridge University Press, London.

Thornthwaite, C. W. (1948). An approach toward a rational classification of climate. *Geogr. Rev.*, *38*: 55-94.

Walter, H. (1973). *Vegetation of the Earth in Relation to Climate and the Ecophysiological Conditions*, Springer, Heidelberg.

Walter, H. (1986). The Namib desert. In *Ecosystems of the World*, vol. 12B, *Hot Deserts and Arid Shrublands*, M. Evenari, I. Noy-Meir, and D. W. Goodall (Eds.). Elsevier Science Publishers, Amsterdam, pp. 245-282.

Walter, H., and Box, E. O. (1983a). Caspian lowland biome. In *Ecosystems of the World*, vol. 12A, *Hot Deserts and Arid Shrublands*, M. Evenari, I. Noy-Meir, and D. W. Goodall (Eds.). Elsevier Science Publishers, Amsterdam, pp. 9-41.

Walter, H., and Box, E. O. (1983b). Semi-deserts and deserts of central Kazakhstan. In *Ecosystems of the World*, vol. 12A, *Hot Deserts and Arid Shrublands*, M. Evenari, I. Noy-Meir, and D. W. Goodall (Eds.). Elsevier Science Publishers, Amsterdam, pp. 43-78.

Walter, H., and Box, E. O. (1983c). Middle Asian deserts. In *Ecosystems of the World*, vol. 12A, *Hot Deserts and Arid Shrublands*, M. Evenari, I. Noy-Meir, and D. W. Goodall (Eds.). Elsevier Science Publishers, Amsterdam, pp. 79-104.

Walter, H., and Box, E. O. (1983d). The orobiomes of middle Asia. In *Ecosystems of the World*, vol. 12A, *Hot Deserts and Arid Shrublands*, M. Evenari, I. Noy-Meir, and D. W. Goodall (Eds.). Elsevier Science Publishers, Amsterdam, pp. 161-191.

Walter, H., and Box, E. O. (1983e). The deserts of central Asia. In *Ecosystems of the World*, vol. 12A, *Hot Deserts and Arid Shrublands*, M. Evenari, I. Noy-Meir, and D. W. Goodall (Eds.). Elsevier Science Publishers, Amsterdam, pp. 193-236.

Werger, M. J. A. (1986). The Karoo and southern Kalahari. In *Ecosystems of the World*, vol. 12B, *Hot Deserts and Arid Shrublands*, M. Evenari, I. Noy-Meir, and D. W. Goodall (Eds.). Elsevier Science Publishers, Amsterdam, pp. 283-359.

West, N. E. (Ed.) (1983a). *Ecosystems of the World*, vol. 5, *Temperate Deserts and Semi-Deserts*. Elsevier Science Publishers, Amsterdam.

West, N. E. (1983b). Overview of North American temperate deserts and semi-deserts. In *Ecosystems of the World*, vol. 5, *Temperate Deserts and Semi-Deserts*, N. E. West (Ed.). Elsevier Science Publishers, Amsterdam, pp. 321-330.

West, N. E. (1983c). Great Basin-Colorado Plateau sagebrush semi-desert. In *Ecosystems of the World*, vol. 5, *Temperate Deserts and Semi-Deserts*, N. E. West (Ed.). Elsevier Science Publishers, Amsterdam, pp. 331-349.

West, N. E. (1983d). Western intermountain sagebrush steppe. In *Ecosystems of the World*, vol. 5, *Temperate Deserts and Semi-Deserts*, N. E. West (Ed.). Elsevier Science Publishers, Amsterdam, pp. 351-374.

West, N. E. (1983e). Intermountain salt-desert shrubland. In *Ecosystems of the World*, vol. 5, *Temperate Deserts and Semi-Deserts*, N. E. West (Ed.). Elsevier Science Publishers, Amsterdam, pp. 375-397.

West, N. E. (1983f). Colorado Plateau-Mohavian blackbrush semi-desert. In *Ecosystems of the World*, vol. 5, *Temperate Deserts and Semi-Deserts*, N. E. West (Ed.). Elsevier Science Publishers, Amsterdam, pp. 399-411.

West, N. E. (1983g). Southeastern Utah galleta-threeawn shrub steppe. In *Ecosystems of the World*, vol. 5, *Temperate Deserts and Semi-Deserts*, N. E. West (Ed.). Elsevier Science Publishers, Amsterdam, pp. 413-421.

West, N. E. (1983h). Comparisons and contrasts between the temperate deserts and semi-deserts of three continents. In *Ecosystems of the World*, vol. 5, *Temperate Deserts and Semi-Deserts*, N. E. West (Ed.). Elsevier Science Publishers, Amsterdam, pp. 461-472.

Williams, O. B., and Calaby, J. H. (1985). The hot deserts of Australia. In *Ecosystems of the World*, vol. 12A, *Hot Deserts and Arid Shrublands*, M. Evenari, I. Noy-Meir, and D. W. Goodall (Eds.). Elsevier Science Publishers, Amsterdam, pp. 269-312.

2

Genesis and Classification of Arid Region Soils

DAVID M. HENDRICKS *University of Arizona, Tucson, Arizona*

I. INTRODUCTION

Soils of arid regions possess many characteristics that distinguish them from soils of humid regions. They commonly have low biological activity and a low level of organic matter, are usually slightly acid to alkaline in reaction (pH) in the surface, and commonly contain accumulations of calcium carbonate and sometimes accumulations of soluble salts, gypsum, or free silica. Some arid soils may have properties that are also present in humid soils, such as shallow subsurface zones of clay accumulation. Arid soils show considerable variation in age, ranging from very recent (late Holocene) to early Pleistocene or older. In contrast to earlier concepts of arid soils being a rather homogeneous group, it is now recognized that they show considerable morphological variability (Jackson, 1957; Nettleton and Peterson, 1983). Modern systems of soil classification generally recognize this variability. This chapter describes the major soil classification systems that include a comprehensive classification of arid and semiarid soils. Emphasis is placed on the U.S. soil taxonomy (Soil Survey Staff, 1975). The processes leading to the formation of important features of arid and semiarid soils are also addressed.

Systematic systems of soil classification have been developed to arrange soils in various classes (taxa) that can be easily remembered. Soils can then be compared, and knowledge and experience of soils in one region can be applied elsewhere to soils with similar properties and environmental conditions. Basic to any soil survey program is a soil classification system. To date no common classification and nomenclature of soils have been adopted worldwide. Historically the classification of soils has been a controversial subject, both nationally and internationally, because soil scientists often do not agree on the soil properties to be used to distinguish and to classify soils. Consequently, definitions and nomenclature are different. Different classifications of soils worked out by national schools of soil scientists are used in particular countries or groups of countries. Classification systems vary with regard to their treatment of arid region soils. Obviously, if a country has substantial areas of arid lands its classification system is apt to have a well-developed classification of arid soils. Conversely, a country lacking arid lands is unlikely to classify arid soils unless it has foreign interests in arid regions.

II. SOIL CLASSIFICATION SYSTEMS

A. U.S. Soil Taxonomy

The U.S. Soil Taxonomy (Soil Survey Staff, 1975) is probably the most widely used and/or referred to soil classification in use today. This system was developed over a period of 15 years by the Soil Survey Staff of the Soil Conservation Service in the United States as a comprehensive classification of all soils of the world. The system was developed with the consultation of soil scientists from many nations. At the present time many aspects of the system are being reviewed by several international committees with proposed revisions being developed. The U.S. Soil Taxonomy is a "morphometric" system in that all the properties used to characterize soils can be measured in the field or in the laboratory. Soils are classified according to the presence or absence of and/or the measured quantities of such properties.

Diagnostic Horizons

Certain soil horizons, referred to as diagnostic horizons, are the primary building blocks of the U.S. Soil Taxonomy (Soil Survey Staff, 1975). The diagnostic horizons that are commonly found in arid and semiarid land soils are listed in Table 1.

Epipedons are diagnostic horizons that have formed at the surface and occur nowhere else in the soil, unless the soil is buried under fluvial, eolian, or volcanic deposits. They are defined mostly in terms of soil color, content of organic matter, and base saturation. Of the epipedons currently recognized, only two are widespread in arid and semiarid regions: the mollic and the ochric. The mollic epipedon is generally darker and higher in organic matter than the ochric epipedon. The other two epipedons listed in Table 1 are of limited occurrence in arid soils. The anthropic epipedon is similar to the mollic epipedon except that it is higher in extractable P, presumably due to human activity. The proposed (Osman, 1989) vermic epipedon is characterized by intense biological (faunal) activity or properties reflecting these activities.

The remaining diagnostic horizons form below the soil surface, although they may become exposed if surface horizons are removed. For the most part, subsurface diagnostic horizons develop from materials leached and accumulated from upper horizons, or they exhibit special features that are useful in differentiating soils. Argillic horizons, for example, are accumulations of clay derived from overlying horizons. Cambic horizons, on the other hand, are only slightly altered by soil-forming processes from the parent material. Of the subsurface horizons listed in Table 1, the hypercalcic and hypergypsic have recently been proposed by the International Committee on Aridisols (Osman, 1989).

TABLE 1 Diagnostic Horizons Common in Arid and Semiarid Land Soils

Diagnostic horizons	Characteristics
Surface horizons	
Mollic epipedon	Surface horizon of accumulation of organic matter; dark colored
Ochric epipedon	Surface horizon of limited accumulation of organic matter; light colored
Anthropic epipedon	Surface horizon of accumulation of organic matter; dark colored; relatively high in citric acid-soluble P
Vermic epipedon[a]	Surface horizon dominated by wormholes, wormcasts, or filled animal burrows
Subsurface horizons	
Argillic	Horizon of clay accumulation
Natric	Horizon of clay and sodium accumulation
Cambic	Horizon of pedogenic alteration usually expressed by soil structure or removal of calcium carbonate
Calcic	Horizon of moderate to high levels of carbonate accumulation
Hypercalcic[a]	Horizon of very high levels of carbonate accumulation
Gypsic	Horizon of moderate to high levels of gypsum accumulation
Hypergypsic[a]	Horizon of very high levels of gypsum accumulation
Salic	Horizon of pronounced soluble salt accumulation
Petrocalcic	Hard indurated horizon with calcium carbonate as the dominant cementing material
Petrogypsic	Hard indurated horizon with gypsum as the dominant cementing material
Duripan	Hard indurated horizon with silica as the dominant cementing material
Albic	Light-colored horizon in which clay and organic matter have been significantly removed by leaching
Oxic	Horizon having a residual concentration of kaolinitic clay and free sesquioxides with very low cation-exchange capacity (relict feature of arid soils)

[a]Proposed diagnostic horizons (Osman, 1989).

Soil Moisture and Temperature Regimes

Soil moisture and soil temperature regimes are also important diagnostic features used to classify soils. The soil moisture regimes described in the U.S. Soil Taxonomy (Soil Survey Staff, 1975) are based on the time and season that the soil moisture control section is saturated, moist, or dry. The definitions of these soil moisture regimes were summarized by Buol et al. (1980) as follows:

Aquic moisture regime: These soils are water saturated for at least enough time, usually several days, so that anaerobic conditions exist. Soil mottling normally is indicative of this condition.

Aridic or torric moisture regime: These soils are both dry more than half the time when not frozen and never moist more than 90 consecutive days when soil temperatures are above 8°C at 50 cm depth.

Udic moisture regime: In most years these soils are not dry for as long as 90 cumulative days.

Ustic moisture regime: In most years these soils are dry for more than 90 cumulative days but less than 180 days. In temperate climates, they are usually moist at least 45 consecutive days in the 4 months after the winter solstice and not dry 45 consecutive days in the 4 months after the summer solstice.

Xeric moisture regime: These soils are only in temperate areas where summers are dry and winters moist. These soils usually are dry more than 45 consecutive days in the summer and moist more than 45 consecutive days in the winter.

Soils with aridic (torric) moisture regimes are dominant in arid regions; soils with ustic and xeric moisture regimes occur in semiarid regions. Aquic soils may occur in topographically low areas that have poor drainage and limited runoff in arid and semiarid regions.

The soil temperature regimes are characterized by the mean annual soil temperature and by the difference between the mean summer and winter soil temperatures measured at a depth of 50 cm. The soil temperature regimes are given in Table 2.

Categories

The U.S. Soil Taxonomy has six categories. These are from top to bottom order, suborder, great group, subgroup, family, and series. The highest categories have the fewest classes and criteria separating classes; the lowest categories have the most classes and criteria.

In the order level there are 10 classes. The criteria used to differentiate orders are highly generalized and based more or less on the kinds and degrees of soil-forming processes. Mostly these criteria include properties that reflect major differences in the genesis of soils. An eleventh soil order, the Andisols, has recently been proposed (Leamy, 1988). Of the 11 soil orders (original 10 plus the Andisols), 8 occur in arid and semiarid regions (see Table 3).

The suborder category is a subdivision of an order within which genetic homogeneity is emphasized. Soil characteristics used to distinguish suborders within an order vary from order to order. For example, soil

TABLE 2 Soil Temperature Regimes[a] of the U.S. Soil Taxonomy

Class	Mean annual soil temperature (°C)	Summer-winter difference (°C)
Pergelic	<0	—
Cryic	0-8	—
Frigid[b]	0-8	>5
Mesic	8-15	>5
Thermic	25-22	>5
Hyperthermic	>22	>5
Isofrigid	0-8	<5
Isomesic	8-15	<5
Isothermic	15-22	<5
Isohyperthermic	>22	<5

[a]Soil temperature at 50 cm.

[b]The frigid regime has higher summer temperatures than the cryic regime.

Source: Soil Survey Staff (1975).

moisture and temperature are the important factors that differentiate the suborders of the Alfisols. The occurrence of different diagnostic horizons separates the suborders of the Aridisols.

The great group category is a subdivision of a suborder. The great group classes are distinguished from one another in most suborders primarily by the kind and sequence of soil horizons. All soils belonging to one of the suborders of Aridisols have argillic horizons. They may also have additional diagnostic horizons, such as a petrocalcic, as well as several others. Soils having these additional horizons are placed in separate great groups. Soil temperature and moisture regimes, however, are employed to differentiate the great groups within suborders of the Entisols.

Great group categories are divided into three kinds of subgroups: typic, intergrade, and extragrade. A typic subgroup represents the basic concept of the great group from which it derives. An intergrade subgroup contains soils of one great group but has some properties characteristic of soils in another great group or other class. These properties are not developed or expressed well enough to include the soils within the great group toward which they grade. Extragrade subgroup soils have aberrant properties that do not intergrade to any known soil.

A soil family contains a group of soils within a subgroup that has similar physical and chemical properties that affect response to management and manipulation. The principal characteristics most widely used to differentiate soil families are texture, mineralogy, and temperature. Other family

TABLE 3 Orders and Their Relations to Arid and Semiarid Lands

Alfisols	Soils with a horizon of clay accumulation of high base saturation outside arid lands and without a mollic epipedon; occur in semiarid lands
Andisols[a]	Soils formed mostly from tephra and related volcanic parent materials and containing volcanic glass and/or short-range-order minerals; occur in arid and semiarid lands
Aridisols	Mineral soils relatively low in organic matter; contain developed soil horizons; extensive in arid lands; occur in some semiarid lands
Entisols	Soils without pedogenic horizons; common in arid lands
Histosols	Organic soils; rare in arid lands
Inceptisols	Soils containing some developed horizons other than one of illuvial clay; occur in semiarid lands
Mollisols	Soils with thick, dark surface horizons relatively high in organic matter and base saturation; occur in semiarid lands
Oxisols	Highly weathered soils with an exchange complex of extremely low activity; may occur in arid and semiarid lands as relicts
Spodosols	Soils with a subsoil horizon of aluminum and organic matter or iron accumulation; not in arid or semiarid lands
Ultisols	Soils with a horizon of clay accumulation of low base saturation; rare in semiarid lands
Vertisols	Fine-textured soils that have deep, wide cracks at some time during the year; present in arid and semiarid lands

[a]Proposed soil order (Leamy, 1988).

criteria that are applied to fewer soils include reaction (pH and carbonates), depth, slope, consistence, moisture equivalent, and cracks.

The soil series is the narrowest category in the U.S. Soil Taxonomy. The soil series have a relatively narrow range of precisely defined properties. The soil series allows specific statements to be made about soils, enhances the value of research, and permits effective application of existing knowledge to use soil resources because information is easily and accurately communicated.

Nomenclature

At first glance, the U.S. Soil Taxonomy appears to be complicated and cumbersome because it utilizes many newly coined words, parts of which are combined to form other words. However, once a person becomes familiar with the construction of the nomenclature, he or she soon recognizes the utility and the advantages of the system.

The nomenclature is designed so that each class has a name that is mnemonic, that is, to help the memory, and can note some properties of the soils of each class. The name also places a class in the system so

TABLE 4 Soil Order Names and Formative Elements

Order	Formative element	Derivation	Mnemonicon
Alfisols	alf	Nonsense syllable	Pedalfer
Andisols	and	Modified from Ando	Ando
Aridisols	id	L. *aridus* , dry	Arid
Entisols	ent	Nonsense syllable	Recent
Histosols	is	Gr. *histos*, tissue	Histology
Inceptisols	ept	L. *inceptum*, beginning	Inception
Mollisols	oll	L. *mollis*, soft	Mollify
Oxisols	ox	Fr. *oxide*, oxide	Oxide
Spodosols	od	Gr. *spodos*, wood ashes	Podzol; odd
Ultisols	ult	L. *ultimus*, last	Ultimate
Vertisols	ert	L. *verto*, turn	Invert

that a person can recognize both the category of the class and the classes of the higher categories to which it belongs.

Class names are coined from Greek and Latin roots, for the most part, that are familiar because of their use in many common words. Order names end in "sols" (Table 3). A formative element from each order is the ending of names of the suborders, great groups, and subgroups. The formative elements of orders are listed in Table 4.

Names of suborders have two syllables. The last syllable is the formative element from the order. The first syllable is the suborder formative element that suggests certain characteristics about the soil. A few examples and meanings of suborder formative elements are given in Table 5. One suborder is the Argids. These are Aridisols, dry soils, with argillic horizons (arg + id = Argids).

Continuing in the same manner, great group names are formed simply by placing a great group formative element in front of the suborder name. The prefix is connotative and suggests the unique features of the particular great group named. The great group Durargids, for example, includes Aridisols with argillic horizons and duripans (dur + Argids = Durargids).

Names of subgroups are mostly binomial. Each binomial has an adjective before the name of the great group to which the subgroup belongs. The term "typic" is added to a great group name to form the subgroup name for soils that are typical or modal of the great group. Thus, Typic Durargids is the name of the modal subgroup of the Durargids great group. An intergrade subgroup also carries the name of its great group but is modified by the adjectival form of the name of the class or classes toward which it grades. The adjective is formed by adding "ic" to the class name. For example, Durargids that have some properties of the Haploxerolls great group belong to the Haploxerollic Durargids subgroup. Extragrade subgroup names are formed by adding to the great group name a formative

TABLE 5 Examples of Formative Elements of Suborder[a] Names

Formative element	Meaning
aqu	Aquic moisture regime
arg	Argillic horizon
ochr	Light-colored surface (ochric epipedon)
orth	Most representative
psamm	Sand or loamy sand texture
torr	Aridic (torric) moisture regime
ust	Ustic moisture regime
xer	Xeric moisture regime

[a]Formative elements may be used in more than one order.

element that describes the extragradational feature. Abruptic Durargids, for example, include soils that have properties of Durargids but have in addition an abrupt transition between the A and B horizons characterized by a large difference in the percentage of clay.

Family names are fairly common terms that denote in more detail the features of a particular subgroup. A fine, montmorillonitic thermic soil family is fine textured, contains montmorillonite as the dominant mineral in the clay, and has a warm or thermic mean annual soil temperature.

Soil series are abstract names that in the United States are sometimes taken from some local geographic feature near the site where the series was first established. The Mingus series, for example, is named for Mingus Mountain, Arizona. Other names are variants of place names, and still others are simply coined. Each soil series in the United States is classified with the U.S. Soil Taxonomy nomenclature. Thus, the Mingus series is placed in the clayey, montmorillonitic, thermic Abruptic Durargids family.

Suborders and Great Groups of Arid and Semiarid Lands

The suborders and great groups that are common or have been identified in arid and semiarid lands are listed in Table 5 together with their important diagnostic properties. More detailed descriptions and definitions of these properties are described by the Soil Survey Staff (1975, 1987), Leamy (1988), Comerma (1989), and Osman (1989). Dregne (1976) describes the geographic distribution as well as the properties of the various classes occurring in arid and semiarid regions. Great groups of the listed suborders that do not occur or are rare in semiarid and arid regions are not included in Table 6.

Alfisols

Alfisols are characterized by prominent horizons of eluviation and illuviation in which clay translocation is the dominant soil-forming process (Rust, 1983). Alfisols occur mostly in a humid to subhumid climate in which the soil dries out to some extent during the year. Alfisols also

TABLE 6 Name and Important Properties of Suborders and Great Groups with Soils Common in or Identified in Arid and Semiarid Lands

Order	Suborder	Great group	Important properties
Alfisols	Ulstalfs		Ustic moisture regime
		Durustalfs	Duripan
		Natrustalfs	Natric horizon
		Paleustalfs	Petrocalcic horizon or very strongly developed argillic horizon
		Haplustalfs	Lack features of other Ustalfs great groups
	Xeralfs		Xeric moisture regime
		Durixeralfs	Duripan
		Natrixeralfs	Natric horizon
		Rhodoxeralfs	Red-colored argillic horizon
		Palexeralfs	Petrocalcic horizon or very strongly developed argillic horizon
		Haploxeralfs	Lack features of other Xeralfs great groups
Andisols[a]	Ustands		Ustic moisture regime
		Durustands	Duripan
		Haplustands	Lack duripan
	Vitrands		High in unweathered volcanic material
		Ustivitrands	Ustic moisture regime
	Xerands		Xeric moisture regime
		Vitrixerands	High in unweathered volcanic materials
		Haploxerands	Less unweathered volcanic materials
	Torrands		Aridic (torric) moisture regime
		No great groups	
Aridisols[b]	Argids		Natric or argillic horizon
		Petroargids	Petrocalcic or petrogypsic horizon
		Natrargids	Natric horizon
		Paleargids	Very strongly developed argillic horizon
		Gypsiargids	Gypsic horizon
		Calciargids	Calcic horizon
		Haplargids	Lack features of other Argids great group

TABLE 6 (continued)

Order	Suborder	Great group	Important properties
	Calcids		Calcic or hypercalcic or petrocalcic within 100 cm of soil surface
		Petrocalcids	Petrocalcic horizon
		Hypercalcids	Hypercalcic horizon
		Haplocalcids	Calcic horizon
	Cambids		Cambic horizon
		Aquicambids	Saturated with water during the growing season for at least 1 month
		Petrocambids	Petrocalcic or petrogypsic horizon or duripan between 100 and 150 cm depth
		Anthrocambids	Anthropic or vermic epipedon
		Haplocambids	Lack features of other Cambids great groups
	Durids		Duripan
		Natridurids	Natric horizon above the duripan
		Argidurids	Argillic horizon above the duripan
		Haplodurids	Lack features of other Durids great groups
	Gypsids		Gypsic or hypergypsic or petrogypsic horizon within 100 cm of soil surface
		Petrogypsids	Petrogypsic horizon
		Hypergypsids	Hypergypsic horizon
		Natrigypsids	Natric horizon
		Argigypsids	Argillic horizon
		Calcigypsids	Calcic or hypercalcic horizon
		Haplogypsids	Lack features of other Gypsid great groups
	Salids		Salic horizon
		Aquisalids	Saturated with water for at least 1 month in a year
		Haplosalids	Lack features of the other Salids great group
Entisols	Psamments		Sand or loamy sand texture
		Torripsamments	Aridic (torric) moisture regime
		Quartzipsamments	More than 90% silica and other resistant minerals
		Xeropsamments	Xeric moisture regime

TABLE 6 (continued)

		Ustipsamments	Ustic moisture regime
	Fluvents		Soils in young floodplain and related deposits
		Xerofluvents	Xeric moisture regime
		Ustifluvents	Ustic moisture regime
		Torrifluvents	Aridic (torric) moisture regime
	Arents		Fragments of diagnostic horizons between 25 and 100 cm depth
		No great groups	
	Orthents		Lack features of the other Entisol suborders
		Torriorthents	Arid (torric) moisture regime
		Xerothents	Xeric moisture regime
		Ustorthents	Ustic moisture regime
Inceptisols	Ochrepts		Ochric epipedon
		Durochrepts	Duripan
		Ustochrepts	Ustic moisture regime
		Xerochrept	Xeric moisture regime
Mollisols	Xerolls		Xeric or aridic moisture regime bordering on xeric
		Durixerolls	Duripan
		Natrixerolls	Natric horizon
		Palexerolls	Petrocalcic or a very strongly developed argillic horizon
		Calcixerolls	Calcic or gypsic horizon
		Argixerolls	Argillic horizon
		Haploxerolls	Lack features of other Xerolls great groups
	Ustolls		Ustic or aridic bordering on ustic moisture regime
		Durustolls	Duripan
		Natrustolls	Natric horizon
		Paleustolls	Petrocalcic or very strongly developed argillic horizon
		Calciustolls	Calcic horizon
		Argiustolls	Argillic horizon
		Vermustolls	Vermic epipedon
		Haplustolls	Lack features of other Ustolls great groups

TABLE 6 (continued)

Order	Suborder	Great group	Important properties
Oxisols	Ustox		Ustic soil moisture regime
		Acrustox	Very low cation-exchange capacity
		Eutrustox	Moderate to high base saturation
		Haplustox	Lack features of other Ustox great groups
	Torrox		Aridic (torric) moisture regime
		Acrotorrox	Very low cation-exchange capacity
		Eutrotorrox	Moderate to high base saturation
		Kandiustox	Clay high in kaolinite
		Haplotorrox	Lack features of other Torrox great groups
Vertisols[c]	Xererts		Xeric moisture regime
		Durixererts	Duripan
		Calcixererts	Calcic or petrocalcic horizon
		Haploxererts	Lack features of other Xererts great groups
	Torrerts		Aridic (torric) moisture regime
		Salitorrerts	Salic horizon
		Calcitorrerts	Calcic or petrocalcic horizon
		Gypsitorrerts	Gypsic horizon
		Haplotorrerts	Lack features of other Terrerts great groups
	Usterts		Ustic moisture regime
		Salusterts	Salic horizon
		Calciusterts	Calcic or petrocalcic horizon
		Gypsiusterts	Gypsic horizon
		Haplusterts	Lack features of other Usterts great groups

[a]The andisols represent a new proposed order. The suborders and great groups listed are described by Leamy (1988).

[b]The Aridisols suborders and great groups listed are a proposed modification (Osman, 1989).

[c]The Vertisols suborders and great groups listed are a proposed modification (Comerma, 1989).

occur in drier regions, where they grade toward the Aridisols. The suborders of the Alfisols are separated on the basis of soil temperature and soil moisture regimes. The Alfisols in semiarid regions include the Ustalfs (Alfisols with ustic moisture regimes) and Xeralfs (Alfisols with xeric moisture regimes). The great groups of both suborders include soils with additional (to the argillic horizon) diagnostic horizons, such as duripans, natric horizons, and petrocalcic horizons.

Andisols

Andisols include soils that have developed in volcanic ejecta (such as volcanic ash, pumice, cinders, and lava) and/or in volcaniclastic materials whose colloidal fraction is dominated by short-range-order (sometimes referred to as amorphous) minerals or Al-humus complexes (Leamy, 1988). Short range order is characteristic of x-ray diffraction analysis of minerals lacking distinct x-ray peaks. Evidence of short range order can, however, be detected by other diagnostic techniques. The dominant process in most Andisols is weathering and mineral transformation. Translocation within the soil and accumulation of the translocated components is generally minimal. Weathering has not proceeded so far so that the soils are dominated by glass and/or short-range-order minerals. Andisols have been identified mostly with humid climates but are also thought to include soils formed in semiarid and arid regions (Flach et al., 1980; Leamy et al., 1980). The suborders of the Andisols are separated on the basis of soil moisture and temperature regimes and on the dominance of volcaniclastic materials. Some Xerands, Ustands, and Vitrands occur in semiarid regions, and Torrands are in arid regions. Two great groups of Ustands are recognized—Durustands and Haplustands—based on the presence or absence of a duripan. Of the two great groups of Vitrands, Ustivitrands, and Udivitrands, only Ustivitrands are found in semiarid regions. The Vitrixerands and Haploxerands are the great groups of Xerands.

Aridisols

Aridisols are soils of dry regions, mostly deserts, and have one or more distinctive pedogenic horizons that have been either enriched or depleted of substances by moving water (Nettleton and Peterson, 1983). Aridisols have at least one or more of the following horizons: cambic; argillic or natric; calcic, petrocalcic, or hypercalcic; gypsic, petrogypsic, or hypergypsic; salic; and duripan. Two suborders of the Aridisols are recognized in the U.S. Soil Taxonomy (Soil Survey Staff, 1975, 1987), the Argids and the Orthids. The Argids contain but the Orthids lack an argillic horizon. However, the classes of the Aridisol order are being revised by the International Committee on Aridisols (ICOMID). In version 6.0 (Osman, 1989), six suborders are recognized based on the presence of different diagnostic horizons. The proposed revision differs significantly from the original U.S. Soil Taxonomy, resulting from increased information on Aridisols from the Near East and North Africa. The suborders and great groups of the Aridisols as proposed by ICOMID are listed in Table 6. At the great group level, the degree of expression of the horizons of accumulation and/or the results of other processes considered subordinate to the particular suborder are employed. The defining element is the degree of expression of the diagnostic horizon with associated features. The Argids, for example, contain argillic horizons as the most important feature. In addition, Argids may have other horizons that separate the different great groups of the Argids. Argids with a petrocalcic or petrogypsic horizon whose upper boundary is between 100 and 150 cm of the soil surface are placed in the Petroargids great group. If the petrocalcic or petrogypsic horizon upper boundary occurs within 100 cm of the surface the soils are included with the Petrocalcids or Petrogypsids instead.

Entisols

Entisols include soils that have only limited development of pedogenic horizons. The lack of development may result from youthfulness: soils on steep, actively eroding slopes that lose soil material at a rate greater than the rate of soil horizon formation or soils on floodplains and other areas that receive increments of alluvium at frequent intervals. Some Entisols may consist of quartz and other minerals that resist alteration. Still other Entisols lack development because of their extremely wet or dry environment, which is not conducive to horizonation. Soils resulting from human disturbance due to deep mixing as a result of construction or excavation are also considered Entisols. The five suborders of the Entisols are established based on these conditions of occurrence. The Aquents are associated with high water tables and reducing conditions (aquic moisture regimes), the Arents lack soil horizons because of human disturbance, the Psamments include the sandy Entisols, the Fluvents occur in cumulative environments, and the Orthents include all other Entisols. Since the soils in each suborder of the Entisols occur under a wide range of climatic conditions, soil moisture and temperature regimes are the main differentiating criteria in separating the great groups.

Inceptisols

Although Inceptisols are mostly associated with humid regions, they may have any soil moisture regime except aridic. Thus they occur to a limited extent in semiarid climates. Inceptisols are soils that have undergone modifications by soil-forming processes that are sufficiently great to distinguish them from Entisols but not intense enough to form the kinds of horizons that are required for placement into the other orders (Foss et al., 1983). The main soil-forming processes result in altered horizons from translocation or loss of iron, aluminum, or bases. Inceptisols of semiarid regions are restricted to the Ochrepts suborder. The three great groups of Ochrepts of semiarid regions are the Durochrepts, the Ustochrepts, and the Xerochrepts, characterized by having duripans, ustic, and xeric moisture regimes, respectively.

Mollisols

Mollisols occur mainly in semiarid and subhumid grassland areas where the soils have dark-colored, base-rich (especially Ca and Mg) surfaces, that is, mollic epipedons. The two suborders containing soils of semiarid regions are the Xerolls and Ustolls. The great groups of both suborders are distinguished primarily on the presence of certain diagnostic horizons (Table 6).

Oxisols

Soils of the Oxisols order are intensely weathered and mostly occur in tropical and subtropical regions. They generally occur on old land surfaces or in sediments derived from them (Eswaren and Tavernier, 1980). Oxisols include two genetically very distinct sets of soils: those with an oxic horizon formed under well-drained tropical weathering conditions and those with continuous plinthite within 30 cm of the surface formed under conditions characterized by being periodically saturated with water. Plinthite, when it dries, hardens irreversibly to form a hard indurated layer commonly referred to as laterite. Oxisols form in climatic conditions humid enough to favor the removal of soluble weathering products and produce residual concentrations of sesquioxides and kaolinite clays. Oxisols that currently occur in arid and semiarid regions are considered relict and to have formed in the past during periods of humid climatic conditions. Oxisols of arid and semiarid regions are in the Torrox and Ustox suborders. The great groups of both suborders are differentiated on the basis of certain soil chemical properties. The Acrustox and Acrotorrox contain soils with a very low capacity to retain bases. The lower part of these soils commonly has a net positive charge. The Eutrustox and Eutrotorrox are soils that have an appreciable supply of bases and moderate to high base saturation. The Kandiustox have a high clay content near the surface in which the clay is high in kaolinite.

Vertisols

Vertisols include a group of clay-textured soils that are characterized by a high percentage of clay dominated by smectite minerals. The dominance of the smectite group of minerals leads to expansion and shrinkage on wetting and drying. This results in the formation of deep wide cracks during the dry seasons that close during the wet seasons. The soils of the suborders of the Vertisols are differentiated on the basis of the time during the year and the duration of the opening and closing of the cracks, reflecting the soil moisture regimes. Thus the cracks of the Xererts are closed during the winter and open during the summer; those of the Torrerts remain open throughout the year or close for a short period during the year in a warm season. The soils of the Usterts suborder are distinguished by cracks that remain open for 90 cumulative days or more in most years but are closed for at least 60 consecutive days during the warm season. Two great groups of the Usterts (Chromusterts and Pellusterts) and Xererts (Chromoxererts and Pelloxererts) have been distinguished on the basis of color; the "chrom" great groups have a brighter color (higher chroma) than the "pell" great groups (Soil Survey Staff, 1975). No great groups of the Torrerts have been recognized. However, classification of the Vertisols is undergoing modification by the International Committee on Vertisols (ICOMERT) (Comerma, 1989; Comerma et al., 1988). The great groups of Vertisols given in Table 6 are those defined in the latest proposal of ICOMERT (Comerma, 1989). In this proposal each of the suborders is divided into great groups by the presence of certain diagnostic horizons.

B. USSR Soil Classification System

The classification of the soils of the world developed by Rozov and Ivanova (1967) and the classification of the soils of the USSR developed at the V. V. Dokuchaev Institute of Soil Science (1986) are now used in the Soviet

TABLE 7 Classification of the Soils of the USSR: A Portion of the Subtropical World Group

Ecological-genetic (bioclimatic groups and classes)	Genetic orders	Biophysicochemical orders[a]				
		A	B	C	D	E
Soils of semidesert moderately warm and subtropical regions T >10°C	Automorphic	Soil types of these biophysicochemical orders are not represented in this ecological-genetic class			Sierozems; irrigated sierozems	
Precipitation (annual) 120-350 mm; moisture coefficient <0.3	Weakly hydromorphic Alluvial				Alluvial meadow; sierozem zone	
Weathering-siallitic-carbonate with weak clay formation	Semihydromorphic				Meadow sierozem; irrigated meadow-sierozem	
Biological cycle: very intensive nitrogen-silicon and nitrogen-silicon-calcium	Hydromorphic				Waterlogged meadow; irrigated meadow; sierozem zone; alluvial clayey bog of sierozem zone	Meadow solonetzes of sierozem; hydromorphic solonchaks of sierozem zone

[a]Divisions of the biophysicochemical orders based mainly on type of organic material, adsorption complex, and soluble salts.

Source: Rozov and Ivanova, 1967.

Union. These classifications are based on soil-forming processes and environmental factors as well as soil properties, so that classes are not identified primarily by intrinsic soil characters that can be readily observed or measured. The main reference classes are 110 genetic soil types, conceived at roughly the same level of generalization as the great groups of the U.S. Soil Taxonomy. Each soil type is defined by specified bioclimatic and hydrologic features and by certain morphological characteristics of the soil profile.

The classification groups the soil types upward into orders and classes as higher categories and subdivides them downward into subgroups, genera, species, varieties, and subvarieties. Classification of the lower categories is well developed whereas development of the classification of the higher categories is more tentative (Rozov and Ivanova, 1967).

The genetic soil types are arranged in accordance with three coordinate axes as shown in Table 7, which shows the arrangement for one ecological-genetic group. The three axes are as follows:

1. Ecological-genetic or bioclimatic classes based on climatic indices, weathering cycle, and type of biological cycle.
2. Four genetic soil orders based on moisture regime, with automorphic (atmospheric moistening) soils regarded as zonal and three hydromorphic groups.
3. Five biophysicochemical orders based on characteristics of organic matter decomposition, soluble salts, degree of base saturation and cation composition, and the general makeup of the soil profile. Since the definitions of these orders are somewhat long and complicated, they are replaced by the letters A to E in Table 7. For example, the biophysicochemical order designated D is defined as "humate-fulvate calcium-adsorbed Ca^{2+} and, to some extent, Mg^{2+} in the humus horizon; Na^{+} often found below; humate part bound with Ca; readily soluble salts in the lower and middle part of the profile."

The tabular scheme is given in full by Rozov and Ivanova (1967) and by the Soil Survey Staff (1975).

An example from Table 7 is the sierozem type, a soil of subtropical semidesert, moderately warm regions, which has an automorphic soil moisture regime formed with Ca^{2+}-rich organic matter and Ca dominating the exchange complex with a $CaCO_3$-enriched horizon and the soluble salts generally leached out of the soil surface.

The soil types containing soils of arid and semiarid regions are presented in Table 8 with their approximate U.S. Soil Taxonomy equivalents. This list is taken from V. V. Dokuchaev Institute of Soils (1986) publication, where the soil types are described in detail and their subdivision into the lower categories is given.

C. French Soil Classification System

The classification of soils by the French soil scientists Aubert and Duchaufour (1956) was used to compile soil maps of southern Europe and Africa. The official French system published by the Commission de Pédologie et de Cartographic des Sols (CPSC) (1967) was developed from that of Aubert

TABLE 8 Approximate Correlation Between USSR Soil Types (Dokuchaev Institute of Soil Science, 1986) of Arid and Semiarid Regions and the U.S. Soil Taxonomy (Soil Survey Staff, 1975)

USSR soil type	U.S. Soil Taxonomy
Chestnut soils	Ustolls, Argids, Cambids
Meadow chestnut soils	Ustolls
Semidesert brown soils	Cambids, Calcids, Gypsids
Semidesert meadow brown soils	Cambids, Calcids, Gypsids
Desert gray-brown soils	Cambids, Argids, Salids, Gypsids
Desert takyrlike soils	Torrifluvents, Torriorthents
Takyrs	Torriorthents, Salids, Torrerts
Desert sandy soils	Torripsamments
Meadow desert soils	Torrifluvents, Torriorthents
Sierozems	Torriorthents, Cambids, Xerolls
Meadow-sierozem soils	Cambids, Fluvents
Semidesert and desert meadow soils	Torrifluvents, Torriorthents
Irrigated sierozems	Torriorthents, Cambids, Xerolls
Irrigated meadow sierozem soils	Cambids, Xerofluvents
Irrigated brown soils of the semidesert zone	Cambids, Calcids
Irrigated gray-brown soils of the desert zone	Cambids, Argids
Irrigated takyrlike soils of the desert zone	Torrifluvents, Torriorthents
Irrigated meadow desert soils	Torrifluvents, Torriorthents
Irrigated meadow soils of semi-desert and desert	Torrifluvents, Torriorthents
Irrigated bog soils of semidesert and desert	Histosols, Fluvents
Gray-cinnamon brown soils	Ustalfs, Ustolls, Calcids, Gypsids
Meadow gray-cinnamon brown soils	Ustalfs, Ustolls (aquic)
Cinnamon-brown soils	Ustalfs, Ustolls, Ustochrepts
Bog soils of the semideserts and deserts	Histosols, Fluvents
Solods	Natrustolls, Natrargids
Automorphic solonetzes	Natrustolls, Natrargids
Semihydromorphic solonetzes	Natrargids
Automorphic solonchaks	Haplosalids
Hydromorphic solonchaks	Aquisalids

TABLE 8 (continued)

Calcareous sod desertified alluvial soils	Torrifluvents
Calcareous meadow alluvial soils	Torrifluvents, Ustifluvents, Salids

and Duchaufour (1956). The classes of the French system are not defined precisely. At the highest category the 12 classes are distinguished by (1) degree of profile development; (2) nature of alteration products; (3) type of humus form and distribution of organic matter; and (4) features due to gleying or salinity. Table 9 lists the 12 classes and those subclasses containing soils occurring in arid and semiarid regions of the French system (Commission de Pedologie et de Cartographic des Sols, 1967) and their approximate U.S. Soil Taxonomy equivalent.

D. Australian Soil Classification System

Australia and southern Africa differ from other arid regions in being dominated by ancient landscapes that have not been affected by the glaciations that played such a large role in the Northern Hemisphere. Because of the age of the surficial materials, the influence of past humid climates is much more pronounced than elsewhere. Thus soil morphology (presence, absence, degree of development, and arrangement of horizons) has had much more significance than soil genesis (effect of climate and vegetation) in the classification of Australian soils. The basic classification unit in the system of Stace et al. (1968) is the great soil group. The great soil groups are grouped according to their degree of profile development and degree of leaching (Table 10).

E. FAO/UNESCO Soil Map of the World

The soil legend developed as part of the FAO/UNESCO (FAO, 1974) Soil Map of the World is not a systematic classification of soils, but the named mapping units are sometimes used as taxonomic units. The legend includes two categories of which the higher is roughly equivalent to the great group of the U.S. Soil Taxonomy. Classes are defined by the use of diagnostic horizons, many of which are very similar to those of the U.S. Soil Taxonomy. Nomenclature is of a mixed origin, and the main units have "traditional" names like Chernozems and Podzols, names adopted from North America like Gleysols, Histosols, and Vertisols, and newly coined names like Luvisols and Acrisols. The main general characteristics of each group of major soils that may occur in arid and semiarid lands are given in Table 11 together with the approximate corresponding U.S. Soil Taxonomy class.

TABLE 9 Classes and Subclasses Containing Arid and Semiarid Soils of the French (Commission de Pédogie et de Cartographie des sols, 1967) Soil Classification and Approximate Equivalents in the U.S. Soil Taxonomy (Soil Survey Staff, 1975)

Class	Subclass	U.S. Soil Taxonomy
1. Raw mineral soils	1.1 Erosion or deposition (non-climatic)	Orthents, Fluvents, Psamments
	1.2 Cold deserts	Cryorthents
	1.3 Hot deserts	Torrifluvents, Torriorthents, Torripsamments
2. Weakly developed soils	2.3 Dry climates	Torriorthents, Ustorthents, Xerorthents
	2.4 Erosion or deposition (non-climatic)	Orthents, Fluvents, Psamments
3. Vertisols	3.2 External drainage	Torrerts, Usterts, Xererts
4. Andosols	Rare in semiarid and arid regions	Andisols
5. Calcimorphic	5.1 Calcareous soils	Ochrepts
	5.2 Little or no free carbonate	Xerolls
	5.3 Gypsiferous soils	Gypsids
6. Isohumic soils	6.2 Very cold climates	Ustolls
	6.3 Rain in the cool season	Xerolls, Xerollic Aridisols
	6.4 Rain in the hot season	Ustolls, Ustollic Aridisols
7. Browned soils	Rare in semiarid and arid regions	Mollisols and Alfisols
8 Podzolized soils	Rare in semiarid and arid regions	Spodosols, spodic subgroups
9. Ferruginous soils	Rare in semiarid and arid regions	Inceptisols, Alfisols
10. Ferralitic soil	10.1 High base saturation	Ustox, Torrox
11. Hydromorphic soils	Rare in semiarid and arid regions	Histosols, aquic suborders
12. Halomorphic soils	12.1 Nondegraded structure	Salids
	12.2 Degraded structure	Natric subgroups of Aridisols, Alfisols, and Mollisols

TABLE 10 Australian Great Soil Groups (Stace et al., 1968) of Arid and Semiarid Regions and Their Approximate Equivalents in the U.S. Soil Taxonomy (Soil Survey Staff, 1975)

Australian great soil group	U.S. Soil Taxonomy
No profile differentiation	
Solonchaks	Salids, Salidic subgroups of Ustolls
Alluvial soils	Fluvents
Lithosols	Lithic subgroups Aridisols, Mollisols, and Alfisols
Calcareous sands	Psamments
Siliceous sands	Psamments
Earthy sands	Orthents
Minimal profile development	
Gray-brown and red calcareous soils	Orthents, Cambids
Desert loams	Argids
Red and brown hardpan soils	Durids
Grey, brown, and red clays	Torrerts, Usterts
Dark soils	
Black earths	Usterts, Xererts
Rendzinas	Xerolls, Ustolls
Mildly leached soils	
Solonetz	Natric great groups of Aridisols, Mollisols, and Alfisols
Solodized solonetz and solodic soils	Natric great groups of Mollisols and Alfisols
Solonized brown soils	Calcids, Argids, Ustalfs, Xeralfs
Red-brown earths	Ustalfs, Xeralfs
Noncalcic brown soils	Xeralfs, Ustalfs
Soils with predominately sesquioxide clay minerals	
Calcareous red earths	Cambids, Calcids, Gypsids, Torrox
Red earths	Ustalfs, Xeralfs, Argids
Mildly to strongly acid and highly differentiated	
Great soil groups absent or rare in arid and semiarid regions	
Dominated by organic matter	
Great soil groups absent or rare in arid and semiarid regions	

TABLE 11 General Characteristics of Groups of Major Soils of the FAO/UNESCO (FAO, 1974) Soil Map of the World and Approximate Equivalents in the U.S. Soil Taxonomy (Soil Survey Staff, 1975)

Fluvisols	Soils in alluvium (Fluvents)
Gleysols	Hydromorphic soils in nonalluvial materials (aquic suborders)
Regosols	Weakly developed soils in loamy or clayey unconsolidated deposits other than alluvium (Orthents)
Lithosols	Shallow soils over hard rock (Lithic subgroups)
Arenosols	Soils in unconsolidated sandy deposits other than alluvium (Psamments)
Andisols	Soils, mainly in volcanic ash, with a low bulk density and an exchange complex dominated by noncrystalline clay (Andisols)
Vertisols	Cracking clay soils (Vertisols)
Solonchaks	Saline soils in nonalluvial materials (Salids)
Solentz	Soils with a natric B horizon (Natrargids, Natrustolls, Natrixerolls, Natrustalfs, Natrixeralfs)
Yermosols	Desert soils with a very weak ochric horizon (containing little organic matter, typic Aridisols)
Xerosols	Desert soils with a weak ochric horizon (containing more organic matter than Yermosols, mollic Aridisols)
Kastanozems	Soils with a mollic horizon and a calcic or gypsic horizon or concentration of lime (Ustolls)
Cambisols	Soils with a cambic horizon and without horizons diagnostic of other classes (Cambids, Inceptisols)
Luvisols	Soils with an argillic horizon of medium to high base status (Alfisols)
Ferralsols	Soils with an oxic horizon (Oxisols)

III. SOIL GENESIS

The genesis of soils consists essentially of two major types of phenomena (Simonson, 1978; Pedro, 1983): (1) phenomena corresponding to the weathering of primary (weatherable) minerals and the release of various chemical elements and their possible recombinations to give mineral constituents stable in the weathering or soil environment; (2) phenomena developing during or after weathering related to the arrangement of soil constituents, their organization, and their redistribution into pedological horizons. These phenomena are referred to as soil-forming processes.

A. Weathering

Weathering is the breakdown and alteration of rocks and minerals near the earth's surface to products that are more in equilibrium with the newly imposed physical and chemical conditions of the pressure, temperature, and moisture environment on the earth's surface. Weathering of the initial materials precedes soil formation in hard rocks and accompanies it in soft rocks and unconsolidated materials. Weathering continues as soil formation proceeds until no more reactants (weatherable minerals) are available. It is convenient to distinguish between physical weathering, the breakdown of material entirely by mechanical methods, and chemical weathering, in which chemical reactions accompany the process. Although these two main types of weathering are treated separately, rarely if ever does one of them take place in the absence of the other. In fact, the two types of weathering commonly enhance one another. Physical weathering exposes fresh surfaces where chemical weathering can take place, and the effects of chemical weathering can sometimes result in physical disintegration of a rock or mineral.

Physical Weathering

Physical weathering is generally considered to dominate over chemical weathering in arid regions. It can be brought about by a variety of forces that may originate within the rocks or be applied externally. The resulting stresses lead to strain and eventually to rupture. A number of processes of physical weathering have been described (Ollier, 1975). Only those processes that are or have been considered important are discussed here.

Insolation Weathering

Extreme air temperature and corresponding daily thermal fluctuations of soil and rock surfaces due to solar irradiation in desert regions have been postulated as important casual agents in the breakdown of coarse fragments into smaller particles. According to the classic insolation theory, a thermal gradient is set up between the surface and the interior of a rock when it is irradiated since rocks are poor conductors of heat. The surface of the rock therefore expands more than the inside, setting up stresses that may lead to fracture. Stresses may also arise owing to the uneven volumetric expansion and other differing thermal properties of different minerals in a rock.

A number of experiments, notably those of Blackwelder (1933) and Griggs (1936), caused serious doubts about the effectiveness of the insolation and daily temperature fluctuations in rock breakdown. Despite this, some workers, Ollier (1975), for example, continue to be proponents of the theory based upon the intuitive interpretation of field evidence. Ollier (1975) and Rice (1984) noted two limitations of the experiments. First, relatively small specimens were used in which expansion could take place sideways, if not all around. Larger rocks or exposed rock outcrop, on the other hand, would contain confined expanding patches or rock compressed by neighboring patches. Since these pathces can only expand outward, the stress is much more concentrated and liable to lead to rupture. Second, the fatigue factor cannot be reproduced in rapid experiments. Thus the small stresses applied over long periods of time might lead to permanent strain, whereas under the experimental conditions of Blackwelder and Griggs the rock specimens were heated for only a few moments.

Whether temperature changes alone can cause fracturing is a subject of debate, but it is generally accepted that weathering is enhanced when rapid and repeated high-magnitude temperature changes occur in conjunction with moisture absorption and/or chemical weathering (Kerr et al., 1984; Ravina and Zaslavsky, 1974; Roth, 1965, Verheye, 1975; Smith, 1977).

The thermal expansion and contraction of rocks caused by fire is known to bring about rock fracturing and spalling (Blackwelder, 1933; Emory, 1944). In addition, the rapid increase in temperature would cause any trapped water in the voids to convert to steam and exert large pressures capable of fracturing a rock. The effect of fire is probably minimal in arid regions but might be important in semiarid regions where there is sufficient fuel (biomass) to sustain fires.

Salt Weathering

Rock disintegration due to the formation and presence of salt crystals is considered an important weathering process in arid regions. Three forms of salt weathering are recognized, each of which is thought to be able to induce sufficient stress to exceed the tensile strength of the host rock (Evans, 1970).

The first form involves the growth of salt crystals during the evaporation of saline solutions either within rock pores or in pre-existing cracks and fissures. This mode of salt weathering has been demonstrated by a number of workers (Evans, 1970): crystal growth can take place against the pressure of the confining rock and exert a force in excess of the rock's tensile strength.

The change in volume when anhydrous salts become hydrated is the second form. For example, the hydration of

$$MgSO_4 \cdot H_2O + 6H_2O = MgSO_4 \cdot 7H_2O$$

experiences a volume change of 173% and

$$Na_2SO_4 + 10H_2O = Na_2SO_4 \cdot 10H_2O$$

an expansion of 300%. For this to be effective two criteria must be met (Fahey, 1985). First, the temperature must be equal to or less than the transition point at which an anhydrous salt can be transformed to its hydrated form, and vice versa. Second, at temperatures below the transition point the atmospheric vapor pressure must exceed the vapor pressure at which the salt-hydrate system dissociates. Goudie (1974) demonstrated that Na_2SO_4 is quite effective in disintegrating both a silica-cemented sandstone and chalk, although the experimental conditions were probably optimum for satisfying the two criteria.

The third mode relates to the thermal expansion and contraction of crystallized salts during temperature changes (insolation weathering). Sodium nitrate, with its relatively high coefficient of expansion, should be the most effective of the commonly occurring salts (Cooke and Smalley, 1968). The experimental results of Goudie (1974), however, showed that the thermal expansion of this salt is largely ineffective.

Salt weathering has been proposed as a mechanism for producing silt-sized particles to serve as a source of loess in desert regions (Derbyshire, 1983; Goudie et al., 1979; Pye and Sperling, 1983; Smith et al., 1989).

Frost Weathering

The conventional view of frost weathering relates to the freezing of water, its increase in volume (about 9%) in a confined space, and the resulting pressure on the enclosing rock. However, pore water is probably never completely confined nor fills the total of a void. Thus only part of the expansion is utilized in disrupting the enclosed rock (Reiche, 1950). Frost-weathering experiments have shown that under laboratory conditions the process can be effective in rock disintegration (Fahey, 1985; Roth, 1970; Swantesson, 1985). The combined action of frost and salt (freezing of saline solutions may also be of importance (Fahey, 1985; McGreevy, 1982; Williams and Robinson, 1981). This process would only be effective in cold deserts, especially polar deserts where frequent freeze-thaw cycles would occur. The lack of water available in arid regions would probably minimize this process except possibly in high-latitude coastal areas (Hall, 1988; Trenhaile and Mercan, 1984).

A more recent frost-weathering hypothesis is correlated with the formation of thin films of moisture on mineral surfaces that resist freezing (Konishchev, 1982; White, 1976). Konishchev (1982) observed that quartz grains of sand size are less resistant to frost weathering than feldspars, whereas mica grains are the most resistant. It was concluded that the relative resistance is related to the protective role of the stable film of unfrozen water, which is thickest in the micas. The disintegration is thought to occur when the thickness of the protective unfrozen water film becomes less than the dimensions of fractures and defects that characterize the surface of mineral particles. This mechanism could play a role in the breakdown of sand particles in cold desert regions. Since the formation of dew has been observed in hot deserts (Ravina and Zaslavsky, 1974; Roth, 1965; Verheye, 1975), dew (or frost) would probably be more common in cold deserts. The condensed moisture on the mineral grains could then undergo freezing and thawing except for the thin unfrozen protective layer.

Process of Transportation and Erosion

Erosion and transportation of the eroded products by wind, water, and ice are important means for the physical disintegration of rocks. Of these geological processes, wind and water are important in influencing the nature of eolian and fluvial sediments that serve as parent materials of soils in arid and semiarid regions, as well as humid regions. Fluvial processes in ephemeral arid streams generally operate over brief periods of time, usually as high-energy floods in which large amounts of sediment with particles ranging up to boulder size are eroded and moved. Under these conditions the rock fragments are quite prone to breakdown. The resultant alluvium, which serves as soil parent material, is characteristically poorly sorted.

The effect of wind is most important with sand-sized particles, as in sand dunes. The impact of saltating sand grains during movement and the sand blast abrasion on pebbles are recognized as causing or potentially causing sand and pebble breakdown (Cooke and Warren, 1973).

Chemical Weathering

The processes of chemical weathering result in the breakdown of the mineral structure concomitant with its solution or partial solution and the recombination of the weathered products to form new minerals.

Hydrolysis

Hydrolysis, the chemical reaction between a mineral and the H^+ or OH^- ions of aqueous solutions, is extremely important in the chemical weathering of silicate minerals. A number of laboratory weathering experiments over the past 20 years or so have demonstrated that when a freshly exposed mineral surface is attacked by an acid aqueous solution, a rapid exchange between the H^+ ions of the solution and the cations at or near the mineral surface takes place. This is followed by a slower removal of Si, Al, and other cations. Different mechanisms and dissolution kinetics have been reported by different workers, so that the processes involved have not been completely elucidated (Birkeland, 1984; Colman and Dethier, 1986). Hydrolytic weathering is not as effective in arid regions as in humid regions, for three main reasons. First, weathering is limited to the periods when moisture is available. Second, the weathering system is generally neutral to alkaline in reaction so that H^+ ion activity is relatively low. In sodic soils and weathering systems at high pH, on the other hand, the OH^- ions may combine with Al and Si at the crystal surface, with the compounds formed quite soluble (Birkeland, 1984). Third, the limited leaching does not remove the soluble components that are released so that equilibrium is approached, reducing the weathering rate.

Carbonation

Carbonation, the reaction of CO_3^{2-} or HCO_3^- ions with minerals, is important in the weathering of limestone and other carbonate-containing rocks. Carbon dioxide dissolves in water as follows:

$$CO_2(g) + H_2O = CO_2(aq) + H_2O$$

$$CO_2(aq) + H_2O = HCO_3^- + H^+$$

$$HCO_3^- = CO_3^{2-} + H^+$$

The chemistry associated with this process has been extensively studied and is described in standard textbooks of soil chemistry (Bohn et al., 1984) and geochemistry (Krauskopf, 1967). Although the solubility of $CaCO_3$ is affected by the temperature, pH, CO_2 partial pressure, the presence of other ions, and the area of the solid/liquid interface, the main factor in controlling the rate of dissolution is the rate of removal of weathering products (Weyl, 1958). Thus the effectiveness of carbonation in chemical weathering in arid regions is limited by the lack of leaching to remove the weathering products, although it plays an important role in redistributing carbonate within the soil profile.

Oxidation-Reduction

Oxidation-reduction is the process by which an element loses or gains an electron. The oxidation-reduction potential depends upon the partial pressure of gaseous O_2 and CO_2 in the soil atmosphere and on the pH of the soil solution. Most weathering environments, especially those in arid regions, have a relatively high oxidation-reduction potential that favors the oxidation of such elements as Fe, Mn, and Cr to their higher valence states. Ferrous Fe, for example, when present in a primary mineral, may be converted to the ferric form. To maintain electrostatic neutrality of the crystal structure, some other ion must leave the lattice, causing the lattice to become unstable and more susceptible to other weathering reactions (Ollier, 1975). If the Fe^{2+} is released into solution by hydrolysis, it oxidizes to form an oxide or hydrous oxide of iron (Schwertman, 1985). The result is that many soils in arid and semiarid regions tend to increase in free iron oxides and to develop redder hues with age (McFadden and Hendricks, 1985; Walker, 1967).

Chelation

Chelation refers to the process by which a chemical species (ligand or chelating agent) can occupy two or more bonding sites around a single cation. Such ligands are usually organic compounds and are produced in a soil by biological activity. Both laboratory and field evidence has demonstrated that chelating agents can contribute to chemical weathering. Laboratory experiments have been performed using both synthesized chelating agents (Huang and Keller, 1970; Manley and Evans, 1986; Mast and Drever, 1987) and natural soil organic materials (Baker, 1973; Sotnikova, 1975; Zunino and Martin, 1977). Lichens are known to chemically break down rock surfaces in which chelation weathering plays an important role (Jackson and Keller, 1970; Vidrich et al., 1982). It is expected that chelation weathering is of minor importance in most arid soils. The organic matter content is generally low and is dominated by base-saturated humic acid with limited active chelating agents (Kononova, 1975).

Clay Mineral Genesis

The clay fraction of most arid soils is composed mostly of smectite, with mica and lesser amounts of kaolinite. Palygorskite is also widespread, and chlorite, vermiculite, and mixed layer clays may be present.

Three modes of origin of clay minerals in soils are known (Eberl, 1985; Millot, 1970): detrital inheritance from the parent material, alteration, and synthesis. All three modes operate simultaneously. When physical weathering processes are involved inheritance is of major importance. Under chemical weathering alteration and synthesis dominate.

Detrital inheritance from parent material is considered very important in arid and semiarid lands, where chemical weathering is limited or has not progressed very far. This mode is of particular importance in sedimentary rocks and fluvial and related parent materials containing clay minerals formed in previous cycles of weathering and soil formation. Some igneous rocks contain mica, and metamorphic rocks may contain mica and chlorite that can subsequently be incorporated in the soil. Hydrothermal and deuteric alteration may also form clay minerals in rocks.

Alteration refers to those changes that modify a clay mineral without altering its layered structure. Probably the most easily appreciated example relates to the weathering of micas (Fanning and Keramidas, 1977; Douglas, 1977), which can be expressed as

Mica ⟶ illite ⟶ vermiculite ⟶ smectite

Synthesis takes place when weathering products (ions or colloids) recombine to form new minerals. The type of clay mineral that forms is strongly influenced by the environment, as summarized by Keller (1970).

Kaolinite formation, for example, is favored by strong leaching with removal of Ca^{2+}, Mg^{2+}, Na^+, and Fe^{2+} and partial removal of Si (as H_4SiO_4) and addition of H^+. The formation of smectite minerals, in contrast, is favored in an environment in which the bases (except K) and H_4SiO_4 are retained or continually supplied. Field evidence (Barshad, 1966; Birkeland and Janda, 1971) and mineral stability diagrams based on thermochemical data (Kittrick, 1970, 1971) support Keller's (1970) contentions.

Mica Minerals

Mica minerals are composed of two silica tetrahedral sheets bonded to both sides of an aluminum octahedral sheet with an interlayer ion, usually K^+, bonding these structures together. Mica dominates the clay in some arid regions, but in others it is only a minor component. Mica is the major clay mineral in the hot deserts of central Asia (Gharaee and Mahjoory, 1984; Minashina and Gradisov, 1973; Sosnovskay, 1982) and in arid southern and interior Australia (Norrish and Pickering, 1983). Little mica has been observed in some soils in the Middle East (Al-Rawi et al., 1969; El-Tezhani et al., 1984; Shadfan et al., 1984).

Most, if not all, mica minerals in soils are considered to have been inherited from the parent material (Fanning and Keramidas, 1977). There is some evidence, however, for a pedogenic origin in some arid soils. In many arid soils mica dominates the coarse (2-0.2 μm) clay fraction, indicating that the mica has most likely been inherited (Singer, 1988). In other arid soils mica concentrates in the fine (<0.2 μm) clay fractions (Norrish and Pickering, 1983; Southard and Southard, 1985). In a number of arid soils mica together with soluble and exchangeable K^+ has been reported (Mahjoory, 1974; Norrish and Pickering, 1983; Nettleton and Brasher, 1983) to be highest in the surface and to decrease with depth. These two types of observations are both suggestive of a pedogenic origin of mica. Nettleton et al. (1973) and Mahjoory (1974) hypothesized that biocycling of K played a role in mica formation from smectite. Singer (1988) suggests that the K for mica formation is introduced with eolian dusts.

Vermiculite

Vermiculite is a 2:1 mineral in which the interlayer potassium is replaced by hydrated cations and there is some isomorphous substitution in both the tetrahedral and octahedral layers. Vermiculites occur in variable amounts in diverse environments, including deserts, but seldom as a major component (Douglas, 1977; Norrish and Pickering, 1983). The almost universal opinion is that vermiculite forms by alteration from mica (Douglas, 1977), although there is evidence that it can also form by synthesis (Barshad and Kishk, 1969). Barshad's (1966) data show that vermiculite is more abundant in soils formed in humid than in more arid climates from both acid and basic igneous rocks.

In some cases the interlayer potassium is removed from some interlayers but not others. Where the potassium is retained the basic mica structure is preserved; where the potassium is removed and replaced by a hydrated cation a vermiculite structure is formed. This results in a mixed layer mineral composed of interstratified mica and vermiculite. This and other types of mixed layer clays are common in soils but seem to be more prevalent in humid soils (Mahjoory, 1974).

Smectite

Minerals of the smectite group have a 2:1 structure with considerable isomorphic substitution that results in weakly held interlayer cations. Smectite minerals are important components or dominate the clay fractions of many arid soils (Dregne, 1976; Minashina and Gradisov, 1973; Nettleton and Peterson, 1983; Gal et al., 1974). Smectites, with their high shrink-swell capacities, are of particular significance in Vertisols, contributing to their cracking and related properties (Wilding and Tessier, 1988). Smectite can be inherited from parent materials in arid soils (Akpokodje, 1987), where they are preserved under conditions of limited leaching (Gal et al., 1974). Soils of semiarid areas with a short leaching season maintain the geochemical conditions of high H_4SiO_4, base status, and pH that are optimal for the synthesis of smectite (Folkoff and Meentemeyer, 1987). During the limited leaching season some chemical weathering of minerals takes place, but the weathering products are not easily leached and can concentrate, especially during the dry season. Buol (1965) expressed doubts that smectite could have formed in some arid regions, and Folkoff and Meentemeyer (1985) suggest that more humid past climates were influential in determining its occurrence in arid regions. Singer (1983) reported the occurrence of smectite in the saprolites of diverse igneous rocks that formed by weathering in the Negev Desert in southern Israel. Smectites can also form by alteration when K is removed from the interlayer region of micas accompanied by small changes in the octahedral and tetrahedral sheets. Ismail (1969, 1970) provides laboratory and field evidence that the alteration of biotite yields smectite and vermiculite in arid soils but only vermiculite under humid conditions.

Chlorite

The crystalline structure of chlorite resembles that of mica, vermiculite, and smectite but differs from vermiculite and smectite in that the interlayer region is occupied by Mg, Fe, or Al hydroxides instead of hydrated cations (Barnhisel, 1977). Two distinct chlorites are now recognized as occurring in soils. The chlorite commonly present in low- to medium-grade metamorphic rocks is relatively unstable and generally weathers rapidly. Arid soils commonly contain this type of chlorite, inherited from the parent material, which may persist with limited chemical weathering. The other type of chlorite, with $Al(OH)_3$ occupying the interlayer, tends to occur in highly weathered and leached acid soils. This type is rare or absent in arid soils.

Kaolinite

Kaolinite is a 1:1 mineral that consists of a silica tetrahedral layer bonded to an alumina octahedral layer. Kaolinite occurs in many soils under a wide range of soil-weathering environments and may form from a large number of different minerals (Dixon, 1977). Since strong leaching environments are most favorable for kaolinite formation, its presence, although widespread, is generally small in arid soils (Hendricks, Nettleton, and Grossman, 1985). Kaolinite can be present in high amounts in arid soils formed from kaolinitic parent materials, or the kaolinite may be relict from past more humid conditions, as in arid and semiarid Oxisols (Hendricks et al., 1985).

Palygorskite

Palygorskite, sometimes called attapulgite, does not have a layer structure like the preceding minerals, although its structure resembles that of smectite to some extent. For details of the structure of palygorskite Henin and Caillere (1975) or Zelazny and Calhoun (1977) can be consulted. Palygorskite is a common component of arid and semiarid soils throughout the world (Singer and Galan, 1984). Palygorskite has been described as forming in sediments of normal marine waters, in marine and nonmarine hypersaline waters, and in freshwater lakes (Zelazny and Calhoun, 1977). Zelazny and Calhoun (1977) state that palygorskite occurrences in soils appear to be largely inherited from the parent material. Other workers (Eswaran and Barzanji, 1974; Millot et al., 1969; Paquet et al. (1969; Shadfarn et al., 1985; Singer and Norrish, 1974; Yaalon and Wieder, 1976) postulate and/or provide evidence for a pedogenic origin in the soils they studied. It appears that palygorskite formation is favored by high pH, H_4SiO_4, and Mg^{2+} and low Al activities (Millot et al., 1969; Singer and Norrish, 1974).

B. Soil-Forming Processes

The generalized theory of soil genesis proposed by Simonson (1978) describes four groups of physical, chemical, and biological processes common to all soils: additions of organic and mineral material as solids, liquids, and gases; their removal; their transfer or translocation; and their transformation. The changing balances between these processes differentiate one soil from another. For example, mineralization and humification of plant litter engage more or less the same processes of transformation in all environments, but differing process rates may lead to different end products. In arid environments eolian additions of particulate matter are widespread, and in some areas the addition of soluble salts from shallow groundwater is of particular importance. The removal of materials from arid soils is primarily limited to surface erosion since the removal of soluble material by through-leaching of the profile is generally lacking. The decomposition of organic matter and chemical weathering involve transformations. Of particular importance in soils of arid and semiarid regions is the translocation or redistribution of materials, such as soluble salts, gypsum, carbonates, and silica, and their accumulation in some layer in the soil.

Genesis of Mollic and Ochric Epipedons

The distinctive feature of mollic and ochric epipedons is the buildup of organic matter; the level is higher in the mollic epipedons. The establishment and maintenance of a particular equilibrium level of organic matter depends upon three factors: the relative amount of biomass added, the rates of organic residue and organic matter decomposition, and the capacity of the soil to retain organic matter (Anderson, 1987). Many studies have shown that the organic matter content of soils increases with increased moisture supplies, presumably due to greater production of organic residues (Anderson, 1987; Fenton, 1983; Jenny, 1941, 1980). The available moisture is most dominant in determining vegetative production in semiarid regions (Anderson, 1987). There is general agreement that most soils with mollic epipedons have or have had a grass vegetation at some time (Fenton, 1983). Although strong relationships between aboveground production and soil organic matter content are evident, it is recognized that belowground production is a more important contribution to organic matter levels in grassland soils.

The rate of decomposition of soil organic matter is also affected by temperature. Jenny (1941, 1980) cites several studies that show that the soil organic matter content decreases as the temperature increases under similar moisture conditions. Maximum decomposition rates in arid grasslands, however, combine moist soil conditions and moderate to high soil temperatures (Wildung et al., 1975).

The soil organic matter of arid and semiarid regions is dominated by humic acid relative to fulvic acid (Kononova, 1975) and is calcium saturated or calcium rich (Smith, 1965). Organic matter of this nature seems to be in a fairly stable form and to undergo less rapid decomposition. Interactions between clay and organic colloids also appear to play an important role in stabilizing soil organic matter (Sotnikova, 1975). Several investigators have shown that organic carbon is strongly correlated with the percentage of clay in the thermic and mesic soils in the Great Plains of the United States (Anderson, 1987). Different types of clay minerals vary in their effectiveness in complexing soil organic matter. Smectite and the noncrystalline clays of Andisols (allophane, for instance) seem to form more stable complexes with organic matter than other types of clay (Kononova, 1975; Tate and Theng, 1980).

In arid and semiarid regions mollic epipedons are generally associated with desert grass and steppes and ochric epipedons are in soils with desert shrub and other nongrass plant communities. The more arid soils usually are rather low in organic matter due to low biomass production caused by limited moisture.

Argillic Horizon Genesis

Soil-forming processes in many soils result in the vertical redistribution of clay. The resulting profile characteristically increases in clay content with depth to a maximum in the subsoil and then decreases. The zone of maximum clay accumulation is the argillic horizon. The argillic horizon is defined by the U.S. Soil Taxonomy (Soil Survey Staff, 1975) as "one that contains illuvial layer-lattice clays. This horizon forms below an eluvial horizon but it may be at the surface if a soil has been partially truncated."

Although clay accumulation through translocation to form argillic horizons in soils is now an established fact, there are two other modes of origin that could result in or contribute to the profile clay distribution typical of soils with argillic horizons: differential weathering and stratification of the parent material. Differential weathering was originally proposed as the mechanism for clay accumulation in arid soils (Brown and Drosdorff, 1940; Nikiforoff, 1937). With this hypothesis the "clay bulge" resulted from longer periods of chemical weathering at a depth below the surface, which remained moist longer. The surface soil, on the other hand, dried more rapidly with shorter periods of chemical weathering. Stratification could be a factor in soils formed from certain parent materials, such as alluvium, if the zone where the argillic horizon forms has finer texture than the overlying and underlying strata.

The three sets of processes involved in the formation of the argillic horizon by clay migration are the dispersion, translocation, and accumulation of the clay (Soil Survey Staff, 1975; McKeague and St. Arnaud, 1969). Hence, one of the prerequisites for clay movement is that the clay particles be peptized or dispersed. This can be achieved by lowering the electrolyte content of the soil solution. In Aridisols this can be brought about by repeated leaching of the surface by fresh water. However, the presence of carbonates in the eluvial horizon prevents the clays from dispersing. Wet and dry cycles, which are characteristic of arid soils, may also enhance dispersion since wetting a dry soil seems to lead to disruption of the fabric (Soil Survey Staff, 1975). Once dispersed the clay moves with the percolating water through the noncapillary voids. The presence of cracks that form during drying stages in Aridisols coupled with the occurrence of occasional heavy downpours would seem to favor the translocation of clay. As the clay moves lower in the profile, deposition of the clay takes place by flocculation upon encountering a higher salt and/or a carbonate-enriched zone and by capillary withdrawal upon encountering a drier soil.

The formation of an argillic horizon is considered a slow process so that soils containing them are old and usually date back into the Pleistocene (Soil Survey Staff, 1975). In fact, some Argids have been forming through at least one or more glacial and interglacial periods. Thus, it is commonly suggested that the argillic horizon in some Aridisols may be a relict feature formed in the past in a more humid climate (Nettleton and Peterson, 1983).

Accumulation of Soluble Salts

The accumulation of soluble salts (minerals more soluble than gypsum) is a common feature of many soils when evaporation exceeds precipitation. The mechanism of salt movement and distribution is very important for the utilization of arid soils, and a considerable literature exists on the origin of the salts and their movement. Salt-affected soils occur in all continents as discontinuous patches within other major soil groups, principally in Aridisols and to a lesser extent in Mollisols and Alfisols and other soil orders. The buildup of salts to the level required for salic horizons (horizons 15 cm or more thick having at least 2% soluble salts) occurs in only a very few, not extensive Aridisols (Nettleton and Peterson, 1983).

Salt buildup is most commonly achieved through evaporation of moisture from the surface or shallow depths within the soil. Salts in solution are drawn upward from a shallow water table by capillary action and are deposited as the water evaporates. The salt can be derived from a saline parent material or a salt-rich geological substratum or from weathering of primary minerals and leaching of saline sediments from the rocks in the catchment area.

Salt buildup can also occur in soils in the absence of a water table as a result of the eolian accretion of salts dissolved in rainwater, in sea spray in coastal areas, or as dry fallout. This mode of origin is reported to be important in parts of Israel (Dan and Yaalon, 1982), in Australia (Isbell et al., 1983) and in the Great Basin of western North America (Young and Evans, 1986).

Natric Horizon Genesis

The natric horizon is similar to the argillic horizon in that it represents the accumulation of illuvial clay, but its formation has been influenced by the presence of soluble and exchangeable sodium. Nettleton and Peterson (1983) describe three situations in which Aridisols with natric horizons have different genetic histories. Pleistocene relict soils occur on remnant surfaces. The exchangeable sodium level has been maintained by Holocene and modern eolian salt. Other soils with natric horizons occur in both Pleistocene and Holocene sediments on alluvial flats with water tables shallow enough to supply the sodium for the natric horizon. The third mode of occurrence is in loamy or sandy Holocene sediments with water tables too deep to provide sodium salts. In these soils both the clay for translocation and the sodium for dispersion were added as eolian dust and salt.

The classic model for the genesis of soils that contain what we now call natric horizons was proposed by Gedroitz early in the twentieth century (Ivanova and Bol'shavov, 1972). The essence of this model is as follows. A salt-affected soil (possibly with a salic horizon) forms. When conditions change, such as a drop in the water table or increased rainfall or irrigation that initiates leaching, the soluble salts are easily washed out. If the soil is dominated by Na^+ ions, clay and organic matter disperse and the surface horizons become structureless. The soil at this stage has a high pH brought about by the presence of Na_2CO_3 according to the reaction

$$\text{Na - clay} + H_2O = \text{H - clay} + NaOH$$

$$2NaOH + CO_2 = Na_2CO_3 + H_2O$$

The dispersed clay and humus move down the profile, which leads to the formation of a natric horizon with its characteristic columnar structure. With further leaching the Na^+ is removed from the clay-humus complex and the natric horizon deteriorates.

Gypsic, Hypergypsic, and Petrogypsic Horizon Formation

All warm deserts of the world exhibit surfaces containing gypsum accumulations (gypsum crusts) where there is a source of gypsum and generally less than 250 mm rainfall per year (Watson, 1983). Watson (1988) describes three common types of desert gypsum crusts, two of which are pedogenic or partially pedogenic: shallow water evaporates; subsurface crusts, which are either microcrystalline (gypsum crystals up to 0.5 m diameter) groundwater evaporates or microcrystalline (up to about 1.0 mm diameter) illuvial accretions; and exhumed illuvial crusts currently exposed at the surface.

Soil horizons of gypsum accumulation form pedogenically when gypsum dissolves in the surface horizons and moves down with the percolating water until it reprecipitates at the depth of wetting. In well-drained dryland soils the gypsum apparently precipitates out of solution rapidly since the crystals are sand size or smaller (Nettleton et al., 1982). The illuviation of gypsum begins in the smaller voids and channels in association with capillary moisture movement and then proceeds to larger and still larger voids as the smaller voids are filled. Eventually space for crystal growth becomes limited, causing crystals to interlock.

Most of the gypsic soils in the western United States are associated with gypsum deposits or gypsiferous sediments (Nettleton et al., 1982). Eolian sources of gypsum have been reported by Dan and Yaalon (1982) in Israel. Akpokodje (1984) reported that the gypsum present in parts of the arid zone in Australia was derived from the weathering of pyrite in phyllitic shales.

Calcic and Petrocalcic Horizon Formation

Soils with accumulations of carbonate are very extensive in arid and semiarid regions. The degree of accumulation ranges from slight to high, depending upon the age of the soil and the quantity of carbonate present originally in the parent material and/or added during soil formation. The amount of carbon stored in calcareous soils worldwide has been estimated as 800×10^{15} g, forming a significant pool in the global cycle (Schlesinger, 1982). This amount is larger than the carbon stored in terrestrial vegetation and about 50% of the carbon stored in soil organic matter worldwide (Schlesinger, 1985).

The formation and accumulation of carbonates in surficial deposits, including soils, have been attributed to numerous processes (McKeague and St. Arnaud, 1969). These processes can be grouped into three classes: (1) chemical or biochemical deposition in a lacustrine environment, (2) deposition from the capillary fringe of a shallow water table, and (3) pedogenic deposition from downward infiltrating soil waters. It is now generally accepted that most calcareous soils have been formed by calcium carbonate moving downward in solution during intermittent periods of wetting in a given arid or semiarid environment (McKeague and St. Arnaud, 1969; Gardner, 1972; Gile et al., 1966). Some of the carbonates are released by weathering of the parent material; others are derived from the influx of carbonate-rich eolian dust or carbonates dissolved in rainwater. They are precipitated at a particular depth of wetting by evapotranspiration processes.

Four stages of the development of calcic and petrocalcic horizons are described by Gile et al. (1966). These stages and the modifications by Machette (1985) are summarized in Table 12. For gravelly parent materials thin discontinuous pebble and gravel coatings develop in stage I, continuous coatings with a weakly cemented matrix in stage II, and coatings on all fragments with many of the voids filled with strongly cemented material in stage III, resulting in a mass of relatively strong cementation. For nongravelly parent material, the stages of development are essentially the same, except for the occurrence of nodules of varying hardness in stage II. For all materials, a continuous strongly cemented pan overlaid by a laminar horizon of nearly pure carbonate occurs in stage IV. The buildup of carbonates in the four stages is more rapid in gravelly than in nongravelly material because gravel has less pore space (Gile et al., 1966). Machette (1985) modified this sequence by suggesting six stages. The first four

TABLE 12 Stages of Carbonate Accumulation in Soils[a]

Stage	Soils developed in gravel	Soils developed in sand, silt, or clay
I	Thin, discontinuous coatings of carbonate on underside of clasts	Dispersed powdery and filamentous carbonate
II	Continuous coating all around and, in some cases, between clasts: additional discontinuous carbonate outsiee main horizons	Few to common carbonate nodules and veinlets, with powdery and filamentous carbonate in places between nodules
III	Carbonate forming a continuous layer enveloping clasts: less pervasive carbonate outside main horizon	Carbonate forming a continuous layer formed by coalescing nodules; isolated nodules and powdery carbonate outside main horizon
IV	Upper part of solid carbonate layer with a weakly developed platy or lamellar structure capping less pervasively calcareous parts of the profile	
V	Platy or lamellar cap to the carbonate layer strongly expressed; in places brecciated and with pisolites of carbonate	
VI	Brecciation and recementation, as well as pisoliths, common in association with the lamellar upper layer	

[a]This table includes modifications by Machette (1985) to the scheme proposed by Gile et al. (1966).

stages are similar to those of Gile et al. (1966), but stages V and VI represent more advanced formation by brecciation (fracturing of the pan and recementation of the fragments) and formation of carbonate pisolites (spherical or subspherical accretionary bodies over 2 mm in diameter).

Rabenhorst and Wilding (1986) describe a five-stage sequence in the formation of petrocalcic horizons on solid limestone. The main process involves the in situ dissolution of the parent carbonates in the limestone and their reprecipitation. Other workers (Arkley, 1963; Marion et al., 1985; Mayer et al., 1988) have developed models for calculating the content and redistribution of carbonates based on various factors, such as dust flux, climate, time, soil available water-holding capacity, partial pressure of CO_2 gas, and the initial carbonate content of the soil.

Duripan Formation

Soils with layers containing pedogenic accumulations of and cementation with silica (duripans) are not as widespread as those with carbonates. They may form whenever monosilicic acid is released by weathering and is not subsequently combined into secondary aluminosilicate clays, taken up by plants, or leached from the soil profile (Chadwick et al., 1987; Flach et al., 1969). In the western United States arid soils with duripans are for the most part associated with easily weatherable glass of volcanic origin present initially in the parent material or added with volcanic ash during the course of soil formation (Chadwick et al., 1989; Flach et al., 1969). Some Australian red and brown hardpan soils have formed from colluvium derived largely from older silicified lateritic profiles (Bettenay and Churchward, 1974). Other sources of silica that have been suggested include dust fallout (other than volcanic ash) containing a high proportion of silica and recycling by plants, especially grass species, which are capable of taking up and concentrating considerable quantities of silica in the form of relatively soluble opal phytoliths (Summerfield, 1983a).

The formation of a duripan is similar to the formation of a petrocalcic horizon in some respects (Nettleton and Peterson, 1983). Both types of pans form at about the common depth of wetting and commonly form below or in the lower part of any argillic or natric horizon present. They both have developmental sequences starting from horizons of slight and diffused or widely spaced cement accumulation to horizons continuously cemented on ancient surfaces with continuously indurated horizons with multiple surficial laminae. Micromorphological evidence, however, indicates different modes of illuviation of the two types of cements (Chadwick et al., 1987; Flach et al., 1969). When the two cementing materials coexist in a soil, which is common in arid soils, silica precipitates higher in the soil profile than $CaCO_3$ but their zones of accumulation overlap greatly. On a microscopic scale their primary zones of precipitation tend to be mutually exclusive, with silica distributed throughout the plasma and $CaCO_3$ as homogenous deposits in larger voids. The origin of duripans (hardpans) in Australia has been attributed to episodic flooding and saturation of the soil on smooth slopes and plains, which causes leaching and acidification of the upper horizons. The leached silica is then deposited as a hardpan during subsequent drying (Bettenay and Churchward, 1974).

An indurated siliceous crust on flat-topped erosion remnants, referred to as silcrete, are extensive in the arid and semiarid regions of Australia (Langford-Smith, 1978) and southern Africa (Summerfield, 1981) and have been reported in other areas (Goudie, 1973). Silcretes are relict features that formed sometime in the past and are not now forming in the areas where they occur (Hutton, Twidale, and Milne, 1978). Although silcretes have been widely used as an indicator of arid or semiarid paleoenvironments (Summerfield, 1983b), it is believed that those in Australia formed primarily in the early Tertiary as a result of long-continued pedogenic activity on or just below a stable surface (Hutton et al., 1978). Summerfield (1983) believes that part of the silcretes in southern Africa were formed in humid climates.

Formation of Vertisols

The process by which Vertisols form are commonly associated with their shrink-swell capacities and the opening and closing of deep, wide cracks during drying and wetting cycles. During the dry season soil material from the surface often falls into the cracks. When the soil becomes wet and swells, the cracks close, exerting pressures upward because of the added excess material that falls into the lower horizons. Soil material is then forced toward the surface. With time, these soils invert themselves.

Material originally in the subsoil moves up and becomes the surface soil, and the surface soil becomes the subsoil. This pedoturbation or self-swallowing model is widely referred to (Ahmed, 1983; Buol et al., 1980; Wilding and Tessier, 1988). Wilding and Tessier (1988) have challenged this model as being the mechanism primarily responsible for gilgae relief, slickenside formation, and cyclic horizonation. They offer two alternative models that may offer a better explanation for the formation of certain Vertisol features. The differential loading model provides a hypothesis for gilgae formation whereby clays move from areas of higher to lower areas of confining pressures upon an underlying plastic subsoil or substrata. The mechanism is thus a plastic extrusion process. The soil mechanics model is especially pertinent for the formation of slickensides, which are caused by soil failure along shear planes. This failure occurs whenever the swelling pressures of soil material in a confined system exceed the shear strength of the soil at a specified moisture content.

Buol et al. (1980) offer three views relating to the place of Vertisols in a genetic sequence. They might be old soils, the end product of a developmental sequence of soils with argillic horizons. As soil formation progresses and clay becomes more abundant in the argillic horizon, the shrink-swell capacity increases until the pedoturbation processes become dominant. Another possibility is that a Vertisol containing smectite might progressively weather so that the smectite is converted to a nonswelling clay, reducing and/or eliminating the pedoturbation process. In this case the Vertisol would be considered relatively young. The third interpretation is that Vertisols are in "equilibrium" with their environment and that the smectites are stable and will persist, barring a change in climate.

Formation of Desert Pavement

Desert pavements are armored surfaces composed of angular or rounded fragments, usually one or two stones thick, set on or in matrices of finer material comprising varying mixtures of sand, silt, or clay (Cooke and Warren, 1973). These surfaces are common on many arid soils. The most commonly accepted theory is that desert pavements arise by deflation of fine material by wind or removal by water erosion due to lack of protection of the surface by vegetation (Cooke and Warren, 1973). Other mechanisms have also been involved whereby the gravel particles migrate upward by wetting and drying in hot deserts (Jessup, 1960; Springer, 1958) and freezing and thawing in soils of cold deserts (Claridge and Campbell, 1981). An additional hypothesis has recently been proposed by McFadden et al. (1987) in which gravels for the pavement are formed by in situ physical weathering of larger stones. The physical weathering is enhanced by the influx of salt-rich eolian dust. The dust accumulates slowly below the evolving pavement, promoting development of cumulative soils.

In many soils a light-colored relatively stone-free vesicular sandy loam to loam horizon up to 10 cm thick occurs under the desert pavement. The vesicular nature of this layer is thought to arise from wetting and drying and may develop quite rapidly after wetting or disturbance of the soil surface (Springer, 1958; Evenari et al., 1974).

One common feature of desert pavements is the dark manganese-rich coatings up to 0.1 mm thick, called desert varnish, that occur on exposed rock surfaces (Dorn and Oberlander, 1982). Considerable work has been carried out on the nature and origin of desert varnish (Allan, 1978; Dorn and Oberlander, 1982; Dragovitch, 1988; Potter and Rossman, 1977; Smith and Whalley, 1988; among others), but the mechanisms of formation are still somewhat obscure. The varnish is clearly formed by mechanisms external to the rock surface as the coatings are quite different in composition from that of the substrate and could not have been formed by outward movement of ions released by weathering. The commonly accepted mode of formation is that it is biogenic in origin, formed by the action of bacteria and fungi growing during the evaporation phase of predawn dewfalls on surface rocks and trapping iron, manganese, clay minerals, and so on, present in windblown dust.

IV. CONCLUSIONS

Historically our knowledge about the nature, distribution, and mode of origin of arid soils has lagged behind that of humid region soils. This has been changing in recent years with the increasing recognition that basic knowledge about soils of arid regions is extremely important in providing a conceptual frame of reference for land use planning and for development and management of soils for agriculture, urban, and other uses. That our knowledge is increasing is reflected by the changes being made in the classification of many arid soils. There are still many gaps in our knowledge, as is brought out in this brief review, in which alternative hypotheses and sometimes controversy exist for explaining the origin of soil features.

REFERENCES

Ahmed, N. (1983). Vertisols. In *Pedogenesis and Soil Taxonomy. II. The Soil Orders*, L. P. Wilding, N. E. Smeck, and G. F. Hall (Eds.). Elsevier, Amsterdam, pp. 91-123.

Akpokodje, E. G. (1984). The influence of rock weathering on the genesis of gypsum and carbonate in some Australian Arid Zone soils. *Aust. J. Soil Res. 22*: 243-251.

Akpokodje, E. G. (1987). The mineralogical relationship between some arid zone soils and their underlying bedrocks at Fowlers Gap Station, New South Wales (N.S.W.) Australia. *Roy. Soc. New South Wales J. 120*: 91-99.

Allan, C. C. (1978). Desert varnish of the Sonoran Desert—optical and electron probe microanalysis. *J. Geol. 86*: 743-752.

Al-Rawi, A. H., Jackson, M. L., and Hole, F. D. (1969). Mineralogy of some arid and semi-arid lands of Iraq. *Soil Sci. 107*: 480-486.

Anderson, D. W. (1987). Pedogenesis in the grassland and adjacent forests of the Great Plains. *Adv. Soil Sci. 7*: 53-93.

Arkley, R. J. (1963). Calculation of carbonates and water movement in soil from climatic data. *Soil Sci. 96*: 239-248.

Aubert, G., and Duchaufour, P. (1956). *Project de classification des sols. VI.* Cong. Int. Sci du Sol, Paris, Vol. 97E: 597-604.

Baker, W. E. (1973). The role of humic acids from Tasmanian podzolic soils in mineral degradation and metal mobilization. *Geochim. Cosmochim. Acta 37*: 269-281.

Barnhisel, R. I. (1977). Chlorites and hydroxy interlayered vermiculite and smectite. In *Minerals in Soil Environments*, J. D. Dixon and A. B. Weed (Eds.). Soil Sci. Soc. Am. Madison, Wisconsin, pp. 331-356.

Barshad, I. (1966). The effect of variation in precipitation on the nature of clay mineral formation in soils from acid and basic rocks. *Int. Clay Conf. (Jerusalem), Proc. Vol. 1*: 167-173.

Barshad, I., and Kishk, F. M. (1969). Chemical composition of soil vermiculture clays as related to their genesis. *Contr. Min. Petrol. 24*: 136-155.

Bettanay, E., and Churchward, H. M. (1974). Morphology and stratigraphic relationships of the Saline Hardpan in arid Western Australia. *J. Geol. Soc. Aust. 21*: 73-80.

Birkeland, P. W. (1984). *Soils and Geomorphology*. Oxford University Press, Oxford.

Birkeland, P. W., and Janda, R. J. (1971). Clay mineralogy of soils developed from Quaternary deposits in the eastern Sierra Nevada, California. *Geol. Soc. Am. Bull. 82*: 2495-2514.

Blackwelder, E. (1933). The insolation hypothesis of rock weathering. *Am. J. Sci. 266*: 97-113.

Bohn, H., McNeal, B. L., and O'Connor, G. A. (1984). *Soil Chemistry* (2nd ed.). John Wiley, New York.

Brown, I. C., and Drosdorff, M. (1940). Chemical and physical properties of soils and their colloids developed from granitic materials in the Mohave Desert. *J. Agr. Res. 61*: 335-352.

Buol, S. W. (1965). Present soil-forming factors and processes in arid and semi-arid regions. *Soil Sci. 99*: 45-49.

Buol, S. W., Hole, F. D., and McCracken, R. J. (1980). *Soil Genesis and Classification* (2nd ed.). Iowa State University Press, Ames.

Chadwick, O. A., Hendricks, D. M., and Nettleton, W. D. (1987). Silica in duric soils. I. A depositional model. *Soil Sci. Soc. Am. J. 51*: 975-982.

Chadwick, O. A., Hendricks, D. M., and Nettleton, W. D. (1989). Silicification of Holocene soils in northern Monitor Valley, Nevada. *Soil Sci. Soc. Am. J. 53*: 158-164.

Claridge, G. G. C., and Campbell, I. B. (1981). A comparison between hot and cold desert soils and soil processes. In *Aridic Soils and Geomorphic Processes, Catena Suppl.*, Vol. 1, D. H. Yaalon (Ed.), pp. 1-28.

Colman, S. M., and Dethier, D. P. (1986). An overview of rates of chemical weathering. In *Rates of Chemical Weathering of Rocks and Minerals*, S. V. Colman and D. P. Dethier (Eds.). Academic Press, London, pp. 1-20.

Comerma, J. A. (1989). *International Committee on Vertisols (ICOMERT)*, 5th Circular Letter. (Unpublished Memo.)

Comerma, J. A., Williams, D., and Newman, A. (1988). Conceptual changes in the classification of Vertisols. In *Vertisols, Their Distribution, Properties and Management*, L. P. Wilding and R. Puentes (Eds.).

Texas A&M University and U.S. Dept. Agric. Soil Management Support Services, College Station, Texas, pp. 41-54.
Commission de Pédologie et de Cartographie des sols (1967). *Classification des Sols*. Lab. Geol. Pedol., Ecole Nat. Super. Agron, Grignon, France.
Cooke, R. V., and Smalley, I. J. (1968). Salt weathering in deserts. *Nature 220*: 1226-1227.
Cooke, R. V., and Warren, A. (1973). *Geomorphology in Deserts*. University of California Press, Berkeley.
Dan, J., and Yaalon, D. H. (1981). Automorphic saline soils in Israel. In *Aridic Soils and Geomorphic Processes, Catena (Suppl.)*, Vol. 1, pp. 101-115.
Derbyshire, E. (1983). On the morphology, sediments and origin of the loess plateau of central China. In *Megageomorphology*, R. Gardner and H. Scogin (Eds.). Oxford University Press, Oxford, pp. 172-194.
Dixon, J. B. (1977). Kaolinite and serpentine group minerals. In *Minerals in Soil Environments*, J. B. Dixon and S. B. Weed (Eds.), Soil Sci. Soc. Am., Madison, Wisconsin, pp. 357-404.
Dokuchaev Institute of Soil Science (1986). *Classification and Diagnostics of Soils of the USSR* (translated from Russian). Amerind Publ. Co., New Delhi.
Dorn, R. I., and Oberlander, F. M. (1982). Rock varnish. *Prog. Phys. Geogr. 6*: 317-367.
Douglas, L. A. (1977). Vermiculites. In *Minerals in Soil Environments*, J. B. Dixon and S. B. Weed (Eds.). Soil Sci. Soc. Am., Madison, Wisconsin, pp. 259-292.
Dragovitch, D. (1988). A preliminary electron probe study of microchemical variations in desert varnish in western New South Wales, Australia. *Earth Surf. Proc. Landforms 13*: 259-270.
Dregne, H. E. (1976). *Soils of Arid Regions*. Elsevier, New York.
Eberl, D. D. (1985). Clay mineral formation and transformation in rocks and soils. *Philos. Trans. Rov. Soc. Lond. [A] 311*: 241-257.
El-Tezhani, M. S., Gradnsov, B. B., Rubilina, N. Y., and Chizhikova, N. P. (1984). Chemical and mineral composition of the finely dispersed component and microstructure of some soils in Sudan. *Sov. Soil Sci.*, (1), 75-81.
Emory, K. O. (1944). Brush fires and rock exfoliation. *Am. J. Sci. 242*: 506-508.
Eswaran, H., and Barzanji, A. F. (1974). Evidence for the neoformation of attapulgite in some soils of Iraq. *Trans. 10th Int. Cong. Soil Sci. (Moscow) 7*: 154-161.
Eswaren, H., and Tavernier, R. (1980). Classification and genesis of Oxisols. In *Soils with Variable Charge*, B. K. G. Theng (Ed.). New Zeal. Soc. Soil Sci., Lower Hutt, New Zealand, pp. 427-442.
Evans, I. S. (1970). Salt crystallization and rock weathering: A review. *Rev. Geomorph. Dynamique 4*: 153-177.
Evenari, M., Yaalon, D. H., and Gatterman, V. (1974). Note on soils with vesicular structure in deserts. *Z. Geomorph. 18*: 162-172.
Fahey, B. D. (1983). Frost action and hydration as rock weathering mechanisms on schist: A laboratory study. *Earth Surf. Proc. Landforms 8*: 535-545.

Fahey, B. D. (1985). Salt weathering as a mechanism of rock breakup in cold climates: An experimental approach. *Z. Geomorph. N.F. 29*: 99-111.

Fanning, D. S., and Keramidas, V. Z. (1977). Micas. In *Minerals in Soil Environments*, J. B. Dixon and S. B. Weed (Eds.). Soil Sci. Soc. Am., Madison, Wisconsin, pp. 195-198.

FAO (1974). *Soil Map of the World, 1:5,000,000*, Vol. 1, *Legend*. UNESCO, Paris.

Fenton, T. E. (1983). Mollisols. In *Pedogenesis and Soil Taxonomy. II. The Soil Orders, Dev. Soil Sci. IIB*, L. P. Wilding, N. E. Smeck, and G. F. Hall (Eds.). Elsevier, Amsterdam, pp. 125-163.

Flach, K. W., Nettleton, W. D., Gile, L. H., and Cady, J. G. (1969). Pedocementation: Induration by silica, carbonates and sesquioxides in the Quaternary. *Soil Sci. 107*: 442-453.

Flach, K. W., Holzhey, C. S., DeConick, F., and Bartlett, R. J. (1980). Genesis and classification of Andepts and Spodosols. In *Soils with Variable Charge*, B. K. G. Theng (Ed.). New Zeal. Soc. Soil Sci., Lower Hutt, New Zealand, pp. 411-426.

Folkoff, M. E., and Meentemeyer, V. (1985). Climatic control of the assemblages of secondary clay minerals in the A horizon of United States soils. *Earth Surf. Proc. Landforms 10*: 621-633.

Folkoff, M. E., and Meentemeyer, V. (1987). Climatic control of the geography of clay minerals genesis. *Ann. Assoc. Am. Geogr. 77*: 635-650.

Foss, J. E., Moorman, F. R., and Rieger, S. (1983). Inceptisols. In *Pedogenesis and Soil Taxonomy. II. The Soil Orders, Dev. Soil Sci. IIB*, L. P. Wilding, N. E. Smeck, and G. F. Hall (Eds.). Elsevier, Amsterdam, pp. 355-382.

Gal, M., Amiel, A. J., and Ravikovitch, S. (1974). Clay mineral distribution and origins in the soil types of Israel. *J. Soil Sci. 25*: 79-89.

Gardner, L. L. (1972). Origin of the Mormon Mesa caliche, Clark County, Nevada. *Geol. Soc. Am. Bull. 83*: 143-156.

Gharaee, H. A., and Mahjoory, A. R. (1984). Characteristics and geomorphic relationships of some representative Aridisols in southern Iran. *Soil Sci. Soc. Am. J. 48*: 1115-1119.

Gile, L. H., and Grossman, R. G. (1979). *The Desert Project Monograph*. U.S. Dept. Agric., Soil Conserv. Serv. (Mimeo), USDA, Washington, D.C.

Gile, L. H., Peterson, F. F., and Grossman, R. B. (1966). Morphologic and genetic sequences of carbonate accumulation in desert soils. *Soil Sci. 101*: 347-360.

Goudie, A. (1973). *Duricrusts in Tropical and Subtropical Landscapes*. Clarendon Press, Oxford.

Goudie, A. (1974). Further experimental investigation of rock weathering by salt and other mechanical processes. *Z. Geomorph. N.F. 21*: 1-12.

Goudie, A. S., Cooke, R. V., and Dorrakamp, J. C. (1979). The formation of silt from quartz dune sand by salt-weathering processes in deserts. *J. Arid Environ. 2*: 105-112.

Griggs, D. T. (1936). The factor of fatigue in rock exfoliation. *J. Geol. 44*: 783-796.

Hall, K. (1988). A laboratory simulation of rock breakdown due to freeze-thaw in a maritime antarctic environment. *Earth Surf. Proc. Landforms 12*: 137-149.

Hendricks, D. M., Nettleton, W. D., and Grossman, R. B. (1985). Problems associated with the U.S. System of Soil Taxonomy mineralogy classes in arid regions. In *Mineral Classification of Soils*, J. A. Kittrick (Ed.). Soil Sci. Soc. Am. Spec. Publ. No. 16, Madison, Wisconsin, pp. 41-52.

Henin, S., and Caillere, S. (1975). Fibrous minerals. In *Soil Components*, Vol. 2, *Inorganic Components*, J. E. Gieseking (Ed.). Springer-Verlag, New York, pp. 335-350.

Huang, W. H., and Keller, W. D. (1970). Dissolution of rock-forming silicate minerals in organic acids. *Am. Min. 55*: 1076-1094.

Hutton, J. T., Twidale, C. R., and Milnes, A. R. (1978). Characteristics and origin of some Australian silcretes. In *Silcrete in Australia*, T. Langford-Smith (Ed.). Dept. Geogr., Univ. New England, Australia.

Isbell, R. F., Reeve, R., and Hutton, J. F. (1983). Salt and sodicity. In *Soils, An Australian Viewpoint*. Div. of Soils, CSIRO. Academic Press, London, pp. 107-117.

Ismail, F. F. (1969). Role of ferrous iron oxidation in the alteration on biotite and its effect on the type of clay minerals formed in soils of arid and humid regions. *Am. Min. 54*: 1460-1466.

Ismail, F. F. (1970). Biotite weathering and clay formation in arid and humid regions, California. *Soil Sci. 109*: 257-261.

Ivanova, Y. N., and Bol'shavov, A. F. (1972). Academician K. K. Gedroyts' theory of solonchaks, solonetzes and solods and its subsequent elaboration. *Sov. Soil Sci.*, (4) 156-168.

Jackson, E. A. (1957). Soil features in arid regions with particular reference to Australia. *J. Aust. Inst. Agr. Sci. 25*: 196-208.

Jackson, T. A., and Keller, W. D. (1970). A comparative study of the role of lichens and "inorganic" processes in the chemical weathering of recent Hawaiian lava flow. *Am. J. Sci. 269*: 446-466.

Jenny, H. (1941). *Factors of Soil Formation*. McGraw-Hill, New York.

Jenny, H. (1980). *The Soil Resource, Origin and Behavior*. Springer-Verlag, New York.

Jessup, R. W. (1960). The stony tableland soils of the southwestern portion of the Australia arid zone and their evolutionary history. *J. Soil Sci. 11*: 188-196.

Keller, W. D. (1970). Environmental aspects of clay minerals. *J. Sediment. Petrol. 40*: 788-813.

Kerr, A., Smith, B. J., Whalley, W. B., and McGreevy, J. P. (1984). Rock temperatures from southeastern Morocco and their significance for experimental rock weathering studies. *Geology 12*: 306-309.

Kittrick, J. A. (1970). Synthesis of kaolinite at 25°C and 1 atmosphere. *Clays Clay Min. 18*: 261-267.

Kittrick, J. A. (1971). Montmorillonite and the weathering environment. *Soil Sci. Soc. Am. Proc. 35*: 815-820.

Konishchev, V. N. (1982). Characteristics of cryogenic weathering in the permafrost zone of the European USSR. *Arct. Alp. Res. 14*: 2161-2165.

Kononova, M. M. (1975). Humus of virgin and cultivated soils. In *Soil Components*, Vol. 1, *Organic Components*, J. S. Gieseking (Ed.). Springer, New York, pp. 475-520.

Krauskopf, K. B. (1967). *Introduction to Geochemistry.* McGraw-Hill, New York.
Langford-Smith, T. (Ed.) (1978). *Silcrete in Australia.* Dept. Geogr., Univ. New England, Australia.
Leamy, M. L. (1988). *International Committee on the Classification of Andisols (ICOMAND)*, Circular Letter No. 10. (Unpublished Memo.)
Leamy, M. L., Smith, G. D., Colmet-Daage, F., and Otowa, M. (1980). The morphological characteristics of Andisols. In *Soils with Variable Charge*, B. K. G. Theng (Ed.). New Zeal. Soc. Soil Sci., Lower Hutt, New Zealand, pp. 17-34.
Machette, M. N. (1985). Calcic soils of the southwestern United States. In *Soils and Quaternary Geology of the Southwestern United States, Geol. Soc. Am. Spec. Paper*, D. L. Weide (Ed.), *203*: 1-21.
Mahjoory, R. A. (1974). Clay minerology, physical and chemical properties of some soils in arid regions of Iran. *Soil Sci. Soc. Am. Proc. 39*: 1157-1164.
Manley, E. P., and Evans, L. J. (1986). Dissolution of feldspars by low molecular weight aliphatic and aromatic acids. *Soil Sci. 141*: 106-112.
Marion, G. M., Schlesinger, W. H., and Fonteyn, P. J. (1985). CALDEP: A model for soil $CaCO_3$ (caliche) deposition in southwestern deserts. *Soil Sci. 139*: 468-481.
Mast, M. A., and Drever, J. I. (1987). The effect of oxalate on the dissolution rates of olgioclase and tremolite. *Geochim. Cosmochim. Acta 51*: 2559-2568.
Mayer, L., McFadden, L. D., and Harden, J. W. (1988). Distribution of calcium carbonate in desert soils: A model. *Geology 16*: 303-306.
McFadden, L. D., and Hendricks, D. M. (1985). Changes in the content and composition of pedogenic iron oxyhydroxides in a chronosequence of soils in southern California. *Quat. Res. 23*: 189-204.
McFadden, L. D., Wells, S. G., and Jercinovich, M. (1987). Influences of eolian and pedogenic processes on the origin and evolution of desert pavements. *Geology 7*: 475-488.
McGreevy, J. P. (1982). Frost and salt weathering: Further experimental results. *Earth Surf. Proc. Landforms 7*: 475-488.
McKeague, J. A., and St. Arnaud, R. J. (1969). Pedotranslocation: Eluviation-illuviation in soils during the Quaternary. *Soil Sci. 107*: 428-434.
Millot, G. (1970). *Geology of Clays.* Springer-Verlag, New York.
Millot, G., Paquet, H., and Ruellan, A. (1969). Neoformation l'attapulgite dans les sols é carpaces calcaires de la Basse Moulowy (Maroc Oriental). *C.R. Searces Acad. Sci. [D] 268*: 2771-2774.
Minashina, N. G., and Gradisov, B. P. (1973). The mineralogy of clay in some desert soils. *Sov. Soil Sci.* (7) 459-472.
Nettleton, W. D., and Brasher, B. R. (1983). Correlation of clay minerals and properties of soils in the western United States. *Soil Sci. Soc. Am. J. 47*: 1032-1036.
Nettleton, W. D., and Peterson, F. F. (1983). Aridisols. In *Pedogenesis and Soil Taxonomy. II. The Soil Orders*, L. P. Wilding, N. E. Smeck, and G. F. Hall (Eds.). Dev. Soil Sci. IIB, Elsevier, Amsterdam, pp. 165-215.

Nettleton, W. D., Nelson, R. E., and Flach, K. W. (1973). Formation of mica in surface horizons of dry land soils. *Soil Sci. Soc. Am. J. 37*: 473-478.
Nettleton, W. D., Nelson, R. E., Brasher, B. R., and Derr, P. S. (1982). Gypsiferous soils in the western United States. In *Acid Sulfate Weathering*, J. A. Kittrick, D. S. Fanning, and L. R. Hossner (Eds.). Soil Sci. Soc. Am. Spec. Publ. No. 10, Madison, Wisconsin, pp. 147-168.
Nikiforoff, C. C. (1937). General trends of the desert type of soil formation. *Soil Sci. 43*: 105-131.
Norrish, K., and Pickering, J. G. (1983). Clay minerals. In *Soils, An Australian Viewpoint*. Academic Press, New York, pp. 281-308.
Ollier, C. (1975). *Weathering*. Longman, London.
Osman, A. (1989). *International Committee on Aridisols (ICOMID)*, Draft Report Version 6.0. (Unpublished Memo.)
Paquet, H., Ruellan, A., Tardy, Y., and Millot, G. (1969). Geochimie d'un bassin versant au Maroc Oriental, Evolution des argiles dans les sols des montagnes et des plaines de la Basse Maulouya. *C.R. Searces Acad. Sci. [D] 269*: 1839-1842.
Pedro, G. (1983). Structuring of some basic pedological processes. *Geoderma 31*: 289-299.
Potter, R. M., and Rossman, G. R. (1977). Desert varnish: The importance of clay minerals. *Science 196*: 1446-1448.
Pye, K., and Sperling, C. H. B. (1983). Experimental investigation of silt formation by static breakage processes: The effect of temperature, moisture and salt on quartz dune sand and granite regolith. *Sedimentology 20*: 49-62.
Rabenhorst, M. C., and Wilding, L. P. (1986). Pedogenesis on the Edwards Plateau, Texas. III. New model for the formation of petrocalcic horizons. *Soil Sci. Soc. Am. J. 50*: 693-699.
Ravina, I., and Zaslavsky, D. (1974). The electric double-layer as a possible factor in desert weathering. *Z. Geomorph. N.F. 21*: 13-18.
Reiche, P. (1950). *A Survey of Weathering Processes and Products*, New Mex. Univ. Publ. Geol. 3. Albuquerque, New Mexico.
Rice, A. (1984). Insolation warmed over. *Geology 4*: 61-62.
Roth, A. S. (1970). Frost action in rocks: Some experimental data. *Trans. Inst. Br. Geogr. 49*: 109-124.
Roth, E. S. (1965). Temperature and water content as factors in desert weathering. *J. Geol. 73*: 454-468.
Rozov, N. N., and Ivanova, E. N. (1967). Classification of the soils of the USSR. *Sov. Soil Sci. 2*: 147-156.
Rust, R. H. (1983). Alfisols. In *Pedogenesis and Soil Taxonomy. II. The Soil Orders*, L. P. Wilding, N. E. Smeck, and G. F. Hall (Eds.). Dev. Soil Sci. IIB, Elsevier, Amsterdam, pp. 253-281.
Sawhney, B. L. (1977). Interstratification in layer silicates. In *Minerals in Soil Environments*, J. B. Dixon and S. B. Weed (Eds.). Soil Sci. Soc. Am., Madison, Wisconsin, pp. 405-435.
Schlesinger, W. H. (1982). Carbon storage in the caliche of arid soils: A case study from Arizona. *Soil Sci. 133*: 247-255.

Schlesinger, W. H. (1985). The formation of caliche in soils of the Mojave Desert, California. *Geochim. Cosmochim. Acta 49*: 57-66.

Schwertman, U. (1985). The effect of pedogenic environments on iron oxide minerals. *Adv. Soil Sci. 1*: 172-199.

Shadfan, H., Mashady, A., Eter, A., and Husrin, A. A. (1984). Mineral composition of selected soils in Saudi Arabia. *Z. Pflanz. Bodenkd. 147*: 657-668.

Shadfan, H., Dixon, J. B., and Kippenberg, L. A. (1985). Palygorskite distribution in Tertiary limestone and associated soil of northern Jordan. *Soil Sci. 140*: 206-212.

Simonson, R. W. (1978). A multiple process model of soil genesis. In *Quaternary Soils*, W. D. Mahaney (Ed.). 3rd Symp. Quat. Res., Toronto. Geo. Abstr., Norwich, England, pp. 1-25.

Singer, A. (1983). Clay formation in saprolites of igneous rocks under semiarid to arid conditions, Negev southern Israel. *Soil Sci. 137*: 322-340.

Singer, A. (1988). Illite in aridic soils, desert dusts and desert loess. *Sediment. Geol. 59*: 251-259.

Singer, A., and Galan, E. (1984). *Palygorskite-sepioloite Occurrences, Genesis and Uses*, Dev. Sed. 37. Elsevier, New York.

Singer, A., and Norrish, K. (1974). Pedogenic palygorskite occurrences in Australia. *Am. Min. 59*: 508-517.

Smith, B. J. (1977). Rock temperature measurements from the northeast Sahara and their implications for rock weathering. *Catena 4*: 41-63.

Smith, B. J., and Whalley, W. B. (1988). A note on the characteristics and possible origins of desert vanishes from southeast Morocco. *Earth Surf. Proc. Landforms 13*: 251-258.

Smith, B. J., McGreevy, J. P., and Whalley, W. G. (1989). Silt production by weathering of a sandstone under hot arid conditions: An experimental study. *J. Arid Environ. 12*: 199-214.

Smith, G. D. (1965). Lectures on soil classification. *Pedologie 4*: 41-63.

Soil Survey Staff (1975). *Soil Taxonomy. A Basic System of Soil Classification for Making and Interpreting Soil Surveys.* U.S. Dept. Agr. Handbook 436, USDA, Washington, D.C.

Soil Survey Staff (1987). *Keys to Soil Taxonomy.* Agency for Intern. Dev., and U.S. Dept. Agric. Soil Management Support Service Tech. Monogr. No. 6 (3rd printing), Cornell Univ., Ithaca, New York.

Sosnovskay, V. P. (1982). Mineralogical composition of clay in some soils of the Gissar Valley. *Sov. Soil Sci.* (5) 83-91.

Sotnikova, G. F. (1975). Effect of temperature on the decomposition of minerals by the fulvic acids of red earths during through leaching. *Sov. Soil Sci.* (1) 46-52.

Southard, R. J., and Southard, A. K. (1985). Genesis of cambic and argillic horizons in two northern Utah Aridisols. *Soil Sci. Soc. Am. J. 49*: 67-171.

Springer, M. E. (1958). Desert pavement and vesicular layer of some desert soils in the desert of the Lahonton Basin, Nevada. *Soil Sci. Soc. Am. Proc. 22*: 63-66.

Stace, H. C. F., Hubble, G. D., Brewer, R., Northcote, K. H., Sleeman, J. R., Mulcahy, M. J., and Hallsworth, E. G. (1968). *A Handbook of Australian Soils.* Rellum Tech. Publ., Glenside, South Australia.

Summerfield, M. A. (1981). Distribution, nature and genesis of silcrete in arid and semi-arid southern Africa. *Aridic Soils and Geomorphic Processes, Catena (Suppl. 1)*: 37-65.

Summerfield, M. A. (1983a). Silcrete. In *Chemical Sediments and Geomorphology*, A. S. Goudie and K. Pye (Eds.). Academic Press, London, pp. 59-91.

Summerfield, M. A. (1983b). Silcrete as a paleoclimatic indicator: Evidence from southern Africa. *Paleogeogr. Paleocl. Paleoecol. 41*: 65-79.

Swantesson, J. (1985). Experimental weathering studies of Swedish rocks. *Geograf. Ann. 67A*: 115-118.

Tate, K. R., and Theng, B. K. G. (1980). Organic matter and its interaction with inorganic soil constituents. In *Soils with Variable Charge*, B. K. G. Theng (Ed.). New Zeal. Soc. Soil Sci., Lower Hutt, New Zealand, pp. 225-252.

Thorn, C. E. (1979). Bedrock freeze-thaw weathering regime in an alpine environment, Colorado Front Range. *Earth Surf. Proc. 4*: 211-228.

Trenhaile, A. S., and Mercan, D. W. (1984). Frost weathering and the saturation of coastal rocks. *Earth Surf. Proc. Landforms 9*: 321-331.

Verheye, W. (1975). Nature and impact of temperature and moisture in arid weathering and soil forming processes—A review. *Pedologie 26*: 205-224.

Vidrich, V., Cecconi, C. O., Ristori, G. G., and Fuse, P. (1982). Verwitterung Toskanischer Gesteine unter Mitwerkung von Flechten. *Z. Pflanz. Bodenkd. 145*: 384-389.

Walker, R. F. (1967). Formation of red beds in modern and ancient deserts. *Geol. Soc. Am. Bull. 78*: 353-368.

Watson, W. (1983). Gypsum crusts. In *Chemical Sediments and Geomorphology: Precipitates and Residue in the Near Surface Environment*, A. S. Goudie and K. Pye (Eds.). Academic Press, New York, pp. 133-161.

Watson, A. (1988). Desert gypsum crusts as paleoenvironmental indicators: A micropetrographic study of crusts from southern Tunisia and the central Namib Desert. *J. Arid Environ. 15*: 19-42.

Weyl, P. K. (1958). The solution kinetics of calcite. *J. Geol. 60*: 163-176.

White, S. E. (1976). Is frost action really only hydration shattering? A review. *Arct. Alp. Res. 8*: 1-6.

Wilding, L. P., and Tessier, D. (1988). Genesis of vertisols: Shrink swell phenomena. In *Vertisols: Their Distribution, Properties, Classification and Management*, L. P. Wilding and R. Puentes (Eds.). Texas A&M Univ. Printing Center, College Station, Texas.

Wildung, R. E., Garland, F. R., and Buschbom, R. L. (1975). The interdependent effects of soil temperature and water content on soil respiration rate and plant decomposition in arid grassland soils. *Soil Biol. Biochem. 7*: 373-378.

Williams, R. B. G., and Robinson, D. A. (1981). Weathering of sandstone by the combined action of frost and salt. *Earth Surf. Proc. Landforms 6*: 1-9.

Yaalon, D. H., and Wieder, M. (1976). Pedogenic palygorskite in some arid brown (Calciorthid) soils of Israel. *Clay Min. 11*: 73-80.

Young, J. A., and Evans, R. A. (1986). Erosion and deposition of fine sediments from playas. *J. Arid Environ. 10*: 103-116.

Zelazny, L. W., and Calhoun, F. G. (1977). Palygorskite (attapulgite), sepiolite, talc, pyrophyllite and zeolites. In *Minerals in Soil Environments*, J. B. Dixon and S. B. Weed (Eds.). Soil Sci. Soc. Am., Madison, Wisconsin, pp. 435-436.

Zunino, H., and Martin, J. P. (1977). Metal-binding organic macro-molecules in soil. I. Hypothesis interpreting the role of soil organic matter in the translocation of metal ions from rocks to biological systems. *Soil Sci. 123*: 65-76.

3

Physical Properties of Arid Region Soils

MICHAEL J. SINGER *University of California, Davis, California*

I. INTRODUCTION

Soils of aridic, xeric, and ustic soil moisture regimes occupy 6.5×10^9 ha in Africa, Southwest Asia, Southeast Asia, and South America (Higgins et al., 1981). Physical properties that are suggested to be common to many arid and semiarid region soils include high erodibility, high runoff generation potential, susceptibility to seal and crust formation, poor water-holding capacity, profile hardening, structural instability, low clay content, low activity clay, high bulk density, and high surface temperatures in the summer months (El-Swaify et al., 1984; Guthrie, 1982; Lal, 1985). It is necessary to be cautious in making generalizations about soils that occur throughout the world in a vast array of combinations of soil-forming factors. For example, most of the soils of the arid and semiarid parts of Chile seem to fit these characteristics (Vial and Wright, 1965), although Vial and Wright show that the soils have a wide range of textures and profile properties over the wide range of precipitation and elevations existing in Chile. Surface textures are often sandy, and subsoils have considerable silt and clay. Claridge and Campbell (1982) show that there are many similarities between hot and cold desert soils, including horizon development and the distribution and chemistry of salts. For the purpose of this review, cold arid regions, such as Antarctica, are not considered.

Most of the 10 (soon to be 11) different soil orders occur in arid and semiarid regions. In California, Alfisols, Aridisols, Entisols, Inceptisols, Mollisols, Vertisols, Ultisols, and Histosols are found in arid and semiarid climates. In Australia and Africa, on old land surfaces, Oxisols are found in arid and semiarid climates. In Israel, 6 of the 10 soil orders occur (Dan et al., 1981). Krantz et al. (1978) state that Alfisols and Vertisols are the two most commonly found soil orders in the semiarid tropics. Because of the wide array of soils found in these climate zones, it is difficult to generalize about their properties.

Some common chemical properties, such as low organic matter and the presence of salts, especially Na, result in poor aggregate stability and lead to common problems among arid region soils, such as high water and wind erodibility and high propensity for crusting. Of special importance to soils of arid and semiarid regions, particularly in developing countries, is the problem of structural deterioration and organic matter loss. As Greenland (1981) has remarked, "the less spectacular problems of structural deterioration and organic matter loss frequently reduce yields to well below the real potential." This chapter considers some of the physical properties and problems common to arid and semiarid region soils.

II. SOLID-PHASE PROPERTIES

A. Particle Size Distribution

Definition

Particle size distribution systems place the nearly infinite range of individual, unaggregated mineral particle sizes into discrete groups. The USDA system (USDA, 1983) separates soil particles into a coarse fraction > 2 mm and a fine fraction < 2 mm. The fine fraction is further subdivided into sand (2-0.05 mm), silt (0.05-0.002 mm), and clay (<0.002 mm). Sand, silt, and clay are sometimes further subdivided into fine and coarse fractions (USDA, 1983). A ternary diagram with 100% sand, silt, and clay at each corner (Fig. 1) is used to determine the textural class name.

The Unified and AASHTO (Americal Association of State Highway and Transportation Officials) systems were developed to classify particle size distribution for engineering purposes. Larger size particles are considered in these systems compared to the USDA system, and there are differences in how the three major fine fractions are defined (Table 1).

Another system of particle size classification is used in the USDA system of soil classification to place soils into particle size families (Soil Management Support Services, 1987). Sand, silt, and clay sizes are identical to the USDA particle size classification shown in Fig. 1, but the textural class names are replaced with particle family classifications for purposes of pedon classification (Fig. 2). In Fig. 2, only the major family particle size classes are shown. Both fine and very fine classes are sometimes referred to as "clayey" and the fine loamy, coarse loamy, fine silty, and coarse silty as "loamy."

Examples

Lal (1987) states that soils of the arid and semiarid tropics tend to be characterized by a gravelly horizon and coarse-textured surface horizons. A review of the soils of the semiarid tropics, derived from the FAO/UNESCO soil map of the world, shows that a wide variety of soil textures are found in these areas (Swindale, 1982). Arid regions of Australia are also reported to have surface textures of sand to sandy loam and, rarely, clay loam (Northcote et al., 1975). Of some 1976 samples analyzed by de Oliveira (1968), those of the semiarid areas were predominantly sand, loamy sand, and sandy loam. Very sandy soils are extensive in the semiarid tropics, and medium and fine-textured soils are extensive in eastern Africa, India, and South America (Swindale, 1982).

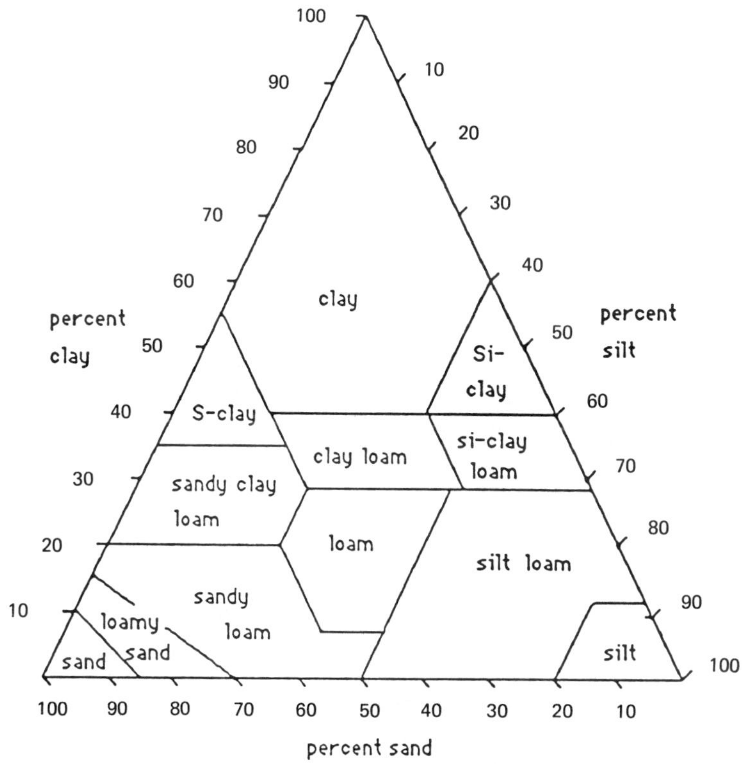

FIGURE 1 USDA textural triangle.

Butler and Hubble (1977) classify the semiarid soils of Australia into five groups: coarse-textured, coherent earths, cracking clays, duplex soils, and shallow stony and gravelly soils. These groups include soils that would be classified as Alfisols, Entisols, Inceptisols, Mollisols, Oxisols, and Ultisols in the USDA system of classification. From this study, it is difficult to generalize about the texture of arid region soils other than to say they occur in a wide range of particle sizes, from the coarse textured with high water intake rates, low plant available water, and high wind and

TABLE 1 Comparison of USDA, Unified, and AASHTO Particle Size Definitions (mm) for Soils

Classification	Gravel	Sand	Silt	Clay
USDA	> 2	2-0.05	0.05-0.002	< 0.002
Unified	> 5	5-0.074	< 0.074[a]	
AASHTO	> 2	2-0.074	< 0.074-0.005	< 0.005

[a]Silt and clay are considered one size separate.

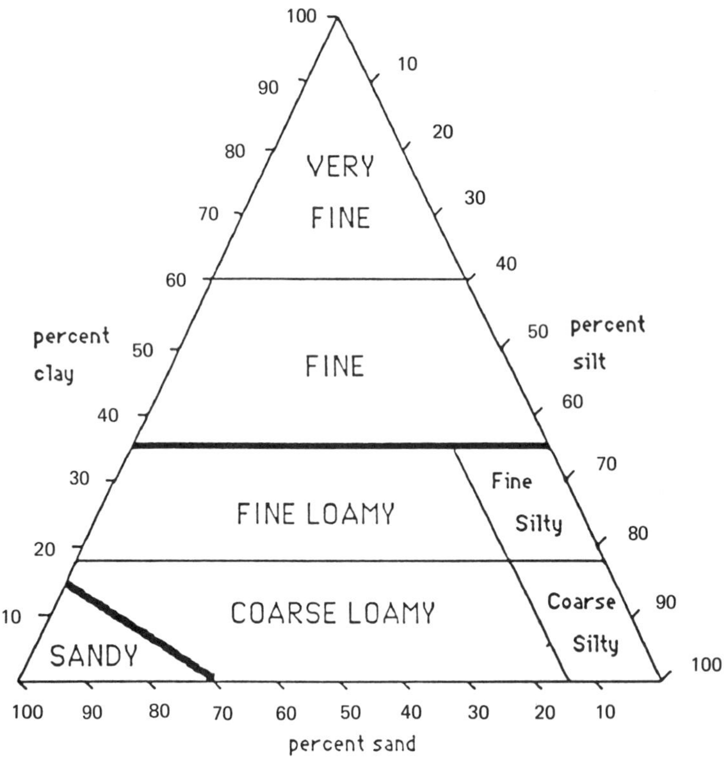

FIGURE 2 Textural triangle with particle size family classes substituted for texture classes.

water erosion potential to the cracking clays, which may have over 86% clay with a wide range of structural properties, low plant available water, low infiltration rates when moist, and generally low erodibilities.

In the United States, approximately 2150 soil series are classified as Aridisols (USDA, 1988). The family particle size classification of these spans the entire range of possibilities, from very fine (>60% clay in the control section) to sandy skeletal (Table 2).

Among the Argids (Table 2), those Aridisols that contain argillic horizons, control sections are dominated by fine and fine loamy particle size classes. The particle size control section for soils with argillic horizons is generally the upper 50 cm of the argillic horizon [see Soil Management Support Services (1987) for a complete definition of control section]. Other major particle size classes are the loamy skeletal and loamy. To be skeletal, a soil must have 35% or more rock fragments larger than 2 mm in diameter in the control section. About 24% of the Argids have skeletal control section particle size classes.

Series within the Orthids (Table 2) have a slightly larger percentage of skeletal control sections (26.9%) and a wide range of particle size families. The "intermediate" textures appear to dominate the family classification, with loamy (12%), coarse loamy (20%), and fine loamy (15%) containing

TABLE 2 Percentage Occurrence of Particle Size Classes in the Orthid and Argid Suborders of the United States

	Occurrence (%)	
Particle size class	Orthids	Argids
Loamy skeletal	23.3	13.8
Sandy skeletal	3.4	—[a]
Clayey skeletal		5.6
Loamy	12.1	11.6
Coarse loamy	20.1	6.7
Fine loamy	15.0	28.6
Clayey		5.0
Fine	6.9	20.3
Very fine		0.3
Silty		3.6
Coarse silty	4.6	
Fine silty	7.9	
Sandy	3.6	0.5
Contrasting		3.9
Others	3.0	

[a]No occurrence.

47.2% of the series (Table 2). These values do not reflect on the surface textures of these soils, which are expected to encompass an equally wide array of textures. It can be concluded from these data that a very wide array of particle size distributions, with and without gravel, occurs in arid and semiarid environments.

B. Mineralogy

The clay mineralogy normally controls soil chemical and physical behavior, even when the amount of clay in a soil is low. As shown in the preceding section, most soils of arid regions contain sufficient clay for clay mineralogy to have an important role in the soil's behavior. American Aridisols are predominantly classified in "mixed" mineralogy families (Table 3). Mixed mineralogy is defined in the U.S. Soil Taxonomy (Soil Survey Staff, 1975) as "the approximate mineralogical composition of selected size fractions of the control section as is used for the particle size classification." Mixed mineralogy is applied to soils that have fragmental, sandy, sandy-skeletal, loamy, or loamy-skeletal particle size classes. It defines mixed mineralogy as <40% of any one mineral other than quartz or feldspars in the 0.02-2 mm fraction.

TABLE 3 Percentage Occurrence of Mineralogy Classes in the Orthid and Argid Suborders of the United States

	Occurrence (%)	
Mineralogy class	Orthids	Argids
Carbonatic	10.3	—[a]
Gypsic	1.9	
Mixed	83.3	72.5
Montmorillonitic	3.5	27.1
Others	0.9	0.4

[a]No occurrence.

Carbonatic, which includes >10% of the Orthids, is defined for any particle size class as more than 40% by weight carbonates plus gypsum, and the carbonates are >65% of the sum of carbonates and gypsum. The particle size used to define carbonatic is the whole soil (particles < 2 mm) or less than 20 mm, whichever has a higher percentage of carbonates plus gypsum.

The montmorillonitic mineralogy class is applied to soils that have a clayey or clayey-skeletal particle size class. It is defined as more than half montmorillonite and nontronite by weight or a mixture that has more montmorillonite than any other one clay mineral. The <0.002 mm size fraction of the control section is used when determining the mineralogy of these particle size classes.

Among the Argids (Table 3), the control section is often dominated by high-activity clays or by a mixture that may include high-activity clays. The Orthids (Table 3) have a much smaller percentage of the families in the montmorillonitic family, in part because many fewer of the soil series have clayey or clayey-skeletal particle size classes. The vast majority are mixed mineralogy, which does not exclude the possibility that the soils contain high-activity clays. According to Eswaran et al. (1981), palygorskite seems to be specific to Aridisols. This does not appear in Table 3.

If one is to generalize mineralogy based on the data in Table 3, it can be said that arid region soils are likely to have a wide variety of clay types, depending on the soil-forming factors. Unlike soils of the humid tropics, in which low-activity clays dominate, the semiarid tropics, other semiarid soils, and arid region soils are likely to have high-activity clays. This should contribute to structural stability. It also leads to the implication that low organic matter, commonly described in arid and semiarid region soils, is the major cause of the poor structural condition of these soils. This conclusion is supported by the work of Norrish and Pickering (1977), who found that a wide range of mineral types existed in the semiarid soils of Australia. Although montmorillonite was much less common than kaolinite, illite, or regularly interstratified clays, all existed in Australian semiarid region soils. Montmorillonite dominated in black earths

TABLE 4 Size Classes of Soil Structure

	Shape of structural unit (mm)			
Size classes	Platy	Prismatic and columnar	Blocky	Granular
Very fine	<1	<10	<5	<1
Fine	1-2	10-20	5-10	1-2
Medium	2-5	20-50	10-20	5-10
Coarse	5-10	50-100	20-50	5-10
Very coarse	>10	>100	>50	>10

(Vertisols) formed on basalt. Illite was dominant where rainfall is highest in the winter and there is a hot dry summer. The exception is where potassium is lacking on the basalts. Kaolinite appears to be in almost all soils.

Eswaran et al. (1981) consider that in many Aridisols the clay fraction has little or no role in the performance of the soil. They state that many of the soluble evaporite minerals, such as the chlorides, sulfates, nitrates, iodates, and borates in addition to the carbonate minerals are more important than the clay minerals in determining the soil chemistry.

C. Structure

Definition

Soil structure describes how individual primary mineral particles are joined to form larger units. The *National Soils Handbook* (USDA, 1983) defines soil structure as the natural organization of soil particles into units. Natural units separated by persistent surfaces of weakness are peds. Peds are described according to their shape, size, and grade (distinctness in place) (Soil Survey Staff, 1951).

Peds are described as platy when the units are flat. They are described as prismatic or columnar when the vertical dimension is longer than the horizontal. Prismatic and columnar peds are flat or slightly rounded in the vertical dimension. Prismatic peds are flat on the horizontal surfaces, and columnar peds are distinctly rounded. Blocky peds are blocklike, bounded by flat or slightly rounded surfaces. If the faces intersect at relatively sharp angles they are described as angular blocky and as subangular blocky if the faces are more rounded. Granular peds are spherical and bounded by curved or very irregular faces.

Peds are described by five size classes: very fine, fine, medium, coarse, and very coarse. The size limits vary for each shape (Table 4). Grade describes the distinctness of peds observed in the field. The determination of grade depends on the ease with which the soil separates into discrete peds and also on the proportion of peds that hold together when the soil is handled. Three classes are used: weak, moderate, and strong.

Emerson (1959) proposed a model of soil aggregates in which organic matter formed linkages between quartz and clay domains. A clay domain or tactoid is a group of oriented clay particles held together primarily by electrostatic forces. The linkages he proposed included quartz-organic matter-quartz, quartz-organic matter-clay domains, clay domains-organic matter-clay domains, and clay domain-clay domain interactions without an organic matter intermediary. Harris et al. (1966) generalize aggregate-forming interactions between mineral particles to be of four types: clay-clay interactions in which a cation forms the linkage between negatively charged clay faces, clay edge-clay face interactions, clay edge or face-clay edge or face interactions in which a polymer forms the linkage, and interactions among uncharged particles linked by silt and organic or inorganic colloids.

Edwards and Bremner (1967) expanded Emerson's model by suggesting that micro- and macroaggregates could be separated at 250 μm. Microaggregates were defined as those aggregates < 250 μm that are difficult to disperse by ultrasonic vibrations. They suggest that these microaggregates are composed of clay and humified organic material linked by polyvalent metals. According to this aggregate concept, any aggregate containing a sand grain is easily disrupted and has little stability. The stability of microaggregates depends on the amount and content of clay and humified organic matter.

Tisdall and Oades (1982) further developed the conceptual aggregate model by dividing aggregates into >2000, 20-250, 2-20, and <2 μm. They proposed that there exists a hierarchical arrangement of soil structure in which individual clay platelets are organized into domains or tactoids bound together by electrostatic forces aided by Fe and organic matter. These are further organized into 2-20 μm aggregates consisting of <2 μm particles bound together by persistent organic molecules. The 20-250 μm aggregates consist of 2-20 μm aggregates bound together by persistent organic materials and crystalline oxides and highly disordered aluminosilicates. The high stability of the <250 μm aggregates is dependent on organomineral interactions that are additive, which is consistent with Emerson's original concepts.

At each level of organization within the hierarchy, different organic binding agents are important. Stabilizing the largest aggregates involves cementing or binding agents, which may be inorganic, organic, or mineral-organic associations. The organic materials involved in stabilizing the largest aggregates are both transient binding agents, such as polysaccharides, and temporary binding agents, such as roots and Vesicular Arbuscular mycorrhizal hyphae (Tisdall and Oades, 1982). Elliot (1986) tested the Tisdall and Oades model and concluded that their concept explained how his soils behaved. He also found that native, uncultivated soils were more stable than cultivated. He concluded that microaggregates are linked together through the action of the temporary (roots and hyphae) and transient (polysaccharides) binding agents. The interaction between the transient and persistent binding agents and clays should be strongly influenced by the surface characteristics of the clay.

In addition to organic matter bonding between particles, iron oxides and hydroxides stabilize soil structure. El Rayah and Rowell (1973) showed that 2.5% Fe_2O_3 or Al_2O_3 on Na-montmorillonite much reduced swelling of the clay. El-Swaify and Emerson (1975) found that Fe and Al oxides were on the surface of clays as amorphous 4 nm spheres. Aluminum precipitated as $Al(OH)_3$ was more effective than Fe as $Fe(OH)_3$ in reducing the slaking of dry illite, kaolinite, or clay mixtures and in increasing the resistance of clay to dispersion in pyrophosphate solutions (El-Swaify and Emerson, 1975). Other authors have found similar results (e.g., Blackmore, 1973; Deshpande et al., 1964). Kaolinite differs from montmorillonite because the kaolinite domains are more rigid and have larger pores (Emerson, 1977). Because the domains are more rigid, the stress induced by entrapped air during wetting causes more rapid slaking of kaolinite than of montmorillonite.

As clays play a critical role in soil structure, soil chemistry that influences clay dispersion and flocculation also affects aggregate stability. The roles of exchangeable sodium percentage (ESP) and electrical conductivity of the soil solution in structural stability have been explained by a number of researchers (e.g., Bresler et al., 1982; Van Olphen, 1977). A complete review is not attempted here as it has been provided elsewhere (Shainberg and Letey, 1984). In general, as ESP increases, the aggregate stability decreases and crusting and erodibility increase because Na causes the clay plates to separate as the swarm of positively charged ions around the clays overlap. The smaller the overlap, the less likely the clays are to disperse. Sodium on the clay exchange complex is hydrated and exerts a swelling pressure on the clay particles. In essence, it forces clay particles apart. The repulsive forces between particles decrease with an increase in salt concentration, thereby reducing the tendency of the Na ions to diffuse outward from the clay and reducing the swelling pressure. The minimum electrolyte concentration that causes flocculation of clay particles is the flocculation value (FV). If the electrical conductivity (EC) of the solution is below the flocculation value of the clay, the clay particles disperse. Oster and Rhoades (1984) illustrate the relationship between ESP and EC. In general, as the EC increases, the ESP necessary to cause dispersion increases. Conversely, at higher electrical conductivity, a higher ESP can be tolerated before soil clays disperse and the management of soil water becomes difficult.

Divalent calcium is more strongly adsorbed to clay surfaces than monovalent sodium, which reduces the tendency of calcium clay to swell. As calcium montmorillonite is aggregated into tactoids (Shomer and Mingelgrin, 1978), the swelling is between the external surfaces of neighboring tactoids rather than between clay platelets. The overall effect is to reduce swelling and dispersion of a Ca-saturated clay compared to a Na-saturated clay and to increase the structural integrity of a Ca soil compared to a Na soil. In reality, soils are neither pure clay nor fully Ca or Na saturated. ESP is a measure of the degree of Na saturation, and soils behave in a manner intermediate between either of the pure clay end members.

The practical consequences of the ESP/EC relationship are that under rain-fed agriculture, the rainfall EC is usually <0.1 dS m^{-1} and clay dispersion, swelling, and structural breakdown are possible even in soils with low ESP. In arid and semiarid soils, where salts accumulate in the profile and ESP is frequently higher than in humid regions, high- or low-quality water can cause serious slow water infiltration and associated runoff and erosion on sloping land.

Examples

There are two horizon structures that may be unique or at least more common in arid and semiarid soils than in soils of humid regimes. These are the vesicular E horizon and the columnar B horizon. The vesicular E horizon is found as a thin surface horizon in some desert soils. The pores are "dead-end" pores: that is, most pores are disconnected. This results in a horizon with a honeycomb appearance and slow permeability.

Formation of the vesicular E under irrigated conditions occurs in nearly saturated soils in which air pressure in pores is sufficient to form a spherical cavity. Upon drying, the pore remains as a vesicle. Additional wetting and drying traps additional air in the soil and forms more vesicles (Miller, 1971). Under nonirrigated conditions, rain seals the surface and prevents the escape of the vapor and vesicles form. Repeated wetting and drying helps to create the structure that Springer (1958) called "unstable and transitory." The unique vesicular structure is found in many desert soils of the United States and other parts of the world (e.g., Australia) (Mott et al., 1979). Vesicular surface horizons have been shown to have lower infiltration rates and more erosion than nonvesicular horizons (Blackburn and Skau, 1974).

Columnar Bt horizons owe their genesis to the presence of Na. The dispersion of clays due to the high ESP leads to coarse or very coarse structure in high-clay B horizons. The surface of the column becomes covered with a salt-rich white cap that is distinctive and not often found in other than arid regions.

D. Consistence

Definition

Soil consistence describes the cohesion among soil particles and the adhesion of soil particles to other materials in the soil. Consistence must be described in relation to soil moisture content because, like soil strength, the behavior of soil depends on the moisture content. When the soil is air dry, consistence is described in terms of the resistance to a breaking force. At the field moisture capacity, soil consistence is described in terms of the force required to deform it, and in the wet state consistence is described in terms of adherence of soil to objects.

Unfortunately, there is no apparent direct relationship between consistence as measured in the field and aggregate stability measured in the laboratory. One would expect soils with hard or firm consistence to have a high degree of aggregate stability. This may not be true, as often soils with high organic matter contents are described as very friable and friable when moist and they have very high percentage of aggregate stability in the laboratory.

Measurement

Consistence is one of the measurements made by field soil scientists when they describe soils. Measurements are made by moistening dry soil to the desired moisture state and then manipulating the soil to determine dry, moist, or wet consistence. Dry consistence is measured by attempting to break a ped, first between thumb and forefinger, followed by more forceful methods. Moist consistence is measured in the same way. Wet consistence is measured by rubbing a wet sample between fingers and measuring the strength necessary to separate the fingers. This is a measure of stickiness. Cohesion is measured by rolling the wet sample into a thin wire. The consistence class is determined by the length and strength of the wire.

These qualitative consistence classes have been correlated to soil strengths (USDA, 1983). Engineers use the Atterberg limits (Casagrande, 1932) to describe the moisture content at which soils change from a brittle solid to a deformable solid that retains its shape, the *lower plastic limit*, and the moisture content at which a soil changes from a deformable solid that holds its shape to one that does not, the *upper plastic* or *liquid limit*. These terms are more closely correlated with actual soil strength than the qualitative USDA terms (McBride and Bober, 1989).

E. Aggregate Stability

Definition

Aggregate stability is a measurement of the resistance of a ped to disruptive force. Kemper and Chepil (1965) define an aggregate as a group of two or more primary particles that cohere to each other more strongly than to surrounding particles. According to Kemper and Rosenau (1986), the stability of aggregates is a function of whether the cohesive forces between particles is greater than the force applied to separate the particles. The sources of disruptive forces are natural (rainfall, wind, and wetting) and human (cultivation and traction). Methods have been developed to assess aggregate stability and to relate the laboratory measurements to the management of soils.

Kemper and Koch (1966) showed that aggregate stability is a function of organic matter, clay, and oxide contents. As organic matter increased from <1% up to 2%, aggregate stability increased rapidly. They also found that as ESP increased from <4 to >20, aggregate stability decreased. Kemper and Rosenau (1984) and Kemper et al. (1987) show that cohesion recovery is primarily a physicochemical process.

The importance of organic matter to aggregate stability is not a simple relationship between total amount of organic carbon and stability because not all the organic compounds in soils are responsible for aggregate stability. As mentioned previously, the model of soil structure of Tisdall and Oades (1982) suggests that different kinds of organic matter are important at different aggregate sizes. Oades (1984) reviewed the role of polysaccharides, cation bridges, and microorganism-clay interactions on aggregate stability.

The stability of pure clay aggregates has been used as a model for determining how organic molecules stabilize aggregates and the interaction between the stabilizing effect and the inorganic chemistry of the soil (e.g., Mortland, 1970). Theng (1982) showed that adsorption of uncharged polymers, such as polyvinyl alcohol, is influenced by the saturating cation and the ionic strength of the solution. Rengasamy et al. (1984) found that the relationship between total cations, sodium adsorption ratio (SAR), and aggregate stability of some Australian red-brown earths varied depending on clay types, clay content, and organic matter. Churchman and Tate (1986) concluded that the variation among aggregate stability of clays from six New Zealand soils was related to differences in clay mineralogy.

Shanmuganathan and Oades (1982) suggest that the dispersible clay content of some Australian soils controls many of the physical properties of the soils. Dispersible clay was determined by shaking a sample by hand for 60 s. An increase in the percentage distribution of 50-250 μm aggregates upon the addition of increasing additions of Fe polycations to dispersed clay reinforces the idea presented earlier of the importance of Fe to aggregate formation and stability.

Measurement

A number of measurement methods have been suggested for measuring aggregate stability. In all cases, only relative stability can be reported because stability can only be considered in terms of the disruptive force. Kemper and Rosenau (1986) review various methods used to determine aggregate stability. Perhaps the most widely used system is that developed by Yoder (1936) in which 50 g of air-dry soil is placed on the topmost sieve in a nest of six sieves. A motor raised and lowered the nest of sieves for a specified time and distance to disrupt the aggregates. The weight of material remaining on each sieve was used to calculate aggregate stability.

Rengasamy et al. (1984) proposed that the dispersive behavior of soils could be determined by measuring the "spontaneous" dispersion followed by dispersion caused by an external force. Spontaneous dispersion was measured by adding distilled water to air-dry aggregates and allowing the combination to stand undisturbed for 12 h. Gentle mixing after 12 h suspends any dispersed clay, which can be measured by standard particle size methods. The same sample is shaken end over end for 1 h and the clay in suspension measured. From these two tests they determined that there were six classes of aggregate stability for the red-brown earths of Australia.

F. Pores and Porosity

Definition

Pores or voids are the spaces between (interaggregate) and within (intraaggregate) aggregates. Pore spaces vary in amount, size, shape, continuity, and tortuosity, all of which influence the rate of fluid and gas flow and the water-holding capacity of soil. Pore size is a continuum from <1 μm to >5 cm. Pore size is sometimes arbitrarily divided into macro- and micropores, although no universally accepted definition for these size classes has been agreed upon. Luxmoore (1981) suggested that a standard pore classification be developed for soils in which micropores are those <10 μm, mesopores are those from 10 to 1000 μm, and macropores are those > 1000 μm. Macropores provide the channels for flow through soil profiles from surface ponding or from perched water tables, and mesopores are most important in the gravitational flow of water during soil drainage. Oades (1984) classifies pores into four groups: <0.2 μm hold "bound" or unavailable water, 0.2-2.5 μm are important for storage of plant-available water, 25-100 μm are important for capillary conduction of water and soil aeration, and those pores > 100 μm are important for aeration, fast drainage of water, and root growth. Lynch and Bragg (1985) suggest that the 0.1-15 μm pores hold water available to plants, the 30-60 μm pores allow water to drain under gravity, and the >60 μm are important in gas exchange. These authors and others agree that soil structure and aggregate stability are important in determining the size distribution of pores and the subsequent water-holding and transmission properties of soils. Norton (1987) found that pore shape and number were the most important properties determining the final infiltration rate in four soils with structural crusts. Reduction in the percentage of planer pores had a good correlation with the final infiltration rate.

Total porosity is a function of the particle density and the dry bulk soil density. The ratio of dry bulk density to particle density is the fraction of the total volume occupied by solids. The pore volume is calculated by subtracting this ratio from 1. If it is assumed that particle density is constant at 2.65 Mg m^{-3}, then total porosity varies inversely with bulk density.

Measurement

Danielson and Sutherland (1986) review the procedures for measuring porosity and pore size distribution. The procedure for measuring pore volume is based on Boyle's law of volume-pressure relationships. If a constant-volume pycnometer is used, the volume of liquid and solid in a soil sample may be obtained from the difference in the volume of gas in the sample chamber with and without the sample in the chamber.

Water desorption and mercury intrusion methods are described by Danielson and Sutherland (1986) as the most common methods for determining pore size distribution. They state that the "evaluation of the size, configuration, and distribution of the soil pores is essentially impossible due to their extremely complicated nature." They state, however, that the size distribution of the larger pores can be measured with some degree of accuracy if one assumes that the volume of drainage water from a saturated sample is proportional to the volume of pores. The size range of pores can be calculated based on the water released from the sample as the sample is equilibrated at higher pressures on a porous plate.

In the mercury intrusion method, pressure is required to force mercury into pores instead of suction to remove water from the saturated sample. The pressure required is dependent on the size and geometry of the pore and the surface tension (Danielson and Sutherland, 1986).

Interpretation

Total porosity and pore size distribution most directly affect water infiltration rates and plant-available water in soils. Both are affected by particle size distribution, soil structure, and forces that disrupt soil structure. Although statements have been written about the low plant-available moisture in arid region soils, it is difficult to see how the varied physical properties of arid region soils can make this generality true. As has already been shown, arid region soils have a wide range of particle size distributions and depths, two main factors influencing porosity and plant-available water. The low organic matter content of arid region soils may contribute to higher bulk density and lower water-holding capacity compared to soils of other temperature and moisture regimes. Organic matter has a much lower specific gravity compared to soil minerals (0.1 versus 2.6 Mg m^{-3}). As the amount of organic matter increases, the total weight per unit volume and bulk density decrease.

Another factor may be more important than the physical properties. Herbel and Gibbens (1987) suggest that for loamy sands and sandy loams in arid rangeland in southern New Mexico, there is insufficient rainfall to wet the profile to any depth to raise the matric potential above -1.5 MPa in most years. Thus, although the soils have the appropriate porosity for rapid intake of water and storage, the environment is such that little plant-available moisture is stored.

III. SPECIAL MORPHOLOGY

A. Surface Horizons

Desert Pavement

Desert pavement is not strictly a soil horizon, but it is unique to arid environments. A desert pavement is formed as a deflation feature. Fine particles are removed from the soil surface, leaving behind coarse fragments. Over time, as more and more fine material is removed, the coarse fragments form an interlocking "pavement" over the surface, protecting it from additional wind erosion. Springer (1958) suggested that coarse fragments are pushed to the soil surface by shrinking and swelling. He based this conclusion on the presence of a coarse fragment-free zone immediately below the desert pavement in many desert soils. Removal or disturbance of the rock pavement can initiate a new cycle of deflation by exposing the underlying soil to wind.

Surface Crusts

Naturally occurring surface crusts include biological and chemical crusts. These are distinguished in this chapter from structural and depositional crusts because of the role management plays in the latter. Biological crusts are considered in another chapter. Chemical crusts, specifically salt incrustations on soil surfaces, are not uncommon in arid regions. They have been shown to decrease wind erodibility by increasing the threshold shear velocity of fine sand (Gillette et al., 1982; Nickling, 1984). The presence of sufficient salt to form a crust and bind particles together generally precludes the use of the soil for agriculture without substantial reclamation. In addition, crusts (salt or other types) tend to be fragile, ephemeral, and subject to disruption by saltating grains once wind erosion has begun. Because of these characteristics, crust stability, although recognized as part of the wind erosion process, is not part of the wind erosion equation (Woodruff and Siddoway, 1965).

B. Subsoil Horizons

In arid and semiarid regions the dominant direction of water flow is upward because evapotranspiration greatly exceeds precipitation. This frequently leads to the accumulation of salts at the soil surface or subsoil. Curiously, in hot and dry environments, cemented and sometimes indurated subsoil horizons are commonly found. This is curious given the initial statement that the flow of moisture is generally upward, not downward. The presence of cemented zones may be interpreted as resulting from the upward flow of salts and subsequent precipitation and desiccation at the upper end of the wetting front or from the downward leaching of salts during extraordinary precipitation episodes followed by desiccation.

Regardless of the genesis of the cemented horizons, several types, distinguished by the cementing compound, are common in arid and semiarid environments. Common are petrocalcic horizons, known by geologists as caliche in which $CaCO_3$ is the dominant cementing agent, petrogypsic horizons cemented with gypsum ($CaSO_4$), and duripans cemented with silica. Iron-cemented pans are less frequent in arid and semiarid climates. Cementation by silica, carbonates, and iron (sesquioxides) was reviewed by Flach et al. (1969).

Petrocalcic

Carbonate as a source of cementing material may originate from within the profile, from calcareous dust added to profiles, and from carbonate-laden groundwater when it rises through a profile. Gile et al. (1965, 1966) proposed four phases of carbonate accumulation in profiles. In phase I, thin discontinuous coatings of carbonate are found in gravelly profiles. In nongravelly profiles, the carbonate is first seen as filaments in pores. In phase II, the pebbles in a gravelly profile have continuous coatings of carbonate and some pores between pebbles are filled with carbonate. Nongravelly soils have thicker and more continuous filaments and few to common nodules. Phase III produces the first recognizable reduction in infiltration because in both gravelly and nongravelly soils there are many interpebble fillings and many nodules and internodual fillings. At this stage, there is nearly complete plugging of the pores.

In the final phase, a true petrocalcic horizon has formed. The pores are completely plugged with carbonate such that little or no water passes through the horizon. This leads to the development of a laminar cemented carbonate layer at the surface of the cemented horizon. This laminar layer is diagnostic for phase IV.

Blumel (1982) separates the process into three stages for soils in which the main source of carbonates is dust. In stage 1 there is genesis of the calcic or petrocalcic horizon as the dust is added to the profile and moved downward with infiltrating water. After formation, there is exhumation and weathering of the cemented zone, with the final stage formation of lamellar structure on the soil (and hardpan) surface. Although these two sequences deal with different phases of the process, both describe the processes similarly. Blumel's stage 1 corresponds to all four phases of the Gile, Peterson, and Grossman model.

The calcareous duricrusts in northwestern Libya were found both at the soil surface and within the soil profile (Ashour and Abd-el-Mogheith, 1983). The location of the crust to a certain extent is determined by the amount of precipitation. If sufficient precipitation occurs the carbonates are washed into the subsoil, where they precipitate.

Petrogypsic

Although there are 850,000 km^2 of gypsiferous soils in the world (Mousli, 1981), much less research has been done on petrogypsic horizons and they appear to be much less common than the carbonate-cemented horizons. It is reasonable to assume that the genesis of petrogypsic and petrocalcic horizons is similar. Watson (1988) identified three main crust types: shallow-water evaporites, macrocrystalline groundwater evaporites, and mesocrystalline illuvial accretions. This last type is a pedogenic crust. The source of gypsum in the crusts from Tunisia was dust. In their review of the desert soils of Israel, Dan and Yaalon (1982) found a good correlation between salt distribution and the stability and age of the pedomorphic surface. They use this correlation to suggest that the major salt sources are airborne, either as dust or as components of rainfall.

The gypsiferous soils of Syria, on the other hand, are formed in an area of rocks very rich in gypsum, and the gypsum may have originated from eolian deposition, transport by water, or dissolution in place (Stoops and Ilaiwi, 1981). The work of these authors suggests that soils with >25% gypsum lack plasticity and cohesion or are cemented. As long as the gypsum accumulations are not continuous, their influence on soil structure is restricted (Stoops and Ilaiwi, 1981).

Duripans

The genesis of silica-cemented horizons has been reviewed by Flach et al. (1969). They report that as little as 10% Si as $Si(OH)_4$ may be sufficient to cement horizons. Frequently, Si and $CaCO_3$ occur together and duripans cannot be distinguished from petrocalcic horizons. As management is the same for all three cemented zones, distinguishing among the cements is not critical.

Management

The presence of subsoil cemented horizons greatly limits the soil volume available for roots and controls the hydrology of the soil profile. Management of a cemented horizon requires that either shallow-rooted plants or plants that can tolerate high water tables be grown. Thus, rice can be grown on soils with a shallow hardpan, but deciduous trees cannot.

In high-intensity agriculture, as in California, cemented horizons are frequently destroyed and mixed with the noncemented portion of the soil by physical and mechanical methods. Large, powerful tractors are employed to drag a steel shank through the soil to disrupt the cemented zone. This has been successfully done to depths up to 2 m, and because the cemented horizon does not re-form for hundreds if not thousands of years, the disruption need only be done once.

IV. MANAGEMENT OF PHYSICAL PROPERTIES

A. Erodibility

Water and wind erosion are both serious problems in arid and semiarid regions. It is not inconsistent with low rainfall to have high soil erosion because the soil is often bare and subject to the direct, although infrequent, impact of raindrops. Wind erosion is a constant hazard for the reason that the dry, often unprotected soil surface is easily moved by wind.

Water Erosion

The amount of soil eroded is a function of the balance between forces detaching and transporting particles and those resisting forces. The process can best be described by separating detachment from transport, knowing full well that the two processes act together. Authors have shown that raindrop impact is most important in detaching soil particles from the main soil mass but that water flowing over the soil surface, particularly when concentrated in rills, can detach additional material (Young and Wiersma, 1973; Meyer, 1971). Overland flow is the agent responsible for most of the transport of soil particles, but net downhill movement due to raindrop impact is small (Morgan, 1978). There are interactions between the depth of overland flow and raindrop size in which the maximum detachment and erosion are coincident with a raindrop size-flow depth ratio of 1 (Ferreira and Singer, 1985; Ferreira et al., 1985). The presence of a thin layer of water maximizes the lifting forces as the crater formed by the raindrop impact closes.

The relative importance of detachment and transport processes is somewhat spatially dependent. Within interrill areas, the relatively flat nonchanneled spaces between rills, splash detachment and transport have been shown to be more important than overland flow transport. Conversely, overland flow transport is the major process in rill areas (Young and Wiersma, 1973; Monke et al., 1977).

The best known relationship between the factors responsible for erosion is the universal soil loss equation (USLE) of Wischmeier and Smith (1965). The USLE states that the long-term average erosion from a field A is a function of six factors; rainfall runoff erosivity factor R, soil erodibility factor K, slope length L and steepness factor S, the amount and "goodness" of vegetation or other cover on the soil surface factor C, and a conservation factor for any erosion control practices instituted to reduce erosion P:

$$A = R \times K \times L \times S \times C \times P \qquad (1)$$

The USLE was developed to unify several regional erosion equations within the United States, hence its name. It has been tested worldwide with varying success, but its universality exists in that the major factors are stated. It has been revised (Wischmeier and Smith, 1978) and is currently under further revision. For the purposes of this chapter, the details of the entire equation are not as important as the properties of arid region soils that influence the soil erodibility factor K.

Soil erodibility has been shown to be primarily a function of particle size distribution and various materials that bind particles together into aggregates and stabilize the aggregates. Wischmeier et al. (1971) developed a graphic solution to an equation for predicting soil erodibility in which increases in silt and very fine sand are predicted to increase soil erodibility. Increasing sand other than very fine sand is also predicted to increase soil erodibility. The other three factors in the nomograph are soil organic matter up to 4%, soil structure, and permeability of the least permeable horizon. The latter two parameters are qualitative inputs into the erodibility relationship. "Better" soil structure and more rapid permeability are related to lower erodibility.

Other researchers have shown that additional factors are important in predicting soil erodibility. These include exchangeable sodium percentage (Singer et al., 1982), amount of iron and aluminum oxides, and kind of clay minerals (Romkens et al., 1977; Trott and Singer, 1983). Singer et al. (1982), using a laboratory rainfall simulator, showed that as little as 2% Na on the exchange complex was sufficient to significantly increase the erodibility of a semiarid region soil. Conversely, small amounts of iron oxides were sufficient to significantly reduce the erodibility of other semiarid region soils (Trott and Singer, 1983).

Imeson and Vis (1982) found that the volcanic ash soils of Colombia had relatively low erodibility due to irreversible dehydration of soils that are normally moist. Sadler (1988) found that the low specific gravity of some volcanic ash soils of Ecuador made them highly susceptible to water erosion. Particle size distribution and aggregate stability were properties closely related to erodibility for most of the soils examined. Erosion of soils from northern Morocco, on the other hand, was closely related to soil crusting as it influenced infiltration rate and time to ponding (Imeson, 1983). Elwell (1977) used the USLE as a basis for developing an erosion model for southern Africa. Texture, permeability of the top 1 m, and soil crusting were factors in the soil erodibility factor.

One can generalize from the literature that in arid and semiarid regions the soil properties contributing to high erodibility are the poor structure, low organic matter, and high crusting potential of the soil. The presence of exchangeable sodium is an important factor in many arid region soils. In humid region soils it is much less important. The low organic matter reported for many arid and semiarid region soils is due to both low organic matter production and rapid organic matter oxidation in the hot, dry climate. Another factor in the high erodibility of these soils is the general lack of stable structure. Hot, dry conditions with few wetting and drying, shrinking and swelling, or freezing and thawing cycles are not conducive to soil structure formation.

Wind Erosion

Throughout the arid and semiarid climate zone, wind erosion is a real or potential problem. Throughout the Middle East, dust storms are a serious problem (Middleton, 1986). In the Arabian Peninsula eolian deposits have been found to adversely affect all agricultural activities (Al-Nakshabandi and El-Robee, 1988). Khalaf and Al-Hashash (1983) measured rates of dust fallout in Kuwait as high as 1002 t km^{-2}. Sand encroachment is one of the major forms of desertification on the African continent. As much as 615 t ha^{-1} soil loss was measured over a 75 day period from Bikaner in India. The loss of organic matter was 1626 kg ha^{-1} during the same period (Gupta et al., 1981).

In the United States, soil erosion by wind is a problem on some 30 million ha of agricultural land (Skidmore, 1982). Most of the problem is in semiarid areas, such as the Great Plains. The effects of wind erosion on yield have not been studied in detail, but the potential impacts of wind erosion are reduction in soil thickness and subsequent loss of moisture and nutrient holding volume. A second serious problem is the abrasion of plants by wind-borne particles. Lyles (1977) estimated average annual reductions in wheat yields from the Great Plains to range from 0.8 to >118 kg ha^{-1} $year^{-1}$ for a wheat-fallow rotation and 2.4 to >136 kg ha^{-1} $year^{-1}$ for a wheat-sorghum-fallow rotation. The highest values in both cases were from West Texas on the most erodible soils.

Downes et al. (1977) reviewed the literature on the effect of wind erosion on crop plants. They state that wind erosion decreases yields or the quality of crops, requiring reseeding, delaying harvest, or completely destroying the crop. In their study they found that 6-day-old seedlings of cabbage, carrots, onions, and peppers were adversely affected by the increasing total kinetic energy of sandstorms. This one example illustrates that the problem of wind erosion is more than the deflation of the land surface.

The earliest work on the physics of wind erosion was by Bagnold (1941). Factors important in predicting wind erosion have been combined into a wind erosion equation (Woodruff and Siddoway, 1965). The present wind erosion equation was developed from the work of Bagnold and Chepil (Chepil and Woodruff, 1963).

The wind erosion equation combines 11 primary field and climatic factors to predict the annual potential erosion losses by wind from a given field.

$$E = f(I',K',C',L',V') \tag{2}$$

where E is the annual soil loss (t ha^{-1} $year^{-1}$), I' is the soil erodibility factor (t ha^{-1} $year^{-1}$), K' is the soil ridge roughness factor, C' is the climatic factor, L' is the width of field factor, and V' is the vegetative cover factor. It is not the purpose of this chapter to review the equation in its entirety; others have done such reviews (e.g., Nickling, 1987). The factors that make wind erosion a particular problem in arid and semiarid areas are the soil erodibility and climatic factors.

Soil erodibility is the product of soil erodibility and knoll erodibility and is defined as the potential annual soil loss from a wide, unsheltered isolated field with a bare, smooth, noncrusted surface. The erodibility index is a function of several soil properties, including texture, structure, and organic matter content. The most important erodibility factor is the percentage of particles greater than 0.84 mm diameter. Soil erodibility increases as the percentage of soil fractions greater than 0.84 mm diameter decreases. Knoll erodibility is a factor that varies with slope and is expressed in terms of percentage slope. The erosion rate for windward slopes longer than 500 feet is about the same as from level land, and knoll erodibility is taken as 100% in this situation (Woodruff and Siddoway, 1965).

Climatic factor is an index of wind erosion as influenced by surface soil moisture and the cube of average windspeed. The index characterizes the wind erosion potential of a particular climate at a given location relative to climatic conditions at Garden City, Kansas. The form of the C' factor is given here to illustrate the importance of the three key variables: wind speed, precipitation, and evapotranspiration.

$$C' = 386 \frac{U_z^3}{[\Sigma_{i=1}^{12} 10(P - E)_i]^2} \tag{3}$$

Average annual wind speed is represented by U_z corrected to 9.1 m, P is monthly precipitation, and E is evapotranspiration. The denominator is the Thornthwaite precipitation-evaporation index and is used instead of actual soil water content because soil moisture data are usually not readily available at most stations. The importance of soil moisture is clear from this equation. The rate of soil movement varies approximately inversely to the square of effective surface soil moisture.

It has been argued (Coote, 1985, as cited by Nickling, 1987) that the use of a monthly estimate of surface soil moisture is a weakness of the model. Regardless of the accuracy of the model, it is important to note that for arid and semiarid areas, surface soil moisture is low during most of the year, and with natural vegetation sparse or nonexistent, the probability of wind erosion is high. Also contributing to high wind erodibility is the small fraction of aggregates larger than the critical 0.84 mm size.

Managing wind erosion in arid and semiarid areas requires careful water management, surface vegetation management, and control of wind speeds with wind breaks. Increasing surface roughness can decrease wind erosion as can the use of windbreaks to reduce the critical wind speed near the ground. Maintaining soil moisture during critical parts of the year with mulches can also contribute to lower wind erosion. For example, Gupta et al. (1983) have shown that a three-row windbreak consisting of *Cassia siamea*, *Albizia lebbeck*, and *C. siamea*, when 4-5 m tall, was most effective in reducing wind velocity, wind erosion, and evaporative losses downwind from the shelterbelt.

B. Crusting and Sealing

Two distinct types of crusting have been observed in semiarid and arid region soils. Each has a particular origin, morphology, and management. Crusts and seals are differentiated based on their moisture condition at the time observed. Seals are moist, and crusts are dry. The first type of crust is the structural crust formed by the combination of raindrop or sprinkler drop impact on a soil surface with physicochemical dispersion of the soil clays. The second is the depositional crust formed by the sedimentation of fine material on the soil surface as turbid water infiltrates into the soil. The source of the turbidity may be either irrigation water flow in a furrow or overland flow from rainfall erosion.

Structural Crust

The structural crust was first described in detail by McIntyre (1958a, b), although the problem was recognized earlier (Hendrickson, 1934; Lowdermilk and Rowe, 1934; Duley, 1939). McIntyre (1958a) observed crusts formed from artificial rainfall on soils packed in columns using thin sections. He described a 0.1 mm thick dense layer without pores and a 0.1-3 mm thick washed-in zone. The top zone he called a skin seal due to compaction.

He found the skin seal and the washed-in zone to have permeability much slower than the bulk soil below these zones (McIntyre, 1958b). Using rainfall between 66 and 99 mm h^{-1} intensity and 5.0 or 5.4 mm drops, he produced crusts with a skin seal that had a permeability of 5×10^{-7} cm s^{-1} and a washed-in zone with permeability of 5×10^{-6} cm s^{-1}.

The physical and chemical conditions that produce crusting have been thoroughly studied subsequent to McIntyre's work. Tackett and Pearson (1965) concluded that raindrop compaction produced a thin, dense, oriented clay layer 1-3 mm thick. Thompson and James (1985) showed that the hydraulic resistance of the seal formed on the surface of a silt-loam soil increased as the water droplet kinetic energy per unit soil surface area increased. Ferry and Olsen (1975) considered clay orientation a critical factor in crust formation and crust properties, particularly strength. Shainberg (1985) reviewed the effect of soil chemistry on soil crust formation. He showed that both ESP and the electrical conductivity of the soil solution control clay swelling and dispersion and the ultimate infiltration rate of the soil crust.

Shainberg and Singer (1988) showed an interaction between raindrop impact energy and soil ESP. As the ESP of two soils increased, they became more susceptible to crust formation at lower rainfall energy levels. Soils with ESP 5 and 10 were more susceptible to sealing by low-energy rain than were ESP 0 soils. For the two soils tested, clay content appeared to be a more important determinant of sealing than organic matter content. Others, including Tarchitzky et al. (1984), Gal et al. (1984), and Onofiok and Singer (1984), have confirmed and expanded upon the earlier findings of McIntyre.

Such crusts are not uncommon in arid and semiarid areas. Charreau and Nicou (1971) found that crusts in the Sahel reduce sorghum yields. Nicou and Charreau (1980) found that this was a major physical problem that increases exponentially with increasing clay content up to about 18% clay if the clay contains 2:1 minerals.

Depositional Crust

Depositional crusts are thicker than structural crusts and do not require rainfall or irrigation drops as part of their genesis. In addition, clay swelling and dispersion are not necessary prerequisites to depositional crust formation. Shainberg and Singer (1985) have shown that the characteristics of depositional crusts are affected by ESP, but they can form in any environment where fine-textured material is suspended in water and allowed to settle. The hydraulic conductivity of soil columns with depositional crust increased by two orders of magnitude as the electrical conductivity of the suspension increased from 0.1 to 0.55 dS m^{-1}. The concentration of solids in the suspension also influences the depositional crust and hydraulic conductivity (Shainberg and Singer, 1986). As the concentration of suspension increases, the thickness of the depositional crust increases and the hydraulic conductivity decreases. Southard et al. (1988) observed the micromorphology of depositional crusts formed in the laboratory with a petrographic light microscope and a scanning electron microscope. Crusts formed from soil material suspended in 0.002 M Cl^- solution had higher bulk density than crusts produced from suspension of soil material in distilled water or 0.005 M Cl^-. The increased bulk density was a function of the degree of orientation of the soil particles. Particles settling from distilled water produced crusts with strongly oriented clay layers.

Moore et al. (1989) concluded that depositional crusting was the major cause of slow water penetration in a prune orchard in which the ESP was <3 dS m^{-1}. Irrigation water entered each irrigation check from a standpipe 30 cm above the soil surface. The force of the water as it spilled from the standpipe onto the soil disturbed the soil surface and entrained clay and silt that was deposited on the soil surface as the irrigation water infiltrated.

Management of depositional crusts depends on elimination or reduction of suspended sediment. In flood irrigation, reduction of water energy at the water inlet to the field can reduce the suspended sediment. Permanent cover in furrows can reduce erosion and sediment production in the furrow. Such amendments as gypsum or phosphogypsum, a by-product of phosphorus fertilizer manufacturing, can be used to increase the electrical conductivity of the irrigation water and increase the flocculation of clays, thereby decreasing the density and increasing the porosity of depositional crusts (Southard et al., 1988). Tillage can be used to disrupt depositional crusts if other methods fail to reduce their formation.

V. CONCLUSIONS

Although generalizations have been made about the physical properties of arid and semiarid region soils, to properly assess soil management problems soil surveys are required. It is clear that soils of arid and semiarid regimes have a wide range of particle size distributions and mineralogy. Because of the climate under which the soils form, the sparse vegetation, and the rapid decomposition of organic matter, two soil physical properties of general concern are poor aggregation and low aggregate stability. These fundamental properties result in a high potential for wind and water erosion. The soils also tend to form slowly permeable seals and crusts that decrease infiltration and increase runoff, further contributing to soil erosion. Sealing and crusting and the low infiltration rate they produce may often be responsible for the low water availability of arid and semiarid region soils. There is little physical reason for the generalization that these soils have low water-holding capacity. The climate, combined with low infiltration characteristics, are more likely reasons for the low plant-available water in many soils.

Two soil crust-forming mechanisms have been recognized, and both occur in arid and semiarid region soils. Structural crusts form from drop impact and depositional crusts from accumulation of sediments on the surface as sediment-laden water infiltrates into the soil profile. Protection of the soil surface from drop impact is the primary mechanism for preventing structural crusts. Depositional crusting is reduced by reducing the amount of suspended sediment in water. Vegetation, either as growing plants or mulch, is effective in reducing both kinds of crusting.

Shallow soil depth to a cemented horizon is another property of arid and semiarid region soils that is partly responsible for low plant-available water. These kinds of horizons are widely distributed in regions where evaporation exceeds precipitation and weathering products accumulate in the solum rather than being leached from the soil profile. Management to increase crop yields or to make the soil suitable for cultivation usually requires disruption of the cemented horizon.

Soil erosion by wind and water is a serious problem in arid regions. Both forms of erosion can be controlled by careful choice of farming systems that are designed to maintain maximum cover on the soil surface during the periods of the year with the highest soil erosion potential.

REFERENCES

Al-Nakshabandi, G. A., and El-Robee, F. T. (1988). Aeolian deposits in relation to climatic conditions, soil characteristics and vegetative cover in the Kuwait desert. *J. Arid Environ. 15*: 229-243.

Ashour, M. M., and Abd-el-Mogheith, S. M. (1983). Calcareous duricrusts in north-western Libya. *J. Arid Environ. 6*: 203-213.

Bagnold, R. R. (1941). *The Physics of Blown Sand and Desert Dunes.* Chapman and Hall, London, p. 265.

Blackburn, W. H., and Skau, C. M. (1974). Infiltration rates and sediment production of selected plant communities in Nevada. *J. Range Manag. 27*: 476-480.

Blackmore, A. V. (1973). Aggregation of clay by the products of iron(III) hydrolysis. *Aust. J. Soil Res. 77*: 75-82.

Blumel, W. D. (1982). Calcretes in Namibia and SE-Spain, relations to substratum soil formation and geomorphic factors. In *Aridic Soils and Geomorphic Processes*, D. H. Yaalon (Ed.). Catena Supplement 1. Cremlingen, FRG, pp. 67-82.

Bresler, E., McNeal, B. L., and Carter, D. L. (1982). *Saline and Sodic Soils: Principles-Dynamics-Modeling.* Springer-Verlag, New York.

Butler, B. E., and Hubble, G. D. (1977). Morphologic properties. In *Soil Factors in Crop Production in a Semi-arid Environment*, J. S. Russel and E. L. Greacen (Eds.). University of Queensland Press, St. Lucia, Queensland, Australia, pp. 9-32.

Casagrande, A. (1932). Research on the Atterberg limits of soils. *Public Roads 13*: 121-130, 136.

Charreau, C., and Nicou, R. (1971). L'amelioration du profil cultural dans les sols sableux et sablo-argileux de la zone tropicale seche ouest-africaine et ses incidences agronomiques. *Agron. Trop. 26*: 209-255.

Chepil, W. S., and Woodruff, N. P. (1963). The physics of wind erosion and its control. *Adv. Agron. 15*: 211-302.

Churchman, G. J., and Tate, K. R. (1986). Aggregation of clay in six New Zealand soil types as measured by disaggregation procedures. *Geoderma 37*: 207-220.

Claridge, G. G. C., and Campbell, I. B. (1982). A comparison between hot and cold desert soils and soil processes. In *Aridic Soils and Geomorphic Processes*, D. H. Yaalon (Ed.). Catena Supplement 1. Cremlingen, FRG, pp. 1-29.

Dan, J., and Yaalon, D. H. (1982). Automorphic saline soils in Israel. In *Aridic Soils and Geomorphic Processes*, D. H. Yaalon (Ed.). Catena Supplement 1. Cremlingen, FRG, pp. 103-115.

Dan, J., Gerson, R., Koyumdjisky, H., and Yaalon, D. H. (1981). *Aridic Soils of Israel, Properties, Genesis and Management.* Agricultural Research Organization Special Publication 190. Bet Dagan, Israel.

Danielson, R. E., and Sutherland, P. L. (1986). Porosity. In *Methods of Soil Analysis*, Part I, *Physical and Mineralogical Methods*, 2nd ed., A. Klute (Ed.). Am. Soc. Agron., Madison, Wisconsin, pp. 443-461.

de Oliveira, L. B. (1968). Observations on particle-size distributions in soils of the North-East Brazil. *Pesquisa Agropecuaia Brasil. 3*: 189-195.

Deshpande, T. L., Greenland, D. J., and Quirk, J. P. (1964). Role of iron oxides in the bonding of soil particles. *Nature (Lond.) 201*: 107-108.

Downes, J. D., Fryrear, D. W., Wilson, R. L., and Sabota, C. M. (1977). Influence of wind erosion on growing plants. *Trans. ASAE 20*: 885-889.

Duley, F. L. (1939). Surface factors affecting the rate of intake of water by soils. *Soil Sci. Soc. Am. Proc. 4*: 60-64.

Edwards, E. P., and Bremner, J. M. (1967). Microaggregates in soils. *J. Soil Sci. 18*: 64-73.

Elliot, E. T. (1986). Aggregate structure and carbon, nitrogen, and phosphorus in native and cultivated soils. *Soil Sci. Soc. Am. J. 50*: 627-633.

El Rayah, H. M. E., and Rowell, D. L. (1973). The influence of iron and the permeability of a Na-soil. *J. Soil Sci. 24*: 137-144.

El-Swaify, S. A., and Emerson, W. W. (1975). Changes in the physical properties of soil clays due to precipitated aluminum and iron hydroxides. I. Swelling and aggregate stability after drying. *Soil Sci. Soc. Am. Proc. 39*: 1056-1062.

El-Swaify, S. A., Walker, T. S., and Virmani, S. M. (1984). *Dryland Management Alternatives and Research Needs for Alfisols in the Semi-arid Tropics.* International Crops Research Institute for the Semi-arid Tropics, Patancheru, P.O. Andhra Pradesh, India.

Elwell, H. A. (1977). *A Soil Loss Estimation System for Southern Africa.* Research Bull. 22, Dept. Conservation and Extension, Causeway, Salisbury, Rhodesia.

Emerson, W. W. (1959). The structure of soil crumbs. *J. Soil Sci. 10*: 235-244.

Emerson, W. W. (1977). Physical properties and structure. In *Soil Factors in Crop Production in a Semi-arid Environment*, J. S. Russell and E. L. Greacen (Eds.). University of Queensland Press, St. Lucia, Queensland, Australia, pp. 78-104.

Eswaran, H., Ilaiwi, M., and Osman, A. (1981). Mineralogy and micromorphology of aridisols. In *Proceedings Third International Soil Classification Workshop*, F. H. Beinroth and A. Osman (Eds.). Arab Center for the Studies of Arid Zones and Dry Lands, Damascus, Syria, pp. 153-174.

Ferreira, A. G., and Singer, M. J. (1985). Energy dissipation for water drop impact into shallow pools. *Soil Sci. Soc. Am. J. 49*: 1502-1507.

Ferreira, A. G., Larock, B. E., and Singer, M. J. (1985). Computer simulation of water drop impact in a 9.6-mm deep pool. *Soil Sci. Soc. Am. J. 49*: 1537-1542.

Ferry, D. M., and Olsen, R. A. (1975). Orientation of clay particles as it relates to crusting of soil. *Soil Sci. 120*: 367-375.

Flach, K. W., Nettleton, W. D., Gile, L. H., and Cady, J. G. (1969). Pedocementation: Induration by silica, carbonates, and sesquioxides in the quaternary. *Soil Sci. 107*: 442-453.

Gal, M., Arcan, L., Shainberg, I., and Keren, I. (1984). Effect of exchangeable sodium and phosphogypsum on crust structure-scanning electron microscope observations. *Soil Sci. Soc. Am. J. 48*: 872-878.

Gile, L. H., Peterson, F. F., and Grossman, R. B. (1965). The K horizon: A master horizon of carbonate accumulation. *Soil Sci. 99*: 74-82.

Gile, L. H., Peterson, F. F., and Grossman, R. B. (1966). Morphological and genetic sequences of carbonate accumulation in desert soils. *Soil Sci. 101*: 347-360.

Gillette, D. A., Adams, J., Muhs, D., and Kihl, R. (1982). Threshold friction velocities and rupture moduli for crushed desert soils for input of soil particles into air. *J. Geophys. Res. 87*: 9003-9015.

Greenland, D. J. (1981). Soil management and soil degradation. *J. Soil Sci. 32*: 301-322.

Gupta, J. P., Aggarwal, R. K., and Raikly, N. P. (1981). Soil erosion by wind from bare sandy plains in western Rajasthan, India. *J. Arid Environ. 4*: 15-21.

Gupta, J. P., Rao, G. G. S. N., Gupta, G. N., and Rao, B. V. R. (1983). Soil drying and wind erosion as affected by different types of shelterbelts planted in the desert region of western Rajasthan, India. *J. Arid Environ. 6*: 53-58.

Guthrie, R. L. (1982). Distribution of great groups of aridisols in the United States. In *Aridic Soils and Geomorphic Processes*, D. H. Yaalon (Ed.), Catena Supplement 1. Cremlingen, FRG, pp. 29-36.

Harris, R. F., Chesters, G., and Allen, O. N. (1966). Dynamics of soil aggregation. *Adv. Agron. 18*: 107-169.

Hendrickson, B. H. (1934). The choking of pore-space in the soil and its relation to run-off and erosion. *Trans. Am. Geophys. Union 15*: 501-505.

Herbel, C. H., and Gibbens, R. P. (1987). Soil water regimes of loamy sands and sandy loams on arid rangelands in southern New Mexico. *J. Soil Water Conserv. 42*: 442-447.

Higgins, G. M., Dudal, R., Pecrot, A., Kassam, A. H., and Christoforides, A. (1981). Distribution of soils with aridic, ustic and xeric moisture regimes in the developing world. In *Proceedings Third International Soil Classification Workshop*, F. H. Beinroth and A. Osman (Eds.). Arab Center for the Studies of Arid Zones and Dry Lands, Damascus, Syria, pp. 27-63.

Imeson, A. C. (1983). Studies of erosion thresholds in semi-arid areas: Field measurements of soil loss and infiltration in northern Morocco. *Catena (Suppl.) 4*: 80-89.

Imeson, A. C., and Vis, M. (1982). Factors influencing the erodibility of soils in natural and semi-natural ecosystems at different altitudes in the Central Cordillera of Colombia. *Z. Geomorph. N.F. (Suppl.) 44*: 91-105.

Kemper, W. D., and Chepil, W. S. (1965). Aggregate stability. In *Methods of Soil Analysis*, Part I, *Physical and Mineralogical Methods*, C. A. Black (Ed.). Am. Soc. Agron., Madison, Wisconsin, pp. 499-510.

Kemper, W. D., and Koch, E. J. (1966). *Aggregate Stability of Soils from Western U.S. and Canada.* USDA Tech. Bull. 1355, U.S. Govt. Printing Office, Washington, D.C.

Kemper, W. D., and Rosenau, R. C. (1984). Soil cohesion as affected by time and water content. *Soil Sci. Soc. Am. J. 48*: 1001-1006.

Kemper, W. D., and Rosenau, R. C. (1986). Aggregate stability and size distribution. In *Methods of Soil Analysis*, Part I, *Physical and Mineralogical Methods*, 2nd ed., A. Klute (Ed.). Am. Soc. Agron., Madison, Wisconsin, pp. 425-442.

Kemper, W. D., Rosenau, R. C., and Dexter, A. R. (1987). Cohesion development in disrupted soils as affected by clay and organic matter content and temperature. *Soil Sci. Soc. Am. J. 51*: 860-867.

Khalaf, F., and Al-Hashash, M. (1983). Aeolian sedimentation in the northwestern part of the Arabian gulf. *J. Arid Environ. 6*: 319-332.

Krantz, B. A., Kampen, J., and Virmani, S. M. (1978). Soil and water conservation and utilization for increased food production in the semi-arid tropics. ICRISAT Journal Article 30.

Lal, R. (1985). Soil surface management. In *Agro-research for the Semi-arid Tropics: North-west Australia*, R. C. Muchow (Ed.). University of Queensland Press, St. Lucia, Queensland, pp. 273-300.

Lal, R. (1987). *Tropical Ecology and Physical Edaphology.* John Wiley and Sons, New York.

Lowdermilk, W. C., and Rowe, P. B. (1934). Still further studies on adsorption of rainfall in its relations to superficial runoff and erosion. *Trans. Am. Geophys. Union 15*: 509-515.

Luxmoore, R. J. (1981). Micro-, meso-, and macroporosity of soil (letter to the editor). *Soil Sci. Soc. Am. J. 45*: 671-672.

Lyles, L. (1977). Wind erosion: Processes and effect on soil productivity. *Trans. ASAE 20*: 880-884.

Lynch, J. M., and Bragg, E. (1985). Microorganisms and soil aggregate stability. *Adv. Soil Sci. 2*: 133-171.

McBride, R. A., and Bober, M. L. (1989). A re-examination of alternative test procedures for soil consistency limit determination. I. A compression-based procedure. *Soil Sci. Soc. Am. J. 53*: 178-183.

McIntyre, D. S. (1958a). Permeability measurements of soil crusts formed by raindrop impact. *Soil Sci. 85*: 185-189.

McIntyre, D. S. (1958b). Soil splash and the formation of surface crusts by raindrop impact. *Soil Sci. 85*: 261-266.

Meyer, L. D. (1971). Soil erosion by water on upland areas. In *River Mechanics*, H. W. Shen (Ed.). Fort Collins, Colorado, pp. 27-1 to 27-25.

Middleton, N. J. (1986). Dust storms in the middle east. *J. Arid Environ. 10*: 83-97.
Miller, D. E. (1971). Formation of vesicular structure in soil. *Soil Sci. Soc. Am. Proc. 35*: 635-637.
Monke, E. J., Marelli, H. J., Meyer, L. D., and DeJong, J. F. (1977). Runoff, erosion, and nutrient movement from interrill areas. *Trans. ASAE 20*: 58-61.
Moore, D. C., Singer, M. J., and Sadler, M. S. (1989). Orchard floow management effects on infiltration and aggregate stability. In *Proc. California Plant and Soil Conf.*, Sacramento, California, pp. 68-73.
Morgan, R. P. C. (1978). Field studies of rainsplash erosion. *Earth Surf. Proc. 3*: 295-299.
Mortland, M. M. (1970). Clay-organic complexes and interactions. *Adv. Agron. 22*: 75-117.
Mott, J., Bridge, B. J., and Arndt, W. (1979). Soil seals in tropical tall grass pastures of northern Australia. *Aust. J. Soil Res. 17*: 483-494.
Mousli, O. F. (1981). Methods of evaluation and classification of gypsiferous soils and their suitability for irrigated agriculture. In *Proceedings Third International Soil Classification Workshop*, F. H. Beinroth and A. Osman (Eds.). Arab Center for the Studies of Arid Zones and Dry Lands, Damascus, Syria, pp. 278-307.
Nickling, W. G. (1984). The stabilizing role of bonding agents on the entrainment of sediment by wind. *Sedimentology 31*: 111-117.
Nickling, W. G. (1987). Recent advances in the prediction of soil loss by wind. In *Proceedings IV International Conference on Soil Conservation* Maracay-Venezuela, I. Pla Sentis (Ed.). Universidad Central de Venezuela, Maracay, Venezuela, pp. 1163-1186.
Nicou, R., and Charreau, C. (1980). Priorities for alleviating soil-related constraints to food production in the tropics. IRRI, Los Banos, Laguna, Phillipines, pp. 371-388.
Norrish, K., and Pickering, J. G. (1977). Clay mineralogic properties. In *Soil Factors in Crop Production in a Semi-arid Environment*, J. S. Russel and E. L. Greacen (Eds.). University of Queensland Press, St. Lucia, Queensland, Australia, p. 327.
Northcote, K. H., Hubble, G. D., Isbell, R. F., Thompson, C. H., and Betlenay, E. (1975). *A Description of Australian Soils.* CSIRO, Canberra.
Norton, L. D. (1987). Micromorphological study of surface seals developed under simulated rainfall. *Geoderma 40*: 127-140.
Oades, J. M. (1984). Soil organic matter and structural stability: Mechanisms and implications for management. *Plant Soil 76*: 319-337.
Onofiok, O., and Singer, M. J. (1984). Scanning electron microscope studies of surface crusts formed by simulated rainfall. *Soil Sci. Soc. Am. J. 48*: 1137-1143.
Oster, J. D., and Rhoades, J. D. (1984). Water management for salinity and sodicity control. In *Irrigation with Reclaimed Municipal Wastewater—A Guidance Manual*, G. S. Pettygrove and T. Asano (Eds.). Report No. 84-1 ws, California State Water Resources Control Board, Sacramento, California, pp. 7-1 to 7-20.
Rengasamy, P., Greene, R. S. B., Ford, G. W., and Meharani, A. H. (1984). Identification of dispersive behavior and the management of red-brown earths. *Aust. J. Soil Res. 22*: 413-432.

Romkens, M. J. M., Roth, C. B., and Nelson, D. W. (1977). Erodibility of selected clay subsoils in relation to physical and chemical properties. *Soil Sci. Soc. Am. J. 41*: 954-960.

Sadler, M. S. (1988). *Factors Affecting the Erodibility of Recent Volcanic Ash-derived Soils from the Central Andes of Ecuador.* MS thesis, University of California, Davis, California.

Shainberg, I. (1985). The effect of exchangeable sodium and electrolyte concentration on crust formation. *Adv. Soil Sci. 1*: 101-122.

Shainberg, I., and Letey, J. (1984). Response of soils to sodic and saline conditions. *Hilgardia 52*: 1-57.

Shainberg, I., and Singer, M. J. (1985). Effect of electrolytic concentration on the hydraulic properties of depositional crust. *Soil Sci. Soc. Am. J. 49*: 1260-1263.

Shainberg, I., and Singer, M. J. (1986). Suspension concentration effects on depositional crusts and soil hydraulic conductivity. *Soil Sci. Soc. Am. J. 50*: 1537-1540.

Shainberg, I., and Singer, M. J. (1988). Drop impact energy-soil exchangeable sodium percentage interactions in seal formation. *Soil Sci. Soc. Am. J. 52*: 1449-1452.

Shanmuganathan, R. T., and Oades, J. M. (1982). Effect of dispersible clay on the physical properties of the B horizon of a red-brown earth. *Aust. J. Soil Res. 20*: 315-324.

Shomer, I., and Mingelgrin, U. (1978). A direct procedure for determining the number of plates in tactoids of smectites: The Na/Ca montmorillonite case. *Clays Clay Min. 26*: 135-138.

Singer, M. J., Janitzky, P., and Blackard, J. (1982). The influence of exchangeable sodium percentage on soil erodibility. *Soil Sci. Soc. Am. J. 46*: 117-121.

Skidmore, E. L. (1982). Soil and water management and conservation: Wind erosion. In *Handbook of Soils and Climate in Agriculture,* V. J. Kilmer (Ed.). CRC Press, Boca Raton, Florida, pp. 371-399.

Soil Management Support Services (1987). *Keys to Soil Taxonomy.* United States Department of Agriculture, Agency for International Development, Cornell University Press, Ithaca, New York.

Soil Survey Staff (1951). *Soil Survey Manual.* Agric. Handb. 18, USDA, U.S. Government Printing Office, Washington, D.C.

Soil Survey Staff (1975). *Soil Taxonomy: A Basic System of Soil Classification for Making and Interpreting Soil Surveys.* Agric. Handb. 436, USDA, U.S. Government Printing Office, Washington, D.C.

Southard, R. J., Shainberg, I., and Singer, M. J. (1988). Influence of electrolyte concentration on the micromorphology of artificial depositional crust. *Soil Sci. 145*: 278-288.

Springer, M. E. (1958). Desert pavement and vesicular layer of some soils of the desert of the Lahanton Basin, Nevada. *Soil Sci. Soc. Am. Proc. 22*: 63-66.

Stoops, G., and Ilaiwi, M. (1981). Gypsum in arid soils morphology and genesis. In *Proceedings Third International Soil Classification Workshop,* F. H. Beinroth and A. Osman (Eds.). Arab Center for the Studies of Arid Zones and Dry Lands, Damascus, Syria, pp. 175-185.

Swindale, L. D. (1982). Distribution and use of arable soils in the semi-arid tropics. In *Managing Soil Resources*. Plenary Session Papers 12th International Congress of Soil Science. New Delhi, India, pp. 67-100.

Tackett, J. L., and Pearson, R. W. (1965). Some characteristics of soil crusts formed by simulated rainfall. *Soil Sci. 99*: 407-413.

Tarchitzky, J., Banin, A., Morin, J., and Chen, Y. (1984). Nature formation and effects of soil crusts formed by water drop impact. *Geoderma 33*: 135-155.

Theng, B. K. G. (1982). Clay-polymer interactions: Summary and perspectives. *Clays Clay Min. 30*: 1-10.

Thompson, A. L., and James, L. G. (1985). Water droplet impact and its effect on infiltration. *Trans. ASAE 28*: 1506-1510, 1520.

Tisdall, J. M., and Oades, J. M. (1982). Organic matter and water stable aggregates in soils. *J. Soil Sci. 33*: 141 163.

Trott, K. E., and Singer, M. J. (1983). Relative erodibility of 20 California range and forest soils. *Soil Sci. Soc. Am. J. 47*: 753-759.

United States Department of Agriculture (1983). *National Soils Handbook*. Supt. Documents, Washington, D.C.

United States Department of Agriculture (1988). *Family Soil Classification Files* (mimeo). Iowa State University Computer Center, Ames, Iowa, pp. 65-84.

Van Olphen, H. (1977). *An Introduction to Clay Colloid Chemistry*, 2nd ed. John Wiley and Sons, New York.

Vial, C. D., and Wright, C. (1965). *Soils of the Arid Zones of Chile.* FAO Soils Bulletin No. 1, Rome, p. 42.

Watson, A. (1988). Desert gypsum crusts as palaeoenvironmental indicators: A micropetrographic study of crusts from southern Tunisia and the central Namib desert. *J. Arid Environ. 15*: 19-42.

Wischmeier, W. H., and Smith, D. D. (1965). *Predicting Rainfall-Erosion Losses from Cropland East of the Rocky Mountains.* USDA Agricultural Research Service Agriculture Handbook no. 282, p. 47.

Wischmeier, W. H., and Smith, D. D. (1978). *Predicting Rainfall Erosion Losses—A Guide to Conservation Planning*. USDA Agricultural Hand-Book 537, p. 58.

Wischmeier, W. H., Johnson, C. B., and Cross, B. V. (1971). A soil erodibility nomograph for farmland and construction sites. *J. Soil Water Conserv. 26*: 189-193.

Woodruff, N. P., and Siddoway, F. H. (1965). A wind erosion equation. *Soil Sci. Soc. Am. Proc. 29*: 602-608.

Yoder, R. E. (1936). A direct method of aggregate analysis of soils and a study of the physical nature of erosion losses. *J. Am. Soc. Agron. 28*: 337-351.

Young, R. A., and Wiersma, J. L. (1973). The role of rainfall impact in soil detachment and transport. *Water Resources Res. 9*: 1629-1636.

4

Chemistry of Arid Region Soils

WILLIAM G. KNIGHT *The Earth Sciences Group and Colorado State University, Fort Collins, Colorado*

I. INTRODUCTION

The rates and extent of both mineral weathering and organic matter addition (and therefore soil development) are interactive functions of time, climate, topography, biological action, and parent material (e.g., Jenny, 1941). Climate (specifically moisture and temperature fluxes) severely restricts pedogenesis in dry environments, and arid region soils are typically characterized as weakly developed. Temperature influences evaporation, biological growth and activities, the rates of chemical weathering reactions and, together with water and slope (topography), governs physical weathering processes. Water serves as the principal chemical reactant in weathering, as the vehicle for removing reaction products from the localized sites of weathering, and, in arid environments, as the single most important agent limiting biological growth and activity. In the absence of water, soil development cannot proceed, and hence, many chronologically old desert soils are pedogenically young. Mature soils developed on arid landscapes, conversely, indicate great age and/or geomorphic stability and may even reflect a period of more intense weathering and biological activity associated with an earlier climate.

The relation of climatic factors to the extent of weathering was generalized by Strakhov (1967) from polar to equitorial latitudes across the meridian. Strakhov's scheme is presented here along with a sequence of idealized soil profiles (after Bohn et al., 1985) to illustrate the more salient effects of climate on weathering rate and soil development (Fig. 1). The latitudinal zone identified for deserts and semideserts depicts both reduced precipitation and the condition of evaporation in excess of precipitation. A great deal of the inorganic character of arid region soils can be explained by consideration of primary mineral weathering alone under these climatic conditions. For example, in the initial stages of weathering the release of Ca^{2+}, Fe^{3+}, and to a lesser extent Na^{+} and K^{+} is emphasized as a consequence of their relative abundance as structural components in the most

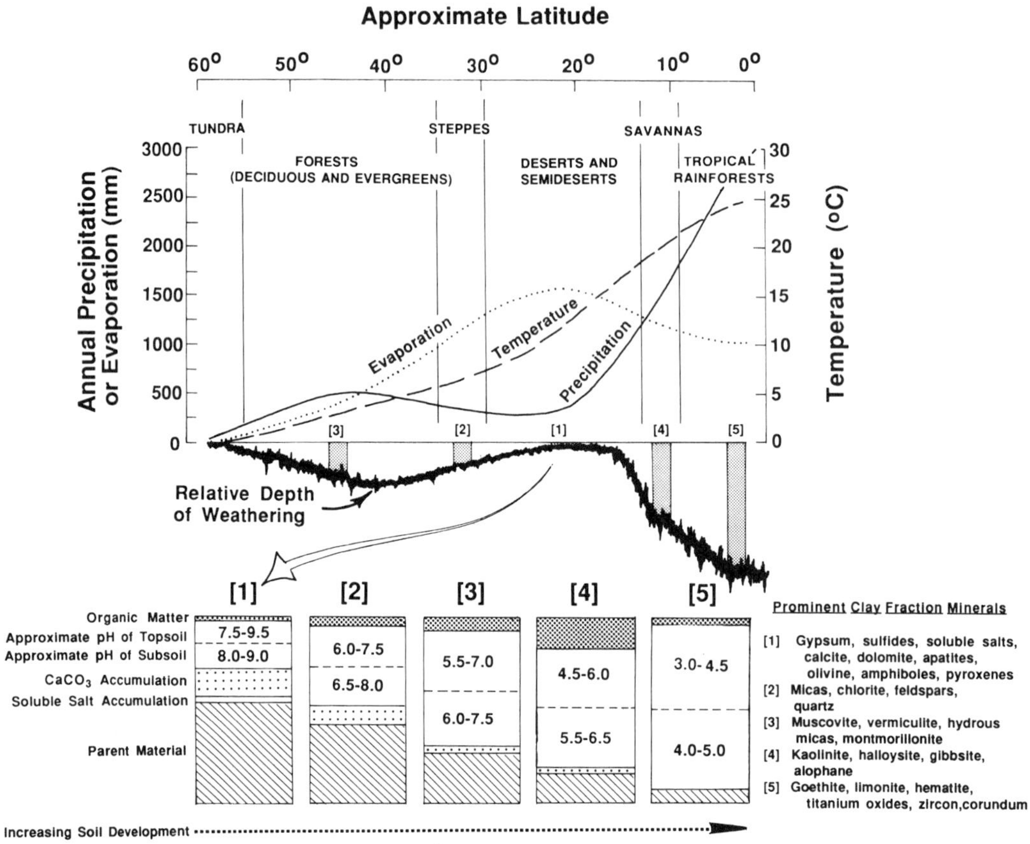

FIGURE 1 Relative depth of weathering in relation to latitude and climatic factors (after Strakhov, 1967), conceptualized soil developmental gradient (adapted from Bohn et al., 1985), and prominent clay fraction mineralogy (adapted from Jackson and Sherman, 1953).

readily weathered (unstable) minerals. These cations are quite soluble and are readily leached, providing adequate rainfall occurs to effect deep percolation. Soils developing in an arid environment, however, tend to retain the alkaline earth and alkali cations to a proportionally greater extent, and subsequently hydroxide bases form and an alkaline pH ensues. It is not only the upward gradient of water movement that promotes the concentration of these cations but also both their retention by the negatively charged surfaces of layer silicates and organic colloids and their reprecipitation as cationic constituents of the nearly diagnostic secondary minerals that form and accumulate in episodically dry soils (i.e., calcite, apatites, gypsum, and halides).

Jackson and Sherman (1953) considered the mineral composition of the clay fraction (including secondary minerals) in soils in relation to the degree of soil development. They delineated 13 groups of minerals whose prominence in the clay fraction reflected the degree of soil maturity.

These are abbreviated here as five groups in Fig. 1 for illustrative purposes. The least developed soils feature accumulations of easily weathered minerals inherited from parent materials, and the highly developed soils contain proportionally more secondary minerals in the clay fraction with a commensurate loss of inherited minerals.

The concentration of the accumulated cations resulting from early weathering and the relatively high pH that results serve to impede continued weathering. To a limited extent the production and dissociation of carbonic acid (reaction of atmospheric or respired CO_2 in water) promotes continued weathering, but in the absence of water the action of CO_2 as a weathering agent is limited. However, the action of acid and complexing agents exuded by pioneering microorganisms on rock (i.e., lichens) may provide for continued weathering in moisture-limited settings.

The genesis of soil begins with the release of the first weathering product from parent rock, and it is no coincidence that such "pioneering" events in soil development are commensurate with the occurrence of pioneering microorganisms. The degree to which biologically mediated mineral weathering is significant in the earliest stages of pedogenesis increases with the increasing harshness of the environment. Thus in very cold (boreal or alpine) or dry (desert) climates, the activities of colonizing organisms can represent a major role in mineral weathering. As the soil develops and plants are established, it is their growth and activities and those of associated microflora that provide for the continued participation of the biosphere in weathering processes, extending the depth of weathering and promoting the formation of new minerals and organomineral complexes.

Accumulation of the organic matter in soils is a rather delicate balance between the rate of addition of biological residues and the rate of their decomposition, with both processes highly dependent on climate. In arid regions, moisture limits the amount of plant growth, the overwhelming contributor of the raw material for humus synthesis. In time, following an initial buildup of soil carbon (commensurate with early soil development), the rates of residue addition versus decomposition are generally held to attain an "equilibrium" reflecting the specific environment in which the soil forms (e.g., Stevenson, 1986). The time required to achieve such equilibrium may be extremely long (as much as 10,000 years for some desert soils developing on wind-blown sand), whereas in more stable, mesic systems organic matter content may reach a steady state (equilibrium implies no net flow of energy) in something of the order of 100 years (e.g., Stevenson, 1986). The organic matter content of desert and arid regions soils (typically Entisols and Aridisols) is characteristically low and conversely is characteristically high in the Mollisols typical of semiarid grasslands. This occurrence may in part be a consequence of the increased moisture and somewhat cooler temperatures of semiarid lands, favoring plant production to a greater extent than microbial decomposition processes (Stevenson, 1986).

In addition to organic matter breakdown, which releases mineral elements (including nitrogen), microorganisms can import, export, store, and alter the chemical states (and therefore properties) of nitrogen in soils (Haynes, 1986; Stevenson, 1986). They can also affect and effect mineral transformations (including those of layer silicates), serve as a temporary "sink" for nutrients (*immobilization*), and release nutrients (*mineralization*) and are the active agents for the synthesis of the complex, high-molecular-weight organic compounds that constitute the humified fraction (Huang and Schnitzer, 1986). Microbial growth and activities are highly dependent on the availability of water, substrates, and cometabolites, and as such they are comparatively retarded in arid region soils. In this regard, characteristic, episodic "flushes" in microbial processes occur commensurate with precipitation events and are critical to the ecology of desert systems.

The ability of the soil to supply essential growth elements is restricted in arid environments not only by the availability of water directly but by the solubility of controlling mineral phases, the rate and extent of organic matter mineralization, adsorption and exchange characteristics, and the physical properties of the soil. Stable mineral or amorphous phases in arid region soils can exert controlling influences on soil pH and the availability of calcium, phosphorus, and iron. In regard to iron and phosphorus in arid region soils, nutritional limits to plant growth are common, and in both these cases organic exudates of plant and microbial origin appear important to the natural remedy of deficiencies. Evaporite (soluble salts) and calcite mineral precipitation and high exchangeable sodium status are consequences of mineral weathering and retarded leaching in an arid climate. These conditions can further compromise the availability of water via deleterious physical effects on the soil that reduce the hydraulic conductivity (caliche deposition, clay swelling, and dispersion) and by reducing the water potential (osmotic effect). The evaporative gradient in arid regions can also promote the accumulation of trace elements toxic to plants and/or animals (i.e., boron, molybdenum, and selenium).

The predominant stable secondary alluminosilicates in arid region soils are smectites (i.e., montmorillonite). These clay minerals exhibit relatively high charge, surface area, and shrink or swell properties and therefore greatly contribute to chemical reactivity and the recalcitrance of humified materials in arid region soils. They also give rise to the prominent cracking features noted for many arid region soils.

II. PRIMARY MINERAL WEATHERING

Soil development proceeds from the physical and chemical weathering of rock under conditions at the earth's surface. These conditions are vastly different from those under which the rocks formed (i.e., high thermal energy and pressure) and feature exposure to water, atmospheric gasses, and the biosphere. Collectively, over time, weathering processes yield unconsolidated, porous materials of lower density, greater volume, and different chemical character than the parent rock. With the exception of a mechanical component associated with the weathering of rock by pioneering organisms, the physical processes that break rocks apart (e.g., erosion, thermal expansion or contraction, and hydrofracture by freezing) are not treated in any depth here (see Bloom, 1978, for more comprehensive information). Rather, the principal contribution of rock disintegration to pedogenesis, as viewed here, is to enhance the exposure of rock mineral surfaces to the far greater influence of chemical weathering. Hunt (1972) gives a useful description of the lithologic factors (also not treated here) that define the physical structure of major rock types in relation to weathering and soil development.

A. Thermodynamic and Kinetic Applications in Weathering

The definition of solution conditions (ionic composition and activities, pH, CO_2, and redox) that favor or restrict the initial or continued weathering of minerals (including secondary mineral formation and dissolution) can be approached theoretically via the employment of thermodynamic principles to electrolyte solutions. Perhaps the best general descriptions of thermodynamics as applied to soil systems currently available are the chapter by Sposito (1986) in *Soil Physical Chemistry* and that given in the text edited by Bolt and Bruggenwert (1976). The monumental works by Lindsay (1979) and Stumm and Morgan (1981) feature the application of chemical equilibria in soils and natural waters as a vehicle for understanding both soil development and the chemical character of soil solutions and are particularly valuable references. Garrels and Christ (1965) give a focused treatment of mineral stability as approached thermodynamically.

Despite their inherent instability, some minerals remain essentially unweathered over relatively long periods of time. The thermodynamic dictum that substances react until their most stable states are achieved is certainly sound, but such ultimate states may not be observed in our laboratory or field experience. Reacting substances may exist in many intermediate forms before (if ever) reaching a stable state and theoretically may arrive at such *metastable* or final, most stable states in a number of ways (*paths*). The study of time as a reaction determinant, together with the study of intermediate products and reaction paths, is *kinetics*. In arid region soils in which the removal of reaction products from the localized sites of weathering is limited, kinetic restrictions to weathering equilibria may be correspondingly important. For a thorough treatment of the application of chemical kinetics to soil systems, including mineral-weathering considerations, the interested reader is referred to the landmark book by Sparks (1988) and his chapter in *Soil Physical Chemistry* (Sparks, 1986), also edited by Sparks.

Free Energy Analysis

The chemical states (valences and structural configurations) of constituent ions in primary minerals are often unstable in the presence of water. Because an ion's stability increases with increasing dilution in the aqueous phase, a strong impetus to dissolve exists. This is *ion dissolution*, and it is the principal mechanism for the chemical weathering of rock. In thermodynamic terms, ion dissolution yields a decrease in the free energy of an ion as it goes from the structured, or coordinated crystalline state to the dissolved state. The free energy change (ΔG) is defined at constant temperature and pressure as

$$\Delta G = \Delta H - T\,\Delta S \tag{1}$$

where ΔH is the change in enthalpy, T is the absolute temperature, and ΔS is the change in entropy. As applied to ion dissolution, enthalpy represents the energy (summed heats of formation) associated with the ion's chemical bonds in the mineral solid. Entropy represents the ion's randomness of motion and position. Changes in ΔH and ΔS then respectively reflect the collective energy released as participant bonds break and the increased freedom (randomness) of an ion in solution compared to an ion in the bound or crystal state.

Mineral weathering reactions are often described using free energy relationships. When the free energies of all chemical species in a given mineral dissolution expression (reaction) are known, the tendency for the reaction to proceed can be predicted. This type of thermodynamic stability evaluation is accomplished by subtracting the sums of the free energies of formations ΔG_f^0 for reactants from summed ΔG_f^0 for reaction products.

$$\Delta G_r^0 = \Sigma\ \Delta G_f^0\,(\text{products}) - \Sigma\ \Delta G_f^0\,(\text{reactants}) \tag{2}$$

When the result is a negative value, the dissolution reaction results in a net decrease in free energy, is *spontaneous*, and may proceed. However, the free energy analysis gives no information about the extent of the reaction because of the restriction that all products and reactants be at unit activity.

The Equilibrium Constant and the Extent of Reaction

Soil water is generally present only as a thin film on particle surfaces, and a concentration-dependent limit to the extent of reaction exists. Since the extent of reaction also depends upon the energy change for the reaction, ΔG_r^0 is related to the equilibrium constant K^0 for the reaction as

$$\Delta G_r = \Delta G_r^0 + RT \ln K^0 \tag{3}$$

At equilibrium, ΔG_r is zero by definition and Eq. (3) becomes

$$\Delta G_r^0 = -RT \ln K^0 \tag{4}$$

The utility of this expression is that it provides a means of developing equilibrium expressions that relate mixed, nonunit activities to free energy relationships. For a given reaction, K^0 is defined as

$$K^0 = \frac{(\text{products})}{(\text{reactants})} \tag{5}$$

where parentheses denote activities. When the activities of all products and reactants are known (measured via ion-selective electrode or calculated via the Davies or similar equations for ionic strength-corrected activity coefficients) and ΔG_r^0 has been determined for a reaction of interest, the equilibrium constant can be found.

As soils dry and water is lost from weathering surfaces, the ions in the reacting solution become increasingly more concentrated and mineral dissolution eventually halts. Conversely, when there is sufficient water to promote the diffusion of reaction products away from weathering sites or when mass flow is in effect, the equilibrium dissolution is favored. In this regard, mineral dissolution in arid environments can be viewed as episodic and proceeds in the intervals between precipitation events to the extent that dilution or removal of weathering products can be achieved.

To some extent, the formation and dissociation of carbonic acid [Eq. (6)] promotes further weathering by neutralizing the bases released from mineral dissolution and by contributing protons to participate in acid hydrolysis reactions.

$$CO_2 + H_2O \longleftrightarrow H_2CO_3 \longleftrightarrow H^+ + HCO_3^- \tag{6}$$

In this regard CO_2 is considered an active weathering agent, but in the absence of water its effect cannot be manifested. Further, in the early stages of soil development and in desert soils, the contribution of CO_2 to the soil atmosphere from biological respiration is limited by the relative paucity of roots and microorganisms.

Activity Ratio Diagrams

It is often desirable to evaluate the relative stabilities of minerals in natural environments as a function of their ionic constituent concentrations and controlling influences (i.e., pH, redox conditions, and P_{CO_2}). This can be accomplished by employing the free energy analysis as abbreviated earlier to independent minerals and then plotting the concentration of a common dissolution species as a function of a common independent variable.

For example, consider the dissolution of low-temperature albite, a principal feldspar mineral, commonly found in igneous rock. The reaction is

$$NaAlSi_3O_{8(\text{low albite})} + 4H^+ + 4H_2O \rightarrow Na^+ + Al^{3+} + 3H_4SiO_4{}^0 \tag{7}$$

Using thermodynamic data compiled by Sadiq and Lindsay (1979), the appropriate stoichiometric coefficients, and Eq. (7), the standard free energy of the reaction (ΔG_r^0) is calculated as

$$\begin{aligned} G_r^0 &= [\Delta G_f^0\,(Na^+) + \Delta G_f^0\,(Al^{3+}) + 3\Delta G_f^0\,(H_4SiO_4{}^0)] \\ &\quad - [\Delta G_f^0\,(NaAlSi_3O_8) + 4\Delta G_f^0\,(H_2O)] \\ &= -15.61 \text{ kJ mol}^{-1} \end{aligned}$$

Using this value and Eq. (4), the equilibrium constant for the reaction is calculated as $10^{2.74}$ at 25°C. Low-temperature albite dissolution at equilibrium can then be characterized via Eq. (5) as

$$10^{2.74} = \frac{(Na^+)(Al^{3+})(H_4SiO_4{}^0)}{(H^+)(NaAlSi_3O_8)(H_2O)} \tag{8}$$

and the extent of reaction at equilibrium and defined temperature-pressure conditions can be evaluated given the activities of solution constituents. Because the activities of water and solids are generally assumed to be unity (not always applicable) (Sposito, 1983) and after taking the log of both sides, Eq. (8) becomes

$$2.74 = \log (Na^+) + \log (Al^{3+}) + \log (H_4SiO_4^0) - 4pH \tag{9}$$

When low-temperature albite weathers in the soil system, (Na^+) may be approximated as "fixed" by exchange equilibria. Substitution of 10^{-3} M as a value for exchangeable Na^+ (Lindsay, 1979) and rearranging reduces Eq. (9) to

$$\log Al^{3+} + 3pH = 5.74 - pH - 3 \log H_4SiO_4^0 \tag{10}$$

The graphic presentation of such equations as ratios of the activities residing on each side of the equation are called *activity ratio stability diagrams*, and their derivation as presented here is based on the *ion association model*. Given appropriate and accurate thermochemical and/or solubility data and defined dissolution reactions, equilibrium stability relationships for any minerals of interest can theoretically be derived and plotted. This has been accomplished for a vast array of minerals in the monumental soil chemistry work by Lindsay (1979). When common terms for the activity ratio diagrams can be derived, relative mineral weathering interpretations are facilitated. To illustrate, stability lines for examples of Ca-, Na-, and K-substituted aluminosilicate primary minerals are presented in comparison to two unsubstituted silicates and gibbsite as Fig. 2 (adapted from Lindsay, 1979). Also included are quartz (SiO_2), soil SiO_2, and amorphous silica to show the $H_4SiO_4^0$ imparted to soil solution by these solid phases. The diagrams (Fig. 2) illustrate that at low solution $H_4SiO_4^0$, the stability of minerals shown follows the order gibbsite > kaolinite > sillaminite > anorthite > microcline > low-temperature albite. At high solution $H_4SiO_4^0$, kaolinite is the most stable mineral shown and sillaminite the least. The increased stability of gibbsite over kaolinite at low $H_4SiO_4^0$ is observed in highly weathered soils from which silica is readily leached as kaolinite. Eventually, $H_4SiO_4^0$ becomes low enough that gibbsite is preferentially precipitated. In arid region soils, silica is largely retained, precluding this condition and favoring the stability of kaolinite. However, as discussed in the treatment of secondary minerals, smectites are the most stable secondary alluminosilicates under conditions typifying arid soils, and their formation is also a function of the retention of high silica concentrations in arid region soils.

Cautions and Limits in Thermodynamic Applications

It should be recognized that the employment of activity ratio-based stability diagrams for making mineral dissolution and transformation predictions is not without handicaps, both theoretical and analytical. The approach itself, as characterized by Sposito (1986), employs the ion association model, which may not have direct or strict thermodynamic significance. This is because the characterization of soil solution composition via molecular speciation relies on extrathermodynamic conventions to obtain single-species activity coefficients. The departure from strict thermodynamic relevance is in the de-emphasis of the macroscopic properties of electrolyte solutions, which are more rigorously described (in thermodynamics) by the total molalities of the stoichiometric components of neutral salts (*specific interaction model*). It is only when activity coefficients can be utilized to produce an accurate macroscopic characterization of electrolyte solutions that the ion association model versus specific interaction model conflict is resolved.

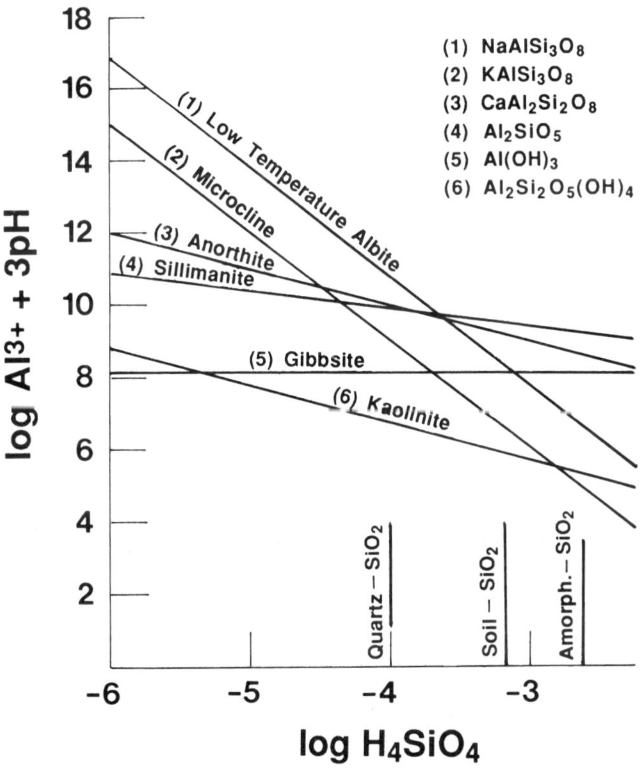

FIGURE 2 Stabilities of some Na-, K-, and Ca-substituted aluminosilicates compared to some unsubstituted aluminosilicates for which equilibrium solution ion activities are 10^{-3} M for Na^+ and K^+ and $10^{-2.5}$ M for Ca^{2+}. (Adapted from Lindsay, 1979.)

The accuracy and reliability of the thermochemical data used to calculate solubility or equilibrium constants is critical to thermodynamic applications. As such, compilations of thermodynamic data for use in describing soil solutions are under continual scrutiny and are subject to revision and updating as new measurements are made and discrepancies resolved. Wagman et al. (1982) give a thorough review of criteria for evaluating published data as well as an excellent compilation themselves. Other particularly good compilations are listed by Sposito (1986).

Other difficulties associated with the application of thermodynamics in general to soil solutions exist and are well elucidated by Morgan (1967) and include determining the presence and extent of kinetic limitations to attaining equilibrium.

Weathering Rate and Kinetics

With the noteworthy exception of Wollast (1967), who evaluated the dissolution of K feldspars, it is only very recently that the kinetics of mineral weathering processes have received concerted research attention. Most of the work with respect to primary minerals has addressed the dissolution of feldspars, amphiboles, and pyroxenes (Wollast, 1967; Lasaga, 1981; Schott et al., 1981; Helgeson et al., 1984; Holdren and Speyer, 1985; Chou and Wollast, 1985). Gibbsite (Bloom and Erich, 1987) and calcite (Frenkel et al., 1983; Amhrein et al., 1985) dissolution have also been investigated recently. The weathering of gibbsite is of little interest here since its occurrence in arid soils is selected against by the relatively high retention of silica. Conversely, calcite ($CaCO_3$) is a prominent mineral in arid region soils, inherited from parent materials and/or formed during pedogenesis in an evaporative environment. It can exert pronounced influences on the chemistry of soils (i.e., pH and solution Ca^{2+} control when in excess of several percent by weight), and its accumulation in lower horizons via reprecipitation results in the widespread "caliche" layers of arid regions.

Types of Kinetic Reaction Controls

Mineral dissolution rates are limited, at a fixed level of undersaturation by *surface reaction-controlled kinetics* involving the detachment of ions or molecules from crystal surfaces and/or by *transport-controlled kinetics*, the transport of dissolved weathering products (*solutes*) away from the reaction site (Berner, 1978). Rate-limiting reactions that are transport controlled or under mixed control in which transport plays an important role are of particular interest in arid region soils as a consequence of diminished system water and flow. The rates of dissolution for most minerals of low solubility, however, are surface reaction controlled where the rate of ion detachment is slow enough to preclude the accumulation of a concentrated layer of ions close to the reacting surface (Dibble and Tiller, 1981). For these minerals, transport processes (diffusion and advection) provide for the movement of dissolved ions into the adjacent undersaturated solution at a rate in excess of their detachment from the mineral surface, and increasing flow rates or stirring does not increase their rate of weathering (Berner, 1978, 1983).

Feldspar, Amphibole, and Pyroxene Dissolution

The overall dissolution of feldspars, amphiboles, and pyroxenes are surface reaction controlled but follow the *parabolic diffusion law*, according to which silica concentration increases with the square root of time (Wollast, 1967; Huang and Kiang, 1972; Luce et al., 1972; Velbel, 1986). The mechanism features a rapid exchange of H^+ for alkali and alkaline earth cations, resulting in a residual thin layer of apparently hydrolyzed aluminosilicate. The formation of this layer explains an observed nonstoichiometric release of Ca^{2+}, Mg^{2+}, Na^+, and K^+ ions relative to Si^{4+} and Al^{3+} (Velbel, 1985). Further dissolution proceeds at the rate of replacement of alkalis by hydrogen at the interface between the residual layer and the mineral surface, after which stoichiometry is restored and the reaction becomes wholly surface controlled (Velbel, 1985).

Calcite Dissolution

Frenkel et al. (1983) determined that the dissolution of calcite in the presence of an exchanger phase occurred by a two-step mechanism. An initial fast reaction (completed after 1 h) was observed to be dependent on the exchanger composition. The following slower reaction was diffusion controlled and depended on both the soil-water suspension ratio and the partial pressure of CO_2. In a later study conducted to further evaluate the nature of the CO_2 dependence, Amrhein et al. (1985) suggested calcite dissolution could be best modeled by three simultaneous surface reactions in which CO_2 attack consisted of an adsorption-hydrolysis sequence featuring a surface-activated complex. This mechanism differs from the notion of simple proton attack on the calcite surface following the production and dissociation of carbonic acid (H_2CO_3). The proposed mechanism, which exhibited very good agreement with measured solution Ca^{2+} for both the high and low calcite surface area cases studied (2.0 and 0.20 $m^2\ L^{-1}$, respectively) is (after Amrhein et al., 1985)

$$CO_2(g) \leftrightarrow CO_2(aq) \tag{11}$$

$$CaCO_3 + CO_2(aq) \leftrightarrow \text{surface complex} \tag{12}$$

$$\text{Surface complex} + H_2O \leftrightarrow Ca^{2+} + HCO_3^- \tag{13}$$

Another important aspect of the study by Amrhein et al. (1985) is that it challenges the long-held assumption of soil solution equilibrium with atmospheric CO_2 [Eq. (11)] and indeed demonstrates that the transfer of $CO_2(g)$ to solution can be an important rate-limiting step in mineral dissolution.

B. Mineral Structure and Stability

A comparison of the ratio of solution cation concentrations in river waters or groundwaters to their abundance in a regional geological terrain provides a relative index of primary mineral weathering. This technique yields a *cation mobility series* and was employed by Goldich (1938) to produce the stability sequence for the mineral constituents of igneous and metamorphic rocks given in Fig. 3. Also shown is another weathering sequence for common minerals of igneous rock (*Bowen reaction series*) developed by comparing dissolution reactions between various powdered minerals in water (Fig. 3, after Blatt et al., 1972). The similarity of the two series emphasizes that in nature as in the laboratory it is the action of water in contact with mineral surfaces that effects weathering. It is no coincidence that the minerals shown in Fig. 3 weather in the same order in which they crystallize. The degree of instability of the primary minerals that make up rock is directly related to the degree of departure from the formative environment of the minerals. The most evident departure is the vastly reduced temperature in the surface environment, and thus those minerals that crystallize first from cooling magma are the most stable at high temperatures and least stable at low earth surface temperatures. Since the structures of minerals (including the types, amounts, and coordinations of structural cations) are a reflection of their formative environment, primary mineral weathering as addressed herein subsequently begins with considerations of structure. Mineral structures and stability relationships are comprehensively addressed in Dixon and Weed (1977) and are well reviewed with respect to weathering considerations by Bohn et al. (1985), from which a good deal of the discussion here is derived.

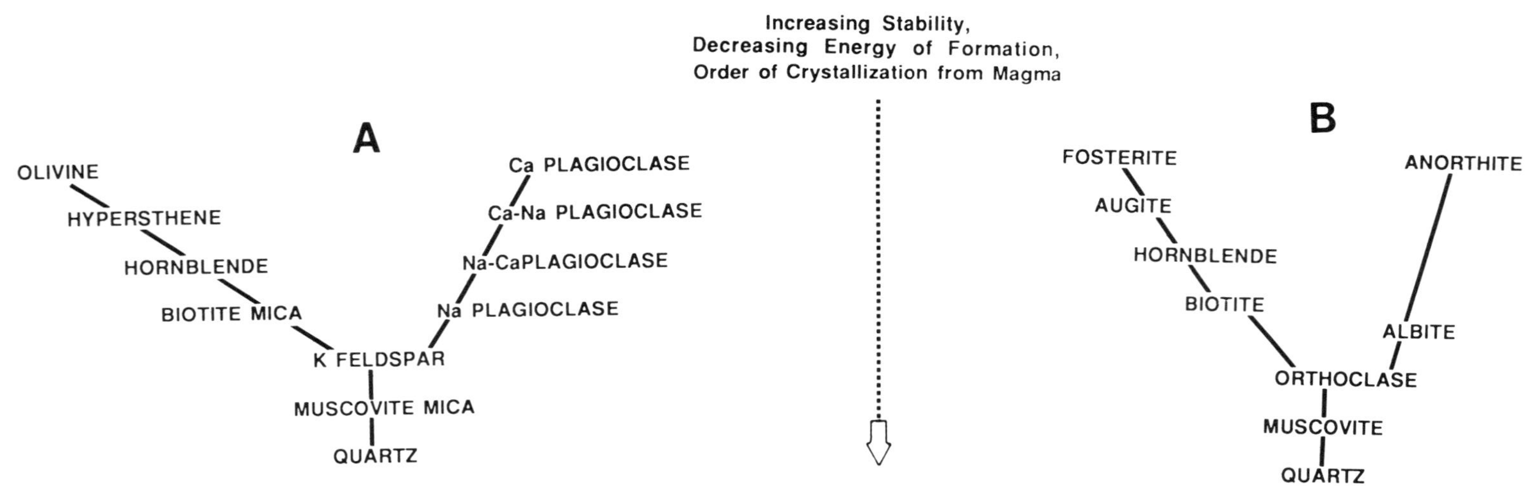

FIGURE 3 Weathering series for common minerals of igneous and metamorphic rocks. (A) Based on cation composition of regional waters in comparison to composition of regional geologic materials. (After Goldich, 1938.) (B) Bowen reaction series based on dissolution of minerals in water. (After Blatt et al., 1972.)

TABLE 1 Silicate Mineral Classification by Structure

Structure	Classification	Structural configuration	Si/Ō ratio	Mineral example
Single tetrahedra	Neosilicates	Single tetrahedra	1:4	Olivine
Disilicate	Sorosilicates	Two tetrahedra sharing one corner	2:7	Hemimorphite
Ring	Cyclosilicates	Closed rings of tetrahedra sharing two oxygens	1:3	Beryl
Single chains	Inosilicates	Continuous single chains of tetrahedra sharing two corners	1:3	Pyroxene
Double	Inosilicates	Continuous double chains of tetrahedra sharing alternately two and three oxygens	4:11	Hornblende
Sheet	Phyllosilicates (layer silicates)	Continuous sheets of tetrahedra each sharing three oxygens	2:5	Montmorillonite
Framework	Tectosilicates	Continuous framework of tetrahedra each sharing all four oxygens	1:2	Quartz

Source: From Dana, 1959, as presented by Bohn et al., 1985.

Crystalline Structures

In a crystal, the number of ions that can be arranged around an oppositely charged central ion (*coordination number*) is a function of the ratio of ionic radii (r_c/r_a, or *radius ratio*) for the central cation (r_c) and a coordinating anion (r_a). From geometric analysis, assuming ions to be rigid spheres, the stable arrangements (packing) of anions around cations can be determined as a function of radius ratio. Coordination numbers for the central cation of 3, 4, 6, 8, and 12 result from radius ratios of 0.15-0.22, 0.22-0.41, 0.41-0.73, 0.73-1.0, and 1.0, respectively (Bohn et al., 1985). The stable arrangements of anions around cations that arise from the given radius ratios are (again respectively) (1) as corners of an equilateral triangle; (2) as corners of a tetrahedron; (3) as corners of an octahedron; (4) as corners of a cube; and (5) as the closest possible packing (Bohn et al., 1985). Since over 90% of the volume of common silicate minerals is attributable to oxygen, it is essentially the packing of oxygen around coordinating cations that governs silicate crystal unit structure. The manner in which the units combine gives rise to the overall structural differences of the silicate minerals (Table 1, after Dana, 1959). Sand- and silt-sized quartz is highly resistant to weathering because each of the four tetrahedral oxygens is shared as a corner of an adjacent tetrahedra, thus imparting a great deal of structural integrity. Olivine neosilicates consist of

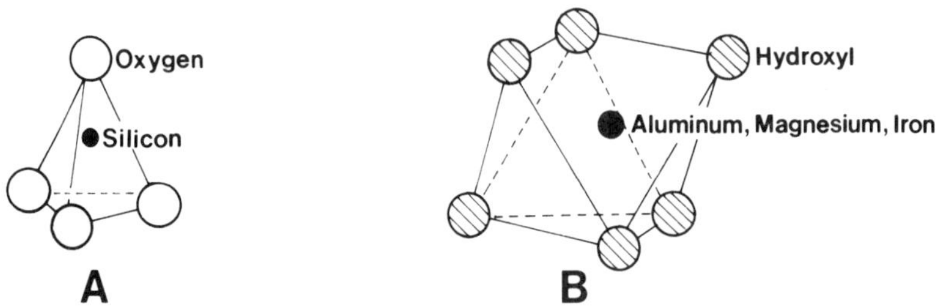

FIGURE 4 Diagrammatic representations of a silicon tetrahedral unit (A) and an aluminum, magnesium, or iron octahedral unit (B).

individual silica tetrahedra whose coordinating oxygens are not shared, pyroxenes are single chains of linked tetrahedra that share two oxygens with adjacent tetrahedra, and amphiboles are double chains of tetrahedra linked together by oxygen atoms in which adjacent tetrahedra share two and three oxygens alternately.

The fundamental *tetrahedral* and *octahedral coordination unit cells* that arise from the coordination of oxygen atoms or hydroxyl ions (OH^-) around central, coordinating Si^{4+} and Al^{3+} cations are pictured in Fig. 4. Both these coordinating cations occur in fourfold tetrahedral coordination with oxygen, but Al^{3+} is most commonly found in the sixfold coordination with hydroxyl ions, which yields an eight-faced octahedral unit (Fig. 4). Cations of like size (*crystal radii*) can replace either Si^{4+} or Al^{3+} in these coordinating positions (*isomorphous substitution*). Crystal radii depend on the number of electrons for a given ion, their attraction to respective nuclei, and the ion's structural position in the crystal. Most often, Mg^{2+}, Fe^{2+}, or Fe^{3+} (crystal radii = 0.066, 0.074, and 0.064 nm, respectively) substitutes for Al^{3+} (0.051 nm) in octahedral coordination and Al^{3+} for Si^{4+} (0.042 nm) in tetrahedral coordination. Substitutions of ions whose crystal radii are in excess of 10-15% of the parent ion are extremely rare.

Isomorphous Substitution and Weatherable Cations

When isomorphous substitution results in a net charge imbalance, electrical neutrality is maintained by the surface retention of ions on the mineral structure or by simultaneous substitutions at other structural locations within the mineral. In a calcium feldspar with half Al^{3+} and half Si^{4+} in tetrahedral coordination positions, the relatively large charge imbalance is satisfied by surface Ca^{2+} between units and leads to a weakened structure. Sodium- and potassium-feldspars, in comparison, feature Na^+ and K^+ charge satisfying ions between units in which Al^{3+} occupies only a quarter of the coordinating tetrahedral positions, resulting in somewhat more stable structures. As in the case of micas, the presence of these charge-satisfying cations between tetrahedra can lead to well-defined cleavage planes that facilitate their breaking apart.

The least stable primary minerals are generally rich in oxidizable (Fe and Mn), water-soluble alkali (Na and K) and alkaline earth (Ca and Mg) cations. The presence of these ions as charge-balancing structural components between crystalline units or in crystal coordination tends to increase the rate of mineral weathering. Olivine neosilicates [$(Fe,Mg)_2SiO_4$)], pyroxenes [$R_2(SiO_2)$], amphiboles [$R_{14}(OH)_4(Si_{16}O_{44})$], and to a lesser extent primary layer silicates, such as biotite [$K(Mg,Fe,Mn)_3Si_3AlO_{10}(OH)_2$], are examples of such relatively unstable minerals. The R in the amphibole and pyroxene formulas is generally Mg^{2+}, Fe^{2+}, Ca^{2+}, or Mn^{2+}, but Na^+ and K^+ are also common. When R is Fe^{2+} or Mn^{2+}, oxidation reactions play an important role in enhancing the rate of weathering because the stable states of iron and manganese under aerobic conditions are Fe(III) and Mn(III-IV). In the absence of oxygen, the transfer of electrons (e^-) between reacting species in soil solution takes on increasing importance as a regulator of chemical reactions. The concentration of e^- (pE) is essentially fixed in aerobic soils, and therefore chemical reactions in soils in which oxygen (the dominant electron acceptor or *oxidizing agent* in nature) is not limited appear more significantly affected by pH. Although anaerobic conditions can arise, they are a transitory exception in arid soils, and reduction reactions in weathering are consequently given muted attention here. However, even in dry soils, "pockets" of O_2-starved locales can evolve within thin films of water where gas diffusion is retarded and microbes are actively consuming oxygen. Further, temporary conditions of significant soil anserobiosis may occur in arid region soils when intense or long-duration rainfall or runoff events saturate soils for more than about 24 h, especially where the soil structure features long diffusion paths (small, tortuous, or discontinuous pores).

C. Role of Organics in Mineral Weathering and Transformations

The interaction of the biosphere with the inorganic precursors of soil has long been recognized as a critical element in mineral weathering and pedogenesis (Slater, 1856; Jenny, 1941; Polynov, 1945; Keller and Fredrickson, 1952; Aristovskya et al., 1969; Fitzpatrick, 1971; Duchaufour, 1982). Specific reviews and papers dealing with biological and biochemical factors in mineral weathering include Silverman and Ehrlich (1964), Zajic (1969), Dommergues and Mangenot (1970), Alexander (1977), Berthelin (1983), and Robert and Berthelin (1986). The latter is a particularly excellent review of the topic and a good deal of the discussion here is drawn from that source.

Microorganisms may be essential adjuncts in the weathering of silicates. Reviews of microbial action in the weathering and transformation of silicate minerals are given by Zajic (1969) and Silver et al. (1986). The role of soil fauna in mineral transformation is not well elucidated, but the action of termites serves to enhance the clay concentration of termitaria over surrounding soil, and conditions for transformations of mica into vermiculite or even smectites exist (Boyer, 1982).

Organic Attack on Rock

The weathering of rock occurs at interfacial microsites between living organisms and the primary minerals that constitute rock (Robert and Berthelin, 1986). The integrity of the interfacial contact is maintained by the strong adhesion of microorganisms to surfaces (Marshall, 1980; Stotzky, 1980) at discrete locales via specific organs, such as the rhizoids of foliaceous lichens or over an entire surface via a polymeric "glue" (generally a polysaccharide gel) in the case of bacteria, algae, or fungi.

A mechanical component of weathering may serve to detach minerals from rock surfaces at microsites where hyphae penetrate rocks or minerals themselves (i.e., by extending between biotite layer packs) (e.g., Robert and Berthelin, 1986). Alternating hydration and dessication of the thallus of foliaceous lichens (Fry, 1924) may provide the mechanical force for the physical disaggregation of minerals and as such would represent an important weathering process in arid environments in which seasonal and sporadic rainfall is followed by periods of drought. Endolithic blue-green algae, lichens, and associated microflora inhabiting fractures and holes in rock may be especially important in rock weathering in desert environments (Friedman, 1971; Friedman and Ocampo, 1976), where their sheltered habitat conserves moisture and affords protection from the desiccating effect of wind. Physical processes appear to be generally (perhaps always) accompanied by biological activity. The frequent observation of calcium oxalate (CaC_2O_4) as a whitish deposit at the base of detached minerals suggests oxalic acid ($H_2C_2O_4$) may be a particularly potent biochemical weathering agent. In fact, a strong mineral weathering role for oxalic acid derived from lichens on rock was suggested over a hundred years ago by Slater (1856), who noted its strong dissolution action on several classes of minerals.

Microbial Uptake of Dissolution Products

Dissolution patterns on quartz under the thallus of the lichen *Acarospora hospitans* were observed by Robert, Veneau, and Berrier as reported in Robert and Berthelin (1986). These workers also found silica microspheres (a few micrometers in diameter) associated with the thallus, suggesting dissolution was followed by absorption and subsequent excretion of silica. This phenomena was also observed by Ascaso et al. (1976) under *Parmelia conspersa* and by Hallbauer and Johns (1977) under *Dimalanea oreina*. The localized uptake (absorption) of weathering reaction products may be a particularly important phenomenon in desert environments in which the removal of dissolved mineral constituents from weathering surfaces is not immediately facilitated by water. The secretion of silica (e.g., Robert and Berthelin, 1986) or iron (Jackson and Keller, 1970) at hyphal surfaces under lichens places the weathering products away from the active weathering sites, thereby promoting continued dissolution. The uptake and concentration of elements by bacteria and fungi can also affect mineral transformation by shifting elemental equilibria in solutions or suspensions. The transformation of micas into vermiculite as a consequence of fungal uptake of K, for example, was reported by Weed et al. (1969).

Exudates as Weathering Agents

Fungi have been much studied in regard to mineral solubilization (Henderson and Duff, 1963; Webley et al., 1963; Boyle and Voigt, 1973; Aristrovskaya et al., 1969; Silverman and Munoz, 1970, 1971; Arrieta and Grez, 1971; Agbim and Doxtader, 1975; Mehta et al., 1978; Rossi, 1978), and their weathering action is a consequence of the production of low-molecular-weight acids and complexing agents. Bacteria (both heterotrophic and autotrophic) have also been reported to produce metabolic compounds capable of solubilizing silicates. Large amounts of Al, Fe, and Mg were released

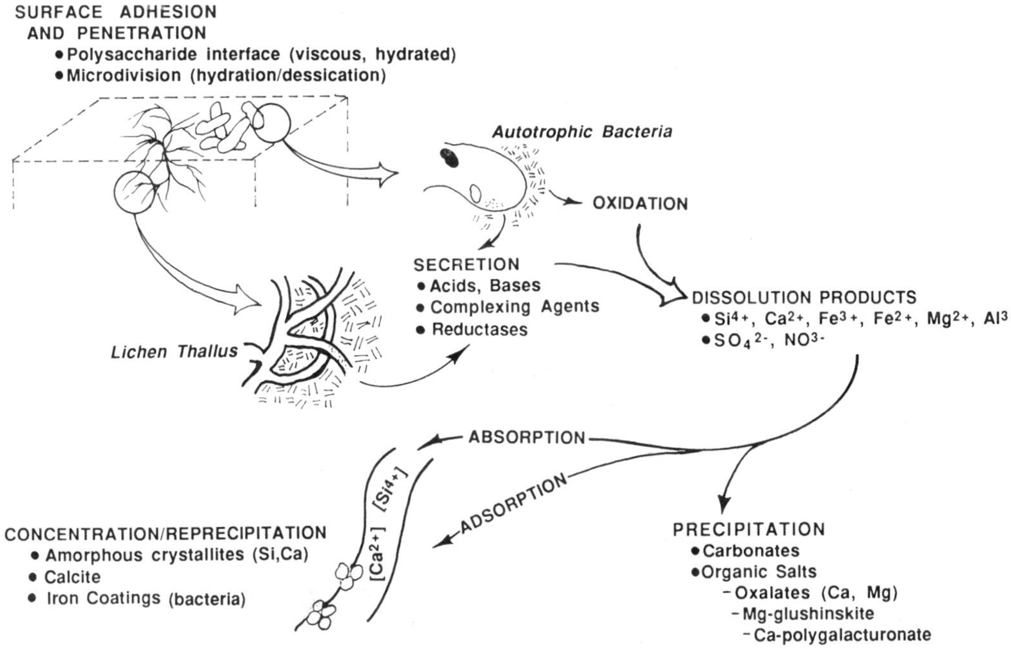

FIGURE 5 Schematic overview of prominent mechanisms and agents in the microbially mediated weathering of primary minerals.

from silicates by some *Pseudomonads* (Berthelin and Belgy, 1979), and *Thiobacillus ferrooxidans* can weather glauconite and microcline (Ivarson et al., 1978). Granite sand (e.g., Robert and Berthelin, 1986) and serpentinized rock (Lebedeva et al., 1978) were attacked by protons (H^+) resulting from the production of nitrate by nitrifying bacteria. The direct reduction of Fe and Mn by reductases of anaerobic and facultative anaerobic bacteria can also serve to weather silicates (glauconite) and is dependent on intimate contact between the bacteria and the mineral surface (e.g., Robert and Berthelin, 1986).

Chelating compounds and siderophores of microbial origin may also be important agents in primary mineral weathering; however to date only the release of Fe and Al from silicates, sulfides, and oxides has been directly demonstrated by chelates (Arrieta and Grez, 1971; Berthelin and Belgy, 1979).

A summary of microbially mediated processes important in the weathering of primary minerals is given as Fig. 5 illustrating the main points presented here.

Alteration of Silicates

One mechanism for the alteration of silicate minerals is metal extraction from their structures (Webley et al., 1960, 1963). The extraction of "fixed" K from muscovite and biotite (e.g., Silver et al., 1986) by fungi may be an important component of mica alteration to albite and vermiculite.

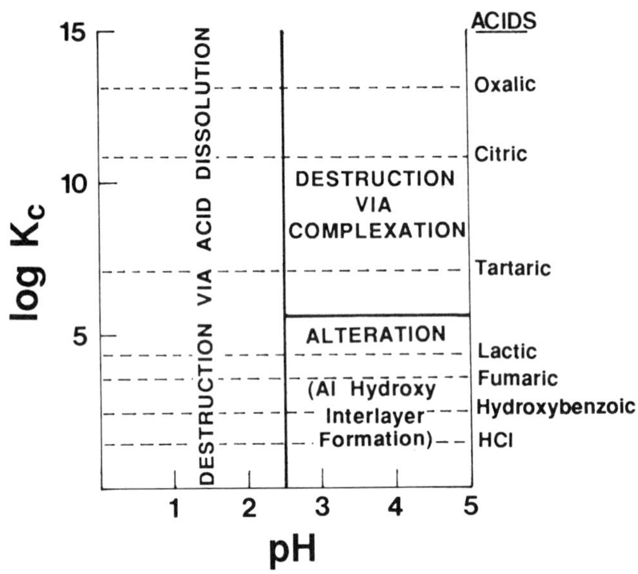

FIGURE 6 Relative weathering or alteration action of various acid and acid-complexing agents on trioctahedral micas as a function of pH. K_c values are stability constants for Al complexes with the acids. [Adapted from Razzaghe, 1976, State Doctoral Thesis, Univ. of Paris VI, as presented by Robert and Berthelin, 1986.]

Several bacterial genera, including *Pseudomonas*, *Bacillus*, and *Arthrobacter*, may be active in the alteration of spodumene to montmorillonite under alkaline conditions (promoted by the microbial production of ammonium and sodium carbonates) and under acid conditions may be followed by the further conversion of montmorillonite to kaolinite (e.g., Silver et al., 1986).

Robert and Berthelin (1986) differentiated three groups of acids with three types of actions according to their extent of attack on micas. In order of the severity of destruction or alteration of the minerals, as witnessed by extraction of elements from the reacting solution (10^{-3} N acids at pH 3-5.5 in contact with the minerals), these were

1. Oxalic, citric, tartaric, and salicylic
2. Quinic, lactic, malonic, formic malic, fumaric, acetic, hydroxybenzoic, and vanillic
3. Succinic, aspartic, alanine, and butyric

The first group exhibited both an acid and a complexing action; the second group demonstrated only an acid influence in weathering, similar to HCl and H_2SO_4 solutions; and the action of the third group was essentially hydrolytic in which the extent of elemental release from mica was similar to that produced by water itself. The action of the first group (acid and complexing) was also observed to release proportionally more Al than Si, whereas the reverse was true for the other two groups. The extent of silicate weathering or alteration as a function of the pH and the type of acid or acid complexing agent involved is summarized in Fig. 6, after Robert and Berthelin (1986).

III. SECONDARY MINERALS, THE CLAY FRACTION, AND SALT-AFFECTED SOILS

The clay fraction of soils is predominantly composed of secondary minerals derived from sedimentary parent materials or formed from the products of primary mineral weathering. To a much smaller extent, some secondary minerals arise from the recrystallization of new minerals entirely in the solid phase. The sand- and, in largest part, silt-sized fractions of soil, in contrast, are mostly primary minerals inherited from parent rock directly or through a sedimentary cycle. Quartz (SiO_2) and mixtures of K^+, Na^+, and Ca^{2+} feldspars are the largest representatives of residual primary minerals in soils, although micas, pyroxenes, amphiboles, olivine, and a host of other, even less abundant minerals may be present. In weakly weathered soils characteristic of dry regions, primary minerals may also occur in mentionable amounts in the clay fraction.

The layer silicates (chlorites, vermiculite, smectites, and kaolins), as they are formed, become the principal sink for the large amounts of silica (Si^{4+}) and to a lesser extent aluminum (Al^{3+}) that are also released as primary mineral weathering products. Montmorillonite (especially Ca montmorillonite) is the featured stable secondary aluminosilicate in arid soils. Its presence and predominance in the clay fraction imparts well-known shrink-swell and low water permeability properties as well as promotes the formation of the surface crusts often associated with arid soils.

With the exception of highly weathered soils (atypical of dry regions) dominated by allophane and hydrous oxides, soils have a net negative charge resulting from the negative charges of layer silicate minerals and organic matter. This property of negative charge gives rise to the *cation-exchange capacity* (CEC) of soils and, together with sruface area, profoundly affects the chemical reactivity of soils. In this regard, smectites exhibit high permanent charge and large reactive surface areas and therefore represent a valuable contribution to the nutrition of plants.

The cationic composition of solutions extracted from arid and semiarid soils is dominated by Na^+, Ca^{2+}, Mg^{2+}, and K^+. The major anionic constituents are Cl^-, $SO_4{}^{2-}$, $HCO_3{}^-$, and $CO_3{}^2$. As many as 40 possible minerals can form during the evaporation of soil solutions featuring these ionic constituents. These include the carbonates (not formally evaporites), sulfates, and halides. Carbonate salts of calcium, magnesium, sodium, and to a limited extent potassium are most commonly found in the soils and subsoils of deserts and semideserts, although those of low solubility, particularly calcite ($CaCO_3$), may persist even in humid environments. The chlorides (NaCl, $CaCl_2$, and $MgCl_2$) are characterized by high solubility (and therefore high toxicity) and, together with the sulfates (particularly gypsum, $CaSO_4$), contribute most to the formation of saline soils.

The stable solid phases of calcium minerals are particularly important to understanding the chemical character of arid region soils. These minerals can serve to control soil pH and the activities and concentrations of

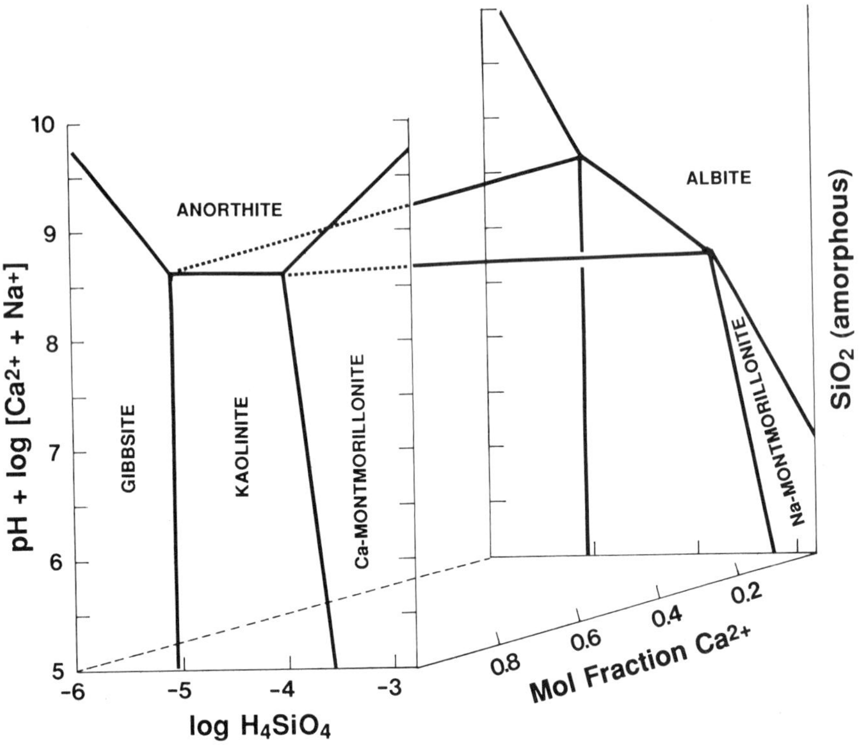

FIGURE 7 Equilibrium stability-phase diagram for the Ca-Na-Al-Si-O-H_2O system. The front vertical plane is the Ca system, and the back plane is the Na system. (Adapted from Bohn and Kittrick, 1984.)

calcium, phosphate, and sulfate ions in soil solution. In the low organic matter soils characteristic of arid regions, these manifestations of calcium mineral stability can establish nutritional limits to growth and, together with the effect of high exchangeable sodium, can exert deleterious influences on the physical properties of soil.

The amount and types of the salts present in combination with the amounts of exchangeable cations (particularly Na^+) govern the nature and extent of the physical and chemical properties that define salt-affected soils.

A. Layer Silicates

Although it has been suggested (Buol, 1965) that arid regions are not conducive to the formation of montmorillonite, the conditions of low rainfall appear to preserve smectite minerals and perhaps even promote their formation (Gal et al., 1974). The formation of montmorillonite (especially Ca montmorillonite) is indeed favored in arid soils as it is more stable (compared to kaolinite, for example) under conditions of high alkali and alkaline earth cations, high Si/Al ratios and high pH (Fig. 7, after Bohn and Kittrick, 1984). Further, the smectites are characteristically more isomorphously substituted and less ordered (structural crystallinity) than kaolinite.

In arid soils, a greater degree of isomorphous substitution may result from the kinetically slow precipitation-dissolution reactions of Si and Al, coupled with the rather long residence time of water (diminished flow) in contact with soil solids. Because of their structural similarities, smectite formation has been held to be a direct result of the weathering or alteration of micas via ion exchange between structural and solution cations. This phenomenon, however, is energetically unfeasible, and smectite formation likely proceeds as a precipitation reaction governed by the solution concentrations of constituent ions. Vermiculites and chlorites, also representing less weathered layer silicates, are also prevalent in arid region soils and serve as sinks for K, Mg, and Si resulting from primary mineral weathering. Kaolinite also occurs in the more developed soils of arid and semiarid regions but is of greatest abundance in soils of warm, moist climates and is a large component of equatorial ocean sediments (Dixon, 1977).

Structures and Reactive Surface Area

In addition to organic matter, the layer silicates provide the largest contribution to both surface area and charge in soils and, hence, soil reactivity. Both surface area and charge are direct manifestations of the structures of the layer silicates. The structures of layer silicates result from the number and sequence of tetrahedral and octahedral sheets bonded together by a layer of shared oxygens.

Micas and Chlorites

Micas, with a typical half-unit cell formula of $K[Al_2(Si_3Al)O_{10}(OH)_2]$, may be dioctahedral (i.e., muscovite) or trioctahedral (i.e., biotite) and are precursors for other 2:1 layer silicates, especially vermiculites. Although not known to form in soils, micas are abundant and are inherited from a wide variety of parent rocks (e.g., shales, slates, granites, schists, and sediments). They are the most important natural source of potassium in soils. The interlayer K^+ (charge-compensating cation) is very strongly adsorbed as a result of negative charge from isomorphous substitution being manifested close to the unit layer surface. Potassium ion is thus essentially nonexchangeable (*fixed*) but is slowly released during weathering. Because of "fixed" interlayer K^+, the reactive area is limited to external surfaces ($70\text{-}120 \times 10^3$ m^2 kg^{-1}). When slightly altered, mica contains interlayer water and less K^+, and *hydrous mica*, more commonly called *illite*, is formed. Mica has been used as a model mineral for the study of organic acid effects on mineral weathering and transformations (Robert and Berthelin, 1986).

Chlorites, with an idealized half-unit cell formula of $AlMg_2(OH)_6)_x$-$(Mg_3(Si_{4-x}Al_x)O_{10}(OH)_2$, are 2:1:1 layer silicate minerals and are largely inherited as primary minerals from metamorphic or igneous rocks (or their sedimentary derivatives) or as alteration products of ferromagnesium minerals (e.g., biotite or hornblende). They occur extensively in soils and include three subgroups (dioctahedral, ditrioctahedral, and trioctahedral) differentiated by the chemical composition of interlayer hydroxide and octahedral sheets (Bailey et al., 1971). The basic structure is four sheets of polyhedra in which three of the sheets form a 2:1 layer similar to that of mica (i.e., tetrahedral sheet-octahedral sheet-tetrahedral sheet). An interlayer hydroxide sheet, which does not feature a plane of atoms shared with an adjacent tetrahedral sheet or sheets, completes the unit structure. This hydroxide sheet is analogous to and serves as the interlayer "cation" between negatively charged 2:1 sheets and links the surface area (70-150 × 10^3 m^2 kg^{-1}) to essentially external surfaces by retarding swelling.

Vermiculites and Smectites

Vermiculites, with an idealized half-unit cell formula $[Mg(H_2O)_6]_{n/2}$-$[(Mg,Fe)_3(Si_{4-n},Al_n)O_{10}(OH)_2]$, exhibit less swelling than smectites because of the higher layer charge (-0.7 per unit formula). Vermiculites are also 2:1 layer silicates whose structures reflect the micas from which they are typically derived via hydrothermal alteration or weathering. Replacement of the interlayer K^+ in mica by the charge-compensating, exchangeable hydrated magnesium cation of the vermiculite cell formula $[Mg(H_2O)_6]^{2+}$ typically expands the c spacing from 1.0 to 1.4-1.5 nm. Some internal balancing of residual mica layer charge in some vermiculites may result from the substitution of Fe^{3+} or Al^{3+} for Mg^{2+} and Fe^{2+} in the octahedral sheet. In this case, an excess negative charge (1.3) results from isomorphous substitution in the tetrahedral sheet and an excess positive charge (0.6) results from isomorphous substitution in the octahedral sheet. Sodium ion then balances the net charge (-0.7 per unit formula), and the vermiculite mineral formula is $Na_{0.7}(Al_{0.2}Fe_{0.4}Mg_{2.4})(Si_{2.7}Al_{1.3})$-$O_{10}(OH)_2$. Vermiculite surface areas range from 600 to 800 × 10^3 m^2 kg^{-1} but are considerably reduced when interlayers are saturated with K^+ or $NH_4{}^+$. This phenomenon (*clay fixation*) prohibits swelling (c spacing reduced to 1.0 nm), and the "fixed" cations cannot be exchanged with ordinary salt solution.

Smectites, the most common of which is *montmorillonite*, with a half-unit cell formula of $Na_x(Al_{2-x}Mg_x)Si_4O_{10}(OH)_2$, in contrast, expand freely and their effective surface area (600-800 × 103 m^2 kg^{-1}) includes both external and internal (as much as 80% of total) surfaces. These are 2:1 layer silicate minerals because their structural units feature two tetrahedral sheets bonded by shared apical oxygens to a single interlying octahedral sheet. Al^{3+} is the principal cation in octahedral coordination, but Mg^{2+} and even Fe^{2+} (imperfect isomorphous substitution) may occur. The sodium cation typically compensates for charge by surface exchange. Low layer charges (0.25-0.6 per formula unit) allows free expansion (swelling) from a completely dry c spacing of 0.95-1.0 nm up to tens of nanometers when fully hydrated.

Kaolins

The *kaolin* group, typified by *kaolinite* $[Al_2Si_2O_5(OH)_4]$, features a structural unit with a single tetrahedral and single octahedral sheet (1:1 layer silicate), with Al^{3+} in octahedral coordination positions and silicon as the only cation in the tetrahedral sheet. This composition may also give rise to the poorly crystallized polymorphs of kaolinite, dickite, or nacrite. Antigorite $[Mg_3Si_2O_5(OH)_4]$ results when Mg^{2+} is in octahedral coordination. Hydrogen bonding between oxygen ions of the tetrahedral sheet and hydroxyl ions of the octahedral sheet in the basal plane binds structural units together. Swelling of the kaolin minerals is prevented by the hydrogen bonding, although a type of kaolinite called halloysite results when water is held between structural units and the c spacing (distance between adjacent tetrahedral sheets of H-bonded structural units), when fully hydrated, expands from 0.72 to 1.0 nm. The surface area of kaolins is relatively small, ranging from 10 to 20 × 10^3 m^2 kg^{-1}, is limited to external surfaces only.

Charge Properties of Layer Silicates

Cation-exchange capacity (CEC) ranges for the 2:1 layer silicates that exhibit a relatively minor pH-dependent charge are 800-1200 and 1200-1500 mmol+ kg^{-1} for montmorillonite and vermiculite, respectively. Kaolinite and chlorite exhibit a very high degree of pH-dependent charge but have much smaller overall CEC values (10-100 and 200-400 mmol+ kg^{-1}, respectively), but vermiculite and montmorillonite charges are relatively independent of pH. Mica (CEC range = 200-400 mmol+ kg^{-1}) appears to be intermediate in this regard. For kaolinite, pH-dependent charge may be either positive (acid environment) or negative (basic environment), as a function of ambient pH and is manifested at crystal edges where AlOH groups are exposed (Jackson, 1964).

The net extent of changes in charge are a function of the total area of exposed edge surfaces. The relative edge area of kaolinite is higher than for 2:1 minerals like montmorillonite because kaolinite tends to stack vertically whereas 2:1 layer silicates have small particle dimensions in the vertical compared to those in the horizontal plane. The relatively minor pH dependence of charge on montmorillonite also appears related to the weakly acid nature of edge groups such that they dissociate only at high pH.

B. Carbonates, Phosphates, Sulfates, and Halides

The stable phases of the prominent evaporite minerals common to arid regions, including calcite (not formally an evaporite), are treated by Doner and Lynn in the comprehensive volume, *Minerals in Soil Environments*, edited by Dixon and Weed (1977). Phosphate minerals are discussed by Lindsay and Vlek in a separate chapter of the same reference. Lindsay (1979) gives thermodynamically based arguments for the occurrence and stabilities of all mineral phases presented here.

Carbonates and Calcite

Carbonates are formed from the decomposition products of crystalline rocks, inherited from sedimentary parent materials or deposited (precipitated) from natural waters or groundwaters in which they are dissolved (Szaboecs, 1989). Calcite and dolomite are the most abundant carbonates in sedimentary rocks, but siderite ($FeCO_3$) and iron-bearing dolomites (ankerites) are also often present (Lippman, 1973). However, iron present as Fe(II) is not stable except under reducing conditions and siderite does not persist. Magnesite ($MgCO_3$) is also derived from sedimentary rocks but where found in soils is intimately mixed with calcite as perhaps pedogenically formed Mg calcite and dolomite (Sherman et al., 1962; St. Arnaud and Herbillon, 1973). The formation of dolomite, however, appears restricted to marine conditions and calcite allows little Mg^{2+} structural substitution. Because Mg^{2+} has a higher hydration energy than Ca^{2+}, magnesium carbonates are commonly found as hydrated $MgCO_3$ forms, such as hydromagnesite [$Mg_5(CO_3)_4(OH)_2 \cdot 4H_2O$], whereas hydrated $CaCO_3$ minerals are very rare. Only two forms are known, $CaCO_3 \cdot 6H_2O$ and $CaCO_3 \cdot H_2O$, and only the latter (monohydrocalcite) has been observed under natural conditions. Sodium carbonates may also be inherited from sedimentary parent material and are common to lacustrine sediments (Lippman, 1973).

By far the most prominent, widely studied, and chemically important salt in arid region soils is calcite ($CaCO_3$). The crystalline polymorphs of $CaCO_3$ (aragonite and vaterite) are not known to form in soils, and vaterite has not been identified in sediments (Lippman, 1973). Aragonite is the stable form of $CaCO_3$ only at temperatures in excess of 50°C (Wray and Farrington, 1957) and may occur in animal shells and recent sediments but is not found in tertiary or older sediments. Calcite formation in soils is intimately dependent on temperature, pH, the activity of Ca^{2+} in soil solution, and the partial pressure of CO_2 in the soil atmosphere. Its precipitation can be represented as

$$Ca^{2+} + H_2O + CO_2(g) \Longleftrightarrow CaCO_3 + 2H^+ \tag{14}$$

The relatively high concentrations of Ca^{2+} from weathering and exchange equilibria, in addition to the evaporative environment of arid soils, favors this reaction even at high CO_2. Increasing CO_2 when water is not limited, however, favors calcite dissolution as

$$CaCO_3 + CO_2 + H_2O \Longleftrightarrow Ca^{2+} + 2HCO_3^- \tag{15}$$

The source of CO_2 in soils is both atmospheric and a consequence of biological activity in soils. Biological activity in soils, even in arid soils, may yield CO_2 partial pressures 10-100 times atmospheric depending on the rates of root and microbial respiration and gas diffusion to the atmosphere. Carbon dioxide solubility is highly temperature dependent (Hodgman, 1960, p. 556), and its hydration may represent a kinetic barrier to rapid equilibration with the soil atmosphere (Amrhein et al., 1985). Carbon dioxide dissolved in water leads to the formation of carbonic acid (H_2CO_3), which dissociates to yield protons (H^+), bicarbonate ion (HCO_3^-), and in turn the carbonate ion (CO_3^{2-}), as follows:

$$CO_2(g) + H_2O \Longleftrightarrow CO_2(aq) + H_2O \tag{16}$$

$$CO_2(aq) + H_2O \Longleftrightarrow H_2CO_3 \tag{17}$$

$$H_2CO_3 \Longleftrightarrow H^+ + HCO^{3-} \tag{18}$$

$$HCO_3^- \Longleftrightarrow H^+ + CO_3^{2-} \tag{19}$$

From the first and second dissociation constants (pK values) for carbonic acid (6.37 and 10.33, respectively, from Wagman et al., 1968) and the solubility product for calcite an expression relating Ca^{2+} ion activity in solution to pH and P_{CO_2} as a function of calcite stability can be derived as

$$\log Ca^{2+} + 2pH = 9.76 - \log P_{CO_2} \quad (20)$$

Calcite, when accumulated in excess of several percent by weight, can control both Ca^{2+} ion activity in soil solution and soil pH. The pH of calcareous soils in which exchangeable sodium is low is around 8.3 and can be approximated (corrections for ionic strength effect on activity coefficients are required) via

$$(H^+) = (10^{-13.5} P_{CO_2})^{1/2} \quad (21)$$

which is derived from the solubility product for calcite in equilibrium with water and $CO_2(g)$. The utility of Eq. (21), in which P_{CO_2} is known, is evident but also must be taken as only approximately valid since the effect of ionic strength on the carbonic acid dissociation "constants" are not accounted for in the derivation and since ion pairing [such species as $Ca(HCO_3)^+$ and $Ca(OH)_2$] affects Ca^{2+} ion activity.

Sulfates and Gypsum

Sulfur in soils is generally associated with the organic fraction but as the organic matter content of soils decreases, proportionally more S is present as mineral forms, predominantly gypsum ($CaSO_4$ $2H_2O$) in arid region soils. Sodium sulfate as thenardite (Na_2SO_4) and magnesium sulfate as epsomite ($MgSO_4$ $7H_2O$) also commonly occur in arid soils but because of their high solubilities (262 and 280 g L^{-1}, respectively, at 25°C) are much less extensive and persistent than gypsum.

Gypsum occurs in soils in a variety of forms (nodules, concretions, slabs, or cementing masses or in very dry desert conditions, powdery deposits of calcium sulfate hemihydrate). It is relatively soluble (2 g L^{-1} at 25°C), and when sulfate (SO_4^{2-}) in solution falls below about 10^{-2} M gypsum does not form or persist. Gypsum and calcite may coexist in soils under unique conditions in the pH range between 7.5 and 8.0 at fixed P_{CO_2}.

In addition to contributing nutritionally important sulfur to solution for plant and microbial uptake, gypsum is important as an indigenous mineral agent effective in the amelioration of high sodicity.

Phosphates and Apatite Minerals

Aside from P derived from organic sources, the availability of phosphorus to plants is a function of the solubility of the stable mineral phase present. In some cases this may be extremely restrictive. In calcareous soils, relatively high pH and Ca^{2+} ion activity in soils promotes the formation of moderately to highly insoluble calcium phosphate minerals. The stabilities of hydroxyapatite [$Ca_5(PO_4)_3OH$], fluorapatite [$Ca_5(PO_4)_3F$], octacalcium phosphate [$CaH(PO_4)_3 \cdot 2.5\ H_2O$], β-tricalcium phosphate [$Ca_3(PO_4)_2$], and dicalcium phosphate ($CaHPO_4$) are portrayed in Fig. 8 (after Lindsay, 1979) in relation to the prominent phosphate minerals forming in acidic soils (i.e., strengite and variscite). Interpretation of the diagram is of course subject to the thermodynamic restrictions inherent in its development, but it serves to illustrate the dependence of solution P on the stable forms of mineral phosphorus. For example, from Fig. 8, at pH 8 hydroxyapatite would be

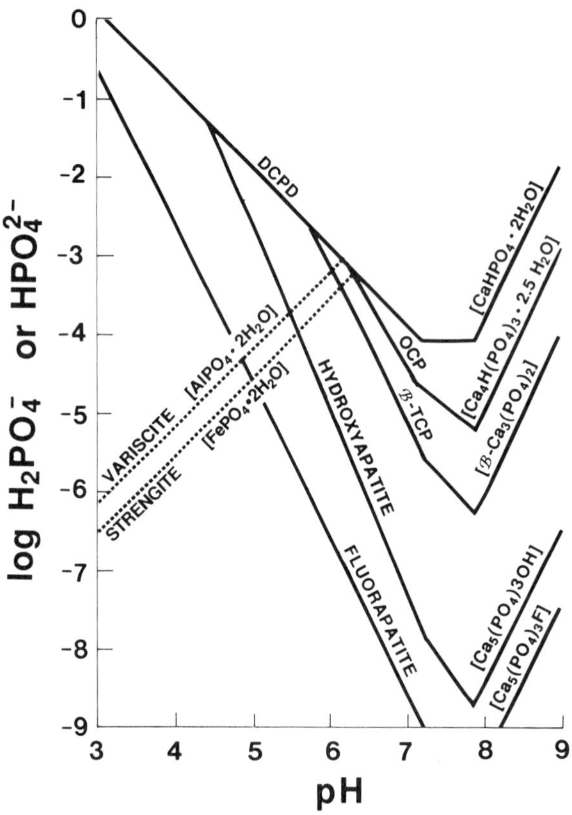

FIGURE 8 Stability of Ca phosphate minerals compared to phosphates of iron and aluminum in equilibrium with solution Ca^{2+} at $10^{-2.5}$ M or as controlled by the solubility of calcite and atmospheric CO_2. (Adapted from Lindsay, 1979.)

the stable mineral P form under equilibrium conditions and the resultant solution phosphorus would be of the order of 10^{-9} M.

Recent work by Jurinak et al. (1986) suggests oxalic acid ($H_2C_2O_4$) and elevated rhizosphere CO_2 may be effective in the enhanced dissolution of apatite minerals in arid soils, thereby promoting increased P availability. Root and microbial respiration can result in pronounced CO_2 differences between the rhizosphere and the bulk soil (Barber and Martin, 1976; Buyanovsky and Wagner, 1983), and in arid region calcareous soils the well-described beneficial influence of mycorrhizae in the P nutrition of infected plants may be in part a function of elevated CO_2 compared to nonmycorrhizal plants (Knight et al., 1989). The prevailing view of mycorrhizae as a more efficient (Cress et al., 1979; Howeler et al., 1982) and additional surface area for P absorption (Sanders and Tinker, 1971) may not in itself fully explain the enhanced P uptake in arid region soils in which P availability is constrained by mineral or organic sinks (Bolan et al., 1984).

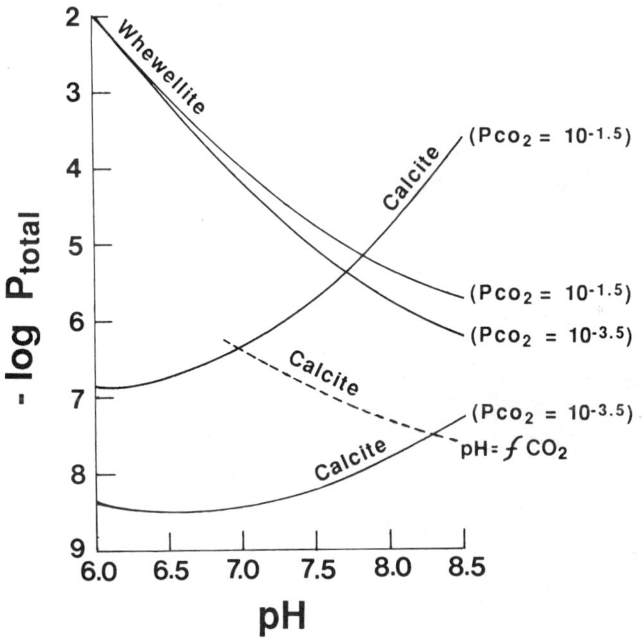

FIGURE 9 The concentration of total soluble phosphorus in equilibrium with hydroxyapatite and calcite or whewellite (calcium oxalate monohydrate). The dashed line represents the field case in which pH is a function of P_{co_2}. Solid lines are when pH is independent of P_{co_2}. (Adapted from Jurinak et al., 1986.)

Mycorrhizae may also increase plant P availability through the differential adsorption of cations and anions, thus modifying rhizosphere pH less directly than by increasing CO_2 (Buwalda et al., 1983) and/or via the synergistic action of P-solubilizing bacteria and mycorrhizae (Azcon et al., 1976). In the case of oxalic acid exuded by plants and/or mycorrhizae, calcium oxalate monohydrate or whewellite (CaC_2O_4) is precipitated, lowering solution Ca^{2+} activity in soil solution and promoting the dissolution of $Ca_5(PO_4)_3OH$. In this event, calcium oxalate, not calcite, becomes the solid phase controlling solution Ca^{2+} ion activity. The effect of whewellite on P availability from apatite is shown in Fig. 9, after Jurinak et al. (1986), along with the effect of enhancing rhizosphere CO_2 from $10^{-3.5}$ to $10^{-1.5}$ atm (100 times ambient atmospheric). When CaC_2O_4 controls solution Ca^{2+} (whewellite lines) and is in equilibrium with hydroxyapatite and fixed atmospheric P_{co_2}, the total P in solution (10^{-6} M) is considerably higher than for calcite in equilibrium with hydroxyapatite under the same conditions ($P_{total} = 10^{-7}$ M). In the whewellite-hydroxyapatite system the effect of increasing P_{co_2} is also marked and further increases P_{total} to 10^{-5}M levels. However, the CO_2 effect is misleading as applied to the calcite-hydroxyapatite system in which pH is a function of P_{co_2}. The dashed line better reflects hydroxyapatite stability with increasing P_{co_2} in the calcite-

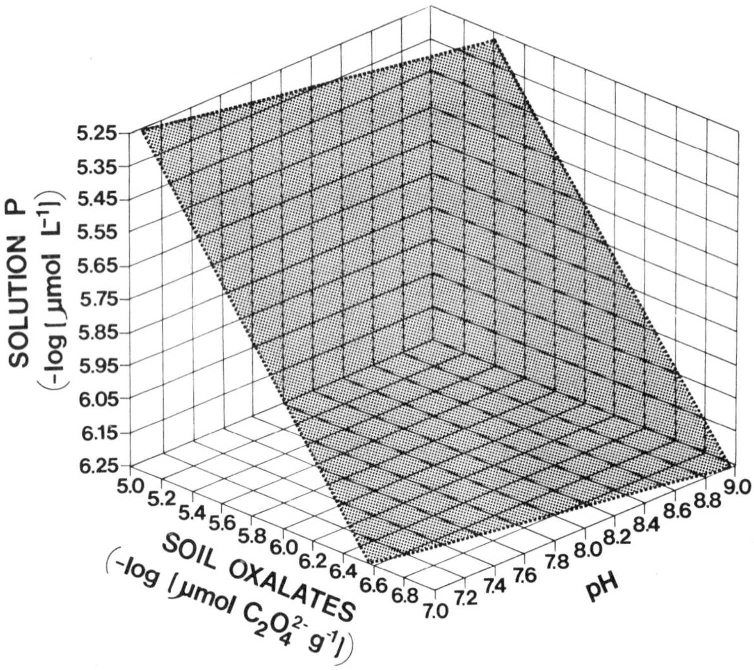

FIGURE 10 The relationship of total solution phosphorus (Ps) to pH and total soil oxalate content in a typic Argixeroll of Wyoming. The multiple linear regression model is $-\log P_s = -2.70 + 0.70 \log$ (soil oxalates) $+ 0.15$ pH; $r^2 = 0.925$. (Unpublished data of Knight.)

hydroxyapatite equilibrium and best represents the field case. The effect of pH as a consequence of increasing P_{CO_2} is much less dramatic in the whewellite-hydroxyapatite system because of the comparatively small pH sensitivity of CaC_2O_4 as opposed to that of $CaCO_3$. Although unable to delineate the stable mineral phase controlling solution P in the soil, unpublished data from the systems studied by Knight et al. (1989) yielded a strong relationship between soil oxalates, pH, and total solution P (Fig. 10).

Halides and Sodium Chloride

Calcium chloride is rarely present in soils because of the preferential precipitation of calcite and gypsum. Magnesium chloride is somewhat more common in arid soils and is more likely found in localized spots of higher humidity, such as some high plateaus of Algeria and southern California. Sodium chloride (NaCl), however, is the most common chloride salt found in salt-affected soils.

C. Salt-Affected Soils

The origins, extents, world geographic distribution, and chemical and physical properties of salt-affected soils are thoroughly addressed in the comprehensive volume by Szabolcs (1989). Salt-affected soils are also well reviewed in selected chapters of the books by James et al. (1982), Bohn et al. (1985), and the volume edited by Bolt and Bruggenwert (1976).

Origins of Salt-Affected Soils

The precipitation and accumulation of mineral salts in situ in arid soils is a consequence of retarded leaching (favoring the retention of ions released during weathering) and the evaporative environment (favoring their concentration), as well as their inheritance from marine sediments (i.e., Mancos shale beds of the upper Colorado River Basin) or sediments formed during the evaporation of saline lacustrine waters and groundwaters. Since arid lands require irrigation for the cultivation of most crops, salts are also imported with irrigation water where evapotranspiration serves to accumulate them over time. When irrigation is poorly managed, salt "loading" of soils can be significant. The application of 1 ha-m of 1000 mg salt L^{-1} irrigation water adds 10 t of salt per hectare annually. Poor irrigation management can also result in the discharge of salt-laden waters to surface waters, compromising the quality of the water available to other users. The Wellton-Mohawk irrigation project of the 1960s is a highly visible example in which irrigation increased the groundwater to a level that intercepted underlying saline deposits in a small Arizona valley. The resultant salinized groundwaters were then discharged to the Gila River and utilized by farmers downstream in Mexico, damaging their crops. The atmospheric deposition of salts may also contribute to their presence in arid region soils as a function of raindrops forming around salt particles. This is particularly appreciable near seacoasts, where the salt concentration of rainfall may be as high as 50-200 mg L^{-1}, decreasing rapidly with distance inland. Even small atmospheric salt deposition over long periods of time, however, may be ultimately substantial, and without leaching in an arid environment, salt accumulation can be pronounced. When affected lands are subsequently managed for salt removal or when particularly high periods of precipitation occur, salts can be flushed and discharged (saline seeps) at other locations. A final mechanism for salt accumulation in arid regions results from a landlocked topography in which waters from surrounding uplands are deposited in a closed basin with no outlet or surface drainage features. Salts dissolved in the contributing waters may then be concentrated during evaporation. The Great Salt Lake of northern Utah is a well-cited example of this phenomenon.

Problems of Salt-Affected Soils

The fundamental concerns associated with salt-affected soils are *salinity* and *sodicity*. High salinity (total soluble salt content) manifests itself in both reduction of the water potential of the soil solution (*osmotic effect*) and the concentration of toxic ions that interfere with normal plant metabolism (*specific ion effect*). The reduction of the osmotic potential component of the total water potential increases the work required to be performed by plants to exclude the salts from uptake, thereby increasing plant stress (*physiological drought*). Common specific ion toxicities in arid regions are associated with chloride, sodium, and boron (e.g., James et al., 1982).

High sodicity (exchangeable Na^+ ion concentration) manifests itself in both the physical deterioration of the soil structure and in the potential for sodium toxicity to plants. In the case of structural deterioration, two mechanisms, swelling and dispersion, appear active (Frenkel et al., 1978). At lower levels of exchangeable sodium (5-15%), Na^+ may be adsorbed on the external edges of clay platelets (*tactoids*) of Ca montmorillonite, where Ca^{2+} ions remain adsorbed on the internal surfaces. Neighboring tactoids repulse one another as a consequence of the diffuse electrical double layer that develops around each tactoid, and dispersion occurs when salinity is low. Dispersed tactoids can migrate and block soil pores, reducing hydraulic conductivity. At higher levels of exchangeable sodium (25-30%), soil swelling becomes prominent as a function of Na^+ ions on the internal surfaces of platelets. The number of monovalent sodium ions that can occupy internal surfaces is in excess of the number of divalent cations (i.e., Ca^{2+}) occupying the same area. An osmotic gradient results that induces water movement into the interplatelet region, causing the clay to swell. When swelling is pronounced pore size is reduced, and again, hydraulic conductivity is impaired. At very high exchangeable sodium percentages (in excess of 50%), swelling becomes so pronounced that the repulsive electrical double-layer forces result in breakdown of the tactoid structure and *deflocculation* of the clay system results.

The presence of salts has an ameliorating effect on the physical consequences of high sodicity (McNeal et al., 1966) due to reduction of the concentration gradient between the bulk soil solution and the internal clay solution and by reducing the thickness of the double layer. The degree of adverse consequences of high sodicity is also a function of the content and mineralogy of the clay fraction. Soils with high clay contents featuring predominantly 2:1 type, swelling layer silicates, such as montmorillonite, are particularly susceptible, whereas coarse-textured soils or soils whose clay fractions are predominantly kaolinite and/or illite are relatively insensitive to the ESP.

Diagnostic Criteria and Classification

The classification of salt-affected soils reflects the types and amounts of salts present in the soil and hence the nature of resultant limitations to plant growth and land use.

The electrical conductivity determined for a saturation extract (EC_e) at 25°C is a measure of the total soluble salt content of the soil water (saturation extract) and thus serves as an index of *salinity*. Its use is based on the dissociation of salts (electrolytes) in water. Since ions carry electrical current, the conductivity of a solution reflects the salt content such that the greater the conductivity, the greater the salt content.

The defining criteria used for classification of sodicity in soils is the *exchangeable sodium percentage* (ESP) defined as (NaX/CEC) 100, where NaX is exchangeable sodium measured in mEq per 100 g and CEC is the cation-exchange capacity of the soil, also in mEq per 100 g. Most often, in practice, the ESP is approximated by determination of the technically less demanding *sodium adsorption ratio* (SAR) via measurement of the solution concentrations of Ca^{2+}, Mg^{2+}, and Na^+ in a saturation extract of the soil. The SAR (where brackets denote concentrations in mEq L^{-1}) is defined as

$$SAR = \frac{[Na^+]}{[(Ca^{2+} + Mg^{2+})/2]^{1/2} \ (mmol\ L^{-1})^{1/2}} \tag{22}$$

and was developed for use with saturation extract measurements by the U.S. Salinity Laboratory (Richards, 1954) based on the work of Gapon (Kelley, 1948), who studied calcium-sodium exchange equilibria in soils; that is,

$$Ca_{1/2}X + Na^+ \Longleftrightarrow NaX + \frac{1}{2} Ca^{2+} \tag{23}$$

At equilibrium, the selectivity coefficient (k_g or the Gapon "constant") for Eq. (23) is

$$k_g = \frac{[Ca^{2+}]_{1/2} NaX}{[Na^+] Ca_{1/2}X \ (mol\ L^{-1})^{1/2}} \tag{24}$$

where brackets denote concentrations in mol L^{-1}. The Gapon equation is arrived at by taking the ratio of exchangeable Na^+ to exchangeable Ca^{2+} in Eq. (24):

$$\frac{NaX}{CaX} = k_g \left(\frac{[Na^+]}{[Ca^{2+}]^{1/2}} \right) \tag{25}$$

The SAR derivation (Richards, 1954) assumes that the chemistry of Ca^{2+} and Mg^{2+} is sufficiently similar to be additive and gives

$$\frac{NaX}{(Ca + Mg)X} = k'_g \left(\frac{[Na^+]}{[(Ca^{2+} + Mg^{2+})/2]^{1/2}} \right) \tag{26}$$

where the additive $Ca^{2+} + Mg^{2+}$ term in the denominator on the right side of the equation is divided by 2 to convert units from mEq L^{-1} to mmol L^{-1} and k'_g is the modified Gapon constant. Since in arid regions Na^+, Ca^{2+}, and Mg^{2+} are by far the dominant exchangeable cations, they are collectively held to reflect the total CEC. This assumption and the expression for SAR [Eq. (22)] then reduces Eq. (26) to

$$\frac{NaX}{CEC - NaX} = k'_g SAR \tag{27}$$

where the left side of the equation is termed the *exchangeable sodium ratio*. Widespread employment of the SAR began with publication of a regression equation between SAR and the exchangeable sodium ratio for 59 salt-affected soils as (after Richards, 1954)

$$ESR = 0.01475(SAR) - 0.0126 \tag{28}$$

and the empirically determined value for k'_g is 0.01475 (mmol $L^{-1})^{-1/2}$. Using the definition for ESP, then, SAR and ESP are related as

$$\frac{ESP}{100 - ESP} = k'_g (SAR) \tag{29}$$

Salt-affected soils are commonly classified as saline (*solonchak* in the USSR classification system) sodic (*solonetz* in the Soviet Union), or saline-sodic as a function of the values measured for the diagnostic criteria. The historical term "white alkali" soils was applied to saline soils because of the frequent (but not ubiquitous) presence of a white salt crust at their surface. Similarly, the dissolution of organic matter in the high-alkaline plus high-Na^+ environment of sodic soils often leads to the formation of dark-colored surface deposits (via evaporation), and the term "black alkali" was applied to sodic soils in the early salinity literature.

As classified by the U.S. Salinity Laboratory (Richards, 1954), saline soils are defined by a saturation extract EC_e in excess of 4 dS m^{-1} and an SAR less than 13. Sodic soils exhibit an SAR greater than 13 with an EC_e below 4 dS m^{-1}, and saline-sodic soils have EC_e and SAR values above 4 dS m^{-1} and 13, respectively. Normal soils are classified by EC_e values below 4 dS m^{-1} and an SAR less than 13. The pH of saline soils is less than 8.5, and the composition of the soil solution is dominated by Ca^{2+} and Mg^{2+} cations and Cl^- and $SO_4{}^{2-}$ anions with smaller amounts of the bicarbonate ion ($HCO_3{}^-$). The pH of sodic soils can be as high as 10 as a function of the formation of NaOH from hydrolysis of exchangeable Na^+ followed by reaction with dissolved CO_2 to form Na_2CO_3. The resultant high concentration of carbonate ions in soil solution favors the precipitation of Ca^{2+} and Mg^{2+} carbonates and leaves Na^+ to dominate the cationic composition of the soil solution. The bicarbonate and carbonate ($CO_3{}^{2-}$) ions then become the predominant anionic species (from dissolution of highly soluble Na^+-salts) with proportionally less Cl^- and $SO_4{}^{2-}$ in soil solution.

IV. ACCUMULATION AND LOSSES OF MINERAL ELEMENTS

Mechanisms for the retention of ions in the soil system include cation exchange, precipitation as secondary minerals or amorphous forms, and binding by humic substances. Consideration of plant uptake and microbial immobilization as temporary sinks for mineral elements in soil rather than as complete losses defines them as retention mechanisms as well. Collectively, retention mechanisms and retarded leaching in arid region soils favor the accumulation of mineral elements released from weathering and organic matter mineralization.

The predominance of iron, calcium, magnesium, sodium, and potassium as chemical constituents of igneous and sedimentary parent materials (Table 2, from Szaboecs, 1989), combined with their favored release from minerals in which they exist as structural cations, suggest that they, along with silica and aluminum, would dominate the soil solution of arid region soils. However, although the release of silica and aluminum is indeed featured in primary mineral weathering, their presence as ions in soil solution is diminished by the precipitation of secondary aluminosilicates. In highly weathered soils aluminosilicates eventually disappear and Al^{3+} accumulates to become a major exchangeable cation. Its retention is favored relative to that of Si^{4+} because the latter is more readily leached. Since the solubility of iron is largely controlled by Fe(III) oxides (Lindsay, 1979), which precipitate following the release of iron from mineral weathering (principally

TABLE 2 Average Chemical Composition (%) of Igneous and Sedimentary Rocks

	Igneous rocks			Sedimentary rocks		
Component	Granite	Grandiorite	Basalt	Sandstone	Clay stone	Limestone
SiO_2	70.77	65.69	51.55	70.63	61.16	5.24
TiO_2	0.39	0.57	1.48	0.15	0.68	0.06
Al_2O_3	14.59	16.11	14.95	4.85	16.21	0.82
Fe_2O_3	1.58	1.76	2.55	1.09	4.23	0.55
FeO	1.79	2.68	9.10	0.31	2.58	0.55
Fe	2.50	3.31	8.86	0.99	4.96	0.43
MnO	0.12	0.07	0.20	Trace	Trace	Trace
MgO	0.89	1.93	6.63	1.18	2.57	7.96
CaO	2.01	4.47	10.00	5.59	3.27	42.97
Na_2O	3.52	3.74	2.35	0.46	1.37	0.05
K_2O	4.15	2.78	0.89	1.33	3.42	0.33
P_2O_5	0.19	0.20	0.30	0.08	0.18	0.04
CO_2				5.11	2.77	41.93

Source: From Szaboecs (1989).

ferromagnesium minerals, such as olivine), its presence in soil solution is also diminished and in calcareous soils it is diminished to the extent of representing a nutritional limit to plant growth. The activity of specific chelating compounds of microbial origin (i.e., hydroxamate siderophores) represents an important mechanism for facilitating the adequate supply of iron to plants in Fe-limited settings.

Calcium, magnesium, sodium, and potassium ions, because of their interaction with exchange surfaces in soils and/or their precipitation as relatively and highly soluble secondary minerals, dominate the soil solution and are not nutritionally restrictive to plant growth in arid region soils.

Despite the slow rate of weathering in arid environments and very low concentrations in primary minerals, trace elements may very slowly accumulate in soils over geological time. In the absence of sufficient leaching to remove them, the accumulation may reach levels deleterious to plant growth or cause injury to animals who graze plants that have concentrated them in their tissues. In arid region soils, boron, molybdenum, and selenium represent such cases and are discussed here.

Although leaching is a transient exception rather than a general occurrence in arid region soils, during seasonal and/or episodic high-precipitation events or rapid snowmelt, leaching losses of highly mobile elements may be facilitated. In some cases leaching losses may be accelerated by anion exclusion phenomena, by metal complexation to more mobile organic moieties or by conversion of ions to move soluble forms (i.e., under reducing conditions).

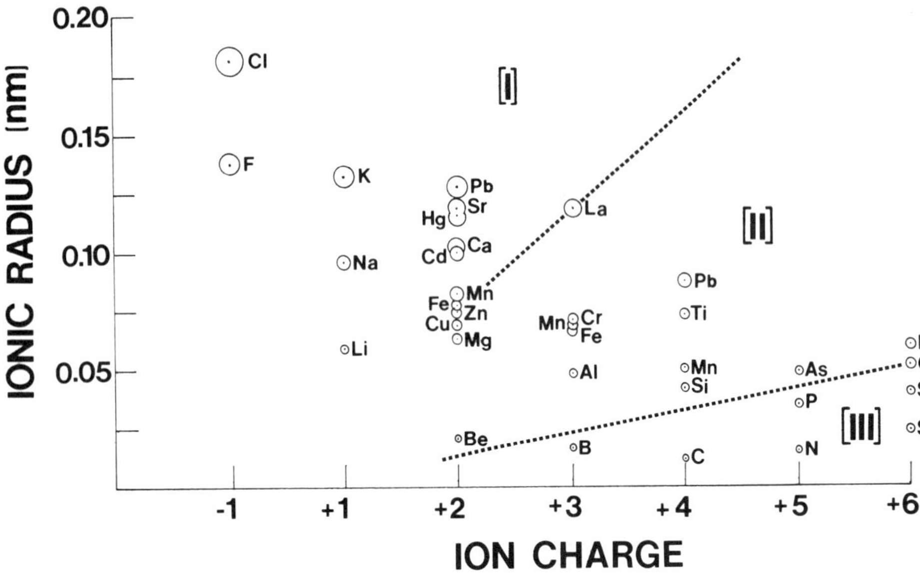

FIGURE 11 Ionic potential for important ions. Relative ion sizes as depicted are not to scale. (After Bohn et al., 1985.)

A. Ion Mobility and Retention

The *ionic potential* (based on the work of Goldrich, 1938) is the ratio of the crystal radius of an ion to its charge (Fig. 11, after Bohn et al., 1985) and is a relatively useful index of the tendency of ions to be removed (leached) from soils over geological time. Large univalent and divalent ions (category I, Fig. 11), including halides (Cl^-, F^-) and the alkali and alkaline earth cations (K^+, Na^+, Ca^{2+}, and Mg^{2+}), are highly soluble, easily weatherable, and readily removed from soil, provided there are sufficient moisture availability and flow to effect these processes. In arid regions, the water gradient is dominantly toward the surface, and therefore despite their high solubility these ions tend to accumulate in developing arid soils. However, it is not only the evaporative gradient that serves to preclude leaching losses but also their retention as exchangeable cations and their precipitation as secondary minerals as discussed in Sec. III.

Ions of high charge with low ionic radii (small ionic potential) repel protons of associated water molecules and form oxyanions that are moderately to highly soluble and also tend to leach. Oxyanions of P, S, N, B, and C (phosphates, sulfate, nitrate, borates, and carbonates) are members of this group (category III, Fig. 11), although orthophosphate mobility is characteristically low in soils, due, in part, to specific adsorption and mineral precipitation phenomena (Enfield and Ellis, 1983). The stable forms and relative concentration distribution of P, S, and C oxyanions in aerobic soils are well defined as a function of pH (Lindsay, 1979) and in the neutral to alkaline range characteristic of arid region soils are predominantly $H_2PO_4^-$ and HPO_4^{2-} (orthophosphates), SO_4^{2-} (sulfate), and the bicarbonate ion (HCO_3^-). Carbonates and sulfates also participate in reactions that form stable secondary minerals in arid region soils. Sulfates and orthophosphates are taken up by plants and microorganisms as essential growth elements and incorporated in the organic fraction of soils. Nitrate is highly subject to leaching but is also readily taken up by plants and microorganisms and is subject to conversion to other N forms with varying degrees and mechanisms for retention and loss from soils. N forms and chemistry are discussed in Sec. V, on organic matter and nitrogen. Boron as the borate ion ($H_2BO_3^-$) is readily leached, and B limitations to plant growth are often associated with highly weathered soils. However, borate leaching is selected against in arid regions and Na and Ca borate salts may even precipitate at the surface of some soils. Boron toxicities are common to many salt-affected soils. Molybdenum and selenium may also accumulate in arid region soils. Their oxyanions increase in availability with increasing pH (Lindsay, 1979), and they are readily taken up by arid zone plants and frequently concentrated to levels toxic to foraging animals (i.e., *molybdenosis* and *selenosis*).

The behavior of ions of intermediate charge and ionic radii (category II, Fig. 11) are somewhat more difficult to predict. Included in this group are Al^{3+}, Fe(III), and Mn(III) and Mn(IV). These ions commonly precipitate as insoluble hydroxides or oxides by repeling H^+ and polarizing associated water molecules and thus tend to accumulate in soils over time. Oxides of aluminum, iron, and manganese among others are commonly abundant in highly weathered soils and are not as appreciable in the characteristically undeveloped soils of dry regions. The formation of the insoluble hydroxides of iron and zinc, however, in calcareous soils of arid regions limits their availability and results in common deficiencies. The action of specific complexing transport agents of microbial origin (i.e., hydroxamate siderophores) have been elucidated as important to the nutrition of plants and microorganisms under Fe-limiting conditions.

Cation Retention and Replaceability

Negative charge manifested on the surfaces of clay minerals and associated with functional groups in organic matter retains cations in the soil. This phenomenon is of profound importance to the elemental fertility of soils since it serves to retard leaching losses of oppositely charged cations (principally Ca^{2+}, Mg^{2+}, Na^+, and K^+) of nutritional importance to plants and microorganisms and hold them in the soil solution (water films) near colloidal surfaces where they are available for uptake. Since these and other cations can exchange for other cations in the aqueous phase they are called *exchangeable cations* and are defined as those released from soil following treatment with neutral salt solutions. Generally, the sum of exchangeable Ca^{2+}, Mg^{2+}, K^+, Na^+, and Al^{3+} is numerically equivalent to the cation-exchange capacity of a soil, although the relative contribution of cationic constituents can differ. Al^{3+} tends to be more accumulated in acid soils, for example, whereas Na^+ is a dominant cation in the sodic soils often encountered in arid regions.

The dilution of an equilibrium solution favors retention of more highly charged species and is called the *valence dilution effect*. The preference or selectivity of exchange surfaces for oppositely charged ions is a function of both valence and the hydrated size of the ion. Because smaller hydrated ions can approach the colloidal surface much closer than larger ions of like charge, they are bound more "tightly." Valence is a much more important factor, however, and can be illustrated qualitatively via Coulombs' law:

$$F = \frac{qq'}{Dr^2} K \tag{30}$$

where F is the force of attraction or repulsion (newtons), q and q' are charges (coulombs), r is the distance of charge separation (m), D is the dialectric constant (78 for water at 25°C), and K is the proportionality constant (8.9×10^9). Changing the charge of q' representing the valence of a single ion from 1+ to 2+, where q is the charge of the colloid surface, doubles F.

The relative replaceability of ions of varying valences in a series have been experimentally determined and is called a *lyotrophic series*. From dehydrated crystal radii and percentage NH_4^+ or K^+ data the following lyotrophic series results, in order of replaceability (after Bohn et al., 1985):

$$Li^+ = Na^+ > K^+ + NH_4^+ > Rb^+ > Cs^+ + Mg^{2+} > Ca^{2+} > Sr^{2+}$$
$$= Ba^{2+} > La^{3+} = Al^{3+} > Th^{4+}$$

Anion Exclusion

Under air-dry conditions, the residence of exchangeable ions on oppositely charged surfaces constitutes a double layer (i.e., a layer of negative charge associated with the colloidal surface and a layer of positive charge associated with the tightly bound cations), called a *Helmholtz double layer*. In the presence of water, diffusion of the cations in the aqueous phase away from the surface tends to decrease their concentration with increasing distance from the surface. The layer of counterions in solution which neutralizes the surface charge, is diffuse and therefore the double layer is termed the *diffuse double layer* (DDL). DDL theory was independently developed by Gouy and Chapman in the first decade of this century and is commonly attributed to them as the Gouy-Chapman DDL theory. The thickness of the DDL is a function of the concentration C_i of the bulk solution (outside the influence of the colloidal charge) such that increasing cation concentration C_c shrinks the DDL by decreasing the diffusion of positive charge away from the charged surface. Since the negative charge of colloidal surfaces also tends to repel like charged ions (*anion repulsion* or *exclusion*), the "cation-rich" nature of the DDL is emphasized. One important consequence of anion exclusion is that it allows, for example, NO_3^- movement in excess of that predicted by convection theory (Thomas and Swodoba, 1970; Krupp et al., 1972). However, bulk solution anion concentration C_a also tends to promote diffusion of anions into the double layer with increasing C_a and thus shrinks DDL thickness. The overall thickness of the DDL is small (0.05-10 nm), depending on C_i and the amounts of monovalent versus divalent cations compared to the size of soil pores (1000-50,000 nm), and is roughly the same size as water film thicknesses in dry soils.

Organometal Complexes

The role of metal complexation by organic matter in the chemistry of trace elements is of importance because it increases the availability of micronutrients via reduction in the amounts that would otherwise precipitate as more insoluble mineral forms and by enhancing their mobility. The concentrations of toxic elements may also be reduced to nontoxic levels by OM complexation, especially to the relatively recalcitrant high-molecular-weight components (i.e., humic acid). Conversely, some organic complexing agents may serve to enhance the bioavailability of toxic metals and/or increase their mobility and thus facilitate their movement to groundwater.

Organic acids and complexing agents lead to the formation of one type of organometal complex, and metal-humate complexes represent another type. The ability of humic and fulvic acids to form stable complexes with metals is attributed to their high content of oxygen-containing functional groups (i.e., -COOH, phenolic, alcoholic, and enolic -OH and C=O). The amounts of metals that can be bound to humic substances appears approximately equal to the content of functional acid groups, principally -COOH. Using typical exchange acidities for humic substances of between 1.5 and 5.0 mEq g^{-1}, a retention of 48-160 mg Cu^{2+} g^{-1} humic acid can be calculated (Stevenson, 1986). Ionic strength, pH, molecular weight, and degree of saturation affect the binding capacity of humified materials.

The solubility of humic acid-complexed metals may be increased or diminished as a function of the solubility of the complexing agent. Fulvic acid-metal complexes, owing to their lower molecular weights and higher contents of acid functional groups, are more soluble than humic acid-metal complexes. Adsorption of humic materials to clays limits their solubilization. High pH and high trivalent cation concentrations tend to favor their precipitation.

The availability of redox-sensitive elements may be strongly affected by humic substances, which can reduce Fe^{3+} to Fe^{2+}, MoO_4^{2-} to MoO^{5+}, VO_3^- to VO^{2+}, and Hg^{2+} to Hg^0.

B. Iron and Micronutrients (Toxicities and Deficiencies)

The geographic distribution, chemistry, cycling, and nutritional importance of trace metals and micronutrients are reviewed in Mortvedt et al. (1972), Davies (1980), and Robb and Pierpoint (1983). The micronutrients of principal nutritional interest and toxicity-related concern in arid region soils are iron (Fe), zinc (Zn), manganese (Mn), copper (Cu), boron (B), molybdenum (Mo), and selenium (Se).

Iron and Hydroxamate Siderophores

The stable forms of iron in calcareous soils are the Fe(III) hydrolysis species $Fe(OH)_3^0$, $Fe(OH)_4^-$, and $Fe(OH)_2^+$. In the pH range 7-9, soluble inorganic forms of Fe in total do not exceed 10^{-10} M (Lindsay, 1979). This level of iron is insufficient for the normal growth of most plants and leads to the condition commonly described as *lime-induced iron chlorosis*. Plants growing in calcareous soils may obtain their Fe requirement naturally by effecting localized (rhizoplane) changes in redox or from chelated forms via the action of plant and microbial exudates or humic substances. Most introduced crops must be artificially supplied with chelated Fe fertilizers.

A group of low-molecular-weight (500-1200 daltons), Fe^{3+}-specific, peptide derivative chelating agents called *hydroxamate siderophores* (HSs) have received increasing attention as particularly important in the iron nutrition (and competition) of microbes (Murphy et al., 1976; Emery, 1977; Neilands, 1979; Kloepper et al., 1980) and plants (Powell et al., 1980, 1982; Cline et al., 1984) under limiting conditions. The effectiveness of HSs as Fe chelates is related to their stability over a wide range of pH (Powell et al., 1980) as evidenced by the example of Desferrioxamine B (DFOB), a siderophore produced by *Streptomyces pilosus* important to the Fe nutrition of sorghum (Cline et al., 1984). The organic acids are ineffective in all but acid media, and the synthetic chelates, although effective at high pH, are less stable than the natural HS, DFOB. Hydroxamate siderophores appear to be present in many soils at significant levels and are concentrated in the rhizosphere and produced by rhizobacteria and mycorrhizal and nonmycotrophic fungi (Powell et al., 1980). The idealized action of HSs produced by a hyphal segment of a mycorrhizal fungi is illustrated in Fig. 12. HSs concentration in soil solution is highly modified by clay content, especially organic matter content, being readily adsorbed on humic acid (Powell et al., 1980).

Zinc, Manganese, Copper, Boron, Molybdenum, and Selenium

The stable mineral phases of zinc that may control the activity of Zn^{2+} are not defined. A useful reference solubility given by Lindsay (1979) is

$$\log Zn^{2+} = 5.8 - 2pH \tag{31}$$

which is based on the experimentally measured solubility of soil Zn. From Eq. (31) it is readily calculated that in a calcareous soil with a pH of 8.3, Zn^{2+} concentration in soil solution is of the order of 10^{-11} M. The potential for zinc limitations to plant growth, as with iron, is therefore enhanced with increasing pH and, in turn, a function of the calcium carbonate content and exchangeable sodium status of the soil.

Manganese deficiencies would also be indicated as a function of the controlling solid phase (MnO_2) predicted at high pH under aerobic conditions (i.e., -log (Mn^{2+}) = 10^{-14} - 20^{-18} M at pH 7-9) (Lindsay, 1972). However, Mn deficiencies are uncommon in calcareous soils, suggesting that other, unidentified solid phases (perhaps coprecipitates of iron and/or other trace metals) may be present and governing Mn^{2+} activity at higher pH.

Copper availability in soils is also highly dependent on pH and markedly decreases to dificient levels in alkaline soils in the absence of organic matter. However, Cu^{2+} is strongly complexed with organic matter in soils, and limitations are much less common than for iron and, to a lesser extent, zinc. A useful reference for estimating changes in Cu^{2+} activity with pH [Eq. (33)] follows from the reaction involving soil and copper (after Lindsay, 1972):

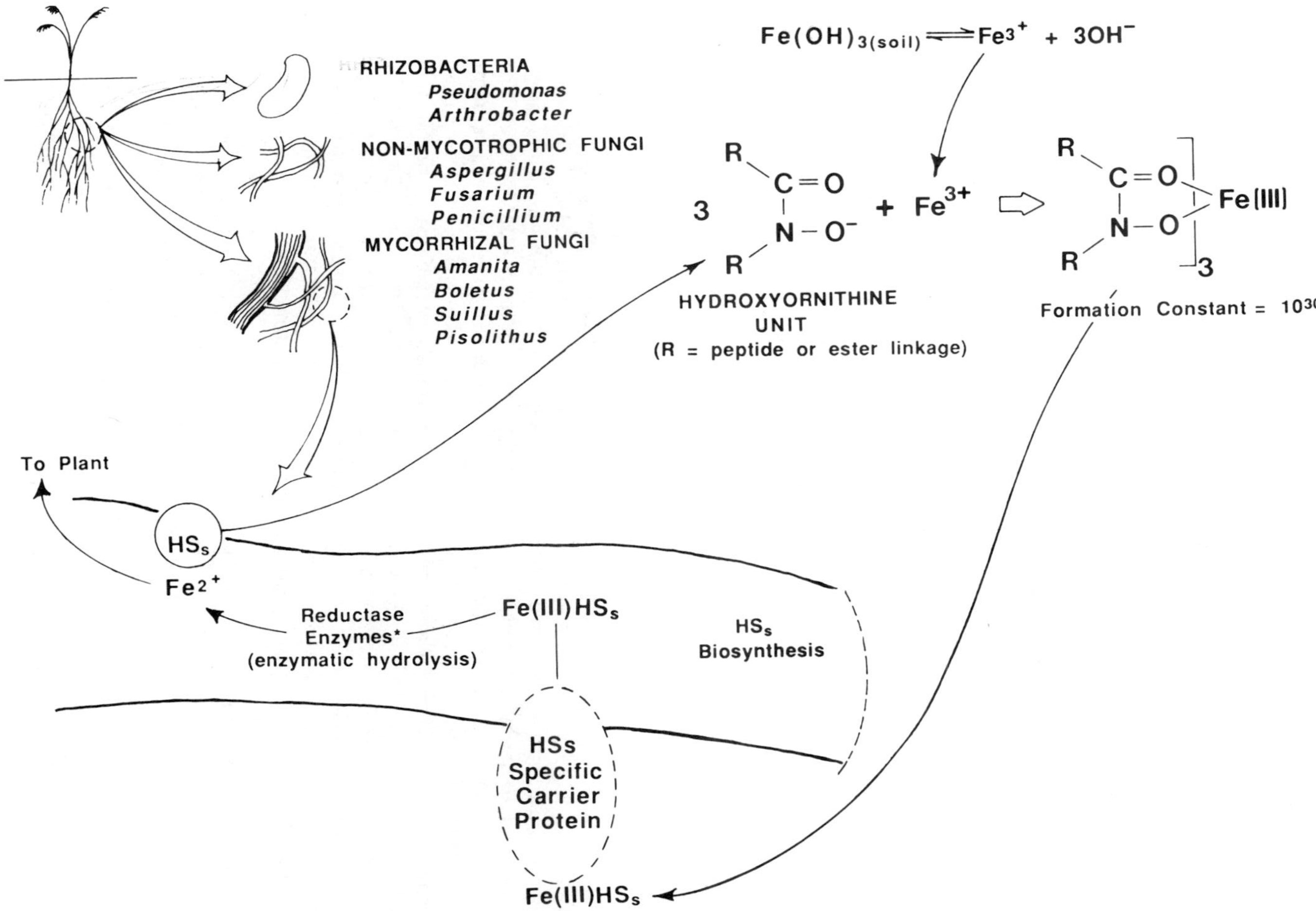

FIGURE 12 Conceptual representation of the mode of action of hydroxamate siderophores in the iron nutrition of plants.

$$Cu^{2+} + \text{soil} \Leftrightarrow \text{Cu-soil} + 2H^+ \quad (32)$$

as

$$(Cu^{2+}) = 10^{3.2}(H^+) \quad (33)$$

The chemistry of boron mineral phases is not well understood, although highly insoluble fluorine-containing borosilicate (tourmaline), Ca and Mg borate, and Fe and Al complexes have been identified and, along with organic matter, may contribute to boron retention in soils. Boron toxicities, however, are relatively common in arid zone soils in which H_2BO_3 leaching is selected against, Na and Ca borate salts accumulate at the soil surface, and B-rich irrigation waters are applied.

Molybdenum and selenium also commonly accumulate to toxic levels in plants in arid regions and result in the conditions of molybdenosis and selenosis in foraging animals. The toxic levels of molybdenum in arid region soils is due to the increasing availability of MoO_4^{2-} with increasing pH (Lindsay, 1972). A reference value for MoO_4^{2-} can be obtained from the following expression (after Lindsay, 1972):

$$(MoO_4^{2-}) = \frac{10^{-20.5}}{(H^+)^2} \quad (34)$$

V. ORGANIC MATTER AND NITROGEN

Soil organic matter is composed of and derived from the tissues and metabolic products of both living and decaying organisms in or on the developing soil. The process by which the essential plant nutrients of nitrogen, phosphorus, and sulfur are liberated from organic matter and subsequently converted to inorganic forms (NH_4^+, NO_3^-, PO_4^{3-}, and SO_4^{2-}) is called *mineralization* and is facilitated by microorganisms. As such, those factors that influence the growth and activity of microorganisms in soil (including temperature, pH, and water potential) also influence mineralization. *Immobilization* refers to the uptake and subsequent conversion of the available mineral forms to organic forms, also via the growth and activity of microorganisms. In arid soils, where sporadic rainfall is punctuated by periods of drought, these processes are largely episodic, with characteristic "flushes" of decomposition and mineral release commensurate with precipitation events.

The participation of N, P, and S mineral forms in inorganic reactions often serves to obscure their release during organic matter breakdown. For example, NH_4^+ may be "fixed" in soil clays and/or lost as gaseous ammonia. Since phosphates do not readily leach, the total P content of soils is normally quite stable. The availability of P, however, as dictated by the solubility of secondary minerals or by its release from organic matter (mineralization), varies widely. For this reason soil P has been categorized as fractions residing in various "pools" according to their availability (e.g., Stevenson, 1986). Both organic and inorganic pools have stable (recalcitrant or insoluble) phases and more readily available (labile) phases.

Adsorption isotherms have been widely employed to evaluate the loss of added P from soil (e.g., Enfield and Ellis, 1983). After a given time to allow equilibration, both the Freundlich and Langmuir equations have been found to successfully describe the disappearance of added P in soils (Berkheiser et al., 1980). This is not surprising in that a multiple-layer Langmuir model has been shown to be equivalent to the Freundlich equation (Brunauer and Copeland, 1967).

Sulfur is generally associated with the organic fraction of aerobic soils, but as the organic matter content of soils decreases in arid soils, relatively more S is present as mineral forms, such as gypsum and epsomite, with generally much smaller amounts of the pyritic minerals chalcopyrite ($CuFeS_2$) or pyrite (FeS_2) inherited from parent materials. The oxidation of pyrite yields SO_4^{2-} which is then incorporated in organic matter via plant and microbial uptake or precipitated with Ca^{2+} as gypsum. Gypsum solubility is independent of pH over its stability range as governed by the activity of SO_4^{2-} in soil solution and can coexist with calcite. Sulfate and phosphate may be precipitated as stable mineral phases (i.e., gypsum or apatites, respectively).

The activities, death, and decay of resident plants, animals, and microorganisms in and on the soil provide food for other living organisms, reactive compounds for chemical interaction with the inorganic component of soils, and precursors for the synthesis of the complex, stable organic compounds collectively called *humified organic matter*. On a mass basis humified substances are the most reactive components of soils, and their contribution to the chemistry of soils is pronounced. These resistant (humified) materials (mean residence time in soil = 250-1000 years) themselves participate in a host of mineral-weathering, transformation, and complexation reactions (Dixon and Weed, 1977; Huang and Schnitzer, 1986). They are a major contributor to the CEC of soils, promote aeration and permeability via aggregation, and enhance the water-holding capacity of soils. The effect of the inorganic component of soils (particularly the clay mineral fraction) on the humified fraction is equally important since it is principally responsible for the recalcitrance of humified materials.

The action of metabolic exudates of microorganisms and nonhumified decay products exhibits varying degrees of chemical reactivity and importance under differing soil conditions. In this regard, each of the processes and materials noted previously influence (and/or are influenced by) the conditions characteristic of the soil, particularly pH, texture, and the type and degree of base saturation.

The topics of this section are well reviewed in selected chapters of the recent books by Paul and Clark (1989), Stevenson (1986), Haynes (1986), and Bohn et al. (1985).

A. Inputs, Decomposition, and Mineralization

The amount, variety, and composition of plant material in concert with the intensity of microbial activity in essence dictate the organic matter content of soils. These factors in turn are principally dependent on climate. Dry matter inputs of dead plant material in arid soils are of the order of zero to 1 kg C m^{-2} $year^{-1}$. Despite the low annual input of litter, organic matter may very slowly accumulate in arid regions until a characteristic "steady-

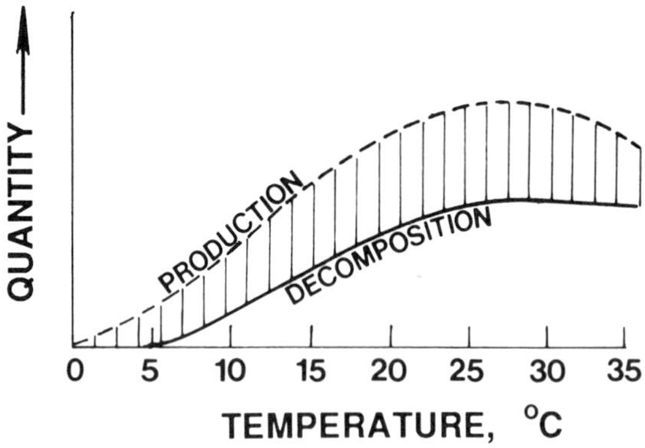

FIGURE 13 Relative effect of temperature on primary production versus organic matter breakdown in an aerobic environment. (Adapted from Senstius, 1958, *Am. Sci. 46*: 355, as presented by Stevenson, 1986.)

state" level is achieved. Stevenson (1986) attributes this steady-state condition, in part (in all soils), to the production of highly resistant colloidal compounds (e.g., humic acids), limitations in the availability of cometabolites (N, P, and S) necessary for the microbial degradation of organic matter, and the protection of humus via its interaction with mineral matter (i.e., formation of organomineral complexes). The range of organic matter contents of arid region soils is zero to about 3% (by mass), but values are more typically less than 1%. The high organic matter content (as much as 5-6% by mass in the upper 15 cm) of grassland soils in semiarid regions (Mollisols) was noted earlier and may be attributable (after Stevenson, 1986) to

1. Greater production of humus precursors produced under grass
2. Inhibited nitrification (i.e., conservation of N)
3. More extensive rhizosphere (where humus synthesis occurs) under grass
4. Poorer soil aeration contributing to preservation of organic matter
5. Enhanced NH_3 fixation by lignin (i.e., conservation of N) as promoted by the high base status of grassland soils

Temperature exerts a differential effect on plant primary production processes and the activities of microorganisms. As applied to organic matter production and decomposition, the concept is illustrated in Fig. 13 (after Stevenson, 1986) and may also help explain the high organic matter content of the grassland soils characteristic of semiarid regions (i.e., Mollisols).

Soil N accumulation during succession has been reviewed by Stevens and Walker (1970), Major (1974), and Jenny (1980). In harsh environments (i.e., deserts and tundra) and during the earliest stages of soil development, it is the activity of primary colonizing microorganisms that imports N and provides for its accumulation. Cyanobacteria are particularly important in this regard (Fogg et al., 1973). As soil development proceeds and conditions become increasingly conducive to secondary succession, N_2-fixing plant species become established and characteristically dominate (Fogg et al., 1973; Burns and Hardy, 1975; Granhall, 1981).

The uptake of N by plants in natural desert systems is considerably reduced compared to that in more mesic environments where growth is more vigorous (Haynes, 1986). Since high temperature and moisture favor microbial activity (including decomposition processes), the rate of litter breakdown is considerably higher in warm, mesic environments than in the moisture-limited soils of arid regions (Olson, 1963; Jenkinson and Ayanaba, 1977).

Decomposition

The larger scale breakup of plant and animal residues is accomplished by mammals, earthworms, and termites following or commensurate with the action of saprophytic plants and decomposing microorganisms. The oxidation of organic carbon in plant, animal, and microbial remains provides energy for the metabolism and growth of *heterotrophic* microorganisms in soil. Residue attack by these organisms is initially rapid where oxygen, water, temperature, and nutrients (*cometabolites*) are not limiting. In arid soils water availability markedly limits biological activities (Knight and Skujins, 1981), both directly and by limiting access of microorganisms to dissolved cometabolites (i.e., N and P). For a review of the effects of limited water on the growth, activities, and ecology of soil microorganisms, see Parr et al. (1981).

Organic materials, as derived from plant, animal, and microbial structures, are both highly variable and highly ordered. The breaking of structural bonds requires overcoming activation energy barriers via catalysts. The catalysts of biological systems are enzymes, and fundamentally, microbial attack on organic matter is enzymatic. The complex and varied composition of organic matter dictates a myriad specific enzymes active in its complete breakdown. For a listing and discussion of soil enzymes and the reactions they catalyze, see Stevenson (1986).

Since decomposing organisms more readily degrade the least resistant compounds (i.e., sugars, cellulose, and proteins) than the more recalcitrant compounds (i.e., lignin), the initially rapid rate of decay slows. Proteins, starches, and cellulose appear especially susceptible to breakdown by molds and spore-forming bacteria. Despite the presence of the more resistant fractions, 50-80% of freshly added organic matter is lost in a year from most temperate region soils. The principal products of the initial decomposition stage in aerobic soils are CO_2, H_2O, NH_3, SO_2, organic acids, and other incompletely oxidized substances. As decomposition continues, new biomass is formed with incorporation into microbial tissue of some of the initial decomposition products and further loss of CO_2 via the continued decomposition of others. In the latter phase of decomposition, the resistant lignaceous materials are slowly metabolized, with the continued formation of new biomass and release of CO_2. The fungi and actinomycetes play a major role in this stage of decomposition and may be particularly important in arid soils owing to their ability to continue metabolic processes at reduced water potentials in soils.

Mineralization

Nitrogen, phosphorus, and sulfur are associated with the organic fraction of soils, and their availability is largely a consequence of microbially mediated mineralization and immobilization. Mineralization and immobilization can be presented as

$$R\text{-}N,P,S\ (\text{immobilized}) \Longleftrightarrow (\text{mineralized})\ NH^{4+},\ NO^{3+},\ PO_4^{3-},\ SO_4^{2-} \qquad (35)$$

where R- is an organic moiety and the arrows depict opposing processes but do not imply equilibrium.

The retention time of immobilized nutrients is of the order of several months to a few years, and the microbial biomass thus serves as a temporary reservoir for N, P, and S. The humified materials have mean residence times of the order of 250-1000+ years and serve as a more "permanent" sink for these nutrients in the sense that their slow decomposition to release nutrients is matched by the synthesis of new humified materials.

The rates of mineralization of N, P, and S are not the same, and they are not generally released in the same ratio as they occur in organic matter. This may be principally because they occur in different compounds and fractions of organic matter whose degradation is more or less readily facilitated by decomposing organisms. The ratios of carbon to the nutrients (in decomposing materials) serve to regulate mineralization-immobilization processes. According to Stevenson (1986), ratio values that tend to favor net mineralization are, for C/N, C/P, and C/S, <20:1, <200:1, and <200:1, respectively. Net immobilization of N, P, and S is favored for C/N, C/P, and C/S ratios of the order of 30:1, 300:1, and 400:1, respectively.

B. Humification

The portion of soil organic matter that occurs in relatively stable form is called *humified* and is synthesized by soil microorganisms from nonhumic decomposition products and metabolic exudates.

The major classes of nonhumified, nonliving organic materials in soils are carbohydrates (principally sugars, hemicellulose, and cellulose) and lipids (fatty acids, fats, waxes, and resins). In addition to serving as foodstuffs for microorganisms and precursors for humified materials, lipids and carbohydrates can also more directly influence soil properties and plant growth. Lipids, such as aldehydes, ketones, phenolic acids, glycosides, and coumarins, are of importance because of their variable toxicity to plants. Other lipids, such as the B vitamins, conversely can stimulate plant growth. The ability of complex polysaccharides to create stable soil aggregates via binding of inorganic soil particles can be important to soil structure and may even be detrimental when microbial gums and/or mucilages accumulate to block soil pores. However, this occurrence is generally restricted to submerged soils and is not of consequence in arid soils. To a minor extent, carbohydrates may supplement the CEC via the charge contribution of the COOH-groups of uronic acid. Carbohydrates can also form complexes with metal ions, often increasing their mobility and bioavailability. The contributions of humified materials to soil aggregation, cation-exchange capacity, and organometallic complex formation are far greater, however, and their formation, structures, and properties are emphasized here.

Collectively, carbohydrates and lipids may represent from about 5% to as much as 30% of the organic matter of most soils. Humified materials, in

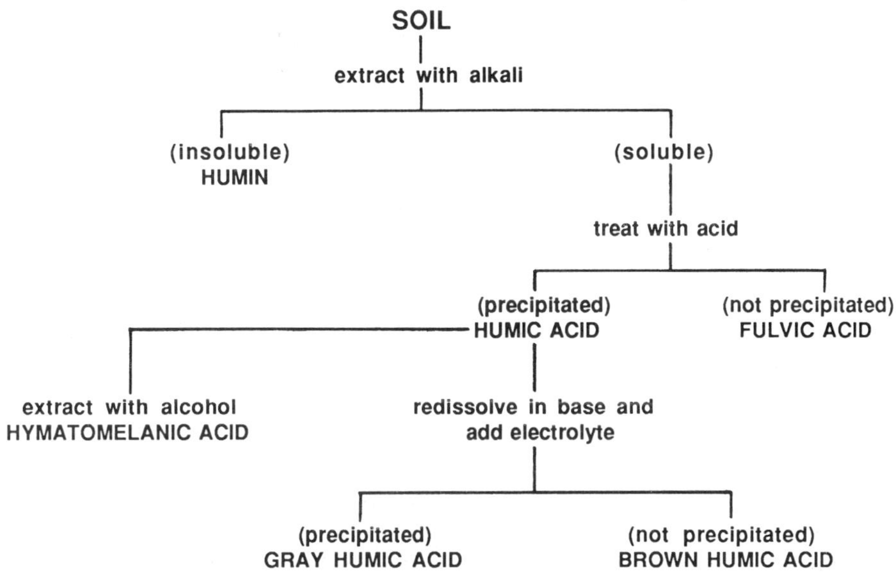

FIGURE 14 Simplified scheme for the extraction and fractionation of humified materials in soil. (After Stevenson, 1986.)

comparison, are by far the largest component of organic matter in inorganic soils (70-80%) and principally consist of the complex compounds humic acid (HA) and fulvic acid (FA) and the hydroxybenzoic acid derivatives formed via oxidative degradation of HA and FA.

Extraction and Characterization of Humified Materials

Humified materials are highly acidic, yellow to black, high-molecular-weight polyelectrolytes formed by secondary synthesis reactions and characterized as a function of their extractability from soils with highly alkaline NaOH and/or neutral pyrophosphate solutions followed by sequential treatments with acid, base, and alcohol (Fig. 14). Upon treatment, the fraction that is not soluble in alkali is called *humin*. The fraction that reprecipitates upon acidification of the soluble extract remaining is called *humic acid*, and the soluble fraction left is called *fulvic acid*. *Hymatomelanic acid* is the fraction that is extractable in alcohol from the precipitated humic acid. After redissolving the humic acid in base and treating with electrolyte two more fractions are obtained. One is *gray humic acid*, which precipitates, and the other is *brown humic acid*, which does not. The gray humic acids are characteristic of the humic acids associated with Mollisols and differ from those associated with peat and Spodosols (i.e., brown humic acids). Further, mollisols appear to contain proportionally higher amounts of humic acids as opposed to the higher relative amounts of fulvic acids found in forest soils (Alfisols, Spodosols, and Ultisols).

Formation and Structures

Humified materials appear to be formed via polymerization of monomers resulting from the decomposition of plant components (including lignin) that have been metabolized by microorganisms with a subsequent increase in soil biomass (Fig. 15). The polymerization stage is commensurate with repeated recycling of the biomass C and N, formation of more biomass, and microbial production of polyphenols (Kononova, 1966).

Fulvic acids are contrasted with humic acid by lower molecular weights, decreased C contents, and both increased O contents and exchange acidity. As conceptualized by Kononova (1966), this reflects a decreased level of polymerization. The structures of humic and fulvic acids remain incompletely defined, but "types" have been postulated that exist in varying degrees of consistency with compositional data and formational theory (Stevenson, 1986).

Properties

The properties of the humified materials differ from the plant, animal, and/or microbial compounds from which they are derived (including lignaceous biopolymers of plants).

Functional groups, such as hydroxyl (-OH), carboxyl (-COOH), amine ($-NH_2$), and phenolic (C_6H_4OH), are present on the surfaces of solids, including layer silicates, organic matter, oxides, and hydrous oxides. When these groups are protonated (gain H^+) or deprotonated (lose H^+), surface charge is correspondingly increased or decreased. The direction and degree of protonation or deprotonation is principally a function of the pH of the soil solution, and thus resultant changes in charge are called pH dependent. On organic matter, charge development is exclusively pH dependent and principally a result of H^+ dissociation from phenolic and carboxylic groups. The extent of pH-dependent charge on organic matter and resultant CEC is dramatic. This is illustrated by the data of Helling et al. (1964) from a survey of 60 Wisconsin soils. For pH values of 3.5, 6.0, and 8.0, the mean CEC (n = 60) in mmol+ kg^{-1} for soil organic matter were 730, 1310, and 2130, respectively. These values, compared to the total soil cation-exchange capacities, resulted in increased organic matter contributions to total soil CEC with increasing pH of 28, 36, and 45%. In arid region soils, then, the relatively low organic matter may be compensated, in part, by the increased exchange capacity associated with higher pH.

The formation of stable aggregates by humate and/or carbohydrates and clays contributes to good soil structure and thus imparts erosion resistance and favorable water movement through soils. Soil aggregate stability is well related to the amount of humus materials extractable in a neutral pyrophosphate solution but is highly correlated with the amounts extractable in a pyrophosphate-NaOH sequential extraction (Chaney and Swift, 1984). The latter procedure yields more of the high-molecular-weight, less oxidized humus compounds and therefore suggests they are the substances most important in soil aggregation. Clay-humate complexes in soils appear to arise from polyvalent cation ($M=Ca^{2+}$, Fe^{3+}) ionic bonding, van der Waals bonding with sugar groups, and ligand exchange with hydrous oxides. The greater surface area of smectites (2:1) clays, characteristic of arid region soils, compared to the kaolin (1:1) clays enhances the stabilizing effect of

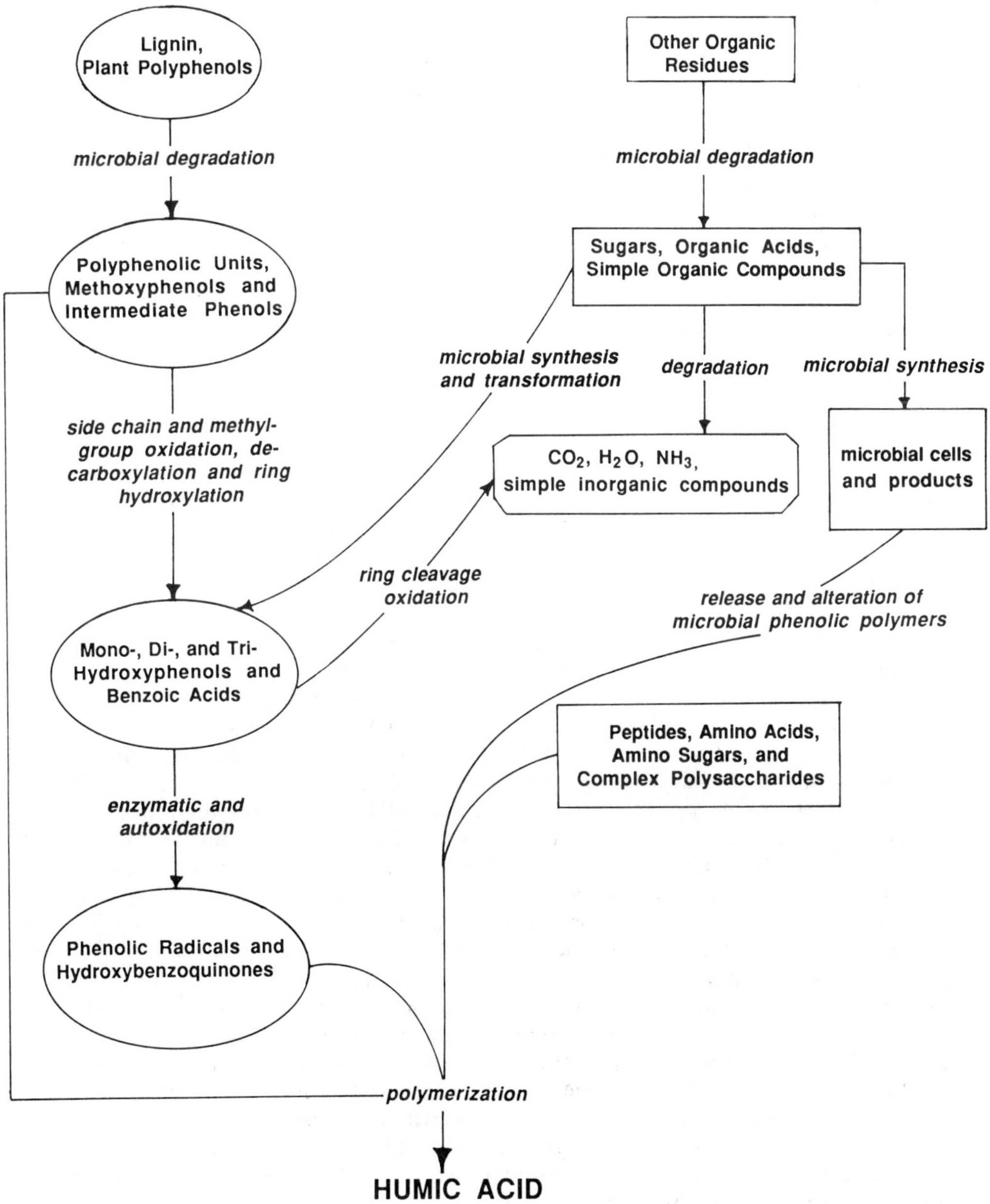

FIGURE 15 Conceptualized humic acid formation paths and precursors. (Adapted from Martin and Focht, 1977.)

clays on organic matter (Sorensen, 1972), probably because of the formation of interlayer complexes (Theng, 1974). The sorption and complexation of organics by clays tend to retard their biodegradation (McLaren and Peterson, 1965; Theng, 1979; Stevenson, 1982).

C. Nitrogen Transformations, Sinks, and Losses

Nitrogen is by far the most widely and comprehensively studied nutritional element in soils. Excellent and comprehensive reviews on the chemistry, biochemistry, and fertility aspects of nitrogen in relation to soils and agriculture are numerous. Particularly noteworthy sources are the books by Haynes (1986) and Stevenson (1986). Also see Skujiņš (1981) for a more directed treatment of N cycling in arid systems. Since microorganisms facilitate most of the N reactions in soil systems, the recent soil microbiology and biochemistry text by Paul and Clark (1989) is also a valuable reference.

Redox Interpretations

Nitrogen chemistry in soil solutions can be approached qualitatively via the study of electrochemical equilibria because the important transformations of N are oxidation and reduction reactions in which microbes and their enzymes serve as catalysts. Within the redox stability range of water, a number of N oxidation states are feasible. However, nitrate (NO_3^-), ammonia (NH_3), and dinitrogen gas (N_2), along with amino acid N forms (i.e., alanine and CH_3-$CHNH_2$-COOH), are the stable states (Bohn et al., 1985; Lindsay, 1979). The predominance of N_2 as the most stable nitrogen species explains the observation that most of the world's nitrogen exists as atmospheric dinitrogen. Nitrate represents only a small fraction and is stable only under oxidizing conditions at pH > 3, and NH_3 is stable only under reducing conditions. Ammonia is generally scarce compared to the amino acid forms of N because of the enhanced stability of amino acids to oxidation. However, when the carbon of amino acids is rapidly oxidizing, NH_3 may be liberated (*ammonification*) at rates in excess of the rates of conversion of NH_4^+ to nitrate (*nitrification*) and ammonia may accumulate.

The high pH, low pE conditions typical of arid to semiarid regions suggest a system dominated by N_2, with a variable ratio of NO_3^- to NH_4^+ that increases with increasing pH and pE. This qualitative evaluation fits well with Ellenberg's (1971) general classification of ecosystem types (based on field nitrification potential estimates) as a NH_4^+/NO_3^- type contrasted with solely NH_4^+ types or NO_3^- types. Although nitrification is generally held to play a minor role in the N cycle of natural ecosystems (e.g., Haynes, 1986) because NO_3^- is normally present in such small concentrations (Clark, 1977; Woodmansee et al., 1978; Vitousek et al., 1979), deserts may be episodic exceptions (Skujiņš, 1981).

The quantitative application of electrochemical equilibria to N transformations in soils is limited by disequilibrium itself. N reactions are dominantly (and fortunately) irreversible. At equilibrium, almost all the world's oxygen would be consumed in the production of nitric acid as

$$N_2 + 5/2O_2 + H_2O \longrightarrow 2HNO_3 \quad (36)$$

The reaction is hindered by a rather large activation energy barrier, however, which is overcome to an extent by lightning discharge in the atmosphere. Similarly, most N transformations in soils are kinetically limited and are dependent on enzymatic catalysis to proceed unidirectionally. The presence of the unstable N compounds (i.e., NO_2^- or N_2O) is a direct consequence of the propensity for disequilibrium.

Kinetic Models

Kinetically based mathematical models and computer simulations have evolved of late as a means of integrating the historically accumulated, large database on nitrogen forms and reactions in soils. They provide a useful, conceptual framework for discussion of the major N-transformation reactions in soils. Tanji (1982) gives an excellent review of N cycle modeling, and Sparks (1988) lists kinetic N models published to date and also gives an abbreviated presentation of the model of Tanji and Mehran (1979). This model was extended from that of Mehran and Tanji (1974) to include nitrogen uptake as affected by variable NO_3^- and NH_4^+ absorption coefficients. The parent model of Mehran and Tanji (1974) features irreversible first-order kinetics for nitrification, denitrification, mineralization, immobilization, and plant uptake. Reversible first-order kinetics were assumed for NH_4^+ ion exchange. In brief, for a given nitrogen species of interest N_c,

$$\frac{d(N_c)}{dt} = -\Sigma\, k_i[N_c] + \Sigma\, k_j[N_m] \tag{37}$$

where N_m is the concentration of other N species, k_i and k_j are first-order rate coefficients for i = 1, ..., n sinks and j = 1, ..., n source mechanisms, respectively, and t is time.

N-Transformation Controls

The optimal pH for nitrification processes is in the range 7-9, and they are generally favored in arid region soils. However, NH_4^+ may accumulate and lead to accumulated, toxic levels of NH_3 at pH values in excess of 7.5 and consequently inhibit the activity of *Nitrobacter* (Morrill and Dawson, 1967; Broadbent et al., 1957; Stojanovic and Alexander, 1958). The NH_4^+-oxidizing bacteria are less sensitive to high NH_4^+ than are *Nitrobacter*, and thus NO_2^- may accumulate at higher pH (Jones and Hedlin, 1970; Nakos and Wolcott, 1972).

Phosphorus availability can limit nitrification and in arid soils may represent a deterrent to the N nutrition of plants. Purchase (1974a, b) reported that NO_2^- oxidizers were inhibited by low P availability in a savanna grassland, and Melillo (1981) reported much higher nitrification rates in Hubbard Brook forest floor materials with added NH_4^+ + P fertilization than with added NH_4^+ alone.

At water potentials in soils of the order of -1000 to -5000 kPa, the total microbial biomass becomes moisture limited and activity is significantly or completely impaired. The fungal and actinomycete populations appear less affected than bacteria, which become inactive between -800 and -1500 kPa (Chen and Griffin, 1966; Wilson and Griffin, 1975). Net ammonification may occur to a lower limit of -4000 kPa (Myers et al., 1982), although values as low as minus 5000 kPa have been reported (Wetselaar, 1968).

N Losses and Sinks

Leaching losses of nitrate can be large and important in many natural and cultivated systems where sufficient moisture to promote their movement out of the soil system exists. In arid regions this event is not of predominant concern as a mechanism for N loss and is not treated further here. The reader is rather referred to Haynes (1986), who gives a comprehensive and well-organized review of NO_3^- leaching considerations.

Erosional losses of nitrogen from arid region soils, however, can be significant owing to the action of wind and water across lands with incomplete and inadequate vegetative cover (Skujiņš, 1981). The data of Fletcher et al. (1978) as reported by Haynes (1986) give ranges of 0.2-2.5 (mean = 1.6) and 0.3-5.1 (mean = 3.4) kg N ha^{-1} $year^{-1}$ for erosional losses of N via wind and water erosion, respectively, for the Great Basin Desert. For the Sonoran Desert, mean wind erosion losses of N were 1.2 (range = 0.2-3.0) and mean water erosion losses were 1.7 (range = 1.0-5.6) kg N ha^{-1} $year^{-1}$.

Gaseous losses of N from arid region soils can also be substantial and occur via denitrification and ammonia volatilization processes. Denitrification processes are well described elsewhere (Payne, 1973, 1981; Firestone, 1982; Knowles, 1982; Fillery, 1983). The pathway of denitrification can be represented as

$$NO_3^-(+5) \rightarrow NO_2^-(+3) \rightarrow NO(+2) \rightarrow N_2O(+1) \rightarrow (N_2O) \quad (38)$$

where values in parentheses denote oxidation states of N, dependent on specific reductase enzymes at each step. The environmental condition of most significance to the rates and extent of denitrification in O_2 such that in reduced conditions, denitrification is emphasized and, conversely, retarded under aerobic conditions.

The volatilization of NH_3 is of importance in arid region soils because it has been well correlated with $CaCO_3$ content (Lehr and Van Wesemael, 1961; Fenn and Kissel, 1975). It appears to be stimulated following the addition of ammoniacal fertilizers (Freney et al., 1981). As suggested by Nelson (1982), addition of $(NH_4)_2SO_4$ to a calcareous soil may result in the reaction sequence

$$CaCO_3 + (NH_4)_2SO_4 \rightarrow Ca^2 + 2OH^- + CaSO_4 + 2NH_4HCO_3 \quad (39)$$

$$2NH_4HCO_3 \rightarrow 2NH_3 + 2CO_2 + 2H_2O \quad (40)$$

Upon hydrolysis of $CaCO_3$,

$$Ca^{2+} + 2OH^- + (NH_4)_2SO_4 \rightarrow CaSO_4 + 2NH_3 + 2H_2O \quad (41)$$

The summarized overall reaction is

$$CaCO_3 + (NH_4)_2SO_4 \rightarrow 2NH_3 + CO_2 + H_2O + CaSO_4 \quad (42)$$

Ammonia volatilization yields a net acidification of the system (Avnimelech and Laher, 1977) as

$$NH_4^+ \rightarrow NH_3 + H^+ \quad (43)$$

This representation implies that pH can exert a controlling influence on the concentrations of NH_4^+ and NH_3 such that NH_3 losses should increase as the soil pH increases (e.g., Lyster et al., 1980). However, substantial NH_3 volatilization can occur in acid soils (Blasco and Cornfield, 1966). It now appears that solution pH at the microlocale of a urea or NH_4^+ salt granule is more important than the bulk soil pH in determining NH_3 losses (Nelson, 1982; Sherlock and Goh, 1984; Black et al., 1984, 1985a, b).

In addition to $CaCO_3$ content and pH, the climatic factors of moisture and temperature also strongly affect NH_3 loss from soils. Conditions in arid regions are particularly conducive to ammonia volatilization because of wetting-drying extremes and high temperatures. The rate and extent of NH_3 loss from fertilizers increases with increasing temperature (Watkins et al., 1972; Lyster et al., 1980). The temperature T dependence of NH_3 loss from soils is a consequence of the effect of T on the equilibrium constants for successive reactions that convert NH_4^+(aq) in soil solution to NH_3(aq) and then to NH_3(g) (Vlek and Stumpe, 1978; Hales and Drewes 1979; Vlek and Craswell, 1981; Sherlock and Goh, 1984, 1985).

Ammonium and potassium ions can migrate between structural layers of 2:1 clay minerals (e.g., montmorillonite and vermiculite) when in an expanded state. When the layers contract, the ions are held within the layers in a nonexchangeable or "fixed" state. Ammonium fixation (despite the term "fixed") is not irreversible, and NH_4^+ is slowly released and replaced from interlayers (Nommik and Vahtras, 1982). The extent of fixation is affected by the total concentration of ammonium in soil solution, its concentration relative to K^+, which competes with NH_4^+ for interlayer positions, and the degree of organic interaction with clays, which may block entry to interlayers (e.g., Haynes, 1986). Ammonium fixation also tends to increase with increasing soil pH (Nommik, 1965) and is greatly increased with the frequency and intensity of wetting and drying events (Blasco and Cornfield, 1966). The presence of montmorillonite in arid region soils, their relatively high pH, and extremes of wetting and drying suggest ammonium fixation can be a large sink for N in desert settings. Indeed in some arid soils of Israel and Australia, between 3 and 1076 $\mu g\ g^{-1}$ of ammonium, representing between 2 and 90% of the total soil N, was reported as fixed (e.g., Young and Aldag, 1982).

REFERENCES

Agbim, N. N., and Doxtader, K. G. (1975). Microbial degradation of zinc silicates. *Soil Biol. Biochem. 7*: 275.

Alexander, M. (1977). *Introduction to Soil Microbiology*. John Wiley, New York.

Amrhein, C., Jurinak, J. J., and Moore, W. M. (1985). Kinetics of calcite dissolution as affected by carbon dioxide partial pressure. *Soil Sci. Soc. Am. J. 49*: 1933.

Aristovska, T. V., Daragan, A. Y., and Kutuzova, R. S. (1969). Microbial factors in the movement of some mineral elements in the soil. *Sov. Soil Sci. 5*: 538.

Arrieta, L., and Grez, R. (1971). Solubilization of iron-containing minerals by soil microorganisms. *Appl. Microbiol. 22*: 487.

Ascaso, C., Galvan, J., and Ortega, C. (1976). The pedogenic action of *Parmelia conspersa, Rhizocarpon geographicum* and *Umbilicaria postulata. Lichenologist 8*: 151.

Avnimelech, Y., and Laher, M. (1977). Ammonia volatilization from soils: Equilibrium considerations. *Soil Sci. Soc. Am. J. 41*: 1080.

Azcon, R., Barea, J. M., and Hayman, D. S. (1976). Utilization or rock phosphate in alkaline soils by plants inoculated with mycorrhizal fungi and phosphate-solubilizing bacteria. *Soil Biol. Biochem. 8*: 135.

Bailey, S. W. (Chairman), Brindley, G. W., Johns, W. D., Martin, R. T., and Ross, M. (1971). Summary of national and international recommendations on clay mineral nomenclature by 1969-70 Clay Min. Soc. Nomenclature Committee. *Clays Clay Min. 19*: 129.

Barber, D. A., and Martin, J. K. (1976). The release of organic substances by cereal roots in soil. *New Phytol. 76*: 68.

Berkheiser, V. E., Street, J. J., Rao, P. S. C., and Yuan, T. L. (1980). Partitioning of inorganic orthophosphate in soil water systems. *CRC Crit. Rev. Environ. Control 10*: 179.

Berner, R. A. (1978). Rate control of mineral dissolution under earth surface conditions. *Am. J. Sci. 278*: 1235.

Berner, R. A. (1983). Kinetics of weathering and diagenesis. *Mineral. 8*: 111.

Berthelin, J. (1983). Microbial weathering processes. In *Microbial Geochemistry*, W. E. Krumbein (Ed.). Blackwell Scientific, Oxford, pp. 223-263.

Berthelin, J., and Belgy, G. (1979). Microbial degradation of phyllosilicates during simulated podzolization. *Geoderma 21*: 297.

Black, A. S., Sherlock, R. R., Smith, N. P., Cameron, K. C., and Goh, K. M. (1984). Effect of previous uring application on ammonia volatilization from 3 nitrogen fertilizers. *N.Z. J. Agr. Res. 27*: 413.

Black, A. S., Sherlock, R. R., Cameron, K. C., Smith, N. P., and Goh, K. M. (1985a). Comparison of three field methods for measuring ammonia volatilization from urea granules broadcast onto pasture. *J. Soil Sci. 36*: 271.

Black, A. S., Sherlock, R. R., and Smith, N. P. (1985b). Ammonia volatilization from nitrogenous fertilizers broadcast onto pastures: Effects of application time and rate. *N.Z. J. Agr. Res. 28*: 469.

Blasco, M. L., and Cornfield, A. H. (1966). Volatilization of nitrogen as ammonia from acid soils. *Nature (Lond.) 212*: 1279.

Blatt, H., Middleton, G. V., and Murray, R. C. (1972). *Origin of Sedimentary Rocks*. Prentice-Hall, Englewood Cliffs, New Jersey.

Bloom, A. L. (1978). *Geomorphology*. Prentice-Hall, Englewood Cliffs, New Jersey.

Bloom, P. R., and Erich, M. S. (1987). Effect of solution composition on the rate and mechanism of gibbsite dissolution in acid solutions. *Soil Sci. Soc. Am. Proc. 27*: 294.

Bohn, H. L., and Kittrick, J. A. (1984). Stability diagram of Ca-Na-K-plagiociase with authigenic minerals. *Chem. Geol. 43*: 181.

Bohn, H. L., McNeal, B. L., and O'Connor, G. A. (1985). *Soil Chemistry*, 2nd ed. John Wiley, New York.

Bolan, N. S., Robson, A. D., Barrow, N. J., and Alymore, L. A. G. (1984). Specific activity of phosphorus in mycorrhizal and nonmycorrhizal plants in relation to the availability of phosphorus to plants. *Soil Biol. Biochem. 16*: 299.

Bolt, G. H., and Bruggenwert, M. G. M. (Eds.) (1976). *Soil Chemistry. A. Basic Elements*. Elsevier, Amsterdam.

Boyer, P. (1982). Quelques aspects de l'action des termites du sol sur les argiles. *Clay Min. 17*: 453.

Boyle, J. R., Voigt, G. K., and Sawhney, B. L. (1967). Biotite flakes: Alteration by chemical and biological treatment. *Science 155*: 193.

Broadbent, F. E., Tyler, K. B., and Hill, G. N. (1957). Nitrification of ammoniacal fertilizers in some California soils. *Hilgardia 27*: 247.

Brunauer, S., and Copeland, L. E. (1967). Surface tension, adsorption, part 5. In *Handbook of Physics*, E. U. Condon and H. Odishaw (Eds.). McGraw-Hill, New York, pp. 90-115.

Buol, S. W. (1965). Present soil-forming factors and processes in arid and semi-arid regions. *Soil Sci. 99*: 45.

Burns, R. C., and Hardy, R. W. F. (1975). *Nitrogen Fixation in Bacteria and Higher Plants*. Springer-Verlag, Berlin.

Buwalda, J. G., Stribley, D. P., and Tinker, P. B. (1983). Increased uptake of anions by plants with VA mycorrhizas. *Plant Soil 71*: 463.

Buyanovsky, G. A., and Wagner, G. H. (1983). Annual cycles of carbon dioxide level in soil air. *Soil Sci. Soc. Am. J. 47*: 1139.

Chaney, K., and Swift, R. S. (1984). The influence of organic matter on aggregate stability of some British soils. *J. Soil Sci. 35*: 223.

Chen, A. W., and Griffin, D. M. (1966). Soil physical factors and the ecology of fungi. V. Further studies in relatively dry soils. *Trans. Br. Mycol. Soc. 49*: 419.

Chou, L., and Wollast, R. (1985). Steady-state kinetics and dissolution mechanisms of albite. *Am. J. Sci. 285*: 963.

Clark, F. E. (1977). Internal cycling of ^{15}N in shortgrass prairie. *Ecology 58*: 995.

Cline, G. R., Powell, P. E., Szaniszlo, P. J., and Reid, C. P. P. (1984). Comparison of the abilities of hydroxamic, synthetic, and other natural organic acids to chelate iron and other ions in nutrient solution. *Soil Sci. Soc. Am. J. 45*: 1158.

Cress, W. A., Throneberry, G. O., and Lindsay, D. L. (1979). Kinetics of phosphorus absorption by mycorrhizal and nonmycorrhizal tomato roots. *Plant Physiol. 64*: 484.

Dana, J. D. (1959). *Dana's Manual of Mineralogy* (rev. by C. S. Hurlbut, Jr., Wiley, New York.

Davies, B. E. (Ed.) (1980). *Applied Soil Trace Elements*. Wiley, New York.

Dibble, W. E., Jr., and Tiller, W. A. (1981). Non-equilibrium water/rock interactions. 1. Model for interface controlled reactions. *Geochim. Cosmochim. Acta 45*: 79.

Dixon, J. B. (1977). Kaolinite and serpentine group minerals. In *Minerals in Soil Environments*, J. B. Dixon and S. B. Weed (Eds.). Soil Sci. Soc. Am., Madison, Wisconsin, pp. 357-403.

Dixon, J. B., and Weed, S. B. (1977). *Minerals in Soil Environments.* Soil Sci. Soc. Am., Madison, Wisconsin.

Dommergues, Y., and Mangenot, F. (1970). *Ecologie Microbienne du Sol.* Masson, Paris.

Duchafour, P. (1982). *Pedology* (English edition translated by T. R. Patton). Allen and Unwin, London.

Ellenberg, H. (1971). Nitrogen content, mineralization and cycling. In *Productivity of Forest Ecosystems*, P. Duvigneaud (Ed.). UNESCO, Paris, pp. 504-514.

Emery, T. (1977). The storage and transport of iron. In *Metal Ions in Biological Systems*, Vol. 7, *Iron in Model and Natural Compounds*, H. Sigel (Ed.). Marcel Dekker, New York, pp. 77-125.

Enfield, C. G., and Ellis, R., Jr. (1983). The movement of phosphorus in soil. In *Chemical Mobility and Reactivity in Soil Systems*, D. W. Nelson, D. E. Elrick, and K. K. Tanji (Eds.). Soil Sci. Soc. Am., Madison, Wisconsin, pp. 93-107.

Fenn, L. B., and Kissel, D. E. (1975). Ammonia volatilization from surface applications of ammonium compounds to calcareous soils. IV. Effect of calcium carbonate content, *Soil Sci. Soc. Am. Proc. 39*: 631.

Fillery, I. R. P. (1983). Biological denitrification. In *Gaseous Losses of Nitrogen from Plant-Soil Systems*, J. R. Freney and J. R. Simpson (Eds.). Martinus Nijhoff/Dr. W. Junk, The Hague, The Netherlands, pp. 33-64.

Firestone, M. K. (1982). Biological denitrification. In *Nitrogen in Agricultural Soils*, F. J. Stevenson (Ed.). Am. Soc. Agron., Madison, Wisconsin, pp. 289-318.

Fitzpatrick, E. A. (1971). *Pedology*. Oliver and Boyd, Edinburgh.

Fletcher, J. E., Sorenson, D. L., and Porcella, D. B. (1978). Erosional transfer of nitrogen in desert ecosystems. In *Nitrogen in Desert Ecosystems*, N. E. West and J. Skujiņš (Eds.). Dowden, Hutchinson & Ross, Stroudsburg, Pennsylvania, pp. 171-181.

Fogg, G. I., Stewart, W. D. P., and Walsby, A. E. (1973). *The Blue-Green Algae*, Academic Press, New York.

Freney, J. R., Simpson, J. R., and Denmead, O. T. (1981). Ammonia volatilization. In *Terrestrial Nitrogen Cycles: Processes, Ecosystem Strategies and Management Impacts*, F. E. Clark and T. Rosswall (Eds.). Ecological Bulletins, No. 33-NFR, Stockholm, pp. 291-302.

Frenkel, H., Goertzen, J. O., and Rhoades, J. D. (1978). Effect of clay type and content, exchangeable sodium percentage and electrolyte concentration on clay dispersion and soil hydraulic conductivity, *Soil Sci. Soc. Am. J. 42*: 32.

Frenkel, H., Amrhein, C., and Jurinak, J. J. (1983). The effect of exchangeable cations on soil mineral weathering. *Soil Sci. Soc. Am. J. 47*: 649.

Friedman, I. (1971). Light and scanning electron microscopy of the endolithic desert algal habitat. *Thycologia 10*: 411.

Friedman, I., and Ocampo, R. (1976). Endolithic blue-green algae in the dry valleys: Primary producers in the Antarctic Desert ecosystem. *Science 193*: 1247.
Fry, E. J. (1924). A suggested explanation of the mechanical action of lithophytic lichens on rocks (shale). *Ann. Bot. 37*: 175.
Gal, M., Amiel, J., and Ravikovitch, S. (1974). Clay mineral distribution and origin in the soil types of Israel. *J. Soil Sci. 25*: 79.
Garrels, R. M., and Christ, C. L. (1965). *Solutions, Minerals and Equilibria.* Harper and Row, New York.
Goldich, S. S. (1938). Study in rock weathering. *J. Geol. 46*: 17.
Granhall, U. (1981). Biological nitrogen fixation in relation to environmental factors and functioning of natural ecosystems. In *Terrestrial Nitrogen Cycles: Processes, Ecosystem Strategies and Management Impacts*, F. E. Clark and T. Rosswall (Eds.). Ecological Bulletins, No. 33, NRF, Stockholm, pp. 131-144.
Hales, J. M., and Drewes, D. R. (1979). Solubility of ammonia at low concentrations. *Atmos. Environ. 13*: 1133.
Hallbauer, D. K., and Johns, H. M. (1977). Attack of lichens on quartzitic rock surfaces. *Lichenologist 9*: 119.
Haynes, R. J. (1986). *Mineral Nitrogen in the Plant-Soil System*, Academic Press, Orlando, Florida.
Helgeson, H. C., Murphy, W. M., and Aagaard, P. (1984). Thermodynamic and kinetic constraints on reaction rates among minerals and aqueous solution. II. Rate constants, effective surface area, and the hydrolysis of feldspar. *Geochim. Cosmochim. Acta 48*: 2405.
Helling, C. S., Chesters, G., and Corey, R. B. (1964). Contribution of organic matter and clay to soil cation exchange capacity as affected by the pH of the saturating solution. *Soil Sci. Soc. Am. Proc. 28*: 517.
Henderson, M. E. K., and Duff, R. B. (1963). The release of metallic and silicate ions from minerals, rocks and soils by fungal activity. *J. Soil Sci. 14*: 236.
Hodgman, C. E. (Ed.-in-chief) (1960). *Handbook of Chemistry and Physics*, 41st ed. Chemical Rubber, Cleveland, Ohio.
Holdren, G. R., and Speyer, P. M. (1985). pH-dependent changes in the rates and stoichiometry of dissolution of an alkali feldspar at room temperature. *Am. J. Sci. 285*: 994.
Howeler, R. H., Asher, C. J., and Edwards, D. G. (1982). Establishment of an effective mycorrhizal association on cassava in flowing solution culture and its effects on phosphorus nutrition. *New Phytol. 90*: 229.
Huang, W. H., and Kiang, W. C. (1972). Laboratory dissolution of plagioclase feldspars in water and organic acids at room temperature. *Am. Min. 57*: 1849.
Huang, P. M., and Schnitzer, M. (Eds.) (1986). *Interactions of Soil Minerals with Natural Organics and Microbes.* Special Publication No. 17, Soil Sci. Soc. Am., Madison, Wisconsin.
Hunt, C. B. (1972). *Geology of Soils.* Freeman, San Francisco.
Ivarson, K. C., Ross, G. J., and Miles, M. M. (1978). Alterations of micas and feldspars during microbial formation of basic ferric sulfate in the laboratory. *Soil Sci. Soc. Am. J. 42*: 518.

Jackson, M. L. (1964). Chemical composition of soils. In *Chemistry of the Soil*, F. E. Bear (Ed.). Reinhold, New York, pp. 71-141.

Jackson, T. A., and Keller, W. D. (1970). A comparative study of the role of lichens and "inorganic" processes in the chemical weathering of recent Hawaiian lava flows. *Am. J. Sci. 269*: 446.

Jackson, M. L., and Sherman, G. D. (1953). Chemical weathering of minerals in soils. *Adv. Agron. 5*: 221-317.

James, D. W., Hanks, R. J., and Jurinak, J. J. (1982). *Modern Irrigated Soils*. John Wiley, New York.

Jenkinson, D. S., and Ayanaba, A. (1977). Decomposition of carbon-14 labelled plant material under tropical conditions. *Soil Sci, Soc. Am. J. 41*: 912.

Jenny, H. (1941). *Factors of Soil Formation*. McGraw-Hill, New York.

Jenny, H. (1980). *The Soil Resource Origin and Behaviour*. Springer-Verlag, Berlin.

Jones, R. W., and Hedlin, R. A. C. (1970). Nitrite instability in three Manitoba soils. *Can. J. Soil Sci. 50*: 339.

Jurinak, J. J., Dudley, L. M., Allen, M. F., and Knight, W. G. (1986). The role of calcium oxalate in the availability of phosphorus in soils of semiarid regions: A thermodynamic study. *Soil Sci. 142*: 255.

Keller, W. D., and Fredrickson, O. (1952). Role of plants and colloidal acids in the mechanisms of weathering. *Am. J. Sci. 250*: 594.

Kelley, W. P. (1948). Cation exchange in soils. ACS Monograph No. 109.

Kloepper, J. W., Leong, J., Teintze, M., and Schroth, M. N. (1980). Enhanced plant growth by siderophores produced by plant-promoting rhizobacteria. *Nature 286*: 885.

Knight, W. G., and Skujiņš, J. (1981). ATP concentration and soil respiration at reduced water potentials in arid soils. *Soil Sci. Soc. Am. J. 45*: 657.

Knight, W. G., Allen, M. F., Jurinak, J. J., and Dudley, L. M. (1989). Elevated carbon dioxide and solution phosphorus in soil with vesicular-arbuscular mycorrhizal western wheatgrass. *Soil Sci. Soc. Am. J. 53*: 1075.

Knowles, R. (1982). Denitrification. *Microbiol. Rev. 46*: 43.

Kononova, M. M. (1966). *Soil Organic Matter*. Pergamon Press, Oxford.

Krupp, H. K., Biggar, J. W., and Nielsen, D. R. (1972). Relative flow rates of salt and water in soils. *Soil Sci. Soc. Am. Proc. 36:* 412.

Lasaga, A. C. (1981). Rate laws of chemical reactions. In *Kinetics of Geochemical Processes*, A. C. Lasaga and R. J. Kirkpatrick (Eds.). Reviews in mineralogy, Vol. 8., Mineralogical Soc. Am., Washington, D.C.

Lebedeva, E. V., Lyalikova, N. N., and Bugelsky, Y. Y. (1978). Participation of nitrifying bacteria in weathering of serpentized ultrabasic rocks. *Mikrobiologiya 47*: 238.

Lehr, I. J., and Van Wesemael, J. C. (1961). The volatilization of ammonia from lime rich soils. *Landbouwk, Tijdschr. 73*: 1156.

Lindsay, W. L. (1972). Inorganic phase equilibria of micronutrients in soils. In *Micronutrients in Agriculture*, J. J. Mortvedt, P. M. Giordano, and W. L. Lindsay (Eds.). Soil Sci. Soc. Am., Madison, Wisconsin, pp. 41-57.

Lindsay, W. L. (1979). *Chemical Equilibria in Soils.* John Wiley, New York.
Lippman, F. (1973). *Sedimentary Carbonate Minerals.* Springer-Verlag, New York.
Luce, R. W., Bartlett, R. W., and Parks, G. A. (1972). Dissolution kinetics of magnesium silicates. *Geochim. Cosmochim. Acta 36*: 35.
Lyster, S., Morgan, M. A., and O'Toole, P. (1980). Ammonia volatilization from soils fertilized with urea and ammonium nitrate. *J. Life Sci. 1*: 167.
Major, J. (1974). Nitrogen accumulation in succession. In *Vegetation Dynamics,* R. Knapp (Ed.). Dr. W. Junk, The Hague, The Netherlands, pp. 207-213.
Marshall, K. C. (1980). Bacterial adhesion in natural environments. In *Microbial Adhesion to Surfaces,* R. C. W. Berkeley et al. (Eds.). Soc. Chem. Ind., Chichester, UK, pp. 187-196.
Martin, J. P., and Focht, D. D. (1977). Biological properties of soils. In *Soils for Management of Organic Wastes and Waste Waters,* L. F. Elliot and F. J. Stevenson (Eds.). Am. Soc. Agron., Madison, Wisconsin, pp. 114-169.
McLaren, A. D., and Peterson, G. H. (1965). Physical chemistry and biological chemistry of clay mineral-organic nitrogen complexes. In *Soil Nitrogen,* W. W. Bartholomew and F. E. Clark (Eds.). Am. Soc. Agron., Madison, Wisconsin, pp. 261-284.
McNeal, B. L., Norvell, W. A., and Coleman, N. T. (1966). The effect of solution composition on the swelling of extracted soil clays. *Soil Sci. Soc. Am. Proc. 30*: 313.
Mehran, M., and Tanji, K. K. (1974). Computer modeling of nitrogen transformations in soils. *J. Environ. Qual. 3*: 391.
Mehta, A. P., Torma, A. E., and Murr, L. E. (1978). Biodegradation of aluminum bearing rocks by *Penicillium simplicissimum. IRCS Med. Sci. 6*: 416.
Mellilo, J. M. (1981). Nitrogen cycling in deciduous forests. In *Terrestrial Nitrogen Cycles: Processes, Ecosystem Strategies and Management Impacts,* F. E. Clark and T. Rosswall (Eds.). Ecological Bulletins, No. 33, NFR, Stockholm, pp. 427-442.
Morgan, J. J. (1967). Applications and limitations of chemical thermodynamics in natural water systems. In *Equilibrium Concepts in Natural Water Systems,* W. Stumm (Ed.). Am. Chem. Soc., Washington, D.C., p. 1.
Morrill, L. G., and Dawson, J. E. (1967). Patterns observed for the oxidation of ammonium to nitrate by soil organisms. *Soil Sci. Soc. Am. Proc. 31*: 757.
Mortvedt, J. J., Giordano, P. M., and Lindsay, W. L. (Eds.) (1972). *Micronutrients in Agriculture.* Soil Sci. Soc. Am., Madison, Wisconsin.
Murphy, T. P., Lean, D. R. S., and Nalewajako, C. (1976). Blue-green algae: The excretion of iron-selective chelators enables them to dominate other algae. *Science 192*: 900.
Myers, R. J. K., Campbell, C. A., and Weier, K. L. (1982). Quantitative relationship between net nitrogen mineralisation and moisture content of soils. *Can. J. Soil Sci. 62*: 111.
Nakos, G. G., and Wolcott, A. R. (1972). Bacteriostatic effect of ammoni-

um on *Nitrobacter agilis* in mixed culture with *Nitrosomonas europaea*. *Plant Soil 36*: 521.

Neilands, J. B. (1979). Biomedical and environmental significance of siderophores. In *Trace Metals in Health and Disease*, N. Kharasch (Ed.). Raven Press, New York, pp. 27-41.

Nelson, D. W. (1982). Gaseous losses of nitrogen other than through denitrification. In *Nitrogen in Agricultural Soils*, F. J. Stevenson (Ed.). Am. Soc. Agron., Madison, Wisconsin, pp. 327-363.

Nommik, H. (1965). Ammonium fixation and other reactions involving a nonenzymatic immobilization of mineral nitrogen in soils. In *Soil Nitrogen*, W. V. Bartholomew and F. E. Clark (Eds.). Am. Soc. Agron., Madison, Wisconsin, pp.200-251.

Nommik, H., and Vahtras, K. (1982). Retention and fixation of ammonium and ammonia in soils. In *Nitrogen in Agricultural Soils*, F. J. Stevenson (Ed.). Am. Soc. Agron., Madison, Wisconsin, pp. 123-126.

Olson, J. S. (1963). Energy storage and balance of producers and decomposers in ecological systems. *Ecology 44*: 322.

Parr, J. F., Gardner, W. R., and Elliot, L. F. (Eds.) (1981). *Water Potential Relations in Soil Microbiology*. Special Publication No. 9, Soil Sci. Soc. Am., Madison, Wisconsin.

Paul, E. A., and Clark, F. E. (1989). *Soil Microbiology and Biochemistry*. Academic Press, New York.

Payne, W. J. (1973). Reduction of nitrogenous oxides by microorganisms. *Bacteriol. Rev. 37*: 409.

Payne, W. J. (1981). *Denitrification*. Wiley, New York.

Polynov, B. B. (1945). Premiers stades de la formation des sols sur roches massives cristallines. *Trad. Pochvovedenie 7*: 327.

Powell, P. E., Cline, G. R., Reid, C. P. P., and Szaniszlo, P. J. (1980). Occurrence of hydroxamate siderophore iron chelators in soils. *Nature 287*: 833.

Powell, P. E., Szaniszlo, P. E., Cline, G. R., and Reid, C. P. P. (1982). Hydroxamate siderophores in the iron nutrition of plants. *J. Plant Nutr. 5*: 653.

Purchase, B. S. (1974a). Evaluation of the claim that grass root exudates inhibit nitrification. *Plant Soil 41*: 527.

Purchase, B. S. (1974b). The influence of phosphate deficiency on nitrification. *Plant Soil 41*: 541.

Richards, L. A. (1954). Diagnosis and improvement of saline and alkali soils. U.S.D.A. Handbook No. 60. USDA, Washington, D.C.

Robb, D. A., and Pierpoint, W. S. (1983). *Metals and Micronutrients*. Academic Press, New York.

Robert, J., and Berthelin, J. (1986). Role of biological and biochemical factors in soil mineral weathering. In *Interactions of Soil Minerals with Natural Organics and Microbes*, P. M. Huang and M. Schnitzer (Eds.). Special Publ. No. 17, Soil Sci. Soc. Am., Madison, Wisconsin, pp. 453-495.

Rossi, G. (1978). Potassium recovery through leucite bioleaching possibilities and limitation. In *Metallurgical Applications of Bacterial Leaching and Related Microbiological Phenomena*, l. e. Murr et al. (Eds.). Academic Press, New York, pp. 297-319.

Sadiq, M., and Lindsay, W. L. (1979). Selection of standard free energies of formation for use in soil chemistry. Colorado State Univ. Tech. Bull. No. 134, Fort Collins, Colorado.

Sanders, F. E., and Tinker, P. B. (1971). Mechanism of absorption of phosphate from soil by *Endogone* mycorrhizas. *Nature (Lond.) 233*: 278.

Schott, J., Berner, R. A., and Lennart-Sjoberg, E. L. (1981). Mechanism of pyroxene and amphibole weathering. I. Experimental studies of Fe-free minerals. *Geochim. Cosmochim. Acta 45*: 2123.

Sherlock, R. R., and Goh, K. M. (1984). Dynamics of ammonia volatilization from simulated urine patches and aqueous urea applied to pasture. I. Field experiments. *Fertil. Res. 5*: 181.

Sherlock, R. R., and Goh, K. M. (1985). Dynamics of ammonia volatilization from simulated urine patches and aqueous urea applied to pasture. II. Theoretical derivation of a simplified model. *Fertil. Res. 6*: 3.

Sherman, G. D., Schultz, F., and Alway, F. J. (1962). Dolomite in soils of the Red River Valley, Minnesota. *Soil Sci. 94*: 304.

Silver, M., Ehrlich, H. L., and Ivarson, K. C. (1986). Soil mineral transformation mediated by soil microbes. In *Interactions of Soil Minerals with Natural Organics and Microbes*, P. M. Huang and M. Schnitzer (Eds.). Special Publication No. 17, Soil Sci. Soc. Am., Madison, Wisconsin, pp. 497-519.

Silverman, M. P., and Ehrlich, H. L. (1964). Microbial formation and degradation of minerals. *Adv. Appl. Microbiol. 6*: 153.

Silverman, M. P., and Munoz, E. F. (1970). Fungal attack on rock: Solubilization and altered infrared spectra. *Science 169*: 985.

Silverman, M. P., and Munoz, E. F. (1971). Fungal leaching of titanium from rock. *Appl. Microbiol. 22*: 923.

Skujiņš, J. J. (1981). Nitrogen cycling in arid ecosystems. In *Terrestrial Nitrogen Cycles: Processes, Ecosystem Strategies and Management Impacts*, F. E. Clark and T. Rosswall (Eds.). Ecological Bulletins, No. 33. NFR, Stockholm, pp. 477-491.

Slater, J. W. (1856). On some reactions of oxalic acid. *Chem. Gaz., April 1*: 13-14.

Sorensen, L. H. (1972). Stabilisation of newly formed amino acid metabolites in soil by clay minerals. *Soil Sci. 114*: 5.

Sparks, D. L. (Ed.) (1986). Kinetics of reactions in pure and mixed systems. In *Soil Physical Chemistry*. CRC Press, Boca Raton, Florida, pp. 83-145.

Sparks, D. L. (1988). *Kinetics of Soil Chemical Processes*. Academic Press, New York.

Sposito, G. (1983). The chemical forms of trace metals in soil. In *Applied Environmental Geochemistry*, I. Thornton (Ed.). Academic Press, New York, p. 123.

Sposito, G. (1986). Thermodynamics of the soil solution. In *Soil Physical Chemistry*, D. L. Sparks (Ed.). CRC Press, Boca Raton, Florida, pp. 147-178.

St. Arnauld, R. J., and Herbillon, A. J. (1973). Occurrence and genesis of secondary magnesium-bearing calcites in soils. *Geoderma 9*: 279.

Stevens, P. R., and Walker, T. W. (1970). The chronosequence concept and soil formation. *Q. Rev. Biol. 45*: 333.

Stevenson, F. J. (1982). *Humus Chemistry, Genesis, Composition, Reactions.* Wiley, New York.
Stevenson, F. J. (1986). *Cycles of Soil.* John Wiley, New York.
Stojanovic, B. J., and Alexander, M. (1958). Effect of inorganic nitrogen on nitrification. *Soil Sci. 86*: 206.
Stotzky, G. (1980). Surface interactions between clay minerals and microbes, virus and soluble organics and the probable importance of these interactions to the ecology of microbes in soil. In *Microbial Adhesion to Surfaces,* R. C. W. Berkely et al. (Eds.). Soc. Chem. Ind., Chichester, UK, pp. 231-247.
Strakhov, N. M. (1967). *Principles of Lithogenesis*, Vol. 1 (translation by J. P. Fitzsimmons). Oliver and Boyd, Edinburgh.
Stumm, W., and Morgan, J. J. (1981). *Aquatic Chemistry*, 2nd ed. John Wiley, New York.
Szabolcs, I. (1989). *Salt-Affected Soils.* CRC Press, Boca Raton, Florida.
Tanji, K. K. (1982). Modelling of the soil nitrogen cycling. In *Nitrogen in Agricultural Soils,* F. J. Stevenson (Ed.). Am. Soc. Agron., Madison, Wisconsin, pp. 721-770.
Tanji, K. K., and Mehran, M. (1979). Conceptual and dynamic models for nitrogen in irrigated croplands. In *Nitrate in Effluents from Irrigated Lands*, P. F. Pratt (Ed.). Final Report to National Science Foundation, University of California, Riverside, pp. 555-646.
Theng, B. K. G. (1974). *The Chemistry of Clay-Organic Reactions.* Hilger, London.
Theng, B. K. G. (1979). *Formation and Properties of Clay-Polymer Complexes.* Elsevier, Amsterdam.
Thomas, G. W., and Swoboda, A. R. (1970). Anion exclusion effects on chloride movement in soil. *Soil Sci. 110*: 163.
Velbel, M. A. (1985). Geochemical mass balances and weathering rates in forested watersheds of the Southern Blue Ridge. *Am. J. Sci. 285*: 904.
Velbel, M. A. (1986). Influence of surface area, surface characteristics, and solution composition on feldspar weathering rates. *Am. Chem. Soc. Ser. 323*: 615.
Vitousek, P. M., Gosz, J. R., Grier, C. C., Melillo, J. M., Reiners, J. M., and Todd, R. L. (1979). Nitrate losses from disturbed ecosystems. *Science 204*: 469.
Vlek, P. L. G., and Craswell, E. T. (1981). Ammonia volatilization from flooded soils. *Fertil. Res. 2*: 227.
Vlek, P. L. G., and Stumpe, J. M. (1978). Effect of solution chemistry and environmental conditions on ammonia volatilization losses from aqueous systems. *Soil Sci. Soc. Am. J. 42*: 416.
Wagman, D. D., Evans, W. H., Parker, V. B., Halow, I., Bailey, S. M., and Shumm, R. H. (1968). Selected values of chemical thermodynamic properties. Natl. Bur. Stand. Tech. Notes, Washington, D.C., pp. 270-273.
Wagman, D. D., Evans, W. H., Parker, V. B., Shumm, R. H., Halow, I., Bailey, S. M., Churney, K. L., and Nutall, R. L. (1982). The NBS tables of chemical thermodynamic properties. *J. Phys. Chem. Ref. Data 11* (Suppl. 2).

Watkins, S. H., Strand, R. F., De Bell, D. S., and Esch, J. (1972). Factors influencing ammonia losses from urea applied to northwestern soils. *Soil Sci. Soc. Am. Proc. 36*: 354.

Webley, D. M., Duff, R. B., and Mitchell, W. A. (1960). A plate method for studying the breakdown of synthetic and natural silicates by soil bacteria. *Science 188*: 766.

Webley, D. M., Henderson, E. K., and Taylor, F. (1963). The microbiology of rocks and weathered stones. *J. Soil Sci. 14*: 102.

Weed, S. B., Davey, C. B., and Cook, M. G. (1969). Weathering of mica by fungi. *Soil Sci. Soc. Am. Proc. 33*: 702.

Wetselaar, R. (1968). Soil organic nitrogen mineralization as affected by low soil water potentials. *Plant Soil 29*: 9.

Wilson, J. M., and Griffin, D. M. (1975). Water potential and the respiration of microorganisms in the soil. *Soil Biol. Biochem. 7*: 199.

Wollast, R. (1967). Kinetics of the alteration of K-feldspar in buffered solutions at low temperature. *Geochim. Cosmochim. Acta 31*: 635.

Woodmansee, R. G., Dodd, J. L., Bowman, R. A., Clark, F. E., and Dickinson, C. E. (1978). Nitrogen budget of a shortgrass prairie ecosystem. *Oecologia (Berl.) 34*: 363.

Wray, J. L., and Farrington, D. (1957). Precipitation of calcite and aragonite. *J. Am. Chem. Soc. 79*: 2031.

Young, J. L., and Aldag, R. W. (1982). Inorganic forms of nitrogen in soils. In *Nitrogen in Agricultural Soils*, F. J. Stevenson (Ed.). Am. Soc. Agron., Madison, Wisconsin, pp. 423-448.

Zajic, J. E. (1969). *Microbial Biogeochemistry*. Academic Press, New York.

5

Biology of Arid Region Soils—Faunal Components

YOSEF STEINBERGER *Bar Ilan University, Ramat Gan, Israel*

And in that day I will make a covenant for them, with the beasts of the field, and with the fowls of heaven, and the creeping things of the ground (Hosea 2:20)

I. INTRODUCTION

This chapter concerns the faunal component of arid soils. The two main groups, the vertebrates and invertebrates, exert a separate ecological impact on arid regions. In arid regions, biological activity is telescoped downward into the subterranean part of the biosphere. This effect is marked when we consider the vertebrates, most of which are adapted to burrowing. The invertebrate fauna also has a considerable effect on arid regions. Both vertebrates and invertebrates may function as primary decomposers and as the distributors of organic mineral.

II. VERTEBRATES

The main group of burrowing vertebrates in many arid soils is small mammals belonging to the order Rodentia. In the Australian deserts, marsupials are particularly common. They comprise one of the most varied groups, ranging from the marsupial "mice" to "rats" which weigh a few grams, to the great kangaroo, whose weight is almost 100 kg (Troughton, 1954). Other groups of vertebrates that burrow in the soil, using it only as a refuge, include various amphibians, such as the salamander (Delson and Whitford, 1973), burrowing toads such as *Scaphiopus* (Ruibal et al., 1969), and various groups of reptiles, such as testodiens, lizards, and snakes. In this section we concentrate mainly on the effects on soil caused by rodents.

Much of the information available about desert rodents concerns the North American fauna. In this part of the world, rodents belonging to the Heteromyidae, such as kangaroo rats; Sciuridae, such as ground squirrels, and Cricetidae, such as gerbils are the most common.

In other deserts of the world, different rodents achieve prominence. In the Australian deserts, for example, murids predominate (Newsome and Corbett, 1975), and in the deserts of South America, the ctenomyids, such as the tuco-tucos and the viscacha (*Lagostomus*) (Weir, 1974), construct extensive burrow systems.

A more completely subterranean component of the soil rodent fauna is provided by members of the North American Geomyidae—gophers—and Bathyergidae.

The extensive burrow systems of these small mammals frequently contribute to the instability of the soil structure. Still, these burrows contribute to aeration of the soil and water infiltration, and because these animals store food material collected above ground in these subterranean locations, they actively promote the movement of nutrients into the soil.

The existence of a feeding "guild" among desert rodents is well established (Eisenberg, 1975). Four main types of feeding activity can be identified: carnivores, insectivores, granivores, and omnivores. Of these, the granivore niche, from an ecological point of view, is the most important. North American Heteromyidae are, par excellence, seed feeders ("granivores"), and different methods of seed collecting permit an allocation of resources that leads to high-level species diversity. For example, the kangaroo rat, *Dipodomys meriomi*, selects for high resource densities, "cloisters of seeds," but the pocket mouse, *Perognatus amplus*, does not show any preference for clumped or dispersed seeds. These two species select discharged seeds, whereas the banner tail kangaroo rat, *Depidomys spectabolis*, harvests intact seed pods (Schroder, 1979).

Seeds represent a food resource that is essentially unlimited in deserts. However, environmental conditions at the surface may limit the extent of foraging activities during certain times of the year or may cause these foragers to select alternative items of diet. One solution to the former problem is to store seeds in the subterranean nests ("caching"). Species that cache seeds, such as *Dipodomys* species, tend to maintain this feeding habit throughout the year. In the Old World deserts, the most common rodents are jerboas (family Dipodidae) (Fig. 1). These rodents resemble the American kangaroo rat, the gerbil, and *Meriomys* (family Gerbillinae). They cache food erratically and switch their diet from dry food seeds to green plant material when it is available (Chew, 1974; Brown et al., 1979b; Abramsky, 1983).

Food selection and foraging activity may also be influenced by other factors, such as whether the species are active during the day, as is the case of the ground squirrels of the genus *Citellus*, or nocturnal, as is more generally the case among the Heteromyidae and the Cricetidae. Again, Heteromyidae forage mainly for seeds present on the surface of the ground, although they can also exploit buried seeds. Unlike the Heteromyidae species of the genera *Reithrodontomys* and *Peromyscus* are pouched specialists and may forage in the aerial part of the vegetation. The last two mentioned genera, along with *Onycomis*, can probably be considered to belong to the insectivore-granivore niche. They switch their diet at different times of the year, exploiting insects as a food source during the summer

FIGURE 1 The jerboa (*Jaculus jaculus*), from the Negev Desert.

and changing to a diet of seeds during the colder months when insect food is in short supply (Brown et al., 1979a).

Herbivory is practiced by few desert rodents as a primary activity. The pock rat, *Neotoma*, feeds extensively on fresh tissues of cacti belonging to the genus *Opuntia*, but little information is available about the contribution of its grazing activity to the turnover of organic matter.

In North American deserts, rodents live sympatrically with rabbits and hares (lagomorphs). The food habitat of lagomorphs shows that during foraging periods, they leave considerable quantities of unconsumed material on the ground. Studies of the food habits of jackrabbits, *Lepus californicus*, in the Nevada, Utah, and Chihuahuan deserts (Hayden, 1966; Westoby, 1980; Steinberger and Whitford, 1983), have shown that these populations utilize different forbs and shrubs during the year. Steinberger and Whitford (1983) have shown that although the number of jackrabbits was only 0.25 individuals per hectare, their contribution of organic matter to the system, by the activity of pruning *Larrea*, was 101 kg ha^{-1} $year^{-1}$. This amount is equal to one-fifth of the estimated yearly total litter infall of the creosote bush.

The consequences of all these feeding activities are transport of seed, fruits, and, in some cases, nonreproductive parts of plants to subterranean levels, where they may be used as food resources and then released as partly decomposed fecal material. These are then incorporated into the decomposer food chain. Not all seed caches are relocated by the forager, and such seeds are then able to germinate when environmental conditions become suitable. Hence the granivore rodents function at two levels: they promote the distribution of seeds, and they also accelerate the process of organic decomposition in the soil.

III. INVERTEBRATES

Both burrowing and nonburrowing interstitial elements in the invertebrate fauna of arid soil can be identified. Groups that habitually produce soil burrows incluee various insects, such as ants, termites, and beetles, such as the tenebrionids and scarabaeids.

Most woodlice (Isopoda) utilize existing soil spaces but do not burrow actively. They utilize soil crevices or the spaces within leaf litter that accumulates at the soil surface. An exception to this is the desert-dwelling wood louse of the North African and Arabian deserts, *Hemilepistus*. The interstitial element is provided by the soil water fauna, which is represented in arid soils principally by protozoa and nematodes. Little information is available about the ecological impact of desert soil protozoa, and here we deal only with nematodes.

Another part of the interstitial fauna lives in the air pockets between the soil and litter particles. These are represented mainly by the microarthropods Collembola and Acarina.

A. Ants

Ants are social insects active on the soil surface during different times of the day. Some species are active during the day, and others are nocturnal. However, this distinction may not always hold and members of the North American genera *Pogonomyrmex* and *Solenopsis*, which are active during the day in the cooler months of the year, may become nocturnal or crepuscular during the warmer months (Brown et al., 1979a).

The social behavior of ants is made cohesive by the development of the nest, which may be completely subterranean, visible on the soil surface as a mound, or totally above ground.

There are striking ecological parallels between the rodents that we have just considered and the ants, despite that the majority of rodents are nocturnal and the majority of ants are diurnal. Both groups are efficient burrowers, and in both groups feeding guilds have been identified.

Carnivory was probably the original habit of the arthropods and is still practiced today by the primitive ants *Ponerinae* and *Dorylinae*, along with many *Myrmicinae*. However, many groups have adapted to a vegetarian habitat. Leaf cutters and "Attii" occur in South America, where they cultivate fungi for food in their underground nests. Some species, notably of the genus *Camponotus*, feed on wood or on "honey dew," which they collect outdoors. However, the most interesting feature of the ecology of desert-dwelling ants is the presence of a grain-feeding guild. (The different deserts of North America, Frag. Africa, and Australia, harder vicariants in this regard.) In North America, the genera *Pogonomyrmex*, *Pheidole*, *Solenopsis*, *Veromessor*, and *Novomessor* fill this granivore niche to a large extent. In Africa and in the Negev Desert, this niche is occupied by *Messor* (Gabbour et al., 1977; Steinberger et al., 1990a, b) and in Australia by *Meranopus* (Pisarski, 1978).

FIGURE 2 Entrance to an active nest of *Messor ebeninus* in the Negev dessert. The annual plants growing within the nest entrance germinated from seeds that were brought by the ants.

The high species diversity of granivorous ants in parts of North America deserts can be attributed to a number of factors that are essentially behavioral in character. These factors include patterns of foraging activity as well as activity periods during the diurnal cycle, seed selection, and the ability of particular species to switch their diets. *Pogonomyrmex*, *Pheidole*, *Solenopsis*, and *Veromessor* are group foragers, for the most part, although some species of *Pogonomyrmex*, such as *P. desertorum*, forage individually (Whitford, 1978). There is also considerable evidence that different species of these harvester ants select different types of seeds. Their choice is determined by their mandible gape, which determines the size of the seed that can be carried. For example, the mandible gape of *Pheidole rougulosa*, a small species, is 0.19 mm, whereas the gape of a much larger *Pogonomyrmex rugosus* us 0.87 mm. These two species show entirely different preferences for seed size selection (Chew and DeVita, 1980). *P. rugosus* and *Novomessor cockerellia* feed on seeds during the daytime, but they have a bimobile activity pattern and switch their diet, feeding on insects at night (Whitford and Ettershank, 1975).

Three *Messor* spp. (*M. rugosus*, *M. ebenius*, and *M. arenarius*) are the most abundant harvester ants in the Israeli Negev Desert. They are abundant and are one of the most important granivores in these ecosystems (Steinberger et al., 1990a). These harvester ant species collect seeds from at least 80 plant species, according to the plant parts found in the chaff piles (Fig. 2), representing 67% of the total number of species found in the area (Steinberger et al., 1990b). Their activity is a combined function of temperature and seed availability. Moisture plays a secondary role in determining the activity of harvester ants in Israel, as opposed to North American and South African harvester ants (Kay, 1978; Marsh, 1985), whose activities are directly affected by moisture.

The coexistence of the three *Messor* spp. in the same habitat is facilitated by the difference in their seed size range and by their daily activity cycles: the smallest ant species *M. rugosus* is active during the daytime and collects only abundant small-scale seeds; the biggest one, *M. arenarius*, is active during the day and collects mainly medium- to large-scale seeds and plant parts; the medium-sized species, *M. ebeninus*, whose food range overlaps with that of the other two species, is active mainly at night (Steinberger et al., 1990a).

Like the rodents, granivorous ants can also cache seeds in their underground nests. The ecological consequences of this behavior are similar to those already described in connection with rodents. That granivorous ants and rodents are sympatric in, for example, the North American deserts and appear as potential competitors raises the interesting question of how such competition can persist. Brown et al. (1979b) suggested that such stable coexistence can be maintained for the following reasons:

1. Ants appear to harvest seed size classes in proportion to their occurrence in the environment. They take smaller seeds, on average, than rodents, which seem to forage selectively for large seeds.
2. Ants are able to cope better than rodents with the unpredictability of resources through their regulation of colony size and foraging activity.
3. Ants are able to gain greater rewards by foraging on lower densities of seeds than rodents are, owing to their smaller body size.
4. Rodents are more efficient than ants in collecting dense clusters of seeds and are more efficient in digging for buried seeds.

The importance of each of these factors may well vary from one desert region to another.

B. Termites

This group of social insects occurs in all the desert regions of the world and, in these arid environments, tends to construct subterranean nests. These nests vary in complexity from simple diffuse networks of galleries to elaborate structures that may extend several meters below ground and whose presence is marked by a surface mound that may have elaborate chimneys associated with it.

Both termite and ant populations have a social structure. Termite populations consist of a number of castes, which may differ from each other morphologically, physiologically, and behaviorally.

Once a granivorous feeding guild is established, feeding responsibilities in the termite colony become the function of the worker caste. In this particular group of arthropods there are xylophage, harvesters, scavengers, humus feeders, and fungi growers. Many of the so-called lower termites feed on wood "xylophage," and this is a primitive condition. Wood feeding in deserts is an important activity, and the activities of the dry wood-feeding *Oteriternes occidentalis* may be responsible for the breaking up of this organic material (Nutting et al., 1975) and the use of it as a food substrate.

Harvester termites are common in the tropics and subtropics. Termites do not forage exclusively on seeds but also collect fragments of leaves and stems. In this regard, their feeding behavior overlaps with that of the scavenger guild (Crawford, 1979; Louw and Seely, 1984). Examples of these include members of the genera *Amitermes*, *Nasutitermes*, and *Tumulitermes*. The habit of feeding on humus is shown by various termites in South America, Africa, and Southeast Asia, but humus feeders apparently do not occur in Australia. Little is known about the ecological effects of this habit.

In the Chihuahuan Desert ecosystems, the population density of termites ranges from 300,000 to 660,000 individuals per hectare (Whitford et al., 1983). Their main period of activity occurs between March and November, when the foragers feed on a variety of plant material. Feeding galleries are constructed between September and November, after the summer rains have wet the soil.

The Negev Desert harvester termite (Fig. 3) *Anacantothermis ubachi* is the most common species in this habitat. Its activity is mainly confined to wetter periods of the year—the rainy winter season and the high dew season. These termites collect about 70 kg dry plant material per hectare per annum and store it underground (Steinberger, personal observation).

FIGURE 3 Harvester termites *Anacantothermis ubachi* during the active period.

It is interesting to note that a great concentration of this termite's nests lies within the boundaries of the ant nest's chaff piles. The termites in this place seem to collect mainly the plant material brought by the ants (Steinberger, personal observations).

The cultivation of fungus gardens is characteristic of the termite subfamily *Nacrotermitinae*. These gardens are prepared and tended by the termites, and the fungi grown here belong to the genus *Termitomyces*. There is a symbiotic relationship between the fungus and the termites, although the exact nature of this relationship is not known. Possibly the fungus provides a direct source of food for the termites. Alternatively, the termite may produce enzymes that can predigest the fungal cone on which termitomycete grows.

Termites also represent a source of nutrition for other soil animals that prey upon them: scorpions, ants, and reptiles.

C. Beetles

Although many coleopterans are associated with desert soil, the members of two families, namely Tenebrionidae and Scarabaeidae, are the most common of surface-active fauna. In the North American deserts, the genus *Eleodes* is common in the Chihuahuan Desert and *Cryptoglossa* species inhabit the sand dunes in the Mohave Desert. These desert dwellers are found in many other deserts. Most Tenebrionidae are flightless, and the legs sometimes provide an impressive digging structure. It is in the dune deserts of Namibia that the Tenebrionidae achieve the most spectacular diversity. Here, species belonging to the genera *Onymacris*, *Jyrosis*, *Calosis*, and *Lepidochora* coexist by virtue of the fact that they have different activity patterns, both circadian and seasonal, and different habitat preferences. *Onymacris plana* is active throughout the year, during the daytime on sand dunes and slip-face bases. *Onymacris laeviceps* is active in the same habitat during twilight in the summer (Wallwork, 1982).

A pattern of activity similar to that shown by *O. plana* is also shown by *Onymacris rugatipennis*, but this species confines itself to riverbed gravel sites. Another inhabitant of sand dune crusts and slope faces is *Lepidochora porti*. This species, like *O. planar* and *O. rugatipennis*, is active throughout the year but differs from the other two species by being nocturnal (Edney, 1971).

In the Namib Desert, as in many other arid environments, the water sources for the Tenebrionidae are irregular and unpredictable in time and space.

These desert Tenebrionidae are well adapted to arid environments, not only in the activity pattern and habitat selection already mentioned, but also in other behavioral and physiological characteristics that permit the conservation of body water. The coastal sand dunes of Namibia, where many of these species live, experience advective fog that comes in from the sea at certain times of the year. This fog supplies a cource of moisture for the beetles in an area where rain may not fall for several years. The habit of "fog basking" allows moisture to condense on their legs and elytra, and this is then absorbed (Seely and Hamilton, 1976; Hamilton and Seely, 1976; Louw and Seely, 1984).

The wing covers, *elytra*, of these insects are fused together and enclose a sublateral space into which the abdominal spharacels open. On these, a high relative humidity is permitted to develop, reducing the vapor pressure deficit across the body surface and allowing the sublateral space to act as a thermal buffer (Ahearn and Hadley, 1969; Hadley, 1970). Therefore we can find species that are active during the day in the hottest and driest periods and others that are nocturnal (Edney, 1971).

Tenebrionidae are for the most part scavengers of wind-blown particles of organic debris. In the Mohave Desert, *Cryptoglossa* spp. feed on dead insects and feces of kangaroo rats that are blown about by the winds and collected in the dunes (Crawford, 1981).

Representatives of the Tenebrionidae and Scarabaeidae are abundant in the Negev and Sinai deserts. Many of these Beetles are active during the day in the hot season and have both physiological and behavioral adaptation for this. *Sepidium korah*, a black tenebrionid, covers itself with fine dust and sand particles that reflect a large proportion of radiation and act as camouflage (Fig. 4).

D. Isopods

The woodlice isopoda are crustaceans that have become remarkably successful in deserts. Members of the genera *Porcellio, Armadillidium, Venezillo,* and *Hemilepistus*, have established themselves in arid and semiarid environments. *Armadillidium* is distributed throughout the New World, and *Porcellio*

FIGURE 4 *Sepidium korah* active during the day, from the Negev desert.

is distributed in the Old and New World deserts. In the Old World desert, members of the *Porcellio* and *Hemilepsius* predominate, particularly *Hemilepisus*, which is distributed at least as far as Afghanistan. *Hemilepisus* is well adapted to the desert environment because it has several suitable physiological and behavioral features. Its ability, unique among isopods, to burrow helps it seek refuge in the soil by creating refuges beneath the soil surface (Warburg, 1965; Edney, 1967; Ghabbour et al., 1982; Wallwork, 1970).

Hemilepsius commences to burrow in the soil of the Negev Desert at the end of the winter season, and at this time, the females are ready to produce a new generation. In these burrows the new generation is produced, and the new individuals participate in the construction of a more extensive burrow system that allows the "family" to go deep enough into the soil to enable them to survive in periods of desiccation at the soil surface (Linsenmair and Linsenmair, 1971; Shachak et al., 1976; Steinberger et al., 1989).

Among the isopods, this familial characteristic is unusual. It requires a degree of recognition between members of the family and is promoted by the production of pherohormonal badges that identify family members (Linsenmair and Linsenmair, 1971; Linsenmair, 1972).

It is easy to underestimate the importance of these arthropods in desert ecosystems. Admittedly they do not feature significantly in the soil ecosystems in the New World. However, they achieve a great importance in Old World deserts, where it has been calculated that *Hemilepisus reaumuri* (Fig. 5)

FIGURE 5 The desert isopod *Hemilepisus reaumuri* and its feces, from the Negev Desert.

TABLE 1 Biomass of Common Herbivores on Loess Plain of the Negev Desert

Species	Biomass (g per 1000 m^2)
Isopods (*Hemilepisus reaumuri*)	1000
Porcupines (*Histrix indica*)	150
Jerboa (*Jaculus jaculus*)	70
Hares (*Lepus europeus*)	2.5

in the Negev Desert has a biomass that exceeds that of its principal "herbivores" competitors, as indicated in the data in Table 1 (Steinberger, 1989).

Isopods in general and *Hemilepisus* species in particular are detritivorous; that is, they feed on organic material that originates from a variety of plants. Isopods from temperature regions have the habit of reingesting fecal materials, which Hassal and Rushton (1982) have interpreted as an external rumen. The fecal feeding habit (coprophagy) apparently allows an interval of time between successive ingestion to enable the growth of microbial populations whose activity releases essential nutrients that then become available to the isopods. Although these events have not been documented for desert isopods, such as *Hemilepisus*, there is no reason to suppose that they cannot participate in this kind of feeding activity. From November to February, when *H. reaumuri* is restricted to the burrow system (Shachak, 1980), such coprophagy (fecofeeding) probably occurs and may contribute to the circulation of essential ions, such as copper (Cu).

The habit of *H. reaumuri* to feed on fungi and algal crusts may also be significant. Fungi and algae may not in themselves be particularly nutritious, but they may provide enzymes that can be utilized by the isopods' digestive system for the degradation of higher plant material. Feeding experience on soil crusts indicated a significant change in the soil fungi found in the isopod feces compared to those in the intact soil crust (Glazer et al., 1983). The soil crust acts as a substrate, not only for fungi, but also for bacterial populations in the digestive system that help to decompose organic matter and release protein from it. The soil particles also act as grinders and mechanically break down ingested plant material (Steinberger, 1989). The evidence for the efficacy of these acquired enzymes remains to be established, but it is known that these enzymes can be effective in the digestive capability of other soil-dwelling saprophates, notably the termites (Martin and Martin, 1979).

In conclusion, in certain deserts, such as those in the Middle East, isopods such as *H. reaumuri* are important components in the soil ecosystem because they represent an important component of the soil biomass and participate in the turnover of organic materials in the soil.

E. Nematodes

One of the most abundant fractions of the soil water fauna in desert soil ecosystems is the nematode fauna. The taxonomic classification of nematodes in general, and soil nematodes in particular, presents formidable problems. However, an ecological classification has been devised that allows us to identify different functional groups of nematodes (Yeates, 1971). Nematodes are ubiquitous and are probably important in all soil ecosystems, but their ability to maintain viable populations in these arid environments provide them with a special significance. Soil nematodes tend to be concentrated around the rooting systems of plants. Freckman and Mankau (1977) have shown that the distribution of soil nematodes is correlated with their proximity to the plant rooting system. Plant roots exude CO_2, amino acids, and various growth factors that may be attractive to members of soil water fauna such as nematodes.

From an ecological point of view, soil nematodes can be classified into four feeding guilds: (1) fungal feeders, (2) omnivore predators, (3) microbial feeders, and (4) plant parasites. In desert soils like those studied by Freckman and Mankau (1977) at Rock Valley in Nevada and by Steinberger et al. (1989) in the Negev Desert, microbial feeders prey particularly on bacterial colonies, and their densities decrease as bacterial populations become diluted with increasing distance from the root zone rhizosphere.

The interaction between nematodes and microbial populations may influence decomposition pathways in the soil. For example, the prevalence of bacterial feeding nematodes like *Acrobeloides* may direct decomposition in a different direction than that which would be generated by fungivorous nematodes such as *Aphelenchus*.

Freckman and Mankau (1977), examining the nematode fauna associated with four different plant species, namely, *Lycium andersonii*, *Larrea tridentata*, *Ambrosia dumosa*, and *Krameria pravifolia*, found that plant parasitic nematodes were more numerous under *L. andersonii* and *L. tridentata* than under *A. dumosa* and *K. pravifolia*. We can only speculate on the reasons for these differences. The prevalence of plant parasites on the *Lycium* and *Larrea* suggests that herbivores, rather than a detritus food chain, may be associated with these rhizospheres, and they are therefore involved in a food chain based on grazing activity rather than on detritus.

The activity of nematodes in the soil is related to the presence of water films surrounding the soil particles, but the persistence of nematode fauna in arid soils depends upon the ability of these animals to survive through periods of desiccation during the year. Soil nematodes have adapted to this situation and show drought tolerance. This tolerance has been identified as anhydrobiosis. Anhydrobiotic nematodes assume a shriveled appearance and to all intents and purposes appear to be dead. In this condition these animals can survive high levels of ionizing radiation and temperature extremes. Their recovery from this inactive stage occurs when moisture levels exceed 3% (Freckman, 1978).

Studies of a single autumn dewfall event (Steinberger et al., 1989) showed rapid changes in soil water content, which was evident to a depth of 30 cm. Oscillation in the density of nematode populations beneath the perennial canopies of *Zygophyllum dumosum* and outside the shrub varied in response to moisture, although the distribution in trophic groups at these sites were identical.

Whitford et al. (1981), who also studied the influence of simulated rainfall, found that the total number of free-living nematodes decreased as the moisture decreased. The results from these studies indicated the importance of moisture for triggering nematode population activity.

Population densities of soil nematodes are measured in terms of millions per square meter. For example, Freckman (personal communication) provides estimates of 1.24×10^6 m^{-2} for the mean annual density at Rocky Valley in Nevada. These figures are somewhat higher (8×10^5 individuals m^{-2}) than those estimated for the Negev Desert (Steinberger et al., 1988). They may overestimate the extent of nematode activity in these soils since a proportion of this fauna may be in an anhydrobiotic stage. Anhydrobiosis can affect all tropic groups, as suggested by Freckman (personal communication).

F. Collembola

Collembola are frequently found in arid and semiarid soils and litter (Greenslade and Greenslade, 1973; Santos et al., 1978; Santos and Whitford, 1983; Steinberger and Whitford, 1984; Kamill et al., 1985; Wallwork et al., 1985). The Collembola are coupled with mites in these areas. Because of the size of these soil animals, the availability of water is essential. With no special respiratory structure and a lack of a tracheal system, the oxygen supply mainly depends on gas exchange through the cuticle.

The group of animals is very sensitive to desiccation, although a degree of tolerance varies from one species to others. Greenslade (1981) reported on some epigeneic Collembola (*Entomobrya* spp.) found in the semiarid regions in Australia that were found active on sand dunes during the hottest time of the year. She also reported on several collembolan genera in the Australian and African deserts that were resistant to desiccation.

In contrast, Loring et al. (1988) found an increase in the abundance of the collembolan population in the Chihuahuan Desert during the warm wet season, with very strong nocturnal tendencies. Some species, such as *Sminturides pumilis* and *Sminturides sexpinnatus*, were found responding to precipitation in both the Chihuahuan and Sudan deserts. The quick response to moisture and a diapause state during the dry season shows a possibility of resembling the anhydrobiotic state found between the wet seasons (Poinsot, 1974). Other representatives from the genera *Folsomides* and *Brachystomella* tend to show the same activity patterns as those already reported. According to these findings, vertical migration of a collembolan population for shorter periods and diurnal movement, as reported for nematodes, should also be taken into consideration.

From an ecological point, the Collembola can be classified as detritivores, feeding on decaying organic material, although some of them feed on fungi and bacteria (Christiansen, 1964). The feeding habitat of collembolan species may be important in promoting and regulating, directly or indirectly, decomposition processes, which are of a great importance in desert systems.

G. Acaridiae: Mites

This group of microarthropods is widely distributed, representing one of the most important groups in soil and litter fauna in arid and semiarid deserts, and comprising up to 90% of all microarthropod populations (Wallwork, 1972). Numerically they take second place after the nematodes, reaching a mean maximum density of 43,000 individuals m^{-2} during most of the year, although during the day in hot seasons the mean density is 1200 individuals m^{-2} (Steinberger and Whitford, 1985). The overall estimate of microarthropod density per year in the North American, Australian, and Negev deserts ranges between 2000 and 8000 individuals m^{-2} (Wood, 1971; Wallwork, 1982; Wallwork et al., 1985; Steinberger and Wallwork, 1985). Both the nematodes and the microarthropods that are present in high numbers have been shown to play an important role in regulating the rates of decomposition, mineralization, and mobilization of nutrients (Elkins et al., 1982; Whitford and Freckman, 1985; Whitford, 1988).

From a systematic viewpoint, the soil Acarina belong to one of four orders: the Prostigmata, Mesostigmata, Astigmata, and the Cryptostigmata. Representatives from these orders can be found feeding on plant detritus, fungi, algae, and bacteria and preying on different groups of soil invertebrates.

Living in a rainfall-regulated ecosystem, animals need to keep "track" to maintain biological functions. Wetting and drying affects the populations of soil microarthropods, resulting in marked changes in numbers and diversity. These changes affected not only the moisture in the soil profiles but also the organic matter on a daily basis. Whitford et al. (1981) and Steinberger and Wallwork (1985) showed that microarthropods from the Chihuahuan (North American) and Negev deserts were triggered by avoidance reaction to low moisture and high temperature by moving down in the soil horizon to a higher soil water content and cooler environment. This immediate response (in abundance) to the daily change has been associated with the movement ability of those populations in the soil profile. The movement of soil Acarina is made possible by a combination of behavioral, physiological, and morphological adaptations. The "waxy bloom" epicuticular layer present on the surface of the integument of many terrestrial desert-dwelling arthropods and found on soil Acarina as well (Steinberger et al., 1986) appears to be related to water loss and reflectivity and may also enable movement between the soil particles without damaging the cuticle.

H. Snails

One of the greatest astonishments to people who visit the desert, particularly the Negev Desert, is the white cover of snails that appears on the soil surface and on vegetation. Similar to other invertebrates, such as isopods, the snails (pulmonate gastropods) belong to an aquatic environment rather than a terrestrial one. According to Noy-Meir (1973, 1974), the definition of a desert ecosystem "is a system regulated by water, where water input is variable and unpredictable." This "commodity" of water, which is in short supply, variable in amount, and unpredictable in time, is the main factor that determines the activity pattern. Therefore, snails, as have many other invertebrates, have developed physiological and behavioral adaptations that enable them to survive without water and food supplies during the long and hot summers and to become active for short periods of time during the rainy winter season (Shachak et al., 1976; Steinberger et al., 1981). Field observations have shown that these snails feed on algae and other organic matter on the soil surface after rainfall. The activity and feeding period, as well as other functions, such as locomotion, mating, egg laying, climbing on bushes, and burrowing in the soil and under stones, takes place during the short, wet period, a total of 5-7% of the year.

During the inactive or estivation period, the remaining 93 to 95% of the year, the shell opening is covered by a membrane, the epiphragm, that contains a large amount of calcium carbonate. This epiphragm isolates snails from the outer environment, creating a minienvironment intershell that decreases water loss and lowers the temperature to above 5°C cooler than the outer temperature on hot days (Schmidt-Nielson et al., 1972).

The ability to feed and complete biological functions during short periods of time and to maintain a low biological activity, that is, estivation for long periods of time, shows the opportunistic character of these desert creatures.

The most common snails found in the Negev Desert are *Trochoidae seetzeni*, estivating during the summer period attached to shrubs (Yom-Tov, 1972); *Sphincterochila zonata (=boissieri)*, which can be found on the soil surface of loess plain areas and at the base of desert shrubs (Shachak et al., 1975); and *Sphincterochila prophetarum* (*=arabica*), which inhabits rocky hillsides (Steinberger et al., 1983).

IV. CONCLUSIONS

This brief and relatively short chapter is based on only part of the accumulated knowledge of desert soil fauna. I apologize for the omission of many aspects that are probably no less important than those I have mentioned. Generally speaking, there is a need to know more about the organisms, their contributions, and their reactions occurring in soil, both aboveground and underground.

Ecologists have mostly been preoccupied with studying the relation between the behavioral and physiological processes of individual organisms and their environment, rather than with their role and functions. This was also the way in which I tried to describe those specific organisms, which according to my judgment are unable to survive in harsh environments. One of the most common features of these environments is "moisture or water availability" from different sources, for example, from fog, dew, or rainfall, that is completely unpredictable in space, time, and quantity. Only those organisms able to develop ways to cope with this unpredictability will be able to survive.

I hope that although this chapter was written in a simple form, it will be beneficial to the understanding of the different faunal components of arid soils.

ACKNOWLEDGMENTS

I would like to express my sincere thanks to Professor J. A. Wallwork, who supported and encouraged me morally and intellectually to write this chapter. He also helped to mold my thinking about desert organisms.

To my family and the many people who contributed time and effort to the development of this chapter, I express my sincere thanks. I would also like to thank the Life Science Department at Bar-Ilan University for their help and support.

REFERENCES

Abramsky, Z. (1983). Experiments on seed predation by rodents and ants in the Israeli desert. *Oecologia (Berl.) 57*: 328-332.

Ahearn, G. A., and Hadley, N. F. (1969). The effects of temperature and humidity on water loss in two desert tenebrionid beetles, *Eleodes armata*, and *Cryptoglossa verrucosa*. *Comp. Biochem. Physiol. 30*: 739-749.

Brown, J. H., Davidson, D. W., and Reichman, O. J. (1979a). An experimental study of competition between seed-eating desert rodents and ants. *Am. Zool. 19*: 1129-1143.

Brown, J. H., Reichman, O. J., and Davidson, D. W. (1979b). Granivory in ecosystems. *Annu. Rev. Ecol. Syst. 10*: 201-227.

Chew, R. M. (1974). Consumers as regulators of ecosystems: An alternative to energetics. *Ohio J. Sci. 74*: 359-370.

Chew, R. M., and DeVita, J. (1980). Foraging characteristics of a desert ant assemblage: Functional morphology and species separation. *J. Arid Environ. 3*: 75-83.

Christiansen, K. (1964). Bionomics of Collembola. *Annu. Rev. Entomol. 9*: 137-149.

Crawford, C. S. (1979). Desert detritvores: A review of life history patterns and trophic roles. *J. Arid Environ. 2*: 31-42.

Delson, J., and Whitford, W. G. (1973). Adaptions of the tiger salamander, *Ambystoma tigrinum*, to arid habitats. *Comp. Biochem. Physiol. 46A*: 631-638.

Edney, E. D. (1967). Water balance in desert arthropods. *Science 156*: 1059-1066.

Edney, E. B. (1971). Some aspects of water balance in tenebrionid beetles and a thysanuran from the Namib Desert of Southern Africa. *Physiol. Zool. 44*: 61-76.

Eisenberg, J. F. (1975). The behavior patterns of desert rodents. In *Rodents in Desert Environments*, I. Prakash and P. K. Ghosh (Eds.). *Monographiae biologicae*, Vol. 28, Junk, The Hague, pp. 189-224.

Elkins, N. Z., Steinberger, Y., and Whitford, W. G. (1982). The role of microarthropods and nematodes in decomposition in a semiarid ecosystem. *Oecologia (Berl.) 55*: 303-310.

Freckman, D. W., and Mankau, R. (1977). Distribution and trophic structure of nematodes in desert soils. *Ecol. Bull. 25*: 511-514.

Freckman, D. W. (1978). Ecology of anhydrobiotic soil nematodes. In *Dry Biological Systems*, J. H. Crowe and J. S. Clegg (Eds.). Academic Press, New York, pp. 345-357.

Ghabbour, S. I., Mikhail, W. Z. A., and Rizk, M. A. (1977). Ecology of soil fauna of Mediterranean desert ecosystems in Egypt. I. Summer populations of soil mesofauna associated with major shrubs in the littoral sand dunes.

Glazer, A., Steinberger, Y., and Dubinsky, Z. (1983). Soil fungi as a component in the diet of the desert isopod *Hemilepistus reaumuri*—a preliminary study. In *Developments in Ecology and Environmental Quality*, Vol. 2, Balaban, International Science Service. Rehovot.

Greenslade, P. (1981). Survival of Collembols in arid environments: Observations in South Australia and the Sudan. *Pedobiologia 4*: 218-228.

Greenslade, P. J. M., and Greenslade, P. (1973). Epicaeic Collembola and their activity in a semiarid locality during summer. *Pedobiologia 13*: 227-235.

Hadley, N. F. (1970). Micrometeorology and energy exchange in two desert arthropods. *Ecology 51*: 434-444.

Hamilton, W. J., III, and Seely, M. K. (1976). Fog basking by the Namib Desert beetle, *Onymacris unquicularis*. *Nature 262*: 284-285.

Hassal, M., and Rushton, S. P. (1982). The role of coprophagy in the feeding strategies of terrestrial isopods. *Oecologia (Berl.) 53*: 374-381.

Hayden, P. (1966). Food habits of black-tailed jack rabbits in southern Nevada. *J. Mammal. 47*: 42-46.

Kamill, B. W., Steinberger, Y., and Whitford, W. G. (1985). Soil microarthropods from the Chihuahuan Desert of New Mexico. *J. Zool. 205*: 273-286.

Kay, C. A. R. (1978). Preferred temperatures of desert honey ants (Hymenoptera, Formicidae). *J. Thermal Biol. 3*: 213-217.

Linsenmair, K. E. (1972). Die Bedeutung familienspezifischer "Abzeichen" für den Familienzusammenhalt bei der sozialen Wustenassel *Hemilepistus reaumuri*. *Z. Tierpsychol. 31*: 131-163.

Linsenmair, K. E., and Linsenmair, C. (1971). Paarbildung und Paarsuzammenhalt bei den monogamen Wustenassel *Hemilepistus reaumuri* (Crustocea, Isopoda, Oniscoidea). *Z. Tierpsychol. 29*: 134-155.

Loring, S. J., Weems, D. W., and Whitford, W. G. (1988). Abundance and diversity of surface active collembola along a watershed in the northern Chihuahuan Desert. *Am. Midland Naturalist 119*: 21-30.

Louw, G. N., and Seely, M. K. (1984). *Ecology of Desert Organisms*. Longman, London.

Marsh, A. C. (1985). Thermal responses and temperature tolerance in a diurnal desert ant, *Ocymyrmex barbiger*. *Physiol. Zool. 58*: 629-636.

Martin, M. M., and Martin, J. S. (1979). The distribution and origins of the cellulolytic enzymes of the higher termite *Macrotermes natalensis*. *Physiol. Zool. 52*: 11-21.

Newsome, A. E., and Corbett, L. K. (1975). Outbreaks of rodents in semi-arid and arid Australia: Causes, preventions and evolutionary considerations. In *Rodents in Desert Environments*, I. Prakash and P. K. Ghosh (Eds.). *Monographiae biologicae*, Vol. 28, Junk, The Hague, pp. 117-153.

Noy-Meir, I. (1973). Desert ecosystems: Environment and producers. *Annu. Rev. Ecol. Syst. 4*: 25-51.

Noy-Meir, I. (1974). Desert ecosystems: Higher trophic levels. *Annu. Rev. Ecol. Syst. 5*: 195-214.

Nutting, W. L., Haverty, M. I., and Lafage, J. P. (1975). Demography of termite colonies as related to various environmental factors: Population dynamics and role in the detritus cycle. US/IBP, Desert Biome Research Memorandum, Ecology Center, Utah State University, Logan, Utah.

Pisarski, B. (1978). Comparison of various biomes. In *Production Ecology of Ants and Termites*, IBP, Vol. 13, M. V. Brian (Ed.). Cambridge University Press, Cambridge, pp. 326-331.

Poinsot, N. (1974). Comportement de certains Collemboles dans les biotopes xerigues mediterraneens: un nouveau cas d'anhydrobiose. *C. R. Acad. Sci. Paris 278*: 2213-2215.

Ruibal, R., Tevis, L., and Roig, V. (1969). The terrestrial ecology of the spadefoot toad, *Scaphiopus hammondi*. *Copeia 1969*: 571-584.

Santos, P. F., and Whitford, W. G. (1983). Seasonal and spatial variation in the soil microarthropod fauna of the White Sands National Monument. *Southewest. Naturalist 28*: 417-421.

Santos, P. E., DePree, E., and Whitford, W. G. (1978). Spatial distribution of litter and microarthropods in Chihuahuan desert ecosystem. *J. Arid Environ. 1*: 41-48.

Schmidt-Nielson, K., Taylor, C. R., and Shkolnik, A. (1972). Desert snails: Problems of survival. In *Comparative Physiology of Desert Animals*, G. M. O. Maloiy (Ed.). Zool. Soc., London, pp. 1-13.

Schroeder, G. D. (1979). Foraging behaviour and home range utilization of the bannertail kangaroo rat (*Dipodomys spectabilis*). *Ecology 60*: 657-665.

Seely, M. K., and Hamilton, W. J. III (1976). Fog catchment sand trenches constructed by tenebrionid beetles, *Lepidochora*, from the Namib Desert. *Science 193*:

Shachak, M. (1980). Energy allocation and life history strategy of the desert isopod *H. reaumuri*. *Oecologia (Berl.) 45:*

Shachak, M., Orr, Y., and Steinberger, Y. (1975). Field observation on the natural history of *Sphincterochilla zonata* (Bourguignat, 1853) (= *S. boissieri* Charpentier, 1847). *Argamon Isr. J. Malac. 6*: 20-46.

Shachak, M., Chapman, E. A., and Orr, Y. (1976). Some aspects of the ecology of the desert snail *Sphincterochilla boissieri* in relation to water and energy flow. *Isr. J. Med. Sci. 12*(8): 887-891.

Steinberger, Y., and Whitford, W. G. (1984). Spatial and temporal relationship on soil microarthropods on a desert watershed. *Pedobiologia 26*: 275-284.

Steinberger, Y., and Wallwork, J. A. (1985). Composition and vertical distribution patterns of the microarthropod fauna in a Negev desert soil. *J. Zool. (Lond.) [A] 206*: 329-339.

Steinberger, Y., and Whitford, W. G. (1983). The contribution of shrub pruning by jack rabbits to litter input in a Chihuahuan desert ecosystem. *J. Arid Environ. 6:* 183-187.

Steinberger, Y., and Whitford, W. G. (1985). Microarthropods of a desert tabosa grass (*Hilarie mutica*), *Am. Midland Naturalist 114*(2): 225-234.

Steinberger, Y. (1989). Energy and protein budgets in the desert isopod *Hemilepistus reamuri*. *Acta Oecol. Oecol. Gener. 10*(2): 117-134.

Steinberger, Y., Grossman, S., and Dubinsky, Z. (1981). Some aspects of the ecology of the desert snail *Sphincterochilla prophetarum* in relation to energy and water flow. *Oecologia (Berl.) 50*: 103-108.

Steinberger, Y., Grossman, S., Dubinsky, Z., and Shachak, M. (1983). Stone microhabitats and the movement and activity of desert snails, *Sphincterochilla prophetarum*. *Malacol. Rev. 16*: 63-70.

Steinberger, Y., MacQuitty, M., Langesame, J., and Wallwork, J. A. (1986). Observation on the fine structure of the cuticle of two desert-dwelling oribatid mites (Acari: Cryptostigmata). *Isr. J. Entomol. 20*: 7-12.

Steinberger, Y., Orion, D., and Whitford, W. G. (1988). Population dynamics of nematodes in the Negev Desert soil. *Pedobiologia 31*: 223-228.
Steinberger, Y., Loboda, I., and Garner, W. (1989). The influence of autumn dewfall on spacial and temporal distribution of nematodes in the desert ecosystem. *J. Arid Environ. 16*: 177-183.
Steinberger, Y., Leschner, H., and Shmida, A. (1990a). Chaff piles of harvester ants (*Messor* spp,) nests in desert ecosystem. *Insectes Sociaux* (submitted).
Steinberger, Y., Leschner, H., and Shmida, A. (1990b). Activity pattern of harvester ants (*Messor* spp.) in the Negev desert ecosystem. *Insectes Sociaux* (submitted).
Troughton, E. (1954). *Furred Animals of Australia*, 5th ed. Angus & Robertson, London.
Wallwork, J. A. (1970). *Ecology of Soil Animals*. McGraw-Hill, New York.
Wallwork, J. A. (1972). Mites and other microarthropods from the Joshua Tree National Monument, California. *J. Zool. (Lond.) 168*: 91-105.
Wallwork, J. A. (1982). *Desert Soil Fauna*. Praeger, New York.
Wallwork, J. A., Kamil, B. W., and Whitford, W. G. (1985). Distribution and diversity patterns of soil mites and other microarthropods in a Chihuahuan desert site. *J. Arid Environ. 9*: 215-231.
Warburg, M. R. (1965). Water relation and internal body temperature of isopods from mesic and xeric habitats. *Physiol. Zool. 38*: 99-109.
Weir, B. J. (1974). The tuco-tuco and plains viscacha. In *The Biology of Hystricomrph Rodents*, I. W. Rowlands and B. J. Weir (Eds.). *Symp. Zool. Soc. Lond. 34*: 13-130.
Westoby, M. (1980). Diets of black-tailed jack rabbits in Curlew Valley, Northern Utah. *J. Wildlife Manag. 44*: 942-948.
Whitford, W. G. (1978). Foraging by seed-harvesting ants. In *Production Ecology of Ants and Termites*, M. V. Brian (Ed.). IBP Vol. 13, Cambridge University Press, Cambridge, pp. 107-110.
Whitford, W. G. (1988). Decomposition and nutrient cycling in distributed arid ecosystems. In *The Reconstruction of Distributed Arid Lands. An Ecological Approach*, E. B. Allen (Ed.). Westview Press, Colorado, pp. 136-161.
Whitford, W. G., and Ettershank, G. (1975). Factors affecting foraging activity in Chihuahuan Desert harvester ants. *Environ. Entomol. 4*: 689-696.
Whitford, W. G., and Freckman, E. W. (1985). The role of soil biota in soil processes in the Chihuahuan Desert. In *Arid Lands Today and Tomorrow*, Proceedings of Int. Research and Development Conference, Tucson, Arizona, 1985, E. Whitehead, C. F. Hutchinson, B. B. Timmermann, and R. G. Varady (Eds.). Westview Press, Colorado, pp. 1063-1073.
Whitford, W. G., Freckman, D., Elkins, N. Z., Parker, L. W., Parmalee, R., Phillips, J., and Tucker, S. (1981). Diurnal migration and responses to simulated rainfall in desert soil microarthropods and nematodes. *Soil Biol. Biochem. 13*: 417-425.
Whitford, W. G., Parker, L. W., Schaefer, D., Santos, P. F., and Steinberger, Y. (1983). The contribution of soil fauna to nutrient cycles in desert systems. In *New Trends in Soil Biology*, P. Lebrun, H. M.

Andre, A. de Medts, C. Gregoir-Wido, and G. Wauthy (Eds.). Proceedings of the VII Int. Colloquium of Soil Zoology, Louvain-La-Neuve (Belgium). Die Brichard, Ottignies-Louvain-la-Neuve, Belgium, pp. 49-59.

Wood, T. G. (1971). The distribution and abundance of *Folsomides deserticola* (Collembola: Isotomidae) and other microarthropods in arid and semiarid soils in Southern Australia, with a note on nematode populations. *Pedobiologia 11*: 446-468.

Yeates, G. W. (1971). Feeding types and feeding groups in plant and soil nematodes. *Pedobiologia 11*: 446-468.

Yom-Tov, Y. (1972). Field experiments on the effect of population density and slope direction on the reproduction of the desert snail. *Trochoidea* (Xerocrassa) *seetzeni*. *J. Anim. Ecol. 41*: 17-22.

6

Temporal and Spatial Organization of Desert Plant Communities

EDITH B. ALLEN *San Diego State University, San Diego, California*

I. INTRODUCTION

Plant communities of semiarid lands and deserts have spatial and temporal patterns that are characteristic of these ecosystems. Climate and topography are the main driving forces that create these patterns, either directly by limiting plant productivity, density, and composition or indirectly by dictating patterns of soil formation and fertility (Fig. 1). The soil and vegetation resources of deserts are closely integrated; soils not only determine plant species, but plants affect nutrient cycling and the spatial characteristics of the soils (Crawford and Gosz, 1982). Because low precipitation typically results in low plant cover, the soils are subject to erosion on any slight slope or if they are coarse textured (Dorn, 1988). Rocky uplands and depositional lowlands create characteristic spatial patterns of plants across landscapes (Shmida et al., 1986). The plants themselves tend to be widely spaced and therefore impose a smaller scale spatial organization nested within the elements of the landscape. Finally, deserts are subject to disturbance, and the consequent plant succession, or temporal changes, are often unique to deserts.

There is a large body of literature, some of it reviewed here, that deals with various aspects of vegetation patterns in deserts. The incentive for measuring pattern has always been to understand functioning. More recently, disturbance followed by restoration or reclamation has produced a new incentive: to reclaim deserts, we must understand their small- and large-scale patterns as well as the functional aspects of these patterns. Ecosystem studies have traditionally been carried out on a large scale (McIntosh, 1985), but the challenge of understanding nutrient cycling at the scale of the individual plant is just beginning to be met (M. F. Allen, 1988a, b). Agronomic approaches to reclamation have typically prevailed, including seeding in rows, fertilizing, and irrigating. An ecologically sound approach is to simulate natural plant patterns and take advantage

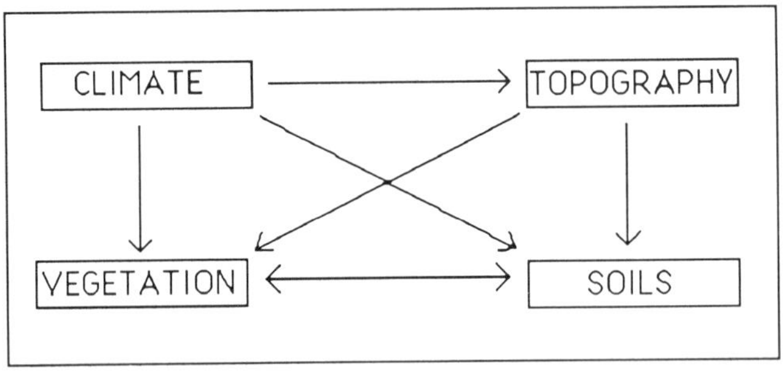

FIGURE 1 Relationships among climate, topography, and soils in structuring plant communities.

of the natural variation in effective moisture and nutrients along topographical gradients (E. B. Allen, 1988). This chapter outlines the spatial and temporal organization of semiarid land and desert plant communities and discusses specifically how this organization is important for reclamation or restoration.

II. SPATIAL ORGANIZATION

A. Horizontal Patterns

Across Landscapes

The high degree of soil erosion in deserts creates characteristic landforms. In mountainous or hilly terrain, erosion causes gentle slopes that are commonly called *bajadas* in the southwestern United States (Dorn, 1988). The vegetation of the bajada grades from xeric upland species to more mesic lowland species (e.g., Wierenga et al., 1987). Soil nutrients change along the gradient. For instance, slopes in southwestern Wyoming had low soil P in the erosional sites and relatively high P in the depositional sites (M. F. Allen, 1988a). Both species composition and productivity change along these gradients (Ludwig, 1986). At the base of the bajada may be an *arroyo* or *wadi* that drains water from the valley or, alternatively, a *playa*, or closed basin that allows no drainage, but rather evaporation typically coupled with salt accumulation. Phreatophytes, often trees, are associated with the larger arroyos, and shorter, salt-tolerant species are in the playas.

Where mountains were not uplifted but rather valleys have been eroded, a *hamada*, or high plain left by erosion, may be found that supports some of the most xeric desert vegetation. This occurs because the hamada is strictly a water runoff region, unlike lower elevational sites (Fig. 2). In regions where the hamadas or mountains are of the appropriate geological substrate, such as sandstone, sand dunes may form downwind. Because of deep infiltration dune areas may also be characterized by phreatophytes.

Kehl (1987) created a schematic representation of landscapes containing oases in Egypt and the Sudan (Fig. 3). In these severely arid climates the species composition and productivity are controlled by groundwater, where available, rather than by surface water, as in Fig. 2. Both the low-lying oases and the dunes surrounding them support phreatophytes.

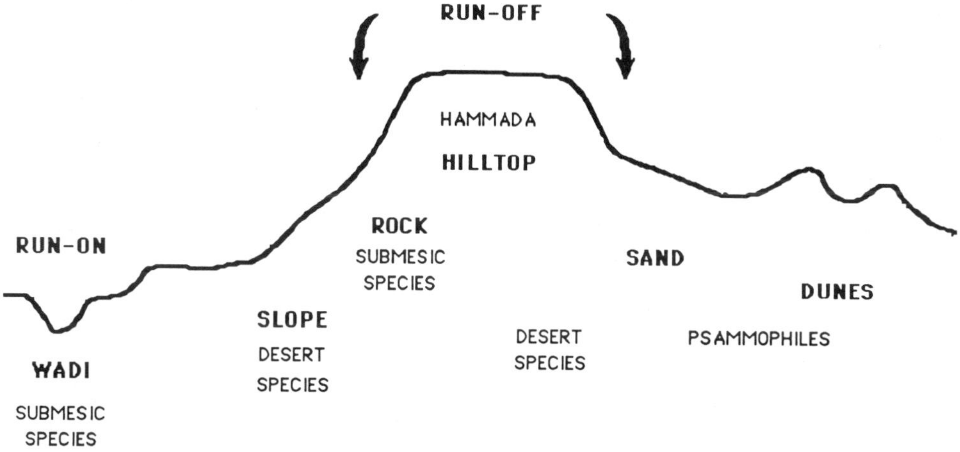

FIGURE 2 A desert landscape showing the main habitat types. (Redrawn from Shmida et al., 1986.)

At a distance from the oasis where the groundwater is deeper the plants become sparse to nonexistent. Localized soil-plant relationships are also important in explaining the distribution of plants across landscapes. For instance, the heavy silty clay soils of the Libyan Plateau preclude plant establishment even where precipitation and salinity may not be limiting (Kehl et al., 1984). In the Sonoran Desert plant species composition is also related to soil particle size distribution and salinity (Phillips and MacMahon, 1978).

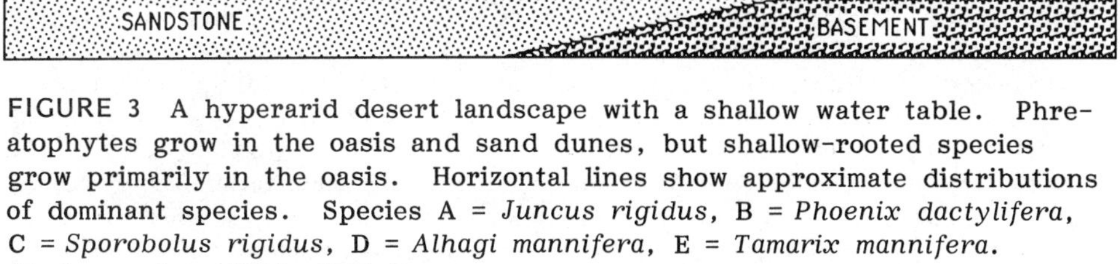

FIGURE 3 A hyperarid desert landscape with a shallow water table. Phreatophytes grow in the oasis and sand dunes, but shallow-rooted species grow primarily in the oasis. Horizontal lines show approximate distributions of dominant species. Species A = *Juncus rigidus*, B = *Phoenix dactylifera*, C = *Sporobolus rigidus*, D = *Alhagi mannifera*, E = *Tamarix mannifera*. (Redrawn from Kehl, 1987.)

TABLE 1 Comparison of Soil Characteristics Under and Between Shrubs of *Artemisia tridentata*

	Under shrub	Interspace
P, mg kg^{-1} (bicarb. extr.)	45 (11)[a]	10 (5)
Organic matter (mg kg^{-1})	94 (5)[a]	54 (8)
VAM spore counts (X 10^4 kg^{-1})	15 (2.4)[a]	9.3 (3.5)
VAM hyphal length (mg kg^{-1})	2.9 (0.7)[a]	0.8 (0.5)
Total hyphal length (mg kg^{-1})	369 (50)[a]	71 (23)

[a]Significantly different at $P > 0.95$ using a Mann-Whitney U test.

Source: Data from Allen and MacMahon (1985).

Islands of Fertility

The plants of the desert communities just described, with the possible exception of moist sites like arroyos, playas, or oases, tend to be widely spaced. They have in common another feature, that the individuals have higher soil nutrients and organic matter under their canopies than the adjacent bare interspaces. These are called islands of fertility, whose function was possibly first described in 1970 by Garcia-Moya and McKell. Since then there have been many detailed descriptions and experimental studies of the structure and function of these islands. Because nitrogen and organic matter are limiting in the desert, the studies have focused on these. Nitrogen is higher under shrubs than in the interspaces or in the deep soil far beneath studied shrubs, ostensibly because of nutrient "mining" by the roots, while organic matter accumulates from litterfall (Charley and West, 1975; West and Klemmedson, 1978; West, 1989). Moisture was higher under than between shrubs in an Australian shrubland, probably because of greater infiltration (Specht, 1981). More recently, mycorrhizal fungi and phosphorus were also found in higher concentrations under than between shrubs (Table 1) (Allen and MacMahon, 1985).

Shrub islands have an impact on smaller, less abundant, and ephemeral plant species. Rather than competing with the shrubs, in many communities herbaceous species tend to be more abundant under than between desert shrubs (e.g., Went, 1942; Barth and Klemmedson, 1978; Garcia-Moya and McKell, 1970; Parker, 1984). Seedlings of *Artemisia herba-alba* were denser at the edge of its canopy than in interspaces, although seedling emergence was reduced immediately at the base of mature plants (Friedman and Orshan, 1975). Conflicting patterns of understory vegetation establishment also occurred under the annual *Salsola kali*, where in harsh environments it acted as a "nurse plant" and facilitated seedling establishment; in more benign environments, where it colonized densely, later seral species were not able to establish as successfully (E. B. Allen, 1989). The conflicting patterns of denser versus sparse plant colonization observed under shrubs are likely due to interactions and relative availability of nutrients, water, and microorganisms and the degree of microclimatic amelioration by the shrubs.

B. Vertical Patterns

Canopy Height

Arid land plant communities are not nearly as diverse in canopy heights as more mesic communities, yet some important patterns have been observed. First, as implied by the studies on colonization of plants under and between shrubs, distinct canopies and understories do exist. Down a bajada gradient in the Sonoran Desert the vegetation became increasingly complex, with one vegetation layer in the upland but four to five layers in the arroyo (Vander Wall and MacMahon, 1984). In this study the increasing complexity of the vegetation was related to increasing bird species diversity. Interestingly, the same understory shrub occurred all along the bajada, and the increasing complexity was due to the addition of new species rather than any substitution. This implies there is no dependence of the lower canopy vegetation on the upper canopy species, and perhaps little aboveground interaction. In this case, simulating the structural height diversity of the native community for restoration purposes would be most important for the sake of the birds rather than for the sake of the plants. However, the herbaceous understory may be dependent on the shrub overstory, as already reviewed, and some species cannot be successfully established without first re-creating islands of fertility.

Rooting Depth

Varied rooting depths are known to occur in many ecosystems and are thought to be one aspect of resource partitioning (Parrish and Bazzaz, 1976). During excavations of coarse roots, Cody (1989) observed the many different rooting strategies of desert plants, including shallow, coarsely rooted succulents, fibrous, relatively shallow rooted grasses, deeply taprooted herbs, and deeper rooted shrubs and phreatophytes. The water relations of some co-occurring desert plants have been worked out, but root depths in relation to soil moisture are seldom known with any precision (Drew, 1979; Nilsen et al., 1984). Rooting depth is probably related to phenology to some degree, with shallow-rooted annuals finishing their life cycles and senescing earlier than deeper rooted perennials (Allen and Knight, 1984; Kemp, 1983). To maximize the use of soil water and resources, it is certainly advisable to select species for revegetation that simulate the natural diversity of rooting strategies in any community. In the absence of more detailed information on resource partitioning, such advice may be the most reasonable course of action to follow.

More recently, studies have been undertaken to examine the microorganisms of deeply rooted plants. Mycorrhizal fungi, rhizobia, nematodes, and other microorganisms were found at 4-5 m depth, just above the water table, in *Prosopis juliflora* in the Chihuahuan Desert (Virginia et al., 1986). In an undisturbed *Artemisia* steppe, mycorrhizal fungi were found as deep as 1 m, penetrating cracks in the sandstone. Alternatively, in disturbed *Artemisia* steppe, replanted shrubs did not have mycorrhizal fungi in their deeper roots even after 4 years (M. F. Allen, 1988a). This suggests one of the difficulties of restoring arid lands. The shallow roots of shrub seedlings may limit their establishment, because deep-rooted shrubs rely on deeply stored moisture for survival. When these shallow roots are combined with an even slower growing microflora, the potential for survival may diminish further. Some plants of *Prosopis* that were known to colonize in approximately the past half-century had not yet formed deep microbial associations, suggesting that the microorganisms move downward at a much slower rate than the roots (Freckman and Virginia, 1989 and pers. comm.).

III. TEMPORAL ORGANIZATION

A. Desert Plants and Precipitation

Precipitation and Temperature Patterns

The productivity of deserts and semiarid lands is limited by the abundance and variability of precipitation to a greater extent than in other biomes (Evans and Thames, 1981). Incorporating the concept of effective precipitation may be more important in deserts than in other biomes because of the high evaporative demand due to high temperatures and/or high runoff due to low vegetation cover. A useful world map of temperature and precipitation, which slows both seasonality and abundance of precipitation, was designed by Meigs (1953). A glance at the map shows that the equatorial hot deserts, grasslands, and savannas are characterized by summer precipitation coupled with hot summer temperatures, but the temperate cold deserts, shrub steppes, and Mediterranean shrublands have cool or cold wet winters with dry, warm to hot summers. Intermediate latitudes often have bimodal, winter plus summer precipitation, and may be shrublands or grasslands, depending on the season of prevalent moisture and local adaptations of plants. Climate diagrams (Walter, 1979) give a more detailed account of the effective precipitation in each area and are useful tools for understanding and restoring local vegetation patterns.

Effects of Variability of Precipitation

Plants are adapted not only to the climatic means, but to the extremes in any location. Deserts and semiarid lands are known for their especially high year-to-year coefficient of variation in precipitation (MacMahon, 1980), posing a unique stress on desert plants. Extreme spatial patchiness of precipitation may also occur in deserts that depend on convective precipitation, with areas receiving several centimeters adjacent to others that receive none (e.g., Evenari et al., 1982). Deep-rooted mature plants may survive in extreme deserts that do not receive any precipitation for periods of several years (Alaily et al., 1987; Bornkamm, 1987; Haynes et al., 1987). Alternatively, prolonged drought may cause mortality of existing plants and shifts in vegetation types to more drought-adapted species as during one of the recent droughts in the African Sahel (Breman and Cisse, 1977) and during the great drought of the 1930s in the North American Great Plains (Weaver and Albertson, 1944). In the case of the American drought, Weaver and Albertson documented the demise of all perennial plants in sections of the shortgrass prairie and their replacement by annual weeds during the worst years of the drought.

The year-to-year variability in precipitation is even more stressful to shallow-rooted seedlings than to mature plants. Observations in the Mojave Desert of North America indicated that high-precipitation years were coupled with high seedling establishment and led Went (1949) to document the "pulse phenomenon" of seedling establishment. The soil surface moisture requirements of seedlings during their first few weeks are higher than those of the mature plant. Although mature desert plants may survive in even the driest years, seedlings establish only during the wettest. The moisture requirements of different species vary, so some establish every year but others establish only every decade or even century. Infrequent establishment events were hypothesized for plants of southwestern Egypt, where most of the known 16 plant species (mainly shrubs) were found dead during various surveys (Alaily et al., 1987). Shrub seedling establishment was noted for the first time in modern history in 1978 in the eastern Sahara, an event that was correlated with groundwater recharge and cloud cover detected by satellite imagery (Haynes et al., 1987). Jordan and Nobel (1981) noted an even-aged stand of *Ferocactus acanthoides* seedlings in the Mojave Desert and used photosynthetic modeling to hypothesize that the seedlings were established 4 years previously during a wet winter followed by a wet summer. Weaver and Albertson (1944) noted the establishment of *Bouteloua gracilis* seedlings when the great drought of the 1930s broke in the North American shortgrass prairie, a rare establishment event that present-day researchers have not observed (B. E. Bement, personal observations). West et al. (1979) concluded that the pulse phenomenon may not be as important in the cold semidesert of southern Idaho, as much precipitation comes in the form of snow in the spring and is therefore more effective than rainfall that runs off and evaporates in climates with summer precipitation.

For the purpose of land reclamation, knowledge of both the amount and the variability of precipitation are obviously important. The minimum amount determines what mature plants can be expected to survive, and the variability dictates what seedlings can be established during dry and wet years and what irrigation or water harvesting regimes need to be imposed. Unfortunately, the minimum moisture requirements for plant establishment have been determined for very few species (e.g., Wilson and Briske, 1978), although the plant water potentials of seedlings that are several months old were much lower than expected, down to -130 bar (Oechel, 1988).

Seasonal Response to Precipitation

Plant phenology is tightly coupled to precipitation. In a comprehensive survey of Chihuahuan Desert species, Kemp (1983) documented that certain species could always be observed, with maximum germination and productivity at set times of the year depending on precipitation and temperature. In the shortgrass prairie C_3 annuals and perennials tended to germinate earlier in the growing season than C_4 plants, and the competitive interactions between C_3 and C_4 plants were dependent on temperature (Allen, 1982). Went (1949) noted that cool season annuals tended to germinate and establish during winter precipitation, and warm season annuals established during periods of summer precipitation. Went's observations occurred before the C_3 and C_4 photosynthetic pathways were discovered and correlated with seasonality.

Years after the discovery of the C_4 pathway there is still debate over the adaptive values of the two pathways. The high water use efficiency of C_4 plants make them ideal for arid conditions, yet C_3 plants may be more drought stress tolerant than within-community C_4 plants, as in the case of *Agropyron smithii* (C_3) and *Andropogon hallii* (C_4) in Nebraska sandhill prairie (Barnes and Harrison, 1982). Teeri and Stowe (1976) showed that the highest diversity of C_4 grasses in North America was correlated with high temperatures rather than low precipitation, since wet regions in the southeastern United States had as high a percentage of C_4 plants as dry regions in the Southwest. For the purposes of restoration, selecting a mixture of plants that reproduces the seasonality of the native community would be desirable to maintain productivity throughout the growing season.

Another factor as important as photosynthetic pathway that determines phenology is rooting depth. Phreatophytes are "uncoupled" from the precipitation regime because their roots extend to the water table. There are many examples of phreatophytes in deserts, including plants of oases (Kehl et al., 1984), dunes (Virginia et al., 1986), and washes, wadis, or arroyos (Nilsen et al., 1984). Other species, such as the deep-rooted shrubs found in shrub steppe, may have a longer lasting water supply than the more shallow rooted, co-occurring grasses. The puzzling coexistence of the C_4 *Atriplex confertifolia* with C_3 grasses in cold desert in Utah was solved by Caldwell et al. (1977). Summer precipitation is scarce in this region, so one would predict C_4 plants not to be abundant. However, *A. confertifolia* had optimum temperatures for photosynthesis that were considerably lower than expected for a C_4 plant. It begins growth almost as early as the C_3 grasses and can maintain productivity later during the growing season, presumably because of its deeper roots and water-conserving habit.

Year-to-year variation in precipitation and temperature affects the timing of various phenological events as well as the productivity of plants. The net annual productivity of some desert ecosystems varied by more than an order of magnitude, even between sequential years, if precipitation was markedly variable (Ludwig, 1986). The onset of flowering and seed set were hastened by higher temperatures and low moisture in several Great Basin grass species, and in years of extreme drought many plants fail to produce seeds at all (Sauer and Uresk, 1976; E. B. Allen, personal observations). Not surprisingly, the phenology of roots is also coupled with the precipitation patterns and photosynthetic pathway of plants (Fernandez and Caldwell, 1975). In one instance inoculation with mycorrhizal fungi caused delayed phenology of *A. smithii* in the field, in conjunction with a more favorable water balance (Allen and Allen, 1986).

The phenology of soil microorganisms is, not surprisingly, tightly coupled with the plants. This is especially true in the case of mutualistic microorganisms, such as mycorrhizal fungi. Maximum sporulation of the same species of fungi occurred early in the growing season in the rhizosphere of C_3 grasses and later for C_4 grasses in the Nebraska sandhills (Allen et al., 1984). The mycorrhizae of *Atriplex gardneri* form arbuscules for several weeks in the spring after snowmelt in cold desert, then they form vesicles, and sporulation begins as the soil dries in early summer (M. F. Allen, 1983).

The phenology, rooting depth, and local physiological adaptations of plants must all be considered in restoring deserts and semiarid lands. There are numerous anecdotes about failures in revegetation (unpublished, because failures are seldom published!) that may be related to phenology. For instance, *B. gracilis* (C_4) failed to establish in a revegetated coal mine in southeast Wyoming in an area that receives little summer precipitation, and *A. smithii* (C_3) failed to establish in northern New Mexico in an area that receives little winter and mostly summer precipitation (E. B. Allen, personal communications with reclamation personnel). Deep-rooted shrubs are often more difficult to establish than shallow-rooted grasses, which discourages their use in favor of grasses. However, recent successes in shrub establishment have been reported (McKell, 1989).

B. Succession

Seral Stages

"Autosuccession," the replacement of late successional species by the same species after disturbance, has become synonymous with desert succession (Wells, 1961; MacMahon, 1981). However, autosuccession occurs only in the most arid deserts (e.g., Carpenter et al., 1986; MacMahon, 1981), and even there certain "late seral" species are recognized as superior colonizers that establish at a disturbed site more rapidly (Prose et al., 1987). The introduction of Eurasian weeds, such as *S. kali*, to North America has added a pioneer seral stage to desert ecosystems where there was none, or only a very sparse one (Allen and Knight, 1984). This places *S. kali* and other alien colonizers in a unique strategy according to Grime's (1979) triangle of plant life history strategies, a ruderal that withstands stress.

Unlike the desert, several definitive stages of succession have been observed in semiarid lands. For instance, in *Artemisia* steppe some typical stages after abandonment from agriculture are pioneer annuals, perennial forbs, colonizing perennial grasses, climax perennial grasses, and shrub-grass (Allen and Knight, 1984; E. B. Allen, 1988). Recovery from severe drought resulted in similar stages in the shortgrass prairie, with the exception of the shrub stage (Costello, 1944). Where the disturbance to the soil is not as severe, the return time to the "climax" vegetation has been estimated at 50 years or less in these ecosystems (e.g., Costello, 1944), a much shorter time than that of more mesic forested ecosystems in which climax depends on growth to maturity of large trees. MacMahon (1981) suggested that the time to achieve climax in deserts and semiarid lands is limited by the ability of plants to establish under low and variable precipitation, but once plants are established succession proceeds rapidly through the few stages.

Trajectories of Succession

Not all successions can be expected to result in climax vegetation. Different kinds of disturbances, the imposition of continual stress, and the colonization of alien species may cause changes in the rate, pattern, or trajectory of succession (E. B. Allen, 1988). Different rates of succession (Fig. 4a) are illustrated by observations of grassland succession during the drought of the 1930s, when the annual weed stage was lengthened in duration because of drought (Costello, 1944; Weaver, 1954). The degree of disturbance was also related to the rate of recovery in the Mojave Desert (Webb et al., 1987).

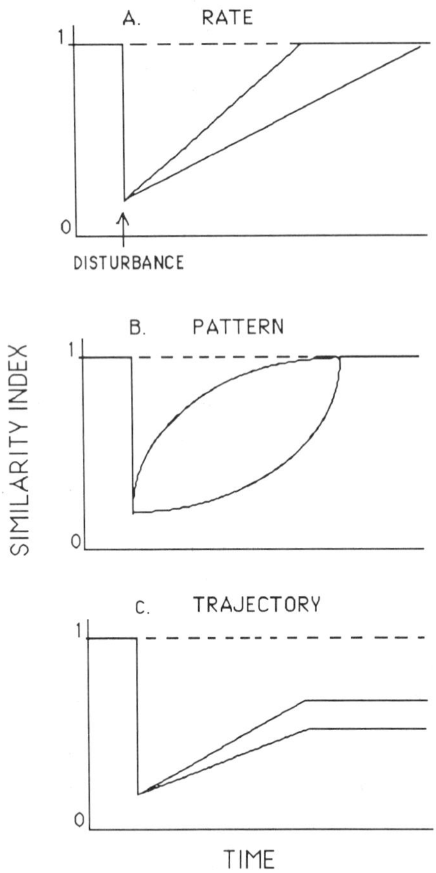

FIGURE 4 Three models showing the course of succession when different rates, patterns, or trajectories are involved. In the first two models a disturbed site resembles the original vegetation with time, but in the third the predisturbance vegetation is never achieved. (Redrawn from E. B. Allen, 1988.)

Different patterns of succession followed by convergence (Fig. 4b) occurred in North American grasslands that were colonized by different species of colonizing annuals, but these were succeeded by the same species of late seral grasses (e.g., Weaver and Albertson, 1944; Weaver, 1954). Inoculation of several disturbed sites with mycorrhizal fungi did not change the species that colonized but did change their relative composition (E. B. Allen, 1989; Allen and Allen, 1986, 1988). The densities of different plant species sometimes increased and sometimes decreased with inoculation, depending on the density of nonmycorrhizal colonizing annuals, soil moisture, position on the slope, and other factors (Allen, 1989).

Different trajectories of succession (Fig. 4c) appear to be common in deserts and semiarid lands. Introductions of the Eurasian weeds *Avena* spp. in California (Heady, 1977) and *Bromus tectorum* in the Great Basin (Hironaka and Tisdale, 1963; Mack, 1981) have caused apparently permanent reversions of perennial native vegetation to annual grasslands. Even the annual *S. kali*, which typically persists for only 3-4 years, caused changes in shrub steppe vegetation that lasted beyond 10 years because it inhibited the establishment of shrubs during the early stages of succession (Allen and Knight, 1984; E. B. Allen, 1988). Purposeful introductions of perennial forage grasses have persisted as nearly pure stands for 40-50 years (Bock et al., 1986; Johnson, 1986). The worldwide problem of desertification is causing some irreversible changes because the stress of human population pressure is not alleviated after vegetation is removed; the vegetation does not return either by natural succession or restoration efforts, and the topsoil is lost (Glantz and Katz, 1987).

Formation of Islands

Large-scale disturbances of the desert tend not only to deplete the soil of nutrients and organic matter but also to homogenize soil resources across communities and landscapes (Allen and MacMahon, 1985; M. F. Allen, 1988a). Questions that have received recent attention are how the islands of fertility form initially and what mechanisms are involved. A recent review by Garner and Steinberger (1989) summarizes the mechanisms for nitrogen transport. These include both biological and physical mechanisms. The plants themselves transport N from interspaces and deep soil and deposit their leaf and root litter primarily at the base of the plant (West, 1989). Animals of all sizes concentrate N near plants in the course of their normal use of individual plants. The physical processes include wind and water.

An experimental approach to island formation was taken on a revegetated surface mine in Wyoming (M. F. Allen, 1988a, b). *Artemisia tridentata* seedlings began to show accumulation of phosphorus and mycorrhizal fungi around their bases after only 4 years. Allen showed experimentally that the shrubs changed the airflow dynamics sufficiently to allow deposition of mycorrhizal fungi and, by implication, other microorganisms, organic matter, and nutrients as well. Thus the all-important process of "self-fertilization" can be restored in a disturbed site if the appropriate species are reintroduced at appropriate densities.

The importance of understanding and simulating natural processes of succession for successful restoration or reclamation has been emphasized (MacMahon, 1983). Restoration is, in fact, simply a form of intervention to hasten the rate of succession. Most often restoration consists of revegetating with late seral species. However, in most situations the soil is disturbed and consists of early seral microorganisms. Where plants are uniformly distributed in the planting process, as in attempts to create pastures from deserts and semideserts, there is no opportunity for fertile islands to re-form. Initial bursts of productivity are often followed by declines (Whitford, 1988). These declines may be attributed to a low average productivity, reductions in soil microorganisms, and decreased decomposition rates (Whitford, 1988; M. F. Allen, 1988a). Such failures can be averted by an understanding of the need for fertile islands to form in arid lands and by implementing management practices to re-create these islands.

IV. SUMMARY

Plant community organization of deserts and semiarid lands is unique in many respects. Spatial organization is characterized by manyfold increases in plant productivity downslope from arid hilltops to washes or wadis with run-on water. Arid lands are known for widely spaced plants, creating islands of fertility around individual plants. Temporal organization is created by variable and patchy precipitation that causes shifting patches of productivity, depending on sites of cloudbursts and on the ability of plants to establish during years of varying precipitation. This pulse phenomenon of plant establishment also dictates the rate of succession. Since seral stages tend to be few in arid lands, succession proceeds rapidly after the initial establishment events. The goal for restorationists and reclamationists is to understand and simulate these spatial and temporal patterns.

ACKNOWLEDGMENTS

I thank Michael F. Allen for reviewing the manuscript. This synthesis was compiled while the author was supported by NSF grants BSR-8818076 and 8818266 and USDA grant 88-37264-4026.

REFERENCES

Alaily, R., Bornkamm, R., Blume, H. P., Kehl, H., and Zielinski, H. (1987). Ecological investigations in the Gilf Kebir (SW-Egypt). *Phytocoenologia 15*: 1-20.

Allen, E. B. (1982). Germination and competition of *Salsola kali* with native C_3 and C_4 species under three temperature regimes. *Bull. Torrey Bot. Club 109*: 39-46.

Allen, E. B. (1988). Some trajectories of succession in Wyoming sagebrush grassland: Implications for restoration. In *The Reconstruction of Disturbed Arid Lands. An Ecological Approach*. E. B. Allen (Ed.). Westview Press, Boulder, Colorado, pp. 89-112.

Allen, E. B. (1989). The restoration of disturbed arid landscapes with special reference to mycorrhizal fungi. *J. Arid Environ. 17*: 279-286.

Allen, E. B., and Allen, M. F. (1986). Water relations of xeric grasses in the field: Interactions of mycorrhizas and competition. *New Phytol. 104*: 559-571.

Allen, E. B., and Allen, M. F. (1988). Facilitation of succession the nonmycotrophic colonizer *Salsola kali* (Chenopodiaceae) on a harsh site: Effects of mycorrhizal fungi. *Am. J. Bot. 75*: 257-266.

Allen, E. B., and Knight, D. H. (1984). The effects of introduced annuals on secondary succession in sagebrush-grassland, Wyoming. *Southwest. Naturalist 29*: 407-421.

Allen, M. F. (1983). Formation of vesicular-arbuscular mycorrhizae in *Atriplex gardneri* (Chenopodiaceae): Seasonal response in a cold desert. *Mycologia 75*: 773-776.

Allen, M. F. (1988a). Belowground structure: A key to reconstructing a productive arid ecosystem. In *The Reconstruction of Disturbed Arid Lands. An Ecological Approach*, E. B. Allen (Ed.). Westview Press, Boulder, Colorado, pp. 113-135.

Allen, M. F. (1988b). Re-establishment of VA mycorrhizas following severe disturbance: Comparative patch dynamics of a shrub desert and a subalpine volcano. *Proc. Roy. Soc. Edinburgh 94B*: 63-71.

Allen, M. F., and MacMahon, J. A. (1985). Impact on disturbance on cold desert fungi: Comparative microscale dispersion patterns. *Pedobiologia 28*: 215-224.

Allen, M. F., Allen, E. B., and Stahl, P. D. (1984). Differential niche response of *Bouteloua gracilis* and *Pascopyrum smithii* to VA mycorrhizae. *Bull. Torrey Bot. Club 111*: 361-365.

Barnes, P. W., and Harrison, A. T. (1982). Species distributions and community organization in a Nebraska Sandhills mixed prairie as influenced by plant/soil-water relationships. *Oecologia (Berl.) 52*: 192-201.

Barth, R. C., and Klemmedson, J. O. (1978). Shrub-induced spatial patterns of dry matter, nitrogen and organic carbon. *Soil Sci. Soc. Am. Proc. 42*: 804-809.

Bock, C. E., Bock, J. H., Jepson, K. L., and Ortega, J. P. (1986). Ecological effects of planting African lovegrasses in Arizona. *Natl. Geogr. Res. 2*: 456-463.

Bornkamm, R. (1987). Growth of accidental vegetation on desert soils in Southwest Egypt. *Catena 14*: 267-274.

Breman, H., and Cisse, A. M. (1977). Dynamics of Sahelian pastures in relation to drought and grazing. *Oecologia (Berl.) 28*: 301-315.

Caldwell, M. M., White, R. S., Moore, R. T., and Camp, L. B. (1977). Carbon balance, productivity and water use of cold-winter desert shrub communities dominated by C_3 and C_4 species. *Oecologia (Berl.) 29*: 275-300.

Carpenter, D. E., Barbour, M. G., and Bahre, C. J. (1986). Old field succession in Mojave Desert shrub. *Madrono 33*: 111-122.

Charley, J. L., and West, N. E. (1975). Plant-induced soil chemical patterns in some shrub-dominated semi-desert ecosystems of Utah. *J. Ecol. 63*: 945-964.

Cody, M. L. (1989). Growth-form diversity and community structure in desert plants. *J. Arid Environ. 17*: 199-209.

Costello, D. F. (1944). Natural revegetation of abandoned plowed land. *Ecology 25*: 312-326.

Crawford, C. S., and Gosz, J. R. (1982). Desert ecosystems: Their resources in space and time. *Environ. Conserv. 9*: 181-194.

Dorn, R. I. (1988). A rock varnish interpretation of alluvial-fan development in Death Valley, California. *Natl. Geogr. Res. 4*: 56-73.

Drew, M. C. (1979). Root development and activities. In *Arid-land Ecosystems*, D. W. Goodall and R. A. Perry (Eds.). Cambridge University Press, London, pp. 573-606.

Evans, D. D., and Thames, J. L. (1981). *Water in Desert Ecosystems*. Dowden, Hutchinson and Ross, Stroudsburg, Pennsylvania.

Evenari, M., Shanan, L., and Tadmor, N. (1982). *The Negev: The Challenge of a Desert*. Harvard University Press, Cambridge, Massachusetts.

Fernandez, O. A., and Caldwell, M. M. (1975). Phenology and dynamics of root growth of three cool semi-desert shrubs under field conditions. *J. Ecol. 63*: 703-714.

Freckman, D. W., and Virginia, R. A. (1989). Plant feeding nematodes in deep-rooting desert ecosystems. *Ecology 70*: 1665-1678.

Friedman, J., and Orshan, G. (1975). The distribution, emergence and survival of *Artemisia herba-alba* in the Negev Desert of Israel in relation to distance from the adult plant. *J. Ecol. 63*: 627-632.
Garcia-Moya, E., and McKell, C. M. (1970). Contribution of shrubs to the nitrogen economy of a desert wash plant community. *Ecology 51*: 81-88.
Garner, W., and Steinberger, Y. (1989). A proposed mechanism for the formation of "fertile islands" in the desert ecosystem. *J. Arid Environ. 16*: 257-262.
Glantz, M. H., and Katz, R. W. (1987). African drought and its impacts: Revived interest in a recurrent phenomenon. *Desertification Control Bull. 14*: 22-30.
Grime, J. P. (1979). *Plant Strategies and Vegetation Processes.* John Wiley, New York.
Haynes, V. C., Megringer, P. J., Johnson, D. L., Haas, H., Muller, A. B., Zaghloul, E. S., Swedan, A., and Wyeman, T. A. (1987). Evidence for the first nuclear-age recharge of shallow groundwater, Arba'in desert, Egypt. *Natl. Geogr. Res. 3*: 431-438.
Heady, H. F. (1977). Valley grasslands. In *Terrestrial Vegetation of California,* G. Barbour and J. Majors (Eds.), John Wiley, New York, pp. 491-514.
Hironaka, M., and Tisdale, E. W. (1963). Secondary succession in annual vegetation in southern Idaho. *Ecology 44*: 810-812.
Johnson, K. L. (Ed.) (1986). Crested wheatgrass: Its values, problems, and myths. Symposium Proceedings, Utah State University Range Science Department, Logan, Utah.
Jordan, P. W., and Nobel, P. S. (1981). Seedling establishment of *Ferocactus acanthodes* in relation to drought. *Ecology 62*: 901-906.
Kehl, H. (1987). Zonation and establishment of vegetation in selected uninhabited Egyptian and Sudanese oases. *Catena 14*: 275-290.
Kehl, H., Stahr, K., and Gauer, J. (1984). Soil-vegetation relationship of a small catchment area on the Libyan Plateau in N.W. Egypt. *Berl. Geowissensch. Abhandl. 50*: 303-324.
Kemp, P. R. (1983). Phenological patterns of Chihuahuan desert plants in relation to the timing of water availability. *J. Ecol. 71*: 427-436.
Ludwig, J. A. (1986). Primary production variability in desert ecosystems. In *Pattern and Process in Desert Ecosystems,* W. G. Whitford (Ed.). University of New Mexico Press, Albuquerque, New Mexico, pp. 5-18.
Mack, R. N. (1981). Invasion of *Bromus tectorum* into western North America: An ecological chronicle. *Agro-Ecosystems 7*: 145-165.
MacMahon, J. A. (1980). Ecosystems over time: Succession and other types of change. In *Fresh Perspectives from Ecosystem Analysis,* R. H. Waring (Ed.). Oregon State University Press, Corvallis, Oregon, pp. 27-58.
MacMahon, J. A. (1981). Successional processes: Comparisons among biomes with special reference to probable roles of and influences on animals. In *Forest Succession,* D. C. West, H. H. Shugart, and D. B. Botkin (Eds.). Springer-Verlag, New York, pp. 277-304.
MacMahon, J. A. (1983). *Nothing Succeeds Like Succession: Ecology and the Human Lot.* Utah State University Press, Logan, Utah, pp. 1-31.

McIntosh, R. P. (1985). *The Background of Ecology.* Cambridge University Press, Cambridge.
McKell, C. M. (Ed.) (1989). *The Biology and Utilization of Shrubs.* Academic press, New York.
Meigs, P. (1953). World distribution of arid and semi-arid homoclimates. In *Review of Research on Arid Zone Hydrology. Arid Zone Programme 1.* UNESCO, Paris.
Nilsen, E. T., Sharifi, M. R., and Rundel, P. W. (1984). Comparative water relations of phreatophytes in the Sonoran Desert of California. *Ecology 65*: 767-778.
Oechel, W. C. (1988). Seedling establishment and water relations after fire in a Mediterranean ecosystem. In *The Reconstruction of Disturbed Arid Lands. An Ecological Approach*, E. B. Allen (Ed.). Westview Press, Boulder, Colorado, pp. 34-45.
Parker, M. A. (1948). Decrease in abundance of *Vulpia octoflora* upon removal of *Gutierrezia microcephala* in a New Mexico grassland. *Southwest. Naturalist 29*: 353-354.
Parrish, J. A. D., and Bazzaz, F. A. (1976). Underground niche separation in successional plants. *Ecology 57*: 1271-1288.
Phillips, D. L., and MacMahon, J. A. (1978). Gradient analysis of a Sonoran Desert bajada. *Southwest. Naturalist 23*: 669-680.
Prose, D. V., Metzger, S. K., and Wilshire, H. G. (1987). Effects of substrate disturbance on secondary plant succession: Mojave Desert, California. *J. Appl. Ecol. 24*: 305-313.
Sauer, R. H., and Uresk, D. W. (1976). Phenology of steppe plants in wet and dry years. *Northwest Sci. 50*: 133-139.
Shmida, A., Evenari, M., and Noy-Meir, I. (1986). Hot desert ecosystems: An integrated view. In *Hot Deserts and Arid Shrublands*, M. Evenari et al. (Eds.). Elsevier, Amsterdam.
Specht, R. L. (1981). The water relations of heathlands: Morphological adaptations to drought. In *Heathlands and Related Shrublands: Quantitative Studies. Ecosystems of the World 9B*, R. L. Specht (Ed.). Elsevier, Amsterdam, pp. 123-129.
Teeri, J. A., and Stowe, L. G. (1976). Climatic patterns and the distribution of C_4 grasses in North America. *Oecologia (Berl.) 23*: 1-12.
Vander Wall, S. B., and MacMahon, J. A. (1984). Avian distribution patterns along a Sonoran Desert bajada. *J. Arid Environ. 7*: 59-64.
Virginia, R. A., Jenkins, M. B., and Jarrell, W. M. (1986). Depth of root symbiont occurrence in soil. *Biol. Fertil. Soil 2*: 127-130.
Walter, H. (1979). *Vegetation of the Earth and Ecological Systems of the Geobiosphere*, 2nd ed. Springer-Verlag, New York.
Weaver, J. E. (1954). A seventeen year study of plant succession in prairie. *Am. J. Bot. 41*: 31-38.
Weaver, J. E., and Albertson, F. W. (1944). Nature and degree of recovery of grassland from the great drought of 1933 to 1940. *Ecol. Monogr. 14*: 393-479.
Webb, R. H., Steiger, J. W., and Turner, R. M. (1987). Dynamics of Mojave Desert shrub assemblages in the Panamint Mountains, California. *Ecology 68*: 478-490.

Wells, P. V. (1961). Succession in desert vegetation on streets of a Nevada ghost town. *Science 134*: 670-671.
Went, F. W. (1942). The dependence of certain annual plants on shrubs in southern California. *Bull. Torrey Bot. Club 69*: 100-114.
Went, F. W. (1949). Ecology of desert plants. *Ecology 30*: 1-13.
West, N. E. (1989). Spatial pattern-functional interactions in shrub-dominated plant communities. In *The Biology and Utilization of Shrubs*, C. M. McKell (Ed.). Academic Press, New York, pp. 283-305.
West, N. E., and Klemmedson, J. O. (1978). Structural distribution of nitrogen in desert ecosystems. In *Nitrogen in Desert Ecosystems*, N. E. West and J. J. Skujiņš (Eds.). Dowden, Hutchinson and Ross, Stroudsburg, Pennsylvania, pp. 1-16.
West, N. E., Rhea, K. H., and Harniss, R. O. (1979). Plant demographic studies in sagebrush grass communities of southeastern Idaho. *Ecology 60*: 376-388.
Whitford, W. G. (1988). Decomposition and nutrient cycling in disturbed arid ecosystems. In *The Reconstruction of Disturbed Arid Lands*, E. B. Allen (Ed.). Westview Press, Boulder, Colorado, pp. 136-161.
Wierenga, P. J., Hendricks, G. H., Nash, M. H., Ludwig, J., and Daugherty, L. (1987). Variation of soil and vegetation with distance along a transect in the Chihuahuan Desert. *J. Arid Environ. 13*: 53-63.
Wilson, A. M., and Briske, D. D. (1978). Drought and temperature effects on the establishment of blue grama seedlings. Proceedings of the First International Rangeland Congress, pp. 359-361.

7

Soil Microbiology in Reclamation of Arid and Semiarid Lands

THOMAS L. KIEFT *New Mexico Institute of Mining and Technology, Socorro, New Mexico*

I. INTRODUCTION

Any consideration of soil ecosystems requires inclusion of the living components. Soil microorganisms have central roles in all aspects of soil processes, including the initial formation of soils from mineral parent material, the establishment of soil structure, and the biogeochemical cycling of elements. It is in this last role that microorganisms most profoundly influence soil processes. The broad metabolic diversity exhibited by soil microbes enables them to participate in the oxidation and reduction reactions that comprise the natural cycles of carbon, nitrogen, sulfur, iron, manganese, and other elements. This microbially mediated cycling of biologically important elements in turn influences the useful properties of soils, such as their ability to support plant life.

These generalizations regarding the importance of soil microorganisms are as valid for arid and semiarid soils as they are for their more mesic counterparts. Desert soils contain all the same functional groups and most, if not all, of the same major taxonomic groups of microorganisms found in soils of more humid regions (Focht and Martin, 1979; Skujiņš, 1984). However, there are qualitative and quantitative differences that set desert soil microbial communities apart from those of other soils. The two factors that are generally most limiting to microbial activities in soil are availability of organic matter and availability of water (Dommergues et al., 1978; Focht and Martin, 1979). As these are often severely limiting in desert soils, lower total numbers and biomass of microbes in desert ecosystems are usually found. Soil microbial biomass is roughly proportional to soil organic matter (Schnurer et al., 1985), and consequently, the mineral soils of arid regions generally support lower masses of microbes. Most of the microbes that are indigenous to desert soils show characteristic physiological and morphological adaptations that enable survival and activity under the harsh conditions of low energy and/or moisture availability. Other stressful conditions in desert ecosystems can include high surface soil temperatures, soil alkalinity, and high salinity. Ecosystems like deserts, where biological processes are controlled more by physical and chemical stresses, such as drought, than by biological interactions, are thought to have a lower species diversity (Noy-Meir, 1985; Atlas and Bartha, 1987). The generalization probably applies to desert soil microbial communities; however, it is difficult to test. In desert soils, as with most other natural environments, only a small fraction of the estimated total number of microbial species can be cultured and identified, making it virtually impossible to determine actual microbial species diversity. It may therefore be more useful to consider the functional diversity of microorganisms. As already stated, all the major functional groups are represented in arid and semiarid soils. However, the physical and chemical conditions may favor some over others, thereby altering the relative contributions or rates of various microbially mediated processes. Finally, qualitative and quantitative differences occur among the microbial communities of the various environments encompassed by the terms "arid" and "semiarid" soils, for example, hot desert soils, Antarctic dry valley soils, saline-sodic soils.

This chapter provides a survey of the various microbial components of soil ecosystems, with specific information given whenever possible on the contributions of these components to desert and semiarid soil processes. Because desert ecosystems have only recently been studied extensively, our understanding of the microbiology of these systems is not always complete. In some cases it is necessary to discuss a particular soil component (e.g., desert soil viruses) with reference to nonarid systems and then attempt to predict the relative contribution of this component. Spatial and temporal patterns of microbial distribution in desert and semiarid systems are presented. Methods for estimating microbial numbers, biomass, and activities in desert soils are discussed. Emphasis is also placed on microbial adaptations to the harsh, even extreme, conditions of arid soils. This chapter does not attempt to review all the scientific literature on the microbiology of desert soils; Skujiņš (1984) can be consulted for a comprehensive review. Certain topics covered in greater detail in later chapters (e.g., desert biological crusts and nitrogen cycling) are only introduced here. Finally, this information is intended to explain microbial processes that may be important in desertification and in the reversal of desertification or the reclamation of disturbed arid lands. In some cases, the microbes may serve as indicators of the current status of a soil or the progress of a reclamation effort; for example, increases in soil microbial biomass may indicate a positive trend toward increased resource availability and plant growth. Increased understanding of the microbiota of desert soils may also lead to the adoption of appropriate environmental manipulations that promote favorable microbial processes, such as microbial stabilization of soil organic matter into humic compounds (Skujiņš and Richardson, 1985).

II. SURVEY OF SOIL MICROORGANISMS

The soil microbiota can be divided into five major taxonomic groups: bacteria, fungi, algae, protozoa, and viruses. The bacteria are the prokaryotic microorganisms (kingdom Monera in the five-kingdom system of Whittaker,

1969), having a relatively simple internal structure. The cyanobacteria (loosely termed blue-green algae) are included among the prokaryotes, although functionally they resemble the eukaryotic algae and are therefore discussed with this group. The actinomycetes are a group of branched, filamentous bacteria that are often treated as distinct from the remainder of the bacteria, based on their morphological and metabolic resemblance to fungi (Alexander, 1977). The eukaryotic soil microbes are the fungi, algae, and protozoa (kingdom Protista). Microscopic metazoan animals can also be considered microorganisms. The viruses are acellular, obligately intracellular parasites of cellular forms of life.

The microbes can also be categorized on a functional basis. The autotrophic microorganisms are those that obtain carbon in inorganic form. These can be further divided into the photoautotrophs, which obtain energy from light (e.g., algae and cyanobacteria), and the chemoautotrophs, which gain energy from the oxidation of inorganic substrates (e.g., nitrifying bacteria and sulfur-oxidizing bacteria). The heterotrophs obtain both energy and carbon from organic matter, and these are the most numerous in soils. Osmotrophs (bacteria, fungi, and some protozoa) transport soluble organic nutrients through their plasma membranes; phagotrophs (most protozoa) engulf particulate matter, including other microbes, for their nutrition.

A further useful classification is that of Winogradsky, who recognized autochthonous and zymogenous populations of soil microbes (Atlas and Bartha, 1987). The autochthonous microbes (originally humus-degrading in Winogradsky's scheme) are those indigenous soil organisms that metabolize the relatively ubiquitous but difficult to degrade polymeric organic matter of soil. Their growth rates on these recalcitrant substrates are generally slow. The zymogenous organisms are opportunists that can grow rapidly on those brief occasions when readily degradable organic matter is present in high concentrations. Autochthonous microbes roughly correspond to K-selected organisms (Andrews and Harris, 1986) and oligotrophs (Poindexter, 1981), which have high affinities for growth-limiting substrates and high starvation survival abilities. Zymogenous microbes correspond to r-selected organisms (Andrews and Harris, 1986) and copiotrophs (Poindexter, 1981), which have high maximum specific growth rates. The majority of desert soil microbes can probably be considered oligotrophs, or at least closer to the K end of an r-K continuum. Bacteria of the genus *Arthrobacter* are oligotrophic microbes adapted to both the low organic carbon and desiccation of arid soils (Cacciari and Lippi, 1987). Bacteria of the genus *Pseudomonas*, which inhabit relatively nutrient-rich rhizosphere habitats, typify the soil copiotrophs.

A. Bacteria

Bacteria (including actinomycetes) are generally the numerically dominant microorganisms in soils (Paul and Clark, 1989), including desert soils (Skujiņš, 1984). The majority of data on densities of bacteria are based on standard plate counts using a limited number of organic growth media. These numbers can therefore be considered underestimates of the total number of heterotrophic bacteria in soil. Estimates of bacterial densities in arid and semiarid soils vary over several orders of magnitude (Table 1).

TABLE 1 Microbial Propagules per Gram of Soil in Desert Surface Soils

Soil	Aerobic bacteria and actinomycetes	Fungi	Reference
Kara-Dum Desert			
Crusty solonetzous soil	8.6×10^4	2.5×10^3	Lobova (1960)
Sandy soil, under vegetation	4.1×10^5	6.3×10^3	Lobova (1960)
PreSaharan desert, Tunisia	$3\text{-}7 \times 10^4$	8×10^3	Skujiņš (1984)
Sahara Desert	$1\text{-}8 \times 10^6$	15	Cameron (1969)
Atacama Desert	<10	0	Cameron (1969)
Sonoran Desert			
Arizona	1.7×10^6	1.6×10^4	Skujiņš (1984)
Baja California	2.5×10^6	4.9×10^3	Cameron (1969)
Chihuahuan Desert	1.7×10^5	2×10^3	Cameron (1969)
Chihuahuan Desert, Jornada	10^6	1.2×10^4	Skujiņš (1984)
Great Basin Desert, Pine Valley	9×10^5	6×10^3	Skujiņš (1984)
Mohave Desert			
Nevada	$0.3\text{-}1.6 \times 10^7$	$0.4\text{-}5 \times 10^4$	Vollmer et al. (1977)
California	9×10^4	3.4×10^5	Cameron (1969)

Direct microscopic counts provide a better estimate of the total number of microorganisms in a soil sample. The major drawback of this method is that no distinction is made between living, actively respiring microorganisms and dead or relatively inactive cells. Modifications of the direct count, for example by preincubation with the electron transport analog, 2-(*p*-iodophenyl)-3-(*p*-nitrophenyl)-5-phenyltetrazolium chloride (INT), and then counting cells that have taken up and reduced the INT to INT-formazan, can give an estimate of the percentage of bacteria that are active (Zimmerman et al., 1978). Using a direct microscopic count method, Bamforth (1984) found $1\text{-}7 \times 10^8$ bacteria g^{-1} soil in the Sonoran Desert of the United States. This is approximately two orders of magnitude higher than plate counts of Sonoran Desert soil bacteria (Cameron, 1969). Several factors may account for this discrepancy. First, a number of the bacteria may be dead or inactive, as is generally the case in natural ecosystems. Normally in soil, the majority of microbes are relatively inactive, and this is undoubtedly especially true for desert soils. The low plate count numbers also support the idea that there are many soil microbes, particularly the starving and the autochthonous or oligotrophic microbes that are resistant to cultivation. Bamforth reported that the majority of bacteria observed in his study were small cocci. These could be starved "dwarf" cells (Bae et al., 1972; Bae and Casida, 1973) and/or inherently small cells (Bakken and Olsen, 1986). Both these groups are typically difficult to isolate, and growth, when it occurs at all, is generally on media of extremely low substrate concentration. Unfortunately, standard plate counts are frequently performed using only rich media. Finally, differences between plate counts in one study and direct counts in another may merely reflect spatial and temporal bacterial variability within desert soils.

The most common bacterial genera cultured from arid soils are *Arthrobacter* (Cacciari and Lippi, 1987), *Corynebacterium, Pseudomonas, Acinetobacter, Bacillus, Proteus,* and *Micrococcus* (Skujiņš, 1984). Endospore-forming bacteria, such as *Bacillus* (aerobic) and *Clostridium* (anaerobic) may constitute a high proportion of desert soil microflora. Reports in the literature range from near zero to over 50% of the soil bacteria being spore formers, predominantly *Bacillus* (Skujiņš, 1984).

Aerobic bacteria probably predominate over anaerobic bacteria in arid soils; however, fewer studies of desert anaerobes have been performed. Anaerobic microsites harboring anaerobic bacteria, for example, within soil aggregates, occur even in highly desiccated soils. Sexstone et al. (1985) found anoxic conditions in the centers of soil aggregates from aerobic soil horizons. Metabolic processes occurring within these sites are fermentation (producing CO_2, H_2, and a variety of organic end products, such as alcohols and volatile fatty acids), denitrification (producing mostly N_2 but also N_2O), sulfate reduction (producing H_2S), and methanogenesis (producing CH_4). Factors that increase soil moisture and soil organic matter can be expected to enhance these anaerobic processes, with possible detrimental effects on soil nutrition (denitrification) and possible benefits (e.g., anaerobic biodegradation of some pollutants that are generally refractory to aerobic microbial processes (Sulflita et al., 1982).

A diversity of myxobacteria (prokaryotes with complex life cycles involving vegetative cells with gliding motility and fruiting structures) have been cultured from arid and semiarid soils (Reichenback, 1970; Larkin and Dunigan, 1973; Brockman, 1976). Brockman isolated 13 species of the genera *Archangium, Cystobacter, Stigmatella, Polyangium,* and *Chondromyces* from arid Mexican soils. The myxobacteria appear to be well adapted to desiccation survival as myxospores (resting cells produced by fruiting bodies) and as microcysts (myxospores coated by slime capsules). A few of the myxobacteria (e.g., *Polyangium*) are cellulolytic, but the majority obtain their nutrients by lysis of other microbes (Reichenbach and Dworkin, 1981).

The autotrophic bacteria comprise the photoautotrophs (i.e., the cyanobacteria discussed later along with the algae) and the chemoautotrophs, which derive energy from oxidation of inorganic compounds. The chemoautotrophs include H_2 oxidizers, sulfur oxidizers, and ammonia and nitrite oxidizers. It can be expected that all these physiological groups of chemoautrophic bacteria exist in arid and semiarid soils, but as with the chemoheterotrophs, their species diversity and their overall numbers and biomass are likely to be lower. As discussed in a later section, the chemoautotrophs are considered as a group to be less drought tolerant than the heterotrophs, and so their contributions to elemental cycles may be somewhat reduced in desert and semiarid soils. However, as in more mesic soils, some of their oxidations may also be carried out by heterotrophs. Specifically, the heterotrophic nitrifiers and heterotrophic sulfur oxidizers may be responsible for more significant proportions of these oxidations in arid systems than in nonarid ones.

B. Actinomycetes

The actinomycetes are bacteria characterized by branched filamentous morphology and the production of desiccation-resistant spores or conidia. In this regard they resemble the fungi, but their filaments are smaller in diameter than fungal hyphae and they are prokaryotic. They function in soil as decomposer organisms with capabilities for degrading relatively recalcitrant polymeric organic compounds, including cellulose, hemicelluloses, lignin, pectin, keratin, and chitin (Alexander, 1977; Williams, 1985). The relatively slow growth of the actinomycetes places them in Winogradsky's autochthonous category. They typically occur in later successional stages following introduction of new organic matter to soil (Alexander, 1977). Because the actinomycetes are favored over other bacteria by conditions of desiccation and high pH (Alexander, 1977), they can be especially important in desert ecosystems, comprising 50% or more of the total soil bacteria (Skujiņš, 1984). Total propagules of desert soil actinomycetes, based on standard plate counts, range as high as 1.9×10^6 actinomycetes g^{-1}. However, these may be overestimates of the numbers of actively respiring actinomycetes, since the majority of actinomycetes in soil may be present as metabolically inactive spores that germinate on plate count media (Alexander, 1977). Bamforth (1984), reporting on direct counts of actinomycetes in Sonoran Desert soils, found total hyphal lengths of 0-90 m g^{-1}. Rao and Venkateswarlu (1983) reported *Streptomyces*, *Micromonospora*, and *Thermoactinomyces* as the dominant actinomycete genera in Indian deserts. Of 50 actinomycete strains isolated from a Russian aridisol, 49 were found to be *Streptomyces* species (Teplyakova and Maksimova, 1957). Although the actinomycetes appear to constitute a major portion of the microbiota of most desert and semiarid soil microbes, low percentages of actinomycetes have also occasionally been reported (Skujiņš, 1984).

The actinomycetes are widely known for their antibiotic production, with as many as 75% of soil populations producing one or more secondary compounds inhibitory to other microorganisms (Alexander, 1977). Although it has often been pointed out that the importance of these antibiotics in nature has yet to be clearly demonstrated (Atlas and Bartha, 1987), there is evidently strong selective pressure to retain the genetic capability for their production. The widespread occurrence of antibiotic resistance among naturally occurring bacterial strains also testifies to the prevalence of this type of microbial warfare or "interference competition" in natural habitats (Wicklow, 1986). (The same arguments can also be applied to antibiotic production in fungi.) A very large number of clinically useful antibiotics, including streptomycin, tetracycline, erythromycin, and chloramphenicol, are produced by various species of *Streptomyces* (Brock and Madigan, 1988). The search for new antibiotic-producing strains continues; actinomycete strains isolated from unusual or extreme habitats, such as arid and semiarid soils, may be especially useful (Labeda, 1986).

Nitrogen-fixing actinomycetes of the genus *Frankia*, which are in symbiotic association with the roots of nonleguminous plants, have received widespread attention in recent years (LeChevalier, 1986). These actinomycete-plant associations are especially favored in nitrogen-poor environments.

C. Fungi

Soil fungi generally account for the majority of the soil microbial biomass and may outweigh the bacteria by factors of 3-10 (Anderson and Domsch, 1973, 1975; Shields et al., 1973; Foster, 1988), although they may be fewer in number than the bacteria when enumerated by standard plate counts (Paul and Clark, 1989). The fungi are a diverse group of eukaryotic, heterotrophic, generally osmotrophic microorganisms that include the slime molds, yeasts, and filamentous forms and often have complex reproductive structures. Many of the soil fungi, like the actinomycetes, are favored under drought conditions as a consequence of their desiccation-resistant spores and the ability of hyphae to grow through air-filled soil pores. Numbers of fungi in soil measured by the standard plate count are generally one to two orders of magnitude below those of the bacteria in desert soils (Table 1). Fungal populations are also generally lower than those for nondesert soils (Skujiņš, 1984). Plate count data for desert fungi range from 0 fungal propagules g^{-1} soil in the Uzboi Desert in southwestern Asia (Lobova, 1960) to 3.4×10^5 propagules g^{-1} in the Mohave Desert in California (Cameron, 1969). Other estimates for fungal numbers in the southwestern United States are 1.2×10^5 propagules g^{-1} for surface soil in the Great Basin Desert during the rainy season and 3.6 to 50×10^4 propagules g^{-1} in Nevada (Skujiņš, 1984). As with the bacterial numbers, desert sites other than those in southwestern North America often show lower total numbers of fungi.

As with the actinomycetes, standard plate counts of fungi may reflect high numbers of relatively inactive fungal spores. A better measure is total hyphal lengths per gram of soil, as observed by direct microscopy. Bamforth (1984) reported fungal hyphal lengths ranging from 0-10 m g^{-1} (unstained) and 10-43 m g^{-1} (phenolic aniline blue stained) for arid Lower Sonoran Desert sites to 18-94 m g^{-1} (unstained) and 7-210 m g^{-1} (phenolic aniline blue stained) for semiarid Upper Sonoran Desert sites. Stained-versus-unstained data give an estimate of the viable cells, if it is assumed that dead or injured cells are able to exclude the dye (Kari, 1986). It would be useful to have hyphal length data for other desert ecosystems, but these measurements have not been made. Another method for estimating the proportion of fungal hyphae that are viable and metabolically active involves staining with fluorescein diacetate (FDA) (Soderstrom, 1977). In this method, nonfluorescent FDA is taken up and hydrolyzed by active cells, leaving fluorescein in the cytoplasm. Viable cells can be counted by epifluorescence microscopy. Estimates of the percentage viability of fungi in desert soils measured by this or another method would be particularly enlightening.

The dominant genera of desert soil fungi are generally the same as those found in moister soils. Rao and Venkateswarlu (1983) reported 80% of fungal isolates from Indian deserts to be of the genera *Aspergillus* and *Penicillium*. Also present were *Cladosporium*, *Alternaria*, *Chaetomium*,

Fusarium, Mucor, and *Helminthosporium*. *Alternaria* was reported as the dominant genus in the surface of a pre-Saharan desert in Tunisia (Skujiņš, 1984). Other commonly encountered desert soil fungal genera that have been recorded are *Alternaria, Curvularia, Stemphylium,* and *Mucor* (Nicot, 1955, 1960). Yeasts appear to be significant components of the microflora in unusual desert habitats and in specialized decomposer niches. Yeasts of the genus *Cryptococcus* have been found to be dominant organisms of Antarctic desert soils (Vishniac and Klinger, 1986), and yeasts have been found in significant numbers in decaying cactus tissue (Barker et al., 1987).

D. Algae and Cyanobacteria

Algae and cyanobacteria occur in virtually any environment with sufficient light and moisture, and even the driest desert soils have the requisite moisture, if only for brief intervals, to support algal growth. Algae and cyanobacteria growing at the surface of desert soils can even be responsible for a major portion of the primary production in these ecosystems. Cyanobacterial crusts can also fix considerable amounts of nitrogen. The significance of these soil surface phototrophs to desert soil processes is discussed in a later chapter. Some desert soil phototrophs flourish in protected habitats, for example, on the underside of light-transmitting calcite, where moisture retention is greater than on exposed soil. These sublithic algae are found in hot deserts and also in Antarctic Desert soils (Friedmann, 1971; Wynn-Williams, 1986). Algae and cyanobacteria also occur in desert ecosystems in symbiotic associations with fungi as lichens. The lichens are common pioneer organisms on rock and soil surfaces.

E. Protozoa

Protozoans are unicellular animals. The free-living protozoa are classified according to their means of motility. The flagellates (Mastigophora) are motile by means of flagella; the amebae (Sarcodina) have pseudopodia; and the ciliates (Ciliata) move by cilia. The Sarcodina can be further subdivided into those with tests (hard shells with holes through which pseudopodia extend) and naked amebae without tests. The Sporozoa are entirely parasitic, often with complex life cycles involving multiple hosts, and are not covered here. Other taxonomic groups overlap with the protozoa. The slime molds have ameboid and/or flagellated feeding stages with protozoan features. The photosynthetic euglenoids are also sometimes claimed by protozoologists, as they are flagellated. Protozoans may be phagotrophic and/or osmotrophic. Phagotrophic forms feed primarily on bacteria, but some also consume yeasts, fungal spores, algae, or other protozoa. Most soil protozoa are able to form starvation- and desiccation-resistant cysts, an essential feature for desert soil existence. As much as 100% of soil protozoa may become encysted during periods of low soil moisture (Parker et al., 1984a). The flagellates and the amebae are the most drought tolerant of the protozoa and are therefore favored over ciliates in arid soils (Alexander, 1977; Stout et al., 1982). The flexible nature of the amebae enables them to remain motile in the thin moisture films of desiccated soils. Ciliate populations have been suggested as useful indicator organisms for relatively mesic moisture conditions (Bamforth, 1980). Soil protozoa can be enumerated by direct microscopic examination or by a multiple-tube-dilution most-probable-number method. Protozoal numbers are often quite low (≤100 cells g^{-1}) in desert soils (Bamforth, 1980). However, higher numbers have been reported in a Chihuahuan Desert soil. Parker et al. (1984a) found 2.5×10^4 naked amebae, 4.9×10^3 flagellates, and 7×10^2 ciliates g^{-1}. Direct microscopic counts of protozoa in Upper Sonoran Desert sites by Bamforth (1984) showed a similar dominance of amebae over other protozoa ($0.14\text{-}3.01 \times 10^3$ flagellates, $0.76\text{-}9.40 \times 10^3$ amebae, and $0.11\text{-}0.97 \times 10^3$ ciliates g^{-1}). Soil protozoan numbers increase in a delayed fashion following increases in their bacterial prey numbers in typical predator-prey population patterns (Clarholm, 1981; Parker et al., 1984a; Elliott et al., 1984a). It is generally considered that protozoa increase rates of organic matter biodegradation by re-releasing organic matter bound in bacterial prey and that they serve as important links to higher trophic levels (Elliott et al., 1984a). Experiments in which soil protozoa are eliminated result in reduced remobilization of nutrients and reduced plant growth (Elliott et al., 1979). Protozoan predation may also increase the rates of growth and substrate metabolism in prey bacteria (Elliott et al., 1986). Relative numbers of protozoan groups are monitored in sewage treatment processes as a measure of the overall health and functioning of the system (Pike and Curds, 1971). When more is known of the functioning and taxonomy of the various groups of soil protozoa, they may serve as useful indicators of ecosystem function.

F. Viruses

Viruses are noncellular obligate intracellular parasites of plant, animal, and microbial cells. They consist of a nucleic acid core (DNA or RNA), a protein coat or capsid, and in some animal viruses, a lipid envelope. Viruses infecting bacterial cells are termed bacteriophages, and they can participate in the transfer of genetic information among bacteria of the same species by transduction. Bacteriophages have been detected in a number of soil bacterial genera, including several that can occur in desert soils: *Arthrobacter, Corynebacterium, Brevibacterium* (Trautwetter and Blanco, 1988), *Rhizobium, Bacillus,* and *Pseudomonas* (Trevors et al., 1987).

Sewage sludge, which may harbor high numbers of viruses, has been proposed as an organic matter amendment for reclamation of disturbed soils (Topper and Sabey, 1986). A concern associated with land application of sewage sludge is the survival of human pathogens, including viruses (e.g., poliovirus and hepatitis A virus), and the public health dangers posed by these agents. Viruses may persist in soils by adhesion to soil particles, or they may be transported to underlying groundwater (Pancorbo et al., 1988). In experiments in which soil was amended with sludge along with two enteroviruses (poliovirus 1 and echovirus 1), Bitton et al. (1984) found that virus survival was primarily affected by desiccation and temperature. Warm dry soil conditions resulted in the shortest survival time (no viruses detected after 8 days). This bodes well for sewage application to desert sites, since heat and desiccation appear to accelerate virus inactivation, but further studies with a variety of viruses and with arid and semiarid soils are warranted. Sewage sludge also poses other threats (e.g., heavy metal contamination).

III. MICROBIAL BIOMASS AND ACTIVITY

Assessment of the success of land reclamation efforts is usually made based on measurements of plant cover and primary productivity, but it may be equally important to measure the biomass and activities of soil microorganisms (Klein et al., 1985; Visser, 1985; Anderson and Domach, 1989). It has been suggested that by measuring changes in soil microbial biomass, one can monitor subtle trends in total soil organic matter (Powlson et al., 1987). Methods for measuring the total mass of living microbial cells in soils include direct microscopic counts, extraction and measurement of cell components common to all microbes, and microbial respiration-based assays.

The merits of direct microscopic measurements have been discussed. Procedural details can be found in Schmidt and Paul (1982). Modifications of direct counts using fluorescently labeled antibodies are useful for enumerating specific populations of microbes (Schmidt, 1974).

One of the most widely used cell components for estimating biomass is adenosine triphosphate (ATP) (Jenkinson et al., 1979). A major advantage of this method is the relative ease of the luciferin-luciferase assay for ATP. The ATP assay has been used to assess microbial activities in recovering mine spoils (Hersman and Temple, 1978, 1979). Recent developments in the extraction of ATP from soils (e.g., Webster et al., 1984) have greatly improved the reliability of the method. However, biomass-ATP ratios vary in response to environmental conditions, especially phosphate and nutrient availability (Ahmed et al., 1982; Fairbanks et al., 1984). An extension of the ATP assay involves extraction and measurement of all three phosphorylated adenosine nucleotide compounds: ATP, adenosine diphosphate (ADP), and adenosine monophosphate (AMP). This enables calculation of the adenylate energy charge (AEC):

$$AEC = \frac{[ATP] + 1/2[ADP]}{[ATP] + [ADP] + [AMP]}$$

The AEC is a measure of the physiological state of the soil microbial biomass considered as a whole. Values above 0.8 indicate metabolically active microbes; values around 0.5 indicate relatively inactive, but surviving microbes; and values below 0.5 indicate severe microbial stress (Brookes, 1986). Brookes et al. (1983) measured the AEC in a nonarid soil and found values of 0.85 in freshly collected moist soil and 0.45 in an air-dried sample. We have found AEC values in a grassland soil ranging from 0.8 shortly after wetting to 0.5 after prolonged desiccation, with transient drops to 0.2 occurring during the early stages of desiccation (Rosacker and Kieft, in press). These data show that the physiological state of a soil microbial community varies greatly in response to moisture status, that wetting-drying cycles cause turnovers in soil microbial populations, that the majority of the microbes are only active during brief favorable periods, and that much of the community can survive extended starvation with an AEC around 0.5. The variability in the AEC also indicates that ATP is less than ideal as a measure of microbial biomass. Adenylate energy charge has not yet been measured in any desert soil systems, although it may be an appropriate measure of microbial response in the recovery of disturbed desert soils.

Other soil microbial cell components that can be extracted and assayed as measures of microbial biomass are nucleic acids (Nannipieri et al., 1986), phospholipids (White, 1986), and cell wall components, such as bacterial peptidoglycan (Millar and Casida, 1970). However, each of these has limitations in that the components may exist outside living microbial cells and the assays are more difficult than the ATP assay.

Two commonly used respiration-based assays of soil biomass are the chloroform-fumigation-incubation method (CFIM) (Jenkinson and Powlson, 1976) and the substrate-induced respiration (SIR) assay (Anderson and Domsch, 1978). In the first of these, a soil sample is fumigated with chloroform to lyse all cells, thereby converting microbial biomass to potential microbial substrate. The chloroform is then removed, the soil is reinoculated, and respiration is measured over a 10-day period. After subtraction of the respiration in an unfumigated control soil, the respiration is directly proportional to the amount of soil biomass carbon originally present. Modifications of the CFIM have been used to estimate microbial biomass nitrogen (Shen et al., 1984; Brookes et al., 1985), sulfur (Saggar et al., 1981), and phosphorus (Hedley and Stewart, 1982). Direct extraction of biomass carbon or nitrogen following fumigation is also used (Brookes et al., 1985; Vance et al., 1987; Tate et al., 1988). Although the CFIM is the most commonly used method at present for biomass measurement, it may not be appropriate for desiccated soils (Sparling et al., 1986; Kieft et al., 1987b), thus limiting the applicability of this method in desert and semiarid soils. Organic carbon amendments also appear to interfere with the CFIM (Martens, 1985). A more generally applicable approach may be to analyze the maximal respiratory response of the soil biomass (SIR method). The SIR method involves adding an excess of a readily degradable carbon substrate (usually glucose) and measuring the initial rate of respiration induced by this substrate. The initial rate of respiration (before microbial growth can occur) is directly proportional to the total biomass carbon present in the soil (Anderson and Domsch, 1978; Smith et al., 1985). West and Sparling (1986) have modified this assay for desiccated soils, making it more useful than the CFIM for use in desert soils. The SIR assay has the further advantage of yielding data within 1 day. It should be noted that poisoned controls may be necessary if either of these respiration-based assays is used with calcareous soils, since dissolution of calcium carbonate can result in abiotic CO_2 evolution.

Despite the extensive development of methods for soil microbial biomass that has occurred in the past 15 years, relatively few data exist on microbial biomass in arid and semiarid soils. Nearly all microbial biomass estimates in the literature are of agricultural, forest, or grassland soils (Smith and Paul, 1986). Of these, the grassland soils in semiarid regions are perhaps the most relevant to this discussion. Clark and Paul (1970) estimated the total microbial biomass in a Canadian semiarid grassland soil to be approximately 200 g m^{-2} to a depth of 30 cm. Assuming a soil bulk density of 1.6 and a microbial carbon content of 50%, this amounts to approximately 0.20 mg biomass C g^{-1} soil. This corresponds closely to the original estimate of Jenkinson and Powlson (1976) of 0.17 mg biomass C g^{-1} for a fallow wheatfield soil with high pH (8.0) and low organic C content (0.84%). Measured by the SIR method, biomass C values averaged 0.10 mg g^{-1} in semiarid grassland soil samples from the Sevilleta Wildlife Refuge in central New Mexico during the hot summer months (Kieft, unpublished data). Data from the experiments of Knight and Skujiņš (1981), in which they measured ATP in the arid Great Basin Desert following equilibration to a range of water potentials, can be converted to biomass C data using a biomass C/ATP ratio of 250 (Paul and Clark, 1989). By this analysis, their highest biomass C values were only 5 and 7 μg biomass C g^{-1} soil. The very low biomass values in this experiment may reflect the very arid nature of the sites and that the samples were air dried once prior to the experiment. Using the same conversion factor, ATP values reported in Skujiņš (1984) can be converted to biomass C values ranging from 3 μg g^{-1} for a pre-Saharan desert soil in Tunisia to 112 μg g^{-1} for a Sonoran Desert site in Arizona. Severe soil disturbance can be expected to diminish microbial biomass. For example, Visser (1985) reported a drop in surface soil microbial biomass C from 0.32 to 0.01 mg g^{-1} when a semiarid prairie site in Canada was strip-mined. Monitoring of soil microbial biomass as a soil is reclaimed or in a chronosequence of reclaimed sites in arid and semiarid areas should be especially fruitful.

Biomass measures may be useful in identifying long-term or large-scale patterns of microbial development, whereas measurements of microbial activities are helpful in determining short-term responses to environmental conditions. Overall microbial activity can be measured as basal soil respiration rates, usually measured as rates of CO_2 evolution. These vary in response to environmental parameters, including moisture content. Sometimes the basal respiration is considered in relation to estimates of soil microbial biomass to obtain an estimate of the proportion of the biomass that is active (Insam and Domsch, 1988; West et al., 1988). As discussed, microbial adenylates are often considered better estimates of microbial activity than biomass. Hersman and Temple (1979) found ATP and soil respiration to be correlated in coal strip-mine spoils in eastern Montana. Soil enzymes have also been measured as indicators of microbial activity. Commonly used soil enzyme assays include dehydrogenase, phosphatase, urease, and pectinolyase (Hersman and Temple, 1979; Klein et al., 1985; West et al., 1988). Stroo and Jencks (1982) measured respiration and enzyme activities in chronosequences of recovering nonarid mine spoils. They found respiration rates, amylase, and phosphatase activities to be correlated and to increase with time of recovery. West and coworkers (1988) found ATP and phosphatase activity to be correlated in desiccated soils. Klein et al. (1985) suggested that analysis of various soil enzymes may be effective in predicting the success of a soil reclamation project, in assaying the effects of potentially toxic compounds on recovering soils, and in monitoring the progress of soil recovery.

There is currently no clear choice of a single most appropriate measure of soil microbial biomass or activity for monitoring recovery of the microbial community in a soil reclamation project. Therefore, the best approach is probably to use a combination of several methods, for example, a biomass assay, a soil respiration assay, soil enzyme assays, and possibly standard plate count assays. However, standard plate count assays alone have a very limited utility. Finally, patterns of soil recovery may best be observed by examining ratios of two parameters, such as the ratio of biomass carbon to total organic carbon or the ratio of soil basal respiration rate to soil biomass carbon. These ratios have been shown to be well correlated with macroclimate (Anderson and Domsch, 1989), with land use (Insam et al., 1989), with desiccation stress (West et al., 1988), and with the degree of recovery from soil disturbance (Insam and Domsch, 1988).

IV. MICROBIAL DISTRIBUTION

A. Spatial Patterns

Vertical Distribution

In general, soil microbial numbers as well as organic matter concentrations are highest near the surface and decline as one moves downward in the soil profile (Alexander, 1977; Paul and Clark, 1989). This truism has been applied to desert soils (Noy-Meir, 1974; Skujiņš, 1981, 1984; Evenari, 1985), but it may not be valid in every case. Kovda et al. (1979) stated that desert soil biological activity, as measured by respiration, dehydrogenase activity, and proteolysis, is greatest in the 0-3 cm horizon; the greatest density of microorganisms occurs deeper in the soil profile (5-20 cm). Bamforth (1984) found higher numbers of bacteria and greater actinomycete and fungal hyphal lengths in Sonoran Desert litter layers than in the underlying soils. Skujiņš and West (1973) found 1.2×10^3 fungal propagules at the soil surface of a Great Basin Desert site and none at 1 m depth. Although this pattern is probably typical of most arid soils, in some cases higher microbial numbers occur at depths of approximately 30 cm, where heat and desiccation stress are less severe (Skujiņš, 1984). In Antarctic dry valley soils, highest numbers often occur just beneath the surface, where temperature fluctuations and freeze-thaw cycles are less extreme (Friedmann, 1980; McKay and Friedmann, 1985). Deeply rooted desert plants can have associated populations of microorganisms at great depths. Populations of *Rhizobium* sp. capable of nodulating with mesquite (*Prosopis glandulosa*) have been found in high numbers to depths greater than 10 m in Chihuahuan Desert and Sonoran Desert sites (Virginia et al., 1986; Jenkins et al., 1987, 1989).

Fertile Islands

The patterns of biological activity in desert ecosystems are dominated by the spatial organization of perennial plants, which create more or less stable "fertile islands" (Garcia-Moya and McKell, 1970; Binet, 1981; Phillips and McMahon, 1981; Skujiņš, 1981, 1984). These perennial shrubs enrich the soil organic matter beneath their canopies by means of litter accumulation and root influences. Although in some instances wind may redistribute plant litter, in most cases it accumulates beneath the plant canopy. Lower branches in some plant species serve to trap and concentrate litter. The shrubs, along with associated plant species, also increase faunal activity beneath the canopy, thereby further concentrating organic matter (Binet, 1981). Concomitant with organic matter accumulation in these plant-associated soils are marked concentrations in soil microbial numbers, biomass, and activity. Baker and Wright (1988), working in a sand dune area of southern New Mexico, found higher numbers of bacteria in the canopy-influenced soils beneath mesquite (*P. glandulosa*) than in unvegetated soils between plants. Virginia et al. (1982) found nearly 60 times higher rates of denitrification upon soil wetting beneath *Prosopis* compared to rates in plant interspace soils. In a study of creosote (*Larrea tridentata*) canopy and interspace soils at a Chihuahuan Desert site, organic carbon was found to be 2.6 times higher, microbial biomass was 3.6 times higher, and soil dehydrogenase activity was 5.8 times higher in the canopy soil than in interspace soil (Kieft, unpublished data). We have found twofold biomass concentrations beneath the canopies of both the yucca (*Yucca glauca*) and four-winged saltbush (*Atriplex canescens*) at the semiarid Sevilleta Wildlife Refuge in New Mexico (Kieft, unpublished data). This belowground spatial heterogeneity of microorganisms is a key feature of ecosystem function, and destruction of these plant-microbe associations has been shown to accompany severe disturbance in arid ecosystems (Allen and MacMahon, 1985).

Rhizosphere Effects

The stimulatory effects of desert plants on the abundance and activities of microorganisms is most pronounced in the region immediately surrounding the roots, that is, in the rhizosphere. Root exudates, sloughed-off root cells, and lysed root cells generate microhabitats enriched in organic nutrients surrounding the roots. Steep chemical gradients extend from the rhizoplane surface away from the roots to the bulk soil. Microbial substrates in the rhizosphere include simple sugars, polysaccharides, amino acids, enzymes, organic acids, nucleotides, vitamins, and plant hormones (Elliott et al., 1984b).

Environmental stress on plants, especially desiccation stress, appears to stimulate plant root production and release of mucigel, consisting of hydrated polysaccharides with galactose and galacturonic acid residues (Martin, 1977; Giddens and Todd, 1984), thereby further stimulating microbial populations. The rhizosphere microorganisms may also contribute increased amounts of extracellular polysaccharide material in response to drought, conferring increased moisture retention and desiccation protection on plant roots (Skujiņš, 1984). Mucigel production in the rhizosphere of desert plants can result in a densely cemented sand and mucigel rhizosheath, which acts as a moisture-retentive barrier and can also harbor increased numbers of nonsymbiotic nitrogen-fixing bacteria (Wullstein et al., 1979; Wullstein and Pratt, 1981). Wullstein (1980) has further suggested that rhizosheath-forming grasses should be especially effective in the reclamation of sandy arid lands. Negative impacts of rhizosphere microorganisms can occur as well, as with root pathogens and microbes that compete with plant roots for inorganic nutrients (Giddens and Todd, 1984).

The rhizosphere effect is often measured as a ratio of the number of microorganisms in the rhizosphere to the numbers in the surrounding bulk soil (R/S). Because the bulk soils of deserts are so depauperate in terms of organic matter and microbial populations, R/S values for deserts are typically many times greater than those for nondesert soils. An R/S value of 700 is not unusual for a desert soil (Skujiņš, 1984), compared to a typical R/S value of 5.6 for the rhizosphere of a nondesert lupine plant

(Papavivas and Davey, 1961). Rhizosphere microbes include bacteria, fungi, and protozoa. Some are free-living; others are symbiotic (mycorrhizal fungi and N_2-fixing root nodule bacteria). Nakas and Klein (1980) found bacteria to be more active than fungi in organic matter decomposition in rhizospheres of a semiarid grassland. Species composition is generally markedly different in the rhizosphere from that in surrounding soils. Some microbes, such as the nonsymbiotic nitrogen fixers *Azotobacter* and *Clostridium*, may occur primarily in the rhizosphere. Small gram-negative rod-shaped bacteria, especially pseudomonads, are thought to predominate over other bacteria in the rhizosphere (Paul and Clark, 1989). Nonmycorrhizal rhizosphere fungi include *Alternaria*, *Aspergillus*, and *Fusarium* (Mahmoud et al., 1964).

B. Seasonal Patterns

Just as spatial patterns of microbial numbers and activities are more pronounced in desert soils, the temporal patterns are accentuated as well. Seasonal patterns of microbial numbers and biomass are generally correlated with seasonal patterns of water availability, with active microbial growth usually favored during rainy seasons (Skujiņš, 1984). Fluctuations in microbial populations in response to water availability can occur on a seasonal basis in deserts with predictable rainy seasons; in other deserts, rainfalls and the resulting blooms of microorganisms occur randomly throughout the year. The "pulse and reserve" paradigm for plant activity in deserts (Noy-Meir, 1973) can be applied in many cases to desert soil microorganisms as well. Microbial growth and activity follow immediately after a rainfall as water and nutrients become available. Later, as the soil dries, a portion of this accumulated biomass perishes, but the remaining viable biomass can be considered the reserve biomass, some in the form of desiccation-resistant spores and some in the form of vegetative cells. Microbial storage products (e.g., poly-ß-hydroxybutyrate and glycogen) and other endogenous substrates enable the sustained viability of vegetative microbial cells. Exceptions to the water-driven pulse-reserve pattern are certainly possible, as factors other than water can be limiting to microbial activity in some desert ecosystems. For example, Parker et al. (1984a) found little if any fluctuations in bacterial and fungal numbers when Chihuahuan Desert soil plots were repeatedly wetted and dried. However, protozoan numbers did follow the expected pulse and die-off pattern. Species composition as well as total microbial numbers and biomass can be expected to vary seasonally in desert ecosystems. Taylor (1979) found that the communities of culturable fungi in two semiarid sites in New Mexico changed in response to the seasonal availability of moisture.

V. LIMITING FACTORS

This section deals with important physical and chemical factors (other than nutrient limitation) that strongly influence microbial activities in arid land soils and the physiological responses of desert soil microorganisms to these factors.

TABLE 2 Approximate Limiting Water Potentials for Growth of Some Microorganisms at 25°C

Water potential (MPa)	Water activity (a_w)	Bacteria	Yeasts	Fungi
-1.5	0.989	*Spirillum*		*Sclerotium rolfsii* (sclerotia)
-3.0	0.978	Chemoautotrophs		
-7.1	0.95	Most gram-negative rods	Basidiomycetous yeasts	Basidiomycetes
-15	0.90	Most cocci, *Lactobacillus*, *Bacillus*	Ascomycetous yeasts	*Rhizopus nigricans*, *Fusarium*, Mucorales, *Aspergillus flavus*
-22	0.85	*Staphylococcus*	*Saccharomyces rouxii* *Debaryomyces*	*Aspergillus niger*, *Paecilomyces varioti*
-39	0.75	Halobacteria, halococci		*Sporendonema sebi*, *Aspergillus* *Wallemia*
-49	0.70			*Chrysosporium*, *Asperigillus ruber*
-59	0.65			*Erotium*
-70	0.60			*Xeromyces bisporus*

Source: Adapted from Brown (1976), Griffin (1972), and Kushner (1978).

A. Water Potential

Water is generally considered the single most limiting factor to biological activities in desert ecosystems (Noy-Meir, 1974; Kovda et al., 1979; Skujiņš, 1981, 1984). The most widely accepted measure of water availability to biological systems is water potential, which is a measure of the free energy of the water in a system relative to the free energy of a reference pool of free water (Papendick and Campbell, 1981). The water potential of soil is measured in pressure units (megapascals, MPa) and is related by a logarithmic function to water activity a_w or to the relative humidity of an atmosphere in equilibrium with the soil:

$$\psi = RT(V_w)^{-1} \ln a_w = RT(V_w)^{-1} \ln \frac{rh}{100}$$

where

ψ = water potential (MPa)
R = gas constant (8.31×10^{-4} m^3 MPa mol^{-1} K^{-1})
T = temperature (K)
V_w = partial molal volume of water (1.8×10^{-5} m^3 mol^{-1})
a_w = water activity
rh = relative humidity (%)

The major components of total water potential in soil are the matric water potential ψ_m and the solute water potential ψ_s:

$$\psi_{total} = \psi_m + \psi_s$$

Matric water potential represents the reduction in water potential caused by interactions of water molecules with solid/liquid interfaces; that is, it is a measure of desiccation in porous media. Solute water potential is the reduction in water potential caused by the presence of solutes (e.g., NaCl). Pure water has a water activity of 1.0 and a corresponding water potential of 0. Desiccation and/or the presence of solutes results in a negative water potential. Water stress in nonsaline desert soils is predominantly a matric water potential phenomenon. Saline soil microorganisms can experience a combined matric and solute water potential stress.

As a group, soil microorganisms are often considered less affected than the plants and animals by water potential stress (Focht and Martin, 1979). This is undoubtedly true in that many microbial species extend the limits for life to extreme water potentials (Table 2), as well as for other physical and chemical parameters. Although it is customary to present data on the lower limits of microbial tolerance in a review of the influence of water potential, it should be noted that distinctions are not made between solute and matric water potential stresses, and several of the microorganisms listed are adapted for growth in high salt or sugar solutions and are not indigenous to soils. In theory, the responses of microbes to water potential stress should be the same, regardless of whether the stress is dominantly matric or solute (Griffin and Luard, 1979). For the microbes, osmotic adjustment is partly a matter of matching cytoplasmic water potentials to those of their surrounding microhabitats. Actually, cytoplasmic water potentials are generally a bit lower, resulting in a slightly positive turgor pressure. Although it is frequently assumed for convenience that the effects of solute and matric water potentials are equivalent (because solute stress is so much easier to study than matric stress), there are some major differences.

Matric stress results in decreased solute diffusion and concomitant reduction in cell motility as water films on soil solids decrease in thickness. Bacterial motility in porous media is severely restricted at matric water potentials below approximately 0.1 MPa (Wong and Griffin, 1976). Thus, matric water potential stress is frequently accompanied by starvation stress as microbes become physically separated from available substrates. As stated earlier, this favors filamentous growth forms (actinomycetes and fungi) which are able to bridge air-filled soil voids. The shift in importance from bacteria to fungi in soil decompositional processes is thought to occur at matric potentials of about -1.5 MPa (Sommers et al., 1981). Soil fungi generally show optimum respiration rates at slightly negative matric water potentials, and they remain active at water potentials well below the lower limits for most bacteria. Soil matric stress also, of course, favors the formation of desiccation-resistant resting stages, such as bacterial endospores and cysts, actinomycete and fungal spores, cyanobacterial akinetes, slime mold sclerotia, and fungal spores. In the case of the endospore-forming bacteria, sporulation is triggered by nutritional factors, but the resulting endospores are phenomenally desiccation tolerant. Vegetative bacterial cells can also be very desiccation resistant. Although the thick cell walls on the gram-positive bacteria are thought to confer greater drought tolerance (Harris, 1981), many gram-negative bacteria can also persist for months in dry soil (Chen and Alexander, 1973). Production of extracellular polysaccharide slime layers may aid in the dessication tolerance of these organisms (Focht and Martin, 1979; Mary et al., 1986).

When the solutes generating a negative solute water potential (high osmolarity) are permeable to cell membranes, they frequently exert a toxic effect on intracellular function. This is especially the case with NaCl, the dominant solute of saline soils. Sodium ions are excluded (at energetic cost) by virtually all microbial cells, even those growing in saturated salt solutions. In their stead, cells accumulate intracellular compatible solutes (also termed osmolytes or osmoprotectants), which are relatively compatible with intracellular function and may even stabilize proteins in the presence of high salt concentrations (Yancey et al., 1982). Osmolytes of bacteria are commonly amino acids or related compounds, with glutamate prevalent among gram-negative bacteria and proline most common among gram-positive bacteria, including actinomycetes (Measures, 1975; Killham and Firestone, 1984; Csonka, 1989). Glycine betaine and KCl are the dominant organic intracellular solutes of the extremely halophilic bacteria (Imhoff and Rodriguez-Valera, 1984). Eukaryotic microorganisms as a rule synthesize polyols, such as glycerol, in response to salt stress (Luard, 1982a, b). Although intracellular solutes have been the primary focus of osmoregulation studies, there are other necessary cellular adaptations to decreased solute water potentials. For instance, membrane lipid composition is modified in the presence of salts (Hanna et al., 1984).

Although it is generally assumed that soil microbes respond physiologically to desiccation in the same manner as they do to salt stress, this has not been clearly demonstrated. Adebayo and Harris (1971) compared the solute and matric responses of two fungi, one a root pathogen (*Phytophthora cinnamomi*) and the other a soil saprophyte (*Alternaria tenuis*). They found that both these fungi were much more sensitive to matric than to solute water potential stress. However, these fungi were not arid soil isolates. One may hypothesize that fungi from nonsaline arid soils would be better adapted to desiccation than to salts, but this has not been tested. McAneny et al. (1982) revealed that *Arthrobacter crystallopoietes* is less inhibited by matric than by solute water potential decreases and that its optimum matric water potential for growth is -1.0 MPa. Soroker et al. (1987) have found that biomass, respiration, and the AEC responses of the microbial community from a saline semiarid saltpan soil differ markedly when challenged by solute and matric stresses of equal magnitude. Soroker et al. (1988) have tested salt-sensitive mutants derived from salt-tolerant isolates for their abilities to survive and compete under salt or desiccation stress. Loss of salt tolerance did not necessarily confer loss of desiccation tolerance. These data emphasize the differences in microbial physiology in regard to osmoregulation and desiccation tolerance, and they underscore the need for further research into the mechanisms of drought tolerance in microorganisms.

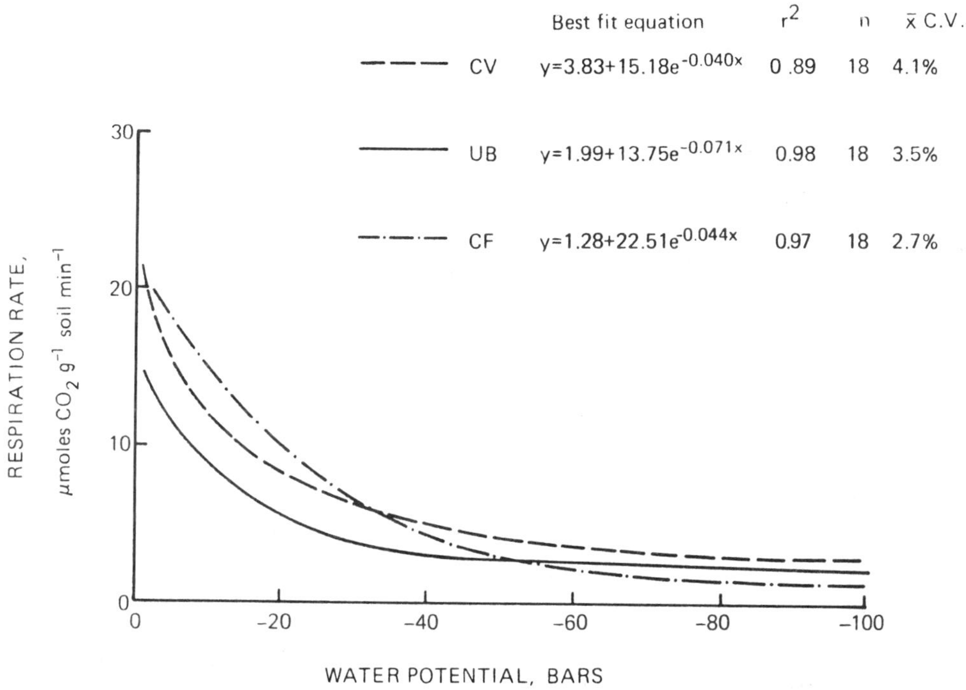

FIGURE 1 Soil respiration rates as a function of water potential in two arid soils (CV and UB) and one nonarid soil (CF). (From Knight and Skujiņš, 1981.)

Numerous studies of nonarid soils have demonstrated that overall microbial activity, as measured by soil respiration rates, is highest at water potentials near saturation and that these rates decline with decreasing moisture availability (Sommers et al., 1981). The same pattern has been seen in two arid Great Basin Desert soils (Fig. 1) (Knight and Skujiņš, 1981). Knight and Skujiņš also compared the microbial ATP contents of the two arid soils and one nonarid soil as they were air dried (Fig. 2). The nonarid forest soil showed highest ATP values at water potentials near saturation, whereas the arid region soils had highest ATP values at approximately -1 MPa. This demonstrates specific adaptations of the desert soil microflora for maintaining viability under stressful water potential conditions.

Water stress in semiarid and arid soils may adversely affect some functional groups more than others. Illustrative of this point, symbiotic nitrogen fixers of the genus *Rhizobium* are notoriously susceptible to drought (Fuhrmann et al., 1986; Mary et al., 1986). This suggests that these microbes have a low probability of survival as free-living cells outside moist, root-associated microhabitats. However, Miller and Pepper (1988) have recently isolated a fast-growing *Rhizobium* from a Sonoran Desert lupine, and they report that this strain has an unusually high capacity for survival in desert soils. Free-living nitrogen fixers of the genus *Azotobacter* are also poorly adapted to low water potentials, showing no respiratory or

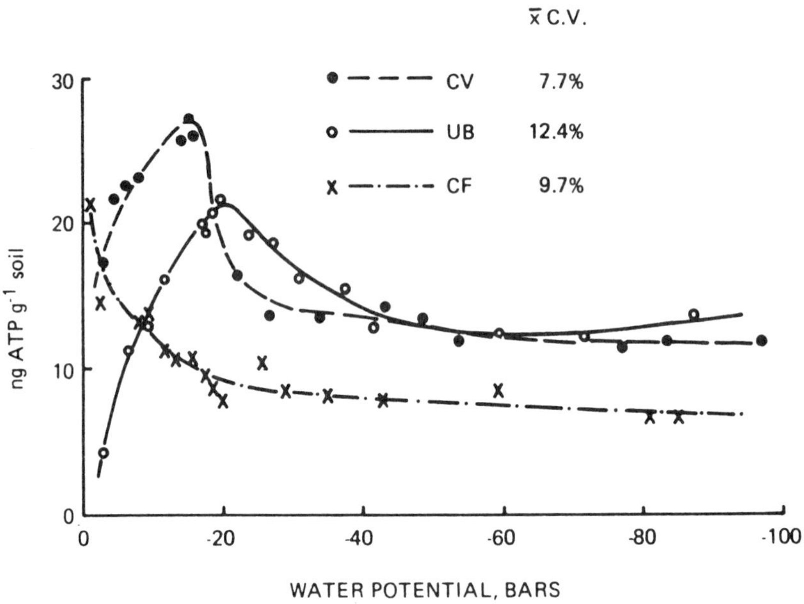

FIGURE 2 Soil ATP concentrations as a function of water potential in two arid soils (CV and UB) and one nonarid soil (CF). (From Knight and Skujiņš, 1981.)

nitrogen-fixing activity below -2.1 MPa (Kroeckel and Stolp, 1984). *Azotobacter* may even be absent altogether from extremely hot desert soils (Skujiņš, 1981, 1984). By favoring fungi over bacteria, dry conditions may be important in freeing fungi from bacterial antagonism and, therefore, in persistence of fungal pathogens of plants (Cook and Duniway, 1981; Griffin, 1981). Certain groups of microbes involved in the decomposition of soil organic matter may be selected against in the dry conditions of deserts, a factor that may lead to the accumulation of a particular organic compound. For instance, carboxymethylcellulose degraders are thought to be poorly drought adapted (Skujiņš 1984). Pesticide-degrading microorganisms may be especially lacking in desert soils, also. Finally, the activities of many, if not all, of the chemolithotrophic bacteria (e.g., nitrifiers and sulfur oxidizers) are generally thought to be restricted to high water potential conditions (> -2 to -3 MPa) (Skujiņš, 1984; Kieft et al., 1987a; Kieft and Spence, 1988).

Arid region microbes are generally better able to utilize very small amounts of moisture than are the higher organisms. Even miniscule rainfalls that are smaller than the "effective rainfalls" required by plants can increase the water potential sufficiently in microhabitats to stimulate microbial activity. Water vapor and dew can furnish adequate moisture for brief bursts of microbial activity (Sala et al., 1981). This has been shown particularly among the desert soil lichens (Kappen, 1988) and desert soil algae (Friedmann, 1971). Individual snowflakes can even provide sufficient moisture to sustain microbial life in Antarctic dry valley soils (Friedmann, 1987).

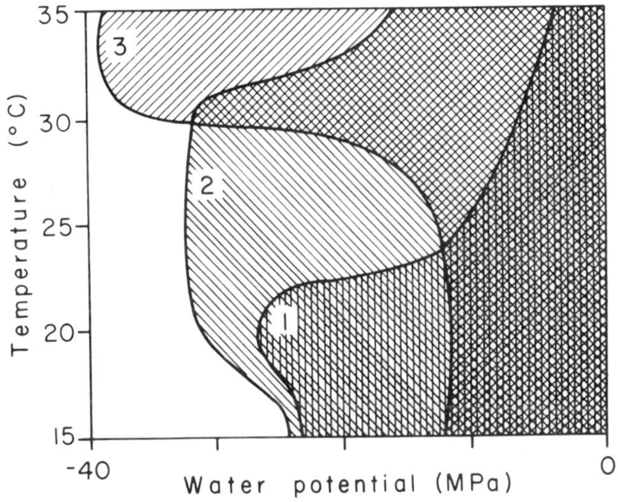

FIGURE 3 Interacting effects of temperature and water potential on selection of microbial communities colonizing hair on a soil surface. Community 1 is dominated by *Penicillium chrysogenum* and *P. frequentans*; community 2 is predominantly *P. citrinum* and *Asperfillus versicolor*; community 3 is predominantly *A. flavus*, *A. niger*, and *A. terreus*. (From Griffin and Luard, 1979.)

It has been known for decades that wetting and drying cycles in soil have major influences on the turnover of carbon and other soil nutrients (Stevenson, 1956; Birch, 1958, 1959; Kieft et al., 1987b; Cortez, 1989), with rates of organic matter degradation greater overall when a soil is alternately wetted and dried than if it is kept constantly moist. Desiccation-induced death of soil microbes releases carbon and other nutrients; rewetting of dried soils results in rapid growth and metabolism by the surviving microbes. Curiously, the rapid wetting of soil may be at least as stressful to microbes as the gradual drying, thereby contributing to the microbial death and recycling of immobilized cell material that occur as a result of wetting and drying cycles (Kieft et al., 1987b).

B. Temperature

The high temperatures reached at the surfaces of hot desert soils constitute another extreme condition to be survived by microorganisms. Extremely high temperatures may select for endospore-forming bacteria. However, many vegetative desert microbes are capable of heat tolerance to at least 70°C. Because the high heat in desert surface soils is generally accompanied by extreme desiccation, there is little possibility for microbial growth at these temperatures and so desert soil microbes are not considered thermophilic (Skujiņš, 1984). However, a few thermophilic microbes can be isolated from nearly any environment, including desert soils and even including Antarctic soils (Horowitz et al., 1972). Temperature is interactive with other abiotic parameters in its effects on soil microbes, as shown for the influence of temperature and water potential on three communities of fungi (Fig. 3).

C. Soil Reaction (pH)

Hydrogen ion concentration in the soil solution is usually measured in a soil sample to which water has been added, thereby estimating the bulk pH, which is an average of the pH of all the microsites in the soil. For arid and semiarid soils the bulk pH is generally alkaline, with values ranging as high as 10. Many generalizations have been made about the effects of pH on the distribution and activity of soil microorganisms. These are generally valid rules of thumb: (1) bacteria generally have a narrower range of pH tolerance than do the fungi; (2) fungi are favored over bacteria in acid environments, for example, forest litter layers; and (3) actinomycetes are favored over fungi and other bacteria in alkaline conditions. These generalizations are supported by a plethora of plate count data, some of which are presented elsewhere in this chapter.

Although bulk pH is a useful parameter, it should be understood that the actual pH levels of the microenvironments in which microbes reside can be very different (Swift et al., 1979). The negative surface charge of clay minerals can concentrate hydrogen ions. Moreover, microbial metabolism can generate or consume hydrogen ions, with acidification the more prevalent activity. These somewhat acidified microsites can explain the presence of neutrophilic or even acidophilic microbes in extremely alkaline desert soils.

Hydrogen ion concentrations can also have an indirect effect on soil microbes by influencing the solubility of nutrients and of metallic compounds. Alkaline conditions may reduce solubilities of Ca, Fe, and K and may also reduce the solubilities of toxic metal ions. Here again, however, microsite effects may differ.

D. Heavy Metals

The presence of heavy metal accumulations is a frequent concern in land reclamation processes. This is particularly true in two types of soil reclamation situations: (1) when sulfide minerals in the form of mine tailings are present (Mills, 1985; Wildung and Garland, 1985), and (2) when sewage sludge is used as fertilizer and organic soil amendment (Day et al., 1988). Sulfide minerals are biologically oxidized by *Thiobacillus* spp. and other chemolithotrophs in a process that releases sulfuric acid and soluble metal ions (Erlich, 1981). The decreased pH associated with this sulfide mineral oxidation further solubilizes soil metals. Municipal sewage sludges are invariably contaminated with heavy metals, particularly if they are derived from heavily industrialized areas. Repeated applications of sludge as a soil conditioner can result in substantial heavy metal accumulations. Heavy metals of major concern include Cu, Zn, Cd, Pb, Ni, and Cr. Accumulation of these metals leads to questions regarding their effects on soil microbial populations and on microbially mediated soil processes. Although considerable information now exists on heavy metal effects in nonarid soils, studies examining metal effects specifically in arid and semiarid sites are fewer in number.

The majority of studies comparing heavy-metal-contaminated soils to noncontaminated control soils have shown some negative influence of metals on microbes. These include reports of decreases in microbial biomass

(Brookes and McGrath, 1984; Brookes et al., 1986), decreases in diversity (Nordgren et al., 1983; Barkay et al., 1985), and decreases in microbial activities (Chang and Broadbent, 1981; Doelman and Haanstra, 1979, 1986; Skujiņš et al., 1986; Reber, 1989). Data from a nonarid agricultural soil in Great Britain show that the biomass of metal-contaminated sludge-amended soils was significantly lower than that of control soils and that the inhibitory effect of metals on the accumulation of microbial biomass persists for decades following abatement of application (Brookes and McGrath, 1984; Brookes et al., 1986). The same patterns were found by these authors for direct microscopic counts and for soil microbial ATP concentrations. That a mesic soil can be so profoundly influenced by sludge application has disturbing implications for desert soils. Microbial biomass, activities, and diversities in arid and semiarid soils are already rather low; further diminishment by sludge metals may significantly slow rates of decomposition and nutrient turnover.

In a few of the studies of sludge-amended soils, heavy metal accumulation appears not to reduce microbial numbers and activities. Boyle and Paul (1989) found an increase in soil microbial biomass and in nitrogen and carbon mineralization rates in a sludge-amended soil in a nonarid environment. This may reflect relatively low metal levels in the sludge and/or a short period of application (8 years), a low availability of metal ions due to soil characteristics (Debosz et al., 1985), or the stimulation of metal-resistant soil microbes (Barkay et al., 1985). Different soil types, especially with different pH values of the soil solutions, and different textures can be expected to vary greatly in their susceptibility to metal-induced stress on soil microflora. It has also been shown that plant and microbial responses to metal perturbation can be opposite to one another (Reber, 1989), indicating the need to monitor both in sludge-amended plots. Further study of heavy metal effects in arid and semiarid systems is obviously needed.

VI. MICROBIAL EFFECTS ON THE SOIL CARBON CYCLE

A. Soil Formation

Microbes are the primary colonizers of mineral surfaces or parent materials that give rise to the formation of soils. As such, microbially mediated processes play an essential role in the early events of pedogenesis. Jenny (1980) has pointed out that in an environment limited to rock, air, and water, only the nitrogen-fixing photoautotrophs (cyanobacteria and purple sulfur bacteria) are self-sufficient enough to generate biomass. The purple sulfur bacteria can be eliminated from consideration, since they are obligately anaerobic and rock surfaces are exposed to oxygen-rich air. Nitrogen may also be available in precipitation, enabling the growth of non-nitrogen-fixing photoautotrophs (algae). These photoautotrophs can also occur as primary colonizers within lichen associations. The lichens are especially important to the process of pedogenesis because they can produce substances (including chelates and phosphatases) that solubilize and weather the parent rock material. Weathering of the minerals leads to the formation of clay minerals (Jacks, 1965). Establishment of a photoautotrophic microbial community leads to an accumulation of organic matter in

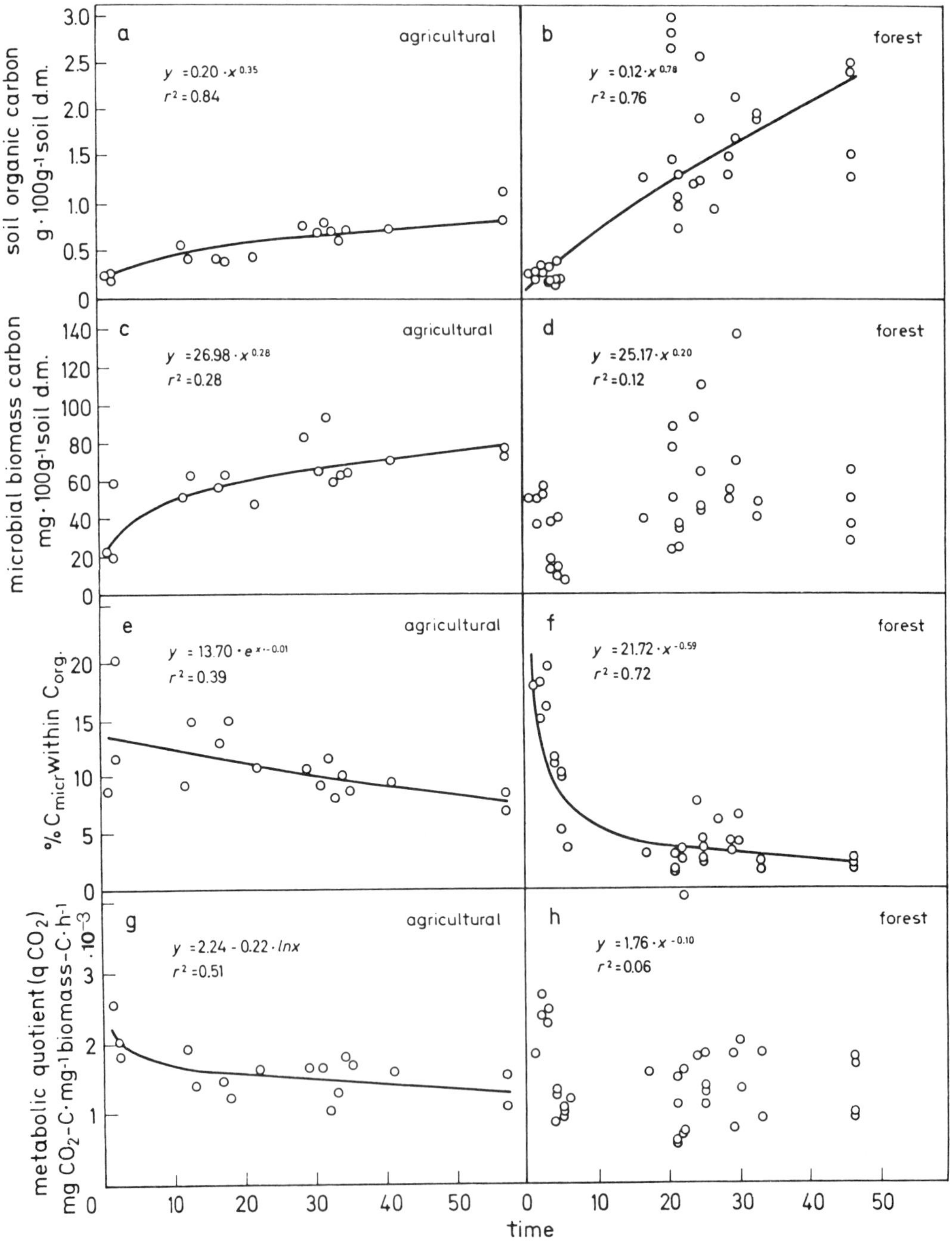

FIGURE 4 Soil organic carbon, biomass, ratio of biomass C to total organic C, and metabolic quotient (ratio of soil respiration and biomass C) for two chronosequences of soils under reclamation following mining. These are two nonarid soils. However, this may also be a very useful approach for monitoring soil recovery in semiarid and desert systems. Steady-state values of one or more of these parameters indicate a stable climax community. (From Insam and Domsch, 1988.)

the form of dead cells and cellular exudates. These in turn serve as substrates for a variety of heterotrophic microbes. Products of heterotrophic metabolism (organic acids and carbonic acid) further contribute to the weathering of mineral surfaces. Eventually, a quantity of stable organic matter sufficient to support higher plants accumulates. Plant litter, dead roots, and root exudates provide further inputs of organic matter to the forming soil. The combined action of the various microbes results in stabilization of a portion of the soil organic matter as humus and in the aggregation of soil particles to form the structure of the soil. Therefore the microbiological contributions to pedogenesis are integral to the process, and they begin with an autotrophic succession and progress to a heterotrophic succession in which decomposition is the primary form of metabolism.

The recovery or reclamation of disturbed soils can be viewed as a process resembling the original stages of soil development. With an extreme disturbance (e.g., strip-mining), parent material may be exposed, resulting in microbial secondary succession being initiated by the photoautotrophic primary colonizers. In less severe cases (e.g., desertification induced by cultivation, grazing, and roadbuilding), the soil and a portion of its microbiota may remain and can thus be viewed as an intermediate stage of secondary succession. Since pedogenesis occurring under natural circumstances in arid systems is inevitably extremely slow, it is often desirable to accelerate the process by the addition of organic matter as plant residues or sewage sludge or in some other form. However, these organic residues do not instantly confer all the beneficial properties of stable soil organic matter; the necessary transformations of organic amendments, most of them microbially mediated, generally require years to tens of years. The process of reclamation can be measured as the accumulation of stable organic matter (Skujiņš and Richardson, 1985), as the degree to which the microbial biomass has developed (Powlson et al., 1987), or as a ratio of those parameters (Fig. 4) (Insam and Domsch, 1988). Soil recovery may be considered complete when one or more of these parameters reach(es) equilibrium. The ratio of soil microbial biomass carbon to total soil organic carbon decreases during soil recovery, as a larger portion of the soil organic matter becomes stabilized into humus; this ratio has been suggested as an especially useful measure of long-term soil reclamation (Insam and Domsch, 1988).

B. Soil Structure

Soil structure is a function of soil aggregation, and this in turn is largely influenced by the activities of soil microorganisms. Aggregation of the soil promotes plant growth by increasing permeability and porosity and by making the soil more amenable to penetration by plant roots. Microbes influence aggregation in the following ways: (1) by the formation of filaments that provide a structural framework for soil aggregates; (2) by production of extracellular polysaccharide material, which acts to cement together clay particles that would otherwise be repelled by their negative surface charges; and (3) by the process of humification, since humic compounds bind clay particles together. Fungal and actinomycete hyphae, along with plant roots, are involved in the temporary formation of relatively large aggregates, whereas stabilization of microaggregates by humic compounds is more or less permanent. Root exudates, dead roots, and plant litter provide the organic substrates for microbial growth and production of polysaccharides that form the cement for aggregation (Forster, 1979; Kovda et al., 1979). Cryptogamic crusts appear to contribute to soil structure by stabilizing the soil surface and by contributing organic matter to the soil (Skujiņš, 1984). Soil crumb formation is usually greatest in fine-textured clay materials and slow to nonexistent in sand. Forster (1979) described the stabilization of sand grains in a maritime dune. Microbially induced aggregation was most important in this system, and it was found that the rates of aggregation were greatest during the winter season when plant residues were available as microbial substrates. Factors involved in soil aggregation are reviewed in more detail by Focht and Martin (1979), Tisdall and Oades (1982), and Tate (1987). Future efforts in arid soil reclamation should include the development of methods for monitoring and optimizing rates of soil aggregation by microbes and microbial products.

C. Decomposition

An age-old tenet of microbial ecology is the principle of microbial infallibility, which states that given favorable conditions, all naturally occurring organic compounds can be decomposed by microorganisms (Alexander, 1965). All plant, animal, and microbial residues are microbially degraded, some quite rapidly and some only very slowly. This principle has held up over the years; however, it cannot necessarily be extended to all synthetic xenobiotic organic compounds.

The role of microorganisms as processors of soil organic matter is summarized in Fig. 5. Components of organic matter inputs, such as plant and animal residues, can be classified into various categories based on their

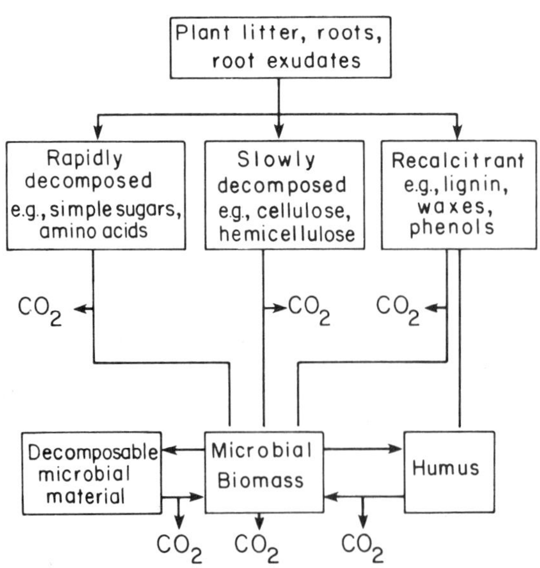

FIGURE 5 Simplified diagram of carbon flow in a soil ecosystem. (After van Veen et al., 1985.)

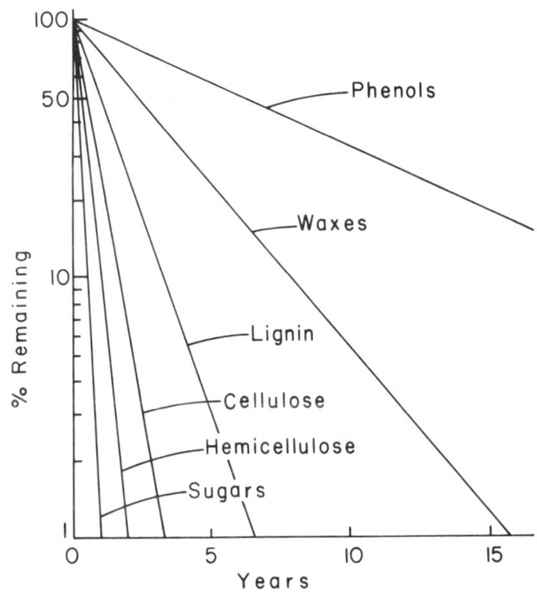

FIGURE 6 Logarithmic decomposition rates of various components of plant litter. These data are from a nonarid soil. Rates in arid and semiarid systems may be somewhat slower, but decomposition rates of individual components relative to each other should be very similar. (After Minderman, 1968.)

biodegradability, for instance in the three-component scheme of van Veen et al. (1985). In the first category are the most readily degraded compounds, including simple sugars, amino acids, and some polysaccharides and proteins. Under favorable conditions, these compounds are utilized in a few hours to a few days. In the second category are polymers, such as cellulose, hemicellulose, and chitin, which are degraded more slowly and may persist in the soil for weeks to months. The most slowly degraded plant compounds are typified by the lignins. Fats, waxes, and tannins are also slowly degraded components of plant and animal residues. The relative rates of decomposition of these components in a nonarid ecosystem are shown in Fig. 6. Heterotrophic microbial activity results in approximately 50% of the organic carbon substrate being mineralized to CO_2 and the remainder being incorporated into microbial biomass. The latter constitutes another important category of soil organic matter, one that serves both as catalyst for biogeochemical cycling and as reservoir of soil nutrients. Turnover of the nutrients that have been immobilized in microbial biomass occurs during predation and as a result of microbial death induced by environmental fluctuations, such as wetting-drying and freeze-thaw cycles.

A portion of the plant and animal residues and a portion of the microbial biomass are converted to humus, which is the dark-colored amorphous, water-insoluble fraction of soil organic matter. Although poorly defined, humus is universally considered the organic fraction in soil that contributes most to the physical and chemical properties of soil that encourage plant life (Cundell, 1977; Skujiņš and Richardson, 1985). Humification is the formation of a highly stable complex mixture of heteropolymeric, phenolic humic substances, including humic acids, fulvic acids, and humin. The process of humification is at least as complex as the vast array of heterogeneous polymers it produces. However, it can be briefly summarized as a depolymerization of plant and microbial residues, followed by a condensation or polymerization into humic and fulvic acids. The first step is brought about by a combination of physical, chemical, and biological factors, with microbially produced enzymes playing a central role; the condensation step is considered primarily chemical. A detailed discussion of humic acid formation is presented by Tate (1987). Rates of humification in desert soils are slower than in more mesic environments, and final equilibrium humus contents are typically low: 1% in desert soils and 3% in semiarid soils (Kovda et al., 1979). Although highly refractile, humic substances can also be biologically degraded. In a stable, climax ecosystem, the rates of formation and decomposition of humic substances should be equal, with a steady-state concentration of humus, whereas disturbance may lead to a decline in humus.

Factors controlling the rates of decomposition in terrestrial systems include the amount of moisture, the quality of the organic substrates (measured by C/N ratio and lignin content, for example), and the sizes and diversity of decomposer populations of organisms. Meentemeyer's (1978) model of litter decomposition rates, which is based on actual evapotranspiration (AET) as an index of macroclimate and on lignin concentration as an index of litter quality, predicts that decomposition should be slow in semiarid and desert systems. According to this model, both the low moisture availability and relatively high lignin content of xerophytes (Skujiņš, 1984) are factors mediating against rapid decomposition in desert systems. Conventional wisdom says that overall patterns of moisture availability result in large accumulations of plant litter, with decomposer activities limited to brief pulses following rainfalls (Noy-Meir, 1974; Skujiņš, 1984). However, patterns of decomposer activity are more complex than this, with extreme spatial as well as temporal variability the rule. As Schimel and Parton (1986) have stressed, one needs to examine microclimatic as well as macroclimatic effects on decomposition in arid systems.

Studies of litter decomposition at the Jornada site in the northern Chihuahuan Desert by Whitford and various coworkers have been particularly enlightening. Their findings indicate that (1) decomposition rates in the Chihuahuan Desert may be equivalent to those of more mesic systems (Schaefer et al., 1985), (2) detritivorous microarthropods are especially important in plant litter decomposition (Elkins et al., 1982; Parker et al., 1984b; Whitford, 1986), (3) rates of decomposition do not vary as a function of rainfall in this system (Schaefer et al., 1985; Whitford, 1986; Whitford et al., 1986; MacKay et al., 1987), and (4) litter buried as a result of animal activity decays at a more rapid pace than surface litter as a result of the more moderate subsurface conditions (Santos et al., 1984). The lack of correlation of moisture and decomposition data may reflect the fact that soil animals, which are not primarily moisture limited, serve crucial functions in organic matter redistribution, in primary decomposition, and in predation on microorganisms (Whitford, 1986).

Most of the available studies are of litter decomposition, but one may assume that control of decomposition of dead roots and root exudates follows the same rules. Rates of decomposition of belowground plant materials should proceed more rapidly since they do not require translocation into lower horizons and the more moderate subsurface temperature and moisture conditions should favor faster degradation.

Disappearance of litter from the soil surface can also occur by purely physical and chemical processes in desert systems. Nonbiological factors that may contribute to the elimination of organics from the litter layer include solubilization and leaching, volatilization, and ultraviolet light damage. Montana et al. (1988) compared litter mass losses in a Chihuahuan Desert site and a nondesert site and concluded that nonbiological processes dominate in arid systems. They propose an aridity gradient that controls whether litter decomposition is primarily biological or abiological, with abiological processes dominant in the most arid desert sites.

Rates of decomposition of straw amendments have been used to test the efficacy of various organic matter amendments in reclaiming semiarid strip-mine spoils (Elkins et al., 1984; Parker et al., 1987) and degraded rangeland soil (Whitford et al., 1989) in New Mexico. In these studies, restoration of soil function was considered to require diverse communities of microflora and microfauna, and these were considered to be restored when straw decomposition rates were similar to those of undisturbed control sites. Soil amendment with recalcitrant organic matter, such as tree bark or wood chips, had the most stimulatory effects on straw decomposition, especially in the strip-mine spoils. Significant differences in decomposition were seen among the various amendments, despite minimal microbial biomass differences. These authors suggest that relatively recalcitrant, high C/N ratio amendments may be especially beneficial in recovery of arid and semiarid soils and that this benefit is derived from a stimulation of the soil microfauna and a concomitant increase in turnover of microbial biomass and soil organic carbon.

It is unfortunate that most of the available data on decomposition in desert systems are restricted to North American hot deserts and that most of those are from the Chihuahuan Desert. Obviously there is a need for comparative studies from other arid and semiarid ecosystems.

VII. MYCORRHIZAE

Mycorrhizae are nearly ubiquitous fungi that exist in mutualistic association with plant roots, providing inorganic nutrients and soil moisture in return for plant photosynthate. Mycorrhizae act as extensions of the plant root systems, expanding the surface area for uptake of water and inorganic ions and, in the process, extending the region of the rhizosphere. This results in a greater radius of redistribution of carbon fixed by the plant. Solubilization and transport of the relatively immobile nutrient phosphorus comprise a primary benefit of the mycorrhizae to their plant symbionts. Mycorrhizae can be divided into the ectomycorrhizae, which do not penetrate the plant roots, and the endomycorrhizae, which do. The former are Ascomycetes and Basidiomycetes, which form associations with woody forest plants and are of only minor importance in arid ecosystems. Of the endomycorrhizae the majority are vesicular-arbuscular (V-A) mycorrhizae, and these are the most widespread of all the mycorrhizae and the most prevalent in arid and semiarid lands (Trappe, 1981). Although they have not been successfully cultured in the laboratory, the V-A mycorrhizae are thought to be Zygomycetes, as indicated by their lack of hyphal septa. Their name is a description of the fungal structures occurring intracellularly in the root cortex. Arbuscules are branched filamentous structures which exchange nutrients between plant and fungus. The vesicles are globular structures that probably act as storage product repositories (Hayman, 1978). Associations of V-A mycorrhizae and plants appear to be rather nonspecific, with a single fungal species able to infect plants from a variety of families.

Besides enhancing the competitive ability of associated plants under conditions of inorganic nutrient limitation, the V-A mycorrhizae have also been suggested to function under arid conditions in moisture uptake and retention (Reid, 1984). The mycorrhizal hyphae may act as conduits for the bulk flow of water. Bulk water flow is driven by plant leaf transpiration and by the resulting moisture gradient extending from the leaf through the plant, through the roots, and finally through the mycorrhizae to the soil. This flow of water through the hyphae is greater than the flow of thin water films of a desiccated soil. The mycorrhizae are also thought to make a better bridge between the soil solution and the root surface than that which can occur in the absence of mycorrhizae. This is especially true in dry soils, where the fungal hyphae are able to penetrate narrow pores and thin water films more effectively than the larger plant root hairs. Of course, these increases in bulk water flow and increases in the effective volume of the soil from which moisture can be obtained also translate into increases in plant nutrients that can be gleaned from the soil solution by root-mycorrhizal systems in arid soils.

Although relatively few studies have been made of the incidence of mycorrhizae in desert and semiarid regions, some data do exist. In a recent study of arid Saudi soils, Malibari et al. (1988) identified a variety of V-A mycorrhizal spores in the rhizospheres of plants, with nearly 50% of these being of one species, *Glomus fasciculatum*. In the same study, infectivity was measured by the degree of infection of onions as indicator plants, and this was found to increase with the age of the plants. Koske et al. (1975) and Tisdall and Oades (1979, 1980) have described the importance of V-A mycorrhyzae in aggregation of sandy soils. Zak and Whitford (1986) reported the occurrence in a Chihuahuan Desert dune system of an ascomycete (*Carbomyces emergens*) that may form ectomycorrhizal associations with shrubs. Surveys of mycorrhizal associations in other arid land ecosystems have also been made (El-Giahmi et al., 1976; Mahmoud et al., 1987, cited in Malibari et al., 1988; White et al., 1989).

The V-A mycorrhizae are found in greatest numbers in undisturbed soils and are not commonly present in severely disturbed sites (Reeves et al., 1979; Miller, 1979; Allen and Allen, 1980; Call and McKell, 1982; Johnson and McGraw, 1988a, b). This phenomenon has been well documented in both arid and nonarid regions. Reeves et al. (1979) compared V-A mycorrhizal associations on a disturbed site (old roadbed) and on an adjacent undisturbed site in a semiarid Colorado sagebrush region. On the undisturbed site, 40 of the 42 plant species were mycorrhizal, whereas only 6 of the 21 species on the disturbed site were mycorrhizal. More significantly, V-A mycorrhizal plants accounted for over 99% of plant cover on the undisturbed site and less than 1% on the disturbed sites. Miller (1979) reported very similar results for disturbed and undisturbed Red Desert sites in Wyoming. Evidently, V-A mycorrhizae are common symbiotic partners of the plants of stable, climax desert communities, but they are only infrequently associated with the plants that are pioneer colonizers of disturbed desert sites. As such, the V-A mycorrhizae can be viewed as "indicator microorganisms" that can indicate the extent of progress of revegetation and recovery in the secondary succession following disturbance of an arid site. The availability of V-A mycorrhizal propagules may also be the major determinant of the rate of succession from weedy, nonmycorrhizal annual plants to stable, climax communities following disturbance (Allen, 1988). Propagules consist of spores, hyphae, and infected root cells. Of these the spores can be expected to survive desiccation and starvation for the longest periods. The mycorrhizal spores are also the most useful form of the fungi for identification purposes (Trappe, 1982). Disturbance may eliminate or drastically reduce the number of viable mycorrhizal propagules in a site, and the nonmycorrhizal primary colonizers of disturbed sites may delay or inhibit the establishment of V-A mycorrhizal-plant associations. Naturally occurring dispersion of V-A mycorrhizal propagules occurs primarily by the actions of wind and animals (Warner et al., 1987). Re-establishment of V-A mycorrhizae may be accelerated in disturbed sites by inoculation with undisturbed soil or by transplanting plants with established V-A mycorrhizal associations (Trappe, 1981; Danielsen, 1985; Skujiņš and Allen, 1986). The development of management techniques for promoting the re-establishment of V-A mycorrhizal-plant associations, including inoculation techniques, is an active area of current research, and efforts in this area should provide significant benefits for the reclamation of disturbed arid lands.

VIII. ALLELOPATHY

The term "allelopathy" refers to the influence exerted by one plant on another by means of chemical exudates. The concept of allelopathy has been extended to include the effects of these plant products on microbial activities, particularly those microbes involved in the biogeochemical cycling of nitrogen (Rice, 1985, 1986).

It has been noted that the losses of nitrogen, especially as nitrate, are lower from ecosystems with stable climax communities than from disturbed ecosystems, early successional stage ecosystems, or agricultural systems. One explanation for this phenomenon is that the climax community plants produce allelopathic substances that influence the uptake of nitrogen by other plants and substances that inhibit the microbial oxidation of ammonia to nitrate (nitrification). Since nitrate is more mobile than ammonia in soil, it is advantageous to the plant for available nitrogen to remain in the more immobile form. Numerous studies have demonstrated low rates of nitrification and/or inhibition of nitrifying bacteria in the presence of various range, forest, and agricultural plants (Rice and Pancholy, 1972, 1973; Baldwin et al., 1983; Rice, 1985). Plant polyphenolic compounds, tannins in particular, are considered the allelochemicals most responsible for inhibition of nitrification (Rice and Pancholy, 1973). The autotrophic ammonia-oxidizing bacteria (*Nitrosomonas*) and nitrite-oxidizing bacteria (*Nitrobacter*) are thought to be especially susceptible to inhibition by polyphenolic plant products. Skujiņš and coworkers have tested the nitrification potential of arid soils under different types of shrubs (*Artemesia tridentata, Ceretoides lanata,* and *Atriplex confertifolia*) and have found a transient accumulation of nitrite, indicating a specific inhibition of nitrite oxidation (Skujiņš, 1981, 1984). Other allelochemical influences on nitrification in arid systems have also been reported (Rixon, 1971; Tiedemann and Klemmedson, 1973; Charley and West, 1977).

Nitrogen fixation by desert soil microorganisms is also susceptible to the allelopathic effects of desert shrubs. Rychert and Skujiņš (1974) compared nitrogen fixation rates beneath shrub canopies to rates in open spaces between plants. Rates of nitrogen fixation (measured as acetylene reduction) were lower by 74% under *A. confertifolia*, 49% under *Eurotia lanata,* and 86% under *A. tridentata.* Rice (1986) has also suggested that denitrification may be inhibited by allelopathy, but this may simply be the indirect effect of nitrification inhibition, resulting in lower availability of nitrate to the denitrifiers.

IX. SALINE-SODIC SOILS

High concentrations of soluble salts occur in saline soils, also called solonchak soils, a condition common in arid and semiarid regions. Sodic soils, also called solonetzic soils, are those in which exchangeable cations are dominated by sodium. When these conditions occur together, the soils are termed saline-sodic. Soil pH values in saline and saline-sodic soils are frequently even more alkaline than those usually found in arid and semiarid soils. High concentrations of soil salts can accumulate naturally at sites in which drainage is inadequate relative to rainfall. Salinity problems can be generated by sustained irrigation in poorly drained areas and/or irrigation with high-salinity water. Sewage sludge application or addition of processed oil shale wastes may also lead to buildup of soil salts. Two common measures of soil salinity are electroconductivity (EC) of a saturation extract and the sodium adsorption ratio (SAR) of the soil solution = $[Na]/[([Ca] + [Mg])/2]^{1/2}$, where all concentrations are in mEq L^{-1}. Electroconductivity values in excess of 4 dS m^{-1} indicate saline soil; SAR values above approximately 15 are indicative of sodic soils (U.S. Salinity Laboratory Staff, 1954; Shainberg and Letey, 1984).

As with most of the extreme conditions that occur in desert systems and that have been discussed thus far, soil salinity has a greater inhibitory effect on plants than on the soil microflora. Plants are inhibited in saline soils by the reduced solute water potentials and by the toxicity of specific ions, especially sodium and chloride ions. Extremely saline sites may be devoid of plant life but still harbor a variety of microbes. Microbes indigenous to saline soils are adapted to growth under conditions of low solute water potentials, as described in the section on water potential. Saline-soil microbes must also be tolerant of rapid changes occurring as saline soil solutions are diluted by precipitation. For this reason, saline soil microbes tend to be halotolerant rather than obligately halophilic. Quesada et al. (1982) cultured mostly halotolerant bacteria along with a few halophilic strains from five saline soils in Spain. Dominant genera were *Pseudomonas, Bacillus, Alcaligenes, Micrococcus, Arthrobacter,* and *Planococcus.* They also found typical rhizosphere effects with 10-100 times more culturable bacteria in the rhizospheres of halophytic plants than in the bulk soil. Gupta and Bajpai (1974) used standard plate count methods to compare microflora of nonsaline, moderately saline, and extremely saline soils in India. They found decreases in fungi with increases in salinity. Numbers of actinomycetes were relatively unaffected by salinity. These authors also reported an inverse relationship between numbers of nitrifying bacteria and salinity, with the nitrite oxidizers the most salt sensitive. Inhibition of *Nitrobacter* may be due to the toxicity of free ammonia under alkaline conditions (Focht and Martin, 1979). The alkaline conditions of sodic soils may be more inhibitory than the salinity; the one soil in the study by Gupta and Bajpai that was highly salt affected (EC = 28 dS m^{-1}) but only moderately alkaline (pH = 7.8) had bacterial and fungal populations comparable to those of their nonsaline soils. Soils with fewest bacteria and fungi were those with combined saline and alkaline conditions (EC = 33-84 dS m^{-1}, pH = 9.4-9.9). Using similar methods in a study of California soils, Killham and Firestone (1984) found relatively low numbers of total bacteria and high populations of actinomycetes in the most saline-sodic soil tested. A moderately halotolerant *Streptomyces griseus* was isolated from one of their saline soils and found to have typical osmoregulatory physiology for a moderately halotolerant bacterium, that is, synthesis of proline as an intracellular compatible solute.

Malik and Azam (1980) compared biomass and rates of decomposition and humification of halophytic plant residues in three Pakistani soils: one nonsaline (EC = 4 dS m^{-1}, SAR = 4.5), one saline (EC = 9.9 dS m^{-1}, SAR = 6.9), and one saline-sodic (EC = 8.6 dS m^{-1}, SAR = 16.9). Microbial biomass C values were 0.040, 0.032, and 0.029 mg g^{-1} for the three soils, respectively. Rates of decomposition and humification followed the same pattern. The halophytic plant of this study was a grass, *Diplachne fusca,* proposed for planting as a first successional stage for the recovery of salt-affected soils. Malik et al. (1979) tested the effects of soil salinity on five species of fungi isolated from saline soils and found that respiration, cellulase activity, and humification were all inhibited by increasing salinity. Of the fungi tested, *Chaetomium globosum* was the most halotolerant.

The saline soils of the Antarctic dry valleys are perhaps the most extreme desert habitat on earth, with combined conditions of low temperature, low nutrient availabilities, low matric water potentials, and low solute water potentials. Despite these extremes, a number of indigenous microbes have been found (Horowitz et al., 1972; Vishniac and Klinger, 1986; Miller and Leschine, 1984; Wynn-Williams, 1986). The diversity of microbial forms in these and other saline deserts testifies to the resistance and adaptability of microorganisms to extreme conditions.

Microorganisms are present in saline and saline-sodic soils (albeit in low numbers) and are therefore presumably actively engaged in their usual functions of biogeochemical cycling. Unfortunately, however, their activities cannot be expected to significantly alter conditions underlying salinity problems, that is, microbial uptake and oxidation-reduction reactions do not eliminate salts from the soil environment. However, organic amendments may stimulate microbial activities that are beneficial to soil structure, possibly increasing permeability characteristics (Sandoval and Gould, 1978). Benefits may also be derived in sodic soils from the microbial production of inorganic and organic acids. The application of elemental sulfur to stimulate microbial production of sulfuric acid has been suggested (Cundell, 1977). Chemical treatments intended to reduce sodium absorption ratios in saline soils, such as gypsum amendments, may have an inhibitory effect on the microflora, thereby reducing populations and diminishing microbially mediated nutrient cycling (Carter, 1986; Giambiagi and Lodeiro, 1988).

X. SUMMARY

Clearly, microorganisms and microbial processes are integral to the soil environment of arid and semiarid lands, and they are of paramount importance in the revitalization of these lands following disturbance. It is hoped that the information presented in this chapter has served to convince those previously uninitiated in soil microbiology while providing some new insights for those readers already possessing a microbiological perspective. In the study of arid soil reclamation, microbes must be considered in several important contexts. First, they function as sensitive indicators of the status of the soil vis-à-vis resource availability. Microbial numbers, diversity, biomass, and activities can all be tremendously useful as monitoring tools for the soil reclamation process. In this context, there is a need for more data on microbial biomass and activities in both undisturbed and disturbed arid lands. Greater use of quantitative biochemical assays of microbes and their various processes will be particularly valuable in arid land research. Second, microbes must be considered for their catalytic roles in soil formation (i.e., organic matter accumulation and stabilization and soil structuring) and in nutrient cycling. If these microbial functions are sufficiently understood, they can be manipulated to nurture beneficial microbial processes while limiting those that are less desirable in the reclamation process. Third, one must consider the direct effects of the microbes growing in close associations with plants and having direct effects on nutrient availability to those plants. Management practices should be designed to promote the formation of mycorrhizal associations and the associations of symbiotic and associative nitrogen-fixing bacteria in the rhizosphere. Fourth, the microbes must be considered sensitive, perturbable living systems. Although as a group the microbes are less sensitive to environmental extremes than are the higher organisms of the system, they cannot be taken for granted. The naturally occurring environmental limitations in desert soils cause the indigenous microflora to function at slow rates compared to the microbes in nonarid soils. Perturbations (e.g., by salinity or heavy metal accumulation) can further diminish microbial numbers and activities. Therefore, the choice of arid land manipulations, such as irrigation or organic matter amendments, must be made with consideration of microbial as well as plant and animal responses. Finally, the diverse adaptations of soil microorganisms to the often extreme conditions of desert soils make them fascinating objects of study in their own right.

REFERENCES

Adebayo, A. A., and Harris, R. F. (1971). Fungal growth responses to osmotic as compared to matric water potential. *Soil Sci. Soc. Am. Proc. 35*: 465-469.

Ahmed, M., Oades, J. M., and Ladd, J. N. (1982). Determination of ATP in soils: Effect of soil treatments. *Soil Biol. Biochem. 14*: 273-279.

Alexander, M. (1965). Biodegradation: Problems of molecular recalcitrance and microbial fallibility. *Adv. Appl. Microbiol. 7*: 35-80.

Alexander, M. (1977). *Introduction to Soil Microbiology*. Wiley, New York.

Allen, E. B. (1988). Some trajectories of succession in Wyoming sagebrush grassland: Implications for restoration. In *The Reconstruction of Disturbed Arid Lands*, E. B. Allen (Ed.). Westview Press, Boulder, Colorado, pp. 89-112.

Allen, E. B., and Allen, M. F. (1980). Natural re-establishment of vesicular-arbuscular mycorrhizae following stripmine reclamation in Wyoming. *J. Appl. Ecol. 17*: 139-147.

Allen, M. F., and MacMahon, J. A. (1985). Impact of disturbance on cold desert fungi: Comparative microscale dispersion patterns. *Pedobiologia 28*: 215-224.

Anderson, J. P. E., and Domsch, K. H. (1973). Quantification of bacterial and fungal contributions to soil respiration. *Arch. Mikrobiol. 93*: 113-127.

Anderson, J. P. E., and Domsch, K. H. (1975). Measurement of bacterial and fungal contributions to respiration of selected agricultural and forest soils. *Can. J. Microbiol. 21*: 314-322.

Anderson, J. P. E., and Domsch, K. H. (1978). A physiological method for the quantitative measurement of microbial biomass in soils. *Soil Biol. Biochem. 10*: 215-221.

Anderson, T.-H., and Domsch, K. H. (1989). Ratios of microbial biomass carbon to total organic carbon in arable soils. *Soil Biol. Biochem. 21*: 471-479.

Andrews, J. H., and Harris, R. F. (1986). r- and K-selection and microbial ecology. *Adv. Microb. Ecol. 9*: 99-148.

Atlas, R. M., and Bartha, R. (1987). *Microbial Ecology: Fundamentals and Applications*. Benjamin Cummings, Menlo Park, California.

Bae, H. C., and Casida, L. E. (1973). Responses of indigenous microorganisms to soil incubation as viewed by transmission electron microscopy of cell thin sections. *J. Bacteriol. 113*: 1462-1473.

Bae, H. C., Cota-Robles, E. H., and Casida, L. E. (1972). Microflora of soil as viewed by transmission electron microscopy. *Appl. Microbiol. 23*: 637-648.

Baker, E. H., and Wright, R. A. (1988). Microbiology of a duneland ecosystem in southern New Mexico, U.S.A. *J. Arid Environ. 15*: 253-259.

Bakken, L. R., and Olsen, R. A. (1986). Dwarf cells in soil. A result of starvation of "normal" bacteria, or a separate population? In *Perspectives in Microbial Ecology*, F. Megusar and M. Gantar (Eds.).

Slovene Society for Microbiology, Ljubljana, Yugoslavia, pp. 561-566.
Baldwin, I. T., Olson, R. K., and Reiners, W. A. (1983). Protein binding phenolics and the inhibition of nitrification in subalpine balsam fir soils. *Soil Biol. Biochem. 15*: 419-423.
Bamforth, S. S. (1980). Terrestrial protozoa. *J. Protozool. 27*: 33-36.
Bamforth, S. S. (1984). Microbial distributions in Arizona deserts and woodlands. *Soil Biol. Biochem. 16*: 133-137.
Barkay, T., Tripp, S. C., and Olson, B. H. (1985). Effect of metal-rich sewage sludge application on the bacterial communities of grasslands. *Appl. Environ. Microbiol. 49*: 333-337.
Barker, J. S. F., Starmer, W. T., and Vacek, D. C. (1987). Analysis of spatial and temporal variation in the community structure of yeasts associated with decaying *Opuntia* cactus. *Microb. Ecol. 14*: 267-276.
Binet, P. (1981). Short-term dynamics of minerals in arid ecosystems. In *Arid-land Ecosystems: Structure, Functioning and Management*, Vol. 2, D. W. Goodall and R. A. Perry (Eds.). Cambridge University Press, Cambridge, pp. 325-356.
Birch, H. F. (1958). Further observations on humus availability and nitrification. *Plant Soil 11*: 262-286.
Birch, H. F. (1959). The effect of soil drying on humus decomposition and nitrogen availability. *Plant Soil 10*: 9-32.
Bitton, G., Pancorbo, O. C., and Farrah, S. R. (1984). Virus transport and survival after land application of sewage sludge. *Appl. Environ. Microbiol. 47*: 905-909.
Boyle, M., and Paul, E. A. (1989). Carbon and nitrogen mineralization kinetics in soil previously amended with sewage sludge. *Soil Sci. Soc. Am. J. 53*: 99-103.
Brock, T. D., and Madigan, M. T. (1988). *Biology of Microorganisms*. Prentice Hall, Englewood Cliffs, New Jersey.
Brockman, E. R. (1976). Myxobacters from arid Mexican soil. *Appl. Environ. Microbiol. 32*: 642-644.
Brookes, P. C. (1986). The adenylate energy charge ratio and ATP concentration of the soil microbial biomass. In *Perspectives in Microbial Ecology*, F. Megusar and M. Gantar (Eds.). Slovene Society for Microbiology, Ljubljana, Yugoslavia, pp. 673-678.
Brookes, P. C., and McGrath, S. P. (1984). Effects of metal toxicity on the size of the soil microbial biomass. *J. Soil Sci. 35*: 341-346.
Brookes, P. C., Tate, K. R., and Jenkinson, D. S. (1983). The adenylate energy charge of the soil microbial biomass. *Soil Biol. Biochem. 15*: 9-16.
Brookes, P. C., Landman, A., Pruden, G., and Jenkinson, D. S. (1985). Chloroform fumigation and the release of soil nitrogen: A rapid direct extraction method to measure microbial biomass nitrogen in soil. *Soil Biol. Biochem. 17*: 837-842.
Brookes, P. C., Heijnen, C. E., McGrath, S. P., and Vance E. D. (1986). Soil microbial biomass estimates in soils contaminated with metals. *Soil Biol. Biochem. 18*: 383-388.
Brown, A. D. (1976). Microbial water stress. *Microbiol. Rev. 40*: 803-846.
Cacciari, I., and Lippi, D. (1987). Arthrobacters: Successful arid soil bacteria, a review. *Arid Soil Res. Rehab. 1*: 1-30.

Call, C. A., and McKell, C. M. (1982). Vesicular-arbuscular mycorrhizae a natural revegetation strategy for disposed processed oil shale. *Reclamation Reveg. Res. 1*: 337-347.

Cameron, R. E. (1969). Cold desert characteristics and problems relevant to other arid lands. In *Arid Lands in Perspective*, W. G. McGinnies and B. J. Goldman (Eds.). University of Arizona Press, Tuscon, pp. 167-205.

Carter, M. R. (1986). Microbial biomass and mineralizable nitrogen in solonetzic soils: Influence of gypsum and lime amendments. *Soil Biol. Biochem. 18*: 531-537.

Chang, F. H., and Broadbent, F. E. (1981). Influence of trace metals on carbon dioxide evolution from a yolo soil. *Soil Sci. 132*: 416-421.

Charley, J. L., and West, N. E. (1977). Micropatterns of nitrogen mineralization activity in soils of some shrub-dominated semi-desert ecosystems in Utah. *Soil Biol. Biochem. 9*: 357-365.

Chen, M., and Alexander, M. (1973). Survival of soil bacteria during prolonged desiccation. *Soil Biol. Biochem. 5*: 213-221.

Clarholm, M. (1981). Protozoan grazing of bacteria in soil-impact and importance. *Microb. Ecol. 7*: 343-350.

Clark, F. E., and Paul, E. A. (1970). The microflora of grassland. *Adv. Agron. 22*: 375-435.

Cook, R. J., and Duniway, J. M. (1981). Water relations in the life-cycles of soilborne plant pathogens. In *Water Potential Relationships in Soil Microbiology*, J. F. Parr, W. R. Gardner, and L. F. Elliott (Eds.). Soil Sci. Soc. Am., Madison, Wisconsin, pp. 119-139.

Cortez, J. (1989). Effect of drying and rewetting on mineralization and distribution of bacterial constituents in soil fractions. *Biol. Fertil. Soils 7*: 142-151.

Csonka, L. N. (1989). Physiological and genetic responses of bacteria to osmotic stress. *Microbiol. Rev. 53*: 121-147.

Cundell, A. M. (1977). The role of microorganisms in the revegetation of strip-mined land in the western United States. *J. Range Manag. 30*: 299-305.

Danielson, R. M. (1985). Mycorrhizae and reclamation of stressed terrestrial environments. In *Soil Reclamation Processes: Microbiological Analyses and Applications*, R. L. Tate and D. A. Klein (Eds.). Marcel Dekker, New York, pp. 173-201.

Day, A. D., Thompson, R. K., and Swingle, R. S. (1988). Sewage sludge as fertilizer for wheat forage in an arid environment. *J. Arid Environ. 15*: 209-214.

Debosz, K., Babich, H., and Stotzky, G. (1985). Toxicity of lead to soil respiration: Mediation by clay minerals, humic acids, and compost. *Bull. Environ. Contam. Toxicol. 35*: 517-524.

Doelman, P., and Haanstra, L. (1979). Effect of lead on soil respiration and dehydrogenase activity. *Soil Biol. Biochem. 11*: 475-479.

Doelman, P., and Haanstra, L. (1986). Short- and long-term effects of heavy metals on urease activity in soils. *Biol. Fertil. Soils 2*: 213-218.

Dommergues, Y. R., Belser, L. W., and Schmidt, E. L. (1978). Limiting factors for microbial growth and activity in soil. *Adv. Microb. Ecol. 2*: 49-104.

El-Giahmi, A. A., Nicolson, T. H., and Daft, M. J. (1976). Endomycorrhizal fungi from Libyan soils. *Trans. Br. Mycol. Soc. 67*: 164-169.

Elkins, N. Z., Steinberger, Y., and Whitford, W. G. (1982). Factors affecting the applicability of the AET model for decomposition in arid environments. *Ecology 63*: 579-580.

Elkins, N. Z., Parker, L. W., Aldon, E. F., and Whitford, W. G. (1984). Responses of soil biota to organic amendments in stripmine spoils in northwestern New Mexico. *J. Environ. Qual. 13*: 215-219.

Elliott, E. T., Coleman, D. C., and Cole, C. V. (1979). The influence of amoebae on the uptake of nitrogen by plants in gnotobiotic soil. In *The Soil Root Interface*, J. L. Harley and R. S. Russell (Eds.). Academic Press, London, pp. 221-229.

Elliott, E. T., Coleman, D. C., Ingham, R. E., and Trofymow, J. A. (1984a). Carbon and energy flow through microflora and microfauna in the soil subsystem of terrestrial ecosystems. In *Current Perspectives in Microbial Ecology*, M. J. Klug and C. A. Reddy (Eds.). American Society for Microbiology, Washington, D.C., pp. 424-433.

Elliott, L. F., Gilmour, C. M., Lynch, J. M., and Tittemore, D. (1984b). Bacterial colonization of plant roots. In *Microbial Plant Interactions*, R. L. Todd and J. E. Giddens (Eds.). Soil Sci. Soc. Am., Madison, Wisconsin, pp. 1-16.

Elliott, E. T., Hunt, H. W., Walter, D. E., and Moore, J. C. (1986). Microcosms mesocosms and ecosystems: Linking the laboratory to the field. In *Perspectives in Microbial Ecology*, F. Megušar and M. Gantar (Eds.). Slovene Society for Microbiology, Ljubljana, Yugoslavia, pp. 472-480.

Erlich, H. L. (1981). *Geomicrobiology*. Marcel Dekker, New York.

Evenari, M. (1985). The desert environment. In *Hot Deserts and Arid Shrublands*, A. M. Evenari, I. Noy-Meir, and D. W. Goodall (Eds.). Elsevier, Amsterdam, pp. 1-22.

Fairbanks, B. C., Woods, L. E., Bryant, R. J., Elliott, E. T., Cole, C. V., and Coleman, D. C. (1984). Limitations of ATP estimates of microbial biomass. *Soil Biol. Biochem. 16*: 549-558.

Focht, D. D., and Martin, J. P. (1979). Microbiological and biochemical aspects of semi-arid agricultural soils. In *Agriculture in Semi-arid Environments*, A. E. Hall, G. H. Cannell, and H. W. Lawton (Eds.). Springer-Verlag, Berlin, pp. 119-147.

Forster, S. M. (1979). Microbial aggregation of sand in an embryodune system. *Soil Biol. Biochem. 11*: 537-543.

Foster, R. C. (1988). Microenvironments of soil microorganisms. *Biol. Fertil. Soils 6*: 189-203.

Friedmann, E. I. (1971). Light and scanning electron microscopy of the endolithic desert algal habitat. *Phycologia 10*: 411-428.

Friedmann, E. I. (1980). Endolithic microbial life in hot and cold deserts. *Origins Life 10*: 223-235.

Friedmann, E. I. (1987). Melting snow in the dry valleys is a source of water for endolithic microorganisms. *Antarct. J. U.S. 13*: 162-163.

Fuhrmann, J., Davey, C. B., and Wollum, A. G. (1986). Desiccation tolerance of clover rhizobia in sterile soils. *Soil Sci. Soc. Am. J. 50*: 639-644.

Garcia-Moya, E., and McKell, C. M. (1970). Contributions of shrubs to the nitrogen economy of a desert-wash plant community. *Ecology 51*: 81-88.

Giambiagi, N., and Lodeiro, A. (1988). Responses of the bacterial microflora of a sodic saline soil to different concentrations of nitrogen and sodium in agar culture. *Soil Biol. Biochem. 21*: 177-178.

Giddens, J., and Todd, R. L. (1984). Rhizosphere microorganisms: Overview. In *Microbial Plant Interactions*, R. L. Todd and J. E. Giddens (Eds.). Soil Sci. Soc. Am., Madison Wisconsin, pp. 51-68.

Griffin, D. M. (1981). Water potential as a selective factor in the microbial ecology of soils. In *Water Potential Relationships in Soil Microbiology*, J. F. Parr, W. R. Gardner, and L. F. Elliott (Eds.). Soil Sci. Soc. Am., Madison, Wisconsin, pp. 141-151.

Griffin, D. M. (1972). *Ecology of Soil Fungi*, Chapman and Hall, London.

Griffin, D. M., and Luard, E. J. (1979). Water stress and microbial ecology. In *Strategies of Microbial Life in Extreme Environments*, M. Shilo (Ed.). Verlag Chemie, Berlin, pp. 49-63.

Gupta, B. R., and Bajpai, P. D. (1974). Some microbiological studies in salt-affected soils. I. Pattern of soil microbial population as affected by salinity and alkalinity. *J. Indian Soc. Soil Sci. 22*: 176-180.

Hanna, K., Bengis-Garber, C., Kushner, D. J., Kogut, M., and Kates, M. (1984). The effect of salt concentration on the phospholipid and fatty acid composition of the moderate halophile *Vibrio costicola*. *Can. J. Microbiol. 30*: 669-675.

Harris, R. F. (1981). Effect of water potential on microbial growth and activity. In *Water Potential Relationships in Soil Microbiology*, J. F. Parr, W. R. Gardner, and L. F. Elliott (Eds.). Soil Sci. Soc. Am., Madison, Wisconsin, pp. 23-95.

Hayman, D. S. (1978). Endomycorrhizae. In *Interactions Between Non-pathogenic Soil Microorganisms and Plants*, Y. R. Dommergues and S. V. Krupa (Eds.). Elsevier, Amsterdam, pp. 401-442.

Hedley, M. J., and Stewart, J. W. B. (1982). Method to measure microbial biomass phosphorus in soils. *Soil Biol. Biochem. 14*: 377-385.

Hersman, L. E., and Temple, K. L. (1978). ATP as a parameter for characterizing strip mine spoils. *Soil Sci. 126*: 350-352.

Hersman, L. E., and Temple, K. L. (1979). Comparisons of ATP, phosphatase, pectinolyase, and respiration as indicators of microbial activity in reclaimed coal strip mined spoils. *Soil Sci. 127*: 70-73.

Hethener, P. (1967). Activité microbiologique des sols à *Cupressus dupreziana* A. Camus au Tassili N'Ajjer (Sahara central). *Bull. Soc. Hist. Nat. Afr. Nord 58*: 39-100.

Horowitz, N. H., Cameron, R. E., and Hubbard, J. S. (1972). Microbiology of the dry valleys of Antarctica. *Science 176*: 242-245.

Imhoff, J. F., and Rodriguez-Valera, F. (1984). Betaine is the main compatible solute of halophilic eubacteria. *J. Bacteriol. 160*: 478-479.

Insam, H., and Domsch, K. H. (1988). Relationship between soil organic carbon and microbial biomass on chronosequences of reclamation sites. *Microb. Ecol. 15*: 177-188.

Insam, H., Parkinson, D., and Domsch, K. H. (1989). Influence of macroclimate on soil microbial biomass. *Soil Biol. Biochem. 21*: 211-221.

Jacks, J. V. (1965). The role of organisms in the early stages of soil formation. In *Experimental Pedology*, E. B. Hallsworth and D. V. Crawford (Eds.). Butterworths, Oxford, pp. 219-226.

Jenkins, M. B., Virginia, R. A., and Jarrell, W. M. (1987). Rhizobial ecology of the woody legume mesquite (*Prosopis glandulosa*) in the Sonoran Desert. *Appl. Environ. Microbiol. 53*: 36-40.

Jenkins, M. B., Virginia, R. A., and Jarrell, W. M. (1989). Depth distribution and seasonal populations of mesquite-nodulating rhizobia in warm desert ecosystems. *Soil Sci. Soc. Am. J. 52*: 1644-1650.

Jenkinson, D. S., and Powlson, D. S. (1976). The effects of biocidal treatments on metabolism in soil. V. A method for measuring soil biomass. *Soil Biol. Biochem. 8*: 209-213.

Jenkinson, D. S., Davidson, S. A., and Powlson, D. S. (1979). Adenosine triphosphate and microbial biomass in soil. *Soil Biol. Biochem. 11*: 521-527.

Jenny, H. (1980). *The Soil Resource: Origin and Behavior.* Springer-Verlag, New York.

Johnson, N. C., and McGraw, A.-C. (1988a). Vesicular-arbuscular mycorrhizae in taconite tailings. I. Incidence and spread of endonaceous fungi following reclamation. *Agr. Ecosyst. Environ. 21*: 135-142.

Johnson, N. C., and McGraw, A.-C. (1988b). Vesicular-arbuscular mycorrhizae in taconite tailings. II. Effects on reclamation practices. *Agr. Ecosyst. Environ. 21*: 143-152.

Kappen, L. (1988). Ecophysiological relationships in different climatic regions. In *Handbook of Lichenology*, Vol. II, M. Galun (Ed.). CRC Press, Boca Raton, Florida, pp. 37-100.

Karl, D. M. (1986). Determination of in situ microbial biomass, viability, metabolism, and growth. In *Bacteria in Nature*, Vol. 2, *Methods and Special Applications in Bacterial Ecology*, J. S. Poindexter and E. R. Leadbetter (Eds.). Plenum Press, New York, pp. 85-176.

Kieft, T. L., and Spence, S. D. (1988). Osmoregulation in *Thiobacillus ferrooxidans*: Stimulation of iron oxidation by proline and betaine under salt stress. *Curr. Microbiol. 17*: 255-258.

Kieft, T. L., Corbett, S., and Jarpe, M. (1987a). Osmoregulation in *Thiobacillus ferrooxidans* and *Thiobacillus thiooxidans*. Abstracts of the 87th Meeting of the American Society for Microbiology, Am. Soc. Microbiol., Washington, D.C., p. 192.

Kieft, T. L., Soroker, E., and Firestone, M. K. (1987b). Microbial biomass response to a rapid increase in water potential when dry soil is wetted. *Soil Biol. Biochem. 19*: 119-126.

Killham, K., and Firestone, M. K. (1984). Salt stress control of intracellular solutes in Streptomycetes indigenous to saline soils. *Appl. Environ. Microbiol. 47*: 301-306.

Klein, D. A., Sorensen, D. L., and Redente, E. F. (1985). Soil enzymes: A predictor of reclamation potential and progress. In *Soil Reclamation Processes: Microbiological Analyses and Applications*, R. L. Tate and D. A. Klein (Eds.). Marcel Dekker, New York, pp. 141-171.

Knight, W. G., and Skujiņš, J. (1981). ATP concentration and soil respiration at reduced water potentials in arid soils. *Soil Sci. Soc. Am. J. 45*: 657-660.

Koske, R. E., Sutton, J. C., and Sheppard, B. R. (1975). Ecology of *Endogone* in Lake Huron sand dunes. *Can. J. Bot. 53*: 87-93.

Kovda, V. A., Samoilova, E. M., Charley, J. L., and Skujiņš, J. (1979). Soil processes in arid lands. In *Arid-land Ecosystems: Structure, Functioning, and Management*, D. W. Goodall, R. A. Perry, and K. M. W. Howes (Eds.). Cambridge University Press, Cambridge, pp. 439-470.

Kroeckel, L., and Stolp, H. (1984). Influences of soil water potential on respiration and nitrogen fixation of *Azotobacter vinelandii*. *Plant Soil 79*: 37-49.

Kushner, D. J. (1978). Life at high salt and solute concentrations: Halophilic bacteria. In *Microbial Life in Extreme Environments*, D. J. Kushner (Ed.), Academic Press, New York, pp. 317-368.

Labeda, D. P. (1986). New Actinomycetes of commercial importance. In *Perspectives in Microbial Ecology*, F. Megušar and M. Gantar (Eds.). Slovene Society for Microbiology, Ljubljana, Yugoslavia, pp. 271-276.

Larkin, J. M., and Dunigan, E. (1973). Myxobacteria from southwestern USA soils. *Soil Sci. Soc. Am. Proc. 37*: 808-809.

Lechevalier, M. P. (1986). Nitrogen-fixing actinomycetes of the genus *Frankia*. In *Perspectives in Microbial Ecology*, F. Megušar and M. Gantar (Eds.). Slovene Society for Microbiology, Ljubljana, Yugoslavia, pp. 253-256.

Lobova, E. V. (1960). *Pochyv pustynnoi zony SSSR*. Akademiya Nauk, Moscow. (Translation: E. V. Lobova, 1967, *Soils of the Desert Zone of the USSR*. U.S. Department of Commerce, Springfield, Virginia.)

Luard, E. I. (1982a). Accumulation of intracellular solutes by two filamentous fungi in response to growth at low steady state osmotic potential. *J. Gen. Microbiol. 128*: 2563-2574.

Luard, E. I. (1982b). Effect of osmotic shock on some intracellular solutes in two filamentous fungi. *J. Gen. Microbiol. 128*: 2575-2581.

MacKay, W. P., Fisher, F. M., Silva, S., and Whitford, W. G. (1987). The effect of nitrogen, water, and sulfur amendments on surface litter decomposition in the Chihuahuan Desert. *J. Arid Environ. 12*: 223-232.

Mahmoud, S. A. Z., Abou-El-Fadl, M., and El Mofty, M. K. (1964). Studies on the rhizosphere microflora of a desert plant. *Folia Microbiol. 9*: 1-8.

Mahmoud, S. A. Z., Ishak, Y. Z., Ramadan, E. M., and Daft, M. J. (1987). Occurrence and infectivity of endomycorrhizas in Egyptian soils. *Egyp. J. Microbiol.* (Special issue): 47-56.

Malibari, A. A., Al-Fassi, F. A., and Ramadan, E. M. (1988). Incidence and infectivity of vesicular-arbuscular mycorrhizas in some Saudi soils. *Plant Soil 112*: 105-111.

Malik, K. A., and Azam, F. (1980). Effect of salinity on ^{14}C-labelled microbial biomass and its contribution to soil organic matter. *Pak. J. Bot. 12*: 117-127.

Malik, K. A., Bhatti, N. A., and Kauser, F. (1979). Effect of soil salinity on decomposition and humification of organic matter by some cellulolytic fungi. *Mycologia 71*: 811-820.

Martens, R. (1985). Estimation of the adenylate energy charge in unamended and amended agricultural soils. *Soil Biol. Biochem. 17*: 765-772.

Martin, J. K. (1977). Effect of soil moisture on the release of organic carbon from wheat. *Soil Biol. Biochem. 9*: 303-304.

Mary, P., Ochin, D., and Tailliez, R. (1986). Growth status of rhizobia in relation to their tolerance to low water activities and desiccation stresses. *Soil Biol. Biochem. 18*: 179-184.

McAneny, K. J., Harris, R. F., and Gardner, W. R. (1982). Bacterial water relations using polyethylene glycol 4000. *Soil Sci. Soc. Am. J. 46*: 542-547.

McKay, C. P., and Friedmann, E. I. (1985). The cryptoendolithic microbial environment in the Antarctic cold desert: Temperature variations in nature. *Polar Biol. 4*: 19-25.

Measures, J. C. (1975). Role of amino acids in osmoregulation of non-halophilic bacteria. *Nature 257*: 398-400.

Meentemeyer, V. (1978). Macroclimate and litter control of decomposition rates. *Ecology 59*: 465-472.

Millar, W. N., and Casida, L. E. (1970). Evidence for muramic acid in soil. *Can. J. Microbiol. 16*: 299-304.

Miller, K. J., and Leschine, S. B. (1984). A halotolerant *Planococcus* from Antarctic dry valley soil. *Curr. Microbiol. 11*: 205-210.

Miller, M. S., and Pepper, I. I. (1988). Survival of a fast-growing strain of lupin rhizobia in Sonoran Desert soils. *Soil Biol. Biochem. 20*: 323-327.

Miller, R. D. (1979). Some occurrences of vesicular-arbuscular mycorrhiza in natural and disturbed ecosystems of the Red Desert. *Can. J. Bot. 57*: 619-623.

Mills, A. L. (1985). Acid mine waste drainage: Microbial impact on the recovery of soil and water ecosystems. In *Soil Reclamation Processes: Microbiological Analyses and Applications*, R. L. Tate and D. A. Klein (Eds.). Marcel Dekker, New York, pp. 35-81.

Minderman, G. (1968). Addition, decomposition, and accumulation of organic matter in forests. *J. Ecol. 56*: 355-362.

Montana, C., Ezcurra, E., Carrillo, A., and Delhoume, A. (1988). The decomposition of litter in grasslands of northern Mexico: A comparison between arid and non-arid environments. *J. Arid Environ. 14*: 55-60.

Nakas, J. P., and Klein, D. A. (1980). Mineralization capacity of bacteria and fungi from the rhizosphere-rhizoplane of a semiarid grassland. *Appl. Environ. Microbiol. 39*: 113-117.

Nannipieri, P., Ciardi, C., Badalucco, L., and Casella, S. (1986). A method to determine soil DNA and RNA. *Soil Biol. Biochem. 18*: 275-281.

Nicot, J. (1955). Remarques sur les peuplements de micromycètes des sables désertiques. *C.R. Searces Acad. Sci. [D] 240*: 2082-2084.

Nicot, J. (1960). Some characteristics of the microflora in desert sands. In *The Ecology of Soil Fungi*, D. Parkinson and J. S. Waid (Eds.). Liverpool University Press, Liverpool, pp. 94-97.

Nordgren, A., Bååth, E., and Söderström, B. (1983). Microfungi and microbial activity along a heavy metal gradient. *Appl. Environ. Microbiol. 45*: 1829-1837.

Noy-Meir, I. (1973). Desert ecosystems: Environment and producers. *Annu. Rev. Ecol. Systematics 4*: 25-51.

Noy-Meir, I. (1974). Desert ecosystems: Higher trophic levels. *Annu. Rev. Ecol. Systematics 5*: 195-214.

Noy-Meir, I. (1985). Desert ecosystem structure and function. In *Hot Deserts and Arid Shrublands*, A. M. Evenari, I. Noy-Meir, and D. W. Goodall (Eds.). Elsevier, Amsterdam, pp. 93-103.

Pancorbo, O. C., Bitton, G., Farrah, S. R., Gifford, G. E., and Overman, A. R. (1988). Poliovirus retention in soil columns after application of chemical- and polyelectrolyte-conditioned dewatered sludges. *Appl. Environ. Microbiol. 54*: 118-123.

Papavivas, G. D., and Davey, C. B. (1961). Extent and nature of the rhizosphere of *Lupinus*. *Plant Soil 14*: 215-236.

Papendick, R. I., and Campbell, G. S. (1981). Theory and measurement of water potential. In *Water Potential Relationships in Soil Microbiology*, J. F. Parr, W. R. Gardner, and L. F. Elliott (Eds.). Soil Sci. Soc. Am., Madison, Wisconsin, pp. 1-22.

Parker, L. W., Freckman, D. W., Steinberger, Y., Driggers, L., and Whitford, W. G. (1984a). Effects of simulated rainfall and litter quantities on desert soil biota: Soil respiration, microflora, and protozoa. *Pedobiologia 27*: 185-195.

Parker, L. W., Santos, P. F., Phillips, J., and Whitford, W. J. (1984b). Carbon and nitrogen dynamics during the decomposition of litter and roots of a Chihuahuan Desert annual, *Lepidium lasiocarpum*. *Ecol. Monogr. 54*: 339-360.

Parker, L. W., Elkins, N. Z., Aldon, E. F., and Whitford, E. F. (1987). Decomposition and soil biota after reclamation of coal mine spoils in an arid region. *Biol. Fertil. Soils 3*: 109-114.

Paul, E. A., and Clark, F. E. (1989). *Soil Biology and Biochemistry*. Academic Press, San Diego.

Phillips, D. L., and McMahon, J. A. (1981). Competition and spacing patterns in desert shrubs. *J. Ecol. 69*: 97-115.

Pike, E. B., and Curds, C. R. (1971). The microbial ecology of the activated sludge process. In *Microbial Aspects of Pollution*, G. Sykes and F. A. Skinner (Eds.). Academic Press, London, pp. 123-147.

Poindexter, J. (1981). Oligotrophy: Fast and famine existence. *Adv. Microb. Ecol. 5*: 63-89.

Powlson, D. S., Brookes, P. C., and Christensen, B. T. (1987). Measurement of soil microbial biomass provides an early indication of changes in total soil organic matter due to straw incorporation. *Soil Biol. Biochem. 19*: 159-164.

Quesada, E., Ventosa, A., Rodriguez-Valera, F., and Ramos-Cormenzana, A. (1982). Types and properties of some bacteria isolated from hypersaline soils. *J. Appl. Bacteriol. 53*: 155-161.

Rao, A. V., and Venkateswarlu, B. (1983). Microbial ecology of the soils of Indian desert. *Agr. Ecosyst. Environ. 10*: 361-369.

Reber, H. H. (1989). Threshold levels of cadmium for soil respiration and growth of spring wheat (*Triticum aestivum* L.), and difficulties with their determination. *Biol. Fertil. Soils 7*: 152-157.

Reeves, F. B., Wagner, D., Moorman, T., and Kiel, J. (1979). The role of endomycorrhizae in revegetation practices in the semi-arid West. I. A comparison of incidence of mycorrhizae in severely disturbed vs. natural environments. *Am. J. Bot. 66*: 6-13.

Reichenbach, H. (1970). A new myxobacterium of the family Sorangiaceae. *Arch. Mikrobiol. 70*: 119-138.
Reichenbach, H., and Dworkin, M. (1981). The order Myxobacterales. In *The Prokaryotes: A Handbook on Habitats, Isolation, and Identification of Bacteria*, Vol. 1, M. P. Starr, H. Stolp, H. G. Truper, A. Balows, and H. G. Schlegel (Eds.). Springer-Verlag, New York, pp. 328-355.
Reid, C. P. P. (1984). Mycorrhizae: A root-soil interface in nutrition. In *Microbial Plant Interactions*, R. L. Todd and J. E. Giddens (Eds.). Soil Soc. Am., Madison, Wisconsin, pp. 29-50.
Rice, E. L. (1985). *Allelopathy*, 2nd ed. Academic Press, Orlando, Florida.
Rice, E. L. (1986). Allelopathic effects on nitrogen cycling. In *Perspectives in Microbial Ecology*, F. Megušar and M. Gantar (Eds.). Slovene Society for Microbiology, Ljubljana, Yugoslavia, pp. 325-331.
Rice, E. L., and Pancholy, S. K. (1972). Inhibition of nitrification by climax ecosystems. *Am. J. Bot. 59*: 1033-1040.
Rice, E. L., and Pancholy, S. K. (1973). Inhibition of nitrification by climax ecosystems. II. Additional evidence and possible role of tannins. *Am. J. Bot. 60*: 691-702.
Rixon, A. J. (1971). Oxygen uptake and nitrification by soil within a grazed *Atriplex vesicaria* community in semi-arid rangeland. *J. Range Manag. 24*: 435-439.
Rosacker, L., and Kieft, T. L. Biomass and adenylate energy charge of a grassland soil during drying. *Soil Biol. Biochem.* (in press).
Rychert, R., and Skujiņš, J. (1974). Nitrogen fixation by blue-green algae-lichen crusts in the Great Basin Desert. *Soil Sci. Soc. Am. Proc. 38*: 768-771.
Saggar, S., Bettany, J. R., and Stewart, J. W. B. (1981). Measurement of the microbial sulphur in soil. *Soil Biol. Biochem. 13*: 493-498.
Sala, O. E., Lauenroth, W. K., Parton, W. J., and Trlica, M. J. (1981). Water status of soil and vegetation in shortgrass steppe. *Oecologia (Berl.) 48*: 327-331.
Sandoval, F. M., and Gould, W. L. (1978). Improvement of saline- and sodium-affected disturbed lands. In *Reclamation of Drastically Disturbed Lands*. American Society of Agronomy, Madison, Wisconsin, pp. 485-504.
Santos, P. F., Elkins, N. Z., Steinberger, Y., and Whitford, W. G. (1984). A comparison of surface and buried *Larrea tridentata* leaf litter decomposition in North American hot deserts. *Ecology 65*: 278-284.
Schaefer, D., Steinberger, Y., and Whitford, W. G. (1985). The failure of nitrogen and lignin control of decomposition in a North American desert. *Oecologia (Berl.) 65*: 382-386.
Schimel, D. S., and Parton, W. J. (1986). Microclimate controls of nitrogen mineralization and nitrification in shortgrass steppe soils. *Plant Soil 93*: 347-357.
Schmidt, E. L. (1974). Quantitative autecological study of microorganisms in soil by immunofluorescence. *Soil Sci. 118*: 141-149.
Schmidt, E. L., and Paul, E. A. (1982). Microscopic methods for soil organisms. In *Methods of Soil Analysis*, Part 2, *Chemical and Microbio-*

logical Properties, 2nd ed., A. L. Page (Ed.). American Society of Agronomy and Soil Science Society of America, Madison, Wisconsin, pp. 803-820.

Schnurer, J., Clarholm, M., and Rosswall, T. (1985). Microbial mass and activity in an agricultural soil with different organic matter contents. *Soil Biol. Biochem. 17*: 611-618.

Sexstone, A. J., Revsbeck, N. P., Parkin, T. B., and Tiedje, J. M. (1985). Direct measurements of oxygen profiles and denitrification rates in soil aggregates. *Soil Sci. Soc. Am. J. 49*: 645-651.

Shainberg, I., and Letey, J. (1984). Responses of soils to sodic and saline conditions. *Hilgardia 52*(2): 1-55.

Shen, S. M., Pruden, G., and Jenkinson, D. S. (1984). Mineralization and immobilization of nitrogen in fumigated soil and the measurement of microbial biomass nitrogen. *Soil Biol. Biochem. 16*: 437-444.

Shields, J. A., Paul, E. A., Lowe, W. E., and Parkinson, D. (1073). Turnover of microbial tissue in soil under field conditions. *Soil Biol. Biochem. 5*: 753-764.

Skujiņš, J. (1981). Nitrogen cycling in arid ecosystems. In *Terrestrial Nitrogen Cycles*, F. E. Clark and T. Rosswall (Eds.). *Ecol. Bull. (Stockh.) 33*: 477-491.

Skujiņš, J. (1984). Microbial ecology of desert soils. *Adv. Microb. Ecol. 7*: 49-92.

Skujiņš, J., and Allen, M. F. (1986). Use of mycorrhizae for land rehabilitation. *MIRCEN J. 2*: 161-176.

Skujiņš, J., and Richardson, B. Z. (1985). Humic matter enrichment in reclaimed soils under semiarid conditions. *Geomicrobiol. J. 4*: 299-311.

Skujiņš, J., and West, N. E. (1973). Nitrogen dynamics in stands dominated by some major cold desert shrubs. US/IBP Desert Biome Res. Memo 73-35, Utah State University, Logan, Utah.

Skujiņš, J., Norstedt, H. O., and Odén, S. (1986). Development of a sensitive method for the determination of low-level toxic contamination in soils. *Swed. J. Agr. Res. 16*: 113-118.

Smith, J. L., and Paul, E. A. (1986). The role of soil type and vegetation on microbial biomass and activity. In *Perspectives in Microbial Ecology*, F. Megušar and M. Gantar (Eds.). Slovene Society for Microbiology, Ljubljana, Yugoslavia, pp. 460-466.

Smith, J. L., McNeal, B. L., and Cheng, H. H. (1985). Estimation of soil microbial biomass: An analysis of the respiratory response of soils. *Soil Biol. Biochem. 17*: 11-16.

Soderstrom, B. E. (1977). Vital staining of fungi in pure cultures and in soil with fluorescein diacetate. *Soil Biol. Biochem. 9*: 59-63.

Sommers, L. E., Gilmour, C. M., Wildung, R. E., and Beck, S. M. (1981). The effect of water potential on decomposition processes in soils. In *Water Potential Relationships in Soil Microbiology*, J. F. Parr, W. R. Gardner, and L. F. Elliott (Eds.). Soil Sci. Soc. Am., Madison, Wisconsin, pp. 97-117.

Soroker, E. F., Waldron, L. J., and Firestone, M. K. (1987). Effects of solute vs. matric water potential on microbial activity and survival in soil. *Agron. Abstr.*, Am. Soc. Agron., Madison, Wisconsin, p. 189.

Soroker, E. F., Caesar, A., Hendson, M., Schroth, M. N., and Firestone, M. K. (1988). The importance of salt-tolerance in the response of bacteria to water-potential stress in soil. *Agron. Abstr.*, Am. Soc. Agron., Madison, Wisconsin, p. 225

Sparling, G. P., Speir, T. W., and Whale, K. N. (1986). Changes in microbial biomass C, ATP content, soil phosphomonoesterase and phosphodiesterase activity following air-drying of soils. *Soil Biol. Biochem. 18*: 363-370.

Stevenson, I. L. (1956). Some observations on the microbial activity in remoistened air-dried soils. *Plant Soil 8*: 170-182.

Stout, J. D., Bamforth, S. S., and Lousier, J. D. (1982). Protozoa. In *Methods of Soil Analysis*, Part 2, *Chemical and Microbiological Properties*, 2nd ed., A. L. Page (Ed.). American Society of Agronomy/Soil Science Society of America, Madison, Wisconsin, pp. 1103-1120.

Stroo, H. F., and Jencks, E. M. (1982). Enzyme activity and respiration in minesoils. *Soil Sci. Am. J. 46*: 548-553.

Suflita, J., Horowitz, A., Shelton, D. R., and Tiedje, J. M. (1982). Dehalogenation: A novel pathway for the anaerobic biodegradation of haloaromatic compounds. *Science 218*: 1115-1117.

Swift, M. J., Heal, O. W., and Anderson, J. M. (1979). *Decomposition in Terrestrial Ecosystems*. University of California Press, Berkeley.

Tate, K. R., Ross, D. J., and Feltham, C. W. (1988). A direct extraction method to estimate soil microbial C: Effects of experimental variables and some different calibration procedures. *Soil Biol. Biochem. 20*: 329-335.

Tate, R. L. (1987). *Soil Organic Matter: Biological and Ecological Effects*. Wiley, New York.

Taylor, E. C. (1979). Seasonal distribution and abundance of fungi in two desert grassland communities. *J. Arid. Environ. 2*: 295-312.

Teplyakova, Z. F., and Maksimova, T. G. (1957). The occurrence of actinomycetes in soils of northern Kazakhstan. *Mikrobiologiya 26*: 327-333.

Tiedemann, A. R., and Klemmedson, J. O. (1973). Effect of mesquite on physical and chemical properties of the soil. *J. Range Manag. 26*: 27-29.

Tisdall, J. M., and Oades, J. M. (1979). Stabilization of soil aggregates by the root systems of ryegrass. *Aust. J. Soil Res. 17*: 429-441.

Tisdall, J. M., and Oades, J. M. (1980). The management of ryegrass to stabilize aggregates of a red-brown earth. *Aust. J. Soil Res. 18*: 415-422.

Tisdall, J. M., and Oades, J. M. (1982). Organic matter and water-stable aggregates in soils. *J. Soil Sci. 33*: 141-163.

Topper, K. F., and Sabey, B. R. (1986). Sewage sludge as a coal mine spoil amendment for revegetation in Colorado. *J. Environ. Qual. 15*: 44-49.

Trappe, J. M. (1981). Mycorrhizae and productivity of arid and semiarid rangelands. In *Advances in Food-Producing Systems for Arid and Semiarid Lands*, J. T. Manassha and E. J. Briskey (Eds.). Academic Press, New York, pp. 581-599.

Trappe, J. M. (1982). Synoptic keys to the genera and species of zygomycetous mycorrhizal fungi. *Phytopathology 72*: 1102-1108.
Trautwetter, A., and Blanco, C. (1988). Isolation and preliminary characterization of twenty bacteriophages infecting either *Brevibacterium* or *Arthrobacter* strains. *Appl. Environ. Microbiol. 54*: 1466-1471.
Trevors, J. T., Barkay, T., and Bourquin, W. (1987). Gene transfer among bacteria in soil and aquatic environments: A review. *Can. J. Microbiol. 33*: 191-198.
U.S. Salinity Laboratory Staff (1954). Diagnosis and improvement of saline and alkali soils. United States Dept. of Agriculture Handbook No. 60, U.S. Government Printing Office, Washington, D.C.
Vance, E. B., Brookes, P. C., and Jenkinson, D. S. (1987). An extraction method for measuring soil microbial biomass. *Soil Biol. Biochem. 19*: 703-707.
van Veen, J. A., Ladd, J. N., and Amato, M. (1985). Turnover of carbon and nitrogen through the microbial biomass in a sandy loam and a clay soil incubated with $[^{14}C(U)]$glucose and $[^{15}N](NH_4)_2SO_4$ under different moisture regimes. *Soil Biol. Biochem. 17*: 747-756.
Virginia, R. A., Jarrell, W. M., and Franco-Vizcaino, E. (1982). Direct measurement of denitrification in a *Prosopis* (mesquite) dominated Sonoran Desert ecosystem. *Oecologia (Berl.) 53*: 120-122.
Virginia, R. A., Jenkins, M. B., and Jarrell, W. M. (1986). Depth of root symbiont occurrence in soil. *Biol. Fertil. Soils 2*: 127-130.
Vishniac, H. S., and Klinger, J. (1986). Yeasts in the Antarctic deserts. In *Perspectives in Microbial Ecology*, F. Megušar and M. Gantar (Eds.). Slovene Society for Microbiology, Ljubljana, Yugoslavia, pp. 46-51.
Visser, S. (1985). Management of microbial processes in surface mined land reclamation in western Canada. In *Soil Reclamation Processes: Microbiological Analyses and Applications*, R. L. Tate and D. A. Klein (Eds.). Marcel Dekker, New York, pp. 203-241.
Vollmer, A. T., Au, F., and Bamberg, S. A. (1977). Observations on the distribution of microorganisms in desert soil. *Great Basin Naturalist 37*: 81-86.
Warner, N. J., Allen, M. F., and MacMahon, J. A. (1987). Dispersal agents of vesicular-arbuscular mycorrhizal fungi in a disturbed arid ecosystem. *Mycologia 79*: 721-730.
Webster, J. J., Hampton, G. J., and Leach, F. R. (1984). ATP in soil: A new extractant and extraction procedure. *Soil Biol. Biochem. 16*: 335-342.
West, A. W., and Sparling, G. P. (1986). Modifications to the substrate-induced respiration method to permit measurement of microbial biomass in soils of differing water contents. *J. Microbiol. Methods 5*: 177-189.
West, A. W., Sparling, G. P., Speir, T. W., and Wood, J. M. (1988). Comparison of microbial C, N-flush and ATP, and certain enzyme activities of different textured soils subject to gradual drying. *Aust. J. Soil Res. 26*: 217-229.
White, D. C. (1986). Quantitative physicochemical characterization of bacterial habitats. In *Bacteria in Nature*, Vol. 2, *Methods and Special Applications in Bacterial Ecology*, J. S. Poindexter and E. R. Leadbetter (Eds.). Plenum Press, New York, pp. 177-203.

White, J. A., Munn, L. C., and Williams, S. E. (1989). Edaphic and reclamation aspects of vesicular-arbuscular mycorrhizae in Wyoming red desert soils. *Soil Sci. Soc. Am. J. 53*: 86-90.

Whitford, W. G. (1986). Decomposition and nutrient cycling in deserts. In *Pattern and Process in Desert Ecosystems*, W. G. Whitford (Ed.). University of New Mexico Press, Albuquerque, pp. 93-117.

Whitford, W. G., Steinberger, Y., MacKay, W., Parker, L. W., Freckman, D., Wallwork, J. A., and Weems, D. (1986). *Oecologia (Berl.) 68*: 512-515.

Whitford, W. G., Aldon, E. F., Freckman, D. W., Steinberger, Y., and Parker, L. W. (1989). Effects of organic amendments on soil biota on a degraded rangeland. *J. Range Manag. 42*: 56-60.

Whittaker, R. H. (1969). *Communities and Ecosystems*. Macmillan, New York.

Wicklow, D. T. (1986). The role of competition in the structuring of microbial communities. In *Perspectives in Microbial Ecology*, F. Megušar and M. Gantar (Eds.). Slovene Society for Microbiology, Ljubljana, Yugoslavia, pp. 131-137.

Wildung, R. E., and Garland, T. R. (1985). Microbial development on oil shale wastes: Influence on geochemistry. In *Soil Reclamation Processes: Microbiological Analyses and Applications*, R. L. Tate and D. A. Klein (Eds.). Marcel Dekker, New York, pp. 107-139.

Williams, S. T. (1985). Oligotrophy in soil: Fact or fiction? In *Bacteria in Their Natural Environments*, M. Fletcher and G. D. Floodgate (Eds.). Academic Press, London, pp. 81-110.

Wong, P. T. W., and Griffin, D. M. (1976). Bacterial movement at high matric potentials. I. In natural and artificial soils. *Soil Biol. Biochem. 8*: 215-218.

Wullstein, L. H. (1980). Nitrogen fixation (acetylene reduction) associated with rhizosheaths of Indian rice-grass used in stabilization of the Slick Rock, Colorado tailings pile. *J. Range Manag. 33*: 204-206.

Wullstein, L. H., and Pratt, S. A. (1981). Scanning electron microscopy of rhizosheaths of *Oryzopsis hymenoides*. *Am. J. Bot. 68*: 408-419.

Wullstein, L. H., Bruening, M. L., and Bolling, W. B. (1979). Nitrogen-fixation associated with sand grain root sheaths (rhizosheaths) of certain xeric grasses. *Physiol. Plant 46*: 1-4.

Wynn-Williams, D. D. (1986). Microbial colonization of Antarctic fellfield soils. In *Perspectives in Microbial Ecology*, F. Megušar and M. Gantar (Eds.). Slovene Society for Microbiology, Ljubljana, Yugoslavia, pp. 191-200.

Yancey, P. H., Clark, M. E., Hand, S. C., Bowlus, R. D., and Somero, G. N. (1982). Living with water stress: Evolution of osmolyte systems. *Science 217*: 1214-1222.

Zak, J. C., and Whitford, W. G. (1986). The occurrence of a hypogeous ascomycete in the northern Chihuahuan Desert. *Mycologia 78*: 840-841.

Zimmerman, R., Iturriaga, R., and Becker-Birk, J. (1978). Simultaneous determination of the total number of aquatic bacteria and the number thereof involved in respiration. *Appl. Environ. Microbiol. 36*: 926-935.

8

Biological Surface Features of Semiarid Lands and Deserts

BLAINE METTING *Cyanotech Corporation, Woodinville, Washington*

I. INTRODUCTION

The lack of water in semiarid and desert regions restricts the development of vascular plant cover. As a result, most of the surface of the ground is exposed to full sunlight and the elements, in response to which a heterogeneous mix of microbiological and cryptogamic communities often occurs. Biological surface communities in semiarid lands and hot and cold deserts show a superficial similarity around the globe. Among factors most influential in determining the kinds of biological surface features and their activities are parent materials (mineral substrates), soil texture, light, source and availability of water, temperature regime, topography, history of use, and time.

Biological communities develop despite harsh climatic conditions because there exist microenvironments suitable for photosynthesis, if only on an ephemeral basis. These communities are variously composed of cyanobacteria, eukaryotic microalgae, lichens, filamentous fungi, and heterotrophic eubacteria. All are dependent upon photosynthesis by free-living or lichenized microalgae or on aerial deposition of organic material.

Friedmann and coworkers (Friedmann et al., 1967; Golubic et al., 1981) offered a descriptive terminology for organisms and microhabitats occupying surfaces in semiarid lands and deserts based on the nature of the substrate. Their scheme is as follows:

1. Edaphic habitats and organisms (*edaphos* = of the soil)
    a. Endedaphic: within or beneath the soil
    b. Epedaphic: upon the soil
    c. Hypolithic: beneath stones, pebbles, or other materials embedded in the surface
2. Lithophytic habitats and organisms (*lithos* = rock)
    a. Chasmolithic: within rock fissures and cracks

b. Cryptoendolithic: within the fabric of rocks; growing in the internal airspace of porous rocks, forming a more or less continuous horizontal layer parallel to the surface
c. Euendolithic: within the rock fabric as a consequence of active penetration of calcareous substrates

Biological surface features often play an important role in the larger context of soil formation and the development and maintenance of vascular plant communities. Surface stabilization against wind and water erosion, together with organic matter accumulation, moisture retention, and biological nitrogen fixation, is important and worthy of greater consideration when managing semiarid lands with productive potential for humankind. This chapter focuses on the terminology and structure of biological surface communities, on descriptive work from selected regions, and on physiological processes of microorganisms responsive to desiccation, excessive light, and extreme temperature. Soil stabilization and biological nitrogen fixation are covered as well. Practical considerations are touched on in the last section of this chapter (Sec. VII.C) and in more depth in Chap. 16.

II. VARIATIONS IN BIOLOGICAL SURFACE FEATURES OF SEMIARID AND ARID LAND

Climate, topography, substrate, and time combine to provide microenvironments and nanoenvironments for colonization by microorganisms in semiarid lands and deserts. Geological surface features gradate from rock to gravel to sand and, ultimately, to incipient soil (see Chap. 2). Microbial and cryptogamic communities are found along the entire gradient, with the exception of actively mobile dune systems. Moving dunes seldom afford sufficient intervals of time when water is available and photosynthetically active light permits the establishment and residence of cryptograms and other microorganisms. With minimal disturbance and sufficient time, essentially all other illuminated microhabitats in both cold and hot semiarid lands and deserts support well-developed communities.

Suitable microenvironments develop in bedrock, gravel (deposited in ephemeral or ancient stream beds), desert pavement, and soil. Gravel-strewn areas, known as desert pavement, cover extensive portions of most arid regions. Indeed, they are more prominent than either sand dunes or exposed bedrock in most hot and cold deserts (Smith, 1968). Desert pavement is often encrusted or cemented with various salts, carbonates, silicates, or gypsum. Typical of central Asia are *takyrs*, meaning in Turkic languages "places without plants." Takyrs are steppe soils with a bare parquetlike surface broken into polygonal aggregates (Kovda et al., 1979).

Less extensive although commonly encountered in arid regions are clay surfaces and salt deposits. By virtue of alluvial transport, these surfaces occupy points of low relief in the landscape. Clay surfaces, called playas in the Western Hemisphere, may be flaky and soft or firm when dry. When wet, they become fluid and sticky and support epedaphic cyanobacteria and microalgae. Subsequent drying results in characteristic "rain crusts," with a typical cracked appearance or a patchwork of dried algal flakes with curled edges (Cameron and Blank, 1966a). Because of their salt content, playa soils sustain relatively low densities of vascular plants, and microbial and cryptogamic communities are quite different from those on other soils. The mineralogy and chemistry of desert soils and clay surfaces is addressed in detail in Chaps. 3 and 4.

Incrustations and thicker salt deposits are sometimes of local significance, varying from smooth and flat to jagged. Endedaphic microalgae and cyanobacteria grow beneath the surface of flat incrustations at depths of a few centimeters where light penetrates, yet the relative humidity of the microenvironment permits their acquisition of water (Cameron and Blank, 1966a).

The dominant biological surface features in semiarid landscapes that support vascular plant communities are cryptogamic crusts. Cryptogams are nonvascular plants, including algae, fungi, lichens, and mosses. Algal crusts are composed of various species of cyanobacteria, eukaryotic microalgae, and filamentous fungi. Lichen crusts are dominated by lichens, which are symbiotic associations of fungi and eukaryotic microalgae or cyanobacteria. Lichen crusts were hypothesized by Cameron and Blank (1966a) to develop from algal crusts, although Killian (1953) reported that the soil algae of the Sahara were never lichenized. Mosses are often associated with algal crusts in the more moist microhabitats of semiarid regions, such as beneath shrubs and in seams or depressions in bedrock. Typical cryptogamic crusts are shown in Figs. 1 and 2.

FIGURE 1 Algal crust in the semiarid Tintic Mountains, Utah, showing a typical pedicled morphology. Pedicles are either with or without lichen caps. (From J. R. Johansen.)

a

FIGURE 2 Cryptogamic crust development in Camp Floyd State Park, Utah. An unburned, undisturbed area (a) is dominated by the moss *Collema tenax* and supports a number of lichens. An algal crust has begun to form on an adjacent area burned 5 years before the photograph (b). The polygonal mats resemble those of Asian takyr crusts. Lichens and mosses have yet to become re-established. (From J. R. Johansen.)

Lithophytic communities are ubiquitous in semiarid and desert regions. Communities develop in response to the hardness and mineral content of rocks, as well as to climate. These communities are less prone to disturbance from livestock and fire than adjacent communities yet are floristically less diverse. Descriptive accounts of cryptogamic crusts, desert varnish, and hypolithic and endolithic microbial communities are subjects of the following sections.

b

FIGURE 2 (continued)

III. CRYPTOGAMIC CRUSTS

A. Algal Crusts

Algal crusts abound on semiarid soils, becoming increasingly rare with increasing aridity. They are most well studied and described from the American Southwest (Cameron, 1960, 1964), the American Great Basin Desert (Table 1), and central Asia (Cameron and Blank, 1966a; Friedmann and Galun, 1974). Good descriptive accounts have also been published from other regions, including China (Reynaud and Lumpkin, 1988), Punjab State, India (Ali and Sandhu, 1972), the western and eastern Sahel of sub-Saharan Africa (Reynaud, 1987; Faurel et al., 1953), and the Kalahari (Skarpe and Henriksson, 1987) and Namibian (Welsh, 1965) deserts of southwest Africa.

The complexity and extent of development of algal crusts varies from thin, flat, ephemeral rain crusts to thickened, pedicled algal crusts.

Alluded to previously, rain crusts appear in small depressions or on broad flat expanses over which water stands for brief periods following rainfall. Microalgal populations develop during periods of free standing water. Single-cell flagellated stages of eukaryotic microalgae often dominate these ephemeral blooms. Desert rain crusts have also yielded the fungi *Rhizopus*, *Mucor*, and *Botrytis* (Fletcher and Martin, 1948).

Upon drying, crusts present a smooth, soil-colored stratum that cracks and curls upward into polygonal fragments. These crusts become pedicled with time owing to erosion around the edges of cyanobacterial mats. The presence of a pedicled crust indicates a potentially erodible soil, and destruction of these crusts can lead to rapid and severe soil loss (Rushforth and Brotherson, 1982). Algal crusts can be differentiated from pedicled rain crusts by their generally darker appearance and common association with lichens or mosses (Fig. 1). In contrast, rain crusts seldom show darkening or association with lichens or mosses. Less well developed crusts exist temporarily, only to become inconspicuous or to completely disappear upon drying (Cameron and Blank, 1966a).

Representing a relatively more permanent and structured epedaphic community, pedicled algal crusts are dominated by filamentous cyanobacteria and fungi. Less abundant but commonly present are filamentous, sarcinoid, and unicellular yellow-green (Xanthophyceae) and green (Chlorophyceae) microalgae, diatoms, and nonfilamentous cyanobacteria. Desert and semiarid regions of Asia include vast expanses irregularly covered by thin, soil-colored, somewhat lustrous algal crusts that occur in 5-25 cm polygonal fragments of uniform size (Bolyshev, 1955). Predominantly composed of filamentous nonheterocystous cyanobacteria, takyr crusts are dominated by species of *Microcoleus*, *Phormidium*, *Lyngbya*, and *Oscillatoria*. The ubiquitous *Nostoc* is commonly present but less abundant. *Oscillatoria* and *Phormidium* predominate in semiarid interior China together with heterocystous *Anabaena*, *Nostoc*, and *Tolypothrix* (Reynaud and Lumpkin, 1988).

Perhaps the most extensively studied algal crusts are those of semiarid North American rangelands and canyonlands. Predominant are the filamentous cyanobacteria. Table 1 lists species encountered on soils of the Great Basin and Colorado plateau. *Microcoleus vaginatus* is by far the most ubiquitous and abundant algal taxon in the deserts of North America and possibly the world. Other predominant species are *Lyngbya limnetica*, *Phormidium tenue*, *Nostoc commune*, *Nostoc muscorum*, and various species of *Oscillatoria* and *Scytonema*. Direct microscopic observations have revealed a rich diatom flora in most algal crusts from semiarid western North America, with *Hantzschia amphioxys*, *Navicula*, *mutica*, and *Pinnularia borealis* most abundant (Table 1). Enrichment for fungi commonly reveals the presence of *Aspergillus*, *Chaetomium*, *Penicillium*, and *Trichoderma* propagules.

An extensive floristic and ecological study of dry algal crusts in Senegal was made over a 10 year period by Reynaud (1987). Principal component analysis identified four distinct ecological groups of microalgae and cyanobacteria among 86 sites. The groups were (1) unicellular plus filamentous green microalgae and diatoms, (2) unicellular diazotrophic and non-N_2-fixing cyanobacteria, (3) homocystous cyanobacteria and *Calothrix* spp., and (4) branched and unbranched heterocystous cyanobacteria. Groups were distinguished on the basis of significant positive or negative correlations

TABLE 1 Most Common Cyanobacteria and Microalgae Occurring on Desert Soils of the Great Basin and Colorado Plateau, USA

	Source							
Species	1	2	3	4	5	6	7	8
Cyanobacteria								
Anabaena variabilis		x	x	x			x	x
Aphanothece castagnei								x
Arthrospira massartii	x							
Calothrix parietina			x					
Chlorogloea fritschii		x						
Chroococcus hansgiri			x					
C. minor								x
C. pallidus				x				
C. rufescens		x			x			
C. turgidis		x	x		x		x	x
Gloeocapsa aeruginosa				x				
G. punctata				x				
G. rupestris								x
Gloeothece linearis				x				x
Lyngbya limnetica	x	x	x					
Merismopedia punctata				x				
Microcoleus vaginatus	x	x	x	x	x	x	x	x
Nostoc commune		x	x					x
N. muscorum	x	x	x				x	x
N. paludosum								x
N. punctiforme			x					x
Oscillatoria agardhii	x							
O. angusta	x				x			
O. boryana					x			
O. germinata	x			x				x
O. lutea						x		
O. tenuis			x				x	
Phormidium minnesotense			x	x				x
P. tenue	x	x			x		x	
Schizothrix calcicola						x		
Scytonema hofmanni			x				x	
S. myochrous		x					x	

TABLE 1 (continued)

Species	Source 1	2	3	4	5	6	7	8
Cyanobacteria								
Stigonema hormoides			x					
Synechococcus aeruginosa	x		x	x		x	x	x
Tolypothrix tenuis			x	x				x
Chlorophyceae								
Chlorococcum			x				x	
Chlorosarcina brevispinosa			x			x		
Klebsormidium		x						x
Palmella miniata			x					
Euglenophyceae								
Euglena gracilis		x		x				
Bacillariophyceae								
Achnanthes exigua			x					
A. lanceolata			x					x
A. linearis		x	x					x
A. microcephala		x						
A. minutissima							x	
Amphora ovalis var. *aphinis*					x			
A. perpusilla					x			
Caloneis bacillum		x	x					
Cocconeis disculus							x	
C. placentula var. *lineata*							x	x
Cyclotella comta		x			x		x	
C. kuetzingiana			x		x			x
C. meneghiniana							x	
C. ocellata					x			
Cymbella naviculiformis							x	
C. turgida		x						
C. ventricosa							x	
Denticula elegans var. *valida*			x	x				x
Diatoma hiemale var. *mesodon*							x	

TABLE 1 (continued)

D. tenue var. *elongatum*							x	
D. vulgare		x					x	
Diploneis oblongella					x			
D. smithii							x	
Epithemia adnata var. *minor*								x
E. turgida			x					x
E. zebra							x	
Fragilaria construens var. *venter*							x	x
F. pinnata								x
F. vaucheriae								x
Gomphonema angustatum var. *producta*							x	
G. olivaceum							x	
G. parvulum			x					
Hantzschia amphioxys	x	x	x	x	x	x	x	x
Melosira granulata							x	
M. italica							x	
M. roseana		x						x
Navicula asellus			x					
N. contenta var. *biceps*								x
N. cryptocephala var. *exilis*							x	
N. cryptocephala var. *veneta*							x	
N. cryptocephaloides			x					
N. cuspidata							x	
N. elginensis var. *rostrata*								x
N. excelsa				x				x
N. lanceolata						x		
N. minima			x					
N. mutica	x	x	x	x	x	x	x	x
N. mutica var. *cohnii*	x	x	x		x			x
N. mutica var. *nivalis*					x		x	x
N. mutica var. *undulata*		x			x			
N. paramutica			x					x
N. pelliculosa					x			
N. peregrina								x

TABLE 1 (continued)

Species	Source 1	2	3	4	5	6	7	8
N. subundulata	x							
N. tenelloides			x					
N. tripunctata		x			x		x	
N. tripunctata var. *schizonemoides*		x						
Nitzschia communis			x					
N. dissipata							x	
N. hantzschiana			x					
N. inconspicua			x					
N. obtusa							x	
N. palea			x				x	
N. paleacea								x
Pinnularia appendiculata		x						x
P. borealis	x	x	x	x	x		x	x
P. brebissonii							x	
P. brebissonii var. *diminuata*							x	
P. microstauron		x						
P. viridis							x	
Rhoicosphaenia curvata							x	x
Rhopalodia gibba		x						x
R. gibberula var. *vanheurckii*			x					x
Stauroneis kriegeri					x			
Stephanodiscus astraea var. *minutula*		x					x	
S. carconensis			x		x			x
S. Niagarae							x	
Surirella ovalis							x	
Synedra fasciculata var. *truncata*								x
S. ulna							x	
Tabellaria fenestrata							x	

Sources: (1) Mills, Utah (St. Clair et al., 1986); (2) northern Arizona (Johansen et al., 1981); (3) Uintah Basin, Utah (Ashley et al., 1985); (4) Tintic Mountains, Utah (Johansen and Rushforth, 1985); (5) Wallsburg, Utah (Johansen et al., 1982); (6) Uintah Basin, Utah (Johansen and Rushforth, 1981); (7) Southern Utah (Anderson and Rushforth, 1976); (8) Camp Floyd, Utah (Johansen and St. Clair, 1986; Johansen et al., 1984).

TABLE 2 Changes in the Floristic Makeup and Standing Crops of Cyanobacterial Crusts Over a 4 Year Period on a Stable Dune Near Thile-Boubacar, Fleuve Region, Senegal[a]

Taxa	July 1979, dune Crest	July 1979, dune Base	February 1985, dune Crest	February 1985, dune Base
Pseudanabaena	3.1	3.6	2.2	17.0
LPP taxa	33.4	23.6	12.3	43.5
Lyngbya	4.7	2.7	1.0	75.0
Scytonema	11.0	11.7	4.9	4.1
Nostoc-Anabaena	0.5	2.8	0.1	2.1
Calothrix	—	0.3	—	0.1
Total biomass	52.7	44.8	20.4	138.2
% N_2-fixing strains	22%	33%	24%	4.5%
Total no. of strains	5	12	8	12

[a]Biomass units are wet mg cm^{-2}.

Source: P. A. Reynaud (unpublished data).

among taxa as well as to edaphic factors, the most important of which were pH and available moisture at the time of sampling. Dry algal crusts from stabilized dunes were similar to crusts from nearby fallow rice fields during the dry season. Both were composed of homocystous and heterocystous cyanobacteria totaling between 10^9 and 10^{10} μm^3 cm^{-2}.

In a separate study of 12 habitats in the Sahel of Senegal, two structurally distinct algal crusts were found (P. A. Reynaud, personal communication). One form consisted of a more or less flat array of interwoven homocystous cyanobacterial filaments, principally species of the LPP (*Lyngbya-Phormidium-Plectonema*) taxonomic complex. Contained at irregularly spaced intervals within the LPP weft were occasional *Nostoc* colonies (*Nostoc punctiforme, N. muscorum,* and *N. commune*). The second structural form of crust was characterized by a cylindrical outer collection of homocystous filamentous species enclosing a core of heterocystous filaments. Crusts were up to 0.2 mm thick (living filaments alone), with aggregated sand particles making the total crust 2-3 mm in thickness.

Four algal crusts from the Sahel were studied in detail (P. A. Reynaud, personal communication). Tables show both floristic and structural differences in the crusts sampled in 1979 and 1985 (Table 2) and their response to rewetting in the laboratory following 4 years of dry storage in the dark (Tables 3 and 4). The sand dune crusts at Thile-Boubacar (Fleuve region, Senegal) were sampled in the dry condition on both dates. During the 6 year interval, the total biomass at the base of the dune increased threefold but that at the crest decreased by a factor of 2.6, as estimated by the serial-dilution method modified for filamentous soil algae (Reynaud and Laloe, 1985). Relative floristic compositions within the

TABLE 3 Changes in Biomass and Floristic Composition of Four Sahelian Algal Crusts from Senegal After 4 Years Dry Storage[a]

Taxa	% Change in wet mg/cm^2			
	Crust A	Crust B	Crust C	Crust D
Pseudanabaena	-1.5	-3.5	+1.1	+4.3
Lyngbya	-4.4	-11.8	-6.9	-10.9
Other LPP taxa	-13.1	-46.5	-30.5	+8.8
Scytonema	+1.0	+13.3	-2.1	-3.0
Nostoc-Anabaena	-0.4	_0.2	+4.1	+2.8
Total biomass	-18.6	-48.3	-35.0	+1.9
% Total biomass change	-35.3	-50.5	-36.3	+4.5
% Change N_2-fixing taxa	+13	+56	+7	-1

[a]Crusts are from Fete-Ole, Thile-Boubacar region (A and B) and Dara and Ferlo, Fleuve region (C and D).

Source: P. A. Reynaud (unpublished data).

TABLE 4 Floristic and Biomass Fluctuations in Four Sahelian Algal Crusts Kept Moist for 3 Weeks Following 4 Years Dry Storage[a]

Taxa	% Change in wet mg/cm^2			
	Crust A	Crust B	Crust C	Crust D
Pseudanabaena	+2.4	+2.4	-5.7	+50.1
Lyngbya	+4.6	+8.7	+26.2	+37.4
Other LPP taxa	-6.1	-3.7	-11.6	+66.5
Scytonema	+13.0	-9.3	+31.0	+1.8
Nostoc-Anabaena	+0.3	0	-0.7	+0.5
Total biomass	+14.1	-1.9	+39.2	+156.3
% Total biomass change	+40	-4	+64	+354
% Change N_2-fixing taxa	+17	-17	+24	-9

[a]Crusts are from Fete-Ole, Thile-Boubacar region (A and B) and Dara and Ferlo, Fleuve region (C and D).

Source: P. A. Reynaud (unpublished data).

crusts did not change with time. Table 3 shows the effects of 4 years of dry storage on biomass and taxonomic composition of the Thile-Boubacar and two other crusts. Table 4 includes data illustrating floristic evolution and biomass fluctuation on the same crusts rewetted in the laboratory over 3 weeks. An interesting note is that the oscillatorian taxa *Oscillatoria* (single trichome per sheath) and *Microcoleus* (many trichomes within a broad diffuse sheath) were not found in Senegalese crusts by Reynaud (1987) (Tables 2, 3, and 4), but they are ubiquitous in arid and semiarid regions in Asia and North America. The *Nostoc-Anabaena* taxa, on the other hand, are common worldwide.

Edaphic fungi have important roles in cryptogamic crusts. Fungi are the principal agents of mineralization of complex organic materials, and in semiarid and arid lands, filamentous fungi contribute to the structural integrity of the soil (Bond and Harris, 1964; Friedmann and Galun, 1974). Durrell and Shields (1960) isolated 41 fungi from an arid Nevada soil. Hunt and Durrell (1966) listed 48 fungi enriched from soils collected in Death Valley, California. Prominent genera were mostly prolific spore-formers (*Aspergillus, Curvularia, Fusarium, Penicillium,* and *Phoma)*, which may or may not reflect the makeup of the fungal flora active in cryptogamic crusts when moisture is available.

B. Lichen Crusts

Lichen crusts are usually more conspicuous than algal crusts and vary in color from dark brown or black to an assortment of white, pink, blue, and green lichens. Deserts harbor a rather diverse flora, with the dominant growth form encountered the crustose lichen. In arid regions, lichen crusts may develop from algal crusts by means of gradual parasitism by fungi. Progressive lichenization of takyrs has been hypothesized to result in complete transformation to a lichen crust, although direct experimental evidence is lacking. Lichen crusts are propagated by fragmentation and by production of vegetative diaspores. Propagules are dispersed by wind or water movement as well as by mitratory animals and livestock. Where fragments of lichen crusts are deposited and lichenization of algal crusts takes place, different patterns of microstratification are possible. Individual crusts can become confluent over more or less extensive areas. As with algal crusts, erosive forces may remove soil from the perimeters of individual crusts, leaving them atop small pedicles extending upward several centimeters or decimeters (Cameron and Blank, 1966a).

Friedmann and Galun (1974) discuss in detail the phytogeography of lichens in semiarid lands and deserts. Most commonly encountered in North America are species of *Collema, Dermatocarpon, Lecanora, Lecidea,* and *Ramalina* (Rushforth and Brotherson, 1982). Semiarid areas of Iran and central Asia support species of *Acarospora, Buellia, Caloplaca, Dermatocarpon, Heppia, Lecanora, Lecidea, Parmelia, Ramalina, Rinodina,* and *Roccella* (Magnusson, 1940-1944; Steiner, 1916; and others cited by Friedmann and Galun, 1974). Frequently reported from dry locales around the world are the terricolous species *Dermatocarpon hapaticum, Collema tenax, Toninia coeruleonigricans, Fulgensia fulgens, Lecides decipens, Squamarina crassa,* and *Squamarina lentigera* (Friedmann and Galun, 1974). Arctic tundras are characterized by the genus *Cladonia* (Reichert, 1953).

Hot deserts support a surprisingly diverse group of lichens as well. For example, Follmann (1965a) listed 153 species from the Atacama Desert, Chile. Among lichens from that very dry region of South America, including coastal Peru, are found species of *Acarospora*, *Buellia*, *Caloplaca*, *Collema*, *Ramalina*, *Solenospora*, and *Teloschites* (Thomson and Iltis, 1968). Faurel et al. (1953) compiled a list of 114 lichen species from the literature pertaining to the Sahara, of which 111 were crustose. One-fifth were in the genera *Dermatocarpon* and *Heppia*, which otherwise represent only 1% of the global lichen flora (Friedmann and Galun, 1974). The Negev Desert of Israel also supports a rich lichen flora, most of which are again of the crustose habit (Galun, 1966; Galun and Reichert, 1960).

Lichens in extreme environments are adapted morphologically for excessive irradiance and periods of desiccation (Herre, 1942; Weber, 1962, 1968). Most obvious is widespread adoption of the crustose habit, which minimizes evaporative surface area and is closely attached to the substrate. Xeromorphic characters include various amorphous exterior layers of necrotic or gelatinous (when wetted) material, or a powdery cover known as the pruina. *Buellia canescens*, for example, has no amorphous upper layer where found in Sweden but is covered by a thick pruina in the Negev Desert (Galun, 1963). Friedmann and Galun (1974) supposed that this superficial layer is of limited protection but may prolong the duration of photosynthetic activity each day by slightly reducing the rate at which water collected during the night as dew is lost.

Bright coloration may also make life more tolerable for desert lichens by reducing the absorption of light (Galun, 1963). Hypolithic growth of lichens beneath a layer of fine sand and dust may act in a similar manner (Follmann, 1965b). Vogel (1955) described a *Buellia* species hypolithic beneath quartz in South Africa that was inverted in the sense that the algal layer was above the fungal cortex. Presumably, quartz replaces the fungal thallus, which in other circumstances would act to ameliorate incoming light.

Ultrastructural investigation of *Caloplaca aurantia* revealed intracellular fungal invasion of the phycobiont (*Trebouxia* sp., Chlorophyceae) in xeric habitats, but those from mesic habitats showed only a superficial association of the partners (Ben-Shaul et al., 1969). Similar observations were reported by Plessl (1963), and Galun et al. (1970). In contrast, intimacy between fungus and cyanobacterium (*Gloeocapsa*) in *Gonohymenia* spp. does not reflect the habitat of origin (Paran et al., 1971).

C. Factors Influencing Cryptogamic Crusts

Substrate, microrelief, and disturbance patterns are the factors that determine the establishment and maintenance of cryptogamic crusts. Physical and chemical properties of substrates are equally important and dictate the primary nature of microhabitats within which cryptogamic communities exist. In turn, metabolic activities contribute to evolution of the substrate and are controlled by interactive microenvironmental parameters, the most important of which are light, water activity, and temperature. In general, lichen crusts are slower to develop and are more stable in the landscape, occupying areas of minimal disturbance, such as rock outcroppings and protected soils. Because they develop more rapidly, algal crusts can occupy microsites more prone to regular disturbance by wind, water, or other forces, such as grazing animals.

Studies of the relationships between cryptogamic crusts and soil type are not numerous. Few include observations taken from contrasting soils or from more than one point in time. Multivariate analysis of cryptogamic crusts has been used on occasion, however, to identify common relationships among comparable samples. Anderson et al. (1982) showed that the extent of crust development at five locations in southern Utah was correlated with texture, electrolytic conductivity, pH, and soil P. Both percentage cover by cryptogamic crusts and their floristic diversity were greater on soils with relatively greater silt content, alkaline pH, and high conductivity. These relationships are contrary to those governing the distribution of vascular plants in the same semiarid habitats. Experimental evidence of a causal effect does not exist, but it is logical to suggest that interception of sunlight and precipitation limits crust development beneath plant canopies and that opportunities for full development of crusts are thus found only where plants are absent, that is, on more saline, alkaline ground or in areas too arid for vascular plant development.

The influence of substrate and relief on the distribution of lichens in dry habitats of the Irano-Turanian and Saharo-Sindian regions and the Negev Desert was reported by Galun (1963). She categorized lichen populations in these habitats as composed of terrestrial species both characteristic and uncharacteristic of the study region, and saxicolous species growing on flint and calcareous stones (Galun, 1963; Friedmann and Galun, 1974).

External perturbations have been shown to profoundly influence the development of cryptogamic crusts. Included are fire, occasional catastrophic storms, and grazing pressure. In all cases, disturbance appears to result in floristic or structural simplification of the crust. In one study, the percentage relative frequencies and absolute densities of diatoms in a sagebrush steppe in Utah were reduced by burning but floristic composition remained comparable (Johansen et al., 1982). In another study, recovery patterns following fire showed that crusts dominated by cyanobacteria and microalgae re-established themselves in 2-3 years (Fig. 2). The more structurally complex lichen crusts, in contrast, had not been re-established after 5 years (Johansen et al., 1984). Uncommon climatic events can have a great influence as well. By direct microscopic examination, Johansen (1984) found that several cyanobacteria and sarcinoid green microalgae increased in abundance but diatom abundance decreased in an algal crust following a rainstorm whose intensity (11 cm precipitation in 7 days) qualified it as a once in 100 year occurrence.

The catastrophic external event most commonly experienced by cryptogamic crusts in semiarid regions is mechanical disturbance, most often trampling by grazing animals. Lichen and lichen-moss crusts can be severely damaged, but algal crusts are more forgiving (Brotherson et al., 1983; Johansen and St. Clair, 1986). Severe grazing pressure probably destroys crusts in some situations to the extent that ecosystem stability and production are compromised in the long term. In sites disturbed by either livestock or fire, the typical pattern of recovery in western North America is

3-8 years for the algal component and 10-15 years for the moss-lichen component (Johansen, 1986b). The importance of balancing acceptable secondary production with ecosystem maintenance is treated in Sec. VII.C.

IV. DIAPHANOUS SUBSTRATES AND HYPOLITHIC MICROALGAE

Diaphanous substrates are translucent and transparent materials beneath which microenvironmental conditions are conducive to algal growth. Embedded in soil or desert pavement, diaphanous materials include white quartz, chalcedony, bone, shell, and other animal fragments, gypsum, dolomite, calcite, salt, and thin sections of mica or feldspar. Ice and snow act as diaphanous substrates in cold deserts and, although not strictly hypolithic, are ecologically equivalent (Cameron and Blank, 1966a). Hypolithic algae have been described from deserts in Australia (Moewus, 1953; Tchan and Beadle, 1955), Chile (Schwabe, 1963), South Africa (Vogel, 1955), and the western United States (Shields and Drouet, 1962; Durrell, 1962; Cameron and Blank, 1966a). The mineralogical and optical properties of diaphanous materials were treated in detail by Cameron and Blank (1966a).

Hypolithic microhabitats differ considerably from adjacent surfaces with respect to insolation, available moisture, and chemical and thermal regimes. Ultraviolet and photosynthetically active radiation penetrate diaphanous materials to varying degrees. With refractive indices greater than 1, some percentage of incident sunlight is reflected. Depending on the nature of the material, sufficient visible wavelength light can penetrate up to a few centimeters, and penetration is greater at higher altitudes. The depth of penetration can be gauged by the accumulation of organic matter on the undersurface of the diaphanous substrate (Cameron and Blank, 1966a).

Berner and Evenari (1978) found floristic differences in hypolithic microalgal communities relative to types of flint in the Negev Desert. Nearly half of all dark flints sampled harbored microalgae in layers up to 40 mm (mostly 5-15 mm) in thickness. Only about 20% of light-colored flints housed microalgae, with, in contrast, algal layers of but 5-10 mm in thickness (maximum of 30 mm). Most likely, the more favorable microhabitats under darker flint were due to the greater attenuation of ultraviolet and excessive visible light.

Short-wave length ultraviolet is absorbed by quartz, a property that affords protection to the hypolithic community. Communities beneath materials more transparent to ultraviolet are proportionally more rich in cyanobacteria; others are predominantly eukaryotic microalgae. Reflection of a portion of the incident sunlight, together with the light color of most diaphanous materials and the increased specific heat capacity afforded by retained moisture, combine in hot deserts to reduce the temperature in the hypolithic microhabitat relative to the soil surface, which can reach 80°C (Cameron and Blank, 1966a).

Except for cryptogamic crusts, organic matter accumulation is greater around and under diaphanous substrates than elsewhere on arid region surfaces devoid of vascular plants. Including both living microorganisms and their debris, the thickness of the organic layer can exceed 1 cm and is largely responsible for retaining moisture against evaporation. The most active microalgal growth is at the organic matter/substrate interface. Heterotrophic bacteria, including diazotrophs (*Azotobacter* and *Clostridium*) can be isolated from the surrounding organic matter, suggesting a situation as complex as that found in algal or lichen crusts (Tchan and Beadle, 1955).

The same microalgae and filamentous cyanobacteria found in cryptogamic crusts are also observed in hypolithic microenvironments. However, coccoid cyanobacteria and green microalgae are more numerous and often dominate hypolithic communities; they include *Anacystis* and *Coccochloris* (cyanobacteria) and the coccoid chlorophytes *Bracteacoccus*, *Stichococcus*, *Spongiochloris*, *Dictyococcus*, *Chlorococcum*, and *Trochisia* (Cameron and Blank, 1966a; Chantanachat and Bold, 1962; Friedmann et al., 1967).

V. ENDOLITHIC MICROBIAL COMMUNITIES

A. Desert Varnish

Desert varnish, or rock varnish, consists of dark-colored deposits of Mn and Fe oxides whose deposition is usually of microbial origin. Biogenic rock varnishes form at microbial weathering fronts on hard surfaces in arid regions and result in the creation of habitable microenvironments. Beneath a layer of brown or black varnish, desert conditions of high irradiance and low water activity are ameliorated to an extent sufficient for the existence of endolithic microorganisms. Cyanobacteria, filamentous fungi, and chemoheterotrophic bacteria that precipitate Mn and/or Fe in pure culture have been isolated from rock varnishes (Krumbein and Jens, 1981).

Rock varnishes are brown, black, or gray, 5 μm to 1 mm thick coatings consisting principally of Mn and Fe oxides with admixtures of quartz, clay, and/or carbonates. Mn can be enriched as much as 10,000-fold over its concentration in the substrate, originating from both the rock and external sources, such as pollen and dust (Krumbein, 1969). Considerable organic material can be present and may contribute to the lacquerlike appearance of some varnishes. Dry cyanobacterial films can contribute to varnishes that appear gray or black (Krumbien and Jens, 1981). Mn is usually enriched to a greater extent than Fe by a still poorly understood process in which the participation of chemoorganotrophic *Pedomicrobium*- and *Metallogenium*-like bacteria are implicated (Dorn and Oberlander, 1981).

Field observations and experiments have provided a good understanding of the developmental progress of rock varnishes along what Krumbein and Jens (1981) termed microbial weathering fronts. The rate of formation and ultimate appearance depend on patterns of precipitation, composition of the rock substrate, and relief. Rock varnishes are peculiar to arid regions. This may reflect the predominance of mineral dissolution and weathering in more humid climates in which evapotranspiration ratios are less than 1. Varnishes are present nearly exclusively on hard surfaces and sandstones. On limestones, soft granites, or gneisses, rock weathering is more rapid than varnish deposition. Microrelief favors varnish formation on hard surfaces in relatively more moist places only where occasional puddling or cracks in the surface do not promote lichenization (Krumbein and Jens, 1981).

Experiments in the field have demonstrated rock varnish formation within 14 years in the Negev Desert. Upon maturation, microbial communities reach a structural equilibrium and maximum thickness that are probably related to attenuation of light and a reduced capacity for photosynthesis and organic matter recycling. Rock varnishes can accumulate in layers in response to periodic availability of above-normal precipitation leading to colonization of the varnish surface followed by new varnish formation. A final property of interest is the longevity of varnish once it is formed. A good indication of longevity comes from examination of ancient human structures and artifacts, including varnish formation on a Neolithic tool dating from between 6000 and 7000 years before the present (Krumbein, 1969).

B. Chasmolithic and Cryptoendolithic Microbial Communities

Chasmolithic microalgae colonize coarse fissures and microscopic cracks in both sedimentary and igneous rock, soft and hard rocks in all biomes, including hot and cold deserts. Cryptoendolithic microbial communities, in particular, are structurally and floristically similar worldwide. In granite, marble, silicified sandstone, and anorthosite, communities are usually chasmolithic, and in porous substrates (e.g., sandstone) the microhabitat and microinhabitants are better described as cryptoendolithic (Bell et al., 1986 and 1988; Friedmann et al., 1967; Friedmann, 1980; Friedmann and Galun, 1974; Friedmann and Ocampo-Friedmann, 1984).

Chasmolithic communities are floristically more diverse among regions; however, both structural and floristic gradients exist between chasmo- and cryptoendolithic microbial communities. Endolithic communities in hot and cold deserts bear superficial resemblance as well, but important differences have been found. These are summarized in Table 5.

Colonization by endolithic microorganisms depends upon the properties of the rock substrate, including color, opacity, and porosity. Endolithic communities are founded on photosynthesis; therefore only light-colored and translucent rocks are suitable. Availability of space is also important, as is a minimum density of the rock (Friedmann and Ocampo-Friedmann, 1984). Within the structures of the microclimate—once again, light, moisture, and temperature are paramount—endolithic communities develop that can include cyanobacteria, eukaryotic microalgae, lichens, heterotrophic eubacteria, yeasts, and filamentous fungi (Table 5).

Quantitative estimates of biomass are comparable between endolithic communities in hot and cold deserts. Organic matter estimates vary between about 32 and 177 g m^{-2}. Organic N varies from 1.8 to 7.1 g m^{-2} in cryptoendolithic lichen communities and from 1.3 to 2.4 g m^{-2} in cryptoendolithic communities of cyanobacteria (Bell and Sommerfeld, 1987; Friedmann, 1980; Friedmann and Ocampo-Friedmann, 1984).

Friedmann et al. (1980) reported estimates of ATP, chlorophyll, and organic N from edaphic and lithic microbial communities in the Antarctic cold, dry valleys of southern Victoria Land. Total organic N in rocks varied from 1.84 to 7.07 g m^{-2} in cryptoendolithic lichen communities and from 1.29 to 2.35 g m^{-2} in cyanobacterial communities. Organic N totaled 3.45 g m^{-2} for chasmolithic algal-bacterial growth in marble. ATP ranged from 0.7 to 15.4 mg m^{-2} for lichens, from 9.0 to 15.7 mg m^{-2} for cyanobacteria, and was 18.4 mg m^{-2} for a chasmolithic community. Chlorophyll ranged from 1.7 to 27.1 mg m^{-2} in lichen-dominated microhabitats and from

TABLE 5 Comparative Floristics of Endolithic Communities from Semiarid Lands and Deserts[a]

Semiarid Colorado sandstone	Hot deserts	Cold deserts
Cryptoendoliths[b]	Chasmoliths	Chasmoliths
Cyanobacteria	Cyanobacteria	Cyanobacteria
Coccoid strains[c]	Coccoid strains	Coccoid strains
Chroococcidiopsis	*Aphanocapsa*	*Chroococcidiopsis*
Gloeocapsa	*Aphanothece*	Chlorophyceae
Gloeothece	CHLOROPHYCEAE	*Hemichloris*
Synechococcus	Coccoid strains	Xanthophyceae
Filamentous strains	*Bracteacoccus*	*Heterococcus*
Anabaena	*Radiosphaera*	LICHENS[e]
Phormidium	*Trochiscia*	*Acarospora*
Chlorophyceae	Sarcinoid strains	*Buellia*
Coccoid strains	*Chlorosarcinopsis*	*Lecidea*
Chlorococcum	*Friedmannia*	Cryptoendoliths
Chlorella	Lichens (*Buellia*)[e]	Cyanobacteria
Coccomyxa	Cryptoendoliths	Coccoid strains
Myrmecia	Cyanobacteria	*Chroococcidiopsis*
Oocystis	Coccoid strains	LICHENS[e]
Chlorophyceae	*Chroococcus*	*Acarospora*
Sarcinoid strains	*Chroococcidiopsis*	*Buellia*
Borodinella	*Gloeocapsa*	*Lecidea*
Chlorosarcinopsis	Chlorophyceae	
Fasiculochloris	Occasional coccoid unicells	
Friedmannia		
Tetracystis		
Filamentous strains		
Klebsormidium		
Stichococcus[d]		

[a]Predominant lifeforms in each microhabitat are capitalized.

[b]Endolithic communities in mesic and semiarid lands vary considerably. Dominant lifeforms differ among substrates and microhabitats.

[c]Taxonomic consideration of coccoid cyanobacteria is difficult and not completely resolved.

[d]*Stichococcus* and, to a lesser extent, *Klebsormidium* are loosely filamentous, commonly occurring as oblong unicells.

[e]Endolithic lichens in hot and cold deserts usually harbor the eukaryotic coccoid green phycobiont *Trebouxia*.

Sources: Bell et al. (1986); Friedmann (1980); Friedmann and Galun (1974); Friedmann and Ocampo-Friedmann (1984).

1.8 to 6.2 mg m^{-2} for cyanobacteria. In the microalgal-bacterial community in marble, chlorophyll concentration was 97.7 mg m^{-2}. In all samples, organic matter was estimated to range from 32 to about 177 g m^{-2} (Friedmann et al., 1980), comparable to the 38-185 g m^{-2} reported from North American and Middle Eastern hot deserts (Friedmann, 1980).

Endolithic conditions in hot deserts impose more severe stress on microorganisms than in cold deserts, and this is reflected in floristic composition (Friedmann, 1980). Except in occasional chasmolithic microsites (Friedmann et al., 1967), hot desert rocks support cyanobacteria almost exclusively and eukaryotes predominate under cold desert conditions.

Hot desert cryptoendolithic microhabitats support cyanobacteria in the genera *Chroococcidiopsis* and, less frequently, *Chroococcus* or *Gloeocapsa*, which form a narrow green zone a few millimeters below and parallel to the rock surface in association with small colorless or orange-pigmented eubacteria (rods or cocci). Cold desert cryptoendolithic communities also include some that fit this description, the most common case a floristically more rich community with eukaryotic microalgae and lichens (Friedmann and Ocampo-Friedmann, 1984). In cold deserts the predominant endoliths are lichens in sandstones (*Buellia* spp.). Typical morphology shows an upper black zone about 1 mm thick, a middle 2-4 mm white zone, and a lowermost green zone. Cellular organization within zones accommodates the particulate and crystalline microtopography of the rock. The lichen is embedded in the rock matrix and covered with a hard rock crust. Thus a true thallus exists, but one unlike the plectenchymous tissue typical of more familiar lichens (Friedmann, 1982).

VI. PHYSIOLOGICAL ECOLOGY OF ARID LAND LICHENS AND MICROALGAE

Moisture availability is the factor most critical in determining the metabolic activity of biological surface communities in semiarid lands and deserts, as well as in landscapes not normally thought to be constrained by insufficient water. The structure and activity of microbial surface communities in the most extreme terrestrial environments, in contrast, are dictated by the simultaneous interactive effects of water activity, light, and temperature (Friedmann, 1980; Friedmann and Ocampo-Friedmann, 1984).

Foremost is the water requirement, whose availability to microorganisms is best reflected in terms of water activity or water potential. Water potential describes the force by which water is held. It is related to the energy that must be used in order that water be assimilated by a microbe in soil. Components of water potential include (1) the osmotic potential due to solutes dissolved in water, (2) matric potential, which includes forces associated with adsorption to surfaces and capillary effects, (3) gravitational potential, (4) pressure potential, resulting from external gas or hydraulic forces, and (5) overburden potential (Papendick and Campbell, 1981). Of these, only osmotic and matric potential are important in most microbial systems, including soils and rocks. Ignoring other components for their small contribution, soil water potential ψ is the sum of matric ψ_m and osmotic ψ_s potentials. For equilibrium conditions, the following thermodynamic relationship applies, with ψ expressed in energy units per mass (pascals; -1 bar = -100 kPa = -0.1 MPa) (Papendick and Campbell, 1981):

$$\psi = \frac{RT}{V_w} \ln a_w$$

where

R = universal gas constant (8.21×10^{-5} m^3 bar mol^{-1} K^{-1})
T = absolute temperature (K)
V_w = partial molal volume of water (1.8×10^{-5} m^3 mol^{-1} at 4°C)
a_w = water activity or the relative humidity of the soil atmospheric fraction ($a_w = P/P_0$, where P = vapor pressure of saturated air at the same temperature)

Lichens and free-living microalgae are remarkably tolerant of long periods of dryness. Cyanobacteria have been revived from herbarium sheets more than a century in age, and records for eukaryotic microalgae are measured in decades (Davis, 1972). Microorganisms are tolerant to very low water potentials. Some ascomycetes and yeasts tolerate solute-controlled stress down to -500 to -600 bar (a_w = 0.69-0.62), and numerous fungi and eubacteria tolerate potentials below -100 bar (Harris, 1981).

A handful of studies have been undertaken to investigate the water relations of lichens and microalgae from deserts. Generally, lichens are sensitive to microclimatic perturbations and are less tolerant to prolonged wet periods than to drought (Peveling, 1987). Morphology is important for water uptake and storage. Dry thalli can imbibe liquid water to saturation in 1-2 min and from the atmosphere in from 3 to 300 min depending on relative humidity. Water is preferentially stored in the cortex and algal layer or in the medulla of the lichen, varying with the habitat in which it is found. The algal component of most lichens comprises less than 10% of the thallus mass (Peveling, 1987).

The desert lichens *Ramalina* and *Evernia* achieve 90% of their maximal activities in water-saturated air at tissue water contents of only 35%. Photosynthesis increases with increased tissue water content to an optimum significantly less than full saturation (Peveling, 1987). Their compensation point at 10,000 lx (luminosity) and 80% relative humidity is at a tissue water potential of -300 bar. Other conditions being suitable, *Leparia* fixes CO_2 at -450 bar (Lange, 1969). Brock (1975a) compared photosynthesis in whole lichens to the separated algal component and found a measure of protection in the symbiosis. Alone, algal photosynthesis was detectable at water activities corresponding to -7 to -145 bar, whereas the whole lichen fixed C from -50 to -307 bar.

Metabolic and biochemical responses by desiccation-tolerant free-living cyanobacteria have been studied in some detail. A wide response range to matric and osmotic stress has been observed and is largely dependent on preconditioning and rate of change in water activity. An edaphic *Microcoleus* from semiarid Idaho proved sensitive to moisture stress, with growth inhibition at -7 bar and cessation at -18 bar. Photosynthesis began to be inhibited at that point and ceased at -28 bar, equivalent to the potential of seawater (Brock, 1975b). Adaptation to water stress by the endolithic coccoid cyanobacterium *Chroococcus* was studied at the ultrastructural level by Potts et al. (1983), who reported shrinkage of the protoplast at -673 MPa without change in overall cell shape or dimension. This resiliency by the gram-negative cyanobacterial envelope together with production of an extracellular mucilagenous sheath has long been supposed to aid in desiccation tolerance (Bewley, 1979; Whitton, 1987).

Physiological studies on *Chroococcus* and four strains of endolithic *Chroococcidiopsis* revealed photosynthetic rates decreasing linearly with both matric and osmotic stress, down to a point. Below -34 bar (a_w = 0.976), all *Chroococcidiopsis* strains showed a sharp drop in activity. In contrast, *Chroococcus* showed optimum photosynthesis between -30 and -40 bar. Absolute $^{14}CO_2$ uptake rates were higher in matric than osmotic systems at the same water potential (Potts and Friedmann, 1981).

Potts and coworkers have published a number of studies of the fundamental aspects of water stress response in the ubiquitous terrestrial heterocystous cyanobacterium *N. commune*. The intracellular ATP pool of *N. commune* resists rapid desiccation to -995 bar (a_w = 0.50). Nitrogenase activity was maintained at -1 bar but was lost with rapid drying (Potts et al., 1984; Potts and Bowman, 1985). Cells in exponential growth continued to manufacture protein for 90 min following desiccation, although neither polysomes nor unique membrane proteins were recovered from desiccated cells (Angeloni and Potts, 1986; Potts, 1986). Certain phycobilisome levels increase initially, but extensive proteolysis takes place at low potentials after 17 days (Potts, 1985). Unique membrane fatty acid profiles have been seen in desiccation-tolerant *N. commune* when compared to profiles reported in the literature for other cyanobacteria (Olie and Potts, 1986). Also, fatty acid and carotenoid profiles differ among strains of diverse geographic origin (Potts et al., 1987).

Water is the ultimate limiting factor for microbial growth in semiarid and desert regions, yet it is important to consider the interactions of other physical factors that determine water availability. The manner in which temperature and light interact with water to influence microbial activity differs between hot and cold deserts.

In temperate semiarid regions and hot deserts it may rain infrequently or not at all so that edaphic, hypolithic, and endolithic lichens and microalgae must imbibe water from the atmosphere. A source of liquid water does exist, however, where significant dew forms during the night and early morning hours. Positive net primary production in these instances can occur for a few minutes or hours early in the morning when light is first available. The period of activity is brief, being limited and finally halted when excessive temperatures are reached and the microenvironment is desiccated for the remaining daylight hours. Should high temperatures persist through the dark hours it is possible for there to be an overall net negative production due to dark respiration (Friedmann and Galun, 1974; Skujiņš, 1984).

Lichen density and species richness in hot deserts vary with exposure and its influence on dew formation. Longer periods of net photosynthesis, more complex thalli, and greater floristic diversity are associated with north-facing habitats in the Negev (Kappen et al., 1980). Interrelationships among light, temperature, and water activity were demonstrated in the laboratory for the Sonoran desert lichens *Parmelia* and *Acaraspora*. Between 40 and 130% thallus water content (weight H_2O/weight thallus) and uniform gas-exchange rates were observed for combinations of four light regimes and five temperatures. At low irradiance gas-exchange rates were depressed at higher temperatures. Temperature optima varied as functions of irradiance from 0 to 10°C at 250 μEin m^{-2} s^{-1} to 20° at 500-750 μEin m^{-2} s^{-1}. Results reflected the fact that the lichens were most active in winter, particularly in early morning hours (Nash et al., 1982). Estimates of CO_2 assimilation of 1.32 mg g^{-1} dry weight and 0.78 mg respiration loss per 24 h period have been reported for lichens from the Negev. These data have been extrapolated to a net production of 8.4% for 200 dew events per annum (Lange et al., 1970).

For many months each year, light is absent in high- (and low-) latitude cold deserts, such as the interior dry valleys of Antarctica. During the summer months, low temperatures limit water activity much of the time when light is available to support endolithic and hypolithic photosynthesis (Kappan and Friedmann, 1983). The relative humidity in cold desert microhabitats varies with temperature and light. On cloudy days, relative humidities up to 74% were measured in cold desert cryptoendolithic microhabitats; the corresponding values on clear days ranged from 16 to 47%. An 80% relative humidity was measured 5 days after a snowfall (Kappen et al., 1981). Of occasional importance, melting snow can provide liquid water in cold deserts (Friedmann, 1978). Together, darkness and low temperatures limit biological activity in cold deserts to fewer than 1000 h $year^{-1}$ (Friedmann et al., 1987). Wind can also be an important factor in determining the temperature and water activity of edaphic and lithic microenvironments (Kappen et al., 1981). A mathematical model constructed using empirical data predicted that microbial colonization of Antarctic soils and rocks should be possible on surfaces with zenith angle less than 15°. At greater angles colonization should be restricted to surfaces with azimuth angles less than 135° or greater than 225° (Nienow et al., 1988). Vertical zonation in rocks appears responsive primarily to photon flux. Maximum irradiance ranges from about 1500 μEin m^{-2} s^{-1} at rock surfaces to less than 1% of that value penetrating to the level of habitation by cryptogams and microorganisms (Kappen et al., 1981).

Experimental work has confirmed the resiliency of cyanobacteria and microalgae to low and high temperatures. Microorganisms are extremely tolerant of low temperatures, as demonstrated by routine lyophilization and storage in liquid N_2. Indigenous strains in soil collected from a hot desert soil survived, grew, and reproduced after extreme treatment in the laboratory (Cameron and Blank, 1966b). Neither abundance nor individual species diversity was affected after samples were held for 1-48 h at -79 and -195°C. Samples evacuated to 55 mmHg and then subjected to diurnal freeze-thaw cycles (15 h at 22°C and 9 h at either -79 or -195°C) for 24-168 h also yielded viable organisms. Species isolated after the treatments were *Microcoleus vaginatus*, *Micrococcus chthonoplastes*, and *Schizothrix calcicola* (cyanobacteria) and the microchaetophoralean chlorophyte *Protococcus* (= *Desmococcus*).

These studies provide data that support the existence of true psychrophily among microalgal populations. Among eukaryotic microalgae isolated from cold deserts, *Hemichloris antarctica* (Chlorophyceae) is a confirmed psychrophile inhabiting the cryptoendolithic zone beneath lichens in porous sandstones. Cells in culture were damaged above 20°C (Tschermak-Woess and Friedmann, 1984). Ecotypes of four antarctic *Heterococcus* species (Xanthophyceae) from edaphic and lithic microhabitats displayed their natural morphology (as seen in the field) only within a restricted temperature range of from 3-18 to 7-22°C. Branching decreased both above and below these ranges, which were generally optimal for growth in culture as well (Darling et al., 1987). As indicated, however, light, not thermal regime, plays the key role in determining zonation within rocks (Nienow et al., 1988).

High temperatures are less well tolerated by microorganisms and are more likely to be detrimental or lethal at greater water activities than under dry conditions. This distinction coupled to greater insolation makes hot deserts more difficult than cold deserts for colonization and habitation by microbes. At high temperatures, the thermal death of microorganisms generally obeys first-order kinetics, with the exception of certain bacterial spores. Metabolizing eukaryotic microalgal populations in culture perish in linear proportion to time spent at elevated temperature, whereas a biphasic response that includes an initial lag has been observed for cyanobacteria (Venkataraman, 1964). High temperatures are better tolerated in the desiccated state and, in particular, in dry soil (Metting, 1981). This has been exploited in India by preservation of cyanobacterial biofertilizer inocula in sand (Venkataraman, 1961). Comparable work with arid land microalgae is lacking.

VII. IMPORTANCE OF SURFACE MICROBIAL COMMUNITIES IN SEMIARID LANDS AND DESERTS

A. Consolidation and Stabilization of Surfaces

Four functional attributes of cryptogamic crusts are often cited as important for ecosystem stability. They include consolidation of surfaces against erosion, improvement of water infiltration and retention, enhancement of seedling establishment, and provision of nutrients by weathering and biological N_2 fixation (Metting et al., 1988). Because it has been subject to much experimentation and in situ observation, the topics of biological N_2 fixation and biological weathering of rock substrates and influence on seedling establishment are treated in the sections that follow (Secs. VII.B and C).

Cryptogams play similar roles both as constituents of mature semiarid and arid ecosystems and as pioneers in primary and secondary plant community succession (Metting, 1981). Free-living cyanobacteria and eukaryotic microalgae are early and vigorous colonizers of volcanic substrates (Henriksson et al., in press; Rayburn et al., 1982) and have for some time been understood to be important in the early stages of secondary succession in temperate habitats. Booth (1941) pioneered study of the relationship between soil algae and erosion control in secondary succession, and his work previewed more recent observations from semiarid lands and deserts. He reported that the loss of soil from dry fallow fields in

Oklahoma was 22 times greater in the absence of an algal crust and that soils with an algal cover retained more water. Early studies in semiarid western North American ecosystems found that soils with cryptogamic crusts contained as much as 300% more organic matter than barren soils and that this was related to moisture retention (Fletcher and Martin, 1948; Shields and Drouet, 1962). Cornet (1981) showed that a *Scytonema* crust in the Sahel of Senegal reduced the loss of moisture to the air nearly 18-fold and that the relative humidity under the first centimeter of a dry soil was 8-9% with a crust and 1.3% without.

Faust (1970) presented experimental evidence that quantified the effect of algal crusts on hydrological properties and erosion for two soils from semiarid Arizona. Using an outdoor rainfall simulator, he reported that significantly greater amounts of suspended sediment were carried in runoff water from the surfaces of denuded plots compared to those encrusted with cryptogams. Tables 6 and 7 show how different physical parameters were influenced by cryptogamic crusts in that study (Faust, 1970). Crusts without algae, referred to as rain crusts, formed following artificial rainfall. For the silt-loam, this was associated with a greater penetration resistance when dry but to reduced resistance for the sandy soil under the same circumstances (Table 6).

Table 7 includes highlights of data for infiltration and runoff from simulated rainfall studies on the two soils. Experiments compared heavy and light rainfall intensities and measured the consequent responses of infiltration rate, total runoff, sediment production rate, and suspended sediment production. Infiltration rates were generally found to be greater with cryptogamic crusts, although their influence was not as great as soil texture.

For all properties measured, the influence of rainfall intensity was great. From the 185 × 370 cm plots, total runoff from an intense rainfall was not much influenced by the presence of a cryptogamic crust but was significantly reduced by crusts in the low-intensity experiments. Likewise, Loope and Gifford (1972) found that dry cryptogamic crusts inhibited infiltration and exacerbated runoff during intense storms. Regardless of soil texture or rainfall intensity, sediment contained in the runoff (the erosiveness index)

TABLE 6 Thickness, Organic C Content, Total N Content, and Penetration Resistance of a Silt-Loam (SL) and Sandy (Sand) Soil With and Without Algal Crusts in Arizona

	+ Algal crust		No algal crust	
	SL	Sand	SL	Sand
Crust thickness, mm	6.45	4.45	5.25	3.70
Organic C, %	0.72	0.37	0.18	0.17
Total N, mg g^{-1} soil	0.58	0.27	0.47	0.15
Penetrometry, kg cm^{-2}	2.46	2.21	1.79	2.56

Source: Faust (1970).

TABLE 7 Influence of Algal Crusts on Infiltration and Erosion from Rainfall Simulation Experiments on a Silt-Loam (SL) and Sandy Soil (Sand) in Arizona: Comparison of Heavy (H) and Light (L) Rainfall Intensities

	+ Algal crust				No algal crust			
	SL	Sand	SL	Sand	SL	Sand	SL	Sand
	H	L	H	L	H	L	H	L
Infiltration rate, mm h^{-1}	11.8	21.7	21.1	24.1	12.9	15.4	20.4	23.4
Total runoff, mm of depth	70.7	5.2	53.4	0.6	68.3	15.1	53.3	2.0
Production rate of settleable sediments, g min^{-1}	6.67	0.63	4.08	0.82	8.42	1.76	7.20	0.96
Suspended sediment production rate, g min^{-1}	3.74	0.39	0.72	0.05	4.23	1.06	0.77	0.10

Source: Faust (1970).

was always greater from bare surfaces (Faust, 1970). In another study, field measurements showed the presence of cryptogamic crusts in a semiarid pinyon-juniper parkland in Arizona to be significantly correlated with the depth of water penetration and decreased runoff (Brotherson and Rushforth, 1983).

B. Biological N_2 Fixation and Mineral Weathering

By means of biological N_2 fixation, the cyanobacterial components of algal and lichen crusts are often the single largest source of N in many semiarid rangelands and desert ecosystems (see also Chaps. 9 and 16). Prior to widespread use of the acetylene reduction assay, Shields et al. (1961) clearly showed that cryptogamic crusts in dry regions contributed significant quantities of N (Table 8). Similar observations have been reported many times in the scientific literature from most semiarid and many desert regions worldwide (Gollerbach and Shtina, 1969; Lynn and Cameron, 1973; Metting, 1981).

The rates and total quantities of N fixed by cyanobacteria in algal and lichen crusts vary widely and depend on the same interactive microenvironmental parameters already discussed: light, available water, and temperature. Variation in time and among locations is great. Well-developed crusts under optimal conditions have been observed to fix at rates of up to 0.18 kg N ha^{-1} day^{-1} when continuously wet and 0.11 kg N ha^{-1} day^{-1} under cycling wet and dry conditions (Mayland et al., 1966). Estimated annual totals range from 1 to 100 kg N fixed ha^{-1} as reported in the literature (Table 9). How much is ultimately available to vascular plants has not been adequately researched, although as much as 70% may be lost by volatilization and denitrification (Skujiņš and Klubek, 1978; West and

TABLE 8 Nitrate and Amino N (ppm) of Desert Algal and Lichen Crusts

	No. samples	Average NO_3^- N	Average amino N
Gypsum sands			
+ Algal crust	20	5.6	212
No crust	41	2.1	0
Lava soils			
+ Algal crust	11	7.2	1504
+ Lichen crust	29	6.1	2019
No crust	10	1.5	866

Source: Shields et al. (1961).

Skujiņš, 1977). Incorporation of biologically fixed N by free-living cyanobacteria and lichens, ammonia excretion, and availability to vascular plants in semiarid zones have been demonstrated by ^{15}N methods (Fuller et al., 1960; Rogers et al., 1966; Mayland et al., 1966; Skujiņš and West, 1974).

In situ measurements of acetylene reduction by cryptogams in semiarid American habitats revealed maximum nitrogenase activity during the early morning hours, when sufficient water was presumably available from dew (Skujiņš and West, 1974). Studies of intact crusts removed to the laboratory showed optimal fixation at 200 μEin m^{-2} s^{-1}, with significant fixation between 10 and 30°C (Rychert et al., 1978). Among habitat types in semiarid western North America, in situ rates ranged from 0.26 to 9.2 ng N_2 cm^{-2} min^{-1} (Rychert et al., 1978), with the most active N_2 fixers crusts dominated by the lichens *Dermatocarpon*, *Grimmia*, and *Peltigera* (Snyder and Wullstein, 1973).

Cyanobacterial crusts from arid African deserts fix comparatively small quantities of N, if such a generalization can be made from the few studies in the literature. Renaut et al. (1975) estimated the in situ nitrogenase activity of continuously wetted cyanobacterial crusts from Morocco at about

TABLE 9 Estimates of N_2 Fixation by Cryptogamic Crusts

Ecosystem	Ecosystem component	Estimated annual N_2 fixation (kg ha^{-1} $year^{-1}$)
Sonoran desert (USA)	Algal crusts	7-11
	Lichen crusts	13-18
Australian arid zone desert	Algal crusts	1.3
Great Basin Desert (USA, semiarid)	Soil crusts	10-100

Source: Rychert et al. (1978).

300 nmol N_2 g^{-1} dry drust h^{-1} with significant inhibition at the high light (>60,000 lx) and temperature (56°C) regime in the field. Estimated N_2 fixation rates by *Scytonema*-dominated algal crusts from five semiarid sites in Nigeria corresponded to approximate annual inputs ranging from 3 to 0.3 g N m^{-2} across a geographic gradient from dry to relatively more dry savanna (Isichei, 1980; Stewart et al., 1978). Skarpe and Henriksson (1987) compared dark- and light-colored cyanobacterial crusts from the western Kalahari (Botswana) for nitrogenase activity in the laboratory. Within 24 h the dark crust, which covered only 1.5% of the collection site, showed a maximum activity corresponding to about 6.8 mmol N_2 fixed m^{-2} h^{-1} compared to 0.6 mmol m^{-2} h^{-1} for the sand-colored crust, which covered nearly 10 times more surface area. Species of *Nostoc* and *Scytonema* were associated with both kinds of crusts (Skarpe and Henriksson, 1987).

Water availability is paramount for fixation of N_2 by free-living or lichenized terrestrial cyanobacteria regardless of their location in wet, semiarid, or arid ecosystems. All terrestrial microorganisms must contend with alternating periods of desiccation and rewetting, a subject that has been the focus of research in the laboratory with edaphic cyanobacteria.

Rehydration of desiccated algal crusts is rapid and is followed, in succession, by respiration, photosynthesis, and nitrogenase activity (Scherer et al., 1984; Potts and Morrison, 1986). Fixation of N_2 by algal crusts is more sensitive to water stress than either photosynthesis or growth. The rate and final degree of desiccation both influence the resumption of N_2 fixation upon rewetting (Rodgers, 1977). Rychert and Skujiņš (1974) found that nitrogenase activity in lichen crusts decreased rapidly upon desiccation to about 2% of optimal activity at -1.3 MPa. Nitrogenase activity was not detected at lower water activities. The surface-to-mass ratio of cyanobacteria is proportional to their rates of water uptake and loss and subsequently, to the response of N_2 fixation to desiccation. This has been demonstrated for three terrestrial species of *Nostoc*. Biotic interactions influence N_2 fixation in cryptogamic crusts as well. Cyanobacterial biomass, but not total N fixed, was decreased by the activities of grazing microfauna in Egyptian desert soil in the laboratory (Ghabbour et al., 1980). The significance of grazing, parasitism, and other in situ biotic interactions is not known.

Heterotrophic N_2 fixation by *Azospirillum, Azotobacter, Frankia, Clostridium,* and other heterotrophic eubacteria in desert soils can be significant in the rhizosphere of vascular plants (see Chaps. 6 and 9). The extent and importance of heterotrophic fixation in surface microbial communities has not been the subject of definitive research. In contrast to the importance of N_2 fixation in cryptogamic crusts, endolithic microbial communities rely to a greater extent on abiotic sources of N conveyed to the microorganisms by wind and water. This appears to be the case for endolithic communities in both hot and cold deserts (Friedmann and Kibler, 1980).

Mineral weathering is accelerated in the presence of cryptogams primarily as a consequence of holding water in contact with rocks longer than would otherwise be the case. Under some circumstances in dry habitats, chasmo- and endolithic lichens may be among the most effective biological weathering agents (Krumbein, 1969). Also, the solution and uptake of rock phosphate by microbes (including cyanobacteria and eukaryotic microalgae) are known to result from acidification of their immediate surroundings, perhaps coupled to siderophore production (Bose et al., 1971; Smith et al., 1978). This attribute is undoubtedly of importance in the colonization of primary substrates, such as volcanic surfaces. On the habitat and ecosystem levels, however, the significance of cryptogamic crusts is probably not as important as physical weathering or the activities of plant roots or mycorrhizal fungi (see Chaps. 3, 6, and 9).

Cryptogamic crusts can be associated with increased levels of other mineral nutrients as well. Significantly greater concentrations of P, K, and organic matter were found in the upper 5 cm of virgin semiarid grassland in Utah compared to an adjacent, lightly grazed area (Kleiner and Harper, 1977). Whether these or other differences that may be found are attributable to cryptogamic crusts is a matter of conjecture, but it would not be surprising to find that positive relationships develop over the long term.

C. Implications for Rangeland Management and Desert Reclamation

The influence of cryptogamic crusts on surface consolidation and soil fertility points to their importance in rangeland ecology and desertification (Metting et al., 1988). Comparative studies within regions have shown how disturbance by fire or grazing can damage crusts and result in erosion (Brotherson et al., 1983; Johansen et al., 1982; Johansen and St. Clair, 1986). Therefore, cryptogamic crusts are quite likely to be of long-term importance in ecosystem stability and primary production. Destruction of crusts may hasten desertification in regions where short-duration intensive grazing is practiced, thus sacrificing renewable production for short-term gains (Johansen, 1986a, b).

Advocates of short-duration intensive grazing (the Savory method), in which animals are concentrated in small areas until most or all forage is consumed, point to increased secondary biomass production and a flush of seedling establishment following the breakup of crusts as sufficient justification (Savory, 1983). There is observational evidence that cryptogamic crusts do inhibit seedling establishment under some circumstances (Dulieu et al., 1977; Savory and Parsons, 1980; Steger, 1982). In contrast, crusts in the Great Basin Desert of Utah were associated with enhanced seedling establishment, a finding corroborated to a degree by field and glasshouse experiments (St. Clair et al., 1984).

Whether it is ecologically or economically feasible to attempt to promote the establishment and maintenance of cryptogamic crusts as a tool in rangeland management or desert reclamation remains to be seen. Preliminary experiments showed that small areas could be stabilized by soil-crust slurries in the field (St. Clair et al., 1986), and under irrigation it has been demonstrated that microalgae can be made to colonize temperate agricultural soils and rice paddies in an economical manner (see Chap. 16). Without further research, prospects for the utilization of cryptogamic crusts for habitat management should not be abandoned.

ACKNOWLEDGMENT

I would like to thank Jeffrey R. Johansen for providing the figures, for Table 1, and for his excellent and most constructive criticism of the manuscript. I thank Pierre A. Reynaud for sharing unpublished information and wish him well in his ornithological adventures.

REFERENCES

Ali, S., and Sandhu, G. R. (1972). Blue-green algae of the saline soils of the Punjab. *Oikos 23*: 228-272.

Anderson, D. C., and Rushforth, S. R. (1976). The cryptogamic flora of desert soil crusts in southern Utah. *Nova Hedwigia 28*: 691-729.

Anderson, D. C., Harper, K. T., and Holmgren, R. C. (1982). Factors influencing development of cryptogamic soil crusts in Utah deserts. *J. Range Manag. 35*: 180-185.

Angeloni, S. V., and Potts, M. (1986). Polysome turnover in immobilized cells of *Nostoc commune* (Cyanobacteria) exposed to water stress. *J. Bacteriol. 168*: 1036-1039.

Ashley, J., Rushforth, S. R., and Johansen, J. R. (1985). Soil algae of cryptogamic crusts from the Uintah Basin, Utah, U.S.A. *Great Basin Naturalist 45*: 432-442.

Bell, R. A., and Sommerfeld, M. R. (1987). Algal biomass and primary production within a temperate zone sandstone. *Am. J. Bot. 74*: 294-297.

Bell, R. A., Athey, P. V., and Sommerfeld, M. R. (1986). Cryptoendolithic algal communities of the Colorado Plateau. *J. Phycol. 22*: 429-435.

Bell, R. A., Athey, P. V., and Sommerfeld, M. R. (1988). Distribution of endolithic algae on the Colorado Plateau in northern Arizona. *Southwest. Naturalist 33*: 315-322.

Ben-Shaul, Y., Paran, N., and Galun, M. (1969). The ultrastructure of the association between phycobiont and mycobiont in three ecotypes of the lichen *Caloplaca aurantia* var. *aurantia*. *J. Microscop. (Paris) 8*: 415-422.

Berner, T., and Evenari, M. (1978). The influence of temperature and light penetration on the abundance of the hypolithic algae in the Negev Desert of Israel. *Oecologia (Berl.)*: 255-260.

Bewley, J. D. (1979). Physiological aspects of desiccation tolerance. *Annu. Rev. Plant Physiol. 30*: 195-238.

Bolyshev, N. N. (1955). *The Origin and Evolution of Takyr Soils*. Moscow University Press, Moscow.

Bond, R. D., and Harris, J. R. (1964). The influence of the microflora on physical properties of soils. I. Effects associated with filamentous algae and fungi. *Aust. J. Soil Res. 2*: 111-122.

Booth, W. E. (1941). Algae as pioneers in plant community succession and their role in erosion control. *Ecology 22*: 38-46.

Bose, P., Nagpal, U. S., Venkataraman, G. S., and Goyal, S. K. (1971). Solubilization of tri-calcium phosphate by blue-green algae. *Curr. Sci. 7*: 165-166.

Brock, T. D. (1975a). Effect of water potential on a *Microcoleus* (Cyanophyceae) from a desert crust. *J. Phycol. 11*: 316-320.

Brock, T. D. (1975b). The effect of water potential on photosynthesis in whole lichens and in their liberated algal components. *Planta 124*: 13-23.

Brotherson, J. D., and Rushforth, S. R. (1983). Influence of cryptogamic crusts on moisture relationships of soils in Navajo National Monument, Arizona. *Great Basin Naturalist 43*: 73-78.

Brotherson, J. D., Rushforth, S. R., and Johansen, J. R. (1983). Effects of long-term grazing on cryptogam crust cover in Navajo National Monument, Ariz. *J. Range Manag. 36*: 479-581.

Cameron, R. E. (1960). Communities of soil algae occurring in the Sonoran desert in Arizona. *J. Ariz. Acad. Sci. 1*: 85-88.

Cameron, R. E. (1964). Terrestrial algae of southern Arizona. *Trans. Am. Microsc. Soc. 133*: 212-218.

Cameron, R. E., and Blank, G. B. (1966a). Desert algae: Soil crusts and diaphanous substrata as algal habitats. *Jet Propulsion Lab. Tech. Rep. 32-971*: 1-41.

Cameron, R. E., and Blank, G. B. (1966b). Soil Studies—desert microalgae. XI. Desert soil algae survival at extremely low temperatures. *Jet Propulsion Lab. Tech. Rep. 37-37 IV*: 174-181.

Chantanachat, S., and Bold, H. C. (1962). Phycological studies. II. Some algae from arid soils. University of Texas Publ. No. 6218, Austin.

Cornet, A. (1981). Le bilan hydrique et son rôle dans la production de la strate herbacée de quelques Phytocenoses Sahéliennes au Sénégal. Languedoc Univ. Sci. Technol., Languedoc, France.

Darling, R. B., Friedmann, E. I., and Broady, P. A. (1987). *Heterococcus endolithicus* sp. nov. (Xanthophyceae) and other terrestrial *Heterococcus* species from Antarctica: Morphological changes during life history and response to temperature. *J. Phycol. 23*: 598-607.

Davis, J. S. (1972). Survival records in the algae, and the survival role of certain algal pigments, fat, and mucilagenous substances. *Biologist 54*: 52-93.

Dorn, R. I., and Oberlander, T. M. (1981). Microbial origin of desert varnish. *Science 213*: 1245-1247.

Dulieu, D., Gaston, A., and Darley, J. (1977). La dégradation des pâturages de la région de N'Djamena (Republique du Tchad) en relation avec la présence de cyanopycees psammophiles: étude préliminaire. *Rev. Elev. Med. Vet. Pays Trop. 12*: 1270-1283.

Durrell, L. W. (1962). Algae of Death Valley. *Trans. Am. Microsc. Soc. 81*: 268-273.

Durrell, L. W., and Shields, L. M. (1960). Fungi isolated in culture from soils of the Nevada test site. *Mycologia 52*: 636-641.

Faurel, L., Ozenda, P., and Schotter, G. (1953). Les lichens du Sahara Algerien. Desert Res. Proc. Int. Symp. Res. Council Israel. Special Publ. *2*: 310-317.

Faust, W. F. (1970). The effect of algal-mold crusts on the hydrologic processes of infiltration, runoff, and soil erosion under simulated rainfall conditions. M.S. Thesis, Unversity of Arizona, Tuscon.

Fletcher, J. E., and Martin, W. P. (1948). Some effects of algae and molds in the rain-crust of desert soils. *Ecology 29*: 95-100.

Follmann, G. (1965a). Fensterflechten in der Atacamawüste. *Naturwissenschaften 14*: 434-435.

Follman, G. (1965b). Eine gesteinsbewohnende Flechtengessellschaft der nordchilenischen Wüstenformationen mit Kennzeichnender *Buellia albula* (Nyl.) Mull. Agr. *Nova Hedwigia 10*: 243-256.

Friedmann, E. I. (1978). Melting snow in the dry valleys is a source of water for endolithic microorganisms. *Antarct. J. U.S. 13*: 162-163.

Friedmann, E. I. (1980). Endolithic microbial life in hot and cold deserts. *Origins Life 10*: 223-235.

Friedmann, E. I. (1982). Endolithic microorganisms in the Antarctic cold desert. *Science 215*: 1045-1053.

Friedmann, E. I., and Galun, M. (1974). Desert algae, lichens, and fungi. In *Desert Biology*, Vol. II, G. W. Brown, Jr. (Ed.). Academic Press, New York, pp. 165-212.

Friedmann, E. I., and Kibler, A. P. (1980). Nitrogen economy of endolithic microbial communities in hot and cold deserts. *Microb. Ecol. 6*: 95-108.

Friedmann, E. I., and Ocampo-Friedmann, R. (1984). Endolithic microorganisms in extreme dry environments: Analysis of a lithobiontic microbial habitat. In *Current Perspectives in Microbial Ecology*, M. J. Klug and C. A. Reddy (Eds.). Am. Soc. Microbiol., Washington, D.C., pp. 177-185.

Friedmann, E. I., Lipkin, Y., and Ocampo-Paus, R. (1967). Desert algae of the Negev (Israel). *Phycologia 6*: 185-195.

Friedmann, E. I., LaRock, P. A., and Brunson, J. O. (1980). Adenosine triphosphate (ATP), chlorophyll, and organic nitrogen in endolithic microbial communities and adjacent soils in the dry valleys of southern Victoria Land. *Antarct. J. U.S. 15*: 164-167.

Friedmann, E. I., McKay, C. P., and Nienow, J. A. (1987). The cryptoendolithic microbial environment in the Ross Desert of Antarctica: Continuous nanoclimate data, 1984 to 1986. *Polar Biol. 7*: 273-287.

Fuller, W. H., Cameron, R. E., and Raica, N. (1960). Fixation of nitrogen in desert soils by algae. *Trans. Cong. Int. Soil Sci. Soc. 3*: 617-624.

Galun, M. (1963). Autecological and synecological observations on lichens of the Negev, Israel. *Isr. J. Bot. 12*: 179-187.

Galun, M. (1966). Additions to the lichen flora of the Negev, Israel. *Isr. J. Bot. 15*: 144-149.

Galun, M., and Reichert, I. (1960). A study of lichens of the Negev. *Bull. Res. Council Isr. [D] 9*: 127-148.

Galun, M., Paran, N., and Ben-Shaul, Y. (1970). An ultrastructural study of the fungus-alga association in *Lecanora radiosa* growing under different environmental conditions. *J. Microsc. (Paris) 9*: 801-806.

Ghabbour, S. I., El-Ayouty, E. Y., Khadr, M. S., and El-Tonsi, A.-M. (1980). Grazing by microfauna and productivity of heterocystous nitrogen-fixing blue-green algae in desert soils. *Oikos 34*: 209-218.

Gollerbach, M. M., and Shtina, E. A. (1969). *Soil Algae*. Acad. Sci., USSR, Moscow.

Goulbic, S., Friedmann, E. I., and Schneider, J. (1981). The lithobiontic ecological niche, with special reference to microorganisms. *J. Sediment Petrol. 51*: 475-478.

Harris, R. F. (1981). Effect of water potential on microbial growth and activity. In *Water Potential Relations in Soil Microbiology*, J. F. Parr, W. R. Gardner, and L. F. Elliot (Eds.). Soil Sci. Soc. Am., Madison, Wisconsin, pp. 23-95.

Henriksson, E., Henriksson, L. E., and Skujiņš, J. (1989). Succession of dinitrogen-fixing terrestrial cyanobacteria on the volcanic island Surtsey, Iceland. *Phykos.*

Herre, A. W. (1942). Additions to and comments upon the lichen flora of the Santa Cruz peninsula. *Am. Midland Naturalist 28*: 752-755.

Hunt, C. B., and Durrell, L. W. (1966). Distribution of fungi and algae. U.S. Geol. Survey, Prof. Paper 509: 55-66.

Isichei, A. O. (1980). Nitrogen fixation by blue-green algal soil crusts in Nigerian savannah. In *Nitrogen Cycling in West African Ecosystems*, T. Rosswall (Ed.). SCOPE/U.N. Environmental Programme, Royal Swedish Acad. Sci., Stockholm, pp. 191-198.

Johansen, J. R. (1984). Response of soil algae to a hundred-year storm in the Great Basin Desert, U.S.A. *Phykos 23*: 51-54.

Johansen, J. R. (1986a). Soil algae and range management. *Appl. Phycol. Forum 3*: 1-2.

Johansen, J. R. (1986b). Importance of cryptogamic soil crusts to arid rangelands: Implications for short duration grazing. In *Short Duration Grazing*, J. A. Tiedman (Ed.). Proc. Short Duration Grazing and Current Issues in Grazing Management Shortcourse, Washington State University, Pullman, pp. 127-156.

Johansen, J. R., and Rushforth, S. R. (1981). Algae of surface waters and soils of areas of eastern Utah. *Nova Hedwigia 34*: 333-339.

Johansen, J. R., and Rushforth, S. R. (1985). Cryptogamic soil crusts: Seasonal variation in algal populations in the Tintic Mountains, Juab County, Utah. *Great Basin Naturalist 45*: 14-21.

Johansen, J. R., and St. Clair, L. L. (1986). Cryptogamic soil crusts: Recovery from grazing near Camp Floyd State Park, Utah, U.S.A. *Great Basin Naturalist 46*: 632-640.

Johansen, J. R., Rushforth, S. R., and Brotherson, J. D. (1981). Subaerial algae of Navajo National Monument, Arizona. *Great Basin Naturalist 41*: 433-439.

Johansen, J. R., Javakul, A., and Rushforth, S. R. (1982). The effects of burning on the algal communities of a high desert soil near Wallsburg, Utah, U.S.A. *J. Range Manag. 35*: 598-600.

Johansen, J. R., St. Clair, L. L., Webb, B. L., and Nebeker, G. T. (1984). Recovery patterns of cryptogamic soil crusts in desert rangelands following fire disturbance. *Bryologist 87*: 238-243.

Kappen, L., and Friedmann, E. I. (1983). Ecophysiology of lichens in the dry valleys of southern Victoria Land, Antarctica. II. CO_2, gas exchange in cryptoendolithic lichens. *Polar Biol. 1*: 227-232.

Kappen, L., Lange, O. L., Schulze, E.-D., Buschbom, U., and Evenari, M. (1980). Ecophysiological investigations on lichens of the Negev Desert. VII. The influence of the habitat exposure on dew imbibition and photosynthetic productivity. *Flora 169*: 216-229.

Kappen, L., Friedmann, E. I., and Garty, J. (1981). Ecophysiology of lichens in the dry valleys of southern Victoria Land, Antarctica.

I. Microclimate of the cryptoendolithic lichen habitat. *Flora 171*: 216-235.

Killian, C. (1953). Symposium discussion. In *Desert Research*. Proc. International Symp. Res. Council Isradel, Special Publ. *2*: 301.

Kleiner, E. F., and Harper, K. T. (1977). Soil properties in relation to cryptogamic groundcover in Canyonlands National Park. *J. Range Manag. 30*: 202-205.

Kovda, V. A., Samoilova, E. M., Charley, J. L., and Skujiņš, J. (1979). Soil processes in arid lands. In *Arid Land Ecosystems*, Vol. 1, D. W. Goodall and R. A. Perry (Eds.). Cambridge University Press, London, pp. 399-470.

Krumbein, W. E. (1969). Über den Einfluss der Mikroflora auf die exogene Dynamik (Verwitterung und Krustenbildung). *Geol. Rundschau 58*: 333-365.

Krumbein, W. E., and Jens, K. (1981). Biogenic rock varnishes of the Negev Desert (Israel), an ecological study of iron and manganese transformation by cyanobacteria and fungi. *Oecologia (Berl.) 50*: 25-38.

Lange, O. L. (1969). Experimentell-ökologische Untersuchungen an Flechten der Negev-Wüste. I. CO_2-Gaswechsel von *Ramalina maciformia* (Del.) Bory unter kontrollierten Bedingungen im Laboratorium. *Flora 158*: 324-359.

Lange, O. L., Schulze, E.-D., and Koch, W. (1970). Experimentall-ökologische Untersuchungen an Flechten der Negev-Wüste. III. CO_2-Gaswechel und Wasserhaushalt von Krusten- und Blattflechten am naturlichen Standort während der sommerlichen Trockenperiode. *Flora 159*: 525-528.

Loope, W. L., and Gifford, G. F. (1972). Influence of a soil microfloral crust on select properties of soils under Pinyon-Juniper in southeastern Utah. *J. Range Manag. 27*: 164-167.

Lynn, R. I., and Cameron, R. E. (1973). The role of algae in crust formation and nitrogen cycling in desert soils. US/IBP Desert Biome Research Memorandum 73-40, Utah State University, Logan.

Magnusson, A. H. (1940-1944). Lichens from Central Asia. Reports from the scientific expedition to the northwestern provinces of China under the leadership of Dr. Sven Hedin, Sino-Swedish Expedition. Publ. 13, *Botany 1*, and Publ. 22, *XI. Botany, 2*. Stockholm.

Mayland, H. F., McIntosh, T. M., and Fuller, W. H. (1966). Distribution of nitrogen fixed in desert algal-crust. *Soil Sci. Soc. Am. Proc. 30*: 56-60.

Metting, B. (1981). The systematics and ecology of soil algae. *Bot. Rev. 47*: 195-312.

Metting, B., Rayburn, W. R., and Reynaud, P. A. (1988). Algae and agriculture. In *Algae and Human Affairs*, C. A. Lembi and J. R. Waaland (Eds.). Cambridge University Press, London, pp. 335-370.

Moewus, L. (1953). About the occurrence of freshwater algae in the semidesert around broken hill (N.S.W., Australia). *Bot. Notiser 4*: 399-416.

Nash, T. H., III, Lange. O. L., and Kappen, L. (1982). Photosynthetic patterns of Sonoran desert lichens. II. A multivariate laboratory analysis. *Flora 172*: 419-426.

Nienow, J. A., McKay, C. P., and Friedmann, E. I. (1988). The crypto-endolithic microbial environmental in the Ross Desert of Antarctica: Light in the photosynthetically-active region. *Microb. Ecol. 16*: 271-189.

Olie, J. J., and Potts, M. (1986). Purification and biochemical analysis of the cytoplasmic membrane from the desiccation-tolerant cyanobacterium *Nostoc commune* UTEX 584. *Appl. Environ. Microbiol. 52*: 706-710.

Papendick, R. I., and Campbell, G. S. (1981). Theory and measurement of water potential. In *Water Potential Relations in Soil Microbiology*, J. F. Parr, W. R. Gardner, and L. F. Elliot (Eds.). Soil Sci. Soc. Am., Madison, Wisconsin, pp. 1-22.

Paran, N., Ben-Shaul, Y., and Galun, M. (1971). Fine structure of the blue-green phycobiont and its relation to the mycobiont in the *Gonhymenia* lichens. *Arch. Microbiol. 76*: 103-113.

Peveling, E. (1987). Lichens. In *Survival and Dormancy of Microorganisms*, Y. Henis (Ed.). John Wiley, New York, pp. 229-266.

Plessl, A. (1963). Über die beziehungen von pilz und alge im flechtenthallus. *Öesterr. Bot. Z. 110*: 194-269.

Potts, M. (1985). Protein synthesis and proteolysis in immobilized cells of the cyanobacterium *Nostoc commune* UTEX 584 exposed to matric water stress. *J. Bacteriol. 164*: 1025-1031.

Potts, M. (1986). The protein index of *Nostoc commune* UTEX 584 (cyanobacteria): Changes induced in immobilized cells by water stress. *Arch. Microbiol. 146*: 87-95.

Potts, M., and Bowman, M. A. (1985). Sensitivity of *Nostoc commune* UTEX 584 (cyanobacteria) to water stress. *Arch. Microbiol. 141*: 51-56.

Potts, M., and Friedmann, E. I. (1981). Effects of water stress on cryptoendolithic cyanobacteria from hot desert rocks. *Arch. Microbiol. 130*: 267-271.

Potts, M., and Morrison, N. S. (1986). Shifts in the intracellular ATP pools of immobilized *Nostoc* cells (cyanobacteria) induced by water stress. *Plant Soil 90*: 211-221.

Potts, M., Ocampo-Friedmann, R., Bowman, M. A., and Tozun, B. (1983). *Chroococcus* S24 and *Chroococcus* N41 (cyanobacteria): Morphological, biochemical and genetic characterization and effects of water stress on ultrastructure. *Arch. Microbiol. 135*: 81-90.

Potts, M., Bowman, M. A., and Morrison, N. S. (1984). Control of matric water potential (ψ_m) in immobilized cultures of cyanobacteria. *FEMS Microbiol. Lett. 24*: 193-196.

Potts, M., Olie, J. J., Nickels, J. S., Parsons, J., and White, D. C. (1987). Variation in phospholipid ester-linked fatty acids and carotenoids of desiccated *Nostoc commune* (cyanobacteria) from different geographic locations. *Appl. Environ. Microbiol. 53*: 4-9.

Rayburn, W. R., Mack, R. N., and Metting, B. (1982). Conspicuous algal colonization of the ash from Mount St. Helens. *J. Phycol. 18*: 537-543.

Reichert, I. (1953). Steppes and deserts in the eastern Holarctic in relation to lichens. Proc. 7th Int. Bot. Cong., p. 676.

Renaut, J., Sasson, A., Pearson, H. W., and Stewart, W. D. P. (1975). Nitrogen-fixing algae in Morocco. In *Nitrogen Fixation by Free-Living Microorganisms*, W. D. P. Stewart (Ed.). Cambridge University Press, London, pp. 229-246.

Reynaud, P. A. (1987). Ecology of nitrogen-fixing cyanobacteria in dry tropical habitats of West Africa: A multivariate analysis. *Plant Soil 98*: 203-220.

Reynaud, P. A., and Laloe, F. (1985). Le méthode des suspensions-dilutions adaptée a l'estimation des populations algales dans une rizière. *Rev. Ecol. Biol. Sol 22*: 161-192.

Reynaud, P. A., and Lumpkin, T. A. (1988). Microalgae of the Lanzhou (China) cryptogamic crust. *Arid Soil Res. Rehabilitation 2*: 145-155.

Rodgers, G. A. (1977). Nitrogenase activity in *Nostoc muscorum:* Recovery from desiccation. *Plant Soil 46*: 671-674.

Rogers, R. W., Lange, R. T., and Nichols, D. J. D. (1966). Nitrogen fixation by lichens of arid soil crusts. *Nature (Lond.) 209*: 96-97.

Rushforth, S. R., and Brotherson, J. D. (1982). Cryptogamic soil crusts in the deserts of North America. *Am. Biol. Teacher 44*: 472-475.

Rychert, R. C., and Skujiņš, J. (1974). Nitrogen fixation by blue-green algae-lichen crusts in the Great Basin Desert. *Soil Sci. Soc. Am. Proc. 38*: 768-771.

Rychert, R. C., Skujiņš, J., Sorensen, J., and Porcella, D. (1978). Nitrogen fixation by lichens and free-living microorganisms in deserts. In *Nitrogen in Desert Ecosystems*, N. E. West and J. Skujiņš (Eds.). Dowden, Hutchinson, and Ross, Stroudsburg, Pennsylvania, pp. 20-30.

Savory, A. (1983). The Savory grazing method or holistic resource management. *Rangelands 5*: 155-159.

Savory, A., and Parsons, S. D. (1980). The Savory grazing method. *Rangelands 2*: 234-237.

Scherer, S., Ernst, A., Chen, T.-W., and Boger, P. (1984). Rewetting of drought-resistant blue-green algae: The time course of water uptake and reappearance of respiration, photosynthesis, and nitrogen fixation. *Oecologia (Berl.) 62*: 418-423.

Schwabe, G. H. (1963). Blaualgen der phototrophen Grenzschicht. Blaualgen und Lebensraum. *Pedobiologia 2*: 132-152.

Shields, L. M. (1957). Algal and lichen floras in relation to nitrogen content of certain volcanic and arid range soils. *Ecology 38*: 661-663.

Shields, L. M., and Drouet, F. (1962). Distribution of terrestrial algae within the Nevada test site. *Am. J. Bot. 49*: 547-554.

Shields, L. M., Mitchell, C., and Drouet, F. (1961). Alga and lichen-stabilized surface crust as a soil nitrogen source. *Am. J. Bot. 44*: 489-498.

Skarpe, C., and Henriksson, E. (1987). Nitrogen fixation by cyanobacterial crusts and by associative-symbiotic bacteria in western Kalahari, Botswana. *Arid Soil Res. Rehabilitation 1*: 55-59.

Skujiņš, J. (1984). Microbial ecology of desert soils. *Adv. Microb. Ecol. 7*: 49-91.

Skujiņš, J., and Klubek, B. (1978). Nitrogen fixation and cycling by blue-green algae-lichen crusts in arid rangeland soils. *Ecol. Bull (Stockh.) 26*: 164-171.

Skujiņš, J., and West, N. E. (1974). Nitrogen dynamics in stands dominated by some major cold desert shrubs. US/IBP Desert Biome Res. Memorandum 74-42, Utah State University, Logan.

Smith, H. T. U. (1968). Geologic and geomorphic aspects of deserts. In *Desert Biology*, Vol. I, G. W. Brown, Jr. (Ed.). Academic Press, New York, pp. 52-100.

Smith, E. A., Mayfield, C. I., and Wong, P. T. S. (1978). Naturally-occurring apatite as a source of orthophosphate for growth of bacteria and algae. *Microb. Ecol. 4*: 105-118.

Snyder, J. M., and Wullstein, L. H. (1973). The role of desert cryptogams in nitrogen fixation. *Am. Midland Naturalist 90*: 257-265.

St. Clair, L. L., Webb, B. L., and Johansen, J. R. (1984). Cryptogamic soil crusts: Enhancement of seedling establishment in disturbed and undisturbed areas. *Reclamation Reveg. Res. 3*: 129-136.

St. Clair, L. L., Johansen, J. R., and Webb, B. L. (1986). Rapid stabilization of fire-disturbed sites using a soil crust slurry: Inoculation studies. *Reclamation Reveg. Res. 4*: 261-269.

Steger, R. E. (1982). Rapid rotation grazing programs in Texas. *Rangelands 4*: 75-77.

Steiner, J. (1916). Aufzahlung der von J. Bornmüller im Oriente gesammelten Flechten. *Ann. Naturhist. Mus. Wien 30*: 24-39.

Stewart, W. D. P., Sampaio, M. J., Isichei, A. O., and Sylvester, R. (1978). Nitrogen fixation by soil algae of temperate and tropical soils. In *Limitations and Potential for Biological Nitrogen Fixation in the Tropics*, J. Döbereiner, R. H. Burris, and A. Hollander (Eds.). Plenum Press, New York, pp. 41-63.

Tchan, Y. T., and Beadle, N. C. W. (1955). Nitrogen economy in semiarid plant communities. 2. The non-symbiotic nitrogen-fixing organisms. *Proc. Linn. Soc. New South Wales 80*: 97-105.

Thomson, J. W., and Iltis, H. H. (1968). A fog-induced lichen community in the coastal desert of southern Peru.

Tschermak-Woess, E., and Friedmann, E. I. (1984). *Hemichloris antarctica*, gen. et sp. nov. (Chlorococcales, Chlorophyta), a cryptoendolithic alga from Antarctica. *Phycologia 23*: 443-454.

Venkataraman, G. S. (1961). A method for preserving blue-green algae for seeding purposes. *J. Gen. Appl. Microbiol. 7*: 96-99.

Venkataraman, G. S. (1964). Thermal resistance and viability of microalgae. *Phykos 3*: 26-28.

Vogel, S. (1955). Niedere "Fensterpflanzen" in der sudafrikanischen Wüste. *Beitr. Biol. Pflanzen 31*: 45-135.

Weber, W. A. (1962). Environmental modification and the taxonomy of the crustose lichens. *Sv. Bot. Tidskr. 56*: 293-333.

Weber, W. A. (1968). A taxonomic revision of *Acarospora*, subgenus *Xanthothallia*. *Lichenologist 4*: 16-31.

Welsh, H. (1965). A contribution to our knowledge of the blue-green algae of southwest Africa and Bechuanasland. *Nova Hedwigia 9*: 131-162.

West, N. E., and Skujiņš, J. (1977). The nitrogen cycle in North American cold-winter semi-desert ecosystems. *Oecol. Plant. 12*: 45-53.

Whitton, B. A. (1987). Survival and dormancy of blue-green algae. In *Survival and Dormancy of Microorganisms*, Y. Henis (Ed.). John Wiley, New York, pp. 109-167.

9

Nutrient Cycling in Soils of Semiarid and Arid Regions

NEIL E. WEST *Utah State University, Logan, Utah*

I. INTRODUCTION

During the International Biological Program (IBP) of the late 1960s to mid-1970s, reviews of nutrient cycling in semiarid and arid environments were commissioned (Charley, 1977; Clark et al., 1978; Jones and Woodmansee, 1979; Kovda et al., 1979; Binet, 1981; West, 1981; Stewart et al., 1983) and new research launched, focusing mainly on nitrogen (N) (West and Skujiņs, 1978; Woodmansee, 1978, 1979; Woodmansee et al., 1981; Bate, 1981; Skujiņs, 1981; O'Connor, 1983). These earlier reviews considered the pre-1980 literature and IBP results then at hand and offered generalizations possible at that time. They also pointed out some remaining problems and researchable topics obvious at that point. Since then, rapid conceptual and technological advances have made possible new ways of investigating nutrient cycles. Improved means of peering at the world and communicating changes in its environment, as well as intensifying land use, have led us now to focus on larger issues (Rosswall et al., 1988). Thus, at the start of the International Geophysical Biological Program (IGBP), it behooves us to re-examine our progress on this topic since the end of IBP, attempt refined generalizations, and point out important unresolved questions to help researchers plan efforts into the 1990s.

Although soils are the focus of this volume, one cannot ignore other aspects of the total ecosystem context within which soils are structured and function. This is because soils involve atmospheric, biological, and mineralogical inputs and outputs. Principles of physics, chemistry, and biology must all be drawn on to explain why the soil functions as it does. Although soil may still be a "black box" to some ecologists and land managers, it cannot remain so if we are adequately to deal with such major issues as desertification, global climatic change, and how to feed, clothe, and shelter the world's burgeoning human population.

Because about 35% of the world's people live in arid to semiarid regions (Brook and Pokshishevsky, 1986) and population growth rates in such environments are generally much higher than in more mesic regions (Brook and Pokshishevsky, 1986), we can only expect exacerbated problems of sustaining production of food, fiber, and fuel from these lands. Better understanding of nutrient cycling limitations to production of these goods and maintenance of services, such as clean water and air, can contribute to dealing with these needs.

First, the whole subject of nutrient limitation to plant growth is now being looked at differently on wildlands than in agricultural contexts (Chapin et al., 1986). This is because wild plants, especially those restricted to infertile soils, respond much less to nutrient addition than do related plants of more fertile soils. Furthermore, natural vegetation is composed of many species and growth forms, all responding differently to available soil nutrient levels and recycling them at different rates. Complex competitive, grazer, and decomposer interactions lead to ecosystem level feedbacks that are not always easy to anticipate (Gosz, 1986). Before such complexities are discussed, I begin with a more reductionist approach.

II. SPATIAL PATTERNS OF NUTRIENTS

Although nutrient cycling is a dynamic or functional property of ecosystems, it is easier to begin with the structural issues. Earlier work on this topic focused on descriptions of spatial distributions of nutrients. The strong vertical patterns of nutrients in desert and grassland areas are still being described in a wider variety of circumstances.

On well-drained sites, soil organic matter (SOM) and constituent organic carbon (C), N, phosphorus (P), sulfur (S), and potassium (K) concentrate near the surface (Romney et al., 1980; Christie, 1981; Barth and Klemmedson, 1982; Schimel et al., 1985b; Bjerregaard et al., 1984). Salts and carbonates leach through the profile but accumulate at relatively shallow depths in deserts compared to the more mesic grasslands (Whitford, 1986a).

On soils with high water tables, salts are wicked upward mainly by capillary action. This salinization process has destroyed the productivity on about twice as much formerly irrigated land as is at present irrigated worldwide (Yensen, 1988). The role of native plants in transpiring water and, thus, lowering the water table and lessening salinization both in situ and downslope is becoming better recognized. For instance, in Australia, former savannas converted to rain-fed agriculture are rapidly leading to salinization down the watershed (Allison and Hughes, 1983). Re-establishment of native perennials on the uplands and planting of drought- and salt-hardy trees and shrubs on the abandoned farmland is now being done to mitigate this process.

Earlier (pre-1980) ecological research, for convenience, usually focused on selected sites thought to typify major land types. The values obtained on such sites were then multiplied by area of the type to obtain estimates of standing crops or dynamics for a larger area. Now that errors of such multiplications are much more serious, we typically study local or regional gradients, catenas, or ecotonal areas instead (Woodmansee and Adamsen, 1983; Swanson et al., 1988; Risser, 1988; Burke, 1989; Burke et al., 1989a) and model in a less discontinuous way (e.g., Schimel et al., 1985b,c;

Parton et al., 1987; Reiners et al., 1989; Joyce et al., 1989; Parton et al., 1989; Burke et al., 1989b).

This recognition of the importance of scale and the geographic context to ecological processes has emerged within the new field of landscape ecology (Forman and Godron, 1986; Gosz, 1986; Cole, 1989). Important examples of its relevance to nutrient cycles in drier regions are the erosional models of Pickup and colleagues (Pickup, 1985; Pickup and Chewings, 1986; Westoby, 1987) used in central Australia, the piospheric reorganization of materials by livestock activities radiating out from watering points (Andrew, 1988), and nutrient cycling in relation also to topography and fences (Senft et al., 1985).

Another recent recognition is the influence of age of landform on the standing crops and labile amounts of nutrients. The usual pattern reported is the increase in SOM and nutrient pools over successional time (e.g., Bush and Van Auken, 1986). Longer time scales must be accounted for as well as shorter. For instance, Lajtha and Schlesinger (1988a) have shown how total soil P decreases with increasing soil age along a desert chronosequence in southern New Mexico. The severe nutrient limitation problems of Australia and East Africa also relate to the extreme age of some soils there. McNaughton (1988) has recently shown how native ungulates in East Africa relate to large spatial heterogeneity in soil fertility and attendant herbage quality.

Low herbage quality leads to slower decomposition. High ratios of C to other nutrients in the litter produced on nutrient-poor sites lead to protracted immobilization of nutrients in microbial biomass and its by-products and therefore slows nutrient release (Chapin et al., 1986; Burke et al., 1989a; Seastedt et al., 1988; Skiles et al., 1989.

In addition to reinforced recognition of vertical anisotropisms and horizontal variations on large scales over long times, additional information about fine-grained horizontal variation is accumulating. Reports by Romney et al. (1980), Cox et al. (1984), Bernard-Reversat and Poupon (1980), Bernard-Reversat (1982), Wikeem and Pitt (1982), Klemmedson and Tiedemann (1986), Tiedemann and Klemmedson (1986), Ludwig et al. (1988a), Hacker (1986), Doescher et al. (1984), Virginia and Jarrel (1983), Whitford (1986b), Lajtha and Schlesinger (1986), Whitford et al. (1987), Schlesinger et al. (1990) add to earlier evidence that "islands of fertility" around the woody plants are common on shrub- and small tree-dominated sites. The deeper, more perennial roots of the woody species gather in nutrients from the less vegetated interspaces and drop litter in higher amounts around themselves. Microbes and animals also focus their activities there (Garner and Steinberger, 1989). This, over time, leads to greater soil tilth and soil moisture infiltration, percolation, and storage next to these plants. Whether the islands of fertility (West, 1989) remain over time as functionally more productive cells within a generally less hospitable matrix of desert to semidesert landscape is an important practical issue (Schlesinger et al., 1990).

Mooney et al. (1980), Richards and Caldwell (1987), Caldwell and Richards (1989), and Caldwell et al. (1990) have shown that the roots of desert shrubs can pump and exude water from wet soils at depths to dry soils within the upper profile during nights of the summer drought period. This allows both the shrubs and nearby herbaceous species to grow over additional time during the warmer months. Mycorrhizae and other soil microbes could also react positively to this water transfer. Since mineral ion diffusivity is reduced by even small reductions in soil moisture, nocturnal flushing of the nutrient depletion zones due to hydraulic lift, even if very limited, may facilitate uptake of nutrient ions in otherwise dry soils (Caldwell et al., 1990).

Although this section focuses on structure, I cannot entirely avoid some mention of function. We can now, having defined the context, stress the main concerns of this review: nutrient cycles within soils. Advances in the basic understanding of the nutrient cycles are considered before we then turn to the implications for environmental problems.

III. NUTRIENT CYCLES

A. Advances in Basic Understanding

Carbon

C is the fundamental building block of life on earth. The evolution of life here has probably involved changing a C-rich, reducing atmosphere into one now dominated by N and O. Various biota have fixed C into SOM, albeit with comparatively less storage in arid to semiarid regions (Schlesinger, 1977). In addition, huge amounts of inorganic carbonates have accumulated in soils of arid to semiarid regions. This latter pool has only recently been identified (Schlesinger, 1982) as a major compartment in the terrestrial cycle. Based on the area of arid soil orders (Dregne, 1976) and mean profile carbonate content of each, Schlesinger (1985a) estimated that there is a global pool of 930×10^{15} g C in calcic and petrocalcic horizons. This estimate does not include calcrete caps. If the ratio of 5.7 times more C stored in inorganic form than in organic (detrital) form in Arizona applies to all deserts, then desert soils could contribute 35% of the world storage of soil C. The combined pool (organic and inorganic) totals over twice the amount of C in the total world biomass (Schlesinger, 1982). Clearly, much more study of the rates of net storage or release now and with projected climatic and vegetational changes are needed (Schlesinger, 1985b; Schlesinger et al., 1989).

Lajtha and Bloomer (1988) have recently shown that $CaCO_3$ appears to be the primary geochemical agent capable of fixation and retention of P within a Chihauhuan Desert chronosequence. Phytotron studies by Lajtha and Schlesinger (1988b) showed $CaCO_3$ directly inhibited the uptake or availability of P to shoots of two Chihuahuan Desert shrubs. Pedogenic carbonate may thus greatly affect both the P availability and the biogeochemical cycle of P in desert ecosystems.

The worldwide buildup of atmospheric CO_2 concentrations from industrial and transportation activity, agricultural land clearing, and drainage has been accelerating over the past 150 years. In addition to its effects on global temperatures, this constitutes a chronic fertilization (Strain, 1988), even in drier regions. Because different plant life forms and physiological modes (C_3, C_4, and CAM) possess differing uptake efficiencies, structural and functional reorganizations are expected, even in drier ecosystems (Smith et al., 1987; Skiles et al., 1989). For instance, Johnson and Lincoln (1989) suggest that the ability of sagebrush (*Artemisia tridentata*) to utilize increased CO_2 may be constrained by the low status of soil nutrients.

Whether any increase in C fixation that results from CO_2 enrichment will result in increased or decreased decomposition is also a crucial question (Mooney et al., 1987). Where other nutrients (especially N) are limiting to growth, plants produce tissues with high C/N ratios when CO_2 levels are elevated. Altered tissue chemistry could change herbivore consumption and litter decomposition rates and consequent nutrient release (Seastedt et al. 1988; Ryan and Rastetter, 1989; Hunt et al., 1989; Johnson and Lincoln, 1989; Kittel et al., 1989). Such feedbacks may alter productivity in the long term.

Hydrogen

H inputs to grassland and desert ecosystems come largely through water. Plants then recombine H, split in the major assimilatory process of photosynthesis, into a wide variety of organic compounds. Animals rearrange more complex organic molecules containing H. Since water is often limiting to both higher plant growth and animal activity, H must be viewed as limiting for a good share of the time. Microorganisms are less restricted by deficiencies of water than higher plants and animals (Kovda et al., 1979; Paul and Clark, 1989; Watwood et al., 1989). This fortunate circumstance usually allows supplies of other critical nutrients to accumulate after light rains, or even dew (Shmida et al., 1986).

Oxygen

O is required for all respiratory activity. It is inhibitory to certain key processes, such as N fixation, denitrification, and heterotrophic nitrification, however (Sprent, 1987a). Even in desert soils, there are times, within certain microsites, that are temporarily lacking in oxygen, allowing these key anaerobic processes to proceed. Methanogenesis is stimulated, however (Watwood et al., 1989).

Nitrogen

If C, H, and O are not limiting, then N usually becomes the next most important element for green plants. The importance of N as a second-order limitation to or regulator of primary and thus all other production in semiarid to desert ecosystems after water limitations are broken has only been recognized for about the past 25 years (Crawford and Gosz, 1982; Whitford, 1986b). Research during IBP did much to summarize our knowledge and accomplish needed new research on this topic. These efforts were thoroughly reviewed by Clark et al. (1978), West and Skujiņš (1978), Woodmansee (1979), Woodmansee et al. (1981), and Skujiņš (1981). Progress since is stressed here.

Because the most abundant form of N, N_2, is very stable, energy-requiring processes are necessary to make it available in forms conducive to life (fixation). In addition to natural electrical discharges, fires, photochemical reactions, and industrial processes, Schrauzer et al. (1979) found that N_2 can be abiotically fixed in titanium-containing sands of desert regions. They speculate that this process could produce fixed N at about one-third the rate that N_2 is oxidized by lightning discharges and about one-tenth the rate at which NH_3 is fixed by biological means.

The major way that atmospheric N is fixed is biologically, through the Mo-mediated enzyme nitrogenase in prokaryotic microorganisms. The energy required comes from chemo- or photosynthesis. Since chemosynthesis has not been proven to be an important pathway and photosynthesis is usually water limited in deserts and semideserts, N fixation rates could be expected to be relatively low there. Instead, N fixation is relatively high (Crawford and Gosz, 1982, 1986). Much N fixation in drier areas is done during short interludes of moisture availability by cyanobacteria and bacteria in soils (Skujiņs, 1984) and in special moist, anaerobic, high C/N microenvironments within animals (e.g., gut of termites) (O'Brien and Slaytor, 1982) and lichens (West, 1990).

Whitford (1986b) and colleagues have found that standing deadwood in the Chihuahuan Desert supports fungi that accumulate a N-rich layer on the wood surface that is then consumed by termites. Snails seem to be doing similar things on standing dead shrubs in Israel (M. Shachak, personal communication, 1989). These observations imply greater physical inputs of N than previously recognized (Whitford, 1986b).

The greatest N fixation in most terrestrial ecosystems takes place in nodules of legumes. Thus, it has always been puzzling that although legumes are commonly found in deserts, only rarely have they been shown to fix much N. A number of drought-tolerant legumes (e.g., *Cercidium*) cannot apparently nodulate, but others (e.g., *Acacia*, *Psorothamus*, and *Prosopsis*), which can nodulate, have not often been shown to do so in drier areas (Sprent, 1987b).

Only *Prosopsis* rooted in permanent, fossil water tables has been shown to fix appreciable N_2 (Rundel et al., 1982; Shearer et al., 1983). Even experimental irrigation of mesquite in sand dunes of the Chihuahuan Desert of northern New Mexico (Virginia et al., 1989) has failed to produce nodulation of *Prosopsis*. Roots of this deep-rooted shrub may be pumping water into near-surface soils (Mooney et al., 1980) where microorganisms are carrying on mineralization, and thus other plants could thrive on the nitrate liberated (Sprent, 1987a). Because there is too little precipitation in grasslands and deserts to leach NO_3^- very far downward (Peterjohn and Schlesinger, 1989), N fixation in the nodules, occurring only much deeper in the soil, is not inhibited.

Free-fixing bacteria, such as *Streptomyces* and *Spirilla*, occur in the rhizosphere of numerous nonleguminous shrubs and grasses found in drier climates. The magnitude of their input has been difficult to assess, however (Paul and Clark, 1989).

Most N fixation in deserts and semideserts probably takes place in symbiotic systems between bacteria or cyano-bacteria and lichens (Sprent, 1985). It was recently recognized, however, that the acetylene reduction method gives results biased on the high side (Weaver, 1986; Paul and Clark, 1989). Accordingly, much of the research using acetylene reduction to obtain absolute rates of N fixation need to be repeated with the more definitive, albeit expensive, ^{15}N technique. Comparisons of relative rates of fixation are still legitimately made via acetylene reduction. Improved means of assessing isotopes of N through mass spectrophotometry (Shearer et al., 1983; Marion, 1987; Shearer and Kohl, 1989) are now allowing more definitive studies of N fixation and water use efficiencies.

Vascular plants take up N, mostly as either NO_3^- or NH_4^+, usually through roots. It is possible that many other soluble organic compounds, especially amino acids, could be utilized directly by plants and microorganisms, thus short-circuiting parts of the main N cycle (Sprent, 1987a). Gaseous NH_3 could also be taken up through aboveground tissues (Mooney et al., 1987). Plants also lose gaseous nitrogen to the air (Farquhar et al., 1983). These alternative pathways have not yet been demonstrated in natural semiarid grasslands or deserts, however.

Whereas NH_4^+ can be affixed to the cation-exchange complex in soil clays or organic matter, NO_3^- is very soluble. The lack of water to move the NO_3^- very far is one probable reason that nitrate can be the most abundant form of combined N found near the soil surface in some grasslands (Schimel and Parton, 1986). In other, usually drier ecosystems (e.g., McNaughton et al., 1988), NH_4^+ may be the dominant form. Ammonia may accumulate in droughty soil because the nitrifiers cannot operate at water potentials at which the ammonifiers are still active.

The enhanced NO_3^- content sometimes observed in the surface layer of a droughty soil may not be due to nitrification but to the upward movement of capillary water carrying NO_3^-. Evaporation at the soil surface leaves the NO_3^- behind (Paul and Clark, 1989).

Whether NO_3^- or NH_4^+ prevails is an important issue since rhizosphere acidification is associated with ammonium nutrition. Lowered pH could affect the solubility and uptake of other nutrients as well.

Even in ecosystems of arid to semiarid regions, the inherent tendency is to "tighten" up the N cycle and focus on the plant-soil part of the cycle (Rosswall, 1976). Organisms and processes that locally conserve N and shift it toward NH_4^+ dominance usually emerge as ecosystems develop via plant succession and soil genesis (Dormaar et al., 1990). This tightening is achieved in a number of ways. The most obvious is internal recycling in the higher plants, especially shrubs. That is, a high proportion of total ecosystem N is retained in the biomass throughout the year. External cycling involves only a small part of the total labile nutrient pool where perennials dominate. This is much less the case where ephemerals prevail (Staaf and Berg, 1981; Reiners, 1981; Wikeem and Pitt, 1982; Rickard, 1985).

If there is little or no net loss of N from an ecosystem, there is little need of fixation. Successional gains go to plants that have particularly efficient means of retaining most N but that scavenge needed increments of N for new growth. This seems to be the case for "climax" shrubs in relatively undisturbed temperate semideserts. For instance, the North American temperate semidesert shrubs *Artemisia*, *Atriplex*, and *Ceratoides*, in addition to high levels of internal recycling, have been shown to repress the growth of cyanobacteria, the major N-fixing component in the systems they dominate (Skujiņš, 1981). Second, litter from these shrubs (Skujiņš, 1981), as in more mesic systems, has been shown to reduce the magnitude of the nitrification pathway and cause consequent losses due to denitrification and leaching. Ammonia volatilization at the higher pH and temperatures is greater, however, unless the necessary cation binding sites are available (Schlesinger and Peterjohn, 1989).

Artemisia, *Atriplex*, and *Ceratoides* species may additionally inhibit the oxidation of nitrite to nitrate during nitrification (Skujiņš, 1981). This can lead to nitrite accumulation, but more often, especially when sufficient organic matter is available, nitrite is diverted to denitrification processes. Thus, it appears that some desert shrubs can have a very adverse effect on soil N accumulation.

Whitford (1988) has shown that the abundance and activity of N-fixing microbes is much higher where grasses have been retained in shrub steppes of New Mexico rather than where shrubs have come to dominance after excessive livestock grazing. The success of shrubs and succulents in taking over degraded shrub steppes may be at least partially explained by their greater tolerance to lower soil nutrient concentrations and superior systems for absorbing, transporting, and recycling vital elements compared to perennial grasses (West et al., 1984).

Dormaar and Smoliak (1985) have confirmed Rice and Pancholy's (1972) observation that nitrification is increasingly inhibited as succession advances in plowed and then abandoned shortgrass prairie. They postulate that both increased aboveground phytomass, caused by the probable higher NO_3-N levels in the revegetated soils over the growing season (i.e., earlier successional stages), and increased rootmass, caused by higher NH_4-N levels over the growing season (i.e., in the "climax" ecosystem stage), are credible observations.

An alternative explanation for increased aboveground phytomass in the revegetated fields is the increased mineralization of N in SOM during the period of cultivation of the prairie soils. Since a large percentage of the SOM is only partially decomposed following plowing (Dormaar, 1975), there remains a potential for considerable mineralization. Cycles of drying and wetting further enhance mineralization (Fisher et al., 1987; Risser, 1988). It is likely that both mechanisms, that is, the effect of NO_3^-/NH_4^+ ratios and increased mineralization due to cultivation, contribute to the increased aboveground productivity.

If mycorrhizal fungi compete successfully for NH_4^+ with other soil microorganisms, the net effect is the same as when plants produce nitrification inhibitors—omission of the nitrification step (Sprent, 1987a). Furthermore, if organic N is taken up directly by mycorrhizae, it is not available for ammonification. The possibility also exists that because mycorrhizae can physically connect different species, direct transfers of nutrients between plants can occur (Francis and Read, 1984; Newman, 1989). These things may be going on among desert shrubs and partially explain their success in either disturbed shrub steppes or truly desert environments.

Plants incorporate NO_3^- or NH_4^+, mostly into proteins. Animals assimilate the proteins. As both plants and animals die, the process is reversed via mineralization. C has been found to have more control over litter decomposition in the Chihuahuan Desert than N (Schaefer et al., 1985). This is because invertebrates can continue activity even when soils dry out since a good share of the decomposition takes place in their gut systems. Recent theoretical (Seastedt et al., 1988) and manipulative studies of invertebrates (Christiansen et al., 1989; Zak et al., 1989) show that trophic relationships rather than faunal densities probably regulate nitrogen dynamics and subsequent plant growth.

Lack of sufficient dung beetles to move livestock feces belowground (Woodmansee, 1979; Savory, 1988) is responsible for substantial losses of NH_3 from rangelands via volatilization from feces and urine (Schimel et al.,

1986). The losses differ enormously, however, at different points along a catena (Schimel and Parton, 1986; Senft et al., 1985).

Vertebrates are also now recognized as important influencers of the N cycle. For instance, Whicker and Detling (1988) have noted that grazing by prairie dogs and other herbivores resulted in higher N concentrations in the shoots of shortgrasses on their colonies than in the same species in adjacent, uncolonized areas. Moderate aboveground herbivory may also enhance root-feeding larvae of scarabeid beetles (Detling, 1988). This may be the result of aboveground grazing producing roots richer in N (Seastedt, 1985). The net result is enhanced rates of N cycling, with concentrations near the soil/air interface (Detling, 1988). McNaughton et al. (1988) and Seagle and McNaughton (1989) have shown how long-term grazing by large herds of ungulates in East Africa has shifted the sward toward dominance by grass species that favor ammonium uptake.

Work during the IBP era (Wetselaar, 1980; Skujiņš, 1981) showed surprisingly high levels of denitrification. Work since (Mosier et al., 1981; Seligman et al., 1985; Feigenbaum et al., 1983; Sorenson and Fresquez, 1985; Peterjohn and Schlesinger, 1989) has reinforced the importance of this phenomenon where soils become compacted or sealed. The required anaerobic conditions are also apparently produced when the microphytic crusts (mosses, liverworts, and algae), so common in arid regions (West, 1990), are saturated by precipitation input. The decaying microphytic crusts also apparently supply the raw materials for these heterotrophic organisms when higher plant litter is in short supply.

Rudaz et al. (1989) have recently shown that both nitrifying and denitrifying bacteria survive in very dry soils associated with annual grasslands and are able to become active producers of N_2O within hours after wetting. Now that improved means of separating the contributions of denitrification and nitrification to total N_2O production are available (Klemmedtsson et al., 1988a, b), experimentation on these processes should be done on some soils of arid to semiarid environments.

The major loss of fixed N from native grasslands comes through volatilization of NH_3 from ungulate excreta and plant and animal residues (Woodmansee, 1979; Bowden, 1986). There may also be undetected volatilization directly from the soil surface and senescing plant parts. Schimel et al. (1986) and Parton et al. (1988a), however, showed that such losses were less than originally expected, especially in heavily grazed grasslands.

Schlesinger and Peterjohn (1989) have recently experimentally shown that additions of water and NH_4Cl stimulated volatilization of ammonia whereas sugar had the opposite effect on three sites in the Chihuahuan Desert. Despite the high rates of nitrification observed, they suggest that the rate of ammonification and the rate of immobilization by soil microbes are important determinants of the potential for the volatile loss of ammonia.

Another dissimilatory manner by which gaseous N may be produced is the abiotic reaction of NO_2^- with the organic constituents of soil. Since high levels of nitrite can be found in the alkaline soils of deserts, this may be a relatively important pathway (Sprent, 1987a). Indeed, Mosier and Parton (1985) suspected that N_2O emissions from shortgrass prairie in Colorado amounted to about 10% of N inputs from atmospheric deposition and N_2 fixation.

Although there is a surprising amount of throughput of N in native ecosystems in arid to semiarid regions, relatively little accumulates there. Crawford and Gosz (1982, 1986) make a case for N actually not being limiting in deserts. They reason that the shrub-centered islands of fertility may occupy as little as 6% and plant litter covers less than one-third of the land surface. When this knowledge is combined with the observation that when soil moisture and/or critical microfauna are present and litter decomposition and plant uptake and growth are as rapid as in more mesic systems, it can be concluded that N deficiency is an artifact of computing averages over the entire land area (whether covered by plants or not) over a year's time.

Limitations of nutrients have traditionally been shown by addition experiments. In the case of N, there have been a number of new N amendment studies in native grasslands, semideserts, and deserts over the past 15 years. For instance, Sharafi et al. (1988) found that N fertilization without additional water resulted in very little enhancement of *Larrea tridentata* aboveground net primary production (ANPP) or root biomass. Fisher et al. (1987, 1988), however, found strong interactions between water and N, even for simulated light rains, in experiments carried out on *Larrea* communities in New Mexico. Whitford (1986b) interpreted these same experiments as showing mainly N control over growth of both shrubs and herbs. Phytotron studies by Lajtha and Klein (1988) showed *Larrea* seedlings responded positively to additions of both N and P, but not significant interactions. Lajtha and Whitford (1989) found age-specific N and water interactions and use efficiencies in *Larrea*. Nobel et al. (1987) showed positive responses of prickly pear cacti (*Opuntia* spp.) to N amendments.

Lauenroth et al. (1978) and Dodd and Lauenroth (1979) have experimentally demonstrated how water and N interact in northeastern Colorado shortgrass prairie. They found that ANPP was more sensitive to alterations in available water than to mineral N, especially for legumes (Lauenroth and Dodd, 1979). However, once water limitations were broken, N availability became limiting and the combined treatment produced 5-13 times as much new growth as the untreated controls. The combined water and fertilizer treatment led to loss of native cacti and invasion of introduced warm season forbs, however. Native arthropods and mammals greatly increased with the treatments involving augmented water. Hunt et al. (1988) found that N fertilization of the same shortgrass prairie in Colorado had a significant effect on ANPP of *Bouteloua gracilis*. N supply also had a small but positive effect on litter decomposition rate and N accumulation in litter.

Carpenter and West (1987) failed to find a strong interaction between water and N additions to mountain big sagebrush twig production, flowering, and fruiting in western Wyoming. Sagebrush, however, is a genetically diverse, relatively slow-growing, long-lived shrub with a very high rate of recycling. Furthermore, the parent materials are richer and atmospheric inputs of N much higher at the Wyoming site than at Dodd and Lauenroth's (1979) site. Elderkin et al. (1986) found *A. tridentata* in northwestern Colorado responded much more strongly to an irrigation and N + P fertilization treatment than only irrigation. Their study did not include soil nutrient status, and thus I am unable to suggest why their results were different.

Gutierrez et al. (1988) found that amendments of water and N to Chihuahuan Desert sites resulted in no effect on total annual plant density but increased density of about eight species. Total aboveground biomass was higher under N amendments only, but not under water alone. There were also no significant N-water interactions on the aboveground phytomass of the annuals. Cornelius and Cunningham (1987) discussed the influence of N enrichment on plant cover and diversity at the same study sites. They found that cover of ephemeral species increased but overall plant species diversity decreased, probably due to changes in availability ratios of limiting resources and decreased heterogeneity in the spatial distribution of N.

Phosphorus

Phosphorus has been attracting more attention since Penning deVries and Krul (1980) and Cole and Heil (1981) pointed out that P positively influences N absorption and fixation in semiarid grasslands. Phytomass is reduced because a P/N ratio of between 0.04 and 0.15 g total P g^{-1} organic N is required in the African mid-Sahel for holding plant production as high as soil moisture allows (Penning deVries and Djiteye, 1982). These authors showed that plants on fertilized rangeland there used water much more efficiently than those on the untreated plots. P is also needed for maximum N mineralization and nitrification (West et al., 1984) and may ultimately be more limiting than N in most terrestrial ecosystems (Gutsick, 1981; McGill and Christie, 1983).

The alkaline, calcic, and gysic soils characteristic of desert ecosystems may tie up P in insoluble forms such that although P could be abundant, it would not be available for plant growth (Lajtha and Schlesinger, 1988b). Lajtha and Schlesinger (1986) offered some indirect evidence that available P was more limiting than N for major shrubs in the northern Chihuahuan Desert. Later direct experiments (Lajtha, 1987) did not confirm this, however, for *Larrea*. Despite the very efficient nutrient resorption or recycling capabilities of desert shrubs, P is not as tightly conserved on the sparsely vegetated Aridisols as in more mesic, acid forest ecosystems (Lajtha and Schlesinger, 1988a). Nobel et al. (1987) showed that prickly pear cacti (*Opuntia* spp.) responded positively to P amendments. Friedel et al. (1980) showed from fertilization experiments in central Australia that P is the most critical element in most grasslands and shrublands there.

Knight (1987) and Jurinak et al. (1986) have recently found that oxalates are abundant in the rhizosphere of *Agropyron smithii*, a widespread xeric grass in western North America. Increased oxalates are correlated with increased concentrations of soluble (plant-available) P. This is a possible mechanism by which P uptake is enhanced by mycorrhizal infection. However, nonmycorrhizal invaders like *Halogeton* and *Salsola* create their own oxalic acid (L. Dudley, personal communication, 1989), possibly explaining some of their success in occupying disturbed rangeland and abandoned fields.

Sulfur

Few studies of the S cycle in noncultivated soils of arid to semiarid environments have been conducted. Since S is required for some amino acids, it is puzzling why this element has not often been directly studied in such contexts. A review of the literature in slightly wetter, Mediterranean annual grasslands (Jones and Woodmansee, 1979) showed that S could often be limiting to maximum plant growth, either directly or through stimulation of legume production.

In a 5 year SO_2 enrichment study on mixed prairie in eastern Montana, Milchunas and Laurenroth (1984) found large increases in the S concentration of aboveground parts of plants, but not large increases in the S concentration in belowground portions of the grasses. There was, however, a cumulative increase in S in the soils. The higher level treatment resulted in a lowering of pH by one unit and a depression of litter decomposition rates. The growth response of the grasses was initially positive but often became negative as time of exposure increased. A defoliation treatment added to some of the SO_2 addition plots indicated a small potential for a negative interaction with SO_2. *A. smithii* has, however, a large capacity to adjust to both SO_2 exposure and defoliation (Lauenroth et al., 1984).

Ludwig et al. (1988b) found that application of a one-time amount of 10 g m^{-2} of S to Chihuahuan Desert shrubland brought no significant response from the perennials or summer ephemerals but did generally reduce soil pH and decrease the diversity of winter ephemerals. Termite activity was also decreased in S treatments (MacKay et al., 1987).

Modeling and literature reviewed by McGill and Cole (1981) and Mitchell and Fuller (1988) leads one to believe that S cycling should follow pathways more similar to N than P. Where anthropogenic sources are near at hand, additional S entering such soils from pollutants may even be making positive influences in deserts through reducing the pH of rainfall and thus making a host of other elements more available (Whitford, 1986b).

Calcium

Ca is usually taken for granted in arid to semiarid regions because of its pronounced accumulation in these poorly leached soils. Native plants have adjusted to its abundance. Lime-induced chlorosis is identified only when humans attempt to plant exotics in such soils (Dennis et al., 1988).

The abundance of filamentous cyanobacteria in the soils of drier regions may be related to their high requirement for Ca (Paul and Clark, 1989). Indeed, the abundance of microphytic soil crusts in general is strongly related to the gypsum content of the parent material (West, 1990).

The acid, low-Ca, yet at least seasonally dry soils on the lateritic peneplains of parts of Australia and East Africa must have selected plants and microbes with low Ca requirements and/or tolerances. This could be an additional problem for those searching for successful replacement species.

Calcite aerosols from desert soils may act to neutralize acid rain in the western United States (Schlesinger and Peterjohn, 1988).

Potassium

Even though Griffith (1978) has found K to be synergistic in N fixation, few have apparently looked at K by itself or in conjunction with other elements in arid to semiarid contexts. Lajtha and Schlesinger (1988a) found no apparent losses of K over a chronosequence in southern New Mexico.

Minor Elements

Examination of other potential nutrients is rarely done for other than irrigated soils in arid to semiarid regions. We may, however, be missing some interesting relationships. For instance, Jensen (1984) has recently found that the K/Mg ratio of the soil showed a negative correlation to grass production and a positive correlation to shrub production for a wide sampling of rangelands in southeastern Idaho. This is probably related to the cation-exchange capacity (CEC) of roots of the two types of plants (Woodward et al., 1984). That is, shrubs have significantly higher root CEC values than grasses. We would thus expect them to absorb Mg from the soil solution more readily than grasses. Conversely, the rate at which shrubs absorb K would be less than that for the grasses, since the shrub exchange sites are selectively attracting divalent cations.

Hunter et al. (1980), Wallace (1987, 1989), and Nelson and Jolley (1989) point to some intriguing questions involving iron (Fe) for plants growing in desert soils. Hunter et al. (1980) describe *L. tridentata* as an Fe-efficient species. Dennis et al. (1988) produced demonstrations of Fe use efficiency by a set of native grasses from the semiarid southwestern United States.

Killingbeck (1989) makes a case for Zn deficiencies limiting N resorption by octotillo (*Fouquieria splendens*) in southern New Mexico. Jarrell and Virginia (1989) suggest that Zn and Cu may limit primary productivity in other Chihuahuan Desert ecosystems.

Excess of such elements as boron, selenium, and arsenic in drier regions continues to cause problems when humans attempt to develop irrigated agriculture there. For instance, drainage water from farmlands in drier reaches of the western United States, when diverted into waterfowl refuges, has disastrous effects on these birds. Irrigation and drainage systems in drier areas act essentially as solar stills concentrating a host of salts. Nobel et al. (1987) have recently shown a positive increase in prickly pear cacti (*Opuntia* spp.) with small amendments of boron.

Nonessential trace elements are not usually examined in dry wildland locales unless there is a need for biogeochemical prospecting (Gough and Erdman, 1983) or to monitor the potential impacts of human developments, especially point sources of pollution, such as power plants (e.g., Rope et al., 1988; Wildung and Garland, 1988).

B. Roles of Animals

There has been increasing attention in recent years to the role of animals in nutrient cycling in semiarid to arid environments. In general, animals greatly increase the rates of nutrient cycling (French, 1979; Seastedt and Crossley, 1984; Parmenter et al., 1987; Ruess, 1987; McNaughton et al., 1988; Seagle and McNaughton, 1989). Under dry and cold conditions, the activity of the microorganisms responsible for decomposition is limited. Higher tropic level consumers may accordingly play a greater role in the reduction of organic matter in such environments (French, 1979). The net result, however, may be rapid pulses of decomposition after each precipitation event (Steinberger and Whitford, 1988).

Homoiothermic herbivores, because they require large amounts of energy to support respiration to maintain body temperature, excrete feces and urine with high concentrations of nutrients available to plants, animals, and microbes in the soil (Chapin et al., 1986). Poikilotherms produce less nutrient-rich excreta in general. Nevertheless, Ingham et al. (1986a, b) found that grazing by soil invertebrates contributed significantly to mineralization in a semiarid grassland. Furthermore, they attributed the stability of their shortgrass prairie system to be at least partially due to the ability of soil predators (mainly nematodes) to consume a wide variety of prey.

Many now agree that invertebrates play an enormous role in cominuting plant litter and moving it belowground for mineralization in both deserts (Whitford and Freckman, 1987) and grasslands (Seastedt and Crossley, 1984; Hunt et al., 1987). Safriel (1987) pointed out that the dietary flexibility and opportunism of many desert herbivores and predators may also contribute to the stability of desert ecosystems. It is conceivable that desert consumer populations can be controlled by plant food quality. Plants low in nutrients may inhibit herbivore reproduction. Stressed plants may produce fewer secondary compounds than normal, resulting in increased herbivory. During wetter times, greater amounts of secondary compounds could be produced, inhibiting the explosion of herbivore populations (Safriel, 1987).

The review by Whitford (1986a) points out that invertebrates play a complex role in the movement and decomposition of litter in the warmer deserts. Their roles cause important deviations from predictability of litter decomposition based solely on precipitation and consequent soil moisture. Cellulolytic breakdown in the gut tract of invertebrates proceeds faster over longer times than in the soil (Crawford and Taylor, 1984).

Termites play a crucial role in moving organic detritus of all kinds into the soil (Nutting et al., 1987; Whitford and Freckman, 1987). In desert grasslands of southern New Mexico, Whitford (1986a) conservatively estimated that subterranean termite biomass exceeded the biomass of domestic livestock at current stocking rates by a factor of 10 and thus considered them a "keystone" group. The removal of termites can temporarily lead to increases in soil N and growth of ephemeral plants (Parker et al., 1982). Termites can, in certain habitats, consume nearly all the aboveground phytomass and relocate the nutrients in their mounds over periods up to about 100 years (Wetselaar, 1980). These are such highly localized deposits of nutrients that broken-up termite mounds have been used as fertilizer in parts of Asia (Sprent, 1987a). Foliar herbivores, such as cattle, may modify N distribution such that there are higher concentrations in roots. This in turn may encourage root herbivores, such as nematodes and scarabs (Seastedt, 1985).

Shachak et al. (1987) have shown how snails grazing on rock-inhabiting algae in the Negev Desert can cause storage of N in desert soils by moving it below the rocks. Isopods in Israel also move surface litter into desert soils (Shachak and Newton, 1985). Thus, the effects of animals on mineral cycles may be complex and not always obvious.

C. Practical Implications

The study of nutrient cycling is inherently more useful than the study of other "aggregate variables," such as energy flow, because nutrients are a finite resource. Thus, questions revolving around stability and the consequences of perturbation (e.g., "desertification") are more tractably addressed through principles of nutrient cycling than other alternatives (Whitford, 1986a).

Nutrients do not influence ecosystem structure and function singly or simply. The interactions between elements and other processes are most important and interesting (Stewart et al., 1983; McGill and Christie, 1983; Reiners, 1983; Gosz, 1986). For instance, Nobel (1989) makes a case for multiplicative indices combining the effects of N, P, K, B, and Na on the productivity of agaves and cacti at two sites in Arizona.

In arid to semiarid climates, the availability of water is usually regarded as most important in determining productivity (Safriel, 1987; Risser, 1988). Sala et al. (1988) have recently shown that the linear relationship of ANPP to precipitation begins to show a reduced upward slope only above 370 mm water input to the native semiarid grasslands of the North American Great Plains. Nutrient regulation is implied thereafter. McNaughton et al. (1988) pointed out that the precipitation-production correlation may be at least partially a mineralization-production cause and effect rather than purely a direct limitation of yield by water. Small rainfalls may contribute more to mineralization than vascular plant growth (Sala and Lauenroth, 1982).

Decomposition of plant litter in deserts is more related to invertebrate activity and number of rainfall events than total precipitation. Rates of decomposition may equal those of more mesic systems (Steinberger and Whitford, 1988).

Within these average, geographic trends there are enormous year-to-year variations. The more arid the region, the greater the fluctuations. In the tropics and subtropics, where temperatures are not usually limiting, plant, invertebrate, and microbial growth takes place anytime there is adequate soil moisture. More N may be available in a grassland soil the year after drought because of less uptake by plants during the drought year but continued microbial activity at lower soil moisture (Risser, 1988). With some years having less effective precipitation, the calculation of annual production and transfer rates, for example, may be deceptive (Crawford and Gosz, 1982, 1986; McGill and Christie, 1983). However, until we can better define the "pulse and reserve" patterns of such systems, we are probably stuck with expressions of mean annual standing crops and activity levels.

In temperate zone deserts and semideserts, especially those at higher elevations, the annual cold season provides further restraints (West, 1983). One must thus be careful in comparing these regions with the lower latitude and altitude situations. Since most research on this and other ecological topics has taken place in temperate and subtropical portions of the Northern Hemisphere, we must be wary of how well that understanding can be extrapolated elsewhere.

Another strong difference between continents that has often been overlooked is age of the landforms. Whereas parent materials are usually young and strongly altered by Ice Age events in Eurasia and North America, Australia and Africa have a considerable portion of their drier reaches on old, well-leached landforms with soils derived from laterites. Much of the vital nutrients were transferred out of these systems long ago. The relict soils can even be acid. The highly developed clays and silicates present special problems, such as surface capping, churning clays, and a greater capacity to fix applied nutrients (Chapin et al., 1986). Although unique plants, and maybe microbes, have evolved to tolerate these more stressful conditions, introductions of exotic organisms and management preferences usually lead to great community and ecosystem reorganization. Such areas are also very sensitive to nutrient additions via atmospheric pollutants.

In addition to pronounced inherent geological and edaphic differences, we must also keep in mind that land use history has been greatly different between and within continents. For instance, there have been millenia to centuries of offtake and spatial redistribution of nutrients via livestock grazing in the Near East and North Africa. In the Sahel (sub-Saharan) region, a long history of livestock grazing is usually combined with some extremely old parent materials. It is thus no wonder that researchers there (Penning deVries and Djiteye, 1982; Breman and deWit, 1983) identified soil fertility as a greater factor than precipitation in limiting rangeland productivity in the middle to southern Sahel. Similarly, Floret et al. (1982) assert that net primary production is limited by nutrients above about 250 mm of total crop year precipitation in pre-Saharan Tunisia. Tadmor et al. (1974) and Thalen (1979) made similar observations in Israel and Iraq, respectively.

The younger, inherently more fertile and less impacted soils of North America may be drawing nutrients stored before the comparatively recent development of the range livestock industry. The long residence time of some vital nutrients in humic materials and affixed to clays means that nutrient depletion could be buffered by SOM over very long times (Woodmansee, 1984). Extensive, low-level nutrient transfers, as done by nomads in subsistence economies (Ellis and Swift, 1988), may not permanently degrade a region. It is more likely that with export economies, beginning with transhumance, nutrient limitation problems begin to arise. Although some (e.g., Savory, 1988) assume original stability and consider degradation must exceed some unspecified threshold, Penning deVries and Djiteye (1982) view degradation in the Sahel as a continuous process beginning with the low levels of exploitation. These alternative models need to be tested elsewhere.

Desertification is a very complex phenomenon involving everything from climatic to social change and their feedbacks. The components of soil change include reductions in SOM and consequent loss of aggregate stability. As such soils puddle or cap over, infiltration and percolation are diminished. This "xerification" process (West, 1986) results in less plant growth and microbial activity. Long-term declines in primary production lead to increased rates of forage utilization as pastoralists attempt to maintain herds. The net transport of nutrients in animal products and increased soil erosion eventually mean lowered soil fertility. This process could be termed "distrophication." If erosion has left only clay "scalds," sand, and/or rocks, the changes can be essentially permanent in the perspective of a human lifetime. Prevention, stabilization, and, especially, reversal of this process are not simple, easy, or cheap. The soils and nutrient cycling aspects must be viewed within a much larger context or any human efforts will be fruitless (Savory, 1988).

Penning deVries and Djiteye (1982) propose that the P/N ratio in grasses can be an indicator of the relative importance of shortages of P or N. In their view, the only way to reverse degradation in the mid-Sahel is through N and P fertilization. In North American grasslands and shrublands, we have found that fertilization lowers the drought resistance of the perennials (Hyder et al., 1976) and favors annuals (Goodman, 1973; Cornelius and Cunningham, 1987; Ludwig et al., 1988b). Which annuals are favored depends on soil type, when soil moisture and nutrients are augmented (Williams and Bell, 1981), and interactions with termites (Gutierrez and Whitford, 1987). Annuals may immobilize N and depress shrub growth (Parker et al., 1984). Annuals are much less efficient users of water and nutrients than perennials (Jackson et al., 1988). Additionally, Kudeyarov (1980) points out that overuse of N fertilizers can inhibit biological N fixation and stimulate "extra" mineralization of soil organic matter.

Savory (1988) attributes the mismanagement of livestock as a major factor in the degradation of semiarid regions. He reasons that extensive long-term use allows the animals to select only the most palatable plants. Furthermore, since most of the nutrient values of forages are lost via abiotic means, such as volatilization, leaching, and wind erosion, we need instead to concentrate livestock on relatively small pieces of ground for relatively short periods of time. This will overcome the usual selectivity to only the most palatable of the forage species and allow time for the vegetation and soils to recover when the animals are moved elsewhere. Predators stimulate the movement of native ungulates. Nomadic pastoralists in a subsistence economy create similar patterns of use. Although this may be so where exports are not within the management objectives (Ellis and Swift, 1988), any move to greater production and exports required to balance imports in a cash economy contradicts these principles. Penning deVries and Djiteye (1982) have shown how the pool of available P in soils of the old landscape within the mid-Sahel can be exhausted within a few decades of such use. Even within generally younger landscapes, differing parent materials lead to sharply different levels of resistance and resilience to perturbation (Schimel et al., 1985a).

Until recently, the influence of fires on nutrient cycling in semiarid environments was largely overlooked. The general view has been that fire has more a stimulatory than degrading effect on grasslands, because very little of the pools of most nutrients is found in the aboveground parts of plants or their litter (Boerner, 1982). The situation is far different, however, where shrubs and/or small trees contain a significant part of the nutrient pools. This seems to be the case in subtropical savannas (Robertson and Rosswall, 1986). In such circumstances, we now have evidence that fires can be more disruptive to nutrient cycles than livestock grazing. For instance, Isichei and Sanford (1980) and Penning deVries and Djiteye (1982) contend that essentially all the aboveground and surface N in plants and litter of the mid-Sahel can be lost through burning. Wetselaar (1980) showed that wildfire in a savannah of the Northern Territory of Australia can lead to losses of >90% of the N and S in aboveground phytomass. Medina (1982) speculates that similar processes are probably going on in the native grasslands of Venezuela. Subsequent pasture growth may be reduced by about 50% until N fixers can successionally replace these losses. Fortunately, filamentous N-fixing cyanobacteria are stimulated by the P released by burning (Paul and Clark, 1989) and some areas can rapidly recover a N pool in soil (Isichei, 1980).

The effects of fires on nutrient cycling in temperate zone shrub steppes and desert shrublands has hardly been examined (Woodmansee and Wallach, 1981). Light ground fires could reduce volatile organics in shrubs or their litter, enhancing N mineralization and nitrification but leaving chemicals inimical to seed germination (Blank and Young, 1990). When fires lead to long-term dominance by introduced annuals [which is often the case in North America (West, 1988)], more frequent and earlier fires and greater erosion follow, reducing nutrient pool sizes. In addition, nutrients and salts could be horizontally diluted, especially at the soil surface, when shrubs are lost. This could further abet the dominance by weedy herbs. Lesser degrees of disturbance, as by herbicidal control of shrubs in a shrub steppe (Burke et al., 1987), result in more subtle changes.

There is so far little evidence that pollution is having direct impacts on nutrient cycling in arid to semiarid regions. Only Woodmansee and Duncan (1980) conclude that significant amounts of N may be added through precipitation in the Central Valley of California. The most likely way that the impacts of pollution will first be noticed is on microphytic soil crusts (West, 1990). Dryland soils are apparently sinks for enhanced methane production (Mooney et al., 1987).

Most savannas and shrub steppes have inherently unstable structure because of a rich mixture of life forms and associated patchiness of form and function (Shmida and Burgess, 1988). They thus need careful management to maintain them since it is very difficult to re-establish the original structure after they are degraded (Young, 1988; West, 1989).

IV. IMPROVEMENTS IN TECHNOLOGY

Since 1980 there have been a number of technological advances that are changing and will further change the way we approach nutrient cycling research in the field and laboratory.

In the past we have had to rely on small, instantaneously taken samples of plants or soils. These were then taken to the laboratory for largely wet chemistry determinations of their nutrient content. The results were then statistically analyzed for central tendencies and inferences made for the plots as a whole (e.g., Jackson et al., 1988). Because of the considerable microscale spatial variation, vertically (in the case of both grasslands and deserts) and horizontally (especially in savannas, shrub steppes, and deserts), and pronounced temporal dynamics, variation was usually so great as to unreliably allow discrimination between site and treatment effects. The use of cation-exchange plates (Penning deVries and Djiteye, 1982) and cation-exchange resins (Lajtha, 1988) now allow us to obtain time-integrated relative differences in liable N, P, and other nutrients between sites and/or treatments. However, we have yet to correlate such results with climate, plant, and microbial activities to see if such approaches are meaningful. We should also keep constantly in mind that any mechanical soil disturbance accelerates microbial activity two to three times that in undisturbed profiles (West, 1981).

Recent advances in mass spectrophotometry have made measurements of stable isotopes, such as ^{15}N (Marion, 1987), ^{13}C, and ^{14}C (Rundel et al., 1989), more feasible. Stable isotope ratio analyses have replaced radiotracers as the main means to track relative uptake and transfer efficiencies. For instance, N fixation can now be more accurately estimated via $^{15}N/^{14}N$ ratios in potentially fixing and nonfixing plants (Shearer et al., 1983; Lajtha and Schlesinger, 1986; Shearer and Kohl, 1989). Thus, the proportion of the N present that is derived from the N in either the atmosphere or soil can be estimated. Isotopes of H and O allow linkages with the hydrological cycle (Allison and Hughes, 1983; Amundson, 1989).

The combination of better measurements with mathematical models (e.g., Penning deVries and Djiteye, 1982; McGill et al., 1981; Parton et al., 1983, 1988; Kittel et al., 1989; Hunt et al., 1989; Skiles et al., 1989) is now allowing us to estimate some regional patterns in turnover and transfers. There remains the problem, however, of extrapolating results from small plots to landscapes or larger units of nature. Gosz et al. (1988) propose to overcome some of these limitations by using Fourier transform infrared spectroscopy (FTIR) to quantify biogenic gas emissions from carbon fixation, respiration, ammonia volatilization, denitrification, nitrification, and methanogenesis and relate them to environmental changes (e.g., precipitation, soil moisture, and temperature) over larger (up to 1 km) scales. This approach should allow avoidance of the problems of additivity of small-scale results yet attain variance diminution with increasing scale. These kinds of nutrient cycling responses can be correlated with soil and atmospheric moisture content over large areas via time domain refractometry (Dalton et al., 1984; Topp and Davis, 1985; Dasberg and Dalton, 1985; Smith et al., 1990) and light detection and ranging (LIDAR) (Matson and Harris, 1988).

Remotely positioned narrowband spectrophotometry has promise for allowing the study of plant nutrient status across realistically large pieces of landscape (Plummer, 1988; Turner et al., 1989). By taking repeat readings and knowing concomitant variation in relevant environmental variables, inference of dynamics should be allowed. Narrowband spectrophotometry (developed first for geobotanical prospecting) has promise even for detecting the amounts of protein or lignin (Westman, 1987).

Livestock carrying capacities could be made more scientific by dividing the amount of biomass with a suitable concentration of N by standardized livestock needs (Penning deVries and Djiteye, 1982).

Satellite imagery (Reiners et al., 1989) and GIS (geographic information systems) allow larger sets of data to be handled to search for landscape-level patterns in fairly sophisticated landform, soils, and vegetation interrelationships, especially in connection with land use and soil erosion issues (Pickup and Chewings, 1988).

V. OPPORTUNITIES FOR NEW RESEARCH

The foregoing shows many leads for further research. Although the importance of further basic explication of the N cycle continues, the importance of P is now clearer and requires more attention. Further research should emphasize multiple-element limitations as well as the development of statistical techniques to explore the variation rather than submerge it in measures of central tendencies (Gosz, 1986). Economic analogies (Bloom et al., 1985) will continue to be useful as water use efficiencies and nutrient ratios become considered in ways more appropriate than Liebig's law of the minimum. Global environmental changes will bring some impacts peculiar to ecosystems of arid to semiarid regions. For instance, how will changing quantities and qualities of precipitation influence carbonates in these soils (Marion et al., 1985)? Acid rain could inadvertently release even more CO_2 from the large, long-fixed pools in these regions. Changes in C/N and other nutrient ratios in new tissue and plant detritus will probably influence herbivory, decomposition, and competitive advantages and, thus, community composition.

Erosional and nutrient cycling phenomena can now be studied over large pieces of landscape via new technology, allowing the collection of vast amounts of data without destructive sampling. Many interesting demonstrations are now underway.

Further work on the functional role of mycorrhizae in nutrient cycling within grasslands, shrub steppes, and deserts seems advisable. For instance, Allen and Allen (1984) have shown that there are definite progressions in the percentages of roots infected by mycorrhizae as abandoned fields or reclaimed mine land recover. Perhaps these fungi aid in tightening nutrient cycles and allowing the maximum amount of phytomass and cover to develop. Allen (1988) advises that we must begin looking at belowground organisms as entities rather than merely functional compartments in a schematic cycle. Mycorrhizae may indeed allow the direct uptake of organic forms of N and P by higher plants and/or the more efficient transfer of NH_4^+ and PO_4^- between higher plant species, cutting down on net losses of N from these systems via volatilization of NH_3 and denitrification of NO_3^-.

Plants are generally more water use efficient when grown on NH_4^+- than NO_3^--rich soils (Raven, 1985). This phenomenon is closely related to the energy required to acquire and assimilate these substances (Sprent, 1987a). McNaughton et al. (1988) have noted that ungulate grazing has shifted preferences of grasses on the Serengeti toward ammonium. Rhizosphere acidification is associated with ammonium nutrition. Thus the increased availability and uptake of ammonium during periods of heavy grazing could affect the solubility and uptake of other nutrients because of altered soil pH. We might see if these trends extend to plants native to more arid regions. If so, this might partially explain the success of certain species following disturbances. It might also be a factor to consider in the selection of new materials in rehabilitation efforts.

Because actinomycetes are more resistant to desiccation than aerobic bacteria, the former are more common in deserts (Skujiņš, 1984). Accordingly, more research attention should be paid to the functions of that group, particularly in heterotrophic nitrification and N fixation via actinorhizal associations (Sprent, 1985).

Examination of the nutrient cycling aspects of proposed new grazing management schemes [e.g., high intensity and short duration (Savory, 1988)] is needed. If Penning deVries and Djiteye's (1982) contention is correct, then there are no synergistic effects of livestock grazing, only a downward trend in soil nutrient pools in the long run. Thus, nutrients will eventually have to be replaced artificially if the productivity of degraded lands is to ever be returned (Tiedemann et al., 1986).

The foregoing principles are being applied to the rehabilitation of both overgrazed and mechanically disturbed lands (Allen, 1988). For instance, the role of C amendments is being actively pursued (Fresquez et al., 1988; Dennis and Fresquez, 1989; Whitford et al., 1989; Watwood et al., 1989). These researchers have shown the importance of adding recalcitrant sources of C so that energy and nutrients are slowly but steadily released. Unlike cultivated, agricultural, or forest environments, where N immobilization by high C/N ratio materials is undesirable, in semiarid rangelands, where moisture availability affects both N immobilization and mineralization, high C/N ratio mulches can be beneficial. This is largely due to the grazing of fungi by microarthropods, whose activity is independent of soil moisture. This invertebrate grazing activity results in mineral N becoming available to plants. Without redevelopment of a complete soil biota, reclamation cannot be considered successful (Whitford, 1988).

Human-induced changes in regional and global climates will probably have enormous influence on arid to semiarid environments in the future. Relatively small changes in precipitation (both amounts and seasonal distributions), temperatures, evapotranspiration, and wind will cause pronounced changes in the balance between plant life forms. This change will in turn influence infiltration, percolation, erosion, and wildfire. Soil organic C, carbonate, various forms of N and P, and their all-important ratios will be altered. Vascular plants, nonvascular plants, and microbes will all be radically influenced. Animal communities will change as plant community structure, energy, and nutrient and allelochemic content change. Both further manipulative studies and modeling efforts are needed to anticipate these changes.

The increasing costs of fossil energy, agricultural amendments, and pesticides, along with greater concern for how those chemicals and increased erosion are influencing water and air quality, are forcing us to address sustainable systems of land use. Better basic understanding of nutrient cycles, including that in the soils of arid to semiarid regions, can contribute to meeting these needs.

VI. CONCLUSION

During the past three decades, there has been lessening of the dominance of water availability as the only interesting controlling factor of productivity and decomposition in ecosystems of arid to semiarid regions. Researchers have also begun to realize that salinization of alluvial soils is not the only interesting topic involving nutrients. We now see some research on the more abundant eolian and skeletal wildland soils. On such sites, nutrient cycling has been identified as more important to questions of ecosystem sustainability, repair, and stability than energy flow (Whitford, 1986a).

Ecologists now recognize the importance of age of landforms, their history of use, and how different portions of the landscape interact in producing dynamic patterns of change on numerous temporal and spatial scales.

The development of more rapid and reliable field and laboratory techniques are allowing us to address more sophisticated questions. Remote sensing of larger pieces of landscapes allow us to detect patterns of static differences and change at scales more appropriate to the needs of land management and assessment of regional and global changes. We also better recognize the inherent fragility of ecosystems of semiarid regions. Long-term rundowns in available forms of nutrients (distrophication), especially of N and P, are now recognized as a part of the complex degradatory process popularly known as desertification.

The concept of islands of fertility has been further demonstrated as an important structural and functional feature of nutrients in desert ecosystems (Garner and Steinberger, 1989). Efforts in monitoring land use and accelerating and directing ecosystem rehabilitation are using this principle (Wallace and Romney, 1980; Wallace et al., 1980; Allen, 1988). The concentration of vital elements and attendant biological activity near the surface of the less horizontally patchy grassland soils reinforce our need to restrict erosive effects (Schlesinger et al., 1990). The roles of fungi in enhancing the uptake and recycling of nutrients in both kinds of systems are now being vigorously researched. Nonvascular, surface-inhabiting microphytes are now better recognized for their possible roles in erosion, water infiltration, N fixation, and gaseous losses, but better experiments are needed to quantify those functions (West, 1990).

The role of animals in spatially concentrating nutrients, accelerating decomposition, and thus controlling nutrient availability and subsequent soil and vegetation development through enhanced nutrient cycling is now a very active area of research.

Use of stable isotopes promises better analyses of uptake, efficiencies, and nutrient interactions across topographical and geographic gradients. Fourier transform infrared spectroscopy may also allow better assessment of gaseous processes on more meaningful spatial and temporal scales. These tools may allow us to finally settle the issue of the relative importance of abiotic versus biotic processes in nutrient cycling within grasslands and deserts. This is especially important because there are those who advocate high-intensity, short-duration livestock grazing as a means to reduce such assumed losses.

Although considerable progress has been made in the computer simulation of nutrient cycles in desert and grassland ecosystems, in both natural and perturbed contexts, such models are only as reliable as the assumptions built into them. Recent experimental findings show that many more important interactions will probably be discovered.

REFERENCES

Allen, M. F. (1988). Belowground structure: A key to reconstructing a productive arid ecosystem. In *The Reconstruction of Disturbed Arid Lands: An Ecological Approach,* AAAS Selected Symposium 109, E. B. Allen (Ed.). Westview Press, Boulder, Colorado, pp. 113-135.

Allen, E. G., and Allen, M. F. (1984). Competition between plants of different successional stages: Mycorrhizae as regulators. *Can. J. Bot. 62*: 2625-2629.

Allison, G. B., and Hughes, M. W. (1983). The use of natural tracers as indicators of soil-water movement in a temperate semi-arid region. *J. Hydrol. 60*: 157-173.

Amundson, R. (1989). The use of stable isotopes in assessing the effect of agriculture on arid and semi-arid soils. In *Stable Isotopes in Ecological Research*, P. W. Rundel, J. R. Ehleringer, and K. A. Nagy (Eds.). Springer-Verlag, New York, pp. 318-341.

Andrew, M. H. (1988). Grazing impact in relation to livestock watering points. *Trends Ecol. Evol. 3*: 336-339.

Barth, R. C., and Klemmedson, J. O. (1982). Amount and distribution of dry matter, nitrogen and organic carbon in soil-plant systems of mesquite and palo verde. *J. Range Manag. 35*: 412-418.

Bate, G. C. (1981). Nitrogen cycling in savanna ecosystems. *Ecol. Bull. (Stockh.) 33*: 463-475.

Bernard-Reversat, F. (1982). Biogeochemical cycle of nitrogen in a semi-arid savanna. *Oikos 38*: 321-332.

Bernard-Reversat, F., and Poupon, H. (1980). Nitrogen cycling in a soil-tree system in a Sahelian savanna: Example of *Acacia senegal*. In *Nitrogen Cycling in West African Ecosystems*, T. Rosswall (Ed.). Royal Swedish Academy of Sciences, Stockholm, pp. 363-369.

Binet, P. (1981). Short-term dynamics of minerals in arid ecosystems. In *Arid-land Ecosystems Structure, Functioning and Management*, Vol. 2, D. W. Goodall and R. A. Perry (Eds.), Cambridge University Press, Cambridge, pp. 325-356.

Bjerregaard, R. S., West, N. E., Caldwell, M. M., and Maryland, H. F. (1984). Standing crops and dynamics of phytomass and minerals in two salt desert shrub communities. *Great Basin Naturalist 44*: 327-337.

Blank, R. R., and Young, J. A. (1990). Chemical changes in the soil induced by fire in a shrub-grass dominated community. Proc. Sixth Wildland Shrub Symposium. Las Vegas (in press).

Bloom, A. J., Chapin, F. S., III, and Mooney, H. A. (1985). Resource limitation in plants—An economic analogy. *Annu. Rev. Ecol. Syst. 16*: 363-392.

Boerner, R. E. J. (1982). Fire and nutrient cycling in temperate ecosystems. *BioScience 32*: 187-192.

Bowden, W. B. (1986). Gaseous nitrogen emissions from undisturbed terrestrial ecosystems: An assessment of their impacts on local and global nitrogen budgets. *Biogeochemistry 2*: 249-280.

Bremen, H., and deWit, C. T. (1983). Rangeland productivity and exploitation in the Sahel. *Science 221*: 1341-1347.

Brooks, S. I., and Pokshishevsky, V. V. (1986). Population problems in the arid and semiarid regions of the world. In *Arid Land Development and the Combat Against Desertification: An Integrated Approach*, United Nations Environment Programme (UNEP), I. P. Gerasimov (Ed.). USSR Commission for UNEP, Moscow, pp. 53-62.

Burke, I. C. (1989). Control of N mineralization in a sagebrush steppe landscape. *Ecology 70*: 1115-1126.

Burke, I. C., Reiners, W. A., Sturges, D. L., and Matson, P. A. (1987). Herbicide treatments effects on properties of mountain big sagebrush soils after fourteen years. *Soil Sci. Soc. Am. J. 51*: 1337-1343.

Burke, I. C., Reiners, W. A., and Schimel, D. S. (1989a). Organic matter turnover in a sagebrush steppe landscape. *Biogeochemistry 7*: 11-31.

Burke, I. C., Yonker, C. M., Parton, W. J., Cole, C. V., Flach, K., and Schimel, D. S. (1989b). Texture, climate, and cultivation effects on soil organic matter content in U. S. grassland soils. *Soil Sci. Soc. Am. J. 53*: 800-805.

Bush, J. K., and Van Auken, O. W. (1986). Changes in nitrogen, carbon and other soil properties during secondary succession. *Soil Sci. Soc. Am. J. 50*: 1557-1601.

Caldwell, M. M., and Richards, J. H. (1990). Hydraulic lift: Water efflux from upper roots improves effectiveness of water uptake by deep roots. *Oecologia (Berl.) 79*: 1-5.

Caldwell, M. M., Richards, J. H., and Beyschlag, W. (1990). Hydraulic lift: Ecological implications of water efflux from roots. In *Plant Root Systems: The Effect on Ecosystem Composition and Structure*, D. Atkinson (Ed.). Blackwell, Oxford (in press).

Carpenter, A. T., and West, N. E. (1987). Indifference of mountain big sagebrush growth to supplemental water and nitrogen. *J. Range Manag. 40*: 448-451.

Chapin, F. S., III, Vitousek, P. M., and Van Cleve, K. (1986). The nature of nutrient limitation in plant communities. *Am. Naturalist 127*: 48-58.

Charley, J. L. (1977). Mineral cycling in rangeland ecosystems. In *Rangeland Plant Physiology*, Range Science Series No. 4, R. E. Sosebee (Ed.). Society for Range Management, Denver, Colorado, pp. 215-256.

Christiansen, T. A., Lockwood, J. A., and Powell, J. (1989). Mediation of nutrient cycling by arthropods in unmanaged and intensively managed mountainbrush habitats. *Great Basin Naturalist 49*: 134-139.

Christie, E. K. (1981). Biomass and nutrient dynamics in a C_3, semiarid Australian grassland. *J. Appl. Ecol. 18*: 907-918.

Clark, F. E., Cole, C. V., and Bowman, R. A. (1978). Nutrient cycling. In *Grasslands, Systems Analysis and Man*, IBP 19, Vol. 2, A. I. Breymeyer and G. M. Van Dyne (Eds.). Cambridge University Press, Cambridge, pp. 659-712.

Cole, M. G. (1989). Landscape ecology: The effect of pattern on process. *Annu. Rev. Ecol. Syst. 20*: 171-147.

Cornelius, J. M., and Cunningham, G. L. (1987). Nitrogen enrichment effects on vegetation of a Northern Chihuahuan Desert landscape. In *Strategies for Classification and Management of Native Vegetation for Food Production in Arid Zones*, E. F. Aldon, G. E. G. Vincent, and W. H. Moir (Eds.). Rocky Mountain Forest and Range Experiment Station, Forest Service, U.S. Department of Agriculture, Ft. Collins, Colorado, pp. 112-116.

Cox, J. R., Parker, J. M., and Stroehlein, J. L. (1984). Soil properties in creosotebush communities and their relative effects on the growth of seeded range grasses. *Soil Sci. Am. J. 48*: 1442-1445.

Crawford, C. S., and Gosz, J. R. (1982). Desert ecosystems: Their resources in space and time. *Environ. Conserv. 9*: 181-195.

Crawford, C. S., and Gosz, J. R. (1986). Dynamics of desert resources and ecosystem processes. In *Ecosystem Theory and Applications*, N.

Polunin (Ed.). John Wiley, New York, pp. 63-88.
Crawford, C. S., and Taylor, E. C. (1984). Decomposition in arid environments: Role of the detritivore gut. *S. Afr. J. Sci. 80*: 170-175.
Dalton, F. N., Herkelrath, W. N., Rawlins, D. S., and Rhoades, J. B. (1984). Time domain reflectometry: Simultaneous assessment of the soil water content and electrical conductivity with a single probe. *Science 224*: 989-990.
Dasberg, S., and Dalton, F. N. (1985). Time domain reflectometry field measurements of soil water content and electrical conductivity. *Soil Sci. Soc. Am. J. 49*: 293-297.
Dennis, G. L., and Fresquez, P. R. (1989). The soil microbial community in a sewage sludge-amended semi-arid grassland. *Biol. Fertil. Soils 7*: 310-317.
Dennis, G. L., Fresquez, P. R., and Johnson, G. V. (1988). Iron uptake by native grasses in a sewage sludge amended semiarid grassland. *J. Plant Nutr. 11*: 1417-1427.
Detling, J. K. (1988). Grasslands and savannas: Regulation of energy flow and nutrient cycling by herbivores. In *Concepts of Ecosystem Ecology*, L. R. Pomeroy and J. J. Alberts (Eds.). Springer-Verlag, New York, pp. 131-148.
Dodd, J. L., and Lauenroth, W. K. (1979). Analysis of the response of a grassland ecosystem to stress. In *Perspectives in Grassland Ecology*, N. R. French (Ed.). Springer-Verlag, New York, pp. 43-58.
Doescher, P. S., Miller, R. F., and Winward, A. H. (1984). Soil chemical patterns under eastern Oregon plant communities dominated by big sagebrush. *Soil Sci. Soc. Am. J. 48*: 659-663.
Dormaar, J. F. (1975). Susceptibility of organic matter of Chernozemic Ah horizons to biological decomposition. *Can. J. Soil Sci. 55*: 473-480.
Dormaar, J. F., Smoliak, S., and Willms, W. D. (1990). Soil chemical properties during succession from abandoned cropland to native range. *J. Range Manag. 43*: 260-265.
Dregne, H. E. (1976). *Soils of Arid Regions*. Elsevier, Amsterdam.
Elderkin, R. L., Jr., Dietz, D. R., Welch, B. L., Ralphs, P. M., and Czarnecki, F. (1986). Fertilization and irrigation of big sagebrush: Effects on nutritional composition, production and use. In *Biology of Artemisia and Chrysothamnus*, Proc. Symp. Provo, Utah, July 9-13 1984. U.S. Dept. Agr., Forest Service, Intermtn. Res. Sta., Gen. Tech. Rep. 200, pp. 140-141.
Ellis, J. E., and Swift, D. M. (1988). Stability of African pastoral systems: Alternate paridigms and their implications for development. *J. Range Manag. 41*: 450-459.
Farquhar, G. D., Wetselaar, R., and Weir, B. (1983). Gaseous nitrogen losses from plants. In *Gaseous Losses of Nitrogen from Plant-Soil Systems*, J. R. Freney and J. R. Simpson (Eds.). Martinus Nijhoff/W. Junk, The Hague, pp. 159-180.
Feigenbaum, S., Seligman, N. G., Benjamin, R. W., and Feinerman, D. (1983). Recovery of tagged fertilizer nitrogen applied to rainfed spring wheat (*Triticum aestivum* L.) subjected to severe moisture stress. *Plant Soil 73*: 265-274.
Feigenbaum, S., Seligman, N. G., and Benjamin, R. W. (1984). Fate of

Nitrogen-15 applied to spring wheat grown for three consecutive years in a semiarid region. *Soil Sci. Soc. Am. J. 48*: 838-843.

Fisher, F. M., Zak, J. C., Cunningham, G. L., and Whitford, W. G. (1988). Water and nitrogen effects on growth and allocation patterns of creosotebush in the northern Chihuahuan Desert. *J. Range Manag. 41*: 387-391.

Fisher, F. M., Parker, L. W., Anderson, J. P., and Whitford, W. G. (1987). Nitrogen mineralization in a desert soil: Interacting effects of soil moisture and N. fertilizer. *Soil Sci. Soc. Am. J. 51*: 1033-1041.

Floret, C. R., Pontanier, R., and Rambel, S. (1982). Measurement and modelling of primary production and water use in a south Tunisian steppe. *J. Arid Environ. 5*: 77-90.

Forman, R. T. T., and Godron, M. (1986). *Landscape Ecology*. John Wiley, New York.

Francis, R., and Read, D. J. (1984). Direct transfer of carbon between plants connected by vesicular-arbuscular mycorrhiza mycelium. *Nature 307*: 53-56.

French, N. R. (1979). Principal subsystem interactions in grasslands. In *Perspectives in Grassland Ecology*, Ecol. Studies 32, N. R. French (Ed.). Springer-Verlag, New York, pp. 173-190.

Fresquez, P. R., Aldon, E. F., and Dennis, G. L. (1988). Carbon dioxide evolution from an organically amended Rio Puerco soil. U.S. Dept. Agr. Forest Service, Res. Note RM-480. Rocky Mtn. Expt. Sta., Fort Collins, Colorado.

Friedel, M. H., Cellier, K. M., and Nicolson, K. P. (1980). Nutrient deficiencies in central Australia semi-desert rangelands with reference to range condition. *Aust. Rangeland J. 2*: 151-161.

Garner, W., and Steinberger, Y. (1989). A proposed mechanism for the formation of "fertile islands" in desert ecosystems. *J. Arid Environ. 16*: 257-262.

Goodman, P. J. (1973). Physiological and ecotypic adaptations of plants to salt desert conditions in Utah. *J. Ecol. 61*: 473-494.

Gosz, J. R. (1986). Biogeochemistry research needs: Observations for Ecosystems Studies Program of the National Science Foundation. *Biogeochemistry 2*: 101-112.

Gosz, J. R., Dahm, C. N., and Risser, P. G. (1988). Long-path FTIR technology for large scale ecological studies. *Ecology 69*: 1326-1330.

Gough, L. P., and Erdman, J. A. (1983). Baseline elemental concentrations for big sagebrush from western U.S.A. *J. Range Manag. 36*: 718-722.

Griffith, W. K. (1978). Effects of phosphorus and potassium on nitrogen fixation. In *Phosphorus for Agriculture—A Situation Analysis*, Potash/Phosphate Institute, Atlanta, pp. 80-94.

Gutierrez, J. R., and Whitford, W. G. (1987). Chihuahuan desert annuals: Importance of water and nitrogen. *Ecology 68*: 2032-2045.

Gutierrez, J. R., DeSilva, O. A., Pagani, M. I., Weems, D., and Whitford, W. G. (1988). Effects of different patterns of supplemental water and nitrogen fertilization on productivity and composition of Chihuahuan Desert annual plants. *Am. Midland Naturalist 119*: 336-343.

Gutsick, V. P. (1981). Evolved strategies in nitrogen acquisition by plants. *Am. Naturalist 118*: 607-637.

Hacker, R. B. (1986). Effect of grazing on chemical and physical properties of an earthy sand in the Western Australian mulga zone. *Aust. Rangeland J. 8*: 11-17.

Hunt, H. W., Coleman, D. C., Ingham, E. R., Ingham, R. E., Elliott, E. T., Moore, J. C., Reid, C. P. P., and Morley, C. R. (1987). The detrital food web in a shortgrass prairie. *Biol. Fertil. Soil 3*: 57-68.

Hunt, H. W., Ingham, E. R., Coleman, D. C., Elliott, E. T., and Reid, C. P. P. (1988). Nitrogen limitations of production and decomposition in prairie, mountain meadow and pine forest. *Ecology 69*: 1009-1016.

Hunt, H. W., Detling, J. K., Elliott, E. T., and Moore, J. C. (1989). Simulation model for the effects of climate change in temperate grasslands (abstract). *Bull. Ecol. Soc. Am. 70*: 148.

Hunter, R. B., Wallace, A., and Romney, E. M. (1980). Field studies of mineral nutrition of *Larrea tridentata*: Importance of N, pH and Fe. *Great Basin Naturalist Memoirs 4*: 162-167.

Hyder, D. N., Huston, W. R., and Burwell, J. B. (1976). Drought resistance of blue grama as affected by atrazine and N fertilizer. *J. Range Manag. 29*: 214-216.

Ingham, E. R., Trofymow, J. A., Ingham, E. R., and Coleman, D. C. (1985). Interactions of bacteria, fungi and their nematode grazers: Effects on nutrient cycling and plant growth. *Ecol. Monogr. 55*: 119-140.

Ingham, E. R., Trofymow, J. A., Ames, R. N., Hunt, H. W., Morley, C. R., Moore, J. C., and Coleman, D. C. (1986a). Trophic interactions and nitrogen cycling in a semiarid grassland soil. I. Seasonal dynamics of the natural populations, their interactions and effects on nitrogen cycling. *J. Appl. Ecol. 23*: 597-614.

Ingham, E. R., Trofymow, J. A., Ames, R. N., Hunt, H. W., Morley, C. R., Moore, J. C., and Coleman, D. C. (1986b). Trophic interaction and nitrogen cycling in a semi-arid grassland soil. II. System responses to removal of different groups of soil microbes or fauna. *J. Appl. Ecol. 23*: 615-630.

Isichei, A. O. (1980). Nitrogen fixation by blue-green algal soil crusts in Nigerian savanna. In *Nitrogen Cycling in West African Ecosystems*, T. Rosswall (Ed.). Royal Swedish Academy of Sciences, Stockholm, pp. 191-198.

Isichei, A. O., and Sanford, W. W. (1980). Nitrogen loss by burning from Nigerian grassland ecosystems. In *Nitrogen Cycling in West African Ecosystems*, T. Rosswall (Ed.). Royal Swedish Academy of Sciences, Stockholm, pp. 325-331.

Jackson, L. E., Strauss, R. B., Firestone, M. K., and Bartolome, J. W. (1988). Plant and soil nitrogen dynamics in California annual grasslands. *Plant Soil 110*: 9-17.

Jarrell, W. M., and Virginia, R. A. (1989). Micro nutrient metals in mesquite dominated Chihuahuan Desert ecosystems (abstract). *Bull. Ecol. Soc. Am. 70*: 155.

Jensen, M. E. (1984). Some edaphic relations of southeastern Idaho wildlands. *Great Basin Naturalist 44*: 265-277.

Johnson, R. H., and Lincoln, D. E. (1989). Basin sagebrush palatability growth and defense; influence of carbon dioxide enrichment and mineral limitation (abstract). *Bull. Ecol. Soc. 70*: 158.

Jones, M. B., and Woodmansee, R. G. (1979). Biogeochemical cycling in annual grassland ecosystems. *Bot. Rev. 45*: 111-144.

Joyce, L. A., Rastetter, E. B., and Parton, W. J. (1989). The effect of texture on decomposition and nutrient release in grassland soils (abstract). *Bull. Ecol. Soc. Am. 70*: 1161.

Jurinak, J. J., Dudley, L. M., Allen, M. F., and Knight, W. G. (1986). The role of calcium oxalate in the availability of phosphorus in soils of semiarid regions: A thermodynamic study. *Soil Sci. 142*: 255-261.

Killingbeck, K. T. (1989). Inefficient nitrogen resorption in a population of octotillo (*Fouquieria splendens* Engelm.): A paradox caused by zinc deficiency? (abstract). *Bull. Ecol. Soc. Am. 70*: 168.

Kittel, T. G. F., Schimel, D. S., Parton, W. J., and Galvin, M. O. (1989). Response of biogeochemical dynamics to climate change: Simulation of the North American central grasslands (abstract). *Bull. Ecol. Soc. Am. 70*: 170.

Klemmedson, J. O., and Tiedemann, A. R. (1986). Long-term effects of mesquite removal on soil characteristics. II. Nutrient availability. *Soil Sci. Soc. Am. J. 50*: 476-480.

Klemmedtsson, L., Svensson, B. H., and Rosswall, T. (1988a). A method of selective inhibition to distinguish between nitrification and denitrification as sources of nitrous oxide in soil. *Biol. Fertil. Soils 6*: 112-119.

Klemmedtsson, L., Svensson, B. H., and Rosswall, T. (1988b). Relationships between soil moisture content and nitrous oxide production during nitrification and denitrification. *Biol. Fertil. Soils 6*: 106-111.

Knight, W. G. (1987). Oxalate and elevated carbon dioxide enhanced phosphorus availability in the rhizosphere of vesicular-arbuscular mycorrhizal *Agropyron smithii*. Ph.D. dissertation, Utah State University, Logan.

Kovda, V. A., Samoilova, E. M., Charley, J. L., and Skujiņš, J. J. (1979). Soil processes in aridlands. In *Arid-land Ecosystems: Structure, Functioning and Management*, Vol. I, D. W. Goodall and R. A. Perry (Eds.). Cambridge University Press, London, pp. 439-470.

Kudeyarov, V. N. (1980). The nitrogen cycle-pedological considerations. In *Nitrogen Cycling in West African Ecosystems*, T. Rosswall (Ed.). Royal Swedish Academy of Sciences, Stockholm, pp. 893-894.

Lajtha, K. (1987). Nutrient reabsorption efficiency and the response to phosphorus fertilization in the desert shrub *Larrea tridentata* (DC) Cov. *Biogeochemistry 4*: 265-276.

Lajtha, K. (1988). The use of ion exchange resin bags for measuring nutrient availability in an arid ecosystem. *Plant and Soil 105*: 105-111.

Lajtha, K., and Bloomer, S. H. (1988). Factors affecting phosphorus adsorption and phosphorus retention in an arid ecosystem. *Soil Sci. 146*: 160-167.

Lajtha, K., and Klein, M. (1988). The effect of varying nitrogen and phosphorus availability on nutrient use by *Larrea tridentata*, a desert evergreen shrub. *Oecologia (Berl.) 75*: 348-353.

Lajtha, K., and Schlesinger, W. H. (1986). Plant response to variations in nitrogen availability in a desert shrubland community. *Biogeochemistry 2*: 29-37.

Lajtha, K., and Schlesinger, W. H. (1988a). The biogeochemistry of phosphorus cycling and phosphorus availability along a desert soil chronosequence. *Ecology 69*: 24-39.

Lajtha, K., and Schlesinger, W. H. (1988b). The effect of $CaCO_3$ on the uptake of phosphorus by two desert shrub species, *Larrea tridentata* (D.C.) Cov. and *Parthenium incanum* H.B.K. *Bot. Gaz. 149*: 328-334.

Lajtha, K., and Whitford, W. G. (1989). The effect of water and nitrogen amendments on photosynthesis, leaf demography, and resource-use efficiency in *Larrea tridentata*, a desert evergreen shrub. *Oecologia 80*: 341-348.

Lauenroth, W. K., and Dodd, J. L. (1979). Response of native legumes to water and nitrogen treatments. *J. Range Manag. 32*: 292-294.

Lauenroth, W. K., Dodd, J. L., and Sims, P. L. (1978). The effects of water and nitrogen-induced stresses on plant community structure in a semi-arid grassland. *Oecologia (Berl.) 36*: 211-222.

Lauenroth, W. K., Milchunas, D. G., and Dodd, J. L. (1984). Response of vegetation. In *The Effects of SO_2 on a Grassland: A Case Study in the United States*, Ecol. Studies 45, W. R. Laurenroth and E. Preston (Eds.). Springer-Verlag, New York, pp. 97-136.

Ludwig, J. A., Cunningham, G. L., and Whitson, P. D. (1988a). Distribution of annual plants in North American deserts. *J. Arid Environ. 15*: 221-227.

Ludwig, J. A., Whitford, W. G., and Cornelius, J. M. (1988b). Effects of water, nitrogen and sulfur amendments on cover, density and size of Chihuahuan Desert ephemerals. *J. Arid Environ. 16*: 35-42.

MacKay, W. P., Fisher, F. M., Silva, S., and Whitford, W. G. (1987). The effects of nitrogen, water and sulfur amendments on surface litter decomposition in the Chihuahuan Desert. *J. Arid Environ. 12*: 223-232.

Marion, G. M. (1987). Use of nitrogen 15 to assess terrestrial nitrogen cycling process. In *Plant Response to Stress*, NATO ASI Series Vol. G 15, J. D. Tenhunen (Ed.). Springer-Verlag, Berlin, pp. 129-155.

Marion, G. M., Schlesinger, W. H., and Fonteyn, P. J. (1985). CALDEP: A regional model for soil $CaCO_3$ (caliche) deposition in southwestern deserts. *Soil Sci. 139*: 468-481.

Matson, P. A., and Harris, R. C. (1988). Prospects for aircraft-based gas exchange measurements in ecosystem studies. *Ecology 69*: 1318-1325.

McGill, W. B., and Christie, E. K. (1983). Biogeochemical aspects of nutrient cycling interactions in soils and organisms. In *The Major Biogeochemical Cycles and Their Interactions*, B. Bolin and R. B. Cook (Eds.). SCOPE, UNESCO, Paris, pp. 271-301.

McGill, W. G., and Cole, C. V. (1981). Comparative aspects of cycling of organic C, N, S and P through soil organic matter. *Geoderma 26*: 267-286.

McGill, W. G., Hunt, H. W., Woodmansee, R. G., and Ruess, J. O. (1981). Phoenix, a model of the dynamics of carbon and nitrogen in grassland soils. *Ecol. Bull. (Stockh.) 33*: 49-115.

McNaughton, S. J. (1988). Mineral nutrition and spatial concentrations of African ungulates. *Nature 334*: 343-345.

McNaughton, S. J., Ruess, R. W., and Seagle, S. W. (1988). Large mammals and process dynamics in African ecosystems. *BioScience 38*: 794-800.

Medina, E. (1982). Nitrogen balance in the *Trachypogon* grasslands of central Venezuela. *Plant and Soil 67*: 303-314.

Milchunas, D. G., and Lauenroth, W. K. (1984). Sulfur deposition, cycling and accumulation. In *The Effects of SO_2 on a Grassland: A Case Study in the Northern Great Plains of the United States*, Ecol. Studies 45, W. K. Lauenroth and E. Preston (Eds.). Springer-Verlag, New York, pp. 61-95.

Mitchell, M. J., and Fuller, R. D. (1988). Models of sulfur dynamics in forest and grassland ecosystems with emphasis on soil processes. *Biogeochemistry 5*: 133-163.

Mooney, H. A. Gulmon, S. L., Rundel, P. W., and Ehleringer, J. (1980). Further observations on the water relations of *Prosopsis tamarugo* of the northern Atacama Desert. *Oecologia (Berl.) 44*: 177-180.

Mooney, H. A., Vitousek, P. M., and Matson, P. A. (1987). Exchange of materials between terrestrial ecosystems and the atmosphere. *Science 238*: 926-932.

Mosier, A. R., and Parton, W. J. (1985). Denitrification in a short grass prairie: A modeling approach. In *Planetary Ecology*, D. E. Caldwell, J. A. Brierly, and C. L. Brierly (Eds.). Van Nostrand-Reinhold, New York, pp. 441-452.

Mosier, A. R., Stillwell, M., Parton, W. J., and Woodmansee, R. G. (1981). Nitrous oxide emissions from a native shortgrass prairie. *Soil Sci. Soc. Am. J. 45*: 617-619.

Nelson, S. D., and Jolley, V. D. (1989). Are wildland plant communities a potential source of understanding biological responses to iron deficiency stress? In *Proc. Symp. Shrub Ecophysiology and Biotechnology*, U.S. Dept. Agr., Forest Service, Intermountain Res. Sta., Gen. Tech. Rep. INT-256, pp. 136-140.

Newman, E. I. (1989). Mycorrhizal links between plants: Their functioning and ecological significance. *Adv. Ecol. Res. 18*: 243-270.

Nobel, P. S. (1989). A nutrient index quantifying productivity of agaves and cacti. *J. Appl. Ecol. 26*: 635-645.

Nobel, P. S., Russell, C. E., Felker, P., Medina, J. G., and Acuna, E. (1987). Nutrient relations and productivity of prickly pear cacti. *Agron. J. 79*: 550-555.

Nutting, W. L., Haverty, M. I., and LaFage, J. P. (1987). Physical and chemical alteration of soil by two subterranean termite species in Sonoran Desert grassland. *J. Arid Environ. 12*: 233-239.

O'Brien, R. W., and Slaytor, M. (1982). Role of microorganisms in the metabolism of termites. *Aust. J. Biol. Sci. 35*: 239-262.

O'Connor, K. F. (1983). Nitrogen balances in natural grasslands and extensively managed grassland systems. *N.Z. J. Ecol. 6*: 1-18.

Parker, L. W., Fowler, H. G., Ettershank, G., and Whitford, W. G. (1982). The effects of subterranean termite removal on desert soil nitrogen and ephemeral flora. *J. Arid Environ. 5*: 53-59.

Parker, L. W., Santos, P. F., Phillips, J., and Whitford, W. G. (1984). Carbon and nitrogen dynamics during the decomposition of litter and roots of a Chihuahuan Desert annual, *Ledidium lasiocarpum*. *Ecol. Monogr. 54*: 339-360.

Parmenter, R. R., Mesch, M. R., and MacMahon, J. A. (1987). Shrub litter production in a sagebrush-steppe ecosystem: Rodent population cycles as a regulating factor. *J. Range Manag. 40*: 50-54.

Parton, W. J., Anderson, D. W., Cole, C. V., and Stewart, J. W. B. (1983). Simulation of soil organic matter formation and mineralization in semiarid agroecosystems. In *Nutrient Cycling in Agricultural Ecosystems*, R. R. Lowrance, R. L. Todd, L. E. Asmussen, and R. A. Leonard (Eds.). University of Georgia, Athens, pp. 533-550.

Parton, W. J., Schimel, D. S., Cole, C. V., and Ojima, D. S. (1987). Analysis of factors controlling soil organic matter levels in Great Plains grassland. *Soil Sci. Soc. Am. J. 51*: 1173-1179.

Parton, W. J., Mosier, A. R., and Schimel, D. S. (1988a). Rates and pathways of nitrous oxide production in a shortgrass steppe. *Biogeochemistry 6*: 45-58.

Parton, W. J., Stewart, J. W. S., and Cole, C. V. (1988b). Dynamics of C, N, P and S in grassland soils: A model. *Biogeochemistry 5*: 109-132.

Parton, W. J., Pielke, R., Burke, I., and Kittel, T. G. F. (1989). Working across scales in grassland ecosystems (abstract). *Bull. Ecol. Soc. Am. 70*: 223.

Paul, E. A., and Clark, F. E. (1989). *Soil Microbiology and Biochemistry*. Academic Press, San Diego.

Penning deVries, F. W. T., and Krul, J. M. (1980). Productivity of Sahelian rangelands in relation to the availability of nitrogen and phosphorus from the soil. In *Nitrogen Cycling in West African Ecosystems*, T. Rosswall (Ed.). Royal Swedish Academy of Sciences, Stockholm, pp. 95-113.

Penning deVries, F. W. T., and M. A. Djiteye (Eds.) (1982). *La Productivité des Pâturages Sahéliens: Une Etude des Sols, des Vegetations et de L'exploitation de cette Resource Naturelle*. PUDOC Centre for Agricultural Publishing and Documentation, Wageningen, The Netherlands.

Peterjohn, W. T., and Schlesinger, W. H. (1989). Nitrogen loss from deserts in the southwestern United States (abstract). *Bull. Ecol. Soc. Am. 70*: 225-226.

Pickup, G. (1985). The erosion cell: A geomorphic approach to landscape classification in range assessment. *Aust. Rangeland J. 7*: 114-121.

Pickup, G., and Chewings, V. H. (1986). Random modelling of spatial variations in erosion and deposition in flat alluvial landscapes in arid Central Australia. *Ecol. Model. 33*: 269-296.

Pickup, G., and Chewings, V. H. (1988). Forecasting patterns of soil erosion in arid lands from Landsat MSS data. *Int. J. Remote Sens. 9*: 69-84.

Plummer, S. E. (1988). Exploring the relationship between leaf nitrogen content, biomass, and the near infrared/red reflectance ratio. *Int. J. Remote Sens. 9*: 177-183.

Raven, J. A. (1985). Regulation of pH and generation of osmolality in vascular plants: A cost-benefit analyses in relation to efficiency of use of energy, nitrogen and water. *New Phytol. 101*: 25-77.

Reiners, W. A. (1981). Nitrogen cycling in relation to ecosystem succession. *Ecol. Bull. (Stockh.) 33*: 507-528.

Reiners, W. A. (1983). Transport processes in the biogeochemical cycles of carbon, nitrogen, phosphorus and sulfur. In *The Major Biogeochemical Cycles and Their Interactions*, SCOPE 21, B. Bolin and R. B. Cook (Eds.). John Wiley, New York, pp. 143-176.

Reiners, W. A., Strong, L. L., Matson, P. A., Burke, I. C., and Ojima, D. S. (1989). Estimating biogeochemical fluxes across sagebrush-steppe landscapes with thematic mapper imagery. *Remote Sens. Environ. 28*: 121-129.

Rice, E. L., and Pancholy, S. K. (1972). Inhibition of nitrification by climax vegetation. *Am. J. Bot. 59*: 752-755.

Richards, J. H., and Caldwell, M. M. (1987). Hydraulic life: Substantial nocturnal water transport between soil layers by *Artemisia tridentata* roots. *Oecologia (Berl.) 73*: 486-489.

Rickard, W. H. (1985). Shoot production and mineral nutrient assimilation in cheatgrass communities. *N.W. Sci. 59*: 169-179.

Risser, P. G. (1988). Abiotic controls on primary productivity and nutrient cycles in North American grasslands. In *Concepts of Ecosystem Ecology*, L. R. Pomeroy and J. J. Alberts (Eds.). Springer-Verlag, New York, pp. 115-129.

Roberts, R. J., and Morton, R. (1985). Biomass of larval Scarabeidae (Coleoptera) in relation to grazing pressures in temperate sown pastures. *J. Appl. Ecol. 22*: 863-874.

Robertson, G. P., and Rosswall, T. (1986). Nitrogen in West Africa: The regional cycle. *Ecol. Monogr. 56*: 43-72.

Romney, E. M., Wallace, A., Kaaz, H., and Hale, J. O. (1980). The role of shrubs on redistribution of mineral nutrients in soil in the Mohave Desert. *Great Basin Naturalist Memoirs 4*: 124-133.

Rope, S. K., Arthur, W. J., III, Craig, T. H., and Craig, E. H. (1987). Nutrient and trace elements in soil and desert vegetation of southern Idaho. *Environ. Monitor. Assess. 8*: 1-24.

Rosswall, T. (1976). The internal nitrogen cycle between micro-organisms, vegetation and soil. *Ecol. Bull. (Stockh.) 22*: 157-167.

Rosswall, T., Woodmansee, R. G., and Risser, P. (Eds.) (1988). *Scales and Global Change; Spatial and Temporal Variability of Biospheric and Geospheric Processes*, SCOPE Report No. 35. John Wiley, New York.

Rudaz, A. O., Davidson, E. A., and Firestone, M. K. (1989). Nitrous oxide production and denitrification immediately following wetting of dry soil (abstract). *Bull. Ecol. Soc. Am. 70*: 249.

Ruess, R. W. (1987). The role of large herbivores in nutrient cycling of tropical savannas. In *Determinants of Tropical Savannas*, Monogr. Ser. 3, B. H. Walker (Ed.). Int. Union Biol. Sci. Geneva, pp. 67-91.

Rundel, P. W., Nilsen, E. T., Sharafi, M. R., Virginia, R. A., Jarrell, W. M., Kohl, D. H., and Shearer, G. B. (1982). Seasonal dynamics of nitrogen cycling for a *Prosopis* woodland in the Sonoran Desert. *Plant and Soil 67*: 343-353.

Rundel, P. W., Ehleringer, J. R., and Nagy, K. A. (Eds.) (1989). *Stable Isotopes in Ecological Research*, Ecol. Studies 68. Springer-Verlag, New York.

Ryan, M. G., and Rastetter, E. G. (1989). Control of carbon and nutrient allocation in biomass (abstract). *Bull. Ecol. Soc. Am. 70*: 250.

Safriel, U. N. (1987). The stability of the Negev Desert ecosystems: Why and how to investigate it. In *Progress in Desert Research*, L. Berkofsky and M. G. Wurtele (Eds.). Rowan and Littlefield, Totowa, New Jersey, pp. 133-144.

Sala, O. E., Parton, W. J., Joyce, L. A., and Lauenroth, W. A. (1988). Primary production of the central grassland region of the United States. *Ecology 69*: 40-45.

Savory, A. (1988). *Holistic Resource Management*. Island Press, Covelo, California.

Schaefer, D., Steinberger, Y., and Whitford, W. G. (1985). The failure of nitrogen and lignin control of decomposition in a North American desert. *Oecologia (Berl.) 65*: 382-386

Schimel, D. S., and Parton, W. J. (1986). Microclimatic controls of nitrogen mineralization and nitrification in shortgrass steppe soils. *Plant Soil 93*: 347-357.

Schimel, D. S., Coleman, D. C., and Horton, K. A. (1985a). Soil organic matter dynamics in paried rangeland and cropland toposequences in North Dakota. *Geoderma 36*: 201-214.

Schimel, D. S., Kelly, E. F., Yonker, C., Aguilar, R., and Heil, R. D. (1985b). Effects of erosional processes on nutrient cycling in semiarid landscapes. In *Planetary Ecology*, D. E. Caldwell et al. (Eds.). Van Nostrand-Reinhold, New York, pp. 571-580.

Schimel, D. S., Stillwell, M. A., and Woodmansee, R. G. (1985c). Biogeochemistry of C, N and P in a soil catena of the shortgrass steppe. *Ecology 66*: 276-282.

Schimel, D. S., Parton, W. J., Adamsen, P. J., Woodmansee, R. G., Senft, R. L., and Stillwell, M. A. (1986). The role of cattle in the volatile loss of nitrogen from shortgrass steppe. *Biogeochemistry 2*: 39-52.

Schlesinger, W. H. (1977). Carbon balance in terrestrial detritus. *Annu. Rev. Ecol. Syst. 8*: 51-81.

Schlesinger, W. H. (1982). Carbon storage in the caliche of arid soils: A case study from Arizona. *Soil Sci. 133*: 247-255.

Schlesinger, W. H. (1985a). The formation of caliche in soils of the Mohave Desert, California. *Geochim. Cosmochim. Acta 49*: 57-66.

Schlesinger, W. H. (1985b). Changes in soil carbon storage and associated properties with disturbance and recovery. In *The Changing Carbon Cycle: A Global Analysis*, D. E. Reichle and J. R. Trabalka (Eds.). Springer-Verlag, pp. 194-220.

Schlesinger, W. H., and Peterjohn, W. T. (1988). Ion and sulfate-isotope ratios in arid soils subject to wind erosion in the southwestern U.S.A. *Soil Sci. Soc. Am. J. 52*: 54-58.

Schlesinger, W. H., and Peterjohn, W. T. (1989). Processes controlling ammonia volatilization in desert ecosystems (abstract). *Bull. Ecol. Soc. Am. 70*: 254.

Schlesinger, W. H., Marion, G. M., and Fonteyn, P. J. (1989). Stable isotope ratios and the dynamics of caliche in desert soils. In *Stable Isotopes in Ecological Research*, P. W. Rundel, J. R. Ehlringer, and K. A. Nagy (Eds.). Springer-Verlag, New York, pp. 309-317.

Schlesinger, W. H., Reynolds, J. F., Cunningham, G. L., Huenneke, L. F., Jarrell, W. M., Virginia, R. A., and Whitford, W. G. (1990). Biological feedbacks in global desertification. *Science 247*: 1043-1048.

Schrauzer, G. N., Guth, D. D., Palmer, M., and Salehi, J. (1979). N-reducing solar cells. In *Solar Energy—Chemical Conversion and Storage*, R. B. King (Ed.). Humana Press, Clifton, New Jersey, pp. 261-269.

Seagle, S. W., and McNaughton, S. J. (1989). Animal nutrition and spatial variability in abundance of Serengeti grazing ungulates (abstract). *Bull. Ecol. Soc. Am. 70*: 258-259.

Seastedt, T. R. (1985). Maximization of primary and secondary productivity by grazers. *Am. Naturalist 126*: 559-564.

Seastedt, T. R., and Crossley, D. A., Jr. (1984). The influence of arthropods on ecosystems. *BioScience 34*: 157-161.

Seastedt, T. R., Ramundo, R. A., and Hayes, D. C. (1988). Maximization of densities of soil animals by foliage herbivory: Empirical evidence, graphical and conceptual models. *Oikos 51*: 243-248.

Seligman, N. G., Feigenbaum, S., Benjamin, R. W., and Feinerman, D. (1985). Efficiency of fallow as a store for fertilizer nitrogen in a semi-arid region. *J. Agr. Sci. (Camb.) 105*: 245-249.

Senft, R. L., Rittenhouse, L. R., and Woodmansee, R. G. (1985). Factors influencing patterns of cattle grazing behavior on shortgrass steppe. *J. Range Manag. 38*: 82-85.

Shachak, M., and Newton, P. G. (1985). The relationship between brood care and environmental unpredictability in the desert isopod *Hemilepistus reaumuri*. *J. Arid Environ. 9*: 199-209.

Shachak, M., Jones, G. G., and Grenot, Y. (1987). Herbivory in rocks and the weathering of a desert. *Science 236*: 1098-1099.

Sharifi, M. R., Meinzer, F. C., Nilsen, E. T., Rundel, P. W., Virginia, R. A., Jarrell, W. M., Herman, D. J., and Clark, P. C. (1988). Effect of manipulation of water and nitrogen supplies on the quantitative phenology of *Larrea tridentata* (creosotebush) in the Sonoran Desert of California. *Am. J. Bot. 75*: 1163-1174.

Shearer, G. D., and Kohl, D. H. (1989). Estimates of N_2 fixation in ecosystems: The need for and basis of the ^{15}N natural abundance method. In *Stable Isotopes in Ecological Research*, P. W. Rundel, J. R. Ehleringer, and K. A. Nagy (Eds.). Springer-Verlag, New York, pp. 342-374.

Shearer, G. D., Kohl, D. H., Virginia, R. A., Bryan, B. A., Skeeters, J. L., Nilsen, E. L., Sharifi, M. D., and Rundel, P. W. (1983). Estimates of N_2-fixation from variation in natural abundance of ^{15}N in Sonoran Desert ecosystems. *Oecologia (Berl.) 56*: 365-373.

Shmida, A., and Burgess, T. L. (1988). Plant growth-form strategies and vegetation types in arid environments. In *Plant Form and Vegetation Structure*, M. J. A. Werger et al. (Eds.). SPB Acad. Publ., The Hague, pp. 211-241.

Shmida, A., Evenari, M., and Noy-Meir, I. (1986). Hot desert ecosystems:

An integrated view. In *Hot Deserts and Arid Shrublands*, Vol. 12B, *Ecosystems of the World*, M. Evenari, I. Noy-Meir, and D. W. Goodall (Eds.). Elsevier, Amsterdam, pp. 379-387.

Skiles, J. W., Kemp, P. R., and Reynolds, J. F. (1989). Simulation analysis of climatic change in a semiarid grassland community (abstract). *Bull. Ecol. Soc. Am. 70*: 265.

Skujiņš, J. (1981). Nitrogen cycling in arid ecosystems. *Ecol. Bull. (Stockh.) 33*: 447-491.

Skujiņš, J. (1984). Microbial ecology of desert soils. *Adv. Microb. Ecol. 7*: 49-91.

Smith, S., Leary, K., Herr, C., and Hokett, S. (1990). Water relations and transpiration of native vegetation in the vicinity of Yucca Mountain, Nevada. *Proc. Sixth Wildland Shrub Symposium*, Las Vegas (in press).

Smith, S. D., Strain, B. R., and Sharkey, T. D. (1987). Effects of CO_2 enrichment on four Great Basin grasses. *Funct. Ecol. 1*: 139-143.

Sorenson, D. L., and Fresquez, P. R. (1985). Patterns of denitrification in reclaimed minesoils in the arid Southwest. In *Proc. Second Ann. Mtg., Am. Soc. Surface Mining and Reclam.*, pp. 327-331.

Sprent, J. I. (1985). Nitrogen fixation in arid environments. In *Plants for Arid Lands*, G. E. Wickens, J. R. Goodin, and D. V. Field (Eds.). George Allen and Unwin, London, pp. 215-229.

Sprent, J. I. (1987a). *The Ecology of the Nitrogen Cycle*. Cambridge University Press, Cambridge.

Sprent, J. I. (1987b). Problems and potentials for nitrogen fixation in deserts. In *Arid Lands: Today and Tomorrow*, E. E. Whitehead et al. (Eds.). Westview Press, Boulder, Colorado, pp. 1049-1062.

Staaf, H., and Berg, B. (1981). Plant litter input to soil. *Ecol. Bull. (Stockh.) 33*: 147-162.

Steinberger, Y., and Whitford, W. G. (1988). Decomposition process in Negev ecosystems. *Oecologia (Berl.) 75*: 61-66.

Stewart, J. W. B., Cole, C. V., and Maynard, D. G. (1983). Interactions of biogeochemical cycles in grassland ecosystems. In *The Major Biogeochemical Cycles and Their Interactions*, B. Bolin and R. B. Cook (Eds.). SCOPE, UNESCO, Paris, pp. 247-269.

Strain, B. R. (1987). Direct effects of increasing atmospheric CO_2 on plants and ecosystems. *Trends Ecol. Evol. 2*: 18-21.

Swanson, F. J., Kratz, T. K., Caine N., and Woodmansee, R. G. (1988). Land form effects on ecosystem patterns and processes. *BioScience 38*: 92-98.

Tadmor, N. H., Eyal, E., and Benjamin, R. (1974). Plant and sheep production and semiarid annual grassland in Israel. *J. Range Manag. 27*: 427-432.

Thalen, D. C. P. (1979). *Ecology and Utilization of Desert Shrub Rangelands in Iraq*. W. Junk, The Hague.

Tiedemann, A. R., and Klemmedson, J. O. (1986). Long-term effects of mesquite removal on soil characteristics. I. Nutrients and bulk density. *Soil Sci. Soc. J. 50*: 472-475.

Tiedemann, A. R., Sanderson, H. R., and Cimon, N. J. (1986). Future site productivity considerations of short-duration, high intensity grazing. In *Short Duration Grazing*, Grazing Mgt. Shortcourse, J. A. Tiedeman (Ed.). Washington State University, Pullman, pp. 139-144.

Topp, G. C., and Davis, J. L. (1985). Measurement of soil water content using time-domain reflectrometry (TDR): A field evaluation. *Soil Sci. Soc. Am. J. 49*: 19-24.

Turner, E. L., Seastedt, T. R., Dyer, M. I., and DeLaney, W. (1989). Effects of grazing and simulated grazing on nitrogen concentrations of tallgrass prairie grasses: Estimation of leaf nitrogen using high spectral resolution reflectance measurements (abstract). *Bull. Ecol. Soc. Am. 70*: 283.

Virginia, R. A., and Jarrel, W. M. (1983). Soil properties in a mesquite-dominated Sonoran Desert ecosystem. *Soil Sci. Soc. Am. J. 47*: 138-144.

Virginia, R. A., Jarrell, W. M., Whitford, W. G., Freckman, D. W., and Jenkins, M. R. (1989). Response of Chihuahuan Desert mesquite dunes to supplemental water. *Bull. Ecol. Soc. Am. 70*: 287-288.

Wallace, A. (1989). Comparative growth of four shrub species in a native desert soil and an amended non-calcareous soil and some unsolved problems in mineral nutrition of desert shrubs. In *Proc. Symp. Shrub Ecophysiology and Biotechnology*. U.S. Dept. Agr., Forest Serv., Intermountain Res. Sta., Gen. Tech. Rep. INT-256, pp. 169-172.

Wallace, A., and Romney, E. M. (1980). The role of pioneer species in revegetation of disturbed desert areas. *Great Basin Naturalist Memoirs 4*: 31-33.

Wallace, A., Romney, E. M., and Hunter, R. B. (1980). The challenge of a desert: Revegetation of disturbed desert lands. *Great Basin Naturalist Memoirs 4*: 216-225.

Watwood, M. E., Dahm, C. N., and White, C. S. (1989). Microbial transformations of benzene and dichloroethane in unsaturated soils of the southwestern United States (abstract). *Bull. Ecol. Soc. Am. 70*: 292.

Weaver, R. W. (1986). Measurement of biological dinitrogen fixation in the field. In *Field Measurement of Dinitrogen Fixation and Denitrification*, Spec. Publ. 18, R. D. Hauck and R. W. Weaver (Eds.). Soil Sci. Soc. Am., Madison, Wisconsin, pp. 1-9.

West, N. E. (1981). Nutrient cycling in desert ecosystems. In *Arid-land Ecosystems: Structure, Functioning and Management*, Vol. 2, D. W. Goodall and R. A. Perry (Eds.). Cambridge University Press, Cambridge, pp. 301-324.

West, N. E. (1983). Overview of North American Temperate Deserts and Semi-deserts. In *Temperate Deserts and Semi-deserts*, Vol. 5, *Ecosystems of the World*, N. E. West (Ed.). Elsevier, Amsterdam, pp. 321-330.

West, N. E. (1986). Desertification or xerification? *Nature 321*: 562-563.

West, N. E. (1988). Intermountain deserts, shrub steppes and woodlands. In *North American Terrestrial Vegetation*, M. G. Barbour and W. D. Billings (Eds.). Cambridge University Press, New York, pp. 209-230.

West, N. E. (1989). Spatial pattern-functional interactions in shrub dominated plant communities. In *The Biology and Utilization of Shrubs*, C. M. McKell (Ed.). Academic Press, San Diego, pp. 283-305.

West, N. E. (1990). Structure and function of microphytic soil crusts in wildland ecosystems of arid to semi-arid regions. *Adv. Ecol. Res. 20*: 179-223.
West, N. E., and Skujiņš, J. (Eds.) (1978). *Nitrogen in Desert Ecosystems*, US/IBP Synthesis Series 9. Dowden Hutchison and Ross, Stroudsburg, Pennsylvania.
West, N. E., Griffin, R. A., and Jurinak, J. J. (1984). Comparison of phosphorus distribution and cycling between adjacent semidesert shrub and cultivated grass-dominated ecosystems. *Plant and Soil 81*: 151-164.
Westman, W. E. (1987). Monitoring the environment by remote sensing. *Trends Ecol. Evol. 2*: 333-337.
Westoby, M. (1987). Soil erosion as a landscape ecology phenomenon. *Trends Ecol. Evol. 2*: 321-322.
Wetselaar, R. (1980). Nitrogen cycling in a semi-arid region of tropical Australia. In *Nitrogen Cycling in West African Ecosystems*, T. Rosswall (Ed.). SCOPE/UNEP, Uppsala, Sweden, pp. 157-169.
Whicker, A. D., and Detling, J. K. (1988). Ecological consequences of prairie dog disturbances. *BioScience 38*: 778-785.
Whitford, W. G. (1986a). Decomposition and nutrient cycling in deserts. In *Pattern and Process in Desert Ecosystems*, W. G. Whitford (Ed.). University of New Mexico Press, Albuquerque, pp. 93-117.
Whitford, W. G. (1986b). Pattern in desert ecosystems: Water availability and nutrient interactions. In *Environmental Quality and Ecosystem Stability*, Vol. III A/B, Z. Dubinsky and Y. Steinberger (Eds.). Bar-Ilan University Press, Ramat Gan, Israel, pp. 109-117.
Whitford, W. G. (1988). Decomposition and nutrient cycling in disturbed arid ecosystems. In *The Reconstruction of Disturbed Arid Lands: An Ecological Approach*, AAAS Selected Symposium 109, E. B. Allen (Ed.). Westview Press, Boulder, Colorado, pp. 136-161.
Whitford, W. G., and Freckman, D. W. (1987). The role of soil biota in decomposition in the Chihuahuan Desert. In *Arid Lands: Today and Tomorrow*, E. E. Whitehead et al. (Eds.). Westview Press, Boulder, Colorado, pp. 1063-1073.
Whitford, W. G., Aldon, E. F., Freckman, D. W., Steinberger, Y., and Parker, L. W. (1989). Effect of organic amendments on soil biota of a degraded rangeland. *J. Range Manag. 42*: 56-60.
Whitford, W. G., Reynolds, J. F., and Cunningham, G. L. (1987). How desertification affects nitrogen limitation of primary production on Chihuahuan Desert watersheds. In *Strategies for Classification and Management of Native Vegetation for Food Production in Arid Zones*, E. F. Aldon, G. E. G. Vincent, and W. H. Moir (Eds.). Rocky Mountain Forest and Range Experiment Station, Forest Service, U.S. Dept. Agric., Fort Collins, Colorado, pp. 143-153.
Wikeem, S. J., and Pitt, M. D. (1982). Soil nitrogen gradients as influenced by sagebrush canopy in southern British Columbia. *N.W. Sci. 56*: 276-286.
Wildung, R. E., and Garland, T. R. (1988). Soils: Carbon and mineral cycling processes. In *Shrub-steppe: Balance and Change in a Semi-Arid Terrestrial Ecosystem*, W. H. Rickard et al. (Eds.). Elsevier, New York, pp. 23-59.

Williams, R. B., and Bell, K. L. (1981). Nitrogen allocation in Mojave Desert winter annuals. *Oecologia (Berl.) 48*: 145-150.
Woodmansee, R. G. (1978). Additions and losses of nitrogen from grassland ecosystems. *BioScience 28*: 448-453.
Woodmansee, R. G. (1979). Factors influencing input and output of nitrogen in grasslands. In *Perspectives in Grassland Ecology*, Ecol. Studies, 32, N. R. French (Ed.). Springer-Verlag, New York, pp. 117-134.
Woodmansee, R. G. (1984). Comparative nutrient cycles of natural and agricultural ecosystems: A step toward principles. In *Agricultural Ecosystems—Unifying Concepts*, B. R. Stinner and G. J. House (Eds.). Wiley Interscience, New York, pp. 145-156.
Woodmansee, R. G., and Adamsen, F. J. (1983). Biogeochemical cycles and ecological hierarchies. In *Nutrient Cycling in Agricultural Ecosystems*, R. R. Lowrance et al. (Eds.). University of Georgia, Athens, pp. 497-516.
Woodmansee, R. G., and Duncan, D. A. (1980). Nitrogen and phosphorus dynamics and budgets in annual grasslands. *Ecology 61*: 873-904.
Woodmansee, R. G., and Wallach, L. S. (1981). Effects of fire regimes on biogeochemical cycles. In *Fire Regimes and Ecosystem Properties*, H. Mooney et al. (Eds.). U.S. Dept. Agr., Forest Service, Gen. Tech. Rep. WO-26, pp. 379-400.
Woodmansee, R. G., Vallis, I., and Mott, J. J. (1981). Grassland nitrogen. *Ecol. Bull. (Stockh.) 33*: 443-462.
Woodward, R. A., Harper, K. T., and Tiedemann, A. R. (1984). An ecological consideration of the significance of cation-exchange capacity of roots of some Utah range plants. *Plant and Soil 79*: 169-180.
Yensen, N. P. (1988). Plants for salty soil. *Arid Lands Newslett. 27*(Fall): 3-10.
Young, J. A. (1988). Seedbeds as selective factors in the species composition of rangeland communities. In *Vegetation Science Applications for Rangeland Analyses and Management*, P. T. Tueller (Ed.). Kluwer Acad. Publ., Dordrecht, pp. 171-188.
Zak, J. C., Freckman, D., and Loring, S. (1989). Trophic relationships within the root region of a desert bunchgrass (abstract). *Bull. Ecol. Soc. Am. 70*: 305.

Part II

RECLAMATION AND IMPROVEMENT OF PRODUCTIVITY

10

Human Activities and Soil Degradation

HAROLD E. DREGNE *Texas Tech University, Lubbock, Texas*

INTRODUCTION

Human-induced soil degradation in the arid region has been going on for thousands of years (Dregne, 1982). Water erosion in uplands and salinization of irrigated lands had a devastating effect on the land and, in some instances, on civilizations in the cultural centers of the Old World. Deforestation around the Mediterranean Sea was followed by accelerated water erosion, beginning at least 2000 years ago. Its effects are still being felt (Mikesell, 1960, 1969). Salinity damage to irrigated land in Mesopotamia (present-day Iraq) contributed significantly to the weakening and ultimate destruction of the hydraulic society that was utterly dependent upon the flow of the Tigris and Euphrates rivers (Jacobsen and Adams, 1958). Erosion and salinization are now global problems. Human-induced salinization is almost always a reversible process, on both irrigated and nonirrigated lands. Water erosion damage may be reversible or irreversible, in human terms, depending upon the kind of soil that is eroded. Wind erosion, although widespread, does not usually bring about permanent damage to the land. As with water erosion, the characteristics of the eroded soil determine the severity of the impact and its reversibility.

II. CAUSES OF SOIL DEGRADATION

There are several direct and indirect causes of soil degradation. Among the direct causes in arid regions are overgrazing, excessive woodcutting, improper soil and water management, land disturbance, cropping lands that are too erodible or too dry, abuse of recreational lands, and inappropriate disposal of the waste products of society. Indirect causes include poverty, increases in human and livestock populations, refusal to believe that a problem exists, absence of environmental awareness, the perception that degradation control is uneconomical, lack of knowledge of how to prevent and control degradation, government policies, and greed. Improper soil and water management and cropping marginal lands are the most important direct causes of accelerated erosion and salinization, whereas poverty and population increases are probably the most significant indirect causes.

Compaction, crusting, pesticide contamination, toxic element accumulation, industrial waste deposition, and loss of organic matter are six other kinds of soil degradation in the arid regions. Compaction of surface and subsurface soil by farm machinery is widespread in cultivated soils, both dryland and irrigated. A survey of irrigated land in the High Plains of Texas concluded that dense compact layers (pans) were present in over 75% of the soils (High Plains Underground Water District, 1984). Excessive soil strength resulting from soil drying, with or without tillage pressure pans, appeared to be responsible for the occurrence of the pans studied in a southern Great Plains project (Taylor et al., 1966).

Surface crusting is a form of compaction that adversely affects water intake and seedling emergence. Crusts may occur on uncultivated soils as well as on cultivated ones. Rainfall characteristics have a key role in crust formation in arid region soils (Hoogmoed and Stroosnijder, 1984). Surface crusts in cultivated land and pans resulting from tillage operations are reversible. Algal (biological) crusts are not compaction crusts and do not reduce infiltration (Fletcher and Martin, 1948).

Toxic element accumulation is a local problem in the arid regions. Heavy metals, boron, and other elements toxic to plants or animals are deposited by smoke, dust, or runoff from industrial and mining operations or are carried in irrigation water (El-Hinnawi and Hashmi, 1982). The affected area is usually small, but the impact can be considerable. Heavy metal accumulation is practically irreversible. Excess boron can be washed out of the soil but at a heavy cost in water.

Pesticide contamination of soils, particularly that associated with the strongly adsorbed chlorinated hydrocarbons and lead and arsenic compounds, is a generally local and low-intensity form of pollution. The heavy metals never degrade, but the hydrocarbons usually do. Most of these pesticides have been banned in the United States. Industrial waste products, such as waste oil and chemicals, are high-intensity pollutants that can sterilize a soil. Fortunately, the affected areas are usually very small.

A pronounced decline in soil organic matter content is usually associated with intensive cropping, summer fallow, or absence of crop rotations with soil-building crops. In rangelands, overgrazing and tree and shrub cutting have a similar effect on organic matter. A soil low in organic matter is almost inevitably lacking in nutrients. Additionally, the soil becomes more susceptible to increased erosion, compaction, crusting, and runoff (Science Council of Canada, 1986). Desirable though crop rotations are, establishing soil-building grasses and legumes in the drylands is frequently difficult unless the rainy season is in the cool period of the year.

III. SOIL DEGRADATION IN A GLOBAL CONTEXT

Good data on the extent and severity of any of the several kinds of soil degradation are very limited or nonexistent. For example, the greatest single threat to the production of crops, globally, is human-induced water

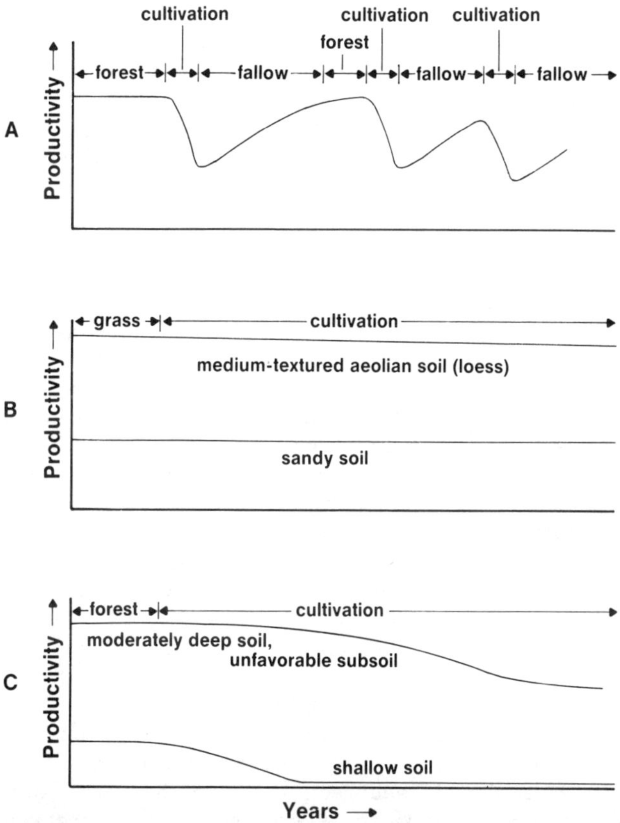

FIGURE 1 Variable effect of erosion on soil productivity. (A) Short-term productivity loss typical of shifting cultivation on deep soils. (B) Slight productivity loss due to moderate to severe erosion on deep, uniform soils. (C) Long-term productivity loss on shallow soils and soils having unfavorable subsoils.

erosion. There is no question but that reduction of soil depth to a thickness of less than 30 cm will permanently—to all practical purposes—lower the production potential of the land. Yet, field data on the amount of land that has reached this threshold or any other threshold are available for only a very few areas of the world (e.g., Papendick et al., 1985). There is no end to the cries about the damage erosion has already done and the threat it poses for the future (Brown, 1984; Dregne, 1982; Eckholm, 1976). What are missing are the data.

Empirical models of water erosion, such as the universal soil loss equation (USLE), have been employed in surveys of water and wind erosion in the United States (U.S. Department of Agriculture, 1981a), Germany (Auerswald and Schmidt, 1986), Ethiopia (Hellden, 1987), and elsewhere. The USLE seems reasonably valid for use in temperate zones but must be modified when used in the tropics. It should be noted that no other known erosion model performs better than the USLE in the tropics.

Of more importance than erosion estimates, using the USLE model or some other means, is the determination of the impact erosion has had on soil productivity. Excessive erosion may have much or little effect on potential productivity, depending upon the thickness and the characteristics of the soil (Fig. 1). A high erosion rate on a deep uniform soil may have no detectable effect on crop yields, whereas a low annual loss could be damaging to a shallow soil (Dregne and Meyer, 1989). To a first approximation, a high erosion rate is more than 50 ton ha^{-1} $year^{-1}$ whereas a low rate would be about 5-10 ton.

Soil productivity losses may be short term (reversible) or long term (irreversible), as illustrated in Fig. 1. Figure 1A represents the short-term soil productivity loss commonly associated with such practices as shifting cultivation. Newly cropped forest or grassland soils that do not receive fertilizers or soil amendments usually suffer a rapid loss of productivity after the initial cultivation. The loss is due to erosion of the surface soil, rapid decomposition of soil organic matter, and leaching of soluble nutrients, or some combination of these factors. After a few years, yields are too low to justify continued cropping and the soil lies fallow for several years. During the fallow period, weeds, shrubs, and trees become established and bring about a restoration of soil organic matter levels and fertility. If the bush fallow period is long enough, full productivity can be restored. If, however, the fallow period is shortened because of a need for cropland, fertility is not restored and yields are reduced even in the first crop year.

Figure 1B illustrates the condition in which soils are deep and have a fairly uniform profile, with little horizon differentiation. Loessial soils found widely in the drylands of China, the Soviet Union, the central United States, and Argentina are typical of such soils, as are the deep sandy soils of the Sahel. Sheet erosion, even when severe, may not reduce yields very much after a moderate loss in fertility during the first few years of cultivation. The short-term loss in fertility is reversible.

The condition to be feared most is depicted in Fig. 1C. Sheet erosion on shallow soils or those with a tight subsoil can cause soil productivity losses that are, for all practical purposes, irreversible, Irreversibility of soil productivity loss does not mean that all productivity is lost. Few soils, no matter how badly eroded they may be, are unable to support some vegetation, however poor it may be. Land too gullied to produce crops may be suitable for grazing or forestry.

Estimates of the consequences of erosion losses are a part of the national resource inventories conducted by the U.S. Department of Agriculture. A model called the erosion-productivity impact calculator (EPIC) is used to calculate how much the erosion estimated by the USLE affects potential productivity. Data requirements for running EPIC in its full configuration are more than most developing countries can supply. However, simplifications have recently been devised that may give useful estimates of the erosion-productivity relation in areas where the database is limited. The EPIC is almost entirely restricted to the United States, despite its usefulness. A major reason for this appears to be the assumption that it is enough to know how much erosion has occurred, in the apparent belief that erosion equates with loss of productivity. This philosophy was

TABLE 1 Erosion-Productivity Studies

Component	Studies	
	Number	Percent
Development stage		
Developed countries	129	85
Developing countries	23	25
Tropical versus nontropical		
Tropical countries	27	18
Nontropical countries	125	82

Source: From Stocking (1985).

prevalent in the United States until the late 1970s, when it became obvious that spending money to control erosion everywhere, regardless of the damage being done, was not acceptable in a time of budget deficits. This was when the policy of focusing on problem soils (targeting) was introduced (Comptroller General, 1983). As the Comptroller General's report said, "damages incurred onsite and offsite are the true costs of erosion and should constitute the basic yardstick for USDA's allocation of conservation resources." In practice, targeting decisions were based on the severity of erosion, not its effect on productivity. The rationale for this decision was that there was not enough known about the relation of erosion to productivity to allow relevant judgments to be made (Park and Sawyer, 1985). It is a condemnation of erosion research that the erosion-productivity relation has been ignored for so long. Real progress has come about only since creation of the National Soil Erosion-Soil Productivity Research Planning Committee (U.S. Department of Agriculture, 1981b). A global review of research on erosion and soil productivity noted the scarcity of experimental data on the relation (Stocking, 1984). The scarcity is demonstrated by the fact that 59% of the experiments Stocking found were conducted in the United States, where only a few experiments had been conducted prior to the 1980s. Table 1 (Stocking, 1985) lists erosion-productivity studies (experimental and nonexperimental) for developed and developing countries as well as for tropical and nontropical zones. The largest number of studies in developing countries was done in Nigeria. That leaves most of the developing countries, which are dominently in the tropics, with little or no data on the impact of erosion on their irreplaceable soil resource. There is no likelihood that the situation will change in the foreseeable future.

A. Wind Erosion

Data on the occurrence of wind erosion are even less available than the small amount of data on where water erosion is a serious problem. Until a few years ago, there was not even a suitable device for capturing wind-eroded particles (Fryrear et al., 1988).

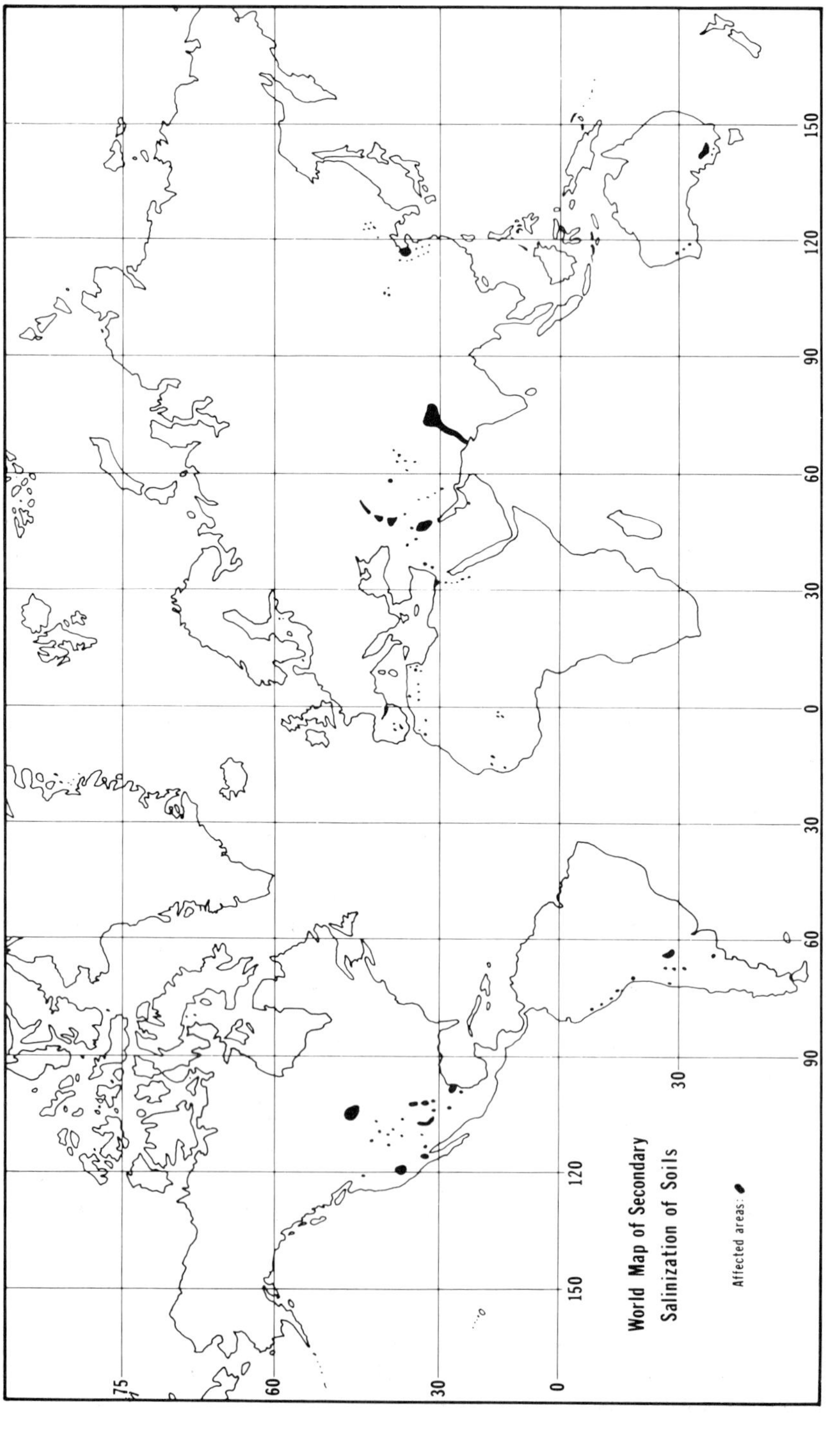

FIGURE 2 Global map of secondary salinization of irrigated and dryland soils.

B. Salinization

Information on the extent of salinization of irrigated land and on dryland saline seeps is much better than for either water or wind erosion. Several estimates have been made of the amount of land having impaired productivity due to salinity. Unfortunately, it is not always clear whether the figures apply to human-induced salinization (secondary salinization) or to the natural salinization (primary salinization) of closed basins, such as chotts, sebkhas, playas, and pans. A World Map of Salt Affected Soils project was initiated by the International Society of Soil Science in the late 1960s. Only two maps have been published, one for Australia (Northcote and Skene, 1972) and one for Europe (Szabolcs, 1974). Vander Pluym (1978) produced a map of the occurrence of dryland saline seeps in the northern Great Plains, and a working party of the Australian Standing Committee on Soil Conservation did the same for Australia (Working Party, 1982). Figure 2 shows several of the places where human-induced secondary salinization has become important. Among the largest areas are the irrigated lands in the North China Plain, the Indus River Valley, Iraq, the west side of the Caspian Sea in the USSR, and the Central Valley of California in the United States. There is a large dryland saline seep area shown along the United States-Canada border. Saline seeps are scattered over much of southwestern Western Australia, southern South Australia, and northwestern Victoria.

Salinization of irrigated land is a common phenomenon. The map of Fig. 2 is too small to show the numerous small irrigated areas affected by salinity in the temperate regions of North and South America, northern Africa, the Soviet Union, northwestern China, and elsewhere. It has been properly stated that salinization is the plague of irrigated agriculture and the major threat to its continuation.

C. Other Types of Degradation

No national, continental, or global maps are known that indicate where compaction, crusting, pesticide and toxic element accumulation, industrial waste deposits, and loss of soil organic matter are problems. Organic matter loss is so prevalent that maps are probably not needed, and compaction by tillage machinery is also common on mechanized farms. Crusting is so widespread in coarse-textured arid region soils and so dependent on soil and crop management that maps of its occurrence would be of doubtful value. Toxic element and industrial waste accumulation is a very localized problem for which local maps would be useful. Mapping pesticide accumulation should also be adequate at a local level for the worst cases. Traces of the pesticides that are degradable only with difficulty can probably be found over much of the earth's cropland where those pesticides have been used in the past. Mapping that condition on a world scale would be very costly.

D. Global Soil Erosion Maps

Small-scale erosion-productivity maps appear to be those that would be most useful for persons searching for general information about where human-induced soil erosion has had a significant adverse impact on the ability of soils to produce crops. Accordingly, a global map has been constructed in an attempt to serve that purpose (Fig. 3). The global

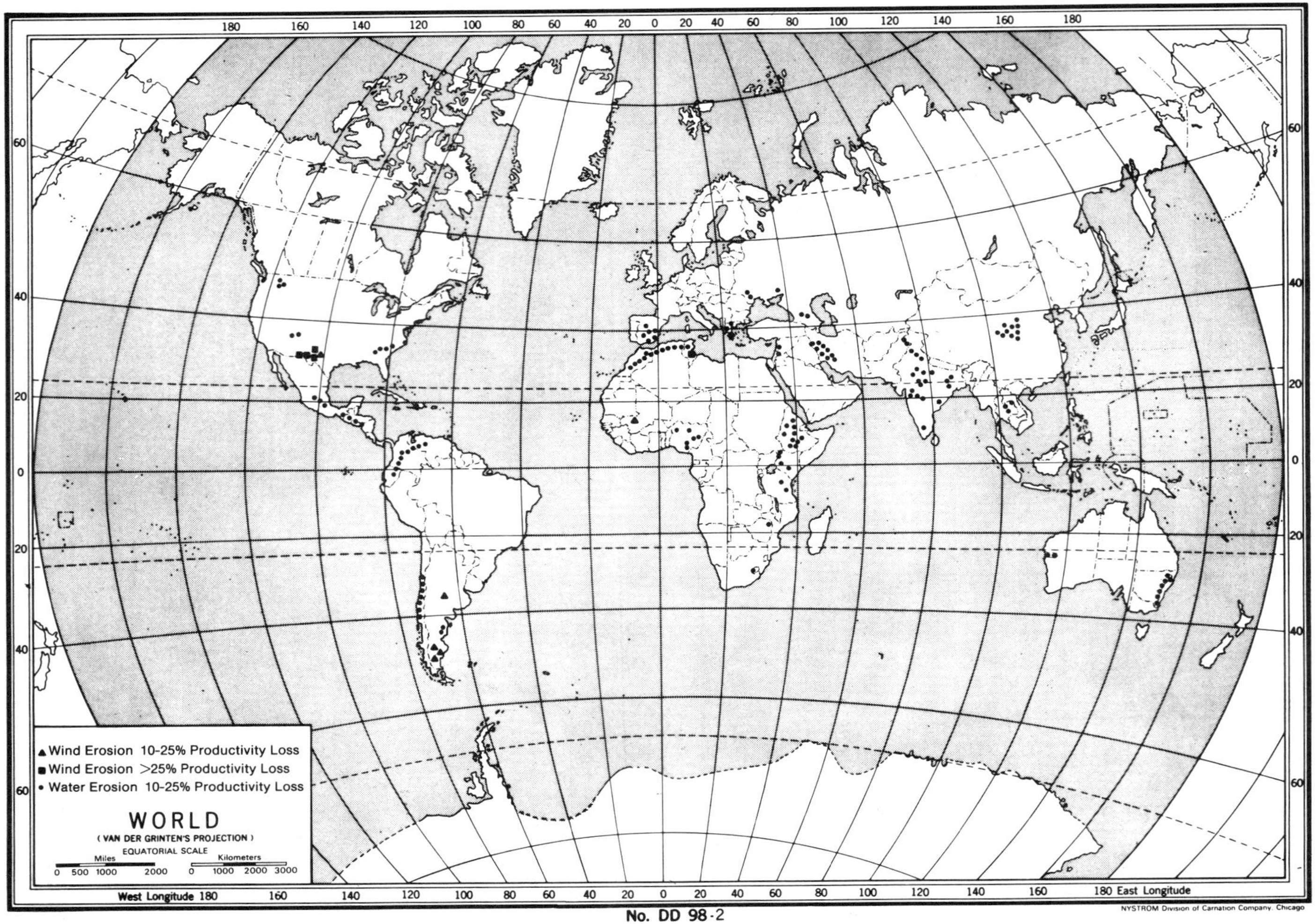

FIGURE 3 Generalized map of the impact of water and wind erosion on soil productivity. Major areas affected by permanent loss of productivity.

map is based, in part, on continental maps, early versions of which have been published (Dregne, 1986).

Given the paucity of geographic information on the extent and severity of accelerated water erosion, recourse in the construction of the global map has had to be taken to anecdotal evidence, qualitative statements in scientific reports, extrapolation of experimental plot data to adjoining areas, my personal observations, and to the few local or national maps that have been prepared. This is a very unsatisfactory basis upon which to draw conclusions on such an important topic as the effect of water erosion on potential soil productivity. It can, at best, be considered an approximation. The "structured informed opinion system" that was proposed by a workshop at Texas Tech University in July 1988 (Dregne and Meyer, 1989) offers a more reliable means of integrating the experience of knowledgeable persons and arriving at a consensus on the extent and severity of the erosion-productivity relation. The system includes a so-called expert system component for use with individuals. The structured informed opinion analysis would be a substitute for the quantitative data on erosion and productivity that are not available now and will not be available for many years, if not decades. The system is in the process of development and testing.

Figure 3 is a conservative estimate of areas in the world where sheet, rill, and gully erosion have reduced potential soil productivity by more than 10%. By the very nature of small-scale maps, each delineation of an affected area invariably includes some soils that are not affected significantly by erosion. Therefore, the estimated productivity loss is for the delineation as a whole, not every hectare of land within the delineation. An example is the Palouse region of Idaho and Washington. This is the only known place where experimental data have clearly demonstrated that erosion has reduced soil productivity over a large area by approximately 22% (Papendick et al., 1985). Yields are reduced below the original potential on the crests and upper slopes but not on the lower slopes, where the topsoil remained deep. Actual crop yields may increase even on the eroded soils if improved technology is employed. However, they will never again be as high as they would have been without erosion.

A conservative approach has been taken in constructing the global map. This means that the likelihood is high that the delineated areas have experienced at least as much productivity loss as is indicated on the map. However, it is also probable that many equally affected areas are not shown because insufficient information was available to allow a reasonably reliable judgment to be made. Moreover, some affected areas may not be shown because they are not sufficiently large to be delineated on this small-scale map.

One observation about the map is that there are many large areas in the world where current knowledge strongly suggests that erosion has reduced potential soil productivity by at least 10%. The extent of the problem is probably much worse. Moreover, it seems likely that the severity of the productivity loss is closer to 20-25% than to 10%.

About half the area delineated in Fig. 3 is in the arid regions, primarily in China, Spain, and North Africa. In China's lesssial plateau, gully erosion on cropland is the principal damaging condition. In Spain, cropland sheet erosion similar to that in the Palouse region of the United States is the cause of the soil degradation. Both sheet and gully erosion have had a harmful effect in North Africa as the result of overgrazing and deforestation as well as cultivating steep lands. In the United States, the water-eroded areas are the Palouse region of the Pacific Northwest, the Navajo Indian Reservation in Arizona and New Mexico, and the Southern Piedmont region in the Southeast.

Wind erosion, notwithstanding its occurrence practically everywhere in the dry regions, does not appear to adversely affect productivity in very many places. In actuality, of course, it may be having a greater effect than anyone can verify. That is a major question for which answers are needed. Certainly, blowouts that expose calcrete, silcrete, or bedrock have a dramatic effect on soil productivity. However, blowouts tend to be small in area and to be localized in their occurrence. The principal large areas where wind erosion has reduced productivity are on the rangelands of the southwestern United States, southern Argentina, and southeastern Tunisia. Other large affected areas are in cultivated lands of western Mali and central Argentina. What is not shown in the map, for either wind or water erosion, is the loss of fertility associated with the removal of fine material from surface soils, the damage done by sandblasting young plants when the harmattan or sirocco winds blow, or the burying of crops when soils erode.

Among the many reports that have been consulted in preparation of the global map, several have been particularly useful. They include Ahn (1977), Bensalem (1977), Craig and Pisone (1988), Das (1977), Darkoh (1986), FAO (1954), Lee (1984), Mikesell (1969), Tikhonov (1960), Trimble (1974), Veblen (1978), Virgo and Munro (1978), Woods (1983), and Zachar (1982).

IV. TYPES OF SOIL DEGRADATION

The immediate cause of most soil degradation is mismanagement of the resource. Mismanagement, however, is seldom willful. Soil erosion, for example, is a self-reinforcing process in that erosion caused by failure to protect the soil leads to reduced yields, which encourages more erosion because less organic matter means less stable aggregates and less resistance to further erosion. Some arid region soils, such as those on steep slopes or very sandy soils, are impossible to cultivate without incurring excessive water or wind erosion. When survival is at stake, conserving resources for the next generation does not enter into land use decisions. The mismanagers are also the victims of soil degradation. In these circumstances, it is futile and unjust to blame the victims for the degradation they have caused. They need socially and economically acceptable help, not condemnation.

A. Erosion

Several factors bring about accelerated soil erosion. Localized high population densities in rural areas has become what is probably the most significant force causing erosion of cultivated lands. High population density, alone, does not necessarily lead to land degradation, as conditions in the Netherlands demonstrate. It is only when people bring about overexploitation of marginal or unsuitable lands that problems arise, and problems may arise even when population densities are low, as can be seen in Australia, Chile, and the United States. As long as land was plentiful, abuse was the accepted practice for the majority of farmers and ranchers because there was always new land waiting to be exploited. A few people had a well-developed conservation ethic, but most did not. The same story has been repeated around the world, and the marks of the abuse are still with us.

Destruction of vegetative cover is the number one contributor to soil erosion. Plowing does the best job of laying the land bare. Overgrazing and excessive tree cutting increase the susceptibility of the land to water and wind erosion but have a minor effect compared to cultivation. In fact, overgrazing and tree cutting seem to have only a modest effect on erosion unless the affected land is on fairly steep slopes. Tree cutting in the Himalayas has a much greater impact than tree cutting in the lower lying rangelands.

B. Salinization

Salinization of irrigated land has plagued civilizations from China to the Middle East to North America, South America, and Australia. Jacobsen and Adams (1958) documented the incidence of salinization and waterlogging over a 3000 year period in Mesopotamia (present-day Iraq). Mesopotamia epitomizes the salinity hazard in irrigation agriculture: flat, low-lying land, high evaporation rates, and slowly permeable soils. Even with low salt levels in the irrigation water, salinization is likely to occur on such land unless periodic flushing of the soil occurs. Because it is a potential hazard in all river basin irrigation systems, it constitutes the greatest single threat to the permanence of irrigated agriculture. Waterlogging due to poor drainage is a contributor to salinization.

Salinization of dry farming land (dryland saline seepage) is an expanding problem in Australia, Canada, and the United States. It affects a few million hectares in each of those three countries, entirely in the semiarid climatic zone. Dryland saline seeps probably occur elsewhere, as well, but have not been recognized or publicized. They are the result of cultivating with shallow-rooted crops on soils that have saline substrata. There is commonly a lag period of a few years between initiation of the responsible cropping system and the appearance of the saline seeps.

C. Organic Matter

Organic matter loss when land is heavily grazed or cultivated for the first time is the usual condition in the arid regions. In the temperate humid regions of North America and Europe, cropping forested land may actually increase soil organic matter levels if a soil-building crop rotation is followed. In the arid regions, this only happens on irrigated land, in practically all cases. Clayey soils naturally have a higher organic matter content than sandy soils in the same agroecological zone.

Organic matter (humus) plays two major agricultural roles in soils. One is to supply nutrients to crops upon mineralization, principally nitrogen, phosphorus, and sulfur. The other is to form stable aggregates that reduce erosion, compaction, and crusting while increasing aeration and

TABLE 2 Crust Effect on Hydraulic Conductivity and Topsoil Moisture Content of a Sandy Soil in Mali

Soil type	Hydraulic conductivity[a]	Topsoil moisture content (%)[b]
Uncrusted soil	100-200 (mm h^{-1})	30-35
Crusted soil	10	15-20

[a]Saturated hydraulic conductivity.

[b]Moisture content of topsoil immediately after rain, percentage weight of water by weight of soil.

Source: From Hoogmoed and Stroosnijder (1984).

biological activity. Humus is a miraculous substance, increasing soil porosity, infiltration rates, aggregate stability, porosity, water-holding capacity, and nutrient-holding capacity, as well as having the other good effects noted previously. Tillage operations hasten humus mineralization. Humus levels in cultivated land in the dry regions typically decline to a low level and stay there.

D. Compaction and Crusting

Both compaction and crusting are natural features of arid region soils as well as human-induced features. Subsoil hardpans are widespread in calcareous soils in the form of calcrete (caliche) and in noncalcareous soils as silcrete (duripan). Surface crusts due to biological activity (algal crusts) and raindrop splash are equally common. In addition, the mere occurrence of strong dehydration of the soil increases compaction and crusting. Drying increases compaction to the depth of the drying, usually 10-40 cm in arid climates. Compaction and crusting, whether natural or human-induced, tend to be greater in soils with a low organic matter content and expanding-type clays. Crusting, especially, is harmful because the crust reduces infiltration rates and increases runoff. Hoogmoed and Stroosnijder (1984) measured permeability rates in crusted and uncrusted sandy soils in the Sahel and noted that permeabilities were at least 10 times less in the crusted soil (Table 2). Reductions of that magnitude induce soil erosion and the edaphic (soil) aridity that exacerbates the moisture restrictions that low rainfall imposes on plant growth. Crusting is common in grazing lands as well as croplands. Livestock trampling contributes to it, as does the loss of plant litter that accompanies overgrazing. Crusting is deliberately encouraged where water harvesting from hillslopes is the objective, as in the ancient Negev wadi agriculture.

The human effect on compaction and crusting is brought about by tillage machinery that packs the soil and causes plow pans to form, by laying the soil bare and exposing more of the surface to the aggregate-destroying impact of raindrops, and by reducing the organic matter content of the soil. Additionally, the use of off-road vehicles on rangelands causes compaction that can reduce the emergence and growth of annual plants in arid regions

(Lathrop and Rowlands, 1983). So does human traffic around tourist attractions such as Ayers Rock in arid central Australia.

Perhaps the worst effect compaction has is to harden the upper part of the soil to the point at which farmers cannot penetrate it with hand implements or animal-drawn tillage instruments. Clay soils are difficult to till even when moderately dry, but the uninitiated do not expect sandy soils to behave similarly. The compaction and crusting problem is particularly troublesome on the sandy soils on the north and south sides of the Sahara. There, farmers customarily cannot do any land preparation prior to the first rains. They must wait for the rain to wet the soil and soften it.

E. Toxic Elements

Heavy metals, such as lead, cadmium, and nickel, are the principal elemental pollutants of soils, primarily because of their toxicity to humans and animals when plants growing on contaminated soils are consumed. Boron is the principal problem element in irrigated soils. It is a constituent of many irrigation well waters, and although it is an essential plant nutrient, there is frequently only a small difference in soil concentration between what is beneficial and what is toxic to crops.

In the arid regions, heavy metal contamination of soils comes from such disparate sources as mining, smelting, and refining; automobiles; industrial waste disposal; and sewage disposal. Some of the pollutants are deposited downwind from sources, some are carried in irrigation and wastewaters, and some are present at land disposal sites. Lagerwerff and Specht (1970) found that concentrations of heavy metals in roadside soil decreased with distance from automobile traffic lanes. Most of the metals stay in the topsoil because of their limited solubility in water and their strong adsorption on soil particles. Sewage sludge applied to agricultural land may carry rather large amounts of heavy metals if the sewage is from industrial sources. Fortunately, the amounts taken up by crops are usually so low that they have little or no adverse effect (Leeper, 1978). This is particularly true if the pH is near neutrality or is alkaline, as is frequently the case in arid regions.

Boron toxicity is expected only in a few dried lake beds in arid regions and where irrigation waters are high in boron. It is a serious local problem, usually associated with soluble salt accumulation in soils.

Heavy metal toxicity is also a local problem associated, in one way or another, with industrial activity. Itai-Itai disease is caused by chronic cadmium poisoning. In Japan, an epidemic in the 1940s was traced to crops grown on soils that had been irrigated with water polluted with cadmium-bearing waste from a zinc mine (El-Hinnawi and Hashmi, 1982).

F. Pesticides

Chlorinated hydrocarbon insecticides have long been known to be strongly adsorbed on soil particles. DDT and its analogs received most attention in the past, before their use was restricted. Those chemicals are decomposed with difficulty, although microbial breakdown ultimately does occur if aerobic conditions prevail. Most organic herbicides, such as 2,4-D, are much more readily decomposable and do not have a long persistence time in soils. Inorganic pesticides of the arsenate type do not degrade in the soil but are strongly adsorbed and release little arsenic to plants.

There remains uncertainty about the fate in the soil of organic pesticides that are not readily water soluble (IAEA, 1986). Some pesticides are held in soil in a form that cannot be extracted by methods commonly used in residue analyses and metabolism studies. In that form, they are called "nonextractable (bound) pesticide residues" (IAEA, 1986). Bound residues may or may not be bioavailable and of toxicological concern.

Pesticide residues were concentrated in cultivated lands until the 1970s, when restrictions were placed on the use of the more persistent pesticides. Since then, residue levels have declined, at least in the developed countries, as biodegradable pesticides came into general use. Local concentrations vary widely, depending upon the type of soil and pesticide, amount applied, method of application, tillage and crop management practices, temperature, and rainfall.

G. Industrial Waste

Pathogenic organisms, heavy metals, and persistent organic chemicals are the primary troublesome pollutants of industrial wastes. Some of the wastes go directly, legally or illegally, into rivers, lakes, and oceans. Others are sent to land disposal sites where the soil and groundwater may or may not be monitored. Much of the waste enters the municipal sewage system.

Harmful industrial wastes going through a sewage treatment plant usually end up in the plant effluent or in the residual sludge. If the wastes are in the effluent, it appears unlikely that they will cause problems whether the effluent is used to irrigate food or nonfood crops (Goldstein, 1977). The sludge, however, can present a problem because some pollutants in municipal wastewaters are much more highly concentrated in sludge than in the effluent (Information Transfers, 1980). Regulations have been formulated to control the composition and marketing of products containing sewage sludge. Heavy metals are the principal health threat, but the threat is minimal unless the sludge is applied in large amounts, as is the case if the sludge is used as a soil conditioner rather than as a fertilizer. As noted earlier, heavy metals are less likely to be a problem in the neutral or alkaline pH of arid region soils.

Treated municipal sewage effluent is safely used for irrigation in hundreds of communities. So are municipal sludges sold commercially by such cities as Milwaukee, Chicago, and Houston. Problems are rare, but vigilance is required to maintain this situation. One major metropolitan district faced a class-action law suit because thousands of households used a high-cadmium sludge on vegetable gardens (Information Transfers, 1980).

V. LAND USE AND SOIL DEGRADATION

The major land uses in arid regions are grazing (80%), dryland farming (5%), and irrigation agriculture (3%). Approximately 12% of the arid regions are too dry (hyperarid) to be used even for significant grazing. The central Sahara is an example of the latter condition. Little or no degradation occurs in the vast majority of the hyperarid lands because the human population is very low.

In the grazing lands, water and wind erosion are the principal forms of soil degradation. Both kinds of erosion plus compaction and crusting as well as some salinization are important dryland farming problems. Irrigated land degradation is primarily due to salinization.

Mining, recreation, and human settlement are three nonextensive land uses that can cause soil degradation in the arid regions. Surface mining frequently leads to water and wind erosion and, sometimes, to pollution of the soil with heavy metals and acid materials. Human settlements, when concentrated, can cause soil erosion on the perimeters of its communities as the vegetative cover is reduced or destroyed. Off-road vehicles are primarily responsible for land disturbance that results in erosion in recreational areas.

VI. DEGRADATION CONTROL

Stopping further degradation of the soil environment in arid regions is possible for all the degradation processes described earlier (for example, water erosion and wind erosion). However, the solutions are costly, in the main, and usually will benefit society only in the long run. Long-term protection of the soil in the interests of future generations is a responsibility every citizen bears. Because society is the benefactor of long-term protection, society—in the form of governments—must assure that such protection is provided, whether with inducements or sanctions, or both. Next to the permanent loss in soil productivity that certain soils experience when eroded excessively, the greatest threat for humankind is probably soil contamination with heavy metals from industrial activity. Heavy metals last forever; how much damage they do to humans is a function of the concentration of metals in the soil and their adsorption characteristics. Rarely is there a direct impact of soil-borne heavy metals on people. Rather, the hazard arises from crop uptake of the metals and later consumption of the crop by humans or animals. Relatively, the soil area affected or potentially affected by harmfully toxic elements, pesticides, or industrial waste accumulation is small, globally. It can be very important locally, however.

Countering the short-term and long-term impact of the crop-related degradation processes of salinization, organic matter loss, compaction, and crusting frequently increases the costs of operation. However, the costs are usually not excessive relative to the benefits obtained. Raising organic matter levels in dryland farming soils is probably most difficult, especially when the cultivated soils are on the dry margin of the semiarid climatic zone. Frequently, sandy soils are common on the uplands in drier regions. Raising their organic matter content to that of the precultivation level is hampered by low crop yields and rapid rates of organic matter decomposition.

A. Erosion

Water and wind erosion control is an almost intractable problem if cropping continues on climatically marginal or highly erodible sloping lands. Wind erosion is the major threat in the drier areas, water erosion on the sloping lands.

Wind Erosion

There are virtually no data on the economic cost of wind erosion. Its on-site damages are due mainly to sand blasting young crops and to covering seedlings with wind-borne soil. In a few instances, as in the southern High Plains in Texas, blowouts are found on cultivated land. Most blowouts occur on rangelands when wind blows away the topsoil, exposes an unproductive subsoil, and piles up sand downwind around shrubs and other barriers and along roads. Off-site wind erosion damage is much greater than on-site damage (Huszar, 1985; Davis and Condra, 1985). Data are largely missing for off-site damage, but it includes the health impairment of dust and dust-borne infectious materials, abrasion of machinery, vision-limiting dust storms that lead to highway accidents, abrasion of paint on buildings, sand piles around buildings, and, perhaps most costly, sand accumulation on highways. The highway problem is taken seriously in Arizona, where warning signs are posted in the most dust-prone areas to warn motorists of the hazard of driving in a dust storm.

The effect of wind erosion on soil productivity is also unknown. Any knowledgeable observer in the southern Great Plains can see that wind erosion has caused organic matter removal and nitrogen deficiency on sandy soils. The deficiency may not show up in normal or dry years, but winter wheat growing in a wet year can show obvious nitrogen deficiency symptoms. Other than a nitrogen deficiency, which can be corrected with fertilizers, there is usually no other evidence that wind erosion affects soil productivity, except, of course, if there are blowouts. When the drought of the Dust Bowl days was ended by good rains, there was practically no indication that the land had suffered, despite horror stories of land being ruined forever.

Many wind erosion control practices have been developed in the United States since the historic studies of Chepil and his colleagues were carried out. Tree windbreaks have been used for centuries. More recently, the value of protective tillage techniques has been recognized, particularly for the drier regions where crop residues are minimal, establishing windbreaks is very difficult or is unacceptable to the farmer, and the soil is highly erodible (King, 1988). Grass strips can be effective and are easier to establish and compete less with the cash crop for water than do tree and shrub windbreaks (Bilbro and Fryrear, 1988). Some soils, however, are simply too erodible to be cultivated without being severely eroded. Those need to be retired and put into grass. Minimum or zero-tillage systems can be effective erosion control practices now that herbicides are available to control weeds. The systems fail or are less effective when crop production is reduced in dry years.

Water Erosion

Water erosion in arid regions appears to take two courses. In the drier areas (less than 250 mm rain, as a first approximation), gullies seem to be more common. This is probably due to the occurrence of intense, short-duration thunderstorms and rapid runoff from degraded rangelands. As the rainfall average increases, sheet and rill erosion seem to be more common. The Palouse semiarid winter wheat area of Washington and Idaho is an example of the latter condition. Rills are numerous and sheet erosion has removed much soil from knobs of hills, but gullies are not common. Certainly, sheet and rill erosion are far more destructive of land than gully erosion, in total, spectacular though the latter be.

How much effect water erosion has on soil productivity in the arid regions (or anyplace else) is largely unknown. The only known region in the world where the erosion-productivity relation has been determined is the Palouse region of the Pacific Northwest in the United States. In that 800,000 ha area, erosion has reduced potential winter wheat yields by 30-40% in 90 years of erosion (Papendick et al., 1985). Erosion not only reduced yields, it also increased costs per unit of production.

Water erosion control practices are well known and are, when properly utilized, effective. This is true for all climatic regions of the world, whether tropical, subtropical, or temperate. Adjustments in such details as vertical and horizontal spacing of terraces depend upon soil, steepness of slope, crops to be grown, and other site-specific factors. Maintaining a permanent vegetative cover is the best way to control either water or wind erosion. Minimum or zero tillage is the next best thing to permanent vegetation. In dryland farming, low levels of plant residue production in the drier climatic zones and during droughts reduce the effectiveness of minimum or zero tillage. Crop management options decrease as the precipitation decreases. Water erosion can be somewhat troublesome in irrigated agriculture, but good water management automatically provides satisfactory erosion control. Water erosion on degraded rangelands is common. Vegetation improvement is usually the only practical procedure to control erosion on rangelands.

B. Obstacles to Degradation Control

Assuming that the magnitude of the erosion problem is known, that acceptable control practices have been devised, and that measures can be taken to minimize soil degradation, the question arises of why the practices and measures are not put into effect globally. Poverty is one reason; lack of concern about environmental deterioration is another; belief that control treatments will not be profitable in the short term is yet another. There may also be a lack of knowledge of how to control the threat or, at least, of where to go to get advice and assistance. The most important, however, is probably our crisis mentality, which says nothing need be done until the problem reaches the crisis stage.

Soil pollution with heavy metals, persistent chemicals, and so on, is in the category of off-site damage in that the pollutants are introduced from the outside, whether the outside is a factory, a sewage sludge plant, or an airplane. As for all off-site damage, it is the land user who suffers, not the producer of the pollutant. Legislation or, perhaps, public pressure are needed to make the producer correct the problem. The producer has little or no incentive to do so otherwise. The public, rightly or wrongly, frequently pays the cost of stopping or repairing the damage. Control of chemical pollutants in arid regions is hampered by the scarcity of large bodies of water or of permanent streams and rivers. Pollutants that could be disposed of by flushing down a stream in the humid regions must be gotten rid of on land in the arid regions. This makes disposal much more difficult.

Other kinds of soil degradation punish the land user, sooner or later. Rehabilitating salinized irrigated land usually becomes profitable immediately or within a few years. The same is true for compaction due to machinery. Soil crusting is more difficult to control and gives positive returns more slowly, as do efforts to restore soil organic matter.

Erosion again comes to the forefront as the problem most likely to be intractable to resolve. The reason is not that control is difficult to implement. Rather, the cost is immediate but the benefits may take years to be realized. Fortunately, water erosion control in the dry regions frequently leads to water conservation, as well, thus carrying the potential to match costs and short-term benefits, which farmers like. Wind erosion control is different. It is sandblasting of young plants, the deposition of sand on top of them, or the blowing away of soil around the roots that causes most on-site wind erosion damage. Such events do not happen every year, ordinarily. If plant damage can be avoided with timely tillage, the returns come that same year. If windbreaks or grass barriers are used to protect the crop, the benefits come over a many-year period, not immediately. Windbreaks serve other purposes, though, an important one of which is aesthetic.

Persuading people to recognize that excessive water erosion is occurring on their land can be surprisingly difficult. Farmers in the Palouse region of the Pacific Northwest often claimed that no serious erosion was occurring on their fields even though extensive rill erosion could be observed from the windows of their homes (Osterman and Hicks, 1988). It is difficult to persuade people to control erosion when they do not see it as a problem.

VII. SOLUTIONS

A. Land Pollution

Solving the land pollution problem in arid regions by on-site control practices is very difficult. Persistent chemical and heavy metal contamination is best prevented by avoiding its occurrence rather than by reclaiming the soil. The latter is virtually impossible to do because of the aridity of the climate, which severely restricts chemical reactions and leaching. Laws and regulations will have to be put into effect to stop further pollution by persistent pesticides, toxic chemicals, and deadly elements.

B. Physical Degradation

On the other hand, reducing the physical degradation of agricultural soils cannot be done by passing laws. Rather, it calls for a combination of actions. First and foremost is the need for research on ways to develop more cost-effective and user-acceptable techniques to increase organic matter levels, reduce soil compaction and crusting, lower soil salinity levels, and minimize soil erosion. Under irrigated and rangeland conditions, improved management can raise organic matter levels and reduce crusting and, in many cases, control salinity levels and erosion. Human-induced compaction can be limited in irrigated soils by judicious changes in the use of machinery. Compaction in rangelands is largely limited to the trampling by livestock around watering points. It is unlikely that this will change.

In addition to research, there is a continuing need to develop the conservation ethic, to educate agriculturalists to recognize the degradation threat, and to provide technical assistance and financial inducements.

The greatest challenge is to halt and reverse the degradation of rainfed croplands in the semiarid regions where rainfall is marginal for crop production. In these drier areas, a lowering of the soil organic matter level is an almost inevitable consequence of cropping. Rotations with soil-building crops are not generally feasible; fertilizer use is uneconomic, by and large; and crop residues are too little to maintain high humus levels and, in turn, reduce crusting and compaction. Grain crop-clean fallow rotations to conserve moisture, as is widely practiced in the wheat-growing areas, increase the susceptibility of soils to erosion in the fallow period. Degradation control, in these circumstances, depends upon tillage, erosion control structures, and water conservation structures and practices instead of vegetation. A major exception is wind erosion control. Shelterbelts can be effective and reasonably long-lasting if the shelterbelts are built with drought-tolerant species and if they can be established successfully. Tillage, however, may be more economical in wind erosion than windbreaks.

C. Marginal and Unsuitable Lands

Some land is simply unsuitable for cultivation without serious degradation. Very sandy soils in the dry regions are in that category when used for dryland farming. They are even hazardous to use for irrigation agriculture and for grazing. These lands should be in native vegetation and only grazed slightly. Moderately to steeply sloping lands are difficult to protect from erosion because a sparse crop cover exposes the soil to splash erosion and rapid runoff. Bench terracing can be an effective erosion control measure on steep slopes, but benches are costly to construct, shallow soils are not suitable for benching, and maintenance is a constant requirement.

Brown (1988) has noted that the Conservation Reserve Program of the U.S. Department of Agriculture reduced soil erosion dramatically in 1 year simply by taking the most erosion-prone land out of cultivation. There is a lesson for the entire world in this action: most erosion is concentrated on a relatively small fraction of cropland. To be effective, erosion control efforts must be targeted to the most erodible soils and those most susceptible to long-term loss in productivity.

D. Land Retirement

As Brown's analysis indicates, in the long run the only practical way to reduce excessive water and wind erosion on marginal lands used for dryland farming is to retire them from cultivation. Conservation of good productive land in more favorable climatic zones can be done by various means that increase production or at least give a positive benefit-cost ratio. If marginal lands are to be removed from cultivation, production must be increased on the better lands and the pressure to break out new land must be reduced. This has happened in the humid eastern United States and Europe, where erosion is probably less and yields greater than ever before. It seems to have good possibilities in China's semiarid loess plateau, too (Postel, 1989). For much of the developing world, however, it will be a massive challenge to find ways to reduce the human and animal pressure that forces people onto ever less productive lands. Management practices that combine erosion control with productivity increases are badly needed. Even more important, probably, is to raise the economic level of the national population, especially the rural population, so that people can find employment in cities and reduce land pressures. Degradation is now a socioeconomic problem as well as a technical problem.

VIII. OFF-SITE DAMAGE

The off-site costs of soil degradation commonly exceed the on-site costs. A study of the economic cost of wind erosion in New Mexico, for example, the first such study ever conducted, showed that off-site damage costs were about 45 times greater than on-site damage costs (Huszar, 1985; Davis and Condra, 1985). Off-farm damages from sheet and rill erosion were estimated, nationally, to be at least twice as great as on-farm damages, amounting to as much as $6 billion per year (Clark et al., 1985). Erosion is by far the principal source of the particulate matter that leaves the place of origin on land and moves off-site by wind or water flow. Particulate matter carries adsorbed pesticides, heavy metals, radioactive elements, and phosphorus, among other things. Water percolating through degraded soils can carry soluble nutrients and pesticides into groundwater supplies or surface water bodies. Nitrates are probably the most studied groundwater pollutants resulting from the agricultural use of land or from landfills.

A. Water Erosion

Sediment, nutrients, and pesticides are the principal off-site pollutants from eroded land (Ribaudo et al., 1989). Sediment deposition on lower lying land can bury crops, change drainage patterns, cover roads, and cause other kinds of local damage. Stream-borne sediment may fill reservoirs, affect aquatic life, reduce the recreational value of lakes, increase flooding, block navigational channels, and cause other damage. Nitrogen and phosphorus carried on soil particles or in solution can be beneficial or harmful. When deposited on land, they can raise the crop production potential. Phosphorus aids fish production when added to nutrient-poor lakes, as in the Andes of Chile, but contributes to the rapid growth of algae and to the eutrophication problem in such water bodies as the Great Lakes. Nitrates seldom affect surface water supplies adversely but can be harmful in groundwater by raising nitrate levels high enough to cause blue-baby disease in infants. Chlorinated hydrocarbon pesticides of the DDT type are adsorbed on soil particles and have caused damage to aquatic life when eroded sediments were deposited in streams and lakes.

Surface coal mines in the arid western United States produce much of the nation's coal. During the last 20 years, states have enacted laws requiring that the mined land be reclaimed when mining ends. Previously, surface mines, with their long rows of spoil, were simply abandoned when the coal had been extracted. The spoil banks eroded easily because no effort was made to revegetate them. Consequently, the abandoned mines were aesthetically unattractive, supplied sediment to nearby streams, and polluted water supplies. Seepage through the spoils picked up soluble salts that contaminated streams and groundwaters in the vicinity. Salinity remains the major off-site pollutant, but land reshaping and the requirement to revegetate the land have done much to improve conditions (Dettmann and Olsen, 1977). Mine seepage in the arid regions is usually alkaline and much less damaging downstream than the acid effluent common in the humid regions.

B. Wind Erosion

Very little is known, apart from the New Mexico study, of the off-site damage done by wind erosion. Sand deposition on roads, fences, and around buildings is a common problem, as are abrasion of machinery and building surfaces. The highest cost found in the Huszar study (Huszar, 1985) was due to damage done to landscaping by sand accumulations. Dust in the home, loss of recreational opportunities, and impaired health followed landscaping in importance. These costs are much less than off-site water erosion costs, largely because windstorms, at least in the United States, tend to be more episodic and less severe than floods.

IX. CONCLUSIONS

Human-induced degradation of arid region soils has been going on for millenia. Wherever people and their animals congregate, degradation follows, whether in China, the Mediterranean, or North America. Water and wind erosion, salinization, loss of organic matter, compaction, crusting, and pesticide contamination are the result of agricultural activities and are by far the most extensive forms of degradation of arid region soils. Toxic element accumulation, principally heavy metals, and industrial waste pollution can be very serious. The latter two kinds of environmental pollutants are generally associated with urban development and mining. The scarcity of large streams and rivers in the arid regions means that using the "solution to pollution is dilution" adage is not viable for water disposal of potential soil contaminants.

Mismanagement of resources is the immediate cause of soil degradation, but the basic causes are likely to be some combination of poverty, ignorance, lack of concern, and greed.

Control measures are known for virtually every kind of soil degradation. Whether control measures are put into practice depends primarily on political pressure, cost-benefit ratios, perception of the severity of the problem, and other nontechnological factors. The availability of site-specific solutions limits the applicability of practices in some areas, particularly in developing countries.

REFERENCES

Ahn, P. (1977). Erosion hazard and farming systems in East Africa. In *Soil Conservation and Management in the Humid Tropics*, D. J. Greenland and R. Lal (Eds.). John Wiley & Sons, New York, pp. 165-176.

Auerswald, K., and Schmidt, F. (1986). *Atlas der Erosionsgefahrdung in Bayern*. GLA-Fachberichte 1, Bayerisches Geologisches Landesamt, Munich.

Bensalem, B. (1977). Examples of soil and water conservation practices in North African countries. Algeria, Morrocco and Tunisia. In Soil Conservation and Management in Developing Countries. FAO Soils Bulletin 33, Rome, pp. 151-260.

Bilbro, J. D., and Fryrear, W. D. (1988). Plant materials for windbarriers in semiarid regions. In *Proceedings, 1988 Wind Erosion Conference*. Texas Tech University, Lubbock, pp. 150-157.

Brown, R. (1984). Conserving soils. In *State of the World 1984*. Worldwatch Institute, Washington, D.C., pp. 53-73.

Brown, R. (1988). Breakthrough on soil erosion. *World Watch 1*(3): 19-24.

Clark, E. H., Haverkamp, J. A., and Chapman, W. (1985). *Eroding Soils: The Off-Farm Impacts*. Conservation Foundation, Washington, D.C.

Comptroller General (1983). Report to the Congress of the United States: Agriculture's soil conservation programs miss full potential in the fight against soil erosion. U.S. General Accounting Office RCED-84-48, Washington, D.C.

Craig, I. A., and Pisone, U. (1988). Overview of rainfed agriculture in northeast Thailand. In *Rainfed Agriculture in Northeast Thailand*, C. Pairintra, K. Wallapapan, J. F. Parr, and C. E. Whitman (Eds.). Office of Agriculture, Science and Technology Bureau, Agency for International Development, Washington, D.C., pp. 24-37.

Darkoh, M. B. K. (1986). Combating desertification in Zimbabwe. *Desertification Control Bull. 13*: 17-28.

Das, D. C. (1977). Soil conservation practices and erosion control in India—A case study. In Soil Conservation and Management in Developing Countries, FAO Soils Bulletin 33, Rome, pp. 11-50.

Davis, B., and Condra, G. D. (1985). The on-site costs of wind erosion in New Mexico. Texas Tech University Department of Agricultural Economics, Lubbock.

Dettman, H., and Olsen, D. (1977). Assessment of water quality impacts of a western coal mine. In *The Reclamation of Disturbed Arid Lands*, R. A. Wright (Ed.). University of New Mexico Press, Albuquerque, pp. 53-67.

Dregne, H. E. (1982). Historical perspective of accelerated erosion and effect on world civilization. In *Determinants of Soil Loss Tolerance*, B. L. Schmidt et al. (Eds.). American Society of Agronomy, Madison, Wisconsin, pp. 1-14.

Dregne, H. E. (1986). Soil and water conservation: A global perspective. *Interciencia 11*: 166-172.

Dregne, H. E., and Meyer, R. E. (1989). Erosion, productivity, and sustainable agriculture in Africa. International Center for Arid and Semiarid Land Studies, Texas Tech University, Lubbock.

Eckholm, E. P. (1976). *Losing Ground.* Worldwide Institute, Washington, D.C.
El-Hinnawi, E., and Hashmi, M.-H. (1982). *Global Environmental Issues.* United Nations Environment Programme, Nairobi, Kenya.
FAO (1954). Soil erosion survey of Latin America. Part II. *J. Soil Water Conserv. 9*: 214-229, 237.
Fletcher, J. E., and Martin, W. P. (1948). Some effects of algae and molds in the rain crust of desert soils. *Ecology 29*: 95-100.
Fryrear, D. W., Stout, J. E., and Gillette, D. A. (1988). Instrumentation for wind erosion. In *Proceedings, 1988 Wind Erosion Conference,* Texas Tech University, Lubbock, Texas, pp. 118-132.
Goldstein, J. (1977). *Sensible Sewage.* Rodale Press, Emmaus, Pennsylvania.
Hellden, U. (1987). An assessment of woody biomass, community forests, land use and soil erosion in Ethiopia. Lund Studies in Geography, Series C, No. 14, Lund University, Lund, Sweden.
High Plains Underground Water District (1984). Hard pans hurt yields. *Cross Section 30*(2): 4.
Hoogmoed, W. B., and Stroosnijder, L. (1984). Crust formation on sandy soils in the Sahel. 1. Rainfall and infiltration. *Soil Tillage Res. 4*: 5-23.
Huszar, P. C. (1985). Off-site economic costs of wind erosion in New Mexico. Colorado State University Department of Agricultural and Natural Resource Economics, Fort Collins.
IAEA (1986). Quantification, nature and bioavailability of bound carbon-14 pesticide residues in soil, plants and food. International Atomic Energy Agency, Vienna.
Information Transfers (1980). *National Conference on Municipal and Industrial Sludge Utilization and Disposal.* Information Transfers, Inc., Silver Spring, Maryland.
Jacobsen, T., and Adams, R. M. (1958). Salt and silt in ancient Mesopotamian agriculture. *Science 128*: 1252-1258.
King, A. D. (1988). Research needs: Clod-farming tillage. In *Proceedings, 1988 Wind Erosion Conference.* Texas Tech University, Lubbock, pp. 37-39.
Lagerwerff, J. V., and Specht, A. W. (1970). Contamination of roadside soil and vegetation with cadmium, nickel, lead, and zinc. *Environ. Sci. Technol. 4*: 583-586.
Lathrop, E. W., and Rowlands, P. G. (1983). Plant ecology in deserts: An overview. In *Environmental Effects of Off-Road Vehicles,* R. H. Webb and H. G. Wilshire (Eds.). Springer-Verlag, New York, pp. 113-152.
Lee, H.-T. (1984). Soil conservation in China's loess plateau. *J. Soil Water Conserv. 39*: 306-307.
Leeper, G. W. (1978). *Managing the Heavy Metals on the Land.* Marcel Dekker, New York.
Mikesell, M. W. (1960). Deforestation in northern Morocco. *Science 132*: 441-454.
Mikesell, M. W. (1969). The deforestation of Mount Lebanon. *Geogr. Rev. 59*: 1-28.

Northcote, K. H., and Skene, J. K. (1972). Australian soils with saline and sodic properties. CSIRO Soil Publication No. 27, Melbourne, Australia.

Osterman, D. A., and Hicks, T. (1988). Highly erodible land: Farmer perceptions versus actual measurements. *J. Soil Water Conserv. 43*: 177-182.

Papendick, R. I., Young, D. L., McCool, D. K., and Krauss, H. A. (1985). Regional effects of soil erosion on crop productivity—the Palouse area of the Pacific Northwest. In *Soil Erosion and Soil Productivity*. American Society of Agronomy, Madison, Wisconsin, pp. 305-320.

Park, W. M., and Sawyer, D. G. (1985). Targeting soil erosion control efforts in a critical watershed. U.S. Department of Agriculture, Economic Research Service, Natural Resource Economics, ERS Staff Report No. AGES850801, Washington, D.C.

Postel, S. (1989). China revives lost land. *Worldwatch 2*(2): 7-8.

Ribaudo, M. O., Colacicco, D., Barbarika, A., and Young, C. E. (1989). The economic efficiency of voluntary soil conservation programs. *J. Soil Water Conserv. 44*: 40-43.

Science Council of Canada (1986). A growing concern: Soil degradation in Canada. Council Statement, September 1986.

Stocking, M. (1984). Erosion and soil productivity: A review. Food and Agriculture Organization, Land and Water Development Division, AGLS Consultant's Working Paper No. 1, Rome.

Stocking, M. (1985). Erosion-induced loss in soil productivity: Trends in research and international cooperation. University of East Anglia Overseas Development Group, Norwich, U.K.

Szabolcs, I. (1974). Salt affected soils in Europe. Martinus Nijhoff, The Hague.

Taylor, H. M., Mathers, A. C., and Lotspeich, F. B. (1964). Pans in the Southern Great Plains soils. 1. Why root-restricting pans occur. *Agron. J. 56*: 328-332.

Tikhonov, A. V. (1960). Soil erosion and its effects on yields in areas of the Volga uplands. *Sov. Soil Sci. 2*: 180-187.

Trimble, S. W. (1974). *Man-Induced Soil Erosion on the Southern Piedmont, 1700-1970.* Soil Conservation Society of America, Ankeny, Iowa.

U.S. Department of Agriculture (1981a). *1980 Appraisal*, Part I, *Soil, Water, and Related Resources in the United States: Status, Condition, and Trends*. U.S. Department of Agriculture, Washington, D.C.

U.S. Department of Agriculture (1981b). Soil erosion effects on soil productivity: A research perspective. *J. Soil Water Conserv. 36*: 82-90.

Vander Pluym, H. S. A. (1978). Extent, causes and control of dryland saline seepage in the northern Great Plains region of North America. In *Dryland-Saline-Seep Control*, H. S. A. Vander Pluym (Ed.). Alberta Agriculture, Lethbridge, Alberta, Canada, pp. 1-48 to 1-58.

Veblen, T. T. (1978). Forest preservation in the western highlands of Guatemala. *Geogr. Rev. 68*: 417-434.

Virgo, K. J., and Munro, R. N. (1978). Soil and erosion features of the Central Plateau region of Tigrai, Ethiopia. *Geogerma 20*: 131-157.

Woods, L. E. (1983). *Land Degradation in Australia.* Australian Government Publishing Service, Canberra.

Working Party (1982). *Salting of Non-irrigated Land in Australia.* Standing Committee on Soil Conservation, Department of Primary Industry, Commonwealth of Australia, Canberra.
Zachar, D. (1982). *Soil Erosion.* Elsevier Scientific, Amsterdam.

11

Grazing Management

CARLTON H. HERBEL and REX D. PIEPER *U.S. Department of Agriculture and New Mexico State University, Las Cruces, New Mexico*

I. INTRODUCTION

A major use of semiarid and arid lands ("rangelands" in this chapter) is grazing or browsing by livestock and wild herbivores. Rangelands also provide habitat for wildlife, recreational opportunities, water, and aesthetic values. Animal production on rangelands reduces the requirements for nonrenewable fossil fuels and utilizes resources not readily usable by other means.

Semiarid and arid rangelands are often fragile and subject to accelerated soil erosion if not managed appropriately. If rangelands are mismanaged so that plants fail to provide sufficient soil cover, the composition of the plant community changes, with the result that productivity is reduced and erosion increases. Continued abuse of the rangeland system can result in severe soil degradation. This does not imply that proper grazing is destructive; some plants produce more biomass with moderate livestock grazing than with protection from livestock grazing. Some native plant communities evolved with grazing use by native animals, whereas other communities evolved with little or no utilization by herbivores. However, native plant communities are not always the most productive under intensive animal grazing (Herbel, 1982). Revegetation with introduced or improved plant species may dramatically increase the productivity of rangelands (Herbel, 1983).

Grazing of livestock is controlled to some degree by range managers and pastoralists around the world. However, many important environmental factors are beyond control of the manager, but these factors need to be considered in planning grazing management and future research on the subject.

Such terms as carrying capacity, animal distribution, management plans, grazing preference, stocking pressure, grazing system, continuous grazing, rotational grazing, grazing utilization, grazing intensity, and flexibility are used to describe various aspects of grazing management (Society for Range

Management, 1974). These terms are addressed in this chapter. *Grazing management* is planned grazing or browsing of plants (native or introduced) by animals (livestock and/or wildlife) to accomplish the objectives of the land manager.

II. PRINCIPLES AND CONCEPTS

Grazing management uses some of the concepts of applied ecology, with emphasis on plant-animal-environment interactions. These involve the temporal and spatial distribution of animals, the kinds and numbers of animals, and animal effects on range plants (Heady, 1971, 1984; Heitschmidt, 1988; Kothmann, 1980). Utilization of plants by animals may affect the quantity and quality of range forage, and the quantity and quality of range forage affect animal production from a unit of rangeland. The environment (soil and climatic components) also affects range forage and, therefore, animal performance.

The performance of grazing animals is determined by (1) animal potential, (2) forage quantity, (3) forage quality, and (4) feeding to supplement range forage. Furthermore, animals often graze the plants and parts of plants having the highest nutrient value (Torell and Hart, 1988).

A. Continuous Grazing

There is a considerable contrast between extensive and intensive grazing management (Table 1). Long-duration or continuous use may involve grazing periods of several months to more than a year—*extensive grazing management*. The periods of deferment from grazing may be zero or extend from several months to more than a year. One or more groups of animals may be used. A relatively small amount of capital expenditure is required for fence and water development. Management input and the risks involved are relatively minor. Decisions on animal numbers and when to move livestock to another range unit are not critical unless the permanent range resource, plants and soils, is in danger of irreversible damage (Herbel, 1982).

McIlvain and Shoop (1969) concluded that a number of rotational schemes they studied were inferior to yearlong continuous grazing at the same stocking rates at Woodward, Oklahoma. They cited some major reasons for the success of continuous yearlong grazing in the southern Great Plains: (1) forage production is primarily dependent on warm-season rainfall, and there can be a 30-fold variation in monthly forage production within a year; (2) most plants are useful to cattle; (3) many of the plants that increase after a drought are excellent grazing plants; (4) cattle grazing competes with forage losses as a plant matures and also with other consumers; (5) young and regrowth forage is more palatable and nutritious than more mature forage; (6) grazed plants conserve soil water for subsequent growth; and (7) favorable precipitation combined with proper grazing management allows rangelands to recover a desirable species composition following drought. Herbel (1973) gave some additional reasons for the success of yearlong continuous grazing: (1) use is light when the plants are actually growing and (2) lighter stocking per unit area means less compaction of the soil by livestock.

TABLE 1 Concept of Grazing Management

Extensive	Intensive
Low risk	High risk
Low cost	High cost
Low management inputs	High management inputs

B. Specialized Grazing Systems

Van Poollen and Lacey (1979) reported that herbage production was 13% higher under specialized grazing system compared to production under continuous grazing on 14 separate studies in U.S. rangeland. On the other hand, gain of individual animals is often reduced by specialized grazing systems since livestock are concentrated on smaller units and lack the opportunity to select a diet as high in nutritive content as they might under continuous grazing (Malechek, 1984; Pieper, 1980). However, in many cases there has been little difference in either vegetational or livestock response between continuous grazing and specialized grazing systems (Driscoll, 1967; Shiflet and Heady, 1971).

C. Short-Duration Grazing

Short-duration grazing use involves one or more groups of domesticated animals that are moved from one unit to an ungrazed unit every 2-28 days (Savory, 1979). Deferment from grazing use varies from a relatively short time during the growing season of plants to a longer period when plants are dormant. This is an intensive system. Short-duration grazing requires a relatively large capital expenditure for fences and water development. Decisions on the movement of livestock must be made daily during the growing season because immature plants are supposed to be grazed only once before they are deferred from grazing. Therefore, the risk and the management inputs are high with the more intensive short-duration grazing system. However, high inputs do not necessarily guarantee high net returns to the operator (Herbel, 1982).

In evaluating grazing systems in South Africa, Roberts (1967) reported: (1) slow rotation systems do not eliminate selective grazing; (2) using 16 units in a high-intensity, short-duration system, 12 are grazed once each for about 2 weeks each 6 months, and the 4 units not in the regular rotation may be used as reserve grazing in droughty years or given a full year's rest if not needed; (3) high-intensity, short-duration grazing combined sufficient rests with efficient grazing use to allow rapid restoration of denuded rangelands but high-intensity, short-duration grazing was less beneficial on rangelands in good condition; (4) restoration of rangelands may be retarded, and intensified abuse may occur, if the stocking rate is increased more rapidly than indicated by range conditions; (5) with variable precipitation, no system can eliminate selective grazing if set grazing periods and stocking rates are maintained, but high-intensity, short-duration reduces selective grazing; and (6) the relatively high capital investment required to implement high-intensity, short-duration grazing can be justified by the increased carrying capacity. Short-duration grazing may also detrimentally influence cryptogamic crusts on many rangelands (Brotherson et al., 1983; Johansen, 1986).

III. GRAZING OPTIMIZATION

Many range workers have noticed that areas protected from grazing by large herbivores often tend to become somewhat stagnant. Early work on grazing influences indicated that grazing was detrimental, despite such observations on vegetation in exclosures. Indeed, Caldwell (1984) has indicated that even low levels of herbivory may reduce the ability of an individual plant to compete successfully in the plant community. Recent research, however, suggests that some degree of defoliation may be beneficial to individual plants in the community.

McNaughton (1979), working in the extremely productive Seregenti Plains in eastern Africa, has studied plant-animal relations in some detail. He developed a series of curves to demonstrate the relationship between the level of defoliation and plant fitness or productivity. Some plants are detrimentally impacted by defoliation, others little impacted until defoliation reaches a certain level and then are detrimentally impacted, but a third group is stimulated by defoliation until a certain level of defoliation is reached, at which point fitness or productivity is reduced until the point of actual death. Others have also expanded on the theory (Hilbert et al., 1981; Dyer et al., 1982; Paige and Whitham, 1987). The basic idea behind the theory is that grazing or herbivory acts as a selective force and that plants continually subjected to herbivory develop adaptations that allow them to survive herbivory. Grasses have long been considered as having special adaptations for grazing, but such a theory was not developed formally. Some workers (Owen and Wiegert, 1976, 1981) have even expressed the view that grasses depend on the herbivores for their survival and evolution. Others have argued against the close relationship between plants and herbivores (Belsky, 1986). Belsky (1986) stated that "None of the grass traits presented as examples of grass mutualism have been shown to encourage herbivory, although they certainly lessen its deleterious effects. In fact, most studies of natural systems show that grazing reduces or has no effect on the productivity of the plants. If this is universally the case, mutualisms could not have evolved."

These arguments have relevance to rangelands in the United States. Bison (*Bison bison* L.) were present in large numbers in the Great Plains prior to settlement. Blue grama [*Bouteloua gracilis* (H.B.K.) Lag. ex Steud.] and buffalo grass [*Buchloe dactyloides* (Nutt.) Engelm.], two dominant grass species of the Great Plains, are very resistant to grazing. Do they owe this resistance to grazing pressure imposed by large concentrations of bison? In contrast are conditions in the Intermountain Region, which supported mainly pronghorn antelope (*Antilocapra americana* Ord) and mule deer (*Odocoileus heminonus* Rafinesque) (Platou and Tueller, 1985), and desert grassland regions of the Southwest, which supported populations of pronghorn antelope but few other large herbivores (Martin, 1968). In these regions some of the major grass species are not resistant to grazing. Notable among these are bluebunch wheatgrass [*Agropyron spicatum*

(Pursh) Scribn. and Smith] in the Intermountain Region (Caldwell et al., 1981) and black grama [*Bouteloua eriopoda* (Torr.) Torr.] in the Southwest. These two grass species may owe their lack of grazing resistance to the absence of grazing pressure before the introduction of domestic livestock into these areas.

IV. STOCKING RATE

Stocking rate is the number of animals grazing on rangeland per unit area. The decision on the stocking rate is the most important made by the land manager. A heavy stocking rate may lead to a change in plant species and increased soil erosion eventually leading to permanent damage to the system. Conversely, a light stocking rate for a number of years often leads to a reduced economic return to the operator. In large units with a poor distribution of grazing animals, it is nearly impossible to prevent areas of heavy utilization near watering points, *sacrifice zones*. In this situation, there is an area of heavy utilization of plants, an area of light utilization at a distance from water, and an area of moderate utilization in between. It is a great challenge to the operator to use techniques that increase the area of moderate use considering a forage crop and economic conditions that vary from year to year. We estimate that proper stocking rates throughout all range units would double animal production in the western United States.

A. Stocking Rate Theory

Proper stocking rate is essential to the successful management of rangeland. To establish the proper stocking rate, Hart (1980) suggested the following determinations:

1. Animal and plant responses.
   a. Achieve maximum gain per animal or per area?
   b. Achieve maximum profit per animal or per area?
   c. Achieve greatest reproductive efficiency of plants and animals?
   d. Maximize use of available forage?
   e. Maximize the improvement of range conditions?
2. Determine the relationship between stocking rate and the desired plant or animal response.
3. Calculate the stocking rate to achieve the desired results.
4. Are there constraints to using the calculated stocking rate?

Answering the questions posed in part 1 will assist in developing goals. Some land managers use experience in developing guides to determine stocking rates. Other land managers determine stocking rate by allocating available forage to animals, using feed requirements for each class of animal.

Under range conditions, gain per grazing animal remains constant below a critical stocking rate and decreases linearly as stocking rate increases above the critical level (Hart, 1978). It is a land manager's duty to determine the critical stocking rate. A guide was developed by Bement

(1969) for grazing rangeland dominated by blue grama in the central Great Plains. Gain per animal and gain per unit area are related to the amount of herbage left ungrazed at the end of the growing season. Maximum returns from yearling cattle were obtained when 336 kg ha^{-1} of herbage was left at the end of the growing season. This figure can be approximated for rangelands grazed only during the growing season by taking 50% of the long-term herbage. For example, if the long-term herbage production on a range unit is 2000 kg ha^{-1}, maximum returns from yearling cattle can be obtained when 1000 kg ha^{-1} of herbage is left at the end of the growing season. The plant and soil resources are maintained, or gradually improved, when 1000 kg ha^{-1} of herbage remains at the end of the growing season. In a droughty year when only 1500 kg ha^{-1} herbage is produced, only 500 kg ha^{-1} herbage should be harvested by the cattle. Conversely, when herbage production is 2500 kg ha^{-1}, 1500 kg ha^{-1} can be harvested. This illustrates the difficulty encountered by land managers because they make decisions on a stocking rate before herbage production is known. In practice, stocking rate decisions are based on the long-term performance of the vegetation on a range unit.

B. Economic Considerations

Economic considerations in determining the stocking rate are important if the land manager wishes to produce the highest profit. If fixed carrying costs per animal (interest, veterinary costs, death loss, and initial costs of land and animals) are high relative to selling price, then profits are greater if gain per animal is maximized. However, if carrying costs are low relative to selling price of animals, profits are greater if gain per unit area is maximized at a reduction in gain per animal (Torell and Hart, 1988; Hart, 1980). In the latter, care must be taken not to abuse the resource.

V. DISTRIBUTION OF GRAZING ANIMALS

Domesticated animals traverse any range unit in response to the number and location of watering places, the distribution of plant species within the unit, topography, the size of the unit, and various other attractants (Herbel and Gould, 1980). Improved distribution of grazing animals over the large units prevalent in the western United States will result in a higher number of animals using range vegetation properly.

A. Water

Domesticated animals must have enough water. Minimum daily requirements for water are as follows: cow with calf, 38 L; ewe with lamb, 4 L; and horse, 45 L (Holechek et al., 1989). Variations in environmental conditions may alter water requirements. In range units with more than one watering point, periodic opening and closing of these points can be used to rotate grazing pressure to different portions of the unit (Martin and Ward, 1970).

In a comparison of wildlife use of water with short-duration and continuous livestock grazing, Prasad and Guthery (1986) showed that water at the center of a short-duration cell may be largely unavailable to white-tailed deer (*Odocoileus virginianus* Zimmermann), collared peccaries

(*Tayassu tajacu* L.), and wild turkeys (*Meleagris gallopavo* Nelson). However, most species of birds made about the same number of visits to water regardless of grazing system. Prasad and Guthery (1986) concluded that peripheral water facilities should be maintained with cell grazing to maintain a maximum diversity of wildlife. The waterers could be fenced to minimize effects on the rotation of livestock even though the outlying water may attract livestock.

B. Location of Plants

Plants that are palatable during a certain part of the year attract animals to the part of the range unit where they occur even though these plant species occur at a distance from water. In the Southwest and other areas, Russian thistle (*Salsola iberica* Sennen and Pau) is palatable and nutritious when young. It can be encouraged to grow with a soil disturbance, such as disking.

C. Topography

In rolling terrain, it is common to see rangelands with bottoms showing the effects of heavy grazing but with light grazing on the adjacent slopes. On a 10% slope, about 75% of the grazing use occurs within 0.8 km of the watering point. Some animals, such as sheep and goats, utilize vegetation on rough terrain better than cattle.

D. Other Structures

Placing salt and supplemental feed away from permanent watering points also helps to distribute domestic animals to areas that otherwise would be lightly utilized (Ares, 1953). Shade and rubbing posts located away from watering points also attract domesticated animals to undergrazed areas. Construction of "drift" fences to prevent livestock from continually grazing in certain favored places within a range unit also helps with distribution problems. Fencing to create smaller units, trails, and roads may be useful in obtaining better livestock distribution under some circumstances.

VI. FLEXIBILITY IN GRAZING

There is a large variation, temporally and spatially, in forage production. Table 2 shows the precipitation and perennial grass yields for three sites on an experimental range in southern New Mexico (Herbel and Gibbens, 1981). The average annual precipitation in this area is 230 mm. The Reagan and Stellar sites, about 50 m apart, are on plains that are occasionally flooded. Both of these sites have clay loam soils. The Cacique site is 13 km from the Reagan and Stellar sites and is a loamy sand soil on a flat, upland area. The average yield of perennial grasses at the Reagan site, primarily burro grass (*Scleropogon brevifolius* Phil.), for 1957-1977 was 950 kg ha^{-1}, but it was 88 kg ha^{-1} in 1960 and 3017 kg ha^{-1} in 1959. The Stellar site was dominated by tobosa [*Hilaria mutica* (Buckl.) Benth.], and the average yield of perennial grasses for 1957-1977 was 1662 kg ha^{-1}; in 1965 it was 15 kg ha^{-1}, and in 1972 it was 3718 kg ha^{-1}.

TABLE 2 Precipitation (mm, October-September) and Dry Weight (kg ha^{-1}) of Perennial Grasses for Selected Years at Three Sites, Jornada Experimental Range, New Mexico

	Reagan[a]		Stellar[b]		Cacique[c]	
Year	Ppt	Yield	Ppt	Yield	Ppt	Yield
1959	216	3017	216	1846	188	428
1960	130	88	130	390	104	63
1963	236	1745	236	1665	122	99
1965	152	69	152	15	140	174
1966	170	719	170	782	229	323
1972	356	1314	356	3718	257	419
1977	203	492	203	541	262	501

[a]Primarily burro grass (*Scleropogon brevifolius*).

[b]Primarily tobosa (*Hilaria mutica*).

[c]Primarily black grama (*Bouteloua eriopoda*) and mesa dropseed (*Sporobolus flexuosus*).

Source: From Herbel and Gibbens, 1981.

Black grama and mesa dropseed [*Sporobolus flexuosus* (Thurb.) Rydb.] dominate the Cacique site and the average yield for 1959-1977 was 321 kg ha^{-1}, but in 1960 it was 63 kg ha^{-1} and in 1977 it was 501 kg ha^{-1}. Table 2 illustrates the tremendous variability that occurs in the forage crop at the same site from year to year and at different sites within the same year (Herbel and Gibbens, 1981). If a constant stocking rate is maintained, it must be low enough to maintain proper utilization when forage production is considerably below average or there will be destructive grazing in drought years (Pieper, 1981). Much of the permanent damage to rangelands has occurred when the stocking rate was considerably higher than indicated by the forage crop. Some flexibility in the stocking rate permits the operator to stock animals according to the variability encountered in the quantity of forage produced.

Most grazing studies have been established at a fixed stocking rate. Downward adjustments in livestock numbers were made only in severe droughts, but adjustments in numbers should be part of the plan for the grazing study (Herbel, 1973). Flexible herd management has been suggested by several workers as the best method for maximizing livestock production without damaging the range resource during droughty periods (Ares, 1952; Boykin, 1967; Jardine and Forsling, 1922; Reynolds, 1954; Stubblefield, 1956). During years when the forage crop is about average, the livestock group is composed of 55-60% breeding animals. The remainder of the herd is composed of young animals. In years of low forage production, adjustments in the size and composition of the group of animals are planned to bring the group within the capacity of the range. For example, in the arid portions of the Southwest, the primary production of perennial grasses occurs in the summer. Therefore, the number of animals relying on this forage can be adjusted in the fall to meet the forage conditions from fall until the following summer. It is an unwise practice to supplement the quantity of forage while the animals are grazing on rangelands. Rather, the surplus animals should be sold or fed in a lot (Herbel et al., 1984). Readily salable animals are marketed, and the breeding-age animals are culled more deeply than normally. In the years with above-average forage production, part or all of the natural increase from the breeding group can be held over until spring or fall, depending on market and forage conditions. Martin (1975) advocated stocking at 90% of estimated carrying capacity to provide a cushion for maximum flexibility.

The cattle herd that grazed on the Jornada Experimental Range in southern New Mexico from 1927 to 1934 averaged 80% cows and 6% yearlings (Paulsen and Ares, 1962). The remaining 14% of the herd was calves and bulls. From 1940 to 1951, when the system of flexible herd management was in effect, cows averaged 61% and yearlings 30% of the total herd. Using the same sale prices for comparison, the average annual gross income was slightly higher in the 1940-1951 period although there were 30% fewer animal units grazing the range than in 1927-1934 (Paulsen and Ares, 1962).

Low forage quality and poor animal growth in Colorado coupled with decreasing prices near the end of the fall grazing season favor the early sale of cattle (Rodriguez and Jameson, 1988). The early sale of domestic livestock and/or harvesting wildlife are also techniques useful in reducing the pressure on rangelands when there is a reduced forage crop. Risk is reduced by using flexible marketing dates.

VII. SOIL PROPERTIES

Johnston (1961) and Johnston et al. (1971) reported that a heavy rate of cattle grazing *Festuca* grasslands in Canada reduced soil organic matter (OM) and total phosphorus (P) content but increased $NaHCO_3$-soluble P compared to a lighter stocking rate. Smoliak et al. (1972) reported that *Stipa-Bouteloua* grasslands in Canada grazed by sheep at a heavy rate had increased soil organic carbon (C) but there was no effect on total and available P and nitrogen (N) content compared to soils with a lighter stocking rate. Organic C and total P content were greater, but total N was lower in soils of ungrazed grassland in the northern Great Plains in central North Dakota when compared to soils of grazed grasslands (Bauer et al., 1987). Bulk densities were highest in surface soils of grazed grassland. Continuous, rotationally deferred, and short-duration rotation grazing systems did not affect soil bulk density in trials in southeastern Wyoming (Abdel-Magid et al., 1987). Van Haveren (1983) found that the degree of soil compaction, as measured by bulk density, depended on the texture, water content, and organic matter.

VIII. GRAZING EFFECTS ON WATERSHEDS

The hydrological impacts of animal grazing result primarily from the interactions of weather, vegetation, soil, topography, and intensity of animal utilization. Blackburn (1984) cited several authors who recognized that heavy continuous grazing accelerated erosion and runoff. Hydrological characteristics are generally not different among ungrazed range units and those that are continuously grazed lightly or moderately.

A. Grazing Systems

Gifford and Hawkins (1976, 1978) found no evidence that any single grazing system consistently increased plant and litter cover on watersheds. Range units grazed under a four-unit, three-herd deferred rotation system in Texas were hydrologically similar to those of a livestock exclosure (Blackburn, 1984; McGinty et al., 1978; Wood and Blackburn, 1981a, b). Range units grazed under a high-intensity, low-frequency grazing system (eight units, one herd with 17 day graze and 119 day rest) were superior or similar hydrologically to moderate continuously grazed units. Conversely, units under a short-duration scheme (14 units, one herd with a 4 day graze and 50 day rest) at double the recommended stocking rate were similar hydrologically to those under heavy continuous grazing (McCalla et al., 1984a, b). Hydrologic conditions under short-duration grazing responded favorably during years of average or higher precipitation; however, during drought the short-duration system rapidly displayed adverse impacts on infiltration rates, roughness, and soil aggregates. After 2 years of above-average precipitation, the hydrological conditions of the short-duration units had not overcome the adverse effects of a single drought year.

Thurow et al. (1988) assessed the hydrological responses of units moderately stocked under continuous and high-intensity, low-frequency grazing systems, and heavily stocked under continuous and short-duration grazing near Sonora, Texas. Moderate stocking maintained infiltration rates, whereas heavy stocking resulted in decreased infiltration rate and increased erosion. The heavy stocking rate and weather, rather than grazing strategy, were the primary factors influencing hydrological responses.

B. Water Quality

Sediment is the major pollutant from rangeland watersheds. Moderate continuous grazing or specialized grazing systems should reduce sediment losses from most watersheds. However, if watersheds have been severely overgrazed, instituting moderate continuous or specialized grazing systems may not reduce sediment losses. Bacteria or nutrients as potential pollutants from livestock grazing are not a serious problem on areas not in riparian zones (Blackburn, 1984; Buckhouse and Gifford, 1976). However, Gary et al. (1983) found that the number of fecal coliform organisms in streamwater were increased when livestock grazed on adjacent rangeland. The numbers of these organisms remain elevated several months after cattle are removed (Stephenson and Street, 1978). Sediments serve as a reservoir of coliform and *Salmonella* organisms (Stephenson and Rychert, 1982). Fecal coliform levels tend to increase as stocking rates increase, as with specialized grazing systems (Tiedemann et al., 1987).

IX. GRAZING EFFECTS ON RIPARIAN ZONES

The direct effects of animal grazing on riparian zones are (1) change, reduction, or elimination of vegetation; (2) higher stream temperatures from a reduction in woody streamside cover; (3) excessive sediment in the channel from streambank and upland erosion; (4) high coliform bacterial counts from upper watershed sources; (5) channel widening from hoof action on streambanks and water erosion; (6) change in the structure of the water column and channel; and (7) elimination of riparian areas by channel degradation and/or lowering of the water table (Skovlin, 1984; Platts and Raleigh, 1984). In a Montana study, there was no correlation between riparian condition and the type of grazing system. Grazing intensity was an important factor in the riparian condition, but not as important as the amount of vegetation utilized by domestic animals during part of the growing season (Platts and Raleigh, 1984). Riparian zones are highly preferred by some classes of domestic animals and must be managed separately from adjacent areas. Specialized systems can protect and enhance riparian habitats, but these have not been identified.

X. GAME RANCHING

Some of the rangelands have been dedicated to producing both wildlife and livestock or, increasingly, wildlife only. Wildlife are used to enhance the beauty of the unit and the opportunity for recreational hunting. In the United States and other countries throughout the world, increased emphasis is placed on wildlife for a source of income (Henderson, 1988).

Generally, sufficient forage remains for wildlife, such as deer or other browsers that have different grazing habits than cattle, when proper utilization of forage species is maintained in the diet of domestic livestock, such as cattle (Smith and Rodgers, 1988). However, dietary overlaps between wildlife and livestock require an allocation of the forage resource. Additionally, proper utilization leaves greater amounts of cover required by ground-nesting birds, such as ducks or grouse.

One or more species of livestock can be used to produce a vegetation community suitable for various species of wildlife. Cattle often have a diet high in grasses with the result that the plant community may have a higher proportion of forbs and shrubs. Deer and pronghorn antelope prefer forbs and shrubs. Upland game birds, such as quail, use forb seeds. Therefore, cattle grazing may improve the habitat for deer, antelope, quail, and other animals that need some forbs and shrubs (Smith and Rodgers, 1988).

Watering locations influence animal selection of foraging areas if preferred vegetation is available. Plant species and location of water wells and drinking troughs can be altered to achieve more uniform distribution of grazing. However, improved distribution can also be obtained by emphasizing cattle in relatively level areas and wildlife in hilly areas and riparian zones.

There are six ways in which wildlife can benefit the land manager increase economic returns (Knight, 1988):

1. Exchange of entry rights for protection from further ingress

2. Exchange of entry rights for the performance of wildlife-enhancing practices
3. Fee areas for nonconsumptive use of wildlife, such as aesthetics
4. Granting entry rights without direct compensation, particularly where wildlife species are a problem
5. Having the wildlife for the land manager's hunting or viewing pleasure
6. Granting hunting rights, as through a lease for a specified purpose

XI. GRAZING SYSTEMS

Environmental conditions vary considerably on rangelands around the world. Consequently, grazing systems should be designed to meet the specific conditions of each area and management condition. In this section, we review grazing studies for major range regions with emphasis on rangelands in the United States.

A. California Annual Rangeland

Studies at the San Joaquin Experimental Range in central California showed a consistent advantage of yearlong continuous grazing over seasonal grazing in breeding cow performance and calf weaning weights (Duncan and Heady, 1969). The best balance of seasonally preferred plant species occurred on ranges grazed continuously yearlong. Similar results with sheep were obtained at Hopland, about 160 km north of San Francisco. Deferred rotation grazing was compared to yearlong continuous grazing. The average weaning weight of lambs was consistently higher with yearlong continuous grazing. The grazing treatment did not affect species composition of plants, density of plants, or herbage production. Heady (1961) concluded that "Yearlong grazing at reasonable stocking rates is the best way to manage the California annual type because it pays primary attention to the day by day animal needs and because yearlong grazing amounts to a partial deferment every year."

B. The Pacific Bunchgrass Region

Hormay and Talbot (1961) described rest rotation grazing for this region as (1) graze a unit all season for maximum livestock production, which may reduce the vigor of the grazed plants; (2) after step 1, rest that unit one or two seasons, until the plants' vigor is restored and litter accumulates; (3) defer grazing in the unit rested in step 2 until seed is ripe, then graze for the remainder of season to trample the seed into the soil and for maximum livestock production; and (4) rest one or two seasons to permit the establishment of new plants. The time required for each step depends on the growth requirements of the major plants. This grazing system reduces selective grazing and improves the reproduction of grazed plants.

Ratliff (1962) used a combination of fencing, salting, water development, and horseback riding to reduce selective grazing at Harvey Valley in northeastern California. After 13 years, Ratcliff and Reppert (1968) concluded that the performance of individual animals was about the same with rest rotation grazing as with season long grazing. In some years, there was an increase in the length of the grazing season but no change in the numbers of livestock with rest rotation grazing. Rest rotation grazing emphasizes plant reproduction by seed and may be less applicable to many ranges where grass reproduction is mostly vegetative. In addition, Hyder and Bement (1977) warn against excessive use during the growing season.

C. Conifer Forest Ranges

Moderate rotation, heavy rotation, and moderate season-long grazing were compared in the Big Horn Mountains near Burgess Junction, Wyoming (D. R. Smith et al., 1967). There was no significant difference in daily animal gain for animals grazing moderate rotation and moderate season long units. Production of the major species, Idaho fescue (*Festuca idahoensis* Elmer), was maintained equally well on the three treatments on soils of granitic origin. On soils derived from sedimentary deposits, the abundance of Idaho fescue declined within the rotation unit grazed heavily.

Deferred rotation and season long grazing at three intensities were compared near La Grande, Oregon (Skovlin and Harris, 1970). Systems of grazing did not cause significant differences in the production of any major plant group or species in the grassland type. In the adjacent forest type, heavy stocking reduced production of the major plant, elk sedge (*Carex geyeri* Boott.), from 455 to 227 kg ha^{-1} after 12 years. There was essentially no difference in the production of elk sedge under deferred rotation and season long grazing with moderate and heavy stocking. Initial management practices, such as fencing, water development, and salt placement, were effective in increasing cattle use in the forest type (Harris, 1954).

D. Southern Africa

Several studies in South Africa indicated that range deterioration was common under continuous grazing (Roberts, 1967; Gammon, 1978a; Booysen and Tainton, 1978). Various rotational systems have been developed to stop or reverse this trend. Some of these rotational schemes have been successful in improving range conditions (Gammon, 1978a; Booysen and Tainton, 1978), but others have showed little difference in vegetation or livestock response to rotational systems, for example, O'Rourke (1978) in Tanzania and Pratchett and Schirvel (1978) in Botswana. Recently, systems involving several paddocks and short grazing periods have become popular (Roberts, 1967; Goodloe, 1969; Howell, 1978). These systems have been designated as short-duration, rapid rotation, or multicamp systems. In a recent review, Skovlin (1987) reported that short-duration grazing at a moderate rate of stocking resulted in range improvement similar to that of conventional schemes. He found numerous instances in which doubling the existing stocking rate in a rapid rotation system on a long-term basis led to severe degradation of the basic range resource. Similar results were obtained on the Charter Estate Grazing Trials in Zimbabwe (Clatworthy, 1984; Gammon, 1984; Parsons, 1984; Worthington, 1984). These trials have been running for a longer period than any short-duration studies in the United States.

E. Intermountain Shrub Region

Season long grazing was compared with deferred rotation grazing in southeastern Oregon (Hyder and Sawyer, 1951). Cow weight gains were greater under season long grazing than deferred rotation grazing. Concentrating the cattle in the deferred rotation system for 2 consecutive years reduced plant vigor.

The important management point on the desert rangelands of southwestern Utah is to occasionally defer key species during late winter (Blaisdell and Holmgren, 1984). Most of this area serves as winter range for sheep. Studies in Nevada also showed that rest rotation grazing failed to improve the sagebrush-grass range, presumably because of heavy grazing during the growing season (Eckert and Spencer, 1986, 1987).

F. Northern Great Plains

Deferred rotation and season long grazing were compared in central North Dakota (Rogler, 1951). Steer gains were not increased with a rotation system when there was sufficient forage for season long grazing. There seemed to be merit in a rotation system for improving range damaged by overgrazing. However, complete deferment until the recovery of range condition was a more rapid method of range improvement (Rogler, 1951). A rotation system can be used when it is necessary to stock at a high rate during occasional years and with older cattle. Younger cattle were less likely to gain weight under a rotation system because they did not use the mature forage in summer and fall as well as older cattle. Lodge (1970) reviewed other grazing studies on native rangelands in the northern Great Plains in Canada and United States and concluded that specialized grazing systems were no better than continuous grazing.

G. Central Great Plains

Hart et al. (1988) compared continuous, four-unit rotationally deferred and eight-unit short-duration grazing for 6 years on a mixed grass range near Cheyenne, Wyoming. The three grazing systems did not affect steer gains differentially. The basal cover of vegetation was not affected by grazing treatment.

H. True Prairie Region

Deferred rotation grazing was compared with season long grazing near Manhattan, Kansas. On the major range site, there was a slightly greater reduction in range condition under deferred rotation grazing than with moderate season long grazing during a 6 year period (Herbel and Anderson, 1959). Moderate season long grazing gave a higher steer gain than deferred rotation grazing (E. F. Smith et al., 1967). Intensive early-season grazing proved beneficial to both livestock and vegetation in the Flint Hills of Kansas (Smith and Owensby, 1978).

I. Southern Great Plains

A number of deferred rotation schemes were studied near Woodward, Oklahoma (McIlvain and Shoop, 1969; McIlvain and Savage, 1951; McIlvain et al. (1955). No rotation system was superior to yearlong continuous grazing.

J. Edwards Plateau

Merrill (1954) compared continuous yearlong grazing at three intensities with deferred rotation grazing at a moderate rate on the Edwards Plateau near Sonora, Texas. After 11 years, the stocking rate in the units in the deferred rotation system was increased 33% (Merrill, 1969). Those units made greater improvement than any of the units grazed continuously. Herbage production was 2458 kg ha^{-1} on the four-pasture, three-herd rotation system compared to 1073 kg ha^{-1} under continuous grazing after 20 years (Reardon and Merrill, 1976).

Near Barnhart, Texas, average annual net returns per animal unit for 7 years were $30.63, $39.03, and $41.71 for continuous, four-unit rotation, and two-unit rotation, respectively (Huss and Allen, 1969). Combination stocking with cattle and sheep; cattle, sheep, and goats; or cattle and goats was more profitable than using cattle or sheep alone (Merrill, 1969; Huss and Allen, 1969).

K. Southern Rolling Plains

Fisher and Marion (1951) compared rotation and continuous grazing at a moderate rate near Spur, Texas, concluding that (1) rotation grazing did not improve the vegetational composition during a 6 year study, (2) rotation grazing increased differential grazing use of the major plants as the season progressed or during drought, and (3) gains of yearling steers grazing on the rotation units were slightly lower than gains of steers grazing on the continuous units. Calf production per animal unit averaged 200, 208, and 221 kg for moderate continuous, two-unit rotation, and a four-unit rotation, respectively, during an 8 year study near Throckmorton, Texas (Kothmann et al., 1970).

Heitschmidt (1986) concluded that production per cow was lower in rotational grazing than moderate continuous grazing at Throckmorton, Texas, because conception rates, weaned calf weights, and calf weaning weights were reduced by rotational grazing. Diet quality and intake were lower with rotation grazing than continuous grazing when forage availability was low, but the opposite occurred when forage availability was high. Overall, sediment production was greater and infiltration rates were lower with rotation grazing than with continuous grazing.

L. Semidesert Grassland of the Southwest

A number of studies near Las Cruces, New Mexico, have contributed to developing the best pasture grazing system and the closely related seasonal suitability grazing system (Paulsen and Ares, 1962; Valentine, 1967; Beck, 1978; Herbel, 1974; Herbel and Nelson, 1969; Holechek and Herbel, 1982).

There were two major vegetation types on the Jornada Experimental Range near Las Cruces 60 years ago, one dominated by black grama and the other dominated by tobosa and burro grass. The grazing system consisted of grazing the tobosa-burro grass type in summer and early fall, and grazing the black grama type for the remainder of the year (Paulsen and Ares, 1962). However, the stand of black grama was depleted because of the severe drought of 1951-1956 (Herbel et al., 1972) and the rapid increase of shrubs (Buffington and Herbel, 1965). The grasses, forbs, and shrubs have a high nutritive value (Nelson et al., 1970). Nearly all species available were grazed to some extent during the year (Herbel and Nelson, 1966). In a 20 year study on a black grama type, Beck et al. (1987) reached the following conclusions:

1. Seasonal suitability grazing system did not improve forage conditions or livestock performance over yearlong continuous grazing.
2. Herbage biomass production is related more to annual and growing season precipitation patterns than to stocking rate.
3. Livestock performance is related more to herbage availability than to either grazing system.

Martin and Severson (1988) compared yearlong grazing with the Santa Rita grazing system for 13 years near Tucson, Arizona. The Santa Rita grazing system is a one-herd, three-unit, 3 year rotation scheme that may improve deteriorated rangelands in the Southwest (Martin, 1978). Martin and Severson (1988) concluded that herbage yields were not related to grazing treatment on rangeland in good condition.

M. North Africa and the Middle East

Grazing studies in northern Africa are limited. Many livestock herders practice some form of nomadic or seminomadic (transhumant) system. These systems usually involve livestock movement away from permanent water during the rainy season and toward permanent water during the dry season (Breman et al., 1978; Holechek et al., 1989; Sandford, 1983). These systems provide maximum flexibility for the herders to meet varying forage and water needs for their livestock.

Heavy stocking on annual grassland rangeland in Niger increased standing plant biomass compared to that under moderate stocking (Wylie et al., 1985). However, animal weight gains were inversely related to stocking rate during the dry season. Heifers lost less weight under a rotational grazing system than under continuous grazing when they were watered on alternate days, but not when watered daily (Wylie and Pieper, 1990). Opportunities for specialized grazing systems appear limited in developing countries, but when animals are under the close control of a herder, some grazing systems could be implemented without extensive fencing and other developments.

In Israel, rotational grazing schemes favored grasses over forbs, which increased under continuous grazing and heavy stocking (Gutman and Seligman, 1979; Seligman and Gutman, 1978). The response of legumes was similar under rotation and continuous stocking with both moderate and heavy stocking. Individual animal performance was also adversely affected by rotational and heavy stocking whereby individual animal selectivity was restricted.

Annual grasslands respond differently than perennial grasslands since seed reproduction is the key rather than the impact of grazing on individual plants. In many cases management suitable for perennial grasslands are misapplied to annual grasslands.

XII. CONCLUSIONS

We have reviewed the literature on grazing management and have reached the following conclusions.

Universal conclusions are inappropriate, because the results of a grazing study are specific to a given site and are closely related to rate of stocking (Gammon, 1978b). The performance of animals on rangelands is related to the quality and quantity of forage available to the animal and the harvesting efficiency of the grazing animal (Heitschmidt, 1988).

Most studies indicate a moderate stocking rate, particularly during the growing season. Grazing at a moderate or lighter rate on key plants provides optimum animal performance. An occasional period of heavy grazing in the dormant season reduces unpalatable or ungrazed plants and promotes more uniform grazing.

Favorable weather is more effective than a specialized grazing system in providing a rapid improvement in species composition.

Fluctuation in the quantity of the forage crop is a frequent occurrence and must be considered in setting a stocking rate.

Any range improvement that increases livestock distribution results in greater livestock production from that unit.

A rapid rotation system may result in improvement in range condition with proper stocking, but once the desired condition has been achieved, a slower rotation or continuous grazing gives the highest animal production.

Ephemeral grasses and forbs may contribute much to animal performance for a brief, but critical, part of the year.

The evolutionary history of a plant species is critical. Some of the major species in the Rocky Mountains, some areas west of the Rocky Mountains, and some desert areas did not evolve under grazing pressure. Species that are not adapted to grazing often do not survive moderate grazing.

Stocking with a variety of animals increases the productivity of rangelands. Rabbits, rodents, insects, and other wildlife, however, consume forages that may be useful to domestic livestock and other animals.

Supplemental feeding to animals grazing range forage should supplement the *quality* of forage, not the *quantity*.

A cell arrangement may improve livestock management (Dahl, 1986).

Heavy trampling may damage beneficial algal crusts and soil microbes.

Intensive grazing requires more management input.

Besides additional research on these principles mentioned, other questions are related to the site-specific importance of (1) seed production, (2) seed planting by trampling, (3) seedling establishment, and (4) trampling effects on soils. Considerable funds have been expended on studying grazing systems, but continuous grazing at a moderate rate will maximize sustained production of animals from most range resources.

CAVEAT

The views expressed in this chapter are those of the authors and do not reflect the official policies of the U.S. Department of Agriculture.

REFERENCES

Abdel-Magid, A. H., Schuman, G. E., and Hart, R. H. (1987). Soil bulk density and water infiltration as affected by grazing systems. *J. Range Manag. 40*: 307-309.

Ares, F. N. (1952). Size and composition of the herd. *Am. Cattle Prod. 34*(7): 14, 16, 18.

Ares, F. N. (1953). Better cattle distribution through the use of meal-salt mix. *J. Range Manag. 6*: 341-346.

Bauer, A., Cole, C. V., and Black, A. L. (1987). Soil property comparisons in virgin grasslands between grazed and nongrazed management systems. *Soil Sci. Soc. Am. J. 51*: 176-182.

Beck, R. F. (1978). A grazing system for semiarid lands. In *Proceedings, First International Rangeland Congress*, D. N. Hyder (Ed.). Society for Range Management, Denver, Colorado, pp. 569-572.

Beck, R. F., McNeely, R. P., and Kiesling, H. E. (1987). Seasonal grazing on black grama grasslands. In *Livestock Research Briefs and Cattle Growers' Short Course*, New Mexico Agr. Exp. Sta., Las Cruces, pp. 85-88.

Belsky, A. J. (1986). Does herbivory benefit plants? A review of the evidence. *Am. Naturalist 127*: 871-891.

Bement, R. E. (1969). A stocking-rate guide for beef production on blue-grama range. *J. Range Manag. 22*: 83-86.

Blackburn, W. H. (1984). Impacts of grazing intensity and specialized grazing systems on watershed characteristics and responses. In *Developing Strategies for Rangeland Management; National Research Council, National Academy of Sciences.* Westview Press, Boulder, Colorado, pp. 927-983.

Blaisdell, J. P., and Holmgren, R. C. (1984). Managing intermountain rangelands—salt-desert shrub ranges. U.S. Department of Agriculture General Technical Report INT 163.

Booysen, P., and Tainton, N. M. (1978). Grassland management: Principles and practice in South Africa. In *Proceedings, First International Rangeland Congress*, D. N. Hyder (Ed.). Society for Range Management, Denver, Colorado, pp. 551-554.

Boykin, C. C. (1967). Profitability and flexibility of two range cattle systems in the Rolling Plains of Texas. *J. Range Manag. 20*: 375-379.

Breman, H., Diallo, A., Traore, G., and Djiteye, M. M. (1978). The ecology of the annual migrations of cattle in the Sahel. In *Proceedings, First International Rangeland Congress*, D. N. Hyder (Ed.). Society for Range Management, Denver, Colorado, pp. 592-595.

Brotherson, J. D., Rushforth, S. R., and Johansen, J. R. (1983). Effects of long-term grazing on cryptogam crust cover in Navajo National Monument, Ariz. *J. Range Manag. 36*: 579-581.

Buckhouse, J. C., and Gifford, G. F. (1976). Sediment production and infiltration rates as affected by grazing and debris burning on chained and seeded pinyon-juniper. *J. Range Manag. 29*: 83-85.

Buffington, L. C., and Herbel, C. H. (1965). Vegetational changes on a semidesert grassland range from 1858 to 1963. *Ecol. Monogr. 35*: 139-164.

Caldwell, M. M. (1984). Plant requirements for prudent grazing. In *Developing Strategies for Rangeland Management; National Research Council, National Academy of Sciences.* Westview Press, Boulder, Colorado, pp. 117-152.

Caldwell, M. M., Richards, J. H., Johnson, D. A., Nowak, R. S., and Dzurek, R. S. (1981). Coping with herbivory: Photosynthetic capacity and resource allocation in two semiarid (*Agropyron*) bunchgrasses. *Oecologia (Berl.) 50*: 14-24.

Clatworthy, J. N. (1984). The Charter Estate grazing trial—results of botanical analysis. *Zimbabwe Agr. J. 81*: 49-52.

Dahl, B. E. (1986). The West Texas experience in short duration grazing. In *Short Duration Grazing*, J. A. Tiedeman (Ed.). Washington State University, Pullman, pp. 103-116.

Driscoll, R. S. (1967). Managing public rangelands: Effective livestock grazing practices and systems for national forests and national grasslands. U.S. Department of Agriculture, AIB 315.

Duncan, D. A., and Heady, H. F. (1969). Grazing systems in the California annual type (abstract). *Am. Soc. Range Manag. 22*: 23-24.

Dyer, M. I., Detling, J. K., Coleman, D. C., and Hilbert, D. W. (1982). The role of herbivores in grasslands. In *Grasses and Grasslands*, J. R. Estes, R. J. Tyrl, and J. N. Brunken (Eds.). University of Oklahoma Press, Norman, pp. 255-295.

Eckert, R. E., Jr., and Spencer, J. S. (1986). Vegetation response on allotments grazed under rest-rotation management. *J. Range Manag. 39*: 166-174.

Eckert, R. E., Jr., and Spencer, J. S. (1987). Growth and reproduction of grasses heavily grazed under rest-rotation management. *J. Range Manag. 40*: 156-159.

Fisher, C. E., and Marion, P. T. (1951). Continuous and rotation grazing on buffalo and tobosa grassland. *J. Range Manag. 4*: 48-51.

Gammon, D. M. (1978a). Patterns of defoliation during continuous and rotational grazing of rangeland by cattle. In *Proceedings, First International Rangeland Congress*, D. N. Hyder (Ed.). Society for Range Management, Denver, Colorado, pp. 603-605.

Gammon, D. M. (1978b). A review of experiments comparing systems of grazing management on natural pastures. *Proc. Grassl. Soc. S. Afr. 13*: 75-82.

Gammon, D. M. (1984). An appraisal of short-duration grazing as a method of veld management. *Zimbabwe Agr. J. 81*: 59-64.

Gary, H. L., Johnson, S. R., and Ponce, S. L. (1983). Cattle grazing impact on surface water quality in a Colorado Front Range stream. *J. Soil Water Conserv. 38*: 124-128.

Gifford, G. F., and Hawkins, R. H. (1976). Grazing systems and watershed management: A look at the record. *J. Soil Water Conserv. 31*: 281-283.

Gifford, G. F., and Hawkins, R. H. (1978). Hydrologic impact of grazing on infiltration: A critical review. *Water Resources Res. 14*: 305-313.

Goodloe, S. (1969). Short duration grazing in Rhodesia. *J. Range Manag. 22*: 369-373.

Gutman, M., and Seligman, N. (1979). Grazing management of Mediterranean foothill range in the Upper Jordan Valley. *J. Range Manag. 32*: 86-92.

Harris, R. W. (1954). Fluctuations in forage of ponderosa pine ranges in eastern Oregon. *J. Range Manag. 7*: 250-255.

Hart, R. H. (1978). Stocking rate theory and its application to grazing on rangelands. In *Proceedings, First International Rangeland Congress*, D. N. Hyder (Ed.). Society for Range Management, Denver, Colorado, pp. 547-550.

Hart, R. H. (1980). Determining a proper stocking rate for a grazing system. In *Proceedings, Grazing Management Systems for Southwest Rangelands Symposium*, K. C. McDaniel and C. Allison (Eds.). New Mexico State University, Las Cruces, pp. 49-64.

Hart, R. H., Samuel, M. J., Test, P. S., and Smith, M. A. (1988). Cattle, vegetation, and economic responses to grazing systems and grazing pressure. *J. Range Manag. 41*: 282-286.

Heady, H. F. (1961). Continuous vs. specialized grazing systems: A review and application to the California annual type. *J. Range Manag. 14*: 182-193.

Heady, H. F. (1971). Management of grazing animals based upon consequences of grazing. In *The Impact of Herbivores on Arid and Semi-arid Rangelands.* Australian Rangelands Society, Perth, Western Australia, pp. 299-309.

Heady, H. F. (1984). Concepts and principles underlying grazing systems. In *Developing Strategies for Rangeland Management; National Research Council, National Academy of Sciences.* Westview Press, Boulder, Colorado, pp. 885-902.

Heitschmidt, R. K. (1986). Short duration grazing at the Texas Experimental Ranch. In *Short Duration Grazing*, J. A. Tiedeman (Ed.). Washington State University, Pullman, pp. 91-102.

Heitschmidt, R. K. (1988). Intensive grazing in arid regions. In *Forage and Grassland Conference, American Forage and Grassland Council*, Baton Rouge, Louisiana, pp. 340-349.

Henderson, F. R. (Ed.) (1988). *Guidelines for Increasing Wildlife on Farms and Ranches.* Wildlife Resources Committee, Great Plains Agricultural Council, Kansas Cooperative Extension Service S-25.

Herbel, C. H. (1973). Grazing systems on native range. In Great Plains Beef Symposium, F. H. Baker (Ed.). Great Plains Agricultural Council Publ. 63, Lincoln, Nebraska, pp. K-1 to K-19.

Herbel, C. H. (1974). A review of research related to development of grazing systems on native ranges of the western United States. In Plant morphogenesis as the basis for scientific management of range resources, K. W. Kreitlow and R. H. Hart (coord.). U.S. Department of Agriculture Misc. Publ. No. 1271, pp. 138-149.

Herbel, C. H. (1982). Grazing management on rangelands. *J. Soil Water Conserv. 37*: 77-79.

Herbel, C. H. (1983). Principles of intensive range improvements. *J. Range Manag. 36*: 140-144.

Herbel, C. H., and Anderson, K. L. (1959). Response of true prairie vegetation on mjaor Flint Hills range sites to grazing treatments. *Ecol. Monogr. 29*: 171-186.

Herbel, C. H., and Gibbens, R. P. (1981). Drought and grazing management decisions. In *Proceedings, International Rancher's Roundup*, L. D. White and A. L. Hoermann (Eds.). Texas Agricultural Extension Servide, Uvalde, pp. 187-192.

Herbel, C. H., and Gould, W. L. (1980). Managing semidesert ranges of the Southwest. New Mexico Extension Circ. 456.

Herbel, C. H., and Nelson, A. B. (1966). Species preference of Hereford and Santa Gertrudis cattle on a southern New Mexico range. *J. Range Manag. 19*: 177-181.

Herbel, C. H., and Nelson, A. B. (1969). Grazing management on semidesert ranges in southern New Mexico. Jornada Exp. Range Rep. No. 1.

Herbel, C. H., Ares, F. N., and Wright, R. A. (1972). Drought effects on a semidesert range. *Ecology 53*: 1084-1093.

Herbel, C. H., Wallace, J. D., Finkner, M. D., and Yarbrough, C. C. (1984). Early weaning and part-year confinement of cattle on arid rangelands of the Southwest. *J. Range Manag. 37*: 127-130.

Hilbert, D. W., Swift, D. M., Detling, J. K., and Dyer, M. I. (1981). Relative growth rates and the grazing optimization hypothesis. *Oecologia (Berl.) 51*: 14-18.

Holechek, J. L., and Herbel, C. H. (1982). Seasonal suitability grazing in the western United States. *Rangelands 4*: 252-255.

Holechek, J. L., Pieper, R. D., and Herbel, C. H. (1989). *Range Management Principles and Practices*. Prentice Hall, Englewood Cliffs, New Jersey.

Hormay, A. L., and Talbot, M. W. (1961). Rest-rotation grazing—a new management system for perennial bunchgrass ranges. U.S. Department of Agriculture Prod. Res. Rep. 51.

Howell, L. N. (1978). Development of multi-camp grazing systems in the southern Orange Free State, Republic of South Africa. *J. Range Manag. 31*: 459-465.

Huss, D. L., and Allen, J. V. (1969). Livestock production and profitability comparisons of various grazing systems, Texas Range Station. Texas Agr. Exp. Sta. Bull. B-1089.

Hyder, D. N., and Bement, R. E. (1977). The status of seasons of grazing and rest in grazing management. In *The Impact of Herbivores on Arid and Semi-arid Rangelands*. Australian Rangelands Society, Perth, Western Australia, pp. 73-82.

Hyder, D. N., and Sawyer, W. A. (1951). Rotation-deferred grazing as compared to season-long grazing on sagebush-bunchgrass ranges in Oregon. *J. Range Manag. 4*: 30-34.

Jardine, J. T., and Forsling, C. L. (1922). Range and cattle management during drought. U.S. Department of Agriculture Bull. 1031.

Johansen, J. R. (1986). Importance of cryptogamic soil crusts to arid rangelands: Implications for short-duration grazing. In *Short Duration Grazing*, J. A. Tiedeman (Ed.). Washington State University, Pullman, pp. 127-136.

Johnston, A. (1961). Comparison of lightly grazed and ungrazed range in the fescue grassland of southwestern Alberta. *Can. J. Plant Sci. 41*: 615-622.

Johnston, A., Dormaar, J. F., and Smoliak, S. (1971). Long-term grazing effect on fescue grassland soils. *J. Range Manag. 24*: 185-188.

Knight, J. E. (1988). Landowner management opportunities. In *Guidelines for Increasing Wildlife on Farms and Ranches*, F. R. Henderson (Ed.). Kansas Cooperative Extensive Service S-25, pp. 1D-31D.

Kothmann, M. M. (1980). Integrating livestock needs to the grazing system. In *Proceedings, Grazing Management Systems for Southwest Rangelands Symposium,* K. C. McDaniel and C. Allison (Eds.). New Mexico State University, Las Cruces, pp. 65-83.

Kothmann, M. M., Mathis, G. W., Marion, P. T., and Waldrip, W. J. (1970). Livestock production and economic returns from grazing treatments on the Texas Experimental Ranch. Texas Agr. Exp. Sta. Bull. B-1100.

Lodge, R. W. (1970). Complimentary grazing systems for the northern Great Plains. *J. Range Manag. 23*: 268-271.

Malechek, J. C. (1984). Impacts of grazing intensity and specialized grazing systems on livestock response. In *Developing Strategies for Rangeland Management; National Research Council, National Academy of Sciences.* Westview Press, Boulder, Colorado, pp. 1129-1158.

Martin, P. S. (1968). The meaning of empty niches. In *AIBS Pleistocene Man-Environment Relationships Symposium*, Columbus, Missouri, pp. 85-94.

Martin, S. C. (1975). Stocking strategies and net cattle sales on semidesert range. U.S. Department of Agriculture Forest Service RM-146.

Martin, S. C. (1978). The Santa Rita grazing system. In *Proceedings, First International Rangeland Congress*, D. N. Hyder (Ed.). Society for Range Management, Denver, Colorado, pp. 573-575.

Martin, S. C., and Severson, K. E. (1988). Vegetation response to the Santa Rita grazing system. *J. Range Manag. 41*: 291-295.

Martin, S. C., and Ward, D. E. (1970). Rotating access to water to improve semidesert cattle range near water. *J. Range Manag. 23*: 22-26.

McCalla, G. R., II, Blackburn, W. H., and Merrill, L. B. (1984a). Effects of livestock grazing on infiltration rates, Edwards Plateau of Texas. *J. Range Manag. 37*: 265-269.

McCalla, G. R., II, Blackburn, W. H., and Merrill, L. B. (1984b). Effects of livestock grazing on sediment production, Edwards Plateau of Texas. *J. Range Manag. 37*: 291-294.

McGinty, W. A., Smeins, F. E., and Merrill, L. B. (1978). Influence of soil, vegetation, and grazing management on infiltration rate and sediment production of Edwards Plateau rangeland. *J. Range Manag. 32:* 33-37.

McIlvain, E. H., and Savage, D. A. (1951). Eight-year comparisons of continuous and rotational grazing on the Southern Plains Experimental Range. *J. Range Manag. 4*: 42-47.

McIlvain, E. H., and Shoop, M. C. (1969). Grazing systems in the southern Great Plains (abstract). *Am. Soc. Range Manag. 22*: 21-22.

McIlvain, E. H., Baker, A. L., Kneebone, W. R., and Gates, D. H. (1955). Nineteen-year summary of range improvement studies at the U.S. Southern Great Plains Field Station, Woodward, Oklahoma. Woodward Prog. Rep. 5506.

McNaughton, S. J. (1979). Grazing as an optimization process. *Am. Naturalist 113*: 691-703.

Merrill, L. B. (1954). A variation of deferred-rotation grazing for use under Southwest range conditions. *J. Range Manag. 7*: 152-154.

Merrill, L. B. (1969). Grazing systems in the Edwards Plateau of Texas (abstract). *Am. Soc. Range Manag. 22*: 22.

Nelson, A. B., Herbel, C. H., and Jackson, H. M. (1970). Chemical composition of forage species grazed by cattle on an arid New Mexico range. New Mexico Agr. Exp. Sta. Bull. 561.

O'Rourke, J. T. (1978). Grazing rate and system trial over 5 years in a medium-height grassland of northern Tanzania. In *Proceedings, First International Rangeland Congress*, D. N. Hyder (Ed.). Society for Range Management, Denver, Colorado, pp. 563-566.

Owen, D. F., and Wiegert, R. G. (1976). Do consumers maximize plant fitness? *Oikos 27*: 488-492.

Owen, D. F., and Wiegert, R. G. (1981). Mutualism between grasses and grazers: An evolutionary hypothesis. *Oikos 38*: 376-378.

Paige, K. N., and Whitham, T. G. (1987). Overcompensation in response to mammalian herbivory: The advantage of being eaten. *Am. Naturalist 129*: 407-416.

Parsons, S. D. (1984). Introduction to the Charter Estate Grazing Trial. *Zimbabwe Agr. J. 81*: 39-40.

Paulsen, H. A., Jr., and Ares, F. N. (1962). Grazing values and management of black grama and tobosa grasslands and associated shrub ranges of the Southwest. U.S. Department of Agriculture Tech. Bull. 1270.

Pieper, R. D. (1980). Impacts of grazing systems on livestock. In *Grazing Management Systems for Southwest Rangelands. A Symposium*, K. C. McDaniel and C. D. Allison (Eds.). Range Improvement Task Force, New Mexico State University, Las Cruces, pp. 133-151.

Pieper, R. D. (1981). The stocking rate decision. In *Proceedings of the 1981 International Rancher's Roundup*, L. D. White and A. L. Hoermann (Eds.). Texas Agricultural Extension Service, Uvalde, pp. 199-204.

Platou, K. A., and Tueller, P. T. (1985). Evolutionary implications for grazing management systems. *Rangelands 7*: 57-61.

Platts, W. S., and Raleigh, R. F. (1984). Impacts of grazing on wetlands and riparian habitat. In *Developing Strategies for Rangeland Management; National Research Council, National Academy of Sciences*, Westview Press, Boulder, Colorado, pp. 1105-1117.

Prasad, N. L. N. S., and Guthery, F. S. (1986). Wildlife use of livestock water under short-duration and continuous grazing. *Wildlife Soc. Bull. 14*: 450-454.

Pratchett, D., and Schirvel, B. (1978). The testing of grazing systems on semiarid rangelands in Botswana. In *Proceedings, First International Rangeland Congress*, D. N. Hyder (Ed.). Society for Range Management, Denver, Colorado, pp. 567-568.

Ratliff, R. D. (1962). Preferential grazing continues under rest-rotation management. Pacific Southwest Forest and Range Exp. Sta. Res. Note 206.

Ratliff, R. D., and Reppert, J. N. (1968). A summary of cattle weights and gains over a 13-year period of rest-rotation grazing (abstract). *Am. Soc. Range Manag. 21*: 55.

Reardon, P. O., and Merrill, L. B. (1976). Vegetative response under various grazing management systems in the Edwards Plateau of Texas. *J. Range Manag. 29*: 195-198.

Reynolds, H. G. (1954). Meeting drought on southern Arizona rangelands. *J. Range Manag. 7*: 33-40.

Roberts, B. R. (1967). Non-selective grazing—an evaluation. *Farmers Weekly (S. Afr.) July 19*: pp. 40-49; *July 26*: pp. 34-39; *August 2*: pp. 34-37.

Rodriguez, A., and Jameson, D. A. (1988). Rainfall risk in grazing management. *Ecol. Model. 41*: 85-100.

Rogler, G. A. (1951). A twenty-five year comparison of continuous and rotation grazing in the northern plains. *J. Range Manag. 4*: 35-41.

Sandford, S. (1983). *Management of Pastoral Development in the Third World.* John Wiley & Sons, New York.

Savory, A. (1979). Range management principles underlying short duration grazing. In *Beef Cattle Science Handbook*, Vol. 16, M. E. Ensminger (Ed.). Agriservices Foundation, Clovis, California, pp. 375-399.

Seligman, N. G., and Gutman, M. (1978). Cattle and vegetation responses to management of Mediterranean rangeland in Israel. In *Proceedings, First International Rangeland Congress*, D. N. Hyder (Ed.). Society for Range Management, Denver, Colorado, pp. 616-618.

Shiflet, T. N., and Heady, H. F. (1971). Specialized grazing systems: Their place in range management. Soil Conservation Service TP-152.

Skovlin, J. M. (1984). Impacts of grazing on wetlands and riparian habitat: A review of our knowledge. In *Developing Strategies for Rangeland Management; National Research Council, National Academy of Sciences.* Westview Press, Boulder, Colorado, pp. 1001-1103.

Skovlin, J. (1987). Southern Africa's experience with intensive short duration grazing. *Rangelands 9*: 162-167.

Skovlin, J. M., and Harris, R. W. (1970). Management of conifer resources for cattle, deer and elk. *Proc. Int. Grassl. Cong. 11*: 75-78.

Smith, D. R., Fisser, H. G., Jefferies, N., and Stratton, P. O. (1967). Rotation grazing on Wyoming's Big Horn Mountains. Wyoming Agr. Exp. Sta. Res. J. 13.

Smith, E. F., and Owensby, C. E. (1978). Intensive-early stocking of Kansas Flint Hills range. *J. Range Manag. 31*: 14-17.

Smith, E. F., Anderson, K. L., Owensby, C. E., and Hall, M. C. (1967). Different methods of managing bluestem pasture. In 54th Annual Livestock Feeders Day, Kansas Agr. Exp. Sta. Bull. 507, pp. 34-38.

Smith, M. A., and Rodgers, J. D. (1988). Grazing management to maintain or enhance wildlife habitat. In *Guidelines for Increasing Wildlife on Farms and Ranches*, F. R. Henderson (Ed.). Kansas Cooperative Extension Service S-25, pp. 9B-14B.

Smoliak, S., Dormaar, J. F., and Johnston, A. (1972). Long-term grazing effects on *Stipa-Bouteloua* prairie soils. *J. Range Manag. 25*: 246-250.

Society for Range Management (1974). *A Glossary of Terms Used in Range Management*, 2nd ed. Range Term Glossary Committee, Society for Range Management, Denver, Colorado.

Stephenson, G. R., and Rychert, R. C. (1982). Bottom sediment: A reservoir of *Escherichia coli* in rangeland streams. *J. Range Manag. 35*: 119-123.

Stephenson, G. R., and Street, L. V. (1978). Bacterial variations in streams from a southwest Idaho rangeland watershed. *J. Environ. Qual. 7*: 150-157.

Stubblefield, T. M. (1956). Greater returns from cow-yearling operations on Southwest ranges. *J. Range Manag. 9*: 8-10.

Thurow, T. L., Blackburn, W. H., and Taylor, C. A., Jr. (1988). Infiltration and interrill erosion responses to selected livestock grazing strategies, Edwards Plateau, Texas. *J. Range Manag. 41*: 296-302.

Tiedemann, A. R., Higgins, D. A., Quigby, T. M., Sanderson, H. R., and Marx, D. B. (1987). Responses of fecal coliform in streamwater to four grazing strategies. *J. Range Manag. 40*: 322-329.

Torell, L. A., and Hart, R. H. (1988). Economic consideration for efficient stocking rates on rangeland. In *Fort Keogh Research Symposium*, R. S. White and R. E. Short (Eds.). Montana Agr. Exp. Sta., pp. 71-76.

Valentine, K. A. (1967). Seasonal suitability, a grazing system for ranges of diverse vegetation types and condition classes. *J. Range Manag. 20*: 395-397.

Van Haveren, B. P. (1983). Soil bulk density as influenced by grazing intensity and soil type on a shortgrass prairie site. *J. Range Manag. 36*: 586-588.

Van Poollen, W. H., and Lacey, J. R. (1979). Herbage response to grazing systems and stocking intensities. *J. Range Manag. 32*: 250-253.

Wood, M. K., and Blackburn, W. H. (1981a). Grazing systems: Their influence on infiltration rates in the Rolling Plains of Texas. *J. Range Manag. 34*: 331-335.

Wood, M. K., and Blackburn, W. H. (1981b). Sediment production as influenced by livestock grazing in the Texas Rolling Plains. *J. Range Manag. 34*: 228-231.

Worthington, D. K. (1984). The Charter Estate Grazing Trial—animal production and management. *Zimbabwe Agri. J. 81*: 41-48.

Wylie, B., and Pieper, R. D. (1990). Influence of watering frequency and grazing system on heifer weight gains in central Nigeria. *World Anim. Rev.*, in press.

Wylie, B., Senock, R., Snyder, L., Roettgen, B., and Porter, B. (1985). *Grazing and Animal-Plant Relationships in Pastoral Development in Central Niger*. U.S. Agency for International Development, Niamey, Niger, pp. 163-224.

12

Reduced Tillage Systems

PAUL W. UNGER *Agricultural Research Service, U.S. Department of Agriculture, Bushland, Texas*

I. INTRODUCTION

Tillage is widely used to control weeds; incorporate soil amendments, nutrients, herbicides, and insecticides; and modify soil physical conditions, thereby improving soil conditions for crop establishment, growth, and yield. Tillage is also used to conserve soil and water.

The number of tillage operations used to produce a crop varies widely and depends on the crop grown; soil and climatic conditions; and the economic level, degree of mechanization used, attitudes, and management capability of the producer. For example, subsistence farmers in Africa use one operation with hand tools to prepare a seedbed and one or two operations after planting, again by hand, to control weeds. In contrast, under highly mechanized conditions in west Texas, 12-15 operations may be used to produce one cotton crop (*Gossypium hirsutum* L.) (Fryrear, 1974). Included are primary and secondary tillage, planting, crust breaking, surface roughening, cultivating, spraying, and harvesting. Although planting, spraying, and harvesting are not tillage operations, they involve traffic on the land that increases soil density, which in turn can retard root growth and reduce crop yields unless the compaction is alleviated by another tillage operation (Fryrear, 1974).

These examples are undoubtedly extremes, because most crop production involves more than the lowest and less than the highest number of operations indicated here. The norm is usually referred to as *conventional tillage*, which is "the combined primary and secondary tillage operations normally performed in preparing a seedbed for a given crop grown in a given geographical area" (SCSA, 1982). In practice, tillage after planting would also be included.

Although conventional tillage is the norm, it may actually be excessive tillage for optimum soil and water conservation and for sustained crop production under the prevailing conditions. Excessive tillage, along with overgrazing by livestock, especially under arid to semiarid conditions, frequently results in decreased soil organic matter, enhanced structure deterioration, increased runoff, low water storage, increased evaporation, increased erosion, and poor crop yields. The effects are interdependent; hence, when one occurs, one or more of the others usually also occurs. This may lead to increasingly unfavorable soil conditions for crop growth and yields. If continued unchecked, the land may become unsuitable for crop production. Such land degradation is of widespread concern, especially for arid to semiarid regions where food production increases have not matched population increases in many cases (Lal, 1988).

In the last 20-30 years, much reduced tillage research has been conducted. The emphasis on reduced tillage systems resulted from such factors as increasing energy costs, high equipment inventories and costs (e.g., tractors and related implements), high labor costs, increased concern about soil erosion, and the need for more efficient utilization of water for crop production (Unger, 1984a). Development of a wide array of herbicides contributed greatly to the development of effective reduced tillage systems for many crops. By controlling weeds with herbicides, tillage for weed control is often no longer required. Hence, reduced tillage systems have the potential for minimizing soil degradation that can result from excessive tillage.

The objectives of this chapter are to illustrate the potentials of reduced tillage systems for sustaining crop yields in arid to semiarid regions, to identify some disadvantages of reduced tillage systems, and to provide information on soil and water conservation practices for situations in which tillage alone is not effective. The systems discussed were developed primarily in North America and Australia, but the principles involved are generally also applicable to crop production in developing countries.

II. REDUCED TILLAGE SYSTEMS

Reduced tillage is a system in which the primary tillage operation is performed in conjunction with special planting procedures to reduce or eliminate secondary tillage operations (SCSA, 1982). The reduction in tillage may be in the number or intensity of operations. A related system, often referred to as no-tillage, involves planting crops with no seedbed preparation other than opening the soil to place seed at the intended depth. This usually involves opening a small slit or punching a hole into the soil. Normally, chemicals rather than cultivation are used for controlling weeds. The system may also be referred to as slot planting or zero tillage (SCSA, 1982). In this chapter, no-tillage is treated as a subset of reduced tillage.

Another system that is widely applicable to arid to semiarid regions is stubble-mulch tillage. This method is widely used in the drier parts of the United States, Canada, Australia, and other countries where it is often the "conventional tillage method" for nonirrigated crop production. Although often considered conventional tillage, stubble-mulch tillage is a reduced tillage method compared to the more intensive tillage that was previously used under the same conditions. The shift in primary tillage methods used during fallow for wheat (*Triticum aestivum* L.) was illustrated by Greb (1979). Changes in soil water storage during fallow and subsequent grain yields were also presented. The results (Table 1) indicate that a new system became conventional tillage in each successive period.

TABLE 1 Progress in a Wheat-Fallow System with Respect to Water Storage and Wheat Yields, Akron, Colorado

		Fallow water storage		
Years	Tillage during fallow[a]	mm	Precipitation (%)	Wheat yield (Mg ha^{-1})
1916-1930	Maximum tillage; plow, harrow (dust mulch)	102	19	1.07
1931-1945	Conventional tillage; shallow disk, rod weeder	118	24	1.16
1946-1960	Improved conventional tillage; begin stubble mulch in 1957	137	27	1.73
1961-1975	Stubble mulch; begin minimum tillage in 1969	157	33	2.16
1975-1990	Projected estimate; minimum tillage; began no-tillage in 1983	183	40	2.69

[a]Based on 14 months of fallow, from mid-July to second mid-September.

Source: Adapted from Greb (1979).

Because stubble-mulch tillage has played such an important role in nonirrigated crop production in many arid to semiarid regions, it is also treated as a subset of reduced tillage. Minimum tillage involving a combination of stubble-mulch tillage and herbicides has become the conventional method at present in many places, especially in the U.S. Great Plains.

A. Systems Involving Residue Incorporation

The introduction in recent years of a wide array of herbicides has permitted the development of reduced tillage systems for numerous crops. Herbicides usually control weeds during at least part of the crop production cycle. However, in contrast to no-tillage systems (discussed in Sec. II.D), most or all of the soil surface is disturbed by tillage for seedbed preparation or at planting (Unger, 1984a).

Reduced tillage studies in the United States were started in the early 1940s with disking or a modified form of plowing being substituted for moldboard plowing. However, corn (*Zea mays* L.) grain yields were lower with reduced tillage because the disking caused soil compaction. In the 1950s, a system that combined plowing and planting into one operation was developed for corn. This system resulted in grain yields comparable to those with conventional tillage and soil erosion was one-sixth of that with conventional tillage. These results and subsequent research showed that any tillage that exceeded a minimum needed to obtain a favorable seedbed was wasteful and often reduced corn yields (Bennett, 1977).

Promising results of early studies, along with the development of suitable herbicides, resulted in a rapid expansion of reduced tillage at numerous

U.S. locations during the 1960s. Results of the studies, conducted mainly in subhumid to humid locations, were summarized by Amemiya (1977), Fisher and Lane (1973), Griffith et al. (1977), Oschwald (1973), and Reicosky et al. (1977). Although the results are from subhumid to humid locations, they are applicable to the wetter parts of semiarid regions and to irrigated crop production in arid and semiarid regions. Hence they are described here briefly.

In the fall (autumn) plow, field-cultivate system, a moldboard plow is used for primary tillage and secondary tillage is reduced to one shallow cultivation with sweeps at planting. A disk or rotary tiller may be substituted for sweep cultivation to produce a finer and firmer seedbed. However, this causes the soil to be more erodible.

In the spring plow, wheel-track-plant system, strip seedbed preparation is performed on soil that was initially plowed only 12-14 h before planting. By planting soon after plowing, major soil drying is avoided and the soil water content at planting is such that the wheels break the clods and firm the seedbed. Seedbed preparation and planting are done in a combined operation. Planted rows may be in tractor or planter wheel tracks. Soil erosion with this system is less than with fall plowing because crop residues remain on the surface until plowing in spring.

The fall (autumn) chisel, field-cultivate system is similar to the fall plow, field-cultivate system, except that moldboard plowing is replaced by chiseling. Because chiseling retains more surface residues than moldboard plowing, the chisel system provides greater erosion control. Variations of this system are chisel-plant, chisel-secondary tillage (with disk, harrow, or cultivator), combination coulter or disk-chisel, and alternate chisel-moldboard plow (Oschwald, 1973).

Tillage in the disk and plant system is accomplished with medium-weight tandem disks operated 8-10 cm deep, by heavy disks operated 15-20 cm deep, or a combination of the two. Usually one disking in fall is followed by one or two diskings in spring before planting. To maintain surface residues for soil and water conservation purposes, disking should be delayed as long as practical and tandem rather than heavy disks should be used because tandem disks do not penetrate as deeply or incorporate as much residues as heavy disks.

Several till-plant systems are available to accomplish tillage and planting in a combined operation. The system developed in Nebraska employs wide sweeps operated 5-8 cm deep on the ridge remaining from the previous crop. This tillage moves old stalks and root clumps into the zone between rows, thus providing a trash-free zone on the ridges for planting. The ridges are formed when the previous crop is cultivated or after harvest with rolling or disk-hiller cultivators, large disk cultivators, or disk bedders. Where heavy disks are used to cut residues and level old ridges, new ridges must be formed each year. The only tillage required where ridges are maintained from year to year is reshaping the ridges with a rolling or disk-type cultivator in the fall or spring.

A variation of this system involves alternating the position of the row and the interrow zone from year to year. In this system, a ridge is initially formed between the rows at planting. At subsequent plantings, the ridges are split and become the planted row. During the ridge-splitting and planting operation, previous crop residues are buried in the new ridge, where they decay during the growing season and help provide a favorable seedbed for the next crop.

In a strip tillage system, only a narrow band of soil is tilled, often with a rotary tiller from which some blades have been removed. A typical strip tillage zone is about 20 cm wide and 5-10 cm deep. When a standard planter is attached to the tiller, tilling and planting can be combined because the tiller chops the residues from the previous crop.

Other reduced tillage systems that maintain crop residues on the surface for at least a part of the crop production cycle include the lister plow-plant, rotary till-plant, and sweep plow-plant systems. These and possibly others are variations of previously discussed systems and differ from them mainly with respect to the implement used. The choice of a system depends on the equipment available, the soil and climatic conditions, the crop to be grown, the size and type of farming operation, and the producer's managerial ability and personal preferences (Griffith et al., 1977).

B. Tillage-Herbicide Combination Systems

When weeds are controlled with herbicides during the interval between crops, the need for tillage is reduced, thus resulting in greater residue retention on the soil surface. Tillage-herbicide combinations have received considerable attention, especially where residue production is low, erosion potential is high, water conservation is important, and troublesome weeds cannot be controlled effectively by either tillage or herbicides alone. There are many tillage-herbicide combination systems. Common practices are to (1) use tillage initially to control existing weeds, loosen the soil, or incorporate some residues, and then use herbicides for subsequent weed control; (2) control weeds initially with herbicides so that more residues can be maintained on the surface during highly erosive periods, and then use tillage to prepare a seedbed as planting time approaches; or (3) use tillage for one crop and herbicides for the second crop in two-crop rotations. The tillage is often performed with sweeps or blades that undercut the surface, thereby controlling weeds and retaining most crop residues on the surface (Phillips, 1969; Smika and Wicks, 1968; Wicks and Smika, 1973; Woodruff, 1972).

Major deterrents to the adoption of tillage-herbicide combination treatments are the high cost and limited availability of some herbicides and the limited availability of suitable herbicide application equipment, especially in developing countries. Of increasing concern worldwide is the effect of increased chemical (including herbicide) usage on the environment. Problems with herbicide use in crop production systems are discussed in greater detail in Sec. II.D.

C. Stubble-Mulch Tillage

Stubble-mulch tillage refers to tilling a soil in such a manner that crop residues are maintained on the soil surface. It is also referred to as mulch farming, trash farming, mulch tillage, and plowless farming (SSSA, 1973). The stubble-mulch tillage and farming system was developed to combat erosion by wind that occurred during a major drought in the Great

TABLE 2 Effect of Tillage Machines on Small Grain Surface Residues Remaining After Each Operation

Tillage machine	Approximate residue maintained (%)
Subsurface cultivators: wide-blade cultivator and rodweeder	90
Mixing-type cultivators: heavy-duty cultivator, chisel, and other types of machines	75
Mixing and inverting disk machines: one-way flexible disk harrow, one-way disk, tandem disk, offset disk	50
Inverting machines: moldboard plow, disk plow	10

Source: From Anderson (1968).

Plains of the United States and Canada during the 1930s. With stubble-mulch tillage, residues remained anchored in the surface soil and thus helped control erosion by wind. The surface residues were soon also found to be effective for controlling erosion by water and for conserving water (Duley and Russel, 1942; McCalla and Army, 1961).

Any tillage that retains plant residues on the soil surface could, in a broad sense, be classified as stubble-mulch tillage (Stallings, 1957). Tillage that undercuts surface residues to loosen soil and control weeds is considered stubble-mulch tillage. This, however, does not preclude using disks or other implements for the first operation to incorporate some residues with soil when unusually large amounts are present (Stallings, 1957; Unger and McCalla, 1980). Such tillage hastens residue decomposition, but sufficient residues are retained on the surface to control erosion. Besides disks, other implements that reduce large amounts of residue are stubble pulverizers and busters (Jacks et al., 1955) and mulch traders, skew-treaders, or spike-tooth harrows used in conjunction with one-way disks and sweep plows (Papendick and Miller, 1977). These implements should be used with caution and in such a manner that sufficient surface residues are maintained to adequately protect the soil against erosion.

Tillage implements that loosen a soil and retain most residues on the surface include V-sweeps (60 cm or wider), straight-blade machines, chisel plows, one-way disk plows (used when large amounts of residue are present), and rodweeders (Unger and McCalla, 1980). Typical amounts of surface residue remaining after each operation with various implements are shown in Table 2. Approximate amounts of surface residues needed to control erosion are given in Table 3 for various soils.

In the Great Plains, initial tillage with sweep or blade inplements is usually 12-15 cm deep with subsequent tillage at successively shallower depths, which helps to provide a firm, well-prepared seedbed for the next crop. Where residue amounts are large, as in the dry-farmed areas of the Pacific Northwest or in more humid regions, the second operation may be deeper than the first to avoid serious clogging. However, subsequent operations again are at shallower depths to provide a good seedbed (Hanway, 1970).

TABLE 3 Approximate Amounts of Residue Needed to Maintain Erosion Below a Tolerable Level of 11 Mg ha^{-1} on Various Types of Soil

Soil texture[a]	Water erosion Flat wheat residue (Mg ha^{-1})	Wind erosion (Mg ha^{-1}) Wheat residue		Growing wheat		Sorghum residue	
		Standing	Flat	Standing	Flat	Standing	Flat
Silts	1.6	0.5	1.0	0.6	0.5	2.0	2.9
Clays and silty loams	2.1	0.9	1.8	1.1	0.9	3.7	5.3
Loamy fine sands	1.0	1.2	2.4	1.3	1.0	4.7	6.9

[a]Silts with 50%, clay and silty clay with 25%, and loamy sand with 10% nonerodible fractions (greater than 0.84 mm diameter).

Source: From Anderson (1968) and Fenster (1973).

Stubble-mulch tillage is adaptable to all soil types, but coarse-textured soils require more surface residues than fine-textured soils to control erosion by wind (Fenster, 1973; Hanway, 1970; McCalla and Army, 1961). Hence, erosion by wind would be more easily controlled by stubble-mulch tillage on fine- than on coarse-textured soils. In addition, where residues amounts are low, a rough, cloddy surface that aids control of erosion by wind usually can be produced by tillage. Such roughened surfaces provide additional protection against erosion by wind in emergency situations.

Control of soil erosion by water with stubble-mulch tillage should be easier on coarse- than on fine-textured soils because water usually infiltrates more rapidly into coarse-textured soils. Hence, less surface residue would be required on coarse-textured soils to keep erosion by water at tolerable levels.

Based on wheat yield ratios with stubble-mulch and clean tillage at various western U.S. locations, the stubble-mulch system is adapted better to arid and semiarid climatic regions than to subhumid and humid regions (Zingg and Whitfield, 1957). Higher yields with stubble-mulch tillage at the drier locations appear related to lower nitrification. This prevents excessive plant growth and, in combination with greater water infiltration, improves the water-fertility balance for the crop (Johnson, 1950; Zingg and Whitfield, 1957). Improved weed control at the drier locations is also undoubtedly a factor.

In that part of the Great Plains where wheat yields were increased by stubble-mulch tillage, plant growth and yields are usually limited by water rather than by nutrient deficiencies. Consequently, lower nitrification early in the growing season when more water may be available limits plant growth and thereby soil water use. As a result, more water is available for crop use during the grain production period (Zingg and Whitfield, 1957). At more humid locations where fertility rather than water is the yield-limiting factor, lower nitrification with stubble-mulch tillage results in lower yields. Increased fertilizer applications can usually overcome the yield depressions. Zingg and Whitfield (1957) reported that the crop response to applied nitrogen at a subhumid location was greater on stubble-mulch tillage than on moldboard-plowed areas.

D. No-Tillage

In ancient times, seed was planted in soil with little or no soil preparation. Tillage intensity increased as the detrimental effects of weeds and poor soil conditions on crops were realized. The trend toward increasing tillage was reversed in many cases in recent years, and probably the first no-tillage experiment in recent times involved overseeding of a legume into an unproductive grass sod without tilling the soil. The grass was closely grazed or burned and the legume was seeded at high rates to reduce competition between the unwanted grass and the legume. Trampling by grazing animals resulted in good seed-soil contact. This experiment showed that no-tillage production was possible if unwanted vegetation could be controlled. A major advance regarding no-tillage occurred when chemicals that destroyed existing vegetation and had little or no residual effect on the crop being established were introduced in the 1950s (Baeumer and Bakermans, 1973). This soon resulted in no-tillage research with field crops being started at several locations in the United States (Baker et al., 1956; Barnes et al., 1955; Phillips, 1954; Wiese et al., 1960). No-tillage research was greatly expanded in the late 1960s and early 1970s after widespread publicity concerning the system. Since then, many aspects of no-tillage crop production have been investigated throughout the world and suitable systems are available for many crops. Much information regarding no-tillage has been published in recent years, which is relied on in this discussion of the advantages and disadvantages of no-tillage systems.

Advantages

The advantages of no-tillage systems compared to clean tillage or even other reduced tillage systems include improved erosion control, increased use of land, increased water conservation, equal or higher crop yields, reduced energy requirements, reduced equipment inventories, reduced equipment wear and tear, reduced labor requirements, and greater net returns. These advantages have been discussed in detail by Phillips and Phillips (1984), Phillips and Young (1973), Sprague and Triplett (1986), and Unger (1984a), and only brief comments are made in this chapter.

Improved control of erosion with no-tillage results from more crop residues being retained on the surface than with other tillage systems. Of course, sufficient residues must be available to achieve maximum effectiveness of no-tillage systems. For controlling erosion by wind, residues shield the soil against the forces of wind, thus reducing wind speeds at the soil surface to below the threshold velocity for erosion to occur. Improved control of erosion by water with no-tillage systems is related to reduced rates and amounts of runoff across the surface and to the surface cover provided by residues, which reduces soil detachment and transport due to raindrop impact and flowing water. Increased land use is related to the improved erosion control resulting from use of the no-tillage system. As a result, lands unsuitable for crop production by conventional methods can often be cropped safely by the no-tillage method (Phillips, 1980).

Improved water conservation with no-tillage results from surface residues that retard the flow of water across the surface, thereby providing more time for water infiltration, and from the surface protection provided by residues. These residues dissipate the energy of falling raindrops, thus minimizing soil aggregate dispersion and surface sealing and, thereby, maintaining favorable rates of water infiltration into soil. Surface residues also reduce soil water evaporation. As a result, more stored water is retained for later crop use. However, reduced runoff, increased infiltration, and reduced evaporation do not assure greater water storage for subsequent use by plants because a soil can retain only a given amount of water. The excess water either percolates or seeps from the soil. Although a soil cannot retain more water than its storage capacity, reduced runoff and evaporation and greater infiltration increase the potential of refilling the soil's storage reservoir after plants have extracted some of the water.

Many factors affect crop yields, and any unfavorable condition can reduce yields. However, if no major problems are encountered, yields with no-tillage are usually comparable to or higher than with other tillage methods. Increased yields with no-tillage often result from improved soil water and/or temperature conditions. The effects of no-tillage on soil water were just discussed. In arid to semiarid regions, high soil temperatures may be detrimental to plants and thus lowering soil temperatures could improve crop yields. Lower soil temperatures (up to 10°C) with no tillage than with clean tillage were reported by Allen et al. (1975) and Rockwood and Lal (1974).

The advantages of reduced equipment inventories, equipment wear and tear, energy requirements, and labor requirements are interrelated. Because less equipment is used in no-tillage systems, equipment inventories are lower. Also, smaller tractors that are used less frequently are often adequate for no-tillage systems. Consequently, wear and tear on the equipment, energy use, and labor requirements are also reduced.

As for crop yields, many factors affect net returns for crop production systems. However, if yields with no-tillage are equal to or exceed those with other tillage systems and if herbicides control weeds at a cost similar to that of tillage, then net returns are usually greater with no-tillage because of the lower equipment inventories, equipment wear and tear, energy use, and labor requirements (Crosson et al., 1986; Harman and Martin, 1987).

Disadvantages

Although no-tillage systems have some major advantages, they also have some disadvantages, including poor adaptability to poorly drained soils, low soil temperatures in cool climates, increased use of chemicals, shift in weed populations, carryover effect of herbicides, adverse effect of herbicides on adjacent crops, poor weed control with herbicides, high cost of herbicides, limited water for spraying, unavailability of suitable equipment (sprayers and planters), potential for greater pest problems (insects, diseases, and rodents), limited residues, increased soil compaction, and the need for greater managerial ability of producers. Fortunately, not all disadvantages apply to all situations. The disadvantages regarding no-tillage systems have been discussed by Phillips and Phillips (1984), Phillips and Young (1973), Sprague and Triplett (1986), and Unger (1984a) and are discussed only briefly in this chapter.

Poor adaptability of no-tillage to poorly drained soils is related to its water-conserving benefits. Because of lower runoff, greater infiltration, and lower evaporation, the already high soil water content and poor aeration conditions of poorly drained soils are aggravated by no-tillage, and the system usually is not recommended for poorly drained soils because yields are often lower with no-tillage under these conditions (Triplett and Van Doren, 1977).

Cooler temperatures with no-tillage are beneficial in hot regions, but they are a disadvantage in cool regions where planting is often delayed until soil temperatures become favorable for rapid seed germination and seedling development (Unger and Stewart, 1976). For cool climates, as in the northern United States, ridging the soil for the previous crop resulted in warmer soil temperatures for the next crop when it was planted on the ridges than when it was flat planted (Radke, 1982).

No-tillage systems depend on chemicals (herbicides) to control weeds; more chemicals may also be used as insecticides and fertilizers. Increased chemical use may increase production costs, which may make no-tillage systems impractical for producers with limited capital. Also, suitable chemicals may not be available. If available and used, greater chemical use may increase the pollution of groundwater or downstream water supplies, especially where readily soluble materials, such as nitrogen (N) fertilizer, are used in large quantities and where phosphorus (P) fertilizers and other chemicals are surface applied (Baker and Johnson, 1979; Kang et al., 1980; McDowell and McGregor, 1980; Triplett et al., 1978).

Less runoff from no-tillage areas usually results in lower chemical losses than from tilled areas, but N losses may be higher because of the greater percolation of water through the profile with no-tillage. Losses can be reduced by making several small applications rather than one large application. Phosphorus losses are usually decreased with no-tillage because of lower losses of soil to which the P is adsorbed. However, McDowell and McGregor (1980) found greater P losses with no-tillage because of incomplete P adsorption by sediments, greater application rates, decreased incorporation, release of P from residues, and possibly a greater P-supplying capacity of sediments in runoff from no-tillage areas. Herbicide and insecticide losses from no-tillage areas are influenced strongly by application rates and the intervals of time between application and runoff events. Herbicide losses were greatest when applied at high rates or when runoff occurred soon after application (Baker and Johnson, 1979; Edwards et al., 1980; Triplett et al., 1978).

Since herbicides control weeds in no-tillage systems, their mode of action and the weeds to be controlled are factors that determine which herbicide can be used in a particular cropping system. The herbicide used must be compatible with crops to avoid damage. If not compatible, the present crop is damaged immediately, future crops may be damaged by herbicide residues in soil, and adjacent crops may be damaged by spray drift. In all cases, herbicides must be used correctly.

Most herbicides control specific weeds or groups of weeds. Excellent results are often achieved, but shifts in weed populations sometimes cause problems because not all weeds are controlled as planned (Phillips, 1969; Richey et al., 1977; Wiese and Staniforth, 1973). In other cases, even adapted herbicides sometimes do not provide good control, especially of troublesome weeds. In such cases, tillage (cultivation) would be beneficial, but that may be difficult because of the surface residues and the unloosened soil. As a result, no-tillage is not usually recommended where troublesome weeds are a problem. The development of improved herbicides could minimize these problems (Richey et al., 1977). Other systems for crop production in these situations are tillage-herbicide combinations or crop rotations. The latter may allow the use of a greater array of effective herbicides.

The high cost of some herbicides is a major disadvantage of no-tillage systems in many cases, especially in developing countries, but also in developed countries. In such cases, the relative costs of herbicides, tillage, or hand labor usually determine which production system is used. Other problems facing producers in developing countries with regard to no-tillage are the limited availability of herbicides, poor quality or limited amounts of water for diluting the herbicides, and lack of suitable sprayers and planters. Major advances in no-tillage planting equipment have been made in recent years, but planting problems sometimes still occur, even in developed countries.

The potential for increased pest problems with no-tillage has been recognized and widely discussed, but no major differences compared with clean tillage are usually encountered. Some insect and disease problems are greater with no-tillage; others are not. For the greenbug (*Schizaphis graminum* Rondani), an aphid that infests wheat and grain sorghum [*Sorghum bicolor* (L.) Moench] in the southern and central Great Plains, populations and damage were much lower on both crops with no-tillage than with clean tillage (Burton et al., 1987). Rodent problems are sometimes greater in no-tillage systems because of the shelter provided by the surface residues.

A major factor that thwarts the success of no-tillage systems in arid and semiarid regions is low residue production by nonirrigated crops. Aggravating the situation is major residue use for fuel, feed, or other purposes, especially in developing countries. In other cases, residues may be destroyed by termites or burned (Lal, 1975; Okigbo and Lal, 1977). Where residues are limited and the soil is eroded or degraded, a no-tillage system is not satisfactory (Charreau, 1977; Lal, 1980a). However, soils can be readied for no-tillage crop production by growing suitable cover crops for several years to improve soil structure and water infiltration (Lal, 1980b).

A major reason for performing tillage is to loosen a soil and improve its condition for crop establishment and growth and to improve water infiltration. Therefore, by inference, no-tillage should result in unfavorable soil conditions. However, except on poorly drained soils, on some sandy soils, under low-temperature conditions, or where weed control is poor, crop yields are generally favorable and water infiltration often is higher on no-tillage than on tilled areas (Unger, 1984a). The generally favorable results with no-tillage are attributed to the fact that soil compaction, as measured by bulk densities, is often not much higher with no-tillage than with other tillage methods after the tilled soils have become settled by rainfall, except where the surface soil has been compacted by equipment traffic (tractor, implement, or harvester) or by grazing animals. Such compaction can usually be alleviated by a sweep or chisel implement operation without destroying a significant part of the surface residues. As a result, water conservation was similar where no-tillage and limited tillage (e.g., one sweep tillage) were used (Unger, 1981; Unger et al., 1971).

As a rule, a higher management level is required for no-tillage than for conventional tillage crop production. Because no-tillage systems are, in many cases, vastly different from conventional tillage systems, farmers have not acquired the experience necessary for successful no-tillage crop production. For success with no-tillage, farmers must know how equipment functions in a given situation, what herbicide to use and what can be done if herbicides fail, and what to do when insect and disease problems become severe. To gain experience, producers should usually try no-tillage on a limited area before devoting their entire resources to this farming technique (Unger, 1984a).

III. WEED CONTROL

Weed control in reduced tillage systems depends on the system being used and, to large extent, on soil water conditions and weed growth stage when the weed control operation is performed. Where residues are incorporated (Sec. II.A), the entire surface is tilled one or more times before planting, thus controlling weeds present when the operation is performed. However, troublesome weeds may regrow and other weeds may emerge after the tillage operation or even after planting the crop. Such weeds can often be controlled with herbicides. In reduced tillage systems where residues are incorporated, weeds can be controlled by cultivation because few residues are present to interfere with the operation. Where weed control is poor, water conservation and crop yields are also poor (Woodruff, 1972), the same as with conventional systems.

In tillage-herbicide combination systems, herbicides are used to control some weeds that normally would be controlled by tillage. To be effective, herbicides must uniformly contact weeds and the exposed soil. Where large amounts of residue are present, nonuniform herbicide application can result in poor weed control (Richey et al., 1977). Poor weed control due to herbicide interception by residues possibly contributed to poor water conservation and crop yields in studies by Phillips (1969) and Wicks and Smika (1973), but complete weed and volunteer wheat control occurred when atrazine [2-chloro-4-(ethylamino)-6-(isopropylamino)-*s*-triazine] was applied to areas having up to 11 Mg ha^{-1} of standing wheat residues on the surface (Unger et al., 1971; Unger and Wiese, 1979).

Weed control with stubble-mulch tillage is highly dependent on soil water content at the time of tillage and the occurrence of rainfall after tillage. Stubble-mulch tillage usually provides good weed control in drier locations, especially if no rain occurs soon after tillage (Zingg and Whitfield, 1957). However, because stubble-mulch tillate severs only the main roots, weeds may survive if secondary roots remain in moist soil or if rainfall occurs before the weeds die (Hanway, 1970). Treaders used in conjunction with stubble-mulch tillage improve weed control, especially when soils are relatively dry when tillage is performed (Hanway, 1970; Zingg and Whitfield, 1957).

Where rainfall is more frequent, weed control is usually more difficult with stubble-mulch tillage than with clean tillage. In the U.S. Great Plains and Pacific Northwest, the weeds most difficult to control with stubble-mulch tillage where wheat is grown are downy brome grass (*Bromus tectorum* L.) and cheat grass (*Bromus secalinus* L.) because these weeds have growing seasons similar to that of wheat. The use of treaders improved control of these weeds (Zingg and Whitfield, 1957). These weeds can also be controlled with herbicides, by growing wheat in rotation with a crop such as sorghum, or by occasional use of clean tillage (Hanway, 1970).

For success with no-tillage, herbicides must satisfactorily control weeds that compete with crops for water, nutrients, space, and light. Some herbicides control only certain types of weeds (broadleaf or grassy); others (contact herbicides) control all vegetation growing at the time of application. Generally good control is achieved when appropriate herbicides are applied to susceptible weeds that are small and growing rapidly. Large weeds are often difficult to control. All weeds are usually more difficult to control when they are under stress due to water deficits or low or high temperatures.

For weed control during the crop's growing season, a pre- or postemergence herbicide can be used. Additional weed control may be required after the crop is established, which often can be achieved by spraying with a suitable herbicide. When the appropriate herbicide would damage the crop, satisfactory control can often be achieved by spot spraying the weeds or by using a rope wick applicator to control weeds that are taller than the crop (Dale, 1980; Wiese and Lavake, 1980). Such herbicides as glyphosate [*N*-(phosphonomethyl)glycine] and paraquat (1,1'-dimethyl-4,4'-bipyridinium ion) are well-adapted to controlling weeds by spot spraying or with a rope wick applicator.

IV. SOIL CONDITIONS AND PROPERTIES

Soil conditions and properties affected by tillage are numerous and include water content, temperature, nutrients, biological regime, structure, erodibility, and salinity. The effects of tillage on soil conditions and properties are discussed jointly, not for each system separately.

A. Water

Semiarid and arid regions receive too little precipitation for good crop production. Sometimes much of the water is lost by runoff or evaporation, and some is used by weeds. Hence, water conservation is usually a primary objective of the tillage system used in semiarid and arid regions.

Surface residues greatly influence water conservation, by influencing both infiltration and evaporation. Water conservation generally increases with increasing amounts of surface residues (Tables 4 and 5) (Greb et al., 1967, 1970; Unger, 1978a). The immediate effect of surface residues with respect to water conservation is greater infiltration (less runoff), as illustrated by data in Tables 6, 7, and 8. Runoff was usually lower with reduced tillage system than with plowing. However, plowing increased infiltration when no residues were involved (Tables 8 and 9) or when surface residue amounts were low (personal communication, O. R. Jones, Bushland, Texas).

TABLE 4 Straw Mulch Effects on Soil Water Storage Efficiency at Sidney, Montana; Akron, Colorado; and North Platte, Nebraska from 1962 to 1965

Mulch rate ($Mg\ ha^{-1}$)	Fallow period precipitation (mm)	Water storage efficiency (%)
0	355	16
1.7	355-549	19-26
3.4	355-648	22-30
6.7	355-648	28-33
10.1	648	34

Source: From Greb et al. (1967).

The value of surface residues for reducing evaporation under field conditions was illustrated by Smika (1976a, shown by Smika and Unger, 1986), who compared evaporation from conventional, minimum, and no-tillage areas following 165 mm rainfall. Water contents were similar on the day after rainfall (Fig. 1A), but after 34 rainless days, water content was lowest with conventional tillage and highest with no-tillage (Fig. 1B). Surface residue amounts were 1.2, 2.2, and 2.7 $Mg\ ha^{-1}$ with conventional, minimum, and no-tillage treatments, respectively.

The combined effect of greater infiltration and lower evaporation is greater soil water storage, provided that weeds are controlled and that

TABLE 5 Straw Mulch Effects on Soil Water Storage During Fallow,[a] Water Storage Efficiency and Grain Sorghum Yield at Bushland, Texas, 1973-1976

Mulch rate ($Mg\ ha^{-1}$)	Water storage[b,c] (mm)	Storage efficiency[b,c] (%)	Grain yield[c] ($Mg\ ha^{-1}$)
0	72 A	22.6 A	1.78 A
1	99 B	31.1 B	2.41 B
2	100 B	31.4 B	2.60 B
4	116 B	36.5 B	2.98 B
8	139 C	43.7 C	3.68 C
12	147 C	46.2 C	3.99 C

[a]Fallow duration of 10-11 months.

[b]Water storage determined to 1.8 m depth. Precipitation averaged 318 mm.

[c]Values followed by the same capital letter are not significantly different at the 5% level (Duncan's multiple-range test).

Source: From Unger (1978a).

TABLE 6 Runoff and Soil Loss as Influenced by Water Applied,[a] Time of Application, Type of Fall Tillage, and Previous Crop (Illinois)

Time[b] (min)	Water applied (cm)	Fall moldboard plow		Disk-chisel		No fall tillage	
		Corn	Soybean	Corn	Soybean	Corn	Soybean
		Runoff (mm)					
60	64	30	39	1	21	23	32
90	95	58	69	8	51	46	60
120	127	86	96	29	83	71	88
		Soil loss (Mg ha^{-1})					
60	64	42	109	0.6	28	4	14
90	95	86	180	4	52	8	26
120	127	127	256	14	75	11	39

[a]Simulated rainfall was applied at a rate of 65 mm h^{-1} after overwinter soil weathering but before any spring tillage.

[b]Cumulative time from start of water application.

Source: From Griffith et al. (1977).

TABLE 7 Runoff and Soil Loss From Bedford Silt-Loam (9% Slope) During 1 h Artificial Rainstorms, 1972[a] (Illinois)

Tillage system	Runoff (mm)		Soil loss (Mg ha^{-1})	
	Hour 1	Hour 2	Hour 1	Hour 2
Spring plow, disk, plant	45	56	23.3	27.1
Spring chisel, field cultivate, plant	11	38	0.7	2.5
Till-plant	37	50	7.4	8.5
No-tillage	26	39	3.6	3.8

[a]Storms of 65 mm h^{-1} were applied within 4 weeks after planting corn in rows that ran across the slope. Data are averages of two replications.

Source: From Griffith et al. (1977).

TABLE 8 Effect of Tillage on Runoff and Soil Losses from Land Cropped to Maize in Nigeria[a]

	Bare fallow		Plowed		No tillage	
Slope (%)	Runoff (%)	Soil loss (Mg ha^{-1})	Runoff (%)	Soil loss (Mg ha^{-1})	Runoff (%)	Soil loss (Mg ha^{-1})
1	18.8	0.2	8.3	0.04	1.2	0.0007
5	20.2	3.6	8.8	2.16	1.8	0.0007
10	17.5	12.5	9.2	0.39[b]	2.1	0.0047
15	21.5	16.0	13.3	3.92	2.2	0.0015

[a]Rainfall was 44.2 mm.

[b]Probably an error.

Source: From Rockwood and Lal (1974).

TABLE 9 Effect of Tillage-Induced Plow Layer Porosity and Surface Roughness on Cumulative Infiltration of Simulated Rainfall

	Potential water storage volume due to		Cumulative infiltration[c] (mm) to		
Tillage treatment[a]	Pore space[b] (mm)	Surface roughness (mm)	Initial runoff	25 mm Runoff	50 mm Runoff
Untilled	81	8	9	21	24
Plow	137	50	171	217	230
Plow-disk-harrow	124	25	53	73	84
Cultivated	97	29	57	83	91
Rotovated	117	15	24	38	41

[a]Plowing and rotovating performed to a 15 cm depth, cultivating to a 7.5 cm depth on untilled soil.

[b]Measured to tillage depth.

[c]Water applied at a 127 mm h^{-1} rate.

Source: From Burwell et al. (1966).

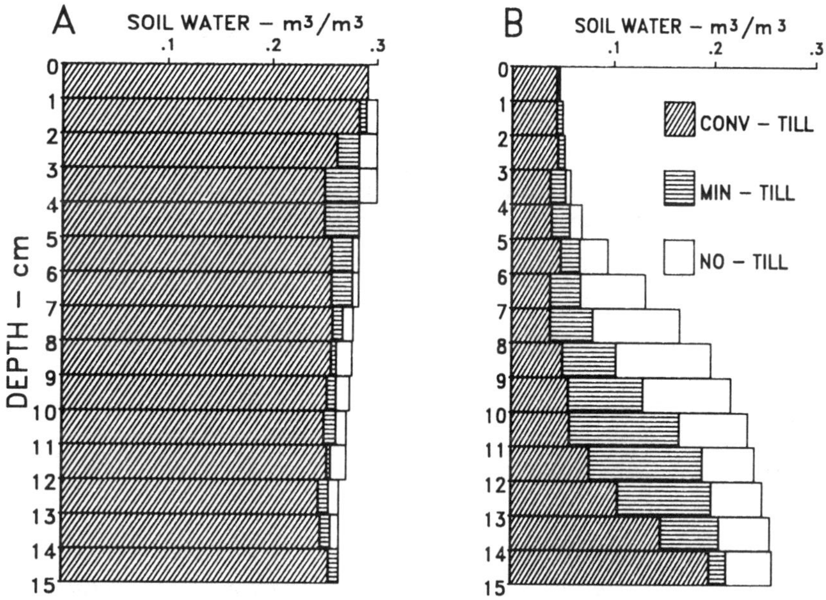

FIGURE 1 Soil water content to a 15 cm depth 1 day (A) and 34 days (B) after 165 mm rainfall as influenced by tillage treatments (CONV-TILL, conventional tillage; MIN-TILL, minimum tillage; NO-TILL, no-tillage). (From Smika and Unger, 1986. Reproduced from *Cropping Strategies for Efficient Use of Water and Nitrogen*, ASA Special Publication 51, 1988, pp. 69-100, by permission of the American Society of Agronomy, Inc., Crop Science Society of America, Inc., and Soil Science Society of America, Inc.)

the soil has the capacity to store the water. The following are examples of reduced tillage effects on soil water storage.

Smika (1976b) summarized the effects of clean and stubble-mulch tillage on soil water gain during fallow at several Great Plains locations (Table 10). At the end of fallow (at wheat planting), the average increase with stubble mulch compared with clean tillage was 27 mm. The progressive improvement in water storage resulting from adoption of improved tillage methods since 1916 in Colorado is shown in Table 1 (Greb, 1979).

In Nebraska, tillage-herbicide combination treatments significantly increased soil water storage and sorghum yields, but not wheat yields, in a 3 year wheat-fallow-sorghum rotation (Table 11). However, tillage-herbicide combination and herbicide only (no-tillage) treatments increased water storage and wheat yields compared with a plow treatment in a 2 year wheat-fallow system (Tables 12 and 13). The yield increases were attributed to increased water storage during fallow (Smika and Wicks, 1968; Wicks and Smika, 1973).

In an irrigated winter wheat-dryland (nonirrigated) grain sorghum rotation (two crops in 3 years), water storage during fallow after wheat, sorghum grain yield, and water use efficiency were highest with no-tillage (Table 14) (Unger and Wiese, 1979). For an irrigated winter wheat-dryland sorghum-dryland sunflower (*Helianthus annuus* L.) rotation, water

TABLE 10 Net Gain in Soil Water During Fallow with Clean and Stubble-Mulch Tillage at Seven Central Great Plains Locations

Location	Years of data (no.)	Tillage method: Clean (mm)	Tillage method: Stubble mulch (mm)
Akron, Colorado	6	142	173
Colby, Kansas	4	115	141
Garden City, Kansas	6	86	90
Oakley, Kansas	4	82	131
North Platte, Nebraska	8	146	203
Alliance, Nebraska	8	29	32
Archer, Wyoming	2	28	42
Weighted average		95	122

Source: From Smika (1976a).

TABLE 11 Effect of Tillage and Herbicide Treatments on Soil Water Content at the End of the Fallow Period[a] and on Wheat and Sorghum Yields in a 3-Year Wheat-Fallow-Sorghum Rotation, Nebraska[a]

Treatment from wheat harvest to sorghum planting: Fall	Treatment from wheat harvest to sorghum planting: Spring	Treatment from sorghum harvest to wheat planting	Soil water gain[b] (mm)	Grain yields ($Mg\ ha^{-1}$): Wheat	Grain yields ($Mg\ ha^{-1}$): Sorghum
Subtillage	Disk	Subtillage (5)[c]	186 A	3.49 B	4.08 A
Subtillage	Atrazine	Subtillate (4)	213 A,B	3.76 B	4.20 A
Atrazine	Atrazine	Subtillage (4)	211 A,B	3.63 B	4.58 A,B
Atrazine	Atrazine	Contact herbicide (4-6)	223 B	3.49 B	4.89 B
Subtillage	Atrazine	Contact herbicide (4-6)	216 A,B	3.63 B	5.02 B

[a]Fallow duration of about 11 months. Values followed by the same capital letter(s) are not significantly different.

[b]Determined to a 3 m depth.

[c]Values in parentheses denote number of operations.

Source: From Smika and Wicks (1968).

TABLE 12 Effect of Tillage and Herbicide Treatments on Soil Water Content at the End of the Fallow Period[a] and on Wheat Yields in a 2 Year Wheat-Fallow Rotation, Nebraska

Operations during fallow: Initial operation following wheat harvest	Operations during fallow: Subsequent operations	Soil water gain[b] (mm)	Grain yield (Mg ha^{-1})
Plow	Subtillage (5)[a]	186 C	3.09 A
Subtillage	Subtillage (5)	238 A	3.36 A,B
Atrazine followed by subtillage	Subtillage (5)	272 A	3.29 A,B
Atrazine	Subtillage (4)	275 A	3.36 A,B
Atrazine	Contact herbicide (4-6)	325 B	3.56 B

[a]Fallow duration of about 14 months. Values followed by the same capital letter(s) are not significantly different.

[b]Determined to a 3 m depth.

[c]Values in parentheses denote number of operations.

Source: From Smika and Wicks (1968).

TABLE 13 Effect of Tillage and Herbicide Treatments on Operations to Control Weeds, Surface Residues, Soil Water Storage During Fallow,[a] and Wheat Yields in a 2 Year Wheat-Fallow Rotation, Nebraska

Treatment	Operations during fallow: Tillage[b] (no.)	Operations during fallow: Herbicide application (no.)	Residues maintained[c] (%)	Soil water gain[d] (mm)	Grain yield (Mg ha^{-1})
Plow	8.5	0.0	0	146	2.69
Stubble mulch	8.7	0.0	21	203	2.88
Atrazine + stubble mulch	7.6	1.4	21	215	2.91
Atrazine + contact herbicide + stubble mulch	5.1	2.8	25	237	3.04
Atrazine + contact herbicide	0.0	6.0	46	274	3.17

[a]Fallow duration of about 14 months.

[b]The plow treatment included one moldboard plowing in the spring. Other tillage was with a sweep implement.

[c]Average amount of residues at start of fallow was 6.6 Mg ha^{-1}.

[d]Determined to a 3 m depth.

Source: From Wicks and Smika (1973).

TABLE 14 Tillage Effects on Water Storage During Fallow After Wheat Harvest, Sorghum Grain Yields, and Water Use Efficiency in an Irrigated Winter Wheat-Fallow-Dryland Grain Sorghum Cropping System, Bushland, Texas, 1973-1977[a]

Tillage method	Water storage (mm)	Grain yield (Mg ha^{-1})	WUE[b] (kg m^{-3})
No-tillage	217 B	3.14 B	0.89 B
Sweep	170 A	2.50 A	0.77 A
Disk	152 C	1.93 C	0.66 C

[a] Values followed by different capital letters are significantly different at the 5% level, based on Duncan's multiple range test.

[b] Water use efficiency based on grain yields, growing season precipitation, and soil water changes.

Source: From Unger and Wiese (1979).

TABLE 15 Effect of Tillage Method on Average Soil Water Storage During Fallow After Irrigated Winter Wheat and On Subsequent Rainfed Grain Sorghum Yields at Bushland, Texas, 1978-1983[a]

Tillage method	Water storage[b] (mm)	Grain yield (Mg ha^{-1})	WUE[c] (kg m^{-3})
Moldboard	89 A	2.56 A,C	0.71
Disk	109 A	2.37 A,D	0.65
Rotary	85 A	2.19 D	0.61
Sweep	114 A,B	1.77 A	0.72
No-tillage	141 B	3.34 B	0.83

[a] Fallow duration of 10-11 months. Values followed by the same capital letters are not significantly different at the 5% level based on Duncan's multiple range test.

[b] Measured to 1.8 m soil depth.

[c] Water use efficiency based on grain yield, growing season precipitation, and soil water changes.

Source: From Unger (1984b).

storage during fallow after wheat and sorghum grain yields again were highest with no-tillage (Table 15) (Unger, 1984b). In both studies, total water use was highest with no-tillage, but this was more than offset by the higher water content at sorghum planting, suggesting that soil water contents were higher with no-tillage throughout the growing season. Such higher water contents allow no-tillage sorghum to respond more favorably to growing season rainfall. Unger et al. (1986) showed that sorghum grain yields increased in response to growing season rainfall as the amounts of surface residues increased.

Surface residues significantly increased overwinter water storage in the Pacific Northwest. At Lind, Washington, water storage was 30 and 50 mm greater with 5 and 11 Mg ha^{-1}, respectively, of wheat residue on the surface from September to March than with no residues (Papendick and Miller, 1977). However, greater evaporative losses during the dry summer with residue treatments than with bare soil negated about half the overwinter gains.

At Pendleton, Oregon, also in the Pacific Northwest, overwinter water storage efficiency was 87% with standing wheat stubble compared with 64% with fall plowing. This resulted in 66 mm more water storage and 25% higher green pea (*Pisum* sp.) yields with the standing stubble treatment (Ramig and Ekin, 1978). In north central Oregon, overwinter water storage was significantly higher with stubble present than where stubble was burned, but there was little or no difference between standing or flailed stubble treatments (Ramig and Ekin, 1984). At Teton, Idaho, 120 mm more water was stored with standing stubble than where the stubble was burned in the fall, with the increase resulting mainly from snow trapping. Winter wheat yields increased about 1.1 kg ha^{-1} for each millimeter of trapped snow water (Massee and McKay, 1979).

Theoretical analyses and field experiences show that surface residues in the Pacific Northwest conserve water mainly during the winter and early spring when the surface is wet much of the time. The increased water conservation results from increased infiltration and reduced first-stage evaporation. Surface residues, however, extend first-stage evaporation, thus resulting in a major loss of stored water during summer (Bristow et al., 1986). Consequently, for maximum water conservation, a cultivation to break capillary continuity in surface layers as soon as possible after major rains cease is important. This hastens the onset of second-stage drying in the surface layer, which greatly reduces the evaporation rate.

B. Temperature

Tillage systems influence soil temperatures mainly through their effect on surface residues, primarily by changing the radiant energy balance of the system (Van Doren and Allmaras, 1978). An insulating influence may also be involved because residue-covered soil is usually warmer than bare soil during cool weather, even during daylight hours (Unger, 1978b). The radiation balance is influenced by reflection of incoming radiation by surface residues, heating of soil and air, and soil water evaporation (Van Doren and Allmaras, 1978). Soil temperatures generally decrease as reflectance increases. The insulating influence of surface residues increases with increasing thickness of the residue layer (Gupta and Gupta, 1983;

Unger, 1978b). Tillage influences the radiation balance and insulating properties of surface residues mainly through its effect on residue amount, distribution, and geometry (standing, flattened, or matted).

Radiation reflectance approaches a maximum as surface residue coverage approaches 100% (Van Doren and Allmaras, 1978). Hence, residue amounts exceeding that needed for 100% coverage should have no effect on soil temperatures if radiation reflectance alone is involved. However, greater amounts do affect soil temperatures (McCalla and Duley, 1946; Unger, 1978b), apparently because of the insulating effect. For example, wheat straw on the surface at 8 or 12 Mg ha^{-1} resulted in lower soil temperatures during a hot period and higher temperatures during a cold period than wheat straw at 4 Mg ha^{-1}, an amount that provided almost 100% coverage (Unger, 1978b). Generally, lower soil temperatures with residues on the surface than with bare soil were also reported by Aston and Fischer (1986), Black and Siddoway (1979), Carter and Rennie (1985), and Wall and Stobbe (1984).

Low soil temperatures under residues are most critical where the growing season is relatively short and where delayed crop establishment may seriously reduce yields. For example, temperatures under heavy residue conditions in the northern United States (latitude N45°) became favorable for planting 6 or 7 days later than in a clean (conventionally tilled) seedbed. At more southerly locations (latitude N35° or less), warming was similar in residue-covered and clean seedbeds (Unger and Stewart, 1976).

C. Nutrients

Adequate nutrients are essential for optimum crop yields in any production system. Because crop residues contain nutrients, the amount available to the next crop may be strongly influenced by the tillage system used. Systems that retain residues on the surface result in slow decomposition and thus slow or limited release of nutrients to the next crop. Also, delayed incorporation of residues, as with the till-plant system, may result in nutrient immobilization and limited release of nutrients to the next crop. To compensate for slow decomposition and nutrient immobilization, more fertilizer may be required for the first few crops after a reduced tillage system is introduced (Davies and Cannell, 1975; Unger and McCalla, 1980). Thereafter, a new equilibrium level should be reached so that nutrient cycling with reduced and clean tillage should be similar.

D. Biological Regime

Soil beneath a crop residue cover is often cooler, wetter, and less aerated than where the residues are plowed under. Hence, the number and distribution of soil organisms with reduced tillage systems may be greatly different from that with clean tillage. Doran (1980) found that the surface layer (0-7.5 cm depth) of most reduced tillage soils had greater microbial populations, higher enzyme activity, and higher total nitrogen (N) and potentially mineralizable N levels than clean tilled soil. Aerobic microorganisms were 10-80% higher and anaerobic bacteria, including denitrifiers, were 60-300% higher with reduced tillage than with clean tillage. At the 7.5-15.0 cm depth, populations of aerobic organisms (especially nitrifiers) were significantly higher with clean tillage than with no-tillage, but populations of facultative anaerobes and denitrifiers were higher with no-tillage at that depth. These results indicate that the biological environment of no-tillage soils is less oxidative than that of clean tillage soils, thus resulting in increases in organic matter and total N in the no-tillage soils. These results explain why reduced tillage soils initially require more fertilizer, as discussed in Sec. IV.C. Under some conditions, phytotoxic substances occur in plant residues or are produced by microorganisms. These substances may inhibit plant growth and may be responsible for the reduced yields that have occurred with reduced tillage in some cases (McCalla and Norstadt, 1974; Elliott et al., 1978).

E. Structure

Soil and water conservation are major goals of reduced tillage systems. To achieve these, soil surface and subsurface conditions must be such that water readily enters the soil and yet keeps it resistant to erosion. Soil structural factors that influence water infiltration and soil erodibility include aggregation, porosity and density, and crusting.

Soil aggregation refers to the cementing or binding together of small soil particles into larger secondary units. When sufficiently water stable, aggregates do not disperse easily and, hence, help to maintain favorable water infiltration rates. They also result in favorable conditions for plant growth. When sufficiently large, stable surface aggregates aid in controlling erosion.

The cementing or binding substances for aggregates may be of mineral or organic origin. Mineral substances are particularly important in tropical and subtropical regions where aggregates cemented by iron are highly stable and allow high infiltration, even after prolonged rainfall (Donahue et al., 1977). Organic substances affecting aggregation are derived from soil organisms (fungi, bacteria, actinomycetes, and earthworms) through their feeding and other actions on plant materials. Plants affect aggregation through exudates from roots, leaves, and stems; leachates from decomposing residues; canopies and surface residues that protect aggregates against raindrop impact, abrasion by wind-borne soil, and dispersion by flowing water; and root action that promotes aggregate formation. If stable aggregates are maintained on the surface, infiltration is higher than for an intensively cultivated, poorly aggregated soil (Black and Siddoway, 1979; Donahue et al., 1977). When aggregates are more than 0.84 mm in diameter, they are considered nonerodible by wind. For reduced tillage systems, surface residues are especially beneficial for preventing aggregate breakdown and maintaining favorable infiltration rates because residues dissipate the energy of falling raindrops, thus minimizing soil dispersion, which could result in surface sealing and reduced infiltration.

Soil porosity and bulk density are inversely related to each other, with any practice affecting one also affecting the other. Tillage methods per se had little affect on soil density where residue amounts were low (Johnson, 1950; McCalla, 1959; Unger, 1969), but densities decreased when increased amounts of residues were involved (Black, 1973; Black and Siddoway, 1979; Juo and Lal, 1977; Koshi and Fryrear, 1973).

Compaction increases soil density and decreases its porosity and is usually undesirable with respect to water conservation and crop production. Compaction generally results from traffic on the land and is often a serious problem under mechanized farming conditions, especially where large equipment is used. Serious natural compaction may occur, especially on sandy soils for which drying rate plays a major role (Gerard, 1965; Taylor et al., 1966).

Natural compaction usually occurs on soils low in organic matter, and loosening is required to overcome it and to enhance subsequent crop growth. Reduced tillage that loosens compacted soil would be suitable for improving such soil, but no-tillage usually would not be satisfactory. However, any reduced tillage system, including no-tillage, may be suitable for subsequent crop production, provided that adequate residues are available to minimize recompaction. Reduced tillage also has potential for minimizing recompaction on such soils and for minimizing compaction on any soil becuase it generally results in fewer trips across the field, often with lighter equipment. Compaction may also be reduced by using a tillage rotation, with moldboard or disk plowing rotated with reduced tillage systems (Dickey et al., 1983; Mulvaney and Paul, 1984). Using controlled traffic in conjunction with reduced tillage systems may even further reduce compaction (Unger and Stewart, 1988).

Soil fauna, especially earthworms, may greatly increase soil porosity, but there generally are few earthworms in intensively cultivated soils. In contrast, earthworm activity was four to five times greater in no-tillage than in plowed soil in the tropics (Lal, 1976). Earthworm channels increase porosity, are highly stable, and permit rapid water infiltration (Hopp and Slater, 1961).

Soil crusts can affect seed germination, seedling emergence, soil aeration, and water infiltration and, consequently, crop production in general. Crusts result from aggregate dispersion and soil particle sorting and rearrangement during rainstorms, followed by soil drying. To minimize crusting, the surface should be protected against raindrop impact that disperses the soil and against flowing or ponded water that is involved in sorting and rearrangement of soil particles. As for minimizing aggregate breakdown, reduced tillage systems that maintain surface residues are important for minimizing crusting (Johnson, 1950; Juo and Lal, 1977; Lal, 1976; Unger, 1984c).

V. SOIL EROSION

A. Water

Examples of reduced tillage effects on erosion by water were included in Tables 6, 7, and 8. Runoff was reduced as surface residues were increased, with soil losses usually reduced more than runoff. Reduction in erosion by water due to surface residues results from reduced dispersion of aggregates, which results in less soil being available for transport, and in less surface sealing, which results in less runoff to transport detached particles across the surface. Surface residues also retard the flow rate of water across the surface, thereby providing more time for water infiltration and more time for sediments to settle from the flowing water. Reduced flow rates also decrease soil detachment due to flowing water and, thereby, further decrease the potential for particle transport across the surface. Where surface residues are limited or not available, tillage that ridges the soil on the contour or creates surface depressions for temporary storage of water to provide more time for infiltration can result in greater infiltration (Burwell et al., 1966) and, hence, less erosion by water.

A complete cover of surface residues provides maximum protection against runoff and erosion. Small grain straw at about 5.6 Mg ha^{-1} or chopped corn stalks at about 8.3 Mg ha^{-1} result in about 95% surface cover (Wischmeier, 1973). Although complete surface cover provides the most protection, wheat straw at 1.1 and 2.2 Mg ha^{-1} reduced runoff velocities to 50 and 40%, respectively, and soil losses to 31 and 18%, respectively, compared with that from bare soil (Meyer et al., 1970).

Residue types affect runoff and soil loss primarily through the degree of surface coverage that they provide for a given amount of residues (by weight). Wheat and soybean (*Glycine max* L.) residues that were left on the surface at harvest and compared on a dry weight basis were equally effective for controlling erosion on a silt-loam soil with 5% slope, but corn stover at 4.5 Mg ha^{-1} was only half as effective as an equal amount of wheat straw on a weight basis (Wischmeier, 1973).

B. Wind

Conditions conducive to soil erosion by wind include a loose, dry, reasonably finely divided soil; a smooth soil surface with limited or no vegetative cover; a large enough field; and wind strong enough to move soil (Skidmore and Siddoway, 1978). These conditions often prevail under arid and semiarid conditions in some areas, with sandy soils particularly susceptible to erosion by wind because of limited coherence between particles, rapid drying, and small particle sizes. Other soils, however, are also subject to erosion by wind when they are loose and dry or when they have been finely divided by tillage, raindrop impact, or freezing and thawing.

Tillage methods strongly influence soil surface relief and residues, with rough or residue-covered surfaces most effective for minimizing erosion by wind. The greatest surface roughness often results from moldboard, lister, or chisel plowing, and greatest surface residue retention results from reduced tillage systems, especially stubble-mulch and no-tillage, provided adequate residues are available. Such operations as disking, harrowing, and land planing that smooth the surface may leave the soil highly susceptible to erosion by wind and generally should not be used on dry land. Surface smoothing may also result from raindrop impact, freezing and thawing, and erosion itself.

Soil movement (erosion) begins at relatively low windspeeds and progressively increases as windspeed and turbulence increase (Chepil and Woodruff, 1963; FAO, 1978). To prevent erosion, windspeed at the soil surface must be reduced to below the threshold speed at which erosion begins (Skidmore and Siddoway, 1978). When sufficient amounts are present, surface residues keep windspeeds below the threshold value and absorb the wind's force, thereby decreasing its force on the soil.

The effectiveness of residues for controlling erosion by wind depends on their kind, amount, texture, height, and orientation. Residues should also be anchored in soil because loose residues may be blown away, thus becoming ineffective for controlling erosion (McCalla and Army, 1961; Woodruff and Siddoway, 1973; Zingg, 1954). When crop residues are available, some type of reduced tillage that retains surface residues is generally more suitable than clean tillage for controlling erosion by wind. No-tillage is usually the most effective because residues remain anchored in the soil. Stubble-mulch tillage is usually also effective because it undercuts the surface and leaves the residues partially anchored in soil. Stubble-mulch tillage also roughens the surface, with roughening greater on fine-textured (clays and loams) than on coarse-textured (sands) soils.

In some cases, few or no residues may be available because of low production or use for other purposes. In other cases, residues have limited usefulness for controlling erosion (e.g., cotton and sunflower). Low residue production is frequently a problem under nonirrigated conditions in arid and semiarid regions and may be especially critical for crops like cotton for which the residues are of limited value for controlling erosion. When residues are limited, regardless of the reason, alternative measures must be relied on to control erosion.

When residues are limited, soils should be kept rough and cloddy to help control erosion by wind. Some roughness results from normal crop production operations. Additional roughness can be obtained by using tillage and planting operations that cover or replace erodible surface materials with materials that are less erodible (Fryrear, 1981). The most effective roughness height for controlling erosion by wind is 50-130 mm (Armbrust et al., 1964; Woodruff and Lyles, 1967). Surface roughness can be maintained by using tillage that minimizes soil pulverization and smoothing and by planting small grains with deep furrow or hoe drills that produce roughness in the 50-130 mm height range (Woodruff and Siddoway, 1973). Even greater protection against erosion by wind is achieved by using these drills when some residues are present.

The amount of cloddiness that tillage produces depends on soil texture, the implement used and its speed of operation, and soil water content at the time of tillage (Woodruff and Siddoway, 1973). Few clods that resist erosion may be formed by tillage in sandy soils. However, if they contain more than about 8% clay, cultivation increases their cloddiness (Harper and Brensing, 1950). Performing plowing, cultivation, and planting operations on sandy soils while they are moist also increases their resistance to erosion by wind. A variety of tillage methods used within a wide range of soil water contents usually produces sufficient cloddiness on fine-textured soils to control erosion by wind. On such soils, even chiseling as an emergency measure is usually effective for controlling erosion by wind when the soil is quite dry (Lyles and Tatarko, 1982).

VI. CROP ESTABLISHMENT, GROWTH, AND YIELD

Limited precipitation and limited stored soil water often limit crop yields under arid and semiarid conditions without irrigation. Consequently, where water conservation is improved as with reduced tillage systems, the yield potential is also increased.

As discussed previously (Sec. IV.A) and shown in Tables 1, 5, 11, 12, 13, 14, and 15, crop yields generally increase as soil water storage increases because of the greater maintenance of surface residues with reduced tillage systems. However, to achieve the favorable yield responses, crop establishment and growth must be satisfactory. Unfortunately, crop establishment and growth factors were generally not discussed, but the favorable yields obtained with reduced tillage systems imply that they were favorable. Although not discussed, personal experiences and some literature reports indicate that crop establishment is often better with reduced than with clean tillage because of improved soil water conditions near the surface. Exceptions are in regions where soil temperatures may be lower with surface residues than with clean tillage. Satisfactory crop establishment, however, also depends on using satisfactory planting equipment. A wide range of suitable planting equipment is now available for use under reduced tillage conditions, even no-tillage conditions, but problems still occur under unusually high residue conditions, especially when the soil is wet. Once satisfactorily established, crop growth with reduced tillage is generally satisfactory because of the improved soil water content and, in some cases, more favorable (lower) soil temperatures in hot climates (Allen et al., 1975; Rockwood and Lal, 1974).

As shown in the previously mentioned tables, treatments involving herbicides alone or sweep tillage-herbicide combination treatments generally resulted in similar yields of wheat and grain sorghum. Yields were similar with the two systems because sweep tillage reduces surface residues only about 10% during each operation. Hence, one or two sweep-tillage operations usually did not incorporate enough residues to affect water storage or crop yields (Phillips, 1969; Unger, 1981; Unger et al., 1971).

VII. PROBLEMS WITH REDUCED TILLAGE SYSTEMS

Much progress has been made in recent years to make reduced tillage systems adaptable for crop production. However, various problems remain for which further research is needed to solve the problems or to find suitable alternatives, especially where the use of reduced tillage systems is essential for soil and water conservation purposes.

A. Continuing Problems

Continuing problems that exist under some conditions include limited or ineffective residues, use of poor residue management practices, unsatisfactory weed control, inadequate equipment, poor cropping sequences, and limited adoption of reduced tillage systems.

Residues may be limited because of low production or use for other purposes. In either case, the full benefits of reduced tillage systems, especially those resulting from maintenance of surface residues for conserving soil and water, may not be obtained. Development of crops with residues that are more resistant to decay or practices that retard residue decay are helpful. When adequate residues cannot be maintained, such practices as contouring, strip cropping, and terracing may be required for conserving soil and water resources (Unger, 1984a). Residues that have little or no value for conserving soil and water are those that provide limited resistance to the forces of wind, falling raindrops, or flowing water. Such crops as cotton, sunflower, green pea, and peanut (*Arachis hypogaea* L.) are in this category because generally only the stems remain after crop harvest.

Herbicides are relied upon for weed control in many reduced tillage systems. However, less than adequate control is obtained in some cases because the herbicides are effective mainly at certain growth stages or environmental conditions, or suitable herbicides are not compatible with the crop being grown. Herbicides are needed that control specific weeds or types of weeds under a wide range of growth stages and environmental conditions.

Vast improvements have been made in recent years regarding equipment suitable for operation under high residue conditions, but equipment problems remain. Equipment is needed that operates without trouble under high-residue conditions, provides good seed-soil contact, causes minimal destruction of surface residues, and places seed at a uniform depth.

Improved cropping sequences are needed that result in more effective timing among the periods of water availability, soil water storage, and crop growing season.

Reduced tillage systems have been developed and tested and have been found adaptable for economical crop production and for effective soil and water conservation in many cases. Yet, most producers do not adopt such systems. Efforts are needed to apprise producers of the resource conserving, crop production, and economical benefits of reduced tillage systems and to assist them in adapting suitable reduced tillage systems to their particular crop production conditions.

B. Potential Problems

As with any new crop production system, all problems may not yet have been encountered. Others may have been expected but have not become widespread. Two of these potential problems are pests and water pollution. Pests (mainly insects and diseases) have long been recognized as potential problems with reduced tillage. Indeed, some are greater problems with reduced tillage, others are less. Consequently, producers using reduced or any other tillage system should monitor their crops for the presence of pests and initiate control measures when warranted.

The potential for water pollution has also long been recognized, and the chemical concentrations in surface runoff water and groundwater have been determined. Again, variable results have been obtained. Usually no greater chemical concentrations were found when fertilizers and pesticides were applied at recommended rates and rainfall did not occur soon after application. However, soluble fertilizers (e.g., nitrogen) may move with infiltrating or runoff water, and split applications of such fertilizers may be required to reduce the potential for water pollution. In all cases, the effects of chemical use on the environment must be monitored so that possible adverse effects on the environment are detected at an early stage.

VIII. CONCLUSIONS

In the last 20-30 years, much research has been conducted to develop effective reduced tillage crop production systems, mainly because of increasing energy costs, high equipment inventories and costs, high labor costs, increased concern about soil erosion, and limited utilization of water resources when conventional tillage systems are used. In reduced tillage systems, tillage intensity or frequency is reduced. In some reduced tillage systems, the entire surface is tilled at least once and all residues are incorporated. In others, the entire area is tilled but most residues are retained on the surface. In yet others, the surface is disturbed only enough to place seeds in the soil, with all residues retained on the surface (no-tillage systems).

Each system is applicable to certain conditions. However, in arid and semiarid regions, water and soil conservation are highly important for favorable and sustained crop production. The residue-retaining systems are especially effective in controlling soil erosion and have substantially increased soil water storage in many cases by reducing runoff (increasing infiltration) and evaporation. Moreover, satisfactory weed control has been achieved with herbicides in many cases, which has further enhanced water conservation.

Although favorable results have been obtained with reduced tillage systems in many cases, these systems are not without problems. One of the major problems in arid and semiarid regions is limited crop residue production without irrigation. This may be aggravated by residue use for feed and fuel, which is common in developing countries. Where residues are limited, the soil and water conservation benefits of residues are not achieved and alternative practices are required to achieve satisfactory resource conservation. Some alternative practices that aid water conservation and control of erosion by water include contouring, terracing, forming depressions for temporary water storage on the surface, and strip cropping. Strip cropping also aids the control of erosion by wind, which can also be reduced by tillage that produces a rough, cloddy surface. Even where these alternative crop production systems must be used for effective resource conservation, reductions in tillage frequency and intensity are still possible under many conditions because herbicides are available for controlling weeds. Reduced tillage intensity or frequency under low residue conditions helps to maintain surface depressions and contour ridges and minimizes surface pulverization, thus reducing the potential for runoff and surface sealing and thereby maintaining favorable water infiltration rates and aiding in the control of erosion by water. Reduced tillage under low residue conditions aids the control of erosion by wind by helping to maintain a rough, cloddy surface.

At present, available reduced tillage systems for semiarid to arid regions were developed primarily for crop production in North America and Australia. These systems may not be applicable without modification for crop production conditions in developing countries. However, the principles involved are applicable to any conditions and should be introduced to developing countries, albeit with caution. The prevailing soil, climatic, social, and economic conditions of the region must be considered when reduced tillage systems are introduced in any region or country.

ACKNOWLEDGMENT

The views expressed here are those of the author and may not reflect the policies of the U.S. Department of Agriculture.

REFERENCES

Allen, R. R., Musick, J. T., Wood, F. O., and Dusek, D. A. (1975). No-till seeding of irrigated sorghum double cropped after wheat. *Trans. Am. Soc. Agr. Eng. 18*: 1109-1113.

Amemiya, M. (1977). Conservation tillage in the western corn belt. *J. Soil Water Conserv. 32*: 29-36.

Anderson, D. T. (1968). Field equipment needs in conservation tillage systems. In *Conservation Tillage in the Great Plains, Proceedings of a Workshop*, Lincoln, Nebraska, February 1968. Great Plains Agricultural Council, University of Nebraska, Lincoln, Publ. 32, pp. 83-91.

Armbrust, D. V., Chepil, W. S., and Siddoway, F. H. (1964). Effects of ridges on erosion of soil by wind. *Soil Sci. Soc. Am. Proc. 28*: 557-560.

Aston, A. R., and Fischer, R. A. (1986). The effect of conventional cultivation, direct drilling and crop residue on soil temperatures during the early growth of wheat at Murrumbateman, New South Wales. *Aust. J. Soil Res. 24*: 49-60.

Baeumer, K., and Bakermans, W. A. P. (1973). Zero-tillage. *Adv. Agron. 25*: 77-123.

Baker, J. L., and Johnson, H. P. (1979). The effect of tillage systems on pesticides in runoff from small watersheds. *Trans. Am. Soc. Agr. Eng. 22*: 554-559.

Baker, L. O., Krall, J. L., Aasheim, T. S., and Hartman, T. P. (1956). Chemical summer fallow in Montana. *Down Earth 11*(4): 21.

Barnes, O. K., Bohmont, D. W., and Rauzi, F. (1955). Effect of chemical and tillage summer fallow upon water-infiltration rates. *Agron. J. 47*: 435-436.

Bennett, O. L. (1977). Conservation tillage in the Northeast. *J. Soil Water Conserv. 32*: 9-12.

Black, A. L. (1973). Soil property changes associated with crop residue management in a wheat-fallow rotation. *Soil Sci. Soc. Am. Proc. 37*: 943-946.

Black, A. L., and Siddoway, F. H. (1979). Influence of tillage and wheat straw residue management on soil properties in the Great Plains. *J. Soil Water Conserv. 34*: 220-223.

Bristow, K. L., Campbell, G. S., Papendick, R. I., and Elliot, L. F. (1986). Simulation of heat and moisture transfer through a surface residue-soil system. *Agr. For. Meteorol. 36*: 193-214.

Burton, R. L., Jones, O. R., Burd, J. D., Wicks, G. A., and Krenzer, E. G., Jr. (1987). Conservation tillage reduces greenbug populations and damage in grain crops. In *Proceedings of the National Symposium on Conservation Systems, Optimum Erosion Control at Least Cost.* American Society of Agricultural Engineers, St. Joseph, Michigan, pp. 351-362.

Burwell, R. E., Allmaras, R. R., and Sloneker, L. L. (1966). Structural alternation of soil surfaces by tillage and rainfall. *J. Soil Water Conserv. 21*: 61-63.

Carter, M. R., and Rennie, D. A. (1985). Soil temperature under zero tillage systems for wheat in Saskatchewan. *Can. J. Soil Sci. 65*: 329-338.

Charreau, C. (1977). Some controversial technical aspects of farming systems in semi-arid west Africa. In *Proceedings of the Symposium on Rainfed Agriculture in Semi-Arid Regions, Riverside, California, April 1977*, Glen H. Cannell (Ed.). University of California, Riverside, pp. 313-360.

Chepil, W. S., and Woodruff, N. P. (1963). The physics of wind erosion and its control. *Adv. Agron. 15*: 211-302.

Crosson, P., Hanthorn, M., and Duffy, M. (1986). The economics of conservation tillage. In *No-tillage and Surface Tillage Agriculture*, M. A. Sprague and G. B. Triplett (Eds.). John Wiley & Sons, New York, pp. 409-436.

Dale, J. E. (1980). Control of Johnson grass by wick-applicated glyphosate (abstract). *Proc. South. Weed Sci. Soc. 33*: 356.

Davies, D. B., and Cannell, R. Q. (1975). Review of experiments on reduced cultivation and direct drilling in the United Kingdom, 1957-1974. *Outlook Agr. 8*: 216-220.

Dickey, E. C., Peterson, T. R., Gilley, J. R., and Mielke, L. N. (1983). Yield comparisons between continuous no-till and tillage rotations. *Trans. Am. Soc. Agr. Eng. 26*: 1682-1686.

Donahue, R. L., Miller, R. W., and Shickluna, J. C. (1977). *Soil—An Introduction to Soils and Plant Growth*, 4th ed. Prentice-Hall, Englewood Cliffs, New Jersey.

Doran, J. W. (1980). Microbial changes associated with residue management with reduced tillage. *Soil Sci. Soc. Am. J. 44*: 518-524.

Duley, F. L., and Russel, J. C. (1942). Effect of stubble mulching on soil erosion and runoff. *Soil Sci. Soc. Am. Proc. 7*: 77-81.

Edwards, W. M., Triplett, G. B., Jr., and Kramer, R. M. (1980). A watershed study of glyphosate transport in runoff. *J. Environ. Qual. 9*: 661-665.

Elliott, L. F., McCalla, T. M., and Waiss, A., Jr. (1978). Phytotoxicity associated with residue management. In *Crop Residue Management Systems*, W. R. Oschwald (Ed.). American Society of Agronomy, Madison, Wisconsin, Special Publ. 31, pp. 131-146.

FAO (Food and Agriculture Organization of the United Nations) (1978). Soil erosion by wind and measures for its control on agricultural lands. FAO Agr. Dev. Paper 71, Rome.

Fenster, C. R. (1973). Stubble mulching. In *Conservation Tillage, The Proc. of a National Conference*. Soil Conservation Society of America, Ankeny, Iowa, pp. 202-207.

Fisher, W. F., and Lane, D. R. (1973). Till-planting. In *Conservation Tillage, The Proc. of a National Conference*. Soil Conservation Society of America, Ankeny, Iowa, pp. 187-194.

Fryrear, D. W. (1974). Dryland farming sandy soils in a semiarid region. In *Symposium on Frontiers of Semi-Arid World*. Texas Tech University, Lubbock, Texas, October 1974.

Fryrear, D. W. (1981). Tillage influences erodibility of dryland sandy soils. In *Proceedings of a Conference on Crop Production with Conservation in the 80s*, Chicago, Illinois, December 1980. American Society of Agricultural Engineers, St. Joseph, Michigan, pp. 153-163.

Gerard, C. J. (1965). The influences of soil moisture, soil texture, drying conditions, and exchangeable cations on soil strength. *Soil Sci. Soc. Am. Proc. 29*: 641-645.

Greb, B. W. (1979). Reducing drought effects on croplands in the West-Central Great Plains. U.S. Dep. Agr. Info. Bull. 420, U.S. Government Printing Office, Washington, D.C., pp. 1-31.

Greb, B. W., Smika, D. E., and Black, A. L. (1967). Effect of straw-mulch rates on soil water storage during summer fallow in the Great Plains. *Soil Sci. Soc. Am. Proc. 31*: 556-559.

Greb, B. W., Smika, D. E., and Black, A. L. (1970). Water conservation with stubble mulch fallow. *J. Soil Water Conserv. 25*: 58-62.

Griffith, D. R., Mannering, J. V., and Moldenhauer, W. C. (1977). Conservation tillage in the eastern corn belt. *J. Soil Water Conserv. 32:* 20-28.

Gupta, J. P., and Gupta, G. N. (1983). Effect of grass mulching on growth and yield of legumes. *Agr. Water Manag. 6*: 375-384.

Hanway, D. G. (1970). Cultural practices in mechanized dryland farming. In *Proceedings of an International Conference on Mechanized Dryland Farming*, Moline, Illinois, August 1969, W. C. Burrows, R. E. Reynolds, F. C. Stickler, and G. E. Van Riper (Eds.). Deere and Co., Moline, Illinois, pp. 102-122.

Harman, W. L., and Martin, J. R. (1987). Economics of conservation tillage research in Texas. In *Conservation Tillage: Today and Tomorrow*, T. J. Gerik and B. L. Harris (Eds.). Texas Agr. Exp. Sta. Misc. Publ. MP-1634, Texas A&M University, College Station, pp. 24-37.

Harper, J., and Brensing, O. H. (1950). Deep plowing to improve sandy land. Oklahoma Agr. Exp. Sta. Bull. B-362, Oklahoma State University, Stillwater.

Hopp, H., and Slater, C. S. (1961). The effect of earthworms on the productivity of agricultural soil. In *The Challenge of Earthworm Research*, R. Rodale (Ed.). The Soil and Health Foundation, Emmaus, Pennsylvania, pp. 67-83.

Jacks, G. V., Brind, W. D., and Smith, R. (1955). Mulching. Commonwealth Bur. Soil Sci. (England) Tech. Comm. 49, Commonwealth Agr. Bur., Farnham Royal, Bucks, England.

Johnson, W. C. (1950). Stubble-mulch farming on wheatlands of the Southern High Plains. U.S. Dep. Agr. Circ. 860, U.S. Government Printing Office, Washington, D.C.

Juo, A. S. R., and Lal, R. (1977). The effect of fallow and continuous cultivation on the chemical and physical properties of an Alfisol in Western Nigeria. *Plant Soil 47*: 567-584.

Kang, B. T., Moody, K., and Adesina, J. O. (1980). Effects of fertilizer and weeding in no-tillage and tilled maize. *Fertil. Res. 1*: 87-93.

Koshi, P. T., and Fryrear, D. W. (1973). Effect of tractor traffic, surface mulch, and seedbed configuration on soil properties. *Soil Sci. Soc. Am. Proc. 37*: 758-762.

Lal, R. (1975). Role of mulching techniques in tropical soil and water management. Int. Inst. Trop. Agr. Tech. Bull. 1, Int. Inst. Trop. Agric., Ibadan, Nigeria.

Lal, R. (1976). No-tillage effects on soil properties under different crops in Western Nigeria. *Soil Sci. Soc. Am. J. 40*: 762-768.

Lal, R. (1980a). Crop residue management in relation to tillage techniques for soil and water conservation. In *Organic Recycling in Africa.* FAO Soils Bull. 3, Rome, pp. 72-78.

Lal, R. (1980b). No-tillage farming in the topics. In *No-tillage Research: Research Reports and Review,* R. E. Phillips, G. W. Thomas, and R. L. Blevins (Eds.). University of Kentucky, Lexington, pp. 103-151.

Lal, R. (1988). Soil degradation and the future of agriculture in sub-Saharan Africa. *J. Soil Water Conserv. 43*: 444-451.

Lyles, L., and Tatarko, J. (1982). Emergency tillage to control wind erosion: Influences on winter wheat yields. *J. Soil Water Conserv. 37*: 344-347.

Massee, T., and McKay, H. (1979). Improving dryland wheat production in eastern Idaho with tillage and cropping methods. Idaho Agr. Exp. Sta. Bull. 581, University of Idaho, Moscow.

McCalla, T. M. (1959). Microorganisms and their activity with crop residues. Nebraska Agr. Exp. Sta. Bull. No. 453, University of Nebraska, Lincoln, pp. 1-31.

McCalla, T. M., and Army, T. J. (1961). Stubble mulch farming. *Adv. Agron. 13*: 125-196.

McCalla, T. M., and Duley, F. L. (1946). Effect of crop residues on soil temperature. *J. Am. Soc. Agron. 38*: 75-89.

McCalla, T. M., and Norstadt, F. A. (1974). Toxicity problems in mulch tillage. *Agr. Environ. 1*: 153-174.

McDowell, L. L., and McGregor, K. C. (1980). Nitrogen and phosphorus losses in runoff from no-till soybeans. *Trans. Am. Soc. Agr. Eng. 23*: 643-648.

Meyer, L. D., Wischmeier, W. H., and Foster, G. R. (1970). Mulch rates required for erosion control on steep slopes. *Soil Sci. Soc. Am. Proc. 34*: 928-931.

Mulvaney, D. L., and Paul, L. (1984). Rotating crops and tillage—both sometimes better than just one. *Crops Soils 36*(7): 18-19.

Okigbo, B. N., and Lal, R. (1977). Role of cover crops in soil and water conservation. In *Soil Conservation and Management in Developing Countries,* FAO Soils Bull. 33, Rome, pp. 97-108.

Oschwald, W. R. (1973). Chisel plow and strip tillage systems. In *Conservation Tillage, The Proceedings of a National Conference.* Soil Conservation Society of America, Ankeny, Iowa, pp. 194-202.

Papendick, R. I., and Miller, D. E. (1977). Conservation tillage in the Pacific Northwest. *J. Soil Water Conserv. 32*: 49-56.

Phillips, R. E., and Phillips, S. H. (Eds.) (1984). *No-tillage Agriculture, Principals and Practices.* Van Nostrand-Reinhold, New York.

Phillips, S. H. (1980). No-tillage, past and present. In *No-tillage Research: Research and Reviews,* R. E. Phillips, G. W. Thomas, and R. L. Blevins (Eds.). University of Kentucky, Lexington, pp. 1-6.

Phillips, S. H., and Young, H. M., Jr. (1973). *No-tillage Farming.* Reiman Associates, Milwaukee, Wisconsin.

Phillips, W. M. (1954). The use of dalapon and 2,4-D for maintaining summer fallowed land in a weed free condition. In *Research Report, Eleventh North Central Weed Control Conference,* N. E. Shafer (Ed.). Fargo, North Dakota, December 1954, pp. 44.

Phillips, W. M. (1969). Dryland sorghum production and weed control with minimum tillage. *Weed Sci. 17*: 451-454.

Radke, J. K. (1982). Managing early season soil temperatures in the Northern Corn Belt using configured soil surfaces and mulches. *Soil Sci. Soc. Am. J. 46*: 1067-1071.

Ramig, R. E., and Ekin, L. G. (1978). Soil water storage as influenced by tillage and crop residue management. Oregon Agr. Exp. Stn. Prog. Rep. SM 78-4, Oregon State University, Corvallis, pp. 65-68.

Ramig, R. E., and Ekin, L. G. (1984). Effect of stubble management in a wheat-fallow rotation on water conservation and storage in eastern Oregon. Oregon Agr. Exp. Stn. Spec. Rep. 713, Oregon State University, Corvallis, pp. 30-33.

Reicosky, D. C., Cassel, D. K., Blevins, R. L., Gill, W. R., and Naderman, G. C. (1977). Conservation tillage in the Southeast. *J. Soil Water Conserv. 32*: 13-19.

Richey, C. B., Griffith, D. R., and Parsons, S. D. (1977). Yields and cultural energy requirements for corn and soybeans with various tillage-planting systems. *Adv. Agron. 29*: 141-182.

Rockwood, W. G., and Lal, R. (1974). Mulch tillage: A technique for soil and water conservation in the tropics. *Span 17*: 72-79.

SCSA (Soil Conservation Society of America) (1982). *Resource Conservation Glossary.* Ankeny, Iowa.

Skidmore, E. L., and Siddoway, F. H. (1978). Crop residue requirements to control wind erosion. In *Crop Residue Management Systems,* W. R. Oschwald (Ed.). American Society of Agronomy, Madison, Wisconsin, Spec. Publ. 31, pp. 17-33.

Smika, D. E. (1976a). Seed zone soil water conditions with reduced tillage in the semiarid Central Great Plains. *Proceedings of the Seventh Conference of the International Soil Tillage Research Organization,* Uppsala, Sweden, June 1976. International Soil Tillage Research Organization, Wageningen, The Netherlands, pp. 37.1-37.6.

Smika, D. E. (1976b). Mechanical tillage for conservation fallow in the semiarid Central Great Plains. In *Conservation Tillage, Proceedings of the Great Plains Workshop,* Fort Collins, Colorado, August 1976. Great Plains Agricultural Council, Fort Collins, Colorado, Publ. 77, pp. 78-91.

Smika, D. E., and Unger, P. W. (1986). Effect of surface residues on soil water storage. *Adv. Soil Sci. 5*: 111-138.

Smika, D. E., and Wicks, G. A. (1968). Soil water storage during fallow in the Central Great Plains as influenced by tillage and herbicide treatments. *Soil Sci. Soc. Am. Proc. 32*: 591-595.

Sprague, M. A., and Triplett G. B. (Eds.) (1986). *No-Tillage and Surface-Tillage Agriculture.* John Wiley & Sons, New York.

SSSA (Soil Science Society of America) (1973). *Glossary of Soil Science Terms.* Madison, Wisconsin.

Stallings, J. H. (1957). *Soil Conservation.* Prentice Hall, Englewood Cliffs, New Jersey.
Taylor, H. M., Roberson, G. M., and Parker, J. J., Jr. (1966). Soil strength-root penetration relations for medium- to coarse-textured soil materials. *Soil Sci. 102*: 18-22.
Triplett, G. B., Jr., and Van Doren, D. M., Jr. (1977). Agriculture without tillage. *Sci. Am. 236*(1): 28-33.
Triplett, G. B., Jr., Conner, B. J., and Edwards, W. M. (1978). Transport of atrazine and simazine in runoff from conventional and no-tillage corn. *J. Environ. Qual. 7*: 77-84.
Unger, P. W. (1969). Physical properties of Pullman silty clay loam as affected by dryland wheat management practices. Texas Agr. Exp. Sta. Misc. Publ. MP-933, Texas A&M University, College Station.
Unger, P. W. (1978a). Straw-mulch rate effect on soil water storage and sorghum yield. *Soil Sci. Soc. Am. J. 42*: 486-491.
Unger, P. W. (1978b). Straw-mulch effects on soil temperatures and sorghum germination and growth. *Agron. J. 70*: 858-864.
Unger, P. W. (1981). Tillage effects on wheat and sunflower grown in rotation. *Soil Sci. Soc. Am. J. 45*: 941-945.
Unger, P. W. (1984a). *Tillage Systems for Soil and Water Conservation.* FAO Soils Bull. 54, Rome.
Unger, P. W. (1984b). Tillage and residue effects on wheat, sorghum, and sunflower grown in rotation. *Soil Sci. Soc. Am. J. 48*: 885-891.
Unger, P. W. (1984c). Tillage effects on surface soil physical conditions and sorghum emergence. *Soil Sci. Soc. Am. J. 48*: 1423-1432.
Unger, P. W., and Stewart, B. A. (1976). Land preparation and seedling establishment practices in multiple cropping systems. In *Multiple Cropping,* R. I. Papendick, P. A. Sanchez, and G. B. Triplett (Eds.). American Society of Agronomy, Madison, Wisconsin, Special Publ. 27, pp. 255-273.
Unger, P. W., and Stewart, B. A. (1988). Conservation techniques for vertisols. In *Vertisols: Their Distribution, Properties, Classification and Management,* L. P. Wilding and R. Puentes (Eds.). Soil Management Support Services, Tech. Monograph 18. Texas A&M University Printing Center, College Station, Texas, pp. 165-181.
Unger, P. W., and Wiese, A. F. (1979). Managing irrigated winter wheat residues for water storage and subsequent dryland grain sorghum production. *Soil Sci. Soc. Am. J. 43*: 582-588.
Unger, P. W., Steiner, J. L., and Jones, O. R. (1986). Response of conservation tillage sorghum to growing season precipitation. *Soil Tillage Res. 7*: 291-300.
Unger, P. W., Allen, R. R., and Wiese, A. F. (1971). Tillage and herbicides for surface residue maintenance, weed control, and water conservation. *J. Soil Water Conserv. 26*: 147-150.
Unger, P. W., and McCalla, T. M. (1980). Conservation tillage systems. *Adv. Agron. 33*: 1-58.
Van Doren, D. M., Jr., and Allmaras, R. R. (1978). Effect of residue management practices on the soil physical environment, microclimate, and plant growth. In *Crop Residue Management Systems,* W. R. Oschwald (Ed.). American Society of Agronomy, Madison, Wisconsin, Special Publ. 31, pp. 49-83.

Wall, D. A., and Stobbe, E. H. (1984). The effect of tillage on soil temperature and corn (*Zea mays* L.) growth in Manitoba. *Can. J. Plant Sci. 64*: 59-67.

Wicks, G. A., and Smika, D. E. (1973). Chemical fallow in a winter wheat-fallow rotation. *J. Weed Sci. Soc. Am. 21*: 97-102.

Wiese, A. F., and Lavake, D. E. (1980). Experiences with a rope wick applicator. *Proc. South. Weed Sci. Soc. 33*: 353-355.

Wiese, A. F., and Staniforth, D. W. (1973). Weed control in conservation tillage. In *Conservation Tillage, The Proceedings of a National Conference,* Soil Conservation Society of America, Ankeny, Iowa, pp. 108-114.

Wiese, A. F., Bond, J. J., and Army, T. J. (1960). Chemical fallow in the southern Great Plains. *Weeds 8*: 284-290.

Wischmeier, W. H. (1973). Conservation tillage to control water erosion. In *Conservation Tillage, The Proc. of a National Conference.* Soil Conservation Society of America, Ankeny, Iowa, pp. 133-141.

Woodruff, N. P. (1972). Wind erosion as affected by reduced tillage systems. *Proceedings of the No-Tillage Systems Symposium,* Ohio State University, Columbus, pp. 5-20.

Woodruff, N. P., and Lyles, L. (1967). Tillage and land modification to control wind erosion. In *Tillage for Greater Crop Production.* American Society of Agricultural Engineers, St. Joseph, Michigan, Publ. 168, pp. 63-67, 70.

Woodruff, N. P., and Siddoway, F. H. (1973). Wind erosion control. In *Conservation Tillage, The Proceedings of a National Conference.* Soil Conservation Society of America, Ankeny, Iowa, pp. 156-162.

Zingg, A. W. (1954). The wind erosion problem in the Great Plains. *Trans. Am. Geophys. Union 35*: 252-258.

Zingg, A. W., and Whitfield, C. J. (1957). A summary of research experience with stubble-mulch farming in the Western States. U.S. Dep. Agr. Tech. Bull. 1166, U.S. Government Printing Office, Washington, D.C.

13

The Importance of Mutualism in Succession

F. B. REEVES and E. F. REDENTE *Colorado State University, Fort Collins, Colorado*

I. INTRODUCTION

Disturbance on a worldwide basis is becoming more intense, and impacts are being felt on a global scale. The accumulation of CO_2 in the atmosphere, acid deposition, desertification resulting from overgrazing and cropping submarginal farmland, clear cutting of tropical rain forests, and drastic disturbances associated with mining are disturbances that will affect the quality of life well into the next century.

For thousands of years the rate of disturbance throughout the world has been slow but steady. However, during the past 50-100 years, disturbance has increased exponentially (Brinck et al., 1988). This increase has been a result of a tremendous rise in human activity causing profound alteration in the environment. For example, it has been established that 850 million ha of natural vegetation has been converted to arable land over the past 120 years, which accounts for nearly 60% of the total cultivated land that existed worldwide in the year 1979 (Richards, 1986). This conversion of natural vegetation to agricultural land makes the land more susceptible to soil erosion. Many of the deserts of the world are, in fact, examples of less successful attempts to turn natural areas into arable land with catastrophic erosion and degradation as a consequence (Brinck et al., 1988).

Humans frequently disrupt the biological resource base, resulting in degraded ecosystems where the biological components do not function normally. Accordingly, we define what is a functional system from our intended use. In some cases, a premium is placed on keeping systems in a successionally immature state (agricultural systems). Such systems are ultrasensitive to disturbances and must be continuously maintained with supplemental sources of energy (Brinck et al., 1988).

As a result of human activity, there is a desperate need to return degraded areas to natural ecosystems. Disturbances that result in degraded areas that have a reduction or a destruction of biota can be mitigated through proper rehabilitation procedures. The extent to which these disturbances can be rehabilitated is dependent upon geological, climatic, ecological, social, and economic conditions of the area disturbed.

In the reclamation process, succession plays a critical role. The severity of the disturbance dictates whether primary or secondary succession will operate. Primary succession occurs when the substrata produced by the disturbance have not previously supported plant growth. This form of succession is relatively slow and involves many seral stages. Secondary succession is the invasion by plants onto areas that have previously supported vegetation. Secondary succession is more rapid than primary succession, but the time required to reach climax is dependent on the type of climax and the quality of the soil being colonized. The rate and direction of succession, whether primary or secondary, is influenced by many belowground processes, including plant-microbial interactions. In natural ecosystems, an equilibrium is established between soil microbes and plant roots; this equilibrium is altered by the growth of the plant and seasonal changes. In managed ecosystems, such as monoculture in agricultural or forest systems, the microbe-root equilibrium is constantly altered with fertilizer and biocide treatments and frequent harvests. Most of our knowledge of rhizosphere-plant interactions is based on such monoculture systems. Thus, care must be exercised in extrapolating the results from such studies to natural systems (Whipps and Lynch, 1986). This chapter addresses some of the important interactions, specifically the importance of fungal-root mutualisms known as mycorrhizae and how mycorrhizae alter resource competition.

II. RESOURCE COMPETITION AND SUCCESSION

The ability to compete for limited resources is an essential attribute for organisms that live in close association with other organisms. Recent studies (Mueller, 1988) support the assumption that natural selection favors those organisms with an increased competitive ability for resources that are limited. Mueller's results lend credence to the primary role this theory has assumed in evolutionary ecology.

In all biomes, the dynamics of plant growth, the potential for differential resource competition, the establishment of complex belowground food webs, and the physicochemical factors of the environment all lead to shifts in available soil resources as a function of higher plant growth and changes in the rhizosphere biota. Observational and experimental data gathered during recorded history support the principle that relative soil fertility determines the relative success of plant species. Indeed, modern agriculture is based on this principle (Tilsdale et al., 1985). Such changes in plant population dynamics are well illustrated during secondary succession following disturbance.

A. Succession

Levels of essential plant nutrients in the soil can alter vegetation composition. Box (1961), Snaydon (1962), Pigott and Taylor (1964), and Beadle (1966) demonstrated that different levels of nitrogen (N), phosphorus (P), and calcium (Ca) can alter species composition in different ecosystems. Meller (1979) and Bornkamm and Hennig (1982) linked changes in vegetation composition during succession to changes in N supply and nutrient cycling. Harcombe (1977) showed that inputs of N, P, and potassium (K) can retard succession in tropical forests and significantly alter species composition. Bazzaz (1979), Parrish and Bazzaz (1982b), and Tilman (1986) found different response patterns to nutrient gradients by species belonging to different seral stages. Tilman (1982, 1984, 1988) emphasized how resource availability and resource supply ratios may influence both vegetation composition and the direction of succession.

The relationships between nutrient cycling and succession were first linked in a conceptual model by Odum (1969). Odum theorized that changes in both nutrient pools and nutrient cycling take place in the course of succession. The model predicted that with the advance of succession:

1. Most of the nutrients become immobilized in organic compounds, and the nutrient supply changes from extrabiotic to intrabiotic.
2. Nutrient cycling becomes more closed and nutrient exchange rates are reduced.
3. The role of detritus in nutrient regeneration becomes more important.

These predictions are consistent with the behavior of some ecosystems, but not with others (MacMahon, 1980; Connell and Slatyer, 1977). Vitousek and Reiners (1975) found increases in nutrient retention with increased net primary productivity (NPP). With increases in NPP, immobilization dominates over mineralization (Gorham et al., 1979; Reiners, 1981). Nutrient cycling rates, soil biological activity, and nutrient retention in biomass are maximized in the intermediate stages of succession and decline as the ecosystem reaches maturity (Titlyanova, 1982). When an ecosystem reaches maturity, NPP stabilizes; equilibrium is reached between nutrient inputs and outputs; and immobilization equals mineralization (Gorham et al., 1979; Reiners, 1981). Woodmansee (1978) has challenged the concept that nutrient input-output ratios approach unity and come to a state of dynamic stability as ecosystems mature, and he suggested that in the climax or near-climax grassland ecosystem, N cycling is "tight," nitrogen accumulates, and the system is not balanced. However, not all ecosystems exhibit these changes in nutrient cycling. A fertile forest site may have large nitrate losses after disturbance, but a lower quality forest site may lose little nitrate after disturbance or during succession (Vitousek et al., 1979).

Changes in the internal cycling of nutrients by plants during succession have also been examined. Reiners (1981) and Vitousek (1982) proposed that an increase in the internal recycling of nutrients may be a mechanism by which late successional species compensate for the reduced availability of mineralized nutrients in the soil. Observations by Hopkinson and Schubauer (1984) on *Spartina alternifolia* support this theory. Woodmansee et al. (1978) estimated that 26% of the N is internally recycled in shortgrass prairie plants.

These studies support the concept that secondary succession is related to changes in plant utilization patterns and nutrient cycling. Although there are theoretical considerations of microbial succession and growth strategies, few experiments have addressed the role of soil biota in succession (Swift and Heal, 1986). Studies suggest that changes in microbial and faunal activity can have an important impact on succession by affecting the patterns of nutrient cycling. For example, changes in nitrification potentials, N_2 fixation, and soil enzymatic activities during succession have been documented in many ecosystems (Pancholy and Rice, 1973; Rice, 1974, 1984; Kapustka and Rice, 1976; Rice and Mallick, 1977; Lodhi, 1979; Montes and Christensen, 1979; Lamb, 1981; Robertson and Vitousek, 1982; Wheeler and Donaldson, 1983; Redente et al., 1984; Klein et al., 1985; Biondini et al., 1985; Metzger et al., 1986).

Carbon cycling and the potential to form mycorrhizae are positively correlated with the dominance of middle and late successional plant species in both natural and induced secondary succession (Redente et al., 1984; Biondini et al., 1984, 1985). Two indices of soil microbial activity, phosphatase and acetylene reduction activity (Speir and Ross, 1978; Sorensen, 1982), sharply decline with the advance of succession (Klein et al., 1985) and are negatively correlated with the intensity of the original disturbance (Redente et al., 1984; Biondini et al., 1984, 1985).

These relationships among nutrient cycling, soil biota, and succession are substantiated by a number of studies that demonstrate that changes in microbial and faunal populations alter nutrient cycling. The early work of Starkey (1929), Lochhead and Chase (1943), and Chan and Katznelson (1961) showed a definite difference in biological activity and substrate utilization between free soil (nonrhizosphere) microorganisms and rhizosphere microorganisms. Heal and Dighton (1986) noted that disturbance can lead to an uncoupling of primary production and decomposition. These workers suggested that changes in the decomposer subsystem can determine whether secondary succession recapitulates previous successional stages. Bowen (1980) emphasized that the plant is the main driving force for the system and that the distribution of the roots and microbial responses to roots are essential considerations in evaluating the development of microbial populations in the rhizosphere.

B. Competition

Competition in arid ecosystems has been questioned (Shreve, 1942; Went, 1955). However, in natural communities plants compete (Braakhekke, 1985), and a recent summary indicates that competition is common and readily detected in both deserts and semiarid grasslands (Fowler, 1986). The acquisition and utilization of essential minerals depend on a complex of factors, including root growth, initial nutrient concentrations in soil (ranging from insufficient to toxic levels), nutrient availability (modified by temporal fluctuations and replacement, symbiotic and free-living root-microbial associations, soil moisture levels, and diffusion coefficients of individual ions), demand, uptake kinetics, transport in the plant, and interactions among elements. The ultimate success of a given plant depends on maintaining at least minimal concentration levels of essential minerals in the living tissue.

Investigators agree that immediately following a severe disturbance, when there are few plants present, there is little competition. At the beginning of secondary succession, a pulse of available nutrients is associated with the initial lack of competition following disturbance. This pulse is the result of the decomposition of plant material and the relatively high moisture level in the soil. Of the major nutrients, increase in the available levels of both N and P have been measured following disturbance (Gorham et al., 1979; Vitousek et al., 1979). Because few plants are initially present following disturbance, N and P accumulate and plants experience luxury resource levels. Rapidly growing competitive and ruderal species are characteristic of such fertile habitats (Chapin, 1980, 1983), and such plants characterize disturbed semiarid soils (Doerr et al., 1984). However, after initial disturbance and colonization by pioneer plants, soil resource availabilities rapidly change. In arid and semiarid environments both N and P precipitate. Nitrate anions are nonspecifically adsorbed, whereas phosphate anions are specifically adsorbed (Mott, 1981). Nitrogen becomes available the next year with moisture; however, phosphorus may be fixed in highly insoluble forms in clay lattices in alkaline soils. As succession proceeds and soil moisture and nutrient availability decrease in semiarid soils, there is a rapid shift in the relative availability of P compared to N. A relative P deficiency can develop in dry environments. Of particular importance for competitive relationships in plant communities is the ability to compete for P, which is one of the major macronutrients required by all plants. Phosphorus may be limited in many environments, and many experiments indicate that phosphorus additions or mycorrhizal colonization substantially improve plant growth. Some plants, when mycorrhizal and grown under limited phosphorus levels, can exhibit 100-1000 times increased growth when compared to nonmycorrhizal plants.

A recent series of important and significant experiments (Hetrick et al., 1989) demonstrated that the addition of phosphorus alone to phosphorus-deficient soil did not increase grass seedling biomass and that the addition of nitrogen increased biomass in mycorrhizal but did not increase biomass in nonmycorrhizal seedlings. These experiments confirm that nitrogen is the most limiting element when mycorrhizae are present, but phosphorus is the most limiting element when mycorrhizae are absent. The implications of these experiments are important for agriculture, for revegetation, and for basic ecological studies. First, the experiments confirm an important role for mycorrhizal fungi in grass seedling growth, and second, the experiments suggest that a reinvestigation of some conclusions that nitrogen is the only or the primary limiting element in grassland systems should be undertaken.

In both tropical soils and in alkaline soils of the western United States, phosphate availability is often low. Low phosphate availability may be a function of fixation of phosphate by iron and aluminum in the soil, and phosphorus fixation is exacerbated in soils with elevated pH conditions. Alkaline iron and aluminum soils result in extremely low levels of available phosphate. Since phosphorus availability is generally low in such soils, the potential importance of mycorrhizae for plant nutrition becomes obvious.

An understanding of the changes in the availabilities of N and P in arid and semiarid environments is essential for understanding competition on such soils and the importance of mycorrhizae formation under such conditions. As environmental conditions change, as root densities change, and as competition for limited resources (moisture and nutrients) changes, belowground conditions are drastically altered. Models, simulations, and observations indicate that root growth into unexploited soil is extremely important in acquiring nutrients that diffuse slowly, and root growth becomes critical as the soil water potential decreases. Root geometry is of minor importance for nutrients that diffuse rapidly (e.g., SO_2, NO_3^-, and NH_4^+), but competition for relatively immobile ions (e.g., PO_4^{3-}, Zn^{2+}, and perhaps K^+) presents alternative problems for the plant (Clarkson, 1985).

III. BACKGROUND FOR MYCORRHIZAL RELATIONSHIPS

Putative mutualism, that is, spores and hyphae of mycorrhizal fungi in subterranean plant structures, is found in the most primitive land plants (*Rhynia* spp. from the Devonian Rhynie chert flora of Scotland) (Kidston and Lang, 1921). Thus mycorrhizal fungi are associated with the first primary succession, the establishment of primitive land plants on previously uninhabited soil. The continuous, 350 million year history of this association, with contemporary fungal representatives that appear to be similar in morphology to the fungi found in the Devonian, Carboniferous, and Triassic (Butler, 1939; Stubblefield et al., 1987), supports the antiquity of this mutualism. The ubiquity of the mycorrhizal symbiosis in extant vascular plants (Trappe, 1987) implies that the relationship is fundamental in the biology of many land plants and that exceptions to the symbiosis are derived rather than primitive. The suggestion has been made that the establishment of mycorrhizal relationships may have been pivotal for the success of a land flora (Pirozynski and Malloch, 1975). By extrapolation, we suggest that symbiotic mycorrhizal associations remain critical for the establishment, development, and maintenance of today's natural ecosystems and for the recovery of severely disturbed ecosystems.

A. Types of Mycorrhizae

There are at least seven major types of mycorrhizal associations, namely, vesicular-arbuscular (V-A) mycorrhizae, ectomycorrhizae, ectendomycorrhizae, ericoid mycorrhizae, orchid mycorrhizae, arbutoid mycorrhizae, and monotropoid mycorrhizae (Harley, 1989). The most common association is the V-A mycorrhizal symbiosis, and ectomycorrhizae are probably the next most common types of symbiosis. Although V-A micorrhizae and ectomycorrhizae differ greatly in their morphology, their physiological functions and effects on hosts are quite similar. Both types of mycorrhizae can occur on the same host (e.g., eucalyptus and leguminous trees). The primary benefits of the symbiosis in most documented cases appear to be associated with an increase in the absorbing surface area of the fungus-root system (Mikola, 1987).

B. Function of Mycorrhizal Associations

Because mycorrhizal associations increase the effective root system of the host, many of the more obvious benefits, such as large increases in host plant biomass, are associated with plants that have relatively thick rather than fibrous root systems (Baylis, 1972). However, some species in the Ericales with fine roots have extensive mycorrhizal colonization (Harley and

Smith, 1983). The significance of such mycorrhizal associations is not fully understood but may be correlated with low soil nitrogen levels (Read, 1987). In general, mycorrhizae offer the potential for plants to increase their ability to capture soil resources. Various species and combinations of species of mycorrhizal fungi and hosts have been studied with reference to the uptake and relative content of P, N, Ca, Mg, Fe, S, Na, Mn, B, Cu, Zn, Si, Br, Cl, I, Pb, Co, Ni, and Cd (Ames and Bethlenfalvay, 1987). Most research has concentrated on changes in P and N uptake in mycorrhizal versus nonmycorrhizal plants. Many experiments demonstrate the enhanced uptake of poorly mobile ions, such as phosphate, copper, zinc, and molybdenum, in mycorrhizal plants when contrasted with nonmycorrhizal plants (Smith, 1980; Hayman, 1983).

In addition to enhancing the uptake of certain poorly mobile elements when in low concentrations in the soil, mycorrhizal colonization is associated with increases in chlorophyll concentrations (Allen et al., 1981a) and rates of photosynthesis (Johnson, 1984), reduced stomatal resistance (Allen et al., 1981b), altered root exudation (Graham et al., 1981), altered hormone content of certain tissues (Allen et al., 1980), and increased drought tolerance (Allen et al., 1981b; Allen and Boosalis, 1983). However, recent research indicates that for selected plants, some of these phenomena, especially changes in water relations, are a secondary consequence of improved phosphorus nutrition and root growth in contrast to direct effects due to mycorrhizal colonization (Koide, 1985; Fitter, 1988). Indeed, Nelsen (1987) suggests that "most changes in water relations reported thus far are likely to be secondary responses due to improved nutrition."

IV. BENEFITS OF MUTUALISM TO HIGHER PLANTS

A. Plant Nutrition

Without argument, when certain plants are grown in soils with deficient levels of essential nutrients, many mycorrhizal plants exhibit substantially and significantly improved mineral nutrition when contrasted with nonmycorrhizal plants of the same species.

In addition to improving P nutrition, recent reviews conclude that the primary role of mycorrhizae in plant N nutrition is to enhance ammonium uptake (Alexander, 1983) or to improve phosphate access, which as a secondary consequence improves N concentrations in host plants (Rousseau and Reid, 1985). Because mycorrhizal hyphae extend some distance from the root, these hyphae can exploit organic and inorganic N sources not directly available to the root (Ames and Bethlenfalvay, 1987). Read (1987) has clearly demonstrated that both ericoid and ectomycorrhizal plants can assimilate protein N. He concluded that mycorrhizal colonization plays a key role in nitrogen nutrition and, further, that the absorption of organic N could lead to a supplemental carbon supply for both the fungus and the host. That certain mycorrhizal fungi can release carbon from lignin, holocellulose, and lignocellulose (Trojanowski et al., 1984) and that certain mycorrhizal fungi have proteolytic ability (Read, 1987) support the suggestion that the organic carbon of the soil can serve as a nutrient source for these heterotrophs.

Interpretations of phenomena vary, but experimental evidence clearly shows that in plants colonized by mycorrhizal fungi phosphorus nutrition is often improved, uptake of other minerals may be improved, and, whether as a direct or indirect result, nitrogen levels and water relations can be altered. Since there are such strong interactions between climate, organisms, topography, parent material (soil and nutrients), and time (Jenny, 1980) and since relative levels of essential nutrients, light, moisture, and hormones profoundly affect plant interactions (Newman, 1983), it is not surprising that the colonization of roots by mycorrhizal fungi can greatly alter the success of plants. Certainly, improvement of a deficient level of a single essential nutrient elicits a chain reaction of events within the plant. During this chain reaction many aspects of overall metabolism may be significantly altered. If colonization of a plant by mycorrhizal fungi improves any of the limiting conditions imposed by the environment, then colonization benefits the plant.

In arid and semiarid environments the major limiting factor for plant survival is often water. Correlated with limited water, however, are limited nutrient levels in the plant. The movement of nutrients in the soil is dependent on sufficient moisture levels, and plant uptake of all nutrients is dependent on plants absorbing nutrients in solution. Further, the diffusion coefficient of some elements, such as phosphorus, is so low that even solubilizing and movement in solution cannot account for the amount of phosphorus needed for plant survival (Reeves, 1988).

Mycorrhizae improve plant phosphorus nutrition in three ways: (1) the hyphae of the fungus extend into the surrounding soil and more effectively mine the limited phosphorus supply than can the root and root hairs alone; (2) the hyphae absorb water from the soil and the nutrients dissolved in the water, thus increasing phosphorus available to the plant; and (3) some mycorrhizal fungi produce extracellular phosphatases that can release phosphorus from the organic materials in the soil.

Autotropism in plants, in part, depends on their ability to mine (for most essential nutrients) or otherwise obtain (e.g., nitrogen fixation) nutrients from the environment. In plants and certain microbes, this strategy involves acquisition of mineral nutrients directly from the adjacent external medium (Goldstein et al., 1989). Typically, there are direct costs associated with nutrient extraction. Unless there are abundant supplies of minerals (and water), all plants must invest substantive fixed carbon in new root tissue. Since "about 95 percent of the world's present species of vascular plants belong to families that are characteristically mycorrhizal" (Trappe, 1977), additional incurred costs are expected for supporting the fungal symbionts. For nitrogen-fixing associations, costs in addition to supporting mycorrhizal fungi must be supported.

Recent work (O'Neill et al., 1987) demonstrates that increased (approximately doubled) atmospheric CO_2 levels lead to increased pine and oak seedling biomass, root biomass, mycorrhizal density, and rates of colonization and proliferation. The authors suggest that increased photosynthesis at elevated CO_2 levels compensates for the increased carbon requirement of the mycorrhizal fungus and that greater mycorrhizal density and increased root growth can facilitate nutrient acquisition, which in turn supports additional biomass increases at elevated CO_2 levels. Clearly there are interactions among carbon acquisition and utilization, mycorrhizae formation and development, and nutrient limitation in such short-term experiments. It remains to be determined whether the observed increases in biomass and mycorrhizal density would continue for long periods of time. Based on the response in this experiment, CO_2 appears to be a limiting "nutrient;" in semiarid and arid environments other limiting factors may be much more important than CO_2 levels.

B. Other Benefits

The question of what constitutes benefit is not resolved. Traditionally, benefit from mycorrhizal colonization has been measured in improved biomass or yield and quantified as a mycorrhizal dependency value for a given species (Gerdemann, 1975). Recent arguments point out that the mycorrhizal response(s) does not always involve significant increases in biomass or yield (Reeves, 1987, 1988; Buckalew, 1989; Harley, 1989), that "mycorrhizal responsiveness" is a more appropriate and inclusive term that describes increases in both biomass and yield as well as other benefits (Reeves, 1987, 1988), and that "we can only judge benefit to the host or to the fungus by determining the selective advantage to them of any traits or characters that we measure" (Harley, 1989).

Since there are obvious and cryptic potential benefits to plants that result from colonization by mycorrhizal fungi, the function of mycorrhizal fungi in competitive interactions between plants and in the establishment and development of plants have been addressed with limited success. Many reviews of mycorrhizal fungi include the equivalent of "the ecology of mycorrhizal fungi" in their content. In reality, such headings typically deal with the ecology of higher plants and the supposed interactions of mycorrhizal fungi and include little experimental evidence to support the hypotheses and conclusions.

Despite the relative lack of data, during the last quarter century there has been an increasing interest in the role of mycorrhizae in controlling, directing, and influencing the establishment of individual plants and natural communities (Schramm, 1966; Reeves et al., 1979; Miller, 1979, 1987a; Janos, 1980, 1988; Mooney, 1984). Walker and Chapin (1987) include formation of mycorrhizae as one of the ten major successional processes that determine change in species composition during succession. MacMahon (1987) points out that studies of patterns of succession on mine spoils "has been the key to elucidating the overall importance of mycorrhizae in all of our ecological studies."

V. MUTUALISM AND DISTURBANCE

A survey of recent literature that deals with ecological relationships and mycorrhizal fungi indicates that mutualism, especially mycorrhizal associations, is recognized as an important and, at times, essential component in re-establishing plant communities on disturbed sites (MacMahon, 1987; Miller, 1987a; Allen, 1988; Janos, 1988; Reeves, 1988). In an excellent review of the mycorrhizal literature associated with grasslands and shrublands, Miller (1987b) includes an extensive discussion of mycorrhizae, disturbance, and succession. He emphasizes that "much of what we have learned from agricultural systems about mycorrhizae may not be relevant to these more stressed environments when maximum biomass production may not be advantageous." In examples in which the plant-soil link is destroyed, as in heavily disturbed or mined lands (Reeves et al., 1979; Miller, 1979; Janos, 1980) or certain clear-cuts in productive forests, attempts at revegetation may not be successful, presumably because the plant-soil links were not successfully re-established (Perry and Rose, 1983; Amaranthus and Perry, 1987; Perry et al., 1989). Several early studies indicated that mycorrhizal fungi appear to be significant components in the revegetation of disturbed land (Daft and Nicolson, 1974; Daft et al., 1975; Khan, 1978; Reeves et al., 1979; Miller, 1979). Mycorrhizal mutualism may be a key element in maintaining successful communities in arid and semiarid soils (Miller, 1987a; Allen, 1988). Perry et al. (1989) emphasize that mutualisms (both nitrogen fixers and mycorrhizae) are especially important when water or nutrients are in short supply. They further point out that seasonal constraints become important in those environments that have a narrow window of opportunity for the re-establishment of plants after disturbance. Mycorrhizal fungi alter succession and community composition by altering the relative success of plants, both in the seedling stage and during later development (Reeves et al., 1979; Miller, 1979, 1987a; Allen and Allen, 1984, 1986, 1988; Williams and Allen, 1984; Caldwell et al., 1985; Grime et al., 1987; Hetrick et al., 1989).

Various studies indicate that V-A mycorrhizal fungal populations decrease when soils are disturbed (Moorman and Reeves, 1979; Jasper et al., 1987; Evans and Miller, 1988), that populations decrease with increasing soil depth in semiarid soils (Schwab and Reeves, 1981), and that soil disturbance reduces V-A mycorrhizal colonization and hence nutrient absorption (Fairchild and Miller, 1988). Several studies indicate that storing topsoil for extended periods of time results in a decrease in both microbial biomass and mycorrhizal inoculum potential (Rives et al., 1980; Gould and Liberta, 1981; Ross and Cairns, 1981; Visser et al., 1983; Abdul-Kareem and McRae, 1984; Miller et al. 1985; Harris et al., 1987). Recent studies indicate that individual populations of V-A mycorrhizal fungi may have narrow ranges of tolerance to environmental conditions (Stahl et al., 1988) and that different species of V-A mycorrhizal fungi have different survival strategies that result in major differences in species distributions in semiarid soils that exhibit large changes in wet-dry cycles (McGee, 1989).

Much evidence indicates that succession on arid and semiarid lands is influenced in part by the relative density of the mycorrhizal inoculum. However, vegetation patterns and mycorrhizal dynamics may be closely associated with small patches or shrub islands (Allen, 1987a). Shrub islands could initialize the establishment of additional areas of mycorrhizal inoculum, and these islands could alter wind patterns and cause the patterns to concentrate wind-dispersed mycorrhizal inoculum. Before obligately mycorrhizal plants grow on soils devoid of mycorrhizal fungi, the fungi must migrate from outside the area or from residual survival pockets, such as shrub islands (Allen, 1987b).

VI. DIRECTIONS FOR PRACTICAL APPLICATIONS

An excellent summary of the potential practical use of mycorrhizae in agricultural systems is available (Jeffries, 1987), as is a summary of the methods and principles used in mycorrhizal research (Schenck, 1982). The major theses presented in the review of agricultural systems are applicable to the recovery of natural systems, with the proviso that caution must be exercised when extrapolating results from managed monoculture(s) to complex, potentially heterogenous plant polycultures under stress conditions.

In general, many native and crop plants show significant responses to mycorrhizal colonization. Results from pot or greenhouse experiments, however, may not be reliable indicators of potential results from field experiments. However, some controlled experiments have been excellent indicators of potential field results. Perhaps the most widely known example is that of the utilization of ectomycorrhizal fungi on container-grown conifer seedlings (Marx et al., 1978). The entire forest industry now recognizes the absolute essential requirement of mycorrhizal fungi for normal growth in conifer species.

Practical applications of V-A mycorrhizal fungi to relatively sterile soils presents a much greater problem. First, it is relatively difficult to produce large quantities of viable and clean inoculum that can be easily manipulated. Second, as a general rule, indigenous mycorrhizal fungi are often superior to imported mycorrhizal fungi in terms of native plant performance. Thus, for most situations, the preferred protocol of ensuring an adequate inoculum of V-A mycorrhizal fungi is to preserve those fungi already present at a particular site (by using careful topsoil storage techniques or by establishing plant species able to host the fungi present at the site), to establish shrub islands (Allen, 1987a) as sites of migration, or to employ natural wind barriers, analogous to snow fences, that aid in the collection of mycorrhizal propagules (Miller, 1987a; Allen, 1987a).

Obviously, additional basic research on the nature of succession in arid and semiarid environments must be completed before the most cost-effective and efficient methods of reinoculating soil with V-A mycorrhizal fungi can be proposed. Just as plants and mycorrhizal fungi benefit from mutualism, so, too, studies in basic and applied ecology can benefit from studies of mutualism, especially mycorrhizal relationships in native and restored plant communities.

VII. CONCLUSIONS

The strong correlation of the appearance of mycorrhizal fungi with increased plant competition, with tightened nutrient cycling, and with altered plant community composition supports the essential role of these symbiotic fungi in controlling and directing plant community patterns. Further, in the exceptional cases in which these fungi are absent, data support their essential role in re-establishing stable and productive communities in stressful environments. Biologists should remain cognizant that many of the benefits resulting from mycorrhizal associations may be subtle, not expressed as obvious increases in plant biomass. Over the millennia, nature has completed many experiments and natural selection has continued to operate. The result is that the mycorrhizal condition is the rule rather than the exception. Clearly, the association offers benefits, especially when one considers that many plants cannot exist in nature without the fungi. Our challenge is to determine what the benefits are and how best to exploit, replace, or maintain the benefits if we are to maintain stable vegetation patterns in competitive environments.

ACKNOWLEDGMENTS

The authors wich to acknowledge the U.S. Department of Energy (DE-FG02-87ER60612) for support during the writing of this chapter.

REFERENCES

Abdul-Kareem, A. W., and McRae, S. G. (1984). The effects on topsoil of long-term storage in stockpiles. *Plant Soil 76*: 357-363.

Alexander, I. J. (1983). Ectomycorrhizas in the nitrogen cycle. In *Nitrogen as an Ecological Factor*, J. A. Lee, S. McNeil, and I. H. Rorison (Eds.). Blackwell, Oxford, pp. 69-93.

Allen, E. B., and Allen, M. F. (1984). Competition between plants of different successional stages: Mycorrhizae as regulators. *Can. J. Bot. 62*: 2625-2629.

Allen, E. B., and Allen, M. F. (1986). Water relations of xeric grasses in the field: Interactions of mycorrhizas and competition. *New Phytol. 104*: 559-571.

Allen, M. F. (1987a). Ecology of vesicular-arbuscular mycorrhizae in an arid ecosystem: Use of natural processes promoting dispersal and establishment. In *Mycorrhizae in the Next Decade*, D. M. Sylvia, L. L. Hung, and J. H. Graham (Eds.). Institute of Food Agricultural Science, University of Florida, Gainesville, pp. 133-135.

Allen, M. F. (1987b). Re-establishment of mycorrhizas on Mount St. Helens: Migration vectors. *Trans. Br. Mycol. Soc. 88*: 413-417.

Allen, M. F. (1988). Belowground structure: A key to reconstructing a productive arid ecosystem. In *The Reconstruction of Disturbed Arid Lands: An Ecological Approach*, E. B. Allen (Ed.). AAAS Selected Symposia Series, Westview Press, Boulder, Colorado, pp. 113-135.

Allen, E. B., and Allen, M. F. (1988). Facilitation of succession by the non-mycotrophic colonizer *Salsola kali* (Chenopodiaceae) on a harsh site: Effects of mycorrhizal fungi. *Am. J. Bot. 75*: 257-266.

Allen, M. F., and Boosalis, M. G. (1983). Effects of two species of VA mycorrhizal fungi on drought tolerance of winter wheat. *New Phytol. 93*: 67-76.

Allen, M. F., Moore, T. S., Jr., and Christensen, M. (1980). Phytohormone changes in *Bouteloua gracilis* infected by vesicular-arbuscular mycorrhizae. I. Cytokinin increases in the host plant. *Can. J. Bot. 58*: 371-374.

Allen, M. F., Sexton, J. C., Moore, T. S., Jr., and Christensen, M. (1981a). Influence of phosphorus source on vesicular-arbuscular mycorrhizae of *Bouteloua gracilis*. *New Phytol. 87*: 687-694.

Allen, M. F., Smith, W. K., Moore, T. S., Jr., and Christensen, M. (1981b). Comparative water relations and photosynthesis of mycorrhizal and non-mycorrhizal *Bouteloua gracilis* (H.B.K.) Lag. ex Steud. *New Phytol. 88*: 683-693.

Amaranthus, M. P., and Perry, D. A. (1987). The effect of soil transfers on ectomycorrhizal formation and the survival and growth of conifer seedlings on old, nonreforested clearcuts. *Can. J. For. Res. 17*: 944-950.

Ames, R. N., and Bethlenfalvay, G. J. (1987). Mycorrhizal fungi and the integration of plant and soil nutrients dynamics. *J. Plant Nutr. 10*: 1313-1321.

Baylis, G. T. S. (1972). Fungi, phosphorus and the evolution of plant roots. *Search 3*: 257-258.

Bazzaz, F. A. (1979). The physiological ecology of plant succession. *Annu. Rev. Ecol. Syst. 10*: 351-372.

Beadle, N. C. W. (1966). Soil phosphate and its role in molding segments of the Australian flora and vegetation with special reference to xeromorphy and sclerophylly. *Ecology 47*: 992-1007.

Biondini, M. E., Bonham, C. D., and Redente, E. F. (1984). Relationships between induced successional patterns and soil biological activity of reclaimed areas. *Reclamation Reveg. Res. 3*: 323-342.

Biondini, M. E., Bonham, C. D., and Redente, E. F. (1985). Secondary successional patterns in a sagebrush (*Artemisia tridentata*) community as they relate to soil disturbance and soil biological activity. *Vegetatio 60*: 25-36.

Bornkamm, R., and Hennig, V. (1982). Experimental ecological study of succession of ruderal plant communities on different soils. 1. Florisitic composition of vegetation. *Flora (Jena) 172*: 267-316.

Bowen, G. D. (1980). Misconceptions, concepts and approaches in rhizosphere biology. In *Contemporary Microbial Ecology*, D. D. Ellwood, J. N. Hedger, J. M. Lynch, and J. H. Slater (Eds.). Academic Press, New York, pp. 283-304.

Box, T. W. (1961). Relationship between plants and soils for four range plant communities in south Texas. *Ecology 42*: 794-810.

Braakhekke, W. G. (1985). The significance of competition for plant diversity. *Z. Zool. Syst. Evolut.-forsch. 23*: 315-327.

Brinck, P., Nilsson, L. M., and Svedin, U. (1988). Ecosystem redevelopment. *Ambio 17*: 84-89.

Buckalew, D. W. (1989). The effects of VA mycorrhizal colonization on concentrations of phosphorus and nitrogen in grasses. Ph.D. dissertation, Colorado State University, Fort Collins.

Butler, E. J. (1939). The occurrences and systematic position of the vesicular-arbuscular type of mycorrhizal fungi. *Trans. Br. Mycol. Soc. 22*: 274-301.

Caldwell, M. M., Eissenstat, D. M., Richards, J. H., and Allen, M. F. (1985). Competition for phosphorus: Differential uptake from dual-isotope-labeled soil interspaces between shrub and grass. *Science 229*: 384-386.

Chan, E. C. S., and Katznelson, H. (1961). Growth interactions of *Arthrobacter globiformis* and *Pseudomonas* sp. in relation to the rhizosphere effect. *Can. J. Microbiol. 7*: 759-767.

Chapin, F. S., III (1980). The mineral nutrition of wild plants. *Annu. Rev. Ecol. Syst. 11*: 233-260.

Chapin, F. S., III (1983). Patterns of nutrient absorption and use by plants from natural and man-modified environments. In *Disturbance and Ecosystems*, H. A. Mooney and M. Godron (Eds.). Springer-Verlag, New York, pp. 175-187.

Clarkson, D. T. (1985). Factors affecting mineral nutrient acquisition by plants. *Annu. Rev. Plant Physiol. 36*: 77-115.

Connell, J. H., and Slatyer, R. O. (1977). Mechanisms of succession in natural communities and their role in community stability and organization. *Am. Naturalist 111*: 1119-1144.

Daft, M. J., and Nicolson, T. H. (1974). Arbuscular mycorrhizas in plants colonizing coal wastes in Scotland. *New Phytol. 73*: 1129-1138.

Daft, M. J., Hacskaylo, E., and Nicolson, T. H. (1975). Arbuscular mycorrhizas in plants colonizing coal spoils in Scotland and Pennsylvania. In *Endomycorrhizas*, F. E. Sanders, B. Mosse, and P. B. Tinker (Eds.). Academic Press, New York, pp. 561-580.

Doerr, T. B., Redente, E. F., and Reeves, F. B. (1984). Effects of soil disturbance and levels of mycorrhizal fungi in a sagebrush-grassland community. *J. Range Manag. 37*: 135-139.

Evans, D. G., and Miller, M. H. (1988). Vesicular-arbuscular mycorrhizas and the soil-disturbance-induced reduction of nutrient absorption in maize. I. Causal relations. *New Phytol. 110*: 67-74.

Fairchild, G. L., and Miller, M. H. (1988). Vesicular-arbuscular mycorrhizas and the soil-disturbance-induced reduction of nutrient absorption in maize. II. Development of the effect. *New Phytol. 110*: 75-84.

Fitter, A. H. (1988). Water relations of red clover *Trifolium pratense* L. as affected by VA mycorrhizal infection and phosphorus supply before and during drought. *J. Exp. Bot. 39*: 595-603.

Fowler, N. (1986). The role of competition in plant communities in arid and semiarid regions. *Annu. Rev. Ecol. Syst. 17*: 89-110.

Gerdemann, J. W. (1975). Vesicular-arbuscular mycorrhizae. In *The Development and Function of Roots*, J. G. Torrey and D. T. Clarkson (Eds.). Academic Press, New York, pp. 491-575.

Goldstein, A. H., Baertlein, D. A., and Danon, A. (1989). Phosphate starvation stress as an experimental system for molecular analysis. *Plant Mol. Biol. Rep. 7*: 7-16.

Gorham, E., Vitousek, P. M., and Reiners, W. A. (1979). The regulation of chemical budgets over the course of terrestrial ecosystem succession. *Annu. Rev. Ecol. Syst. 10*: 53-84.

Gould, A. B., and Liberta, A. E. (1981). Effects of topsoil storage during surface mining on the viability of vesicular-arbuscular mycorrhiza. *Mycologia 73*: 914-922.

Graham, J. H., Leonard, R. T., and Menge, J. A. (1981). Membrane mediated decrease in root exudation responsible for phosphorus inhibition of vesicular-arbuscular mycorrhizae formation. *Plant Physiol. 68*: 548-552.

Grime, J. P., Mackey, J. M., Hillier, S. H., and Read, D. J. (1987). Floristic diversity in a model system using experimental microcosms. *Nature 328*: 420-422.

Harcombe, P. A. (1977). The influence of fertilization on some aspects of succession in a humid tropical forest. *Ecology 58*: 1375-1383.
Harley, J. L. (1989). The significance of mycorrhiza. *Mycol. Res. 92*: 129-139.
Harley, J. L., and Smith, S. E. (1983). *Mycorrhizal Symbiosis.* Academic Press, New York.
Harris, J. A., Hunter, D., and Birch, P. (1987). Vesicular-arbuscular mycorrhizal populations in stored topsoil. *Trans. Br. Mycol. Soc. 89*: 600-603.
Hayman, D. S. (1983). The physiology of vesicular-arbuscular endomycorrhizal symbiosis. *Can. J. Bot. 61*: 944-963.
Heal, O. W., and Dighton, J. (1986). Nutrient cycling and decomposition in natural terrestrial ecosystems. In *Microflora and Faunal Interactions in Natural and Agro-Ecosystems,* M. J. Mitchell and J. P. Nakas (Eds.). Martin Nijhoff/Dr. W. Junk, Dordrecht, The Netherlands, pp. 14-73.
Hetrick, B. A. D., Wilson, G. W. T., and Owensby, C. E. (1989). Influence of mycorrhizal fungi and fertilization on big bluestem seedling biomass. *J. Range Manag. 42*: 213-216.
Hopkinson, C. S., and Schubauer, J. P. (1984). Static and dynamic aspects of nitrogen cycling in the salt march graminoid *Spartina alterniflora. Ecology 65*: 961-969.
Janos, D. P. (1980). Mycorrhizae influence tropical succession. *Biotropica 12*(Suppl): 56-64.
Janos, D. P. (1988). Mycorrhiza applications in tropical forestry: Are temperate-zone approaches appropriate? In *Trees and Mycorrhiza,* F. S. P. Ng (Ed.). Forest Research Institute Malaysia, Kuala Lumpur, pp. 133-188.
Jasper, D. A., Robson, A. D., and Abbott, L. K. (1987). The effect of surface mining on the infectivity of vesicular-arbuscular mycorrhizal fungi. *Aust. J. Bot. 35*: 641-652.
Jeffries, P. (1987). Use of mycorrhizae in agriculture. In *Critical Reviews in Biotechnology,* Vol. 5, G. G. Stewart and I. Russell (Eds.). CRC Press, Boca Raton, Florida, pp. 319-357.
Jenny, H. (1980). *The Soil Resource. Ecological Studies,* Vol. 37. Springer-Verlag, New York.
Johnson, C. R. (1984). Phosphorus nutrition on mycorrhizal colonization, photosynthesis, growth, and nutrient composition of *Citrus arantium. Plant Soil 80*: 35-42.
Kapustka, L. A., and Rice, E. L. (1976). Acetylene reduction (N_2 fixation) in soil and old field succession in central Oklahoma. *Soil Biol. Biochem. 8*: 497-503.
Khan, A. G. (1978). Vesicular-arbuscular mycorrhizas in plants colonizing black wastes from bituminous coal mining in the Illawarra region of New South Wales. *New Phytol. 81*: 53-63.
Kidston, R., and Lang, W. H. (1921). On Old Red Sandstone plants showing structure from the Rhynie chert bed. Aberdeenshire. *V. Trans. Roy. Soc. Edinb. 52*: 855-902, Pl. I-X.
Klein, D. A., Sorensen, D. L., and Redente, E. F. (1985). Soil enzymes: A predictor of reclamation potential and progress. In *Soil Reclamation Processes: Microbiological Analyses and Applications,* R. L. Tate, III, and D. A. Klein (Eds.). Marcel Dekker, New York, pp. 141-171.

Koide, R. (1985). The effect of VA mycorrhizal infection and phosphorus status on sunflower hydraulic and stomatal properties. *J. Exp. Bot.* *36*: 1087-1098.

Lamb, D. (1981). Soil nitrogen mineralization in a secondary rainforest succession. *Oecologia (Berl.)* *47*: 257-263.

Lochhead, A. G., and Chase, F. E. (1943). Qualitative studies of soil microorganisms. VI. Nutritional requirements of the predominant bacterial flora. *Soil Sci.* *55*: 185-195.

Lodhi, M. A. K. (1979). Inhibition of nitrifying bacteria, nitrification and mineralization in soil as related to successional stages. *Bull. Torrey Bot. Club* *106*: 284-289.

MacMahon, J. A. (1980). Ecosystems over time: Succession and other types of change. In *Forests: Fresh Perspectives from Ecosystem Analyses*, R. Waring (Ed.). Proceedings of the Oregon State University Biology Colloquium, April 26-27, 1979, Corvallis, pp. 27-58.

MacMahon, J. A. (1987). Disturbed lands and ecological theory: An essay about a mutualistic association. In *Restoration Ecology, A Synthetic Approach to Ecological Research*, W. R. Jordan, III, M. E. Gilpin, and J. D. Aber (Eds.). Cambridge University Press, New York, pp. 221-237.

Marx, D. H., Morris, W. G., and Mexal, J. G. (1978). Growth and ectomycorrhizal development of loblolly pine seedlings in fumigated and nonfumigated soil infested with different fungal symbionts. *For. Sci.* *24*: 193-203.

McGee, P. A. (1989). Variation in propagule numbers of vesicular-arbuscular mycorrhizal fungi in semi-arid soil. *Mycol. Res.* *92*: 28-33.

Meller, H. (1979). The influence of common heather (*Calluna vulgaris*) on the activity of enzymes in the soil demonstrated at the sandy shores of the Baltic Sea coast of Schleswig-Holstein. *Flora (Jena)* *168*: 320-328.

Metzger, W. C., Klein, D. A., and Redente, E. F. (1986). Bacterial physiological diversity in the rhizosphere of range plants in response to related shale stress. *Appl. Environ. Microbiol.* *52*: 765-770.

Mikola, P. (1987). Mycorrhizae under tropical stress. *Angew. Bot.* *61*: 15-23.

Miller, R. M. (1979). Some occurrences of vesicular-arbuscular mycorrhiza in natural and disturbed ecosystems of the Red Desert. *Can. J. Bot.* *57*: 619-623.

Miller, R. M. (1987a). Mycorrhizae and succession. In *Restoration Ecology, a Synthetic Approach to Ecological Research*, W. R. Jordan, III, M. E. Gilpin, and J. D. Aber (Eds.). Cambridge University Press, New York, pp. 205-219.

Miller, R. M. (1987b). The ecology of vesicular-arbuscular mycorrhizae in grass- and shrublands. In *The Ecophysiology of VA Mycorrhizal Plants*, G. Safir (Ed.). CRC Press, Boca Raton, Florida, pp. 135-170.

Miller, R. M., Carnes, B. A., and Moorman, T. B. (1985). Factors influencing survival of vesicular-arbuscular mycorrhiza propagules during topsoil storage. *J. Appl. Ecol.* *22*: 259-266.

Montes, R. A., and Christensen, V. L. (1979). Nitrification and succession in the piedmont of North Carolina, U.S.A. *For. Sci.* *25*: 287-297.

Mooney, H. A. (1984). Progress and promise in plant physiological ecology. In *Trends in Ecological Research for the 1980's*, H. Cooley and F. B. Golley (Eds.). Plenum Press, New York, pp. 5-17.

Moorman, T. B., and Reeves, F. B. (1979). The role of endomycorrhizae in revegetation practices in the semi-arid West. II. A bioassay to determine the effect of land disturbance on endomycorrhizal populations. *Am. J. Bot.*, 14-18.

Mott, C. J. B. (1981). Anion and ligand exchange. In *The Chemistry of Soil Processes*, D. J. Greeland and M. H. B. Hayes (Eds.). John Wiley & Sons, New York, pp. 179-219.

Mueller, L. D. (1988). Evolution of competitive ability in *Drosphila* by density-dependent natural selection. *Proc. Natl. Acad. Sci. USA 85*: 4383-4386.

Nelsen, C. E. (1987). The water relations of vesicular-arbuscular mycorrhizal systems. In *Ecophysiology of VA Mycorrhizal Plants*, G. Safir (Ed.). CRC Press, Boca Raton, Florida, pp. 71-91.

Newman, E. I. (1983). Interactions between plants. In *Physiological Plant Ecology. III. Responses to the Chemical and Biological Environment*, O. L. Lange, P. S. Nobel, C. B. Osmond, and H. Ziegler (Eds.). Springer-Verlag, New York, pp. 681-710.

Odum, E. P. (1969). The strategy of ecosystem development. *Science 164*: 262-270.

O'Neill, E. G., Luxmoore, R. J., and Norby, R. J. (1987). Increases in mycorrhizal colonization and seedling growth in *Pinus echinata* and *Quercus alba* in an enriched CO_2 atmosphere. *Can. J. For. Res. 17*: 878-883.

Pancholy, S. K., and Rice, E. L. (1973). Soil enzymes in relation to old field succession: Amylase, cellulase, invertase, dehydrogenase, and urease. *Soil Sci. Soc. Am. Proc. 37*: 47-50.

Parrish, J. A. D., and Bazzaz, F. A. (1982a). Competitive interactions in plant communities of different successional ages. *Ecology 63*: 314-320.

Parrish, J. A. D., and Bazzaz, F. A. (1982b). Responses of plants from three successional communities to a nutrient gradient. *J. Ecol. 70*: 233-248.

Perry, D. A., and Rose, S. L. (1983). Soil biology and forest productivity: Opportunities and constraints. In *International Union of Forest Research Organizations Symposium on Forest Site and Continuous Productivity*, R. Ballard and S. P. Gessel (Eds.). USDA Forest Service General Technical Report PNW-163, Portland, Oregon, pp. 229-238.

Perry, D. A., Amaranthus, M. P., Borchers, J. G., Borchers, S. L., and Brainerd, R. E. (1989). Bootstrapping in ecosystems. *BioScience 39*: 230-237.

Pigott, C. D., and Taylor, K. (1964). The distribution of some woodland herbs in relation to the supply of nitrogen and phosphorus in the soil. *J. Ecol. 52*(Suppl.): 175-185.

Pirozynski, K. A., and Malloch, D. W. (1975). The origin of land plants: A matter of mycotrophism. *BioSystems 6*: 153-164.

Read, D. J. (1987). In support of Frank's organic nitrogen theory. *Angew. Bot. 61*: 25-37.

Redente, E. F., Stark, J. M., Biondini, M. E., and Oliver, T. A. (1984). Vegetation structure and succession as they relate to soil disturbance and retorted oil shale. In *Ecological Studies of Natural and Established Ecosystems on Energy Related Disturbances in Colorado*, E. F. Redente and C. W. Cook (Eds.). U.S. DOE Research Report No. DE-AC02-76EV04018, Range Sci. Dept., Colorado State University, Fort Collins, pp. 1-35.

Reeves, F. B. (1987). Mycorrhizal responsiveness and ecology of VAM in arid and semiarid ecosystems. In *Mycorrhizae in the Next Decade*, D. M. Sylvia, L. L. Hung, and J. H. Graham (Eds.). Institute of Food Agricultural Science, University of Florida, Gainesville, pp. 136-138.

Reeves, F. B. (1988). Mineral nutrition, mycorrhizal fungi and succession in semiarid environments. In *Trees and Mycorrhiza*, F. S. P. Ng (Ed.). Forest Research Institute of Malaysia, Kuala Lumpur, pp. 33-50.

Reeves, F. B., Wagner, D., Moorman, T., and Kiel, J. (1979). The role of endomycorrhizae in the revegetation practices in the semi-arid West. I. A comparison of incidence of mycorrhizae in severely disturbed vs. natural environments. *Am. J. Bot. 66*: 6-13.

Reiners, W. A. (1981). Nitrogen cycling in relation to ecosystem succession. In *Terrestrial Nitrogen Cycles. Processes, Ecosystem Stragegies and Management Impacts. Ecol. Bull. (Stockh.) 33*: 507-528.

Rice, E. L. (1974). Inhibition of nitrification by vegetation: Increases during succession and pronounced inhibition by climax ecosystems. In *Allelopathy*, E. L. Rice (Ed.). Academic Press, New York, pp. 77-103.

Rice, E. L. (1984). *Allelopathy*, 2nd ed. Academic Press, New York.

Rice, E. L. and Mallick, M. A. B. (1977). Causes of decreases in residual carbohydrate activity in soil during old-field succession. *Ecology 58*: 1297-1309.

Richards, J. F. (1986). World environmental history and economic development. In *Sustainable Development of the Biosphere*, W. E. Clark and R. E. Munn (Eds.). Cambridge University Press, Cambridge, pp. 53-71.

Rives, C. S., Bajwa, M. I., Liberta, A. E., and Miller, R. M. (1980). Effects of topsoil storage during surface mining on the viability of VA mycorrhiza. *Soil Sci. 129*: 253-257.

Robertson, G. P., and Vitousek, P. M. (1982). Nitrification potentials in primary and secondary succession. *Ecology 62*: 376-386.

Ross, D. J., and Cairns, A. (1981). Nitrogen availability and microbial biomass in stockpiled topsoils in Southland. *N.Z. J. Sci. 24*: 127-143.

Rosseau, J., and Reid, C. P. P. (1985). Interaction of nitrogen phosphorus and mycorrhiza inoculation on nutrient content and growth of *Pinus contorta*. In *Proceedings of the Sixth North American Conference on Mycorrhizae*, R. Molina (Ed.). Forest Research Laboratory, Oregon State University, Corvallis, p. 340.

Schenck, N. C. (1982). *Methods and Principles of Mycorrhizal Research*. American Phytopathological Society, St. Paul, Minnesota.

Schramm, J. E. (1966). Plant colonization studies on black wastes from anthracite mining in Pennsylvania. *Trans. Am. Philos. Soc. 56*: 1-94.

Schwab, S., and Reeves, F. B. (1981). The role of endomycorrhizae in the revegetation practices in the semiarid West. III. Vertical distribution of vesicular-arbuscular mycorrhizal inoculum potential. *Am. J. Bot. 68*: 1293-1297.

Shreve, F. (1942). The desert vegetation of North America. *Bot. Rev. 8*: 195-246.

Smith, S. E. (1980). Mycorrhizas of autotrophic higher plants. *Biol. Rev. 55*: 475-510.

Snaydon, R. W. (1962). Micro-distribution of *Trifolium repens* L. and its relation to soil factors. *J. Ecol. 50*: 133-143.

Sorensen, D. L. (1982). Biochemical activities in soil overlying Paraho processed oil shale. Ph.D. Dissertation, Colorado State University, Fort Collins.

Speir, T. W., and Ross, D. T. (1978). Soil phosphatase and sulfatase. In *Soil Enzymes*, R. G. Burnes (Ed.). Academic Press, London, pp. 197-250.

Stahl, P. D., Williams, S. E., and Christensen, M. (1988). Efficacy of native vesicular-arbuscular mycorrhizal fungi after severe soil disturbance. *New Phytol. 110*: 347-354.

Starkey, R. L. (1929). Some influences of the development of higher plants upon the microorganisms of the soil. III. Influence at the stage of plant growth upon some activities of the organisms. *Soil Sci. 27*: 433-444.

Stubblefield, S. P., Taylor, T. N., and Trappe, J. M. (1987). Vesicular-arbuscular mycorrhizae from the Triassic of Antarctica. *Am. J. Bot. 74*: 1940-1911.

Swift, M. J., and Heal, O. W. (1986). Theoretical considerations of microbial succession and growth strategies: Intellectual exercise or practical necessity? In *Microbial Communities in Soil*, J. Jensen, A. Kjoller, and L. H. Sorensen (Eds.). Elsevier Applied Science, London, pp. 115-131.

Tilman, D. (1982). *Resource Competition and Community Structure*. Princeton University Press, Princeton, New Jersey.

Tilman, D. (1986). Nitrogen-limited growth in plants from different successional stages. *Ecology 67*: 555-563.

Tilman, D. (1988). *Plant Strategies and the Dynamics and Structure of Plant Communities*. Princeton University Press, Princeton, New Jersey.

Tilman, G. D. (1984). Plant dominance along an experimental nutrient gradient. *Ecology 65*: 1445-1453.

Tilsdale, S. L., Nelson, W. L., and Beaton, J. D. (1985). *Soil Fertility and Fertilizers*. Macmillan, New York.

Titlyanova, A. A. (1982). Ecosystem succession and biological turnover. *Vegetatio 50*: 43-51.

Trappe, J. M. (1977). Selection of fungi for ectomycorrhizal inoculation in nurseries. *Annu. Rev. Phytopathol. 15*: 203-222.

Trappe, J. M. (1987). Phylogenetic and ecologic aspects of mycotrophy in the angiosperms from an evolutionary standpoint. In *Ecophysiology of VA Mycorrhizal Plants*, G. R. Safir (Ed.). CRC Press, Boca Raton, Florida, pp. 5-25.

Trojanowski, J., Haider, K., and Hutterman, A. (1984). Decomposition of ^{14}C-labelled lignin, holocellulose, and lignocellulose by mycorrhizal fungi. *Arch. Microbiol. 139*: 202-206.

Visser, S., Fujikawa, J., Giffiths, C. L., and Parkinson, D. (1983). Effect of topsoil storage on microbial activity, primary production and decomposition potential. *Plant Soil 82*: 41-50.

Vitousek, P. (1982). Nutrient cycling and nutrient use efficiency. *Am. Naturalist 119*: 553-572.

Vitousek, P. M., and Reiners, W. A. (1975). Ecosystem succession and nutrient retention: A hypothesis. *BioScience 25*: 376-381.

Vitousek, P. M., Gosz, J. R., Grier, C. C., Melillo, J. M., Reiners, W. A., and Todd, R. L. (1979). Nitrate losses from disturbed ecosystems. *Science 204*: 469-474.

Walker, L. R., and Chapin, F. S., III (1987). Interactions among processes controlling successional change. *Oikos 50*: 131-135.

Went, F. W. (1955). The ecology of desert plants. *Sci. Am. 192*: 68-75.

Wheeler, G. L., and Donaldson, J. M. (1983). Nitrification in an upland forest sere. *Soil Biol. Biochem. 15*: 119-121.

Whipps, J. M., and Lynch, J. M. (1986). The influence of the rhizosphere on crop productivity. In *Advances in Microbial Ecology*, Vol. 9, E. C. Marshall (Ed.). Plenum Press, New York, pp. 187-244.

Williams, S. E., and Allen, M. F. (1984). VA mycorrhizae and reclamation of arid and semi-arid lands. University of Wyoming Agr. Exp. Sta. Rep. No. SA1261, Laramie.

Woodmansee, R. G. (1978). Additions and losses of nitrogen in grassland systems. *BioScience 28*: 448-453.

Woodmansee, R. G., Dodd, J. L., Bowman, R. A., and Clark, F. E. (1978). Nitrogen budget of a shortgrass prairie ecosystem. *Oecologia (Berl.) 34*: 363-376.

14

Nitrogen Fixation by Desert Legumes Associated with Rhizobia

I. L. PEPPER and R. P. UPCHURCH *University of Arizona, Tucson, Arizona*

I. THE *Rhizobium*-LEGUME ASSOCIATION

A. Dynamics of the Symbiosis

Soil exists as a discontinuous matrix of organic and inorganic constituents that allow billions of organisms to coexist in soil. Rhizobia are one group of soil organisms with a unique feature: the ability to form a symbiotic association with members of the plant family, Fabaceae. The rhizobia are ubiquitous aerobic, heterotrophic, gram-negative, diazotrophic bacteria of the genus *Rhizobium*. The Fabaceae is one of the largest families of the angiosperms and is currently estimated to consist of approximately 750 genera and 19,000 species. Rhizobia have the ability to infect legume root hairs and cause the production of root nodules, which are subsequently inhabited by the rhizobia. Within the root nodules, rhizobia reduce or fix atmospheric nitrogen into ammonia, which is utilized by the plant. Simultaneously, plants undergo photosynthesis and reduce or fix carbon dioxide into simple sugars, which are transported to the roots and utilized by the rhizobia. Thus, both symbionts are capable of supplying a nutrient, which usually limits its partner's metabolism. The carbon cost of biological nitrogen fixation (BNF) has been estimated at 6.5 g C g^{-1} N (Phillips, 1980). However, ^{14}C pulse-labeling studies have shown that legumes can compensate for this carbon cost by increased photosynthesis (Paul and Kucy, 1981). Nitrogen (N) is typically the nutrient that limits plant growth, and hence, BNF represents an attractive alternative to high-energy-input production agriculture when low inputs of N fertilizer are required and a sustainable agriculture is necessary.

B. Nodule Formation

An excellent review of this process was recently presented by Long (1989) and is summarized here. The first prerequisite for nodule formation is the presence in the soil of both symbionts. Rhizobia are attracted to the plant roots and inhabit the rhizosphere. This chemotactic response may be due to specific compounds exuded by the plant (Bergman et al., 1988). Rhizobia alter the plant root surface and the growth of root hairs, causing root hair curling (Dazzo and Gardiol, 1984). These changes are the result of a series of complex genetic plant-microbe interactions. The *Rhizobium* genes responsible for nodulation (*nod* genes) are generally not expressed in pure culture. Rather, they are induced by the presence of plant exudates. In general, the inducing compounds are flavenoids, and they have been specifically identified as luteolin for alfalfa and clover (Peters and Long, 1988) and the isoflavone, daidzein (Kosslak et al., 1987). The functions of the *nod* genes are unknown, as are many of the plant genes also involved in the infection process. Nodulins are nodule-specific plant gene products involved in the symbiosis.

As the root hairs curl, rhizobia become trapped within them and the cells of the root cortex begin to divide (Newcomb, 1981). The bacterial cells multiply and invade or infect the outermost cells of the root. Each invaded plant cell produces a polysaccharide cell wall sheath that separates the rhizobia from other plant cells (Callaham and Torrey, 1981). This sheath, the so-called infection thread, is important in allowing the rhizobia to evade the normal plant cell's defense. The rhizobia thus advance via the infection thread and continue to invade newly divided plant root cells as the nodule begins to grow. Finally, the rhizobia are released into the plant cytoplasm but remain surrounded by a plasma membrane of plant origin (Robertson et al., 1978). At this point, N fixation and exchange of nutrients by the symbionts is initiated.

C. Classification of Rhizobia

Rhizobia are actually a group of bacteria with differing biochemical and physiological characteristics and one major common feature, the ability to nodulate legumes. The earliest classification of rhizobia was based on the concept of cross-inoculation groups. In this, a given species of *Rhizobium* formed nodules on host plants within a certain cross-inoculation group, with the species name corresponding to the major host plant within that group (Table 1).

More recently, a new system of classification has been accepted based on the characteristics of the bacteria themselves. This latest system separates the root nodule bacteria into two genera on the basis of their relative growth rates (Table 2) and physiological differences. The genus *Rhizobium* is reserved for fast-growing strains (generation time, 2-4 h), whereas slow-growing strains (generation time, >6 h) are designated *Bradyrhizobium*. The physiological differences between the two genera are discussed in the following section.

Tables 1 and 2 illustrate that the rhizobia that nodulate important production agricultural crops, such as alfalfa, clover, peas, and beans, are more similar to each other than to the rhizobia that nodulate soybeans. The rhizobia that were previously placed in the "cowpea miscellany" group are now known as either *Rhizobium loti* (fast growing) or as biovars of *Bradyrhizobium* (slow growing). Currently, less than 10% of the known

TABLE 1 Classification by Cross-inoculation Groups

Species	Host nodulated	Common name
Group I: Two to six peritrichous flagella; rapid growth on yeast extract media		
R. trifolii	*Trifolium*	Clover
R. Leguminosarum	*Pisum*	Pea
R. phaseoli	*Phaseolus*	Bean
R. meliloti	*Medicago*	Alfalfa
Group II: Polar or subpolar flagellum; slow growth on yeast extract media		
R. japonicum	*Glycine*	Soybean
R. lupini	*Lupinus*	Lupin
Rhizobium	*Vigna*	Cowpea miscellany

Source: From Jordan and Allen (1974).

species of legumes have been examined for the presence of root nodules (Jordan, 1984), but many of the organisms nodulating the less common legumes are within the categories of *R. Loti* or *Bradyrhizobium*, indicating that these organisms are less specific with respect to their legume host than are some of the other species of rhizobia.

TABLE 2 Bergey's New Classification

Genus	Species	Previous species
Rhizobium	*R. leguminosarum*	
	Biovar *leguminosarum*	*R. leguminosarum*
	Biovar *trifolii*	*R. trifolii*
	Biovar *phaseoli*	*R. phaseoli*
	R. meliloti	*R. meliloti*
	R. loti	Fast cowpea, *Leucaena*, lotus
	R. fredii	Fast soybean
Bradyrhizobium	*B. japonicum*	*R. japonicum*
		R. lupini
		Slow cowpea et al.

Source: From Jordan (1984).

TABLE 3 Change in pH and Turbidity of Lupin Strains 43 and 96A11 Grown to Stationary Phase in pH-Adjusted YEM Broth (28°C)

	pH 6		pH 7		pH 8	
Parameter	43	96A11	43	96A11	43	96A11
pH	+0.3	+0.3	-0.5	+0.1	-0.8	-0.3
Turbidity[a]	0.37	0.29	1.19	0.29	0.83	0.27
No. of cells mL^{-1}	10^8	10^8	10^9	10^8	10^9	10^8

[a]Absorbance at 650 nm.

Source: From Shlinkert-Miller and Pepper (1988b).

D. Physiological Characteristics of Rhizobia

Growth Rate

The most conspicuous difference between *Rhizobium* and *Bradyrhizobium* spp. is growth rate, with the former group designated as fast growing. Most legumes are nodulated by either fast- or slow-growing strains, for example, alfalfa by fast-growing strains of *Rhizobium leguminosarum*. However, other legumes form effective nodules with both fast- and slow-growing strains. These include *Acacia* (Dreyfus and Dommergues, 1981), soybean (Keyser et al., 1982; Hattori and Johnson, 1984), and hyacinth bean (Trinick, 1980), and lupin (Schlinkert-Miller and Pepper, 1988a). Usually at any one location, either fast- or slow-growing strains for a particular legume predominate (Schlinkert-Miller and Pepper, 1988b).

Acid Production in Pure Culture

Typically, *Rhizobium* spp. grown in pure culture decrease the pH of the medium, whereas *Bradyrhizobium* spp. increase the pH. However, the production of acid or alkaline end products can be dependent on the carbon and N source (Stowers and Elkan, 1983) as well as the initial pH of the medium (Schlinkert-Miller and Pepper, 1988b). Table 3 illustrates the effect of the initial pH of the medium on the production of acidic end products by fast- and slow-growing lupin rhizobia. Although the fast-growing strain (Lupin 43) generally produces more acidic end products than the slow-growing strain (Lupin 96A11), the magnitude is much higher for an initial pH of 8. At the lower initial pH of 6, both strains actually increase the pH of the medium.

Other Characteristics

Strains of *Rhizobium* possess multiple peritrichous flagella, but *Bradyrhizobium* spp. have a single polar or subpolar flagellum. Fast-growing strains can utilize a wider range of carbohydrates than slow growers (Laurie, 1983). Slow-growing strains preferentially utilize nitrogenous compounds as an energy source (Tan and Broughton, 1982). Slow-growing strains are also more resistant to higher concentrations and a wider range of antibiotics than fast-growing strains, particularly for those antibiotics that affect protein synthesis and cell wall permeability (Kremer and Peterson, 1982). In addition, fast-growing rhizobia typically have large plasmids that carry symbiotic genes and are designated pSym. These vary in size from *R. leguminosarum* plasmids (200-300 kilobases, or kb) to the large megaplasmids (1200-1500 kb) of *Rhizobium meliloti* (Long, 1989). These large plasmids are not found in slow-growing strains, where symbiotic genes are located on the chromosome.

These important differences between fast- and slow-growing strains may well be important for a "successful" symbiosis between two symbionts. The relevance of these factors for desert soil strains is discussed in Sec. III.

II. DESERT LEGUMES

A. Scope

The N-fixing ability of legumes makes them an obvious choice for use in arid and semiarid lands where intensive plant production is practiced. Such systems justify considerable investment at the research stage and during ongoing production cycles. This class of plants is even more important for extensively managed systems used as rangelands, as sources of fuel biomass or for production of useful natural products. Under these extensive conditions the investment per hectare is severely limited even when a practice is performed only once, as during seedling establishment. The challenge is even greater where rehabilitation is involved, since the soil base has already been degraded. Despite this, legumes offer a unique attraction because other sources of N are often unavailable. Hence legume development for use in extensively managed arid and semiarid lands should be a high priority. This treatment of desert legumes gives emphasis to this issue.

Nature has already provided us with a variety of adapted legumes for use in the extensive management of arid and semiarid lands. The Fabaceae family seems to have appeared early in the evolution of the plant kingdom, and members of the family have readily adapted to arid as well as to humid ecosystems. Legumes are indigenous to practically all the deserts of the world, and in some they are abundant. It remains for modern science to select existing desert legumes that are specially adapted and to improve them further for use under extensively managed arid land conditions. In the process it should be possible to identify unique physiological and/or morphological features that have evolved in legumes under desert conditions. An improved understanding of these features could speed the process of developing improved cultivars for use in desert settings. Progress will be hastened if we understand more fully the advantages and limitations presented by legumes.

B. Constraints on the Usefulness of Legumes

In lieu of the advantages attributed to legumes for use in arid lands, it seems peculiar that the preponderance of research for these lands has been devoted to the grass family. Scientists may have avoided several important legume genera as research objects because they are widely known for their toxicity to animals. Although this constraint is valid, it is possible to evolve legumes that are nontoxic and edible by human and beast. There could be some cases in which legume toxicity or intractable nature as a food source would be an advantage. If such features prevented consumption in the wild, the legume species involved could be valuable for rehabilitation of degraded areas. The end result would be a vegetated, stabilized area with an unedible legume component. Edible companion plants could benefit from the N source and be available for animal consumption.

The wide array of toxins present in legumes has been documented (Duke, 1981). Duke lists about 100 toxic substances found in a many wide array of leguminous genera. Some of the toxins are present in several to many genera. Among the more toxic substances are albitocin, erysopine, hydrocanic acid, nicotine, and physostigmine, all with oral LD_{50} values of less than 20 mg kg^{-1}.

In some cases the ideal legume that is palatable to wildlife may be difficult to establish because of excessive consumption during the establishment phase. This is another challenge for the scientist. However, since cattle prefer grasses over legumes this argues for a mixed ecosystem, with members of both plant families present. Possibly the greatest constraint to the use of legumes in extensively managed land masses is the relative lack of research on such plants.

C. Positive Features of Legumes

Legumes can provide a source of fixed N, they fit naturally into a mixture of species, and the limiting factor of toxicity can be turned into an advantage. Tree legumes may be particularly important. Once established such trees can tolerate substantial year-to-year variations in stress, including pressure from animals. Properly engineered, such a species would exclude competing plants from its ground space by competition and the production of toxic substances, or conversely, accommodate a companion species. The companion species could be useful in terms of enhanced soil stability and/or the production of vegetation for animal use. The useful product from the tree could be a fruit or the tree itself, which could provide fuel or wood for other uses. Coppice production could be considered. Since pressure to acquire fuel supplies has led to extensive degradation of arid lands, the promise that legume species hold for providing fuel supplies and, at the same time, rehabilitating arid soils should not be overlooked.

Although tree legumes are suggested as the preferred plant form for use in deserts, the potential for perennial shrubs and herbaceous forms should not be overlooked, nor should annuals. If annuals are used, they must be designed to be naturally reseeding types, as annual reseeding by humans is too resource intensive.

D. Distribution of Desert Legumes

The deserts of the world are so vast and complex as to defy comprehension. Even so, an impressive sampling and digesting of relevant literature has been achieved (McGinnies et al., 1968). From this work, one is able to derive the rough distribution of legumes in the deserts of the world and to sense something of their utility (Tables 4 and 5). It should be pointed

TABLE 4 Some of the Principal Legumes in Eastern Hemisphere Deserts: Distribution and Uses

Genera (no. of species)	Uses
Australian	
Acacia (9)	Animal and human food, wood, dye, construction material
Cassia (1)	—[a]
Indigofera (1)	Animal food
Gobi Desert	
Caragana (4)	Sand binding, animal food, wood
Hedysarum (1)	Human and animal food, wood, sand binding
Oxytropic (1)	Animal food
Takla-Makam Desert	
Alhagi (1)	Medicine
Caragana (3)	—
Turkestan Desert	
Alhagi (1)	Medicine
Ammodendron (1)	—
Astragalus (1)	—
Caragana (3)	Animal food, wood, sand binding
Eremosparton (1)	—
Glycyrrhiza (1)	Human food
Hedysarum (1)	Human and animal food, wood, sand binding
Prosopis (1)	—
Thar Desert	
Astragalus (1)	—
Cassia (1)	Medicine
Crotalaria (1)	Fiber
Dalbergia (1)	Construction material
Indigofera (1)	—
Prosopis (1)	—
Tamarindus (1)	Human food, medicine
Iranian Desert	
Alhagi (2)	Medicine, animal food
Dalbergia (1)	—
Glycyrrhiza (2)	Human food, medicine

TABLE 4 (continued)

Genera (no. of species)	Uses
Iranian Desert	
Prosopis (1)	—
Tamarindus (1)	Human food, medicine
Arabian Desert	
Acacia (3)	Animal food, tannin, gum
Alhagi (2)	Human and animal food
Astragalus (1)	—
Cassia (1)	Medicine
Crotalaria (1)	—
Glycyrrhiza (1)	Human food
Prosopis (1)	—
Retama (1)	—
Trigonella (1)	—
Somali-Chalbi Desert	
Acacia (2)	—
Cassia (1)	Medicine
Sahara Desert	
Acacia (5)	Animal food, tannin, gum
Acanthyllis (1)	—
Alhagi (1)	Human and animal food
Astragalus (1)	—
Cassia (1)	Medicine
Coronilla (1)	—
Crotalaria (1)	—
Genista (1)	Animal food
Glycyrrhiza (1)	Human food
Indigofera (1)	—
Prosopis (1)	—
Retama (2)	Sand binding
Trigonella (1)	—
Kalahari-Namib Desert	
Acacia (3)	Animal food, tannin

[a]Uses not specified.

Source: Adapted from McGinnies et al. (1968).

TABLE 5 Some of the Principal Legumes in Western Hemisphere Deserts: Distribution and Uses

Genera (no. of species)	Uses
North American Desert	
Acacia (2)	Wood
Cercidium (3)	Wood
Dalea (2)	Wood, construction material
Parkinsonia	Wood
Prosopis (1)	Wood, animal food
Atacama-Peruvian Desert	
Acacia (1)	Animal food
Astragalus (3)	—
Caesalpinia (1)	—
Cercidium (1)	Gum
Geoffroea (1)	Animal food
Inga (1)	—
Prosopis (3)	Animal food, wood, construction material
Monte-Patagonian Desert	
Acacia (3)	Animal food, wood
Adesmia (4)	—
Cassia (2)	Fiber
Cercidium (1)	Gum
Geoffroea (1)	Animal food
Glycyrrhiza (1)	—
Neocracca (1)	Human food
Prosopidastrum (1)	Animal food
Prosopis (1)	Human and animal food, wood, sand binding, construction material
Ramorinoa (1)	Human food

Source: Adapted from McGinnies et al. (1968).

out that this listing is illustrative only as many additional legumes are known to inhabit these deserts and information on their utility is fragmentary. However, it will be seen that those genera of legumes listed have documented uses, including human and animal foods, wood, construction materials, medicine, dyes, gums, fibers, tannin, and sand binding. We may also confidently say that desert legumes will be found, in the future, that yield other useful industrial products. Some are known to be potential ground covers, and many leguminous species are known to make excellent shade trees.

In looking to the utility of desert legumes it should be borne in mind that species sometimes have their greatest utility when they are transferred from their natural settings to new habitats. When this happens, it is due in large measure to the new habitat not having certain constraints, such as pests, soil, or climatic conditions that existed in the original setting. One concludes that attributes of desert legumes that appear marginally attractive in their natural habitat should not be dismissed until the species are observed under circumstances that may allow the attribute to be more fully expressed.

E. Noteworthy Programs

A program with special promise is the recently formulated International Legume Database and Information Service (ILDIS), whose coordinating center is at the University of Southampton, England. This organization, sponsored in part by The Royal Botanic Gardens at Kew, England, recognizes that there is a need for information exchange. Although the effort is not exclusively related to arid land legumes, it definitely embraces them. If this activity can be perfected it can serve as a strong impetus to any fledgling projects designed to promote the development of legumes for use in arid land rehabilitation.

The work of Dr. Peter Felker on *Prosopis* serves to illustrate the potential for tree legumes in arid lands (Felker and Clark, 1980; Felker, 1976, 1979). His work was carried out at the University of California, Riverside, for a number of years and more recently has been pursued at the Center for Semiarid Forest Resources in Kingsville, Texas. His numerous published studies reveal the immense diversity within this genus and the prospects for developing biotypes for fruit or wood or for other products. It is clear that the germ plasm diversity available in this genus could provide for the enhancement of any particular attribute desired, such as pest resistance, salt tolerance, heat tolerance, or cold tolerance. Here is a plant already admirably adapted to desert condition. It remains for scientists to put the finishing touches on it to create one or more cultigens with enhanced properties.

The Australian program of adapting annual legumes to arid lands has been underway for about a century. The nature of the program and the outstanding progress that has been made and continues to be made has been summarized (Crawford, 1983). Annual medics (*Medicago* spp.) have been adapted to grow in areas with mean annual rainfall of 250-500 mm. The legumes are harvested by animals and leave a residue of N that is used by a subsequent cereal crop. This system is intermediate between the idealized "extensive" and "intensive" systems, but the net result is a stabilized, productive system under arid conditions. The scientists involved portray an immense variety of germ plasm within and between the large number of related leguminous genera that have been accessioned. They indicate that this base of germ plasm can be used for many decades to come for the continued upgrading of the cultivars used in the plant production system that has evolved. Given this very successful evaluation effort, it should be possible to mount similar efforts to rehabilitate and upgrade arid and semiarid lands on other continents. One visualizes that the conditions and objectives for each desert situation should be specific.

III. USE OF LEGUME-*Rhizobium* ASSOCIATIONS

Regardless of whether legumes are grown intensively with irrigation or extensively as indigenous natural resources without irrigation, the major advantage of legumes is that they are not limited by low soil N if they can fix their own N. When indigenous rhizobia are present in soils, they nodulate legumes. If indigenous rhizobia are not present, additional rhizobia can be added with seed as inoculants. There are several desirable characteristics necessary for useful inoculants:

1. Infective: capable of causing the infection process in legume roots with subsequent nodulation.
1. Effective: capable of actively fixing N.
3. Competitive: capable of nodule occupancy in the presence of other strains of rhizobia. The more competitive the inoculant strain, the greater is the percentage nodulation due to that strain.
4. Able to survive when added to a foreign soil. Inoculant strains added to a particular locale have a greater chance of survival if the new locale is environmentally similar to the original origin of the inoculant strain. For example, desert isolates indigenous to one country may be useful as inoculants for another desert. Thus strains physiologically adapted to specific soil environments have a greater chance of being useful or successful inoculants.

The desirable attributes are not necessarily correlated. Hence competitive strains may or may not be effective and can survive differently in different soils. If competitive effective strains for a given legume are indigenous to a particular soil, no inoculation is necessary. On occasion the indigenous populations of particular strains of rhizobia may be replaced by inoculant strains if the same inoculant is repeatedly added to the soil over a period of several years.

There is a wide spectrum of potential uses for legume-*Rhizobium* associations in semiarid lands and deserts. These include legumes in irrigated production agriculture, legumes grown in nonirrigated agriculture, and legumes used in rangeland, which includes tree legumes. Nitrogen fixation may be useful to increase the yields of the legumes themselves to be used to increase the yield of other crops via intercropping. In addition legume residues or prunings can be used to increase the yield of subsequent crops.

A. Irrigated Production Agriculture

Alfalfa or lucerne (*Medicago sativa* L.) has historically been a major agricultural crop in deserts where irrigation is available. The crop was introduced to the Southwest United States from Mexico around 1850, and since then alfalfa production has increased steadily. It can be grown at elevations ranging from near sea level to over 1500 m. In some Southwest U.S. desert areas alfalfa accounts for 20% of the total irrigated cropland, and yields average 17.0 t ha^{-1} compared to the U.S. national average of 7.6 t ha^{-1} (USDA, 1987). In some areas, yields exceed 20 t ha^{-1}.

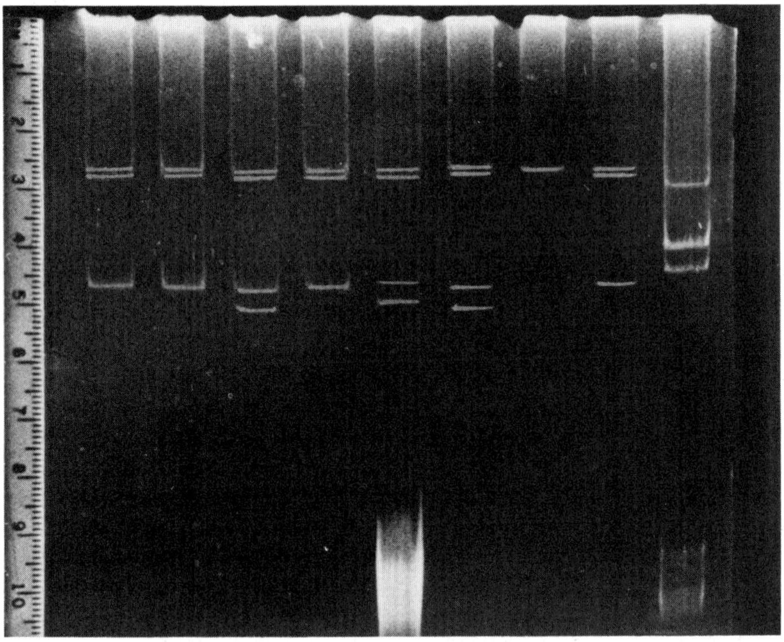

FIGURE 1 Plasmid profile analysis of several field isolates of *Rhizobium meliloti*. (From Shishido and Pepper, 1989.)

Alfalfa is nodulated by *R. meliloti*, and indigenous strains appear to be prevalent where alfalfa is grown. In most instances alfalfa is grown without inorganic N fertilization, and the symbiosis must provide large amounts of fixed N. Estimates of the amount of N that can be fixed by the alfalfa-*R. meliloti* symbiosis are in the range of 200 kg h^{-1} $year^{-1}$ (LaRue and Patterson, 1981). Brown (1976) evaluated six alfalfa cultivars with and without a commercial *R. meliloti* inoculant in the Sonoran Desert region of the United States. No significant differences in forage production, root nodule formation, or percentage protein were found among cultivars or between inoculated and noninoculated treatments. However, neither the cultivars nor the inoculant was adapted to the arid Southwest. A later study by Turley (1980) also demonstrated that cultivars unadapted to the area were inferior, with yield decreases due to high temperature in midsummer.

In a more recent study, Shishido and Pepper (1989) evaluated dominant indigenous *R. meliloti* strains from Sonoran Desert soils. Strains were identified by the use of plasmid profile analyses and intrinsic antibiotic resistance. Chanway and Holl (1986) showed that intrinsic antibiotic resistance alone was not sufficient for strain identification. All dominant Sonoran Desert strains were fast growing and acid producing and contained large megaplasmids. Examples of the plasmid profiles are shown in Fig. 1. Clear differences in profiles can be seen. For example, strains B, C, and G are different strains, whereas strains A, B, D, and H are the same

TABLE 6 Effectiveness of Dominant *R. meliloti* Strains Indigenous to Arizona Soil[a]

Strain	mm Plant^{-1} Nodule weight	mm Plant^{-1} Total N in shoot	Acetylene reduction (nmol plant h^{-1})
AZTCYJ	5.7 (1)	4.80 (1)	635 (1)
AZSC	5.5 (1)	4.77 (1)	609 (1)
AZY	6.4 (1)	4.47 (1)	638 (1)
AZJT	4.7 (1)	4.18 (1)	501 (1)
102F77b	5.2 (1)	4.98 (1)	709 (1)
Control	<0.1 (2)	0.55 (2)	14 (2)
cv, %	21.5	19.9	19.2

[a]Seedlings of alfalfa, cultivar Lew, grown for 6 weeks. Values followed by the same number in parentheses are not significantly different: $p = 0.05$ level, Newman-Keuls range test.

Source: From Shishido and Pepper (1989).

strains. All dominant strains were highly effective. Table 6 shows the effectiveness of these four strains and a commercial strain 102F77b (Nitragin Co., Milwaukee, Wisconsin). All the indigenous strains were as effective as the commercial strain, since no significant differences were found for all parameters evaluated. Strain AZSC was isolated from a soil with an exchangeable sodium percentage of 18.5% and may be useful as an inoculant for alfalfa grown in sodic soils. Strain AZY was isolated from a low-elevation desert area, where summer temperatures are exceedingly high, and was high temperature tolerant.

Both these strains illustrate the physiological adaptation of indigenous strains to the semiarid environment of the desert Southwest United States. Similar physiological adaptation of *R. meliloti* to low-pH soils was shown by Howieson et al. (1988). Strains adapted to the Southwest coupled to cultivars bred for and also well adapted to Southwest conditions allow a highly efficient symbiosis and the successful use of a legume in irrigated production agriculture.

B. Nonirrigated Agriculture

Dry *Phaseolus* beans are another important legume crop that can be grown under irrigated agriculture conditions. However, here we discuss the production of beans under nonirrigated conditions that often prevail in arid deserts, particularly in developing countries. Growth of beans under nonirrigated conditions can be an important source of protein produced by low-energy-input agriculture.

For example, the Hopi, a Native American people of northern Arizona, grow many beans by traditional agricultural methods that utilize no inorganic fertilizer and no irrigation, even though annual rainfall is usually only 22 cm year^{-1}. Beans grown this way require low plant densities to allow better utilization of limited water, and the presence of indigenous bean rhizobia (*R. leguminosarum* biovar. *phaseoli*). Nabhan et al. (1984) showed that species of beans grown by the Hopis included *Phaseolus vulgaris*, *Phaseolus acytifolius*, *Phaseolus coccineus*, and *Phaseolus lunatus*. The crude protein content of these beans ranged from 20.9-26.1%. Pepper et al. (1989) showed that bean rhizobia isolated from a Hopi sand were highly effective.

Bean rhizobia in arid lands also show physiological adaptations and vary widely in the N fixation effectiveness. Ineffective or weakly effective strains of rhizobia are often the case with bean rhizobia, and the bean-rhizobia symbiosis is notoriously weak regardless of location or environment. Reasons for the ineffectiveness of this symbiosis can be due either to symbiont or to the environment in which the beans are grown.

Semiarid lands are often subject to salt and temperature stress that may adversely affect the symbiosis. Hafeez et al. (1988) showed that a soil electrical conductivity (EC_e) of 7.5 dS m^{-1} caused a 60% decline in the dry matter production of mung beans relative to beans grown in similar but nonsaline soils. Nitrogen fixation at this EC_e decreased dramatically even though the rhizobia were salt tolerant. This indicates that both plant and microorganism must be salt tolerant to alleviate the effects of salt stress. Water availability is of course essential for the growth of beans and an effective symbiosis. Venkateswarlu and Rao (1987) recently showed that in the Indian desert, water deficit reduced N fixation CC_2H_2 reduction by mung bean (*Vigna radiata* L.), cluster bean *Cyamopsis tetragonoloba* (L.), and moth bean (*Vigna aconitifolius*). This occurred even for legume varieties adapted to the arid region, with the loss of N fixation activity decreasing from 13 to 100% depending on the level of drought stress.

High-temperature stress can also decrease BNF, although indigenous strains of rhizobia are often physiologically adapted to high temperatures. Josephson and Pepper (1984) showed that bean rhizobia isolated from the Sonoran Desert exhibited increased mean generation times as temperatures increased from 27 to 40°C. At 40°C rhizobia were viable but would not grow. Interestingly, 40°C was the maximum soil temperature normally found at 10 cm depth in the summertime. Eaglesham et al. (1987) showed similar adaptations of *Bradyrhizobium* strains in West Africa. They found strains also to be tolerant of 40°C and desiccation, but such temperature resistance was not correlated with salinity resistance. Karanja and Wood (1988) also selected bean rhizobia strains for use in Kenya on the basis of high-temperature resistance. Piha and Munns (1987) showed that the use of inoculant rhizobia selected for tolerance to high soil temperature may fail because of competition with indigenous strains. They believed that the presence of physiologically adapted indigenous strains was vital for a successful symbiosis at high temperatures.

Under nonstress conditions BNF by the bean-rhizobia symbiosis may be poor because of the characteristics of indigenous strains. Josephson and Pepper (1984) isolated several bean rhizobia from wild desert bean plants that were highly infective on commercial *P. vulgaris* cultivars but totally ineffective. Sometimes rhizobia may not even be infective. For example, Rupela et al. (1987) found poor nodulation of chick-peas (*Cicer arietinum*)

TABLE 7 Competition Between Bean Rhizobia Strains KIM5 and K1

Inoculant (no. of cells)		Strains recovered in nodules (% of total)			
KIM5	K1	KIM5	K1	KIM5 + K1	Unknown
10^8	0	100	0	0	0
0	10^8	0	71.3	27.5[a]	1.2
10^2	10^8	58.3	3.3	36.7	1.7
10^4	10^8	50.0	5.0	43.7	1.3
10^6	10^8	75.0	0	25.0	0
10^8	10^8	85.0	1.2	10.0	3.8

[a]Airborne contamination by KIM5 caused mixed infections.

Source: From Josephson and Pepper (1984).

in India, even though rhizobia populations were high. Other bean rhizobia may be infective and effective but not competitive for nodule occupancy.

The reverse can also be true: many indigenous bean rhizobia are infective and competitive but not effective. Competitiveness is a critical parameter affecting a successful symbiosis. No matter how effective a strain is, if it is not competitive it is not present in many nodules. Competitiveness is clearly not a function of cell numbers in soil or rhizosphere. Table 7 shows the numbers of nodules occupied by bean rhizobia strains KIM5 or K1 when bean seeds were inoculated with variable numbers of both strains. KIM5 is known to be highly competitive and was originally isolated from Kimberly, Idaho. K1 is a desert strain isolated from Tucson, Arizona. KIM5 is extremely competitive and dominates K1 for nodule occupancy even when outnumbered as an inoculant by 10^6:1. However, when beans are inoculated with single-strain inocula in a field study in Tucson, K1 was more effective than KIM5, illustrating that K1 was well adapted to desert conditions.

Thus it can be seen that the choice of inoculant strains used to supersede indigenous strains is critical. Effective strains that are not highly competitive rarely give an inoculation response (as indicated by percentage nodule occupancy due to the inoculant strain) greater than 30%. Inoculation of chick-peas in India by Chandra and Pareek (1985a) gave a variable inoculation response due to strong competition from indigenous strains. Additional studies in India by Dudeja and Khurana (1989) have shown that inoculant strains of rhizobia for pigeon peas (*Cajanus cajan*) did not multiply in the rhizosphere whereas the indigenous strains did.

Other inoculant studies in the desert areas of India have been successful. Chandra and Pareek (1985b, 1987) showed that inoculation response and N fixation by chick-peas could be increased by using increased amounts of inoculant. More recently, Hilali et al. (1989), showed that VIKING 1, a bean rhizobia originally isolated from South America, was effective and competitive in a Moroccan sandy soil. VIKING 1 also persisted in soil for a subsequent planting 1 year later.

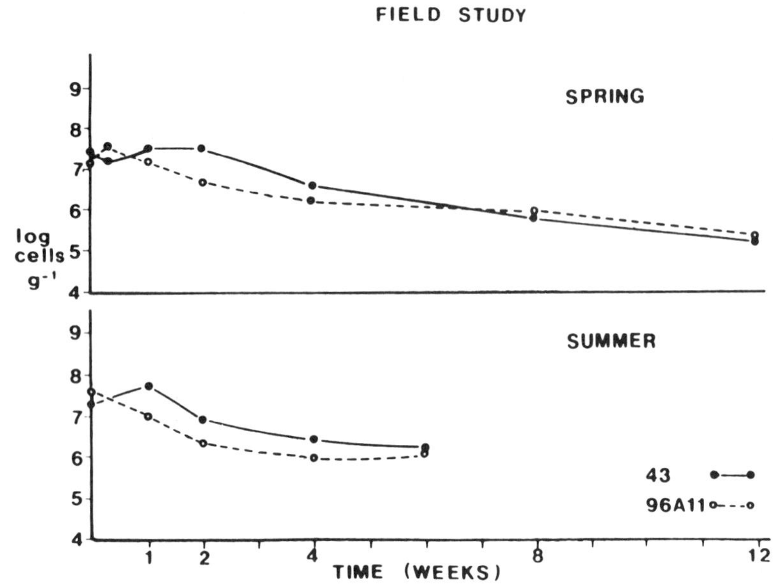

FIGURE 2 Survival of the fast-growing lupin rhizobia strain 43 and the slow-growing strain 96A11 in field soil. Spring months were March through April; summer months were May through June. (From Schlinkert-Miller and Pepper, 1988b.)

It has been shown that the loss of symbiotic properties in bean rhizobia strains may be due to constant genomic rearrangements of reiterated DNA sequences (Martinez et al., 1988). In view of these problems, the bean-rhizobia association is certain to receive continued attention from researchers all over the world as attempts are made to improve bean production without inorganic N fertilizers.

C. Rangeland Legumes

Many indigenous annual and perennial legumes exist in desert ecosystems. These represent important inputs of N, particularly since N typically limits plant growth and desert soils are notoriously low in organic matter and soil N. Although most of the fixed N is used by the legume itself, some N leaks from the root system as exudates that can be utilized by other plants. These indigenous legumes are adapted to semiarid environments and are important sources of nourishment for grazing animals in desert rangelands. Pepper et al. (1980) collected 80 nodulated annual and perennial desert legumes from Arizona, California, and Baja California, Mexico. These genera included *Lupinus*, *Phaseolus*, *Medicago*, *Lotus*, *Melilotus*, *Astragalus*, *Trifolium*, and *Prosopis*. Little information is available on the rhizobia associated with these indigenous species, but there is evidence of the physiological adaptation of desert rhizobia to desert conditions.

Schlinkert-Miller and Pepper (1988a, b) isolated a fast-growing strain of rhizobia from a *Lupinus* species native to Baja California. This strain,

Lupin 43, is different from most lupin rhizobia. Lupin 43 is fast-growing (generation time = 3.6 h), possesses multiple flagella, and produces acid in a defined medium (Table 3). In comparison, slow-growing lupin strains, such as Nitragin 96A11, possess a single flagellum (Fig. 1b) and produce alkaline products in defined media (Table 3). Possession of fast-growing characteristics may have resulted from adaptation to desert conditions, where fast growth rates, multiple flagella, and production of acidic metabolic products may have affected the ability of the strain to survive in arid environments. Figure 2 shows that Lupin 43 survived in higher numbers than 96A11 when individual strains were added to a desert soil under field conditions. Neither strain was indigenous to the desert soil, but the desert strain appeared to be more physiologically adapted to the desert environment than 96A11, which was originally isolated from a nondesert environment. These data illustrate that to be a useful inoculant rhizobia must be able to survive in soil as well as be infective, effective, and competitive. Inoculants can be screened for tolerance to high pH or low pH (Wood and Cooper, 1985; Wood et al., 1988; Yousef et al., 1987). Rangeland legumes are particularly useful for extensively managed system in which inputs of fuel and fertilizer are limited. In developing countries, the introduction of adapted varieties may be feasible to improve the grazing potential of rangelands.

D. Tree Legumes

Nitrogen is generally cited as the most limiting nutrient for plant growth in semiarid environments, and some believe that it may limit plant productivity almost as much as the lack of water (Felker and Clark, 1980). Tree legumes can be important in shading desert soils, which reduces water loss, and also in increasing the N content of soil around the tree legume. Examples of useful tree legumes for semiarid lands include mesquite (*Prosopis*) and *Acacia*, which are found over large areas of semiarid ecosystems.

The potential uses of tree legumes include cattle forage and shade, firewood, living fences, and the enrichment of the N content of desert soils. Seeds of mesquite trees contain up to 69% protein compared to many legume seeds with only 30-45% protein (Felker, 1976). Nilson et al. (1982) estimated that mesquite also has potential as a source of wood fuel, with a plantation productivity of 15,000 kg ha^{-1} $year^{-1}$.

Mesquite can also improve the soil nutrient status of rangeland. Tiedemann and Klemmedson (1973) demonstrated that soil beneath mesquite trees contained almost three times as much organic C and N as that outside the mesquite's foliage cover. Mesquite (*Prosopis glandulosa*) has been shown to significantly increase total N, NO_3^- N, NH_4^+ N, organic C, $NaHCO_3$-extractable PO_3^{3-} P, and saturation extract K^+ in soil beneath the mesquite canopy to a depth of 90 cm (Virginia and Jarrell, 1983).

There is little information available on BNF by tree legumes in semiarid lands, but recent reports have documented nodulation of tree legumes. Field detection of nodules on tree legumes is difficult since the nodules are usually found where soil is moist and therefore may be at depths of 3 m or more (Felker and Clark, 1982). Virginia et al. (1986) showed that mesquite trees developed nodules and fixed N at soil depths greater than 4 m in the California Sonoran Desert. Since N fixation appears to be

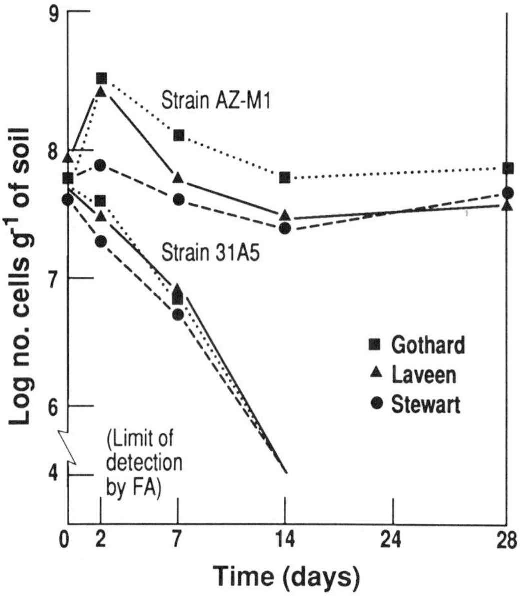

FIGURE 3 Survival of the desert mesquite rhizobia AZM1 and the non-desert isolate 31A5 in three soils. (From Shoushtari and Pepper, 1985b.)

occurring at significant soil depths and yet soil nutrient concentrations increase at the soil surface, then soil nutrients must be received via leaf and pod litter, which is high in protein.

As has been shown in other symbioses, low water or high soil N both inhibit the nodulation of tree legumes. Recently, Miettinen et al. (1988) demonstrated that moisture was needed for the nodulation of mesquite. Shoushtari and Pepper (1985a) also showed that the nodulation of mesquite trees was abundant when moisture was plentiful and also that nodulation was inhibited by high soil N concentrations. Inhibition of nodulation of *Acacia* spp. by N was recently noted by Umali-Garcia et al. (1988). Overall estimates of BNF for tree legumes have ranged from 25-30 kg ha^{-1} $year^{-1}$ (Rundel et al., 1982) to 150 kg ha^{-1} $year^{-1}$ (Felker, 1979). Shearer et al. (1983) used ^{15}N analysis to illustrate that 70% of mesquite N was provided by the symbiosis.

The physiology and effectiveness of mesquite rhizobia isolated from the Sonoran Desert were evaluated by Shoushtari and Pepper (1985a). Desert strains were fast growing and produced acid in defined media. The desert strain AZ-M1 had a mean generation time of 3.5 h and was highly effective when mesquite plants were grown under N-limiting conditions. The presence of applied N significantly inhibited nodulation and N fixation, although total plant N remained relatively constant, illustrating the capacity of N fixation to make N available to plants when soil N is limiting.

The competitiveness of the mesquite strain AZ-M1 and its ability to survive in desert soils was also evaluated and compared to a commercial strain 31A5. Strain AZ-M1 was more competitive than 31A5, but even more striking was its ability to survive in desert soils. The survival of both strains was monitored following the inoculation of three desert soils with individual strains (Fig. 3). The soils varied in texture, salinity, sodium concentration, and pH. The Stewart soil from Safford, Arizona, had a pH of 9.5, an EC_e of 11.2 dS m^{-1}, and an exchangeable sodium percentage of 26.6. Thus these desert soils exhibit characteristics unfavorable to the survival and growth of soil bacteria. Clearly, the desert strain AZ-M1 was able to survive these harsh conditions more successfully than the nondesert strain 31A5, even though neither strain was indigenous to any of the soils. Again, physiological adaptation appears to be implicated and desert rhizobia seem to be well suited to semiarid environments.

Tree legumes represent a potentially attractive new crop for deserts since they are drought tolerant and also fix their own N. Many tree legumes grow profusely in several deserts of the world (Tables 4 and 5), and yet for many tree legumes nothing is known of the organisms that may be fixing N, or even if the trees are nodulated. This is in part due to the depth at which nodules are likely to be present, as well as a lack of funding for research in this area.

E. Intercropping with Legumes

As well as supplying N directly to legumes through the symbiosis, rhizobia can indirectly make N available to other plants. In rangeland this can be via exudation of annual legumes or from leaf and pod litter from legumes that may be returned to the soil. In cropland, legumes can also be intercropped with other non-N-fixing plants. Ruschel et al. (1979) demonstrated increased soil N and uptake of N by maize, due to the *Rhizobium-P. vulgaris* symbiosis. ^{15}N studies in India showed that up to 32% of total N uptake by wheat and maize was due to biological N fixation by cowpeas (*Vigna unguiculate* L.) (Patra et al., 1986). Later they showed that 105 kg N ha^{-1} were obtained from cowpea residues intercropped with maize (Patra et al., 1987).

A more recent innovation, although most widely applied in nonarid areas, has been the concept of "alley cropping." Here a legume is grown in rows and leguminous biomass is harvested periodically and used as a "green manure" for a crop planted between the legume rows, that is, in a "legume alley." Tree legumes, such as *Leucaena leucocephala* and *Acacia albida,* have been widely used in alley cropping systems. Rhizobia from *L. leucocephala* are fast growing and produce acid in defined media, whereas those from *A. albida* are slow growing and alkali producers. Sanginga et al. (1989) isolated several effective indigenous strains from these species in Nigerian soils. *Leucaena* has been estimated to fix 40% of its N requirements (Sanginga et al., 1986). Recently Sanginga et al. (1988b) showed that *Leucaena* could fix 134 kg N ha^{-1} in 6 months following inoculation with an effective rhizobial strain. Additional successful inoculation responses were demonstrated by Roskoski et al. (1987) with *Leucaena* and *Acacia* spp. in Mexico.

Leucaena clippings have been shown to yield 110-580 kg N ha^{-1} annually (Guevarra, 1976). *Leucaena* prunings also increased the yield of maize by 104% in an Nigerian alley cropping system (Mulongoy and van der Meersh, 1988). However, the actual efficiency of N transfer from *Leucaena* clippings to a current crop is usually low, with the main benefit coming from the buildup of organic N, which is subsequently released via mineralization for future crops (Mulongoy and van der Meersch, 1988). *Leucaena* clippings have been shown to increase soil organic C and N (Sanginga et al., 1988). This phenomenon has also been demonstrated in pasture land by Ladd et al. (1981), who showed little immediate transfer of N from legume residues but a buildup of soil organic nutrients available to succeeding crops.

IV. PRIORITIES FOR THE FUTURE

If the potential for using legumes to rehabilitate deserts is to be realized, it is urgent that attention be given to the germ plasm base. This is important, in the first instance, because certain of the species of interest are endangered. In some cases the species is not endangered but certain parts of the germ plasm base for that species may be in jeopardy. The sooner the germ plasm bases are collected, characterized, and preserved, the less chance there is that a valuable genetic trait will be lost. Hence, it is critical that the pace of this work be quickened and that the various activities be linked either formally or informally.

To mount a program for the use of legumes in a specific area, it seems that economics dictates that at least 1 million ha be involved. Given such an area needing rehabilitation or improvement, one can visualize an active search for suitable legume species and for work to augment the genetic traits of the species selected for improvement. This work can follow the traditional approaches of selection among accessions, plant breeding, creation of intraspecific hybrids, and ultimately, the use of genetic engineering techniques. The various steps involved in the demonstration of the use of new species have been described (Upchurch, 1981).

To maximize the usefulness of desert legumes associated with rhizobia, both plant and microbial symbionts must be well suited because strain-by-strain interactions occur (May and Bohlool, 1983). When new cultivars of legumes are introduced into a given area, inoculation may be necessary to maximize N fixation. In some instances, multistrain inoculants may improve the chances of an effective symbiosis. Dorosinskii and Makarova (1977) demonstrated that lupins grown with several inoculant strains gave better yields and higher N contents than those with single-strain inocula. If attention is given to both symbiotic partners, the legume-*Rhizobium* association can improve the productivity of desert and semiarid environments.

ACKNOWLEDGMENTS

Appreciation is expressed to Mr. Horace Miller, who assisted with the literature search for a portion of this chapter.

REFERENCES

Bergman, K., Gulash-Hofee, M., Hovestadt, R. E., Larosiliere, R. C., Ronco, P. G., and Su, L. (1988). Physiology of behavioral mutants of *Rhizobium meliloti*: Evidence for a dual chemotaxis pathway. *J. Bacteriol. 170*: 3249-3254.

Brown, S. C. (1976). The effect of inoculating alfalfa seed with *Rhizobium meliloti*. M.S. thesis, University of Arizona, Tucson.

Callaham, D. A., and Toreey, J. G. (1981). The structural basis for infection of root hairs of *Tritolium repens* by *Rhizobium*. *Can. J. Bot. 59*: 1647-1664.

Chandra, R., and Pareek, R. P. (1985a). Role of host genotype in effectiveness and competitiveness of chickpea (*Cicer arcetinum Rhizobium. Trop. Agr.* (*Trinidad*) *62*: 90-94.

Chandra, R., and Pareek, R. P. (1985b). Role of inoculum dose in strain effectiveness and competitiveness of chickpea (*Cicer arietnum*) *Rhizobium. Indian J. Microbiol. 25*: 7-10.

Chandra, R., and Pareek, R. P. (1987). Effect of inoculum rate on the performance of chickpea (*Cicer arietnum* L.) *Rhizobium* strains in the field. *Biol. Fertil. Soils 5*: 83-87.

Chanway, C. P., and Holl, F. B. (1986). Suitability of intrinsic antibiotic resistance as a method of strain identification in *Rhizobium trifolii. Plant Soil 93*: 287-291.

Crawford, E. J. (1983). Selecting cultivars from naturally occurring genotypes: Evaluating annual *Medicago* species. In *Genetic Resources of Forage Plants*. Part V. *Evaluations*, J. C. McIvor and R. A. Bray (Eds.). CSIRO, East Melbourne, Australia.

Dazzo, F. B., and Gardiol, A. (1984). Host specificity in *Rhizobium*-legume interactions. In *Genes Involved in Microbe Plant Interactions*, D. P. S. Verma and T. H. Hohn (Eds.). Springer, New York, pp. 3-31.

Dorosinskii, L. M., and Makarova, N. M. (1977). Competitive ability of *Rhizobium lupini* strains. *Mikrobiologiya 46*: 143-148.

Dreyfus, B. L., and Dommergues, Y. R. (1981). Nodulation of *Acacia* species by fast- and slow-growing tropical strains of *Rhizobium. Appl. Environ. Microbiol. 14*: 97-99.

Dudeja, S. S., and Khurana, A. L. (1989). Survival and competitiveness of *Bradyrhizobium* sp. in the rhizosphere of pigeon pea (*Cajanus cajan*). *Biol. Fertil. Soils 7*: 63-66.

Duke, J. A. (1981). *Handbook of Legumes of World Economic Importance*. Plenum Press, New York.

Eaglesham, A. R. J., Stowers, M. D., Maina, M. L., Goldman, B. J., Sinclair, M. J., and Ayanaba, A. (1987). Physiological and biochemical aspects of diversity of *Bradyrhizobium* sp. (Vigna) from three West African soils. *Soil Biol. Biochem. 19*: 575-581.

Felker, P. J., and Clark, P. R. (1980). Nitrogen fixation (acetylene reduction) and cross-inoculation in 12 *Prosopis* (mesquite) species. *Plant Soil 57*: 177-186.

Felker, P. J., and Clark, P. R. (1982). Position of mesquite (*Prosopis* spp.) nodulation and nitrogen fixation (acetylene reduction) in 3-m long phraetophytically simulated soil columns. *Plant Soil 64*: 197-305.

Felker, P. J. (1976). Potential utilization of leguminous trees for minimal energy input. Ph.D. dissertation, Michigan State University, East Lansing.

Felker, P. J. (1979). Mesquite, an all purpose leguminous arid land tree. In *New Agricultural Crops*, G. A. Ritchie (Ed.). West View Press, Washington, D.C., pp. 89-132.

Guevarra, A. B. (1976). Management of *Leucaena leucocephala* (Lam) de Wit for maximum yield and nitrogen contribution to intercropped corn, Ph.D. thesis. University of Hawaii, Honolulu.

Hafeez, F. Y., Aslam, Z., and Malik, K. A. (1988). Effect of salinity and inoculation on growth, nitrogen fixation and nutrient uptake of *Vigna radiata* (L.) Wilczek. *Plant Soil 106*: 3-8.

Hattori, J., and Johnson, D. A. (1984). Fast-growing *Rhizobium japonicum* that effectively nodulate several commercial *Glycine max* L. Merrill cultivars. *Appl. Environ. Microbiol. 48*: 234-235.

Hilali, A., Aurag, J., Molina, J. A. E., and Schmidt, E. L. (1989). Competitiveness and persistence of strains of *Rhizobium phaseoli* introduced into a Moroccan sandy soil. *Biol. Fertil. Soils 7*: 213-218.

Howieson, J. G., Ewing, M. A., and Antuono, M. F. (1988). Selection for acid tolerance in *Rhizobium meliloti*. *Plant Soil 105*: 179-188.

Jordan, D. C. (1984). Family III Rhizobiaceae. In *Bergey's Manual of Systematic Bacteriology*, Vol. 1, N. R. Krieg and J. G. Holt (Eds.). Williams & Wilkins, Baltimore, pp. 234-244.

Jordan, D. C., and Allen, O. N. (1974). Family III Rhizobiaceae. In *Bergey's Manual of Determinative Bacteriology*, R. E. Buchanan and N. E. Gibbons (Eds.). Williams & Wilkins, Baltimore, pp. 261-264.

Josephson, K. L., and Pepper, I. L. (1984). Competitiveness and effectiveness of strains of *Rhizobium phaseoli* isolated from the Sonoran Desert. *Soil Biol. Biochem 16*: 651-655.

Karanja, N. K., and Wood, M. (1988). Selecting *Rhizobium phaseoli* strains for use with beans (*Phaseolus vulgaris* L.) in Kenya: Tolerance of high temperature and antibiotic resistance. *Plant Soil 112*: 15-22.

Keyser, H. H., Bohlool, B. B., Hu, T. S., and Weber, D. F. (1982). Fast-growing rhizobia isolated from root nodules of soybean. *Science 215*: 1631-1632.

Kosslak, R. M., Bookland, R., Barkei, J., Paaren, H. E., and Appelbaum, E. R. (1987). Induction of *Bradyrhizobium japonicum* common *nod* genes by isoflavones isolated from *Glycine max*. *Proc. Natl. Acad. Sci. USA 84*: 7428-7432.

Kremer, R. J., and Peterson, H. L. (1982). Nodulation efficiency of legume inoculation as determined by intrinsic antibiotic resistance. *Appl. Environ. Microbiol. 43*: 636-642.

Ladd, J. N., Oades, J. M., and Amato, M. (1981). Distribution and recovery of nitrogen from legume residues decomposing in soils sown to wheat in the field. *Soil Biol. Biochem. 13*: 251-256.

LaRue, T. A., and Patterson, T. G. (1981). How much nitrogen do legumes fix? *Adv. Agron. 34*: 15-36.

Laurie, A. C. (1983). Relationships of rhizobia from native Australian legumes. *Appl. Environ. Microbiol. 45*: 1822-1828.

Long, S. R. (1989). *Rhizobium*-legume nodulation: Life together in the underground. *Cell 56*: 203-214.

Martinez, E., Flores, M., Brom, S., Romero, D., Davila, G., and Palacios, R. (1988). *Rhizobium phaseoli*: A molecular genetics view. *Plant Soil 108*: 179-184.

May, S. N., and Bohlool, B. B. (1983). Competition among *Rhizobium leguminosarum* strains for nodulation of lentils (*Lens esculenta*). *Appl. Environ. Microbiol. 45*: 960-965.

McGinnies, W. G., Goldman, B. J., and Paylore, P. (1968). Vegetation of desert environments. In *Deserts of the World*, W. G. McGinnies, B. S. Goldman, and P. Paylore (Eds.). University of Arizona Press, Tucson.

Miettinen, P., Luukkanen, O., Johansson, S., Eklund, E., and Mulatya, J. (1988). *Rhizobium* nodulation in *Prosopis juliflora* seedlings at different irrigation levels in Eastern Kenya. *Plant Soil 112*: 233-238.

Mulongoy, K., and van der Meersch (1988). Nitrogen contribution by leucaena (*Leucaena leucocephala*) prunings to maize in an alley cropping system. *Biol. Fertil. Soils 6*: 282-285.

Nabhan, G. P., Weber, C. W., and Berry, J. W. (1984). Variation in composition of Hopi Indian beans. *Ecol. Food Nutr. 16*: 135-152.

Newcomb, W. (1981). Nodule morphogenesis. In *International Review of Cytology*, Supplement 13, G. H. Bourne and J. F. Danielli (Eds.). Academic Press, New York, pp. 246-298.

Nilson, E. G., Rundel, P. W., and Sharifi, M. R. (1982). Productivity in native stands of *Prosopis glandulosa* mesquite in the Sonoran Desert of Southern California and some management implications. *Proceedings of the California Riparian Environment Symposium*, September 17-19, Davis, pp. 121-125.

Patra, D. D., Sachdev, M. S., and Subbiah, B. V. (1986). ^{15}N studies on the transfer of legume-fixed nitrogen to associated cereals in intercropping systems. *Biol. Fertil. Soils 2*: 165-171.

Patra, D. D., Sachdev, M. S., and Subbiah, B. V. (1987). Forms of fertilizer nitrogen residues in soil after intercropping of maize-cowpea. *Biol. Fertil. Soils 4*: 155-161.

Paul, E. A., and Kucy, R. M. N. (1981). Carbon flow in plant microbial associations. *Science 213*: 473-474.

Pepper, I. L., Tucker, T. C., Miller, M. S., and Josephson, K. L. (1980). Isolation of nitrogen fixing organisms from Sonoran Desert legumes. *Agron. Abstr.* 158.

Pepper, I. L., Josephson, K. L., Nautiyal, C. P., and Bourque, D. P. (1989). Strain identification of highly competitive bean rhizobia isolated from root nodules: Use of fluorescent antibodies, plasmid profiles and gene probes. *Soil Biol. Biochem. 21*: in press.

Peters, N. K., and Long, S. R. (1988). *Rhizobium meliloti* nodulation gene inducers and inhibitors. *Plant Physiol. 88*: 396-400.

Phillips, D. A. (1980). Efficiency of symbiotic nitrogen fixation in legumes. *Annu. Rev. Plant Physiol. 31*: 29-49.

Piha, M. I., and Munns, D. N. (1987). Sensitivity of the common bean (*Phaseolus vulgaris* L.) symbiosis to high soil temperature. *Plant Soil 98*: 183-194.

Robertson, J. G., Lyttleton, P., Bullivant, S., and Grayston, G. F. (1978). Membranes in lupin root nodules. I. The role of Golgi bodies in the biogenesis of infection threads and peribacteroid membranes. *J. Cell Sci. 30*: 129-149.

Roskoski, J. P., Pepper, I. L., and Pardo, E. (1987). Inoculation of leguminous trees with rhizobia and V.A. mycorrhizal fungi. *For. Ecol. Manag. 16*: 57-68.

Rundel, P. W., Nilson, E. T., Sharifi, M. R., Virginia, R. A., Jarrell, W. M., Kuhl, D. M., and Shearer, G. B. (1982). Seasonal dynamics of nitrogen cycling for a *Prosopis* woodland in the Sonoran Desert. *Plant Soil 67*: 343-353.

Rupela, O. P., Toomsam, B., Miltal, S., Dart, P. J., and Thompson, J. A. (1987). Chickpea *Rhizobium* populations: Survey of influence of season, soil depth and cropping pattern. *Soil Biol. Biochem. 19*: 247-252.

Ruschel, A. P., Salati, E., and Vose, P. B. (1979). Nitrogen enrichment of soil and plant by *Rhizobium-Phaseolus vulgaris* symbiosis. *Plant Soil 51*: 425-429.

Sanginga, N., Mulongoy, K., and Ayanaba, A. (1986). Inoculation of *Leucaena leucocephala* (Lam.) de Wit with *Rhizobium* and its nitrogen contribution to a subsequent maize crop. *Biol. Agr. Hort. 3*: 347-352.

Sanginga, N., Mulongoy, K., and Ayanaba, A. (1988a). Nitrogen contribution of *Leucaena/Rhizobium* symbiosis to soil and a subsequent maize crop. *Plant Soil 112*: 137-141.

Sanginga, N., Mulongoy, K., and Ayanaba, A. (1988b). Nodulation and growth of *Leucaena leucocephala* (Lam.) de Wit as affected by inoculation and N fertilizer. *Plant Soil 112*: 119-135.

Sanginga, N., Mulongoy, K., and Ayanaba, A. (1989). Effectivity of indigenous rhizobia for nodulation and early nitrogen fixation with *Leucaena leucocephala* grown in Nigerian Soils. *Soil Biol. Biochem. 21*: 231-235.

Schlinkert-Miller, M., and Pepper, I. L. (1988a). Physiological and biochemical characteristics of a fast-growing strain of lupin rhizobia isolated from the Sonoran Desert. *Soil Biol. Biochem 20*: 319-322.

Schlinkert-Miller, M., and Pepper, I. L. (1988b). Survival of a fast-growing strain of lupin rhizobia in Sonoran Desert soils. *Soil Biol. Biochem. 20*: 323-327.

Shearer, G., Kohl, D. H., Virginia, R. A., Bryan, B. A., Skeeters, J. L., Nilsen, E. T., Sharifi, M. R., and Rundel, P. W. (1983). Estimates of N_2-fixation from variation in the natural abundance of ^{15}N in Sonoran Desert ecosystems. *Oecologia (Berl.) 56*: 365-373.

Shishido, M., and Pepper, I. L. (1990). Effectiveness of *Rhizobium meliloti* strains indigenous to Sonoran desert soils. *Soil Biol. Biochem. 22*: 11-16.

Shoushtari, N. H., and Pepper, I. L. (1985a). Mesquite rhizobia isolated from the Sonoran Desert: Physiology and effectiveness. *Soil Biol. Biochem. 17*: 797-802.

Shoushtari, N. H., and Pepper, I. L. (1985b). Mesquite rhizobia isolated from the Sonoran Desert: Competitiveness and survival in soil. *Soil Biol. Biochem. 17*: 803-806.

Stowers, M. D., and Elkan, G. H. (1983). The transport and metabolism of glucose in cowpea rhizobia. *Can. J. Microbiol. 29*: 398-406.

Tan, I. P. K., and Broughton, W. J. (1981). Rhizobia in tropical legumes. XIII. Biochemical basis of acid and alkali reactions. *Soil Biol. Biochem. 13*: 389-393.

Tiedemann, A. R., and Klemmedson, J. O. (1973). Nutrient availability in desert grassland soils under mesquite (*Prosopis juliflora*) trees and adjacent open areas. *Soil Sci. Soc. Am. Proc. 37*: 107-111.

Trinick, M. J. (1980). Relationship among the fast growing rhizobia of *Lablab purpureus*, *Leucaena leucocephala*, *Mimosa* spp., *Acacia farnesiana* and *Sesbania grandiflora*. *J. Appl. Bacteriol. 49*: 39-53.

Turley, R. H. (1980). Field studies on the productivity of alfalfa (*Medicago sativa*) grown from seed coated with selected *Rhizobium meliloti* strains. M.S. thesis, University of Arizona, Tucson.
Umali-Garcia, M., Lisbuit, J. S., and Baggayan, R. L. (1988). Effects of *Rhizobium* inoculation on growth and nodulation of *Albizia falcataria* (L.) Fosh. and *Acacia mangium* Willd. in the nursery. *Plant Soil 108*: 71-78.
Upchurch, R. P. (1981). New crops for arid lands. In *Advances in Food Production System for Arid and Semiarid Lands*, J. T. Manassah and E. J. Briskey (Eds.). Academic Press, New York, pp. 773-798.
USDA (1987). *Agricultural Statistics*. United States Department of Agriculture, Washington, D.C.
Venkateswarlu, B., and Rao, A. V. (1987). Quantitative effects of field water deficits on N_2 (C_2H_2) fixation in selected legumes grown in the Indian desert. *Biol. Fertil. Soils 5*: 18-22.
Virginia, R. A., and Jarrell, W. M. (1983). Soil properties in a mesquite-dominated Sonoran Desert ecosystem. *Soil Sci. Soc. Am. J. 47*: 138-144.
Virginia, R. A., Jenkins, M. B., and Jarrell, W. M. (1986). Depth of root symbiont occurrence in soil. *Biol. Fertil. Soil 2*: 127-130.
Wood, M., and Cooper, J. E. (1985). Screening clover and *Lotus* rhizobia for tolerance of acidity and aluminum. *Soil Biol. Biochem. 17*: 493-497.
Wood, M., Cooper, J. E., and Bjourson, A. J. (1988). Response of *Lotus* rhizobia to acidity and aluminum in liquid culture and in soil. *Plant Soil 107*: 227-231.
Yousef, A. N., Al-Nassiri, A. S., Al-Azawi, S. K., and Abdul-Hussain, N. (1987). Abundance of peanut *Rhizobium* as affected by environmental conditions in Iraq. *Soil Biol. Biochem. 19*: 391-396.

15

Use of Actinorhizal Plants in Arid and Semiarid Environments

P. REDDELL *CSIRO, Townsville, Queensland, Australia*

H. G. DIEM and Y. R. DOMMERGUES *Laboratoire de Biotechnologie des Systèmes Symbiotiques Forestiers Tropicaux (ORSTOM/CTFT), Nogent-sur-Marne, France*

I. INTRODUCTION

Actinorhizal plants are nonlegumes that form root nodules in symbiosis with the N_2-fixing actinomycete *Frankia* (Torrey and Tjepkema, 1979). These symbioses occur in a phylogenetically diverse group of plants; at least 24 genera in eight families are reported to nodulate with *Frankia* (Table 1). They include trees, such as *Casuarina* and *Alnus*, shrubs (e.g., *Coriaria* and *Purshia*), and occasionally perennial herbs (*Datisca* spp.). Compared to the vast group of N_2-fixing Leguminosae (Allen and Allen, 1981), there are relatively few species of actinorhizal plants. However, this numerical inferiority is offset by the ability of many to grow in harsh, infertile environments (Bond, 1983).

Under optimal conditions, nodules formed by *Frankia* on the roots of actinorhizal plants are capable of N_2 fixation at rates similar to those found for effectively nodulated legumes (Torrey, 1978). However, in many natural ecosystems and in some plantations, the frequency of nodulation and the significance of N_2 fixation in actinorhizal plants is unknown (Torrey, 1982; Bond, 1983). In this chapter we

1. Describe the occurrence and use of actinorhizal species in arid and semiarid regions
2. Identify likely soil constraints to growth, nodulation, and N_2 fixation in actinorhizal species in arid and semiarid regions
3. Consider approaches to increasing productivity and N_2 fixation in actinorhizal species, both by selection of plant-*Frankia* combinations with superior N_2-fixing potential and by alleviating soil constraints on symbiotic N_2 fixation and plant production

TABLE 1 Actinorhizal Plants

Family	Genus
Betulaceae	*Alnus*
Casuarinaceae[a]	*Allocasuarina*
	Casuarina
	Gymnostoma
	Ceuthostoma
Coriariaceae	*Coriaria*
Datiscaceae	*Datisca*
Elaeagnaceae	*Elaeagnus*
	Hippophae
	Shepherdia
Myricaceae	*Comptonia*
	Myrica
Rhamnaceae	*Ceanothus*
	Colletia
	Discaria
	Kentrothamnus
	Retanilla
	Talguenea
	Trevoa
Rosaceae	*Cercocarpus*
	Chamaebatia
	Cowania
	Dryas
	Purshia

[a]Four genera can be recognized in the family Casuarinaceae: *Casuarina*, *Allocasuarina*, *Gymnostoma*, and *Ceuthostoma*, a small genus of the Melanesian region (Johnson, 1988).

Source: From Benson (1988).

TABLE 2 Casuarinaceae Reputed for Their Drought Tolerance

Species	50th percentile rainfall (mm)	Lowest rainfall (mm)	References
Allocasuarina decaisneana	200-250	40-60	Turnbull (1986)
A. campestris	225-400		Doran and Hall (1983)
A. dielsiana	275-400		Doran and Hall (1983)
Casuarina obesa	200-375	125-275	Turnbull (1986)
C. cristata subsp. *pauper*	250	175	National Research Council (1984)

II. OCCURRENCE AND USE OF ACTINORHIZAL PLANTS IN ARID AND SEMIARID AREAS

Actinorhizal plants known for their tolerance to arid or semiarid environments include a number of species of Casuarinaceae (Table 2) and Rosaceae (e.g., *Purshia tridentata, P. glandulosa, Cowania mexicana*), some South American Rhamnaceae (*Colletia, Kentrothamnus, Retanilla, Talguenea,* and *Trevoa*) and *Elaeagnus angustifolia* (Bond, 1983; Doran and Hall, 2983; El-Lakany, 1983; Zhu, 1985; Righetti et al., 1986). Of this diverse group of plants, the Casuarinaceae appear to have the greatest potential for broad scale rehabilitation of degraded arid and semiarid zone soils.

Currently, only one actinorhizal species, *Casuarina equisetifolia*, has been planted extensively in arid and semiarid regions. This species, grown to stabilize desert and coastal sand dunes in northern and western Africa, India, Pakistan, and southern China, produces wood for fuel and posts as well as providing shelterbelts for agricultural crops and for settlements (Midgley et al., 1983). Plantings are generally nodulated, and N_2 fixation in *C. equisetifolia* has been estimated to contribute 40-60 kg N ha^{-1} $year^{-1}$ in coastal dunes in the dry zone of Senegal (Dommergues, 1963; Gauthier et al., 1985).

Other species of the Casuarinaceae are potentially more tolerant of drought and salinity (e.g., *Casuarina cristata, Casuarina glauca, Casuarina obesa,* and *Allocasuarina decaisneana*) but are not widely tested (Turnbull, 1986; National Research Council, 1984). For example, preliminary trials with *A. decaisneana*, a deep-rooted tree that occurs naturally on sand ridges in central Australia, have been disappointing. This species failed to nodulate and showed poor survival and a mean height growth of only 30-50 cm after 3 years at sites in the arid area of western Africa (Turnbull, 1986; Dommergues and Diem, unpublished). This very poor performance reflects more upon our lack of knowledge of the biology of these plants and their symbiotic requirements than on the intrinsic abilities of the species to grow in dry environments. An adequate assessment of the role of lesser known species of the Casuarinaceae in the future rehabilitation of degraded arid and semiarid zone soils will depend on an increased research effort, both to define their growth requirements and to understand the mechanisms by which they cope with arid conditions.

III. CONSTRAINTS ON GROWTH, NODULATION, AND N_2 FIXATION IN ACTINORHIZAL PLANTS IN ARID, SEMIARID, AND SALINE SOILS

The extent to which soil constraints limit the productivity of actinorhizal plants in arid and semiarid environments is difficult to assess accurately because of a lack of information. The seasonality of growth patterns of actinorhizal plants in these environments is poorly documented, nodulation is often erratic or unknown, and consequently the importance of N_2 fixation in many species is not clear (Reddell et al., 1985; Righetti et al., 1986). For example, in natural populations of the Casuarinaceae in Australia there were large differences in occurrence of nodulation both between species and between and within sampling sites (Reddell et al., 1985). Two species of *Casuarina* (*C. cristata* and *C. obesa*) are found in the warm arid zone of western Australia, often in close proximity. With *C. cristata*, nodules were found at 14% of sites and on only 4% of the individual plants examined. In comparison, nodules were found on 35% of trees of *C. obesa* but nodulated plants were observed at a majority of sites. Greenhouse trials showed that at some sites the absence of infective *Frankia* were responsible for these observations, at others slow plant growth rates (and consequently low N demand) may explain the lack of nodulation. Similar findings have been reported for two actinorhizal genera (*Purshia* and *Cowania*) that occur in semiarid regions of the western United States (Righetti and Munns, 1982; Righetti et al., 1986).

Soil factors could influence the productivity of actinorhizal plants either through direct effects on the stages of symbiotic N_2 fixation (i.e., survival and growth of *Frankia* in the soil and rhizosphere, infection and nodule development processes, and nodule functioning) or through effects on metabolism of the host plant that are independent of the symbiosis. In this section we discuss the soil constraints likely to affect the establishment and growth of actinorhizal plants in arid and semiarid soils. These factors include seasonal moisture deficit, high soil temperatures, deficiencies on nutrients other than N, salinity, and alkalinity. Research on Casuarinaceae-*Frankia* symbioses is emphasized because of the potential importance of this group and the paucity of data available on other actinorhizal plants that occur in arid environments.

A. Soil Moisture Deficit

The effects of water stress on growth and N_2 fixation in *Frankia* were examined experimentally using high-molecular-weight polyethylene glycol solutions (PEG 4000) to simulate soil matric potential (Shipton and Burggraaf, 1982). A 50% reduction in *Frankia* growth was observed at -0.5 to -0.8 M Pa, and growth was negligible at -1.6 MPa. Thus, *Frankia* appears to be less tolerant of moisture stress than some other soil actinomycetes whose limit for growth is in the range of -5 to -10 MPa (Dommergues, 1962). In this experiment, N_2 fixation was also affected by decreased water potential, the highest N_2-fixing activity occurring between -0.3 and -0.7 MPa. Although these data cannot be directly extrapolated to field soils, they suggest that little infection would occur beyond -0.5 to -0.7 MPa due to the greatly depressed *Frankia* growth and the likely limitations on root growth at these potentials. This supports the observations made by Kant and Narayana (1978) that the number and size of nodules on *C. equisetifolia* decreased with increasing severity of water stress.

In addition to these effects on *Frankia* growth, infection, and nodule development, water deficit can also markedly reduce N_2 fixation by nodulated plants. Acetylene reduction activities of nodules of *Allocasuarina verticillata* decline sharply below 25% soil moisture content (Hopmans et al., 1983). However, it is unknown whether this is due to direct effects of nodule dehydration on N_2 fixation processes per se or to indirect effects mediated via the reduced supply of assimilates from the host plant.

Responses of N_2 fixation in actinorhizal plants to the seasonal drought characteristic of many semiarid regions have not been examined. Perhaps the perennial, highly lignified nature of most actinorhizal nodules allows *Frankia* to survive periods when most plant growth and N_2-fixing activity has ceased. In coastal sand dunes in tropical northeast Queensland, active nodule lobes grow from dry woody nodules of *C. equisetifolia* following the first rains of the "wet" season (Reddell, unpublished); similar responses are expected in more arid environments. *Frankia* can probably also survive long drought periods in soil by forming specialized structures, such as sporangia and torulose hyphae (Diem and Dommergues, 1985).

Although it is apparent that growth and N_2 fixation in actinorhizal plants are markedly affected by the soil moisture deficit, there are no studies that compare the relative sensitivity of plants dependent on symbiotic fixation and plants supplied with N fertilizer (i.e., are nodulated plants more or less able to tolerate drought?). Results from a recently established field trial in north Queensland with *C. equisetifolia* showed nodulated seedlings were better able to survive a dry period that occurred soon after planting than were uninoculated seedlings supplied with N fertilizer; the survival rate for nodulated plants was 96%, whereas only 42% of uninoculated plants survived (Reddell and Rosbrook, unpublished.

B. High Soil Temperatures

Temperatures in the surface 20-30 cm of arid zone soils can be in excess of 35°C for significant periods of the year. Despite this, there is no information on the impact of high soil temperatures on the formation and function of actinorhizal nodules.

Most studies with actinorhizal symbioses have defined minimum and optimum temperatures for *Frankia* growth (Burggraaf and Shipton, 1982; Zhang et al., 1986), nodulation processes (Wollum and Youngberg, 1969), and acetylene reduction by detached nodules (Bond and Mackintosh, 1975; Waughman, 1977). These studies show that in general, plant genera and *Frankia* from tropical climates have higher temperature optima (30-35°C) than do those from temperate regions (20-25°C).

One published study has compared the effects of soil temperature between 15 and 30°C on growth of seedlings of *Casuarina cunninghamiana* either supplied with combined N or depending solely on symbiotic N_2 fixation (Reddell et al., 1985). At the higher temperatures there were marked declines in plant growth for both N_2-fixing and N-fertilized plants. For plants dependent on symbiotic N_2 fixation, this reduction in yield was accompanied by a decrease in nodule mass and there was no effect of temperature on the estimated rates of N_2-fixing activity in nodules. This suggests that at the higher temperatures common in arid zone soils, the effects on N_2 fixation in nodulated plants are likely to be mediated via the direct influence of temperature on the metabolism of the host plant.

C. Salinity and Associated Effects

In arid and semiarid regions, plant growth is often not only hindered by drought and high soil temperatures but also the high salt content of some soils. These "salt-affected" soils fall into two main groups that have markedly different properties (Midgley et al., 1984). Saline soils are those in which an excess of NaCl limits plant growth, whereas in alkaline soils Na carbonates and bicarbonates predominate and restrict yield. In the latter group, pH can be in excess of 10 and plants are subjected to the combined effects of high levels of soluble salts and high pH. In both groups of soils depression of plant growth may be caused by osmotic effects on water uptake, nutritional imbalances (Fe, Mn, Cu, and Zn deficiencies), toxicity caused by specific ions (e.g., Na^+, Cl^-, and HCO_3^-) and/or adverse physical properties of the soil (e.g., surface crusting, reduced permeability to water, and increased mechanical resistance to root penetration). The mechanisms of plant tolerance to these two types of problem soils may be different and are considered separately.

Salinity

Our knowledge of the tolerance of actinorhizal plants to NaCl (salinity) is confined to species of *Casuarina.* Two other actinorhizal plants, *Myrica cordifolia* from South Africa and *Hippophae rhamnoides* from Europe, occur in coastal areas (Bond, 1983), but the salt tolerance of these has not been assessed.

Species of the Casuarinaceae differ markedly in their tolerance to NaCl. In screening trials in solution culture with adequate N, *Casuarina* sens. strict. tolerate higher levels of salinity than do species of *Allocasuarina* (*A. verticillata, Allocasuarina littoralis, Allocasuarina torulosa,* and *A. decaisneana*) were killed by NaCl concentrations of 350 mM or less. Differences in tolerance of NaCl between species are reflected in ion uptake and distribution in individual organs of the plant (El-Lakany and Luard, 1982; Luard and El-Lakany, 1984; Aswathappa and Bachelard, 1986; Clemens et al., 1983). Species that survived the highest salinity levels had the lowest concentrations of Na^+ and Cl^- in both roots and shoots.

Field observations confirm the substantial tolerance of some *Casuarina* species to NaCl (Table 3). *C. obesa* occurs as a small tree fringing tidal rivers, swamps, and salt lakes; nodulated plants of this species are found in soils with up to 28 mg Cl^- g^{-1} soil (Reddell et al., 1986). *C. glauca*

TABLE 3 Effect of Increasing Soil NaCl Concentration on Nodule Dry Weight, Biomass of Shoots Expressed in Total N and N_2 Fixation Efficiency (± SEM) in *Casuarina obesa* Seedlings Inoculated with Two Sources of *Frankia* 24 Weeks After Inoculation, 16 Weeks After Commencement of Salinity Treatments

Frankia source	NaCl application (mg NaCl g^{-1} soil)	Nodule dry weight (mg per pot)[a]	Shoot biomass in total N (mg per pot)[a]	N_2 fixation efficiency (mg shoot N mg^{-1} nodule dry weight)
ys (saline soil)	0	550 ± 23	126 ± 5	0.23 ± 0.01
	0.15	562 ± 48	190 ± 4	0.35 ± 0.02
	1.5	292 ± 55	109 ± 32	0.37 ± 0.04
	15	57 ± 14	17 ± 3	0.30 ± 0.04
cb (non-saline soil)	0	203 ± 48	30 ± 5	0.15 ± 0.03
	0.15	176 ± 62	32 ± 6	0.18 ± 0.05
	1.5	123 ± 37	8 ± 3	0.07 ± 0.02
	15	25 ± 2	1	—

[a]Three seedlings per pot.

Source: From Reddell et al. (1985).

grows naturally near tidal estuaries, often directly behind mangroves, but the slightly less tolerant *C. equisetifolia* has enough resistance to grow in soils along seashores where it is exposed to salt spray. The generalization that species found naturally in saline soils have greater salt tolerance should be treated with some caution because of the widespread geographic distribution of some of these species and the dearth of information on the behavior of different provenances. Species or clonal response may also depend on the type of saline environment, for example, coastal sand dunes, areas inundated by seawater, and saline groundwater.

Recent research had demonstrated that strains of *Frankia* isolated from species of actinorhizal plants normally established in saline soils are more tolerant of NaCl in culture than are strains isolated from nonsaline environments (Dawson and Gibson, 1987, and Table 3). Growth and nitrogenase activities of four strains of *Frankia* from the Casuarinaceae (*Casuarina junghuhiana, C. cunninghamiana, A. torulosa,* and *Allocasuarina lehmanniana*) were little affected by NaCl concentrations up to 200 mM, whereas with the exception of an isolate from *Purshia tridentata*, strains isolated from other actinorhizal plants (*Comptonia peregrina, Alnus rubra, Elaeagnus umbellata,* and *Myrica gale*) were severely affected by NaCl concentrations above 50 mM.

Effects of NaCl on growth, nodulation, and N_2 fixation in *Casuarina* have also been assessed (Reddell et al., 1986; Ng, 1987). In both studies, plant dry weights, nodulation, and estimated rates of N_2 fixation either increased or remained unchanged at concentrations of up to 100 mM NaCl. Higher salt levels affected plant growth and nodule mass more than N_2-fixing activity of nodules. One of these papers also raised the possibility that NaCl may more greatly impair NH_4^+ or NO_3^- uptake and/or assimilation than symbiotic N_2 fixation, the sites of which were shown to be physically protected from Cl^- accumulation (Reddell et al., 1986). It is also possible that the salt tolerance of N_2-fixing plants may be enhanced by the products of N_2 fixation, providing the extraosmotic potential needed to prevent plasmolysis of the host cells (Reddell et al., 1986).

These findings suggest that the prospects for selecting salt-tolerant host plant-*Frankia* combinations for use in the reclamation of saline soils are very promising.

Alkalinity

Arid and semiarid zone soils often have alkaline properties. Actinorhizal species tolerant of these soils include *C. cristata* and *C. obesa*, both of which occur and are nodulated on soils of up to pH 9 (Reddell et al., 1986), and *C. equisetifolia* and *H. rhamnoides*, which flourish on sand dunes of pH 8. Other than pH, little is known of effects of properties of alkaline soils on N_2-fixing plants.

The optimum pH for growth of *Frankia* in culture is around 6; growth and N_2 fixation rates decline either side of this value. A pH in the range 6-7 is also optimal for nodulation and N_2 fixation in seedlings of *Casuarina* (Bond, 1957; Coyne, 1973) and *Hippophae* (Bond et al., 1956). The mechanisms by which high pH depresses nodulation and N_2 fixation in actinorhizal plants are unclear.

IV. IMPROVING ESTABLISHMENT AND FUNCTIONING OF ACTINORHIZAL SYMBIOSES IN DRY AND SALT-AFFECTED AREAS

More effective utilization of the Casuarinaceae and other actinorhizal plants in the rehabilitation of degraded arid and semiarid zone soils requires (1) selection of host plant and *Frankia* genotypes with a high potential to fix N_2 that are adapted to the environment, (2) identification of likely edaphic constraints on processes of plant growth, nodulation, and N_2 fixation, and (3) development of appropriate management strategies to reduce the impact of these factors on potential N_2 fixation and productivity. In the previous section we examined some of the soil factors that adversely affect plant growth and symbiotic N_2 fixation; here we consider genetic improvement of the symbionts and site management practices to enhance productivity and N_2 fixation.

A. Inoculation with *Frankia*

Inoculation should be performed using strains of *Frankia* selected on the basis of their ability to nodulate the host plant and fix sufficient N_2 to sustain maximum host plant growth, as well as having specific characteristics (e.g., the ability to enhance the salt tolerance of the host; see Table 3).

TABLE 4 Response of *Casuarina equisetifolia* to Inoculation with Strain ORS021001 Entrapped in a Polymer (Alginate) 1, 2, and 3 Years After Transplantation to the Field[a]

Inoculation	Height (cm)	Dry weight (g per tree)				Total N (g per tree)			
		br	ab	rt	nod	br	ab	rt	nod
July 1985 (1 year after transplantation)									
0	151	163	89	266	0	0.63	1.46	1.24	0
+	182	324	119	309	2.55	1.99	2.17	1.60	0.04
	*	**	**	*		**	**	*	
July 1986 (2 years after transplantation)									
0	291	890	628	507	0	5.64	9.03	2.52	0
+	313	1349	871	626	12.18	9.39	16.24	3.21	0.20
July 1987 (3 years after transplantation)									
0	500	1451	1000	840	0	7.32	15.57	4.74	0
+	575	2220	1356	1031	22.75	12.20	23.35	5.12	0.40
	NS	**	**	**		**	**	NS	

[a]Experiment carried out in a semiarid site in Senegal; br, branches including trunks; ab, assimilatory branchlets; rt, roots; nod, nodules. Significance level, $^{**}P = 0.01$; $^{*}P = 0.05$; NS, nonsignificant.

Source: From Sougoufara et al. (1989b).

For many years, inoculation of seedlings of actinorhizal plants has been performed using suspensions of crushed nodules collected in the field or in the nursery (Torrey, 1982). This technique is still used in many places, including arid and semiarid countries. When practical, crushed nodules should be replaced by pure cultures of selected *Frankia*, not only for their superior performance but also because they substantially reduce the risk of contaminating seedlings or cuttings with pathogenic agents, e.g., *Rhizoctonia solani* and *Pseudomonas solanacearum* (Liang Zichao, 1986).

Since the first successful isolation and culture of *Frankia* in 1978 (Callaham et al., 1978), many strains of *Frankia* have been isolated, some of which are associated with actinorhizal plants growing in arid and semiarid environments. The technology for producing *Frankia* inoculum is now available as the result of recent progress in the in vitro culture of *Frankia* (Diem et al., 1988; Diem and Dommergues, 1990). The use of dried inoculants of *Frankia* cells entrapped in alginate greatly facilitates inoculation work in nurseries. For example, in a field experiment (approximately 1 ha) at Notto, Senegal, to evaluate the response of *C. equisetifolia* to the application of a dried inoculum made of alginate-entrapped cells of *Frankia* strains ORS 021001, the inoculation technique was rated excellent; all inoculated trees nodulated (Table 4). The increase in plant dry weight resulting from inoculation was 40% the second year and 41% the third year; in terms of total N, the inoculation response was 69% the second year and 47% the third. This relative yield improvement was remarkable since the rainfall was only 215 mm in 1985, 375 mm in 1986, and 463 mm in 1987 and, except at the time of planting out, the trees received no supplemental irrigation. These results confirm that *C. equisetifolia* responds to inoculation with *Frankia*, even under drought-stressed conditions.

B. Inoculation with Mycorrhizal Fungi

Endomycorrhizal infection (VAM) is widespread in actinorhizal plants (Gardner, 1986). In P-deficient soil, inoculation with effective endomycorrhizal strains can greatly enhance growth and N_2 fixation (Diem and Gauthier, 1983). Ectomycorrhizal associations are much less frequent than VAM in actinorhizal plants. In the Casuarinaceae, spontaneous true ectomycorrhizae (ECM) are more common in *Allocasuarina* than in *Casuarina* (Reddell et al., 1985).

ECM and VAM fungi can improve the growth of plants placed in water-stressed conditions through two processes (Gianinazzi-Pearson and Diem, 1982).

Increased Water Supply

The water supply could improve if root resistance to water flow decreased, thereby allowing more water to enter the root system (Safir et al., 1972), or if improved P nutrition facilitated water uptake by the plant. The maintenance of a large extramatrical network of endomycorrhizal hyphae may be a feature of particular significance when the soil water potential is low. Both ECM, for example *Pisolithus tinctorius*, and VAM fungi, for example *Glomus mosseae*, which are successful in drought-stressed habitats, produce particularly prolific extramatrical mycelial systems (Read, 1985).

Increased Water Use Efficiency

The drought tolerance of mycorrhizal plants could be improved through more efficient water utilization (Moawad, 1979; Gianinazzi-Pearson and Diem, 1982).

For this reason, actinorhizal plants should be inoculated with mycorrhizal fungi whenever spontaneous mycorrhizal infection does not occur or when the native strains are ineffective. Different inoculation procedures can be used (Schenck, 1982), including the VAM alginate-entrapped inoculants, which have proved successful in the field using VAM on annual legumes (Ganry et al., 1985) and ECM on forest trees (Le Tacon et al., 1986).

C. Improvement of the Host Plant

The N_2-fixing potential of any symbiotic system is determined by both the microsymbiont and the host plant. In the case of actinorhizal plants for the arid and semiarid zones, the following characteristics are important: high potential to fix N_2, ability to form VAM or ECM, and good tolerance of drought, and, if necessary, salinity. More specific parameters which are usually featured in tree-breeding programs, include yield, pest resistance, and the mechanical properties of the wood (El-Lakany, 1986). Different approaches can be adopted to reach these goals.

Species and Provenance Selection

Until recently, investigations of drought-tolerant woody species concentrated essentially on *Eucalyptus, Acacia,* and *Prosopis* (Palmberg, 1985). Selection of provenances of drought-tolerant actinorhizal plants has been undertaken only on a small scale, except in Egypt, where the performance of *C. equisetifolia, C. glauca,* and *C. cunninghamiana* has been compared (El-Lakany, 1983), and in the United States, where interspecific (Hennesey et al., 1985) differences in water stress tolerance of three *Alnus* sp. clones has been established and quantified.

The screening approach based on saline solution cultures followed up by field trials can be used to develop plant material that can thrive in saline environments.

Conventional Plant Breeding

This more elaborate approach is now being applied to *Casuarina* in Egypt. Elite trees have been selected and clonal seed orchards established. Progeny testing is being conducted at different sites in the arid and semiarid regions (El-Lakany, 1983).

Screening Elite Individuals Followed by Vegetative Propagation

This approach is based on screening elite individuals in a population of a given species followed by vegetative propagation to exploit the intraspecific variability of such characteristics as N_2-fixing ability in *C. equisetifolia* (Sougoufara et al., 1990) and the drought resistance in *Alnus glutinosa* (Hennesey et al., 1988).

Most programs of provenance selection and conventional breeding do not consider that the host plant, when associated with *Frankia,* may be able to fix N_2. Preliminary investigations currently underway in Senegal on *C. equisetifolia* are based on this important characteristic, one major screening parameter being the amount of plant N derived from N_2 fixation (Sougoufara et al., 1989).

Somaclonal Variation for Enhancing Salt Tolerance

This approach is based on the somaclonal variation used to enhance salt tolerance, since a callus cultured on a salt-enriched medium produces some cells that exhibit salt tolerance. Through continuous culturing on salt-enriched medium, the callus seems to increase its tolerance (Nabors et al., 1980). The next step is to regenerate the plant from the callus.

Grafting and Micrografting

C. glauca forms many root suckers, which is sometimes a disadvantage. To obtain plants without suckers, *C. glauca* has been grafted onto rootstock of *C. equisetifolia* (Geary, 1983). Grafting could probably also be used to improve the N_2-fixing ability of poorly nodulating Casuarinaceae species.

A more sophisticated technique, micrografting, has been proposed to enable a non-N_2-fixing tree species to nodulate with *Frankia* and, it is hoped, develop the capacity to fix N_2. In this technique, the nonnodulating species is grafted in vitro onto a nodulating (and N_2-fixing) rootstock.

Preliminary results obtained by Kyle and Righetti (1985) with actinorhizal and nonactinorhizal Rosaceae are promising.

Gene Transfer Technology

Applying the ribosomal T-DNA transformation to woody species, Moore et al. (1979) found that apple trees infected with *Agrobacterium rhizogenes* withstood drought better than uninfected trees. Tepfer et al. (1986) have produced genetic grafts with normal aerial shoots attached to transformed, more effective root systems containing foreign genes. Preliminary investigations by E. Duhoux and M. Phelep (unpublished results) indicate that *Casuarina* roots can be transformed by *A. rhizogenes*, which suggests that the Moore or Tepfer approach could also be used with actinorhizal plants.

Much research remains to be done before gene transfer technology combined with the techniques of tissue culture, protoplast fusion, and plant regeneration can be used as standard tools in the development of transgenic actinorhizal plants with the attributes required to tolerate the arid and semiarid environment.

D. Irrigation

Even if both actinorhizal symbionts are highly tolerant to drought and/or salinity, appropriate managerial practices can contribute to reducing the deleterious impact of these constraints on the efficiency of the symbioses. Various practices have been described in other chapters of this book. This section briefly discusses the role of irrigation, which is seldom used on actinorhizal plants, except on the Casuarinaceae. Extensive plantations of *Casuarina* spp. have been developed in Egypt, where they are integrated with agriculture to serve as windbreaks, shelterbelts,* or woodlots.

Trees have long been planted as windbreaks in temperate countries. In arid and semiarid regions, the practice is relatively recent. The positive physical effects of windbreaks on evaporation, wind speed, air and soil temperatures, soil moisture, and humidity, among other factors, have generally been demonstrated, but the biological effects, which are related to stomatal behavior, growth, crop survival and yield, pest attacks, and diseases, have hitherto received little attention (El-Lakany, 1985).

As a case study, consider *C. glauca* and *C. cunninghamiana*, their hybrid, and *C. equisetifolia*, which when inoculated with crushed notules grow satisfactorily and have good biomass yields even in the desert fringes of the Nile Delta under 185 mm annual rainfall, provided that they receive biweekly irrigation during the summer (May through September) and complementary irrigation the rest of the year (El-Lakany, 1983). Irrigation systems include furrow irrigation and sprinkler and trickle irrigation systems.

---

*The terms "shelterbelt" and "windbreak" are frequently used interchangeably. A shelterbelt, however, is usually a wide barrier of trees planted to protect a large area, for example an entire farm, but a windbreak is one or two rows of trees planted to protect a specific field (El-Lakany, 1985).

In Israel, preliminary elimination trials carried out in the Negev under 200-300 mm rainfall have produced promising results using the following Casuarinaceae: *C. cristata*, *C. cunninghamiana*, *C. glauca*, *Casuarina humilis*, *Casuarina inophloia*, *Casuarina luehmannii*, *A. decaisneana*, and *A. torulosa* (Weinstein, 1983). Unfortunately, Weinstein's report does not indicate whether plants were inoculated and does not give any data about the irrigation scheme.

Irrigated plantations of *Casuarina*, as part of agrosystems, have also been tested and proven feasible in a number of other countries, such as India (Kondas, 1983; Singh, 1963) and Burkina Faso (Armitage, 1985).

V. CONCLUDING REMARKS

Actinorhizal plants occur in a wide range of habitats. However, with the exception of some Casuarinaceae, few useful species have been identified and assessed for growth and N_2 fixation in arid and semiarid areas. There is an urgent need to develop techniques to screen and select drought- and salt-tolerant actinorhizal plants and to study their behavior in stressed environments. Research must focus on the ability to fix N_2 (and possibly to form effective associations with mycorrhizal fungi) and other selection criteria, which can be affected by the water regime. In rainfed conditions, selection criteria would be resistance to drought and salinity. In agroforestry systems, which usually include irrigation, the selection criteria would rest on the effectiveness of the trees as windbreaks, their productivity as a source of timber and firewood for the farmers, and their tolerance to salinity.

In both situations, utmost care should be given to establishing symbioses with *Frankia* and with mycorrhizal fungi. This requires (1) the selection of resistant symbiotic strains adapted to drought and/or salinity, and (2) an excellent inoculant and reliable method to apply them.

As with legume symbioses, any program to evaluate and select clones of actinorhizal plants and associated *Frankia* or/and mycorrhizal fungi should provide short-term laboratory or greenhouse screening and long-term field testing.

The potential of actinorhizal plants in arid environments may be better than expected. Selected clones of *C. equisetifolia* inoculated with an effective *Frankia* strain and irrigated throughout the dry season in Senegal fixed 45 g N_2 year^{-1} (Sougoufara et al., 1990). Extrapolations from the experiment in Senegal give figures of 90 and 180 kg N_2 fixed annually ha^{-1} for planting densities of 2000 and 4000 trees ha^{-1}, respectively.

REFERENCES

Allen, O. N., and Allen, E. K. (1981). *The Leguminosae.* The University of Wisconsin Press, Madison.

Armitage, F. B. (1985). *Irrigated Forestry in Arid and Semiarid Lands: A Synthesis.* IDRC, Ottawa.

Aswathappa, N., and Bachelard, E. P. (1986). Ion regulation in organs of *Casuarina* species differing in salt tolerance. *Aust. J. Plant Physiol. 13*: 533-545.

Benson, D. R. (1988). The genus *Frankia*: Actinomycete symbionts of plants. *Microbiol. Sci. 5*: 9-12.

Bond, G. (1957). The development and significance of the root nodules of *Casuarina. Ann. Bot. 21*: 378-380.

Bond, G. (1983). Taxonomy and distribution of non-legume nitrogen-fixing systems. In *Biological Nitrogen Fixation in Forest Ecosystems: Foundations and Applications*, J. C. Gordon and C. T. Wheeler (Eds.). Nijhoff/Junk, The Hague, pp. 55-87.

Bond, G., and Mackintosh, A. H. (1975). Diurnal changes in nitrogen fixation in the root nodules of *Casuarina. Proc. Roy. Soc. Lond [B] 192*: 1-12.

Bond, G., MacConnell, J. T., and McCallum, A. H. (1956). The nitrogen nutrition of *Hippophae rhamnoides. Ann. Bot. 20*: 501-512.

Burggraaf, A. J. P., and Shipton, W. A. (1982). Estimates of *Frankia* growth under various pH and temperature regimes. *Plant Soil 69*: 135-147.

Callaham, D., Del Tredici, P., and Torrey, J. G. (1978). Isolation and cultivation in vitro of the actinomycete causing root nodulation in *Comptonia. Science 199*: 899-902.

Clemens, J., Campbell, L. C., and Nurisjah, S. (1983). Germination, growth and mineral ion concentrations of *Casuarina* species under saline conditions.

Coyne, P. D. (1973). Some aspects of the autoecology of Casuarinas, with particular reference to nitrogen fixation. Ph.D. thesis, Department of Forestry, Australian National University, Canberra.

Dawson, J. O., and Gibson, A. H. (1987). Sensitivity of selected *Frankia* isolates from *Casuarina, Allocasuarina* and North American host plants to sodium chloride.

Diem, H. G., and Dommergues, Y. R. (1985). In vitro production of specialized reproductive torulose hyphae by *Frankia* strain ORS 021001 isolated from *Casuarina junghuhniana* root nodules. *Plant Soil 87*: 17-29.

Diem, H. G., and Dommergues, Y. R. (1990). Current and potential uses and management of Casuarinaceae in the tropics and subtropics. In *The Biology of Frankia and Actinorhizal Plants*, C. R. Schwintzer and J. D. Tjepkema (Eds.). Academic Press, New York, pp. 317-342.

Diem, H. G., and Gauthier, D. (1982). Effet de l'infection mycorhizienne (*Glomus mosseae*) sur la nodulation et la croissance de *Casuarina equisetifolia. C.R. Séances Acad. Sci. Paris [III] 294*: 215-218.

Diem, H. G., Duhoux, E., Simonet, P., and Dommergues, Y. R. (1988). Actinorhizal symbiosis biotechnology: The present and the future. In *Eighth International Biotechnology Symposium Paris 1988 Proceedings*, G. Durand, L. Bobichon, and J. Florent (Eds.). Société Française de Microbiologie, Paris, pp. 984-995.

Dommergues, Y. (1962). Contribution à l'étude de la dynamique microbienne des sols en zone semi-aride et en zone tropicale sèche. *Ann. Agron. 13*: 265-324, 391-468.

Dommergues, Y. (1963). Evaluation du taux de fixation de l'azote dans un sol dunaire reboisé en filao (*Casuarina equisetifolia*). *Agrochemica 7*: 335-340.

Doran, J. C., and Hall, N. (1983). Notes on fifteen Australian casuarina species. In *Casuarina Ecology, Management and Utilisation*, S. J. Midgley, J. W. Turnbull, and R. D. Johnston (Eds.). CSIRO, Melbourne, pp. 19-52.

El-Lakany, M. H. (1983). A review of breeding drought resistant *Casurina* for shelterbelt establishment in arid regions with special reference to Egypt. *For. Ecol. Manag. 6*: 129-137.

El-Lakany, M. H. (1985). Biological effects of shelterbelts and windbreaks in arid regions. In *Sêminaire international sur les brise-vent, Tunis 1983*. IDRC-MR117e, f, CDRI, Ottawa, pp. 104-111.

El-Lakany, M. H. (1986). A decade of research on Casuarina. In *Proceedings of the 18th IUFRO World Congress, Ljubliana, Yugoslavia, Division 1*, Vol. 1, R. Hermann (Ed.). IUFRO, Vienna, pp. 31-37.

El-Lakany, M. H., and Luard, E. J. (1982). Comparative salt tolerance of selected *Casuarina* species. *Aust. For. Res. 13*: 11-20.

Ganry, F., Diem, H. G., and Dommergues, Y. R. (1985). Inoculation with *Glomus mosseae* improves N_2 fixation by field-grown soybean. *Biol. Fertil. Soils 1*: 15-23.

Gardner, I. C. (1986). Mycorrhizae of actinorhizal plants. *MIRCEN J. 2*: 147-160.

Gauthier, D., Diem, H. G., Dommergues, Y. R., and Ganry, F. (1985). Assessment of N_2 fixation by *Casuarina equisetifolia* inoculated with *Frankia* ORS021001 using ^{15}N methods. *Soil Biol. Biochem. 17*: 375-379.

Geary, T. F. (1983). Casuarinas in Florida, USA and in some Caribbean islands. In *Casuarina Ecology, Management and Utilization*, CSIRO, Melbourne, pp. 107-109.

Gianinazzi-Pearson, V., and Diem, H. G. (1982). Endomycorrhizae in the tropics. In *Microbiology of Tropical Soils and Plant Productivity*, Y. R. Dommergues and H. G. Diem (Eds.). Nijhoff/Junk, The Hague, pp. 208-251.

Hennesey, T. C., Bair, L. K., and McNew, R. W. (1985). Variation in response among three *Alnus* clones to progressive water stress. *Plant Soil 87*: 135-141.

Hennesey, T. C., Lorenzi, E. M., and McNew, R. W. (1988). Stomatal conductance and growth of five *Alnus glutinosa* clones in response to controlled water stress. *Can. J. For. Res. 18*: 421-426.

Hopmans, P., Douglas, L. A., Chalk, P. M., and Delbridge, S. G. (1983). Effects of soil moisture, mineral nitrogen, and salinity on nitrogen fixation (acetylene reduction) by *Allocasuarina verticillata* (Lam.) L. Johnson seedlings. *Aust. For. Res. 13*: 189-196.

Johnson, L. A. S. (1988). Notes on Casuarinaceae. III. The new genus *Ceuthostoma*. *Telopea 3*: 133-137.

Kant, S., and Narayana, H. S. (1978). Effect of water stress on growth, nodulation and dinitrogen fixation in *Casuarina equisetifolia*. *Ann. Arid Zone 17*: 216-221.

Kondas, S. (1983). *Casuarina equisetifolia*—a multipurpose tree cash crop in India. In *Casuarina Ecology, Management and Utilization*, J. W. Turnbull and R. D. Johnston (Eds.). CSIRO, Melbourne, pp. 66-76.

Kyle, N. E., and Righetti, T. L. (1985). In vitro micrografting of actino rhizal desert shrubs. In *Nitrogen Fixation Progress*, H. J. Evans, F. J. Bottomley, and W. E. Newton (Eds.). Martinus Nijhoff, Dordrecht, p. 364.

Le Tacon, F., Garbaye, J., and Carr, G. (1986). The use of mycorrhizas in temperate and tropical forests. In *Proceedings of the 18th IUFRO World Congress, Ljubliana, Yugoslavia*, E. Donaubauer (Ed.). IUFRO, Vienna, pp. 513-524.

Liang Zichao (1986). Vegetative propagation and selection of casuarina for resistance to bacterial wilt (in Chinese, summary in English). *Trop. For. (Sci. Technol.) Guangzhou 2*: 1-6.

Luard, E. J., and El-Lakany, M. H. (1984). Effects on *Casuarina* and *Allocasuarina* species of increasing sodium chloride concentrations in solution culture. *Aust. J. Plant Physiol. 11*: 471-481.

Midgley, S. J., Turnbull, J. W., and Johnston, R. D. (Eds.) (1983). *Casuarina Ecology, Management and Utilization*. CSIRO, Melbourne.

Midgley, S. J., Turnbull, J. W., and Hartney, V. J. (1984). Fuelwood species for salt-affected sites. In *ADAD Seminar, Forage and Fuel Production from Salt-Affected Wasteland*. Cunderdin, Western Australia.

Moawad, M. (1979). Ecophysiology of vesicular-arbuscular mycorrhiza in the tropics. In *The Soil-Root Interface*, J. L. Harley and R. Scott Russell (Eds.). Academic Press, London, pp. 197-209.

Moore, L., Warren, G., and Strobel, G. (1979). Involvement of a plasmid in the hairy root disease of plants caused by *Agrobacterium rhizogenes*. *Plasmid 2*: 617-626.

Nabors, M. W., Gibbs, S. E., Bernstein, C. S., and Meis, M. E. (1980). NaCl-tolerant tobacco plants from cultured cells. *Z. Pflanzenphysiol. 97*: 13-17.

National Research Council (1984). *Casuarinas: Nitrogen-Fixing Trees for Adverse Sites*. National Academy Press, Washington, D.C.

Ng, B. H. (1987). The effects of salinity on growth, nodulation and nitrogen fixation of *Casuarina equisetifolia*. *Plant Soil 103*: 123-125.

Palmberg, C. (1985). Selection and genetic improvement of indigenous and exotic multi-purpose tree species for dry zones. *Agrofor. Syst. 4*: 121-127.

Read, D. J. (1985). Non-nutritional effects of mycorrhizal infection. In *Physiological and Genetical Aspects of Mycorrhizae*, V. Gianinazzi-Pearson and S. Gianinazzi (Eds.). INRA, Paris, pp. 169-176.

Reddell, P., and Bowen, G. D. (1985). *Frankia* source affects growth, nodulation and nitrogen fixation in *Casuarina* species. *New Phytol. 100*: 115-122.

Reddell, P., Bowen, G. D., and Robson, A. D. (1985). The effects of soil temperature on plant growth, nodulation and nitrogen fixation in *Casuarina cunninghamiana*. *New Phytol. 101*: 441-450.

Reddell, P., Foster, R. C., and Bowen, G. D. (1986). The effects of sodium chloride on growth and nitrogen fixation in *Casuarina obesa* MIQ. *New Phytol. 102*: 397-408.

Righetti, T. L., and Munns, D. N. (1982). Nodulation and nitrogen fixation in Purshia: Inoculation response and species comparison. *Plant Soil 65*: 383-396.

Righetti, T. L., Chard, C. H., and Backhaus, R. A. (1986). Soil and environmental factors related to nodulation in *Cowania* and *Purshia*. *Plant Soil 91*: 147-160.
Safir, G. R., Boyer, J. S., and Gerdemann, J. W. (1972). Nutrient status and mycorrhizal enhancement of water transport in soybean. *Plant Physiol. 49*: 700-703.
Schenck, N. C. (1982). *Methods and Principles of Mycorrhizal Research*, American Phytopathological Society, Saint-Paul, Minnesota.
Shipton, W. A., and Burggraaf, A. J. P. (1982). *Frankia* growth and activity as influenced by water potential. *Plant Soil 69*: 293-297.
Singh, B. (1968). The first irrigated plantation of Rajasthan. *Indian For. 89*: 690-700.
Sougoufara, B., Duhoux, E., and Dommergues, Y. R. (1989a). Choix des critères de sélection chez un arbre fixateur de N_2: *Casuarina equisetifolia*. *Oecol. Plant. 10*: 215-224.
Sougoufara, B., Diem, H. G., and Dommergues, Y. R. (1989b). Response of field-grown *Casuarina equisetifolia* to inoculation with *Frankia* strain ORS 021001 entrapped in alginate beads. *Plant Soil 118*: 133-137.
Sougoufara, B., Danso, S. K., Diem, H. G., and Dommergues, Y. R. (1990). Estimating N_2 fixation and N derived from soil by *Casuarina equisetifolia* using labelled ^{15}N fertilizer: Some problems and solutions. *Soil Biol. Biochem. 22*: 695-701.
Tepfer, M., Yacoub, A., Lambert, C., Goldmann, A., Rosenberg, C., Dénarié, J., Jung, G., and Slighton, J. (1986). Application of genetic transformation by Ri T-DNA from *Agrobacterium rhizogenes* in plant biotechnology. *Symbiosis 2*: 9.
Torrey, J. G. (1978). Nitrogen fixation by actinomycete-nodulated angiosperms. *Bioscience 28*: 586-823.
Torrey, J. G. (1982). Casuarina: Actinorhizal nitrogen-fixing tree of the tropics. In *Biological Nitrogen-Fixing Technology for Tropical Agriculture*, P. H. Graham and S. C. Harris (Eds.). Centro Internacional de Agricultura Tropical (CIAT), Cali, Colombia, pp. 427-439.
Torrey, J. G., and Tjepkema, J. D. (1979). Symbiotic nitrogen fixation in actinomycete-nodulated plants. *Bot. Gaz. 140(suppl.)*: 1-11.
Turnbull, J. W. (1986). *Multipurpose Australian Trees and Shrubs*. Australian Center for International Agriculture Research, Canberra.
Waughman, G. J. (1977). The effect of temperature on nitrogenase activity. *J. Exp. Bot. 28*: 949-960.
Weinstein, A. (1983). Casuarina introduction trials in Israel. In *Casuarina Ecology, Management and Utilisation*, S. J. Midgley, J.W. Turnbull, and R.D. Johnston (Eds.). CSIRO, Melbourne, pp. 77-79.
Wollum, A. G., and Youngberg, C. T. (1969). Effects of soil temperature on nodulation of *Ceanothus velutinus* Dougl. *Proc. Soil Sci. Soc. Am. 33*: 801-803.
Zhang, Z., Murry, M. A., and Torrey, J. G. (1986). Culture conditions influencing growth and nitrogen fixation in *Frankia* sp. HPFCc13 isolated from Casuarina. *Plant Soil 91*: 3-15.
Zhu, Z. H. (1985). A general situation of protecting forests in China. In *Seminaire international sur les brise-vent, Tunis 1983*. IDRC-MR117e, f, CDRI, Ottawa, pp. 226-235.

16

Microalgal Mass Culture and Forced Development of Biological Crusts in Arid Lands

GJERT KNUTSEN *University of Bergen, Bergen, Norway*

BLAINE METTING *Cyanotech Corporation, Woodinville, Washington*

I. INTRODUCTION

Substantial evidence exists for the important roles that biological crusts have in soil formation, maintenance of soil fertility, and stabilization of soil against erosion by wind and water (see Chap. 8). Nevertheless, forced development of biological crusts for reclamation of disturbed semiarid lands and deserts has received scant attention. Natural establishment of mature plant cover in dry ecosystems is slow. Processes that initiate or hasten desertification or degradation of useful rangeland are numerous and cumulative, and without human intervention the rate of habitat destruction is greater than the rate of repair. Therefore, desert reclamation and efforts to invigorate abused rangelands are important goals toward which the accelerated establishment of biological crusts could make an important contribution.

The usefulness of artificial crust development depends on successful colonization of soil surfaces within a time frame significantly shorter than observed in nature. A few studies suggest that this may be possible on a small scale under certain conditions. Barriers to forced establishment on a large scale include the lack of suitable inocula and the need for the adequate and timely provision of water. Overviews of the microbial ecology of soil algae and the emergence of a microalgal soil-conditioning technology are offered in support of the hypothesis that forced development of biological crusts could be of value in some situations.

Microalgal culture technologies are at present a source of high-quality protein and value-added special chemicals. An introductory overview of microalgal mass culture technology is presented, permitting ready access to the literature for readers wishing to learn more. Simultaneously, a case is made for adoption of microalgal technology by developing nations whose resources include arid lands and sunshine.

Groundwater deposits supply the freshwater needs for many inhabitants of semiarid lands and deserts. Aquifers unfit for human consumption or irrigation of crops underlie many dry regions. Otherwise of little utility, saline and alkaline (mineralized) shallow groundwater supplies could potentially support microalgal biomass technologies whose chief climatic requirements are abundant sunlight, a supply of water, and moderate to high temperatures. Of industrial dimension or integrated with waste treatment at the village level, algal biomass produced with low-quality groundwater could provide the inocula necessary for biological crust formation or, alternatively, protein or products with commercial value.

II. FORCED DEVELOPMENT OF BIOLOGICAL CRUSTS

A. Applications of Microalgae in Agriculture

Microalgae are common inhabitants of terrestrial surfaces, including soils. Cyanobacteria (blue-green algae) and species representing a number of eukaryotic algal groups are encountered in flooded tropical paddies and on soils and on and in rocks in deserts (see Chap. 8 and Metting, 1981). Organic matter input via carbon fixation (photosynthesis), biological nitrogen fixation (N_2 fixation), and surface consolidation are important functional attributes of soil algae. Densities vary from zero to in excess of 10^9 colony-forming units cm^{-2}. On moist soils with adequate sunlight, standing crops can be as large as a few hundred to 1000 dry kg ha^{-1}. In at least one case 24 ton (fresh weight) ha^{-1} of cyanobacterial biomass was reported from a flooded rice soil (Roger and Watanabe, 1986).

Laboratory and field studies have documented productivities approaching 40 dry mg C m^{-2} h^{-1}, but these are neither numerous nor uniform. Nitrogen fixation also depends upon sunlight and available moisture as the factors most important in governing rates and total N fixed on soils. Free-living and symbiotic cyanobacteria are commonly the principal avenue for input of N both to wet tropical soils and as components of biological crusts in semiarid lands and deserts (see Chap. 8).

Communities of microalgae also develop and change in response to soil chemical factors. Most often cited in the literature is the general proclivity of cyanobacteria for neutral to alkaline soils and the predominance of green microalgae at lower pH (Metting, 1990). Shubert and Starks (1985) studied interrelationships among minerals and algal community succession on soils subject to disturbance and reclamation associated with surface mining for coal. They identified and correlated successional patterns related to physicochemical properties on undisturbed soils and adjacent spoils that were similar in composition but variable in age from 1 to 45 years. Microalgal abundance increased with the age of the spoil but was not correlated to any specific chemical property. The availability of nutrient elements varied with pH, and it was difficult to divorce this from positive and negative correlations between mineral levels (Fe, Na, K, and Ca) and microalgal abundance or the presence or absence of specific taxa. Nevertheless, significant relationships at different ages were identified. This suggests that a preliminary survey of undisturbed sites in semiarid and desert regions might reveal important nutrient needs amenable to simple, inexpensive manipulation favoring successful colonization with introduced species or species mixtures.

Microalgae have been exploited as nitrogen biofertilizers for rice cultivation and as polysaccharide-producing soil conditioners in sprinkler-irrigated temperate agriculture (Metting et al., 1988). Algalization is the practice of inoculating flooded paddies with mixtures of cyanobacteria. Inconsistency of performance and because neither the manner nor extent of algalization has changed in over 20 years where it has been championed in India led Roger and Watanabe (1986) to label it an unproven technology. Nevertheless, inoculation with cyanobacteria has been associated with significantly improved crop performance on numerous occasions (Roger and Kulasooriya, 1980). In a small number of studies, introduced cyanobacteria have also been shown to effectively colonize irrigated agricultural soils in temperate climates, although agronomically significant levels of N_2 fixation have not been observed (Metting, 1990; Metting et al., 1988).

Mucilagenous (palmelloid) green microalgae are applied as soil-conditioning agents on a very small scale in the United States. Soil conditioning is any procedure or product application that improves the physical properties of soil for agriculture. Included is improvement of soil structure by genesis and/or stabilization of soil aggregates. Aggregate formation is complex and poorly understood. However, aggregate stabilization is known to be primarily due to adsorption and binding of particulates by polysaccharides of microbial origin together with enmeshment by living microbial filaments (Burns and Davies, 1986). When inoculated onto irrigated sandy soils through center pivot sprinklers, mass-cultured *Chlamydomonas* and *Asterococcus* species (Chlorophyceae) have been shown to significantly improve the integrity of soil aggregates in the face of disruption by wind and slaking in water (Metting, 1988, 1990; Metting et al., 1988).

B. Stabilization of Dry Soils

Experience with algalization and microalgal soil conditioners demonstrates the feasibility of the forced development of algal populations under suitable conditions. Consortia of eukaryotic microalgae, lichens, cyanobacteria, other eubacteria, and filamentous fungi are, after all, successful colonists on rock and soil surfaces in nearly all dry ecosystems (see Chap. 8). In all likelihood, populations of microalgae and cyanobacteria have evolved that are peculiarly adapted to long periods of low water activity. Also, in lieu of long-term experimental evidence from the field, logic suggests that colonization rates in dry habitats are closely related to the availability of water. Therefore, forced development of biological crusts on soils in dry regions depends on inoculum density and quality together with the duration of adequate water activity. Inoculum quality derives from efforts to identify and isolate collections of strains on the basis of growth rates on soil (colonization efficiency), persistence during periods of low water activity, or both. Inoculum density depends on our learning to mass culture appropriate strains. The provision of water depends on climate and the relative value placed on the establishment of biological crusts for soil stabilization in semiarid regions and deserts, where surface water or groundwater is available together with the means for its dispersal to the field.

An intriguing alternative to application of defined species mixtures to semiarid lands was studied by St. Clair and coworkers. Field and greenhouse studies were designed to evaluate the influence of pre-established and introduced biological crusts on the establishment of vascular plant seedlings. Treatments included mechanical disturbance approximate to trampling by grazing animals on burned and unburned plots. A trend was observed toward better seedling establishment on untrampled plots, although differences were significant only under glass (St. Clair et al., 1984). Subsequent experiments showed that slurries prepared from mature crusts could act as effective inocula on dry soils in a semiarid rangeland in Utah (St. Clair et al., 1986). This evidence supports the assumption that it is biologically possible to force the development of biological crusts on semiarid and perhaps even desert soil surfaces. Can we do this on a scale and at rates significant to human needs and in a consistent manner?

III. MICROALGAL MASS CULTURE

A. Exploitation of Mass-Cultured Microalgae

The algae comprise a heterogeneous group of mostly aquatic, nonvascular plants whose only common feature is morphological simplicity (i.e., they are nonarchaegoniate). Algae vary in size from a few micrometers up to 80-90 m and include both microscopic forms and seaweeds. Major algal groups are defined on the basis of pigmentation, cell wall chemistry, biochemical nature of the photosynthate, flagellation, and unique features. Tens of thousands of algal species inhabit the earth, including many thousands of microscopic forms (Bold and Wynne, 1985).

The inherent productivity of microalgal cultures is significantly greater than that of tropical forests, marshlands, or intensive agriculture (Table 1). Thus, microalgal mass culture has been studied for over 40 years as an efficient means for biomass production. Focused upon first by Warburg (1919) as a model system for studying photosynthesis, the unicellular chlorophyte *Chlorella* can grow at rates equivalent to nearly 50 g m^{-2} day^{-1} in the laboratory. Applied research was initially aimed at building a microalgal technology base for protein production for the earth's poor (Burlew, 1953). More recently, economic reality has forced microalgal technology to focus on high-value end products, such as those described here. Cyanobacteria and eukaryotic microalgae that have been commercialized or been subject to extensive applied research are listed in Table 2. Various edaphic, N_2-fixing cyanobacteria are produced for the biofertilization of rice, as are species of *Chlamydomonas* for soil conditioning in the United States (Sec. II.A). Marine microalgae are cultured on a small scale at numerous sites as direct feed for the aquaculture industry. Detailed information on these microalgae and all other aspects of applied microalgal technology can be found by reference to the following texts: Akatsuka (1989), Borowitzka and Borowitzka (1988a), Lembi and Waaland (1988), Richmond (1986a), and Shelef and Soeder (1980).

Chlorella is unique among commercially exploited microalgae because it can be cultured in the dark (heterotrophy) or supplemented with organic C compounds in the light (mixotrophy and facultative photoheterotrophy). All other commercially important microalgae rely on autotrophic growth. Species cultured for large biomass yields are produced in open outdoor systems (Figs. 1 and 2). Microalgae used as sources of high-value

TABLE 1 Productivity of Microalgal Mass Culture Compared to Natural Ecosystems and Agriculture

System	Net primary production ($g\ C\ m^{-2}\ year^{-1}$)
Microalgal mass cultures	
Short-term outdoor yields	8632
Sustained outdoor yields	2354
Seaweed beds and marine reefs	1167
Tropical rainforest	988
Seasonal tropical forest	720
Estuaries	714
Temperate deciduous forest	543
Temperate agriculture	293
Temperate grassland	267

Source: After Woodwell et al. (1978) in Harlin and Darley (1988); Richmond (1986b).

TABLE 2 Microalgae of Commercial or Potential Utility

Microalga	Products and uses
Cyanobacteria (blue-green algae)	
Spirulina	Protein, health foods, α-linolenic acid, water-soluble dyes
Anabaena, Nostoc, Aulosira, other genera	Biofertilizers
Lyngbya, Oscillatoria, Calothrix	Antibiotics, antineoplastic compounds, toxins, phenolics
Phormidium	Emulsifier, polysaccharides
Anabaena, Aphanizomenon, Microcystis	Anatoxin, saxitoxin
Various genera	Phycobilins, restriction endonucleases
Chlorophyceae (green algae)	
Chlorella	Feed protein, wastewater oxygenation, proline, H_2, health foods, pigments
Dunaliella	β-Carotene, glycerol
Scenedesmus	Wastewater oxygenation, feed protein
Botryococcus	Hydrocarbons
Chlamydomonas, Asterococcus	Soil conditioners

TABLE 2 (continued)

Microalga	Products and uses
Neochloris, other genera	Triglycerides
Neospongiococcum, *Haematococcus*	Carotenoids
Bacillariophyceae (diatoms)	
Phaeodactylum	Wastewater oxygenation, palmitic acid
Chaetoceros, other genera	Lipids, ω-3-fatty acids
Thalassiosira, *Navicula*, *Chaetoceros*, other genera	Aquaculture
Cryptophyceae (cryptophytes)	
Cryptomonas, other genera	Carotenoids
Pyrrophyceae (dinoflagellates)	
Gonyaulax, *Gymnodinium*, *Crypthecodinium*, other genera	Saxitoxin, other toxins, carotenoids
Euglenophyceae (euglenoids)	
Euglena	α-Tocopherol
Rhodophyceae (red algae)	
Porphyridium	Polysaccharides, arachidonic acid
Chrysophyceae (golden-grown algae)	
Isochrysis, *Ochromonas*, *Prymnesium*, other genera	Lipids, aquaculture, carotenoids
Eustigmatophyceae	
Nannochloris	Lipids
Xanthophyceae (yellow-green algae)	
Various genera	Lipids, aquaculture

Source: Bold and Wynne (1978); Cohen (1986); DePauw and Persoone (1988); Hoppe et al. (1979); Metting and Pyne (1986); Regan (1988).

chemicals for small markets can be economically cultivated in closed autotrophic systems (photobioreactors).

Today, *Chlorella* species are mass cultured in China and Southwest Asia for sale in Japan and the United States as a protein-rich food. According to Kawaguchi (1980), 46 *Chlorella* factories of varying output were operating in 1979. At least 30 plants in Taiwan together produced about 1000 dry tons annually by 1977 (Soong, 1980). Following the initial *Chlorella* work, renewed interest in microalgal production was provided in the mid-1970s by rediscovery of the non-nitrogen-fixing, filamentous cyanobacterium *Spirulina* as a protein source for human consumption (Switzer, 1980).

According to Durand-Chastel (1980), the French film producer Max Ives Brandily published a popular article recounting *Spirulina* use by people in the Lake Chad region of Africa. It was also shown that the

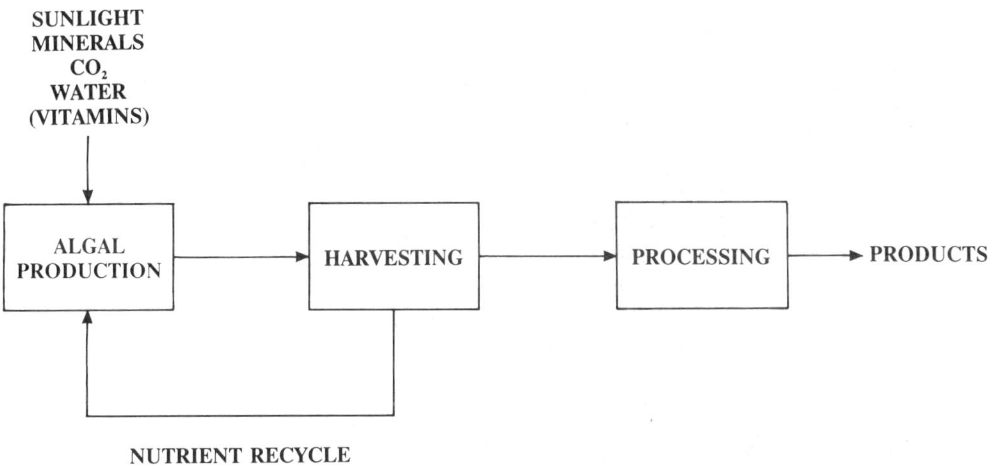

FIGURE 1 Fundamental principle of microalgal biomass production.

Aztec in Mexico traditionally collected natural blooms of *Spirulina* from Lake Texcoco. Today the site of a large soda evaporation project near Mexico City, Sosa Texcoco collects *Spirulina* biomass as a by-product of its chemical operations. Cyanotech Corporation (Hawaii) and DaiNippon Ink & Chemicals (Thailand and Southern California) are currently the largest producers of *Spirulina* biomass for today's marketplace (Table 3).

Spirulina is among the richest sources of plant protein. Having a content up to 70% by weight, *Spirulina* protein quality is nutritionally acceptable for human and animal consumption and readily digestible because of the gram-negative bacterial nature of the cell wall. *Spirulina* biomass contains vitamins and is an excellent source of α-linolenic acid. Carbohydrate and phycobilin content vary among strains and with culture conditions, as do protein and fatty acid profiles (Jassby, 1988). In large measure, the microalga has been successfully mass produced because it tends to dominate outdoor cultures at high alkalinity and pH.

The halotolerant, biflagellate chlorophyte *Dunaliella* is a rich source of both β-carotene and glycerol (Borowitzka and Borowitzka, 1988b). A number of commercial enterprises, some of which are included in Table 3, are engaged in efforts that may one day make production of *Dunaliella* biomass greater than that of any other single microalga. As with other microorganisms, the chemical composition of this alga varies with genotype and culture conditions.

Dunaliella species tolerate hypersaline environments of as much as 20% NaCl by manufacturing glycerol as an internal osmoticum. The predominant commercial value of *Dunaliella* is its content of β-carotene (provitamin A), an antioxidant synthesized as a mixture of isomers in response to high irradiance. Commercial demand for natural β-carotene, which commands perhaps 10% of the total market for β-carotene, has been increasing.

Microalgal cultivation is integral to aquaculture for the production of many crustaceans, shellfish, and fish. De Pauw and Persoone (1988) listed over 30 genera of mostly marine diatoms, chlorophytes, prasinophytes,

FIGURE 2 Aerial view of mass production facility (6 ha) operated by Cyanotech Corporation at Keahole Point, Kona, Hawaii. Dark ponds contain *Spirulina*. Light ponds contain *Dunaliella*. (From Cyanotech Corporation.)

cryptophytes, and haptophytes that include species utilized or of potential utility as aquaculture feeds. In many cases the high value of aquaculture products permits the costly production of microalgae using indoor systems of <100 m^3. Phytoplankton communities in nutrient-enriched natural bodies of water are also fed to animals. However, the improved quality, efficiency, and diversity of aquaculture operations largely depends upon the innovative use of microalgal mass culture technology.

TABLE 3 Location and Approximate Productive Area of Some Commercial *Chlorella*, *Dunaliella*, and *Spirulina* Companies

Company and location	Size (ha)	Microalga
Sosa Texcoco, Mexico[a]	12	*Spirulina*
Siam Algae, Thailand[b]	2	*Spirulina*
Earthrise Farms, California[b]	10	*Spirulina*
Cyanotech Corporation, Hawaii	6	*Spirulina*, *Dunaliella*
Microbio Resources, California	2	*Dunaliella*
Koor Foods, Israel	3	*Dunaliella*
Nippon *Spirulina*, Japan	2	*Spirulina*
Taiwan (four companies)	3-7	*Chlorella*, *Spirulina*
Western Biotechnology, Australia[a]	50	*Dunaliella*

[a]Western Biotechnology and Sosa Texcoco harvest natural populations; other companies' facilities are designed for microalgal production.

[b]Subsidiary companies of DaiNippon Ink & Chemicals (Tokyo).

Source: Ciferri and Triboni (1985); Richmond (1986b); Jassby (1988).

Small markets exist for microalgal products manufactured on a limited scale. These include phycobilins from cyanobacteria used as fluorescent tags in cellular (medical) diagnostic procedures and carotenoids other than β-carotene that have utility as coloring agents. Cyanobacteria yield at least 14 restriction endonucleases commercially sold for use in molecular biology (Cohen, 1986). Microalgae have also been touted as potential sources of other high-value products. Those having received a good deal of attention but from which no economically viable products have as yet emerged include microalgal sources of polysaccharides and ω-3-fatty acids.

ω-3-Fatty acids derive their commercial promise from recent nutritional studies documenting their association with a reduced incidence of heart disease. Current demand is supplied as oil processed from cold-water marine fish, but the compounds themselves are manufactured by microalgae at the bottom of the food chain (Regan, 1988). *Porphyridium*, a genus of microscopic red algae that includes freshwater, edaphic, and marine species, is a source of mucilages with properties of commercial interest (Vonshak, 1988). Finally, microalgae produce compounds with diverse biological activity ranging from antibiosis to plant growth regulatory activity (Metting and Pyne, 1986). Specifically, marine cyanobacteria are being screened by private companies and by subcontractors to the U.S. National Cancer Institute in programs aimed at discovery of novel antineoplastic compounds.

An important application of biomass technology with relevance to developing nations is exploitation of microalgal O_2 production for wastewater treatment. Domestic, municipal, and industrial waste streams have been subject to cleanup with microalgae as integral system components. Oxidation ponds designed to maximize O_2 output by microalgae concomitant with biomass production are called high-rate oxidation ponds (HROP). Conditions in HROPs commonly enrich for species of *Ankistrodesmus*, *Chlorella*, *Coelastrum*, and *Scenedesmus* (nonflagellate chlorophytes). Harvested biomass is suitable as a protein supplement to animal feed. Municipal systems based on this concept are operational in California (Oswald, 1988a) and have also been implemented as village demonstration projects in India and Africa (Venkataraman and Becker, 1985; Fox, 1985, 1988).

B. Physiological Considerations and Productivity

Photosynthesis is the central metabolic process in microalgal production. In photosynthesis, light energy is transformed into chemical energy through reduction of CO_2 to carbohydrate. Light in the 400-700 nm range (photosynthetically active radiation) is captured by algal pigment systems (antenna) associated with reaction centers and electron transport chains, all located in thylakoid membranes within chloroplasts. The absorbed light energy is channeled to the reaction centers, where it drives electrons along electron transport chains from water to NADP. Water is oxidized and NADP is reduced in the process. The products of the light reactions are protons, molecular oxygen, ATP, and reducing power. Reducing power and ATP are subsequently consumed for reduction of CO_2 to carbohydrate by what collectively are termed the dark reactions of photosynthesis. The dark reactions, which can occur in either the light or dark, are dependent on the activity of ribulose 1,5-bisphosphate carboxylase-oxygenase (RUBP carboxylase), the most common enzyme on earth.

Efficient microalgal production depends on understanding photosynthetic mechanisms and how to design systems to maximize exposure of individual cells while minimizing photoinhibition. This is a critical consideration for any system but is of particular relevance if sites in arid equitorial regions are chosen, as is suggested here. Full sunlight has a photon flux density (irradiance) of about 1700 μEin m^{-2} s^{-1}. Photosynthetic rate depends on photon flux according to a saturation function: at low light, photosynthesis increases linearly with increasing irradiance, gradually leveling off to a constant, optimal rate. Above the optimum, the rate of photosynthesis decreases. The phases represent light-limited, light-saturated, and photoinhibited photosynthesis and are influenced by temperature, concentrations of dissolved O_2 and CO_2, nutritional status, and previous growth conditions (Lips and Avissar, 1986).

Long-term exposure to photoinhibitory levels of light can result in pigment and thylakoid membrane degradation followed by cell death. *Dunaliella* and other species known to resist photo-oxidation produce carotenoids, including β-carotene, as protection. An important temperature-dependent process related to photosynthetic efficiency is photorespiration, a factor able to significantly influence net production. Photorespiration is favored by low CO_2 and high dissolved O_2 tensions and results in glycolate production and energy loss by competitive binding of oxygen by RUBP carboxylase. Therefore, understanding the effects of dissolved O_2 is important in microalgal production.

Microalgal productivity is reported as units of biomass produced per area per interval of time. Whereas the minimum quantum requirement and other fundamental physiological limitations to photosynthesis are not agreed upon, maximum theoretical yields extrapolate to more than 500 dry ton ha^{-1} $year^{-1}$. During the past three decades, yields in outdoor culture have risen to sustained short-term levels (weeks to months) of as much as 40 dry g m^{-2} day^{-1} in small raceways (Laws et al., 1986). Estimated commercial yields for production sustained for months in large ponds range from 12 to 15 g m^{-2} day^{-1} (Richmond, 1986b); 1 g m^{-2} day^{-1} = 3.65 t ha^{-1} $year^{-1}$ (Oswald, 1988b).

C. Engineering Considerations in Microalgal Production

Design considerations are paramount for biologically efficient and economically feasible mass culture, harvesting, and processing of desired microalgal species. The first consideration upon which all design parameters rest is the specific application, that is, to process wastewater, foster growth of a specific microalga, or both. Readers wishing to learn more than we present in the following abbreviated account should see Dodd (1986), Oswald (1988b), and Weissman et al. (1987).

Consideration of the following is necessary for the production of microalgal biomass: nutrient requirements (C and inorganic minerals); local climatic regime; site-specific parameters; culture process requirements (retention time, mixing, power demand, and others); and harvesting and processing needs. The past 30 years have witnessed evolution in outdoor culture from meandering (folded) channels and circular lagoons styled upon familiar secondary treatment facilities to the most common designs employed today.

Nutrient requirements include first and foremost the provision of C, which in the end constitutes about half the dry biomass. Because CO_2 is a gas, laws governing its role and availability in aqueous systems are somewhat complex. Equilibrium with the atmosphere is dependent upon pH, alkalinity, and temperature. According to Colman (1988), cyanobacteria and other microalgae employ active transport-dependent concentrating systems for adequate provision of C. Even so, C demand is great in mass cultures approaching maximum photosynthetic efficiency. A major consideration is therefore efficient supply of gaseous CO_2. Loss to the atmosphere (outgassing) is significant because cultures are intentionally shallow and have large surface-volume ratios to make the best use of available light. Carbon dioxide, H_2CO_3, CO_3^-, and CO_3^{2-} exist in aqueous equilibrium. Microalgae use both CO_2 and bicarbonate under most conditions. Removal of dissolved CO_2 by photosynthesis increases pH. Acidity and alkalinity can therefore be controlled by the rate and level of CO_2 input (Kaplan et al., 1986). Indeed, the main C source under alkaline conditions is bicarbonate, and this explains in part the success with *Spirulina* in large-scale mass culture systems built in the 1980s. Bicarbonate salts are, however, about three times more costly than gaseous CO_2 when both are available. Ideally, CO_2 can be made available on site from geological deposits or industrial processes.

Optimal biomass production also depends upon adequate provision of mineral elements, including N, P, K, S, and microelements. Derived from decades of laboratory experimentation, mineral media employed for microalgal species commonly propagated for commercial purposes are given by Borowitzka (1988). Nitrogen is delivered to microalgal mass cultures as inorganic N (NO_3^- and NH_4^+). urea, or undefined organic wastes. Nitrogen gas (N_2) is directly available only to nitrogen-fixing cyanobacteria. *Spirulina*, the microalga (cyanobacterium) commanding today's greatest share of cultivation area, is not capable of biological N_2 fixation. However, cyanobacteria under consideration as biofertilizers and as components of microalgal species mixtures aimed at minimizing rates of desertification commonly perform biological N_2 fixation (i.e., species of *Anabaena* and *Nostoc*; see Chap. 8).

Climate dictates available sunlight and the temperature regime. Climate cannot be controlled; therefore siting is important to ensure maximum solar flux and minimal temperature fluctuation throughout the year. Contemporary mass culture facilities suffer from seasonal effects due to location at temperate latitudes or from monsoons. Typically, plants sited in the continental United States and Israel are unproductive during much of the year because of low temperature and inadequate sunlight. Likewise, plants in Southeast Asia suffer from monsoon rains that dilute mass cultures, making control of pH and nutrient availability difficult. Siting facilities in sun-drenched, equitorial desert regions would obviate these problems, but introduce new ones. Among these is the availability of freshwater and large evaporation rates. Once a region within a country has been chosen, site-specific properties must be identified and dealt with. Principal among these are topography, availability of water and power, wind regime, soil type and permeability, and alternative uses for land and water.

Design considerations vary with scale and depend on targeted yields and any special requirements of the microalga of choice. Smaller units are needed to produce inocula for larger units. Pond area is often scaled up in increments of five, with industrial scale units as large as 5000 m^2. Production ponds are designed to optimize biomass yield by attaining densities that make efficient use of light while minimizing evaporation, outgassing of CO_2, and contamination with competing microalgae, grazers, parasites, and pollutants. Requirements for mixing (turbulence), channel velocity, materials, power needs, and carbonation and how these influence and are influenced by residence time, density, and productivity are subject to detailed explanation by Dodd (1986) and Oswald (1988b).

Dual-channel raceways in widespread use today (Fig. 2) are shallow and constructed to accommodate cultures of 10-50 cm depth. Length-width ratios are from 10 to 40. For example, a common size is 100 $\times$ 10 m. Generally, pairs of raceways share a wall and power stations for operating paddlewheels, the most widely adapted mixing device. The ends of the raceways are curved, and it is at the ends that special construction details are used to minimize stagnation and other flow problems. Mass cultures are circulated to avoid settling and to accommodate illumination, nutrient, and gas transfer. Operational speeds are usually in the range of 10-30 cm s^{-1}. Energy consumption for mixing varies as the third power of the flow rate, as well as with culture depth, mechanical properties of the paddlewheel, and hydraulic head. In smaller systems, airlift and other pumping options can be employed with success.

The choice of wall and bottom materials is important. Most simple is the unlined bottom, which is suitable only when an impermeable soil is available to minimize erosion and suspension of fine particulates. Other options are progressively more expensive, culminating with plastic liners, which are required for food-grade biomass production in the United States and must be replaced at regular intervals. Ponds can be made by excavation or by the use of vertical walls constructed of any number of materials. Integral parts of the plant whose placement must be taken into careful consideration from the beginning include pumps, pipes, wires, vents, control systems, and access roads and paths.

Covered systems have also been devised for microalgal biomass production. These are mostly ponds enclosed within plastic or glass houses. The advantage of covered systems in temperate climates is minimization of temperature loss at night and some extension of production into cooler months. Conversely, overheating becomes a problem during summer months and in hotter regions. Advantages in most situations include reduced evaporative loss and contamination by competitors introduced with the wind. The major disadvantage in all cases is cost and upkeep. That plastics can reduce irradiance at the culture surface up to 50% is another disadvantage, except where photoinhibition and bleaching might otherwise be destructive (Richmond and Becker, 1986).

For special purposes, completely closed reactors offer unique advantages. Closed reactors contain and circulate the culture in transparent plastic tubes or angled flat plexiglass units. Three systems have been described. Torzillo and coworkers in Italy (1986) used both thin-walled polyethylene and rigid polymethyl methacrylate for cultivation of *Spirulina*. Gudin and Thepenier (1986) have successfully cultured mucilagenous species of *Chlamydomonas* and *Porphyridium* in a closed 100 m^2 tubular reactor in southern France. *Porphyridium* was also cultivated in a rectangular flatplate reactor reported by Anderson (1984), in which a 10 m^2 unit was set at an angle and mechanically rotated to follow the angle of the sun in eastern Washington State. In all cases, productivity was acceptable during summer months. Buildup of temperature and O_2 and fouling of the interior walls were reported as problems to be overcome. However, for the production of high-value products, such as special chemicals or aquaculture feed species unable to compete well in the open, closed reactors offer promise.

The harvesting of algal biomass has proved a major barrier to economical production because of the dilute nature of microalgal cultures, whose densities range from 200 to 600 mg L^{-1} (0.02-0.06%), seldom topping 1 dry g L^{-1}. Although harvesting and concentrating are not necessary for the production of filter-feeding animals, it is required for all other uses of the biomass. Compounding the problem is the small size and negative charge and the specific density of microalgae, which is only slightly greater than that of water.

Harvesting alternatives have been described by Mohn (1988). Included are various methods of centrifugation, filtration, microscreening, flotation, settling, and chemical flocculation. Cost varies greatly, with centrifugation generally most expensive. Therefore, the choice of method depends upon the ultimate value of the biomass, whether viability or cellular integrity is desired, and on the degree of concentration required.

TABLE 4 Estimated Costs of Construction and Operation of an Industrial *Spirulina* Facility in California (5 ha) and Pilot-Scale Production of *Scenedemus* in India (Extrapolated to 5 ha)

	U.S. $ × 1000 (% of total)	
Item	California	India
Initial investment and construction		
Buildings	79 (5.1)	174 (21.3)
Site preparation	50 (3.2)	35 (4.3)
Professional costs, permits	88 (5.6)	24 (2.9)
Water and power networks	257 (16.5)	93 (11.3)
Nutrient storage and supply	50 (3.2)	—
Cultivation system	541 (34.7)	175 (21.3)
Total	1558 (100%)	820 (100%)
Annual operating expenses		
Labor	320 (31.4)	45 (7.0)
Nutrients, CO_2, processing	145 (14.2)	256 (39.8)
Other operating expenses	163 (16.0)	50 (7.8)
Cost of capital	156 (15.3)	75 (11.7)
Depreciation	156 (15.3)	185 (28.8)
Administrative	78 (7.7)	32 (5.0)
Total	1018 (100%)	643 (100%)

Source: Jassby (1988); Venkataraman and Becker (1985); Becker and Venkataraman (1980).

D. Economics of Microalgae Production

Several detailed estimates of the cost of microalgal biomass production are given in the literature (Benemann et al., 1987; Jassby, 1988; Tapie and Bernard, 1988; Venkataraman and Becker, 1985). Costs are divided into capital investments, operational costs, and financial costs. Capital and some financial costs are incurred for plant construction and start-up. Operational and other financial costs are those associated with ongoing plant operations. Capital investment costs include land, plant, and equipment. Costs of operation are both fixed and variable. Some labor, overhead, and maintenance costs are fixed, but power, water, nutrients, maintenance, and other labor costs vary with operation and scale.

Table 4 is a breakdown of initial investment costs and annual production costs for a 5 ha *Spirulina* facility with a harvest capacity of 40 dry kg h^{-1} (10 ha). Estimates are based on operational experience at Earthrise Farms, a joint venture with DaiNippon Ink & Chemical (Tokyo). Earthrise Farms is located in the Imperial Valley of Southern California, where a temperate climate restricts production to less than half the year.

Cost estimates from tropical regions are not drastically different. Pilot plant experience with *Spirulina* and *Scenedesmus* in India (Venkataraman and Becker, 1985) and with *Scenedesmus* in Peru (Heussler, 1985) revealed that capital costs were significant. In all cases, harvesting and processing equipment represent the single largest capital expenditure for the industrial-scale production of microalgae, accounting for about one-third. However, major differences in production costs exist between economically developed and lesser developed nations. Mineral nutrients in the form of agricultural fertilizer are less available and, therefore, more expensive in less developed countries. Carbon dioxide, the largest operational cost component, is also likely to be significantly more expensive in the absence of a local source in less developed regions. Parts and supplies for plant maintenance also constitute a greater expense. Conversely, initial land prices and the cost of labor are usually less in developing countries. Water may or may not be comparable. Adequate climatic conditions permitting production throughout the year is the factor most in favor of site development in developing countries.

The best measure of the economics of industrial-scale microalgal mass culture is that a few companies have been able to produce *Spirulina* commercially for years. Commercial microalgae companies understandably do not share details of their markets, prices for their products, or operational costs more revealing than those of Jassby (1988) (Table 4). Nevertheless, our understanding of *Spirulina* operations in California, Hawaii, Israel, and Thailand suggests a production cost of between \$10 and \$12 dry kg^{-1}. Comparable figures extrapolated from pilot-plant operations are \$7.56 kg^{-1} *Scenedesmus* in Peru (Heussler, 1985) and \$11 kg^{-1} *Chlorella* in Taiwan (Kawaguchi, 1980).

IV. OPPORTUNITY FOR SEMIARID LANDS AND DESERTS

A. Industrial or Village-Level Microalgae Production?

We have presented an introduction to the biological aspects of microalgal production together with brief consideration of industrial applications, mostly as they are known to us from operations in countries in temperate zones with mature economies. Nevertheless, it is intuitively clear that microalgal technologies are well suited to arid tropical zones where and whenever land and water resources cannot be used in ways more beneficial to local people and the environment.

The industrial production of microalgal products in semiarid and arid tropical regions will only be feasible where adequate supplies of accessible water exist that cannot be put to better use. Because freshwater is the limiting factor in semiarid lands and deserts, consideration of the industrial production of microalgae rests with supplies of nonpotable surface and groundwaters. Adequate provision of alkaline water could support mass culture of *Spirulina* (protein) for human or animal use. Conceivably, hypersaline resources may support the production of *Dunaliella* for β-carotene and glycerol. Alternatively, resources may support the mass culture of various halotolerant microalgae as aquaculture feed for shellfish, crustacean, or fish production—or any number of high-value chemicals.

Nearly 98% of the freshwater on earth exists as groundwater plus soil moisture (Kalmbach, 1987). Most is freshwater, much being at great depth or otherwise inaccessible by practical means. Distribution of groundwater in deserts is coupled to sedimentation processes, in turn controlled by climate. Groundwater deposits exist adjacent to humid regions or persist from earlier epochs. Most of the important desert aquifers of the southwestern United States and adjacent Mexico, together with those in Chile and parts of Asia are of nonindurated Cenozoic deposits formed by volcanism and block faulting. Erosional processes were responsible for others, including Mesozoic sandstones of Australia's Great Artesian Basin and the Nubian sandstone of the eastern Sahara. Stream channel deposition in basins is the principal mechanism forming desert aquifers. Basin deposition can leave groundwaters whose edges are more fresh. Saline and mineralized groundwaters are often shallow (Thomas, 1987).

We attempted a compilation of quantitative data to reflect knowledge of the extent and quality of groundwater beneath semiarid lands and deserts in Africa and Asia. Unfortunately, the information does not exist in consolidated form. It became clear that such an undertaking was well beyond the time frame, if not the scope of this chapter. A model effort has, however, been made by the Solar Energy Research Institute (Golden, Colorado). Since 1979, the Aquatic Species Biomass Program has focused on the use of saline and mineralized groundwaters of the arid Southwest (Arizona and New Mexico) as a resource for the production of liquid fuels from microalgal mass culture. Efforts have been targeted at screening and selection for halotolerant and halophilic microalgae; physiology, biochemistry, and genetics of lipid production by microalgae; and engineering, land use, and other practical aspects of the subject (Barclay and McIntosh, 1986; Johnson and Sprague, 1987).

Locally, freshwater is available to people in semiarid lands and deserts from rivers or aquifers. Schemes have been proposed and demonstrated for a village-level microalgal production technology. Cultivation of *Spirulina* as part of an integrated village health system was described by Fox (1985) as a viable alternative to industrial-scale microalgae production in lesser developed nations. By his method, microalgal protein is but one by-product of a complete system designed to aid the nutritional and energy self-sufficiency of villages in emerging countries. Fed compost, agricultural residues, and animal and human wastes, a solar-heated biogas generator is the focus of the system. Anaerobic digestion effectively sanitizes wastes and provides biogas for cooking, plus sludge and effluent. Sludge and filtered particulates are used as fertilizer. Residual effluent contains mineral nutrients sufficient for microalgae cultivation. Requirements for mixing are derived from biogas and wind power. Supplemental bicarbonate is periodically required to maintain dominance by *Spirulina*. Filtering through small-mesh cloth is sufficient for collecting the filamentous *Spirulina* biomass. Effluent from the algal production unit is then fed to a ponded area in which aquatic vascular plants are encouraged to grow whose biomass is occasionally collected and composted.

For successful, long-term operation of a rural microalgae production system, Fox (1985) lists 11 points of importance: (1) energy self-sufficiency; (2) low cost; (3) use of village wastes in place of purchased fertilizer; (4) education, that is, about one "expert" for every 50 villages of 100 people; (5) cooperation of local and national political structure (too often difficult to procure); (6) relevant data for local situations, that is, local impact; (7) foster communal involvement; (8) effective recycling of culture medium; (9) limit the size of the microalgal production units to $<500\ m^2$ to effectively use local energy resources; (10) popularization of the effort(s); and (11) retain local control.

B. Summary

Abundant sunlight, high temperatures, and consistent climatic conditions are characteristic of many deserts. These are also the primary requirements, together with provision of water, for microalgal biomass production. An opportunity exists to make productive use of these resources. Indeed, it may be possible to produce inocula composed of soil-binding and N_2-fixing microalgae sufficient for forced development of biological crusts on semiarid and desert soils otherwise threatened by deterioration and destruction. Much work remains to be done.

Alternatively, effective utilization of saline and alkaline shallow groundwaters for microalgal cultivation could foster significant economic advances. Basic research remaining to be performed includes identification of useful groundwater reserves and popularization of these to local and national authorities as significant economic resources. Whether on an industrial or a village scale, the adoption of microalgal mass culture technologies in arid tropical regions is an opportunity we should pursue.

REFERENCES

Akatsuka, I. (Ed.) (1989). *Introduction to Applied Phycology.* SPB Academic, The Hague.

Anderson, D. B. (1984). Polysaccharide production by *Porphyridium creuentum.* M.S. thesis, Washington State University, Pullman.

Barclay, W. R., and McIntosh, R. P. (Eds.) (1986). Algal biomass technologies. An interdisciplinary approach. *Nova Hedwigia 83*: 1-273.

Becker, E., and Venkataraman, L. (1980). Production and processing of algae in pilot plant experience of the Indo-German project. In *Algae Biomass*, G. Shelef and C. Soeder (Eds.). Elsevier-North Holland Press, Amsterdam, pp. 35-50.

Benemann, J. R., Tillett, D., and Weissman, J. C. (1987). Microalgae biotechnology. *Trends Biotechnol. 5*: 47-53.

Bold, H. C., and Wynne, M. J. (1985). *Introduction to the Algae*, 2nd ed. Prentice-Hall, Englewood Cliffs, New Jersey.

Borowitzka, M. A. (1988). Algal growth media and sources of algal cultures. In *Micro-algal Biotechnology*, M. A. Borowitzka and L. J. Borowitzka (Eds.). Cambridge University Press, Cambridge, pp. 456-465.

Borowitzka, M. A., and Borowitzka, L. J. (Eds.) (1988a). *Micro-algal Biotechnology.* Cambridge University Press, Cambridge.

Borowitzka, M. A., and Borowitzka, L. J. (1988b). *Dunaliella.* In *Microalgal Biotechnology*, M. A. Borowitzka and L. J. Borowitzka (Eds.). Cambridge University Press, Cambridge, pp. 27-58.

Burlew, J. D. (Ed.) (1953). *Algal Culture: From Laboratory to Pilot Plant,* Publ. No. 600, Carnegie Institute, Washington, Stanford.

Burns, R. G., and Davies, J. A. (1986). The microbiology of soil structure. *Biol. Agr. Hort. 3*: 95-113.

Ciferri, O., and Tiboni, O. (1985). The biochemistry and industrial potential of *Spirulina*. *Annu. Rev. Microbiol. 39*: 503-526.

Cohen, Z. (1986). Products from microalgae. In *Handbook of Microalgal Mass Culture,* A. Richmond (Ed.). CRC Press, Boca Raton, Florida, pp. 421-454.

Colman, B. (1988). The uptake of inorganic carbon by microalgae. In *Polysaccharides from Microalgae: A New Agroindustry,* J. Ramus and M. C. Jones (Eds.). Duke University Marine Laboratory, Beaufort, North Carolina, pp. 73-82.

DePauw, N., and Persoone, G. (1988). Micro-algae for aquaculture. In *Micro-algal Biotechnology,* M. A. Borowitzka and L. J. Borowitzka (Eds.). Cambridge University Press, Cambridge, pp. 197-220.

Dodd, J. C. (1986). Elements of pond design and construction. In *Handbook of Microalgal Mass Culture,* A. Richmond (Ed.). CRC Press, Boca Raton, Florida, pp. 265-283.

Durand-Chastel, H. (1980). Production and use of *Spirulina* in Mexico. In *Algae Biomass, Production and Use*, G. Shelef and C. D. Soeder (Eds.). Elsevier North-Holland Biomedical Press, Amsterdam, pp. 51-64.

Fox, R. D. (1985). The cultivation of *Spirulina* as part of an integrated village. *Arch. Hydrobiol. Beih., Ergebn. Limnol. 20*: 17-24.

Fox, R. D. (1988). Nutrient preparation and low cost basin construction for village production of *Spirulina*. In *Algal Biotechnology, Proceedings of the Fourth International Meeting SSA Villeneuve d'Ascq,* T. Stadler, J. Mollion, M.-C. Verdus, Y. Karamanos, H. Morvan, and D. Christianen (Eds.). Elsevier Applied Science, Amsterdam, pp. 355-364.

Gudin, C., and Thepenier, C. (1986). Bioconversion of solar energy into organic chemicals by microalgae. *Adv. Biotech. Proc. 6*: 73-110.

Harlin, M. M., and Darley, W. M. (1988). The algae: An overview. In *Algae and Human Affairs*, C. A. Lembi and R. A. Waaland (Eds.). Cambridge University Press, Cambridge, pp. 3-27.

Heussler, P. (1985). Development and results of the Peruvian-German microalgae project. *Arch. Hydrobiol. Beih., Ergebn. Limnol. 20*: 1-8.

Hoppe, H. A., Levring, T., and Tanaka, Y. (Eds.) (1979). *Marine Algae in Pharmaceutical Science*, Walter de Gruyter, Berlin.

Jassby, A. (1988). *Spirulina*: A model for microalgae as human food. In *Algae and Human Affairs,* C. A. Lembi and R. A. Waaland (Eds.). Cambridge University Press, Cambridge, pp. 149-179.

Johnson, D. A., and Sprague, S. (Eds.) (1987). *Aquatic Species Program Annual Report,* Publ. No. SERI/SP-231-3206, Solar Energy Research Institute, Golden, Colorado.

Kalmbach, O. (1987). Principles of modern water-supply technology. In *Encyclopedia Britannica*, 15th ed., Vol. 26, Chicago, pp. 427-431.

Kaplan, D., Richmond, A. E., Dubinsky, Z., and Aaronson, S. (1986). Algal nutrition. In *Handbook of Microalgal Mass Culture,* A. Richmond (Ed.). CRC Press, Boca Raton, Florida, pp. 147-198.

Kawaguchi, K. (1980). Microalgae production systems in Asia. In *Algae Biomass, Production and Use*, G. Shelef and C. J. Soeder (Eds.). Elsevier North-Holland Biomedical Press, Amsterdam, pp. 25-33.

Laws, E. A., Taguchi, S., Hirata, J., and Pang, L. (1986). High algal production rates achieved in a shallow outdoor flume. *Biotechnol. Bioeng. 28*: 191-197.

Lembi, C. A., and Waaland, R. A. (Eds.) (1988). *Algae and Human Affairs*, Cambridge University Press, Cambridge.

Lips, S. H., and Avissar, Y. J. (1986). Photosynthesis and ultrastructure in microalgae. In *Handbook of Microalgal Mass Culture*, A. Richmond (Ed.). CRC Press, Boca Raton, Florida, pp. 43-67.

Metting, B. (1981). The systematics and ecology of soil algae. *Bot. Rev. 47*: 195-312.

Metting, B. (1988). Microalgal soil conditioners. In *Polysaccharides from Microalgae: A New Agroindustry*, J. Ramus and M. C. Jones (Eds.). Duke University Marine Laboratory, Beaufort, North Carolina, pp. 56-61.

Metting, B. (1990). Soil algae. In *The Rhizosphere*, J. M. Lynch (Ed.). Wiley Series in Ecological and Applied Microbiology, Wiley Interscience, New York, pp. 355-368.

Metting, B., and Pyne, J. W. (1986). Biologically active compounds from microalgae. *Enzyme Microb. Technol. 8*: 386-394.

Metting, B., Rayburn, W. R., and Reynaud, P. A. (1988). Algae and agriculture. In *Algae and Human Affairs*, C. A. Lembi and R. A. Waaland (Eds.). Cambridge University Press, Cambridge, pp. 335-370.

Mohn, F. H. (1988). Harvesting of micro-algae. In *Micro-algal Biotechnology*, M. A. Borowitzka and L. J. Borowitzka (Eds.). Cambridge University Press, Cambridge, pp. 395-414.

Oswald, W. J. (1988a). The role of microalgae in liquid waste treatment and reclamation. In *Algae and Human Affairs*, C. A. Lembi and R. A. Waaland (Eds.). Cambridge University Press, Cambridge, pp. 255-281.

Oswald, W. J. (1988b). Large-scale algal culture systems (engineering aspects). In *Micro-algal Biotechnology*, M. A. Borowitzka and L. J. Borowitzka (Eds.). Cambridge University Press, Cambridge, pp. 357-394.

Regan, D. L. (1988). Other micro-algae. In *Micro-algal Biotechnology*, M. A. Borowitzka and L. J. Borowitzka (Eds.). Cambridge University Press, Cambridge, pp. 135-150.

Richmond, A. (Ed.) (1986a). *Handbook of Microalgal Mass Culture*, CRC Press, Boca Raton, Florida.

Richmond, A. (1986b). Outdoor mass cultures of microalgae. In *Handbook of Microalgal Mass Culture*, A. Richmond (Ed.). CRC Press, Boca Raton, Florida, pp. 285-329.

Richmond, A., and Becker, E. W. (1986). Technological aspects of mass cultivation—a general outline. In *Handbook of Microalgal Mass Culture*, CRC Press, Boca Raton, Florida, pp. 245-263.

Roger, P. A., and Kulasooriya, S. (1980). *Blue-green Algae and Rice*, International Rice Research Institute, Los Banos, Philippines.

Roger, P. A., and Watanabe, I. (1986). Technologies for utilizing biological nitrogen fixation in wetland rice: Potentialities, current usage, and limiting factors. *Fertil. Res. 9*: 39-77.

Shelef, G., and Soeder, C. J. (Eds.) (1980). *Algae Biomass, Production and Use*. Elsevier North-Holland Biomedical Press, Amsterdam.

Shubert, L. E., and Starks, T. L. (1985). Diagnostic aspects of algal ecology in disturbed lands. In *Soil Reclamation Processes*, R. L. Tate, III (Ed.). Marcel Dekker, New York, pp. 83-106.

Soong, H. (1980). Production and development of *Chlorella* and *Spirulina* in Taiwan. In *Algae Biomass, Production and Use*, G. Shelef and C. J. Soeder (Eds.). Elsevier North-Holland Biomedical Press, Amsterdam, pp. 96-113.

St. Clair, L. L., Webb, B. L., Johansen, J. R., and Nebeker, G. B. (1984). Cryptogamic soil crusts: Enhancement of seedling establishment in disturbed and undisturbed areas. *Reclamation Reveg. Res. 3*: 129-136.

St. Clair, L. L., Johansen, J. R., and Webb, B. L. (1986). Rapid stabilization of fire-disturbed sites using a soil crust slurry: Inoculation studies. *Reclamation Reveg. Res. 4*: 261-269.

Switzer, L. (1980). *Spirulina*. Proteus Corp., Berkeley, California.

Tapie, P., and Bernard, A. (1988). Microalgae production: Technical and economic evaluations. *Biotechnol. Bioeng. 32*: 873-885.

Thomas, H. E. (1987). The hydrosphere. In *Encyclopedia Britannica*, 15th ed., Vol. 20, Chicago, pp. 770-785.

Torzillo, G., Pushparaj, B., Bocci, F., Balloni, W., Materrasi, R., and Florenzano, G. (1986). Production of *Spirulina* biomass in closed photobioreactors. *Biomass 11*: 61-64.

Venkataraman, L. V., and Becker, E. W. (1985). *Biotechnology and Utilization of Algae—The Indian Experience*, Department Science and Technology, New Delhi, and Central Food Research Institute, Mysore, India.

Vonshak, A. (1988). *Porphyridium*. In *Micro-algal Biotechnology*, M. A. Borowitzka and L. J. Borowitzka (Eds.). Cambridge University Press, Cambridge, pp. 122-134.

Warburg, O. (1919). Uber die Geschwindichkeit der Kohlensaurezersetsung in lebenden. *Biochem. Z. 100*: 230-270.

Weissman, J. C., Tillet, D. M., and Goebel, R. P. (1987). Design and operation of an outdoor microalgae test facility. In *Aquatic Species Program Annual Report*, D. A. Johnson and S. Sprague (Eds.). Solar Energy Research Institute, Publ. No. SERI/SP-231-3206, Golden, Colorado, pp. 231-252.

Woodwell, G. M., Whittaker, R. H., Reiners, W. A., Likens, G. E., Delwiche, C. C., and Botkin, D. B. (1978). The biota and the world carbon budget. *Science 199*: 141-146.

17

Organic Matter Applications

WALLACE H. FULLER *University of Arizona, Tucson, Arizona*

I. INTRODUCTION

The use of organic materials for the rehabilitation of arid lands and deserts and the techniques for arriving at the final product of more stable soil organic matter and humus depend largely on the kind of raw material originally applied, the method of application, and the specific environment under which biodegradation occurs. Climatic conditions in the area largely control the choice(s) of application system that may be selected. The most desirable application system is not always the most economical, requiring a studied choice of methodology depending on the availability of transport and spreading or injection equipment. The alternatives for transport and application to the land vary according to the major conditions just mentioned, which include more specifically (1) the nature of the waste generated for its physical form—solid, slurry, or liquid; (2) the hazardous character and timeliness characteristic of the system; (3) on-site treatment—storage, volume, and removal of undesirable constituents; (4) transportation facilities; (5) treatment required: dewatering, storage, size change, and dilution; and (6) application to the land: mechanical spreading, soil injection, and irrigation; surface, sprinkling, canal channeling, and/or pipe and lining.

Aridity is mostly a comparison between water *supply* and water *need*. Water need refers primarily to potential evapotranspiration and is more difficult to evaluate than supply. Aridity has thus been expressed as some function of temperature and rainfall and is related to observed responses to the environment to climate (Fuller, 1974). Generally, under arid conditions vegetation is sparse and short and develops as individual plants. Organic matter accumulates slowly and often imperceptibly. Erosion, runoff, and evaporation seriously influence the effectiveness of rainfall in supporting plants, biodegradation, and in leaching salts from the surface horizons.

Soils of arid lands do not lack for an active microbial flora (Cameron and Blank, 1965; Fuller, 1969, 1975a). The activity is not inhibited by higher temperatures on the surface but is retarded by a lack of moisture. Arid soils host an abundant and varied microbial population as adequate as that in other soils for biodegradation and transformation of matter when energy sources and moisture become available. Even algae withstand climates of severe heat and dehydration to become a prominent feature of deserts after infrequent rainstorms (Fuller et al., 1960).

II. THEORY OF USE OF ORGANIC MATTER

A. Soil Organic Matter of Arid Lands

Organic matter is an essential component of intensive rehabilitation of arid lands and deserts, and therefore its nature must be related to theories of application of organic residues (Table 1). Organic matter content and composition in soils have received considerable attention because of the disproportionately favorable effect on the chemical, physical, microbiological, and morphological properties of soil compared with the mineral matter (Fuller, 1975a; Tucker and Fuller, 1971; Fuller and Tucker, 1977). Moreover, organic matter is the nitrogen carrier of soils. The rather narrow range of C/N ratios indicates the direct relationship of N with organic matter and the almost complete dependency of N fertility in virgin soils on organic matter. Soil organic matter is defined here as partially rotted residues of the plants that occupied the land, the added organic wastes, and the materials formed by soil organisms. Soil organic matter is not derived wholly from higher plants. Soil organisms represent a substantial if not dominant contribution. The biodegradation and humus synthesis is wholly dependent on the soil microflora, as illustrated in the soil organic matter cycles of Figs. 1 and 2.

The accumulation of humus is low in desert soils compared with those of other climates, but the plant ash residues are higher. The organic matter of desert USSR soils, for example, is reported to be highly mineralized. Ponomareva (1956) emphasizes the importance of humus on desert soil formation as exemplified by her statement: "The mineral forms of biogenic carbon and generally plant ash compounds shape the profile and properties of desert soils."

The content of total organic matter in desert soils of central Asia ranges from 0.3 to 0.8% and seldom exceeds 1.0%, whereas the total nitrogen varies from 0.03 to 0.3% (Lobova, 1967). The gray-brown desert soils have a relatively low C/N ratio (3:5) compared with humid soils, indicating extensive decomposition. Fulvic acids are proportionally higher, being two or three times as abundant in these highly mineralized soils as in other soils. Waxes and resins are also proportionally high among arid soils. Desert plants, the source of organic residues, are abundantly supplied with waxes, resins, and essential oils. In the drier desert soils, the humus content is small and is distributed fairly uniformly throughout the soil profile (Shuvalov, 1966). This is due to the roots of the shrublike desert plants, which do not spread out in the upper horizons; they dry out rapidly but grow more uniformly deep.

Aridisols (hot desert soils) of the Sonoran Desert of North America rarely accumulate more than 1.0% organic matter except in the thin surface (0-2.5 cm) near established vegetation (Fuller, 1974). Winds redistribute fallen leaves, scale, and plant parts, which collect around stationary objects.

TABLE 1 Some Examples of Organic Materials Used on Arid Lands

Material	Nitrogen (%)	P_2O_5 (%)	K_2O (%)
Dried blood	8.0-14.0	0.3-1.5	0.5-0.8
Animal tankage	5.0-10.0	3.0-13.0	Small amounts
Garbage tankage	2.0-4.0	1.0-3.0	0.5-1.5
Process tankages	6.5-10.0	Variable	Small amounts
Fish scrap, dried	6.5-10.0	4.0-8.0	Small amounts
Sewage sludge, orginary	1.6-3.3	1.0	Small amounts
Sewage sludge, activated	4.1-7.5	2.5-4.0	0.75
Cottonseed meal	6.0-9.0	2.0-3.0	1.0-2.0
Bone meals	0.7-5.3	17.0-30.0	—
Castor pomace	4.0-7.0	1.0-1.5	1.0-1.5
Cocoa shell meal	2.5	1.0	2.5
Tobacco stems	1.3-1.6	0.9	4.0-9.0
Sheep or cow manure, dried and pulverized	1.0-2.0	1.0-2.0	2.0-3.0
Poultry manure, dried and pulverized	5.0-6.0	2.0-3.0	1.0-2.0
Bat guano (dry)	1.5-10.0	1.2-4.0	1.0-3.0

Source: From Fuller and Warrick (1985).

The conclusion reached with respect to the quantitative nature of central Asian desert soils (Rozanov, 1961) is that the relatively very low humus content is chiefly due to the activity of a highly biogenic soil process, not exclusively, as previously supposed, to a low amount of vegetative remains entering the soil. In a virgin sierozem (mesic and frigid families of Ustolls and Xerolls of Camborthids, Dur Haplorthids, Orthids, and Paleorthids) of central Asia the total reserve of root residues was found to be 2.13 kg m^{-2} and the organic matter 4.9 kg m^{-2}. Most of the roots were concentrated in the 0-7 cm layer according to Kononova (1961). The proportion of plant growth above the ground compared with the root substance in certain desert soils is distinctly lower than in more humid soils (Table 2).

The presence of a significant amount of organic phosphorus compounds in North American desert soils was first established by Fuller and McGeorge (1951) and Fuller (1974, 1975b). The organic ratio P/C is narrower in desert soils than in most soils of the humid climates.

The mineral nitrogen in central Asian desert soils is almost exclusively in the form of nitrates. According to Rosanov (1961), the amounts are usually low although they vary from season to season depending on rain patterns. Examples reported in North American desert soils indicate some unusually high levels of NO_3^- N as well as low levels (Fuller, 1975a). Very slowly hydrolyzable or stable organic nitrogen compounds account for

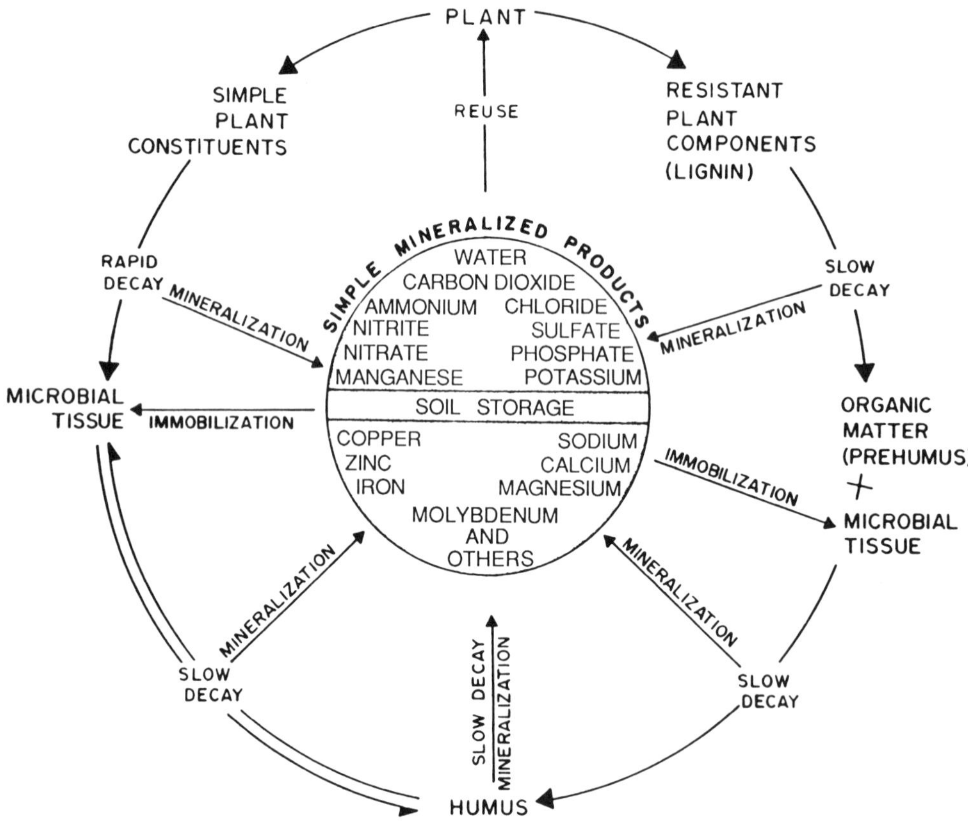

FIGURE 1 The organic matter cycle in soils. (From Fuller and Warrick, 1985.)

nearly 30% of total N in central Asian desert soils are reported by Rozanov (1961). About 55% of organic matter substances containing nitrogen are rather easily hydrolyzable, whereas in the more humid northern prairie soils the amount of this form is much less.

The physicochemical properties of arid soils do not favor the abundant fixation of humus in the USSR (Kononova, 1961). The rapid cycle of the formation and decomposition microbial process, in the short periods when soil moisture is available, accounts for the accumulation of substances of the fulvic acid type. The very rapid rate of decomposition of plant residues in the Sonoran Desert soils of North America is also reported to result in narrow C/N ratios, ranging from 8 to 10 in the upper few centimeters compared with ratios of 10-12 in the more humid soils in the Midwest prairies of the United States (Fuller, 1975a). Even in small amounts, though, organic matter exerts a profound influence on the physical and biological characteristics of the soil. The maintenance of this small amount is essential for maximal plant growth.

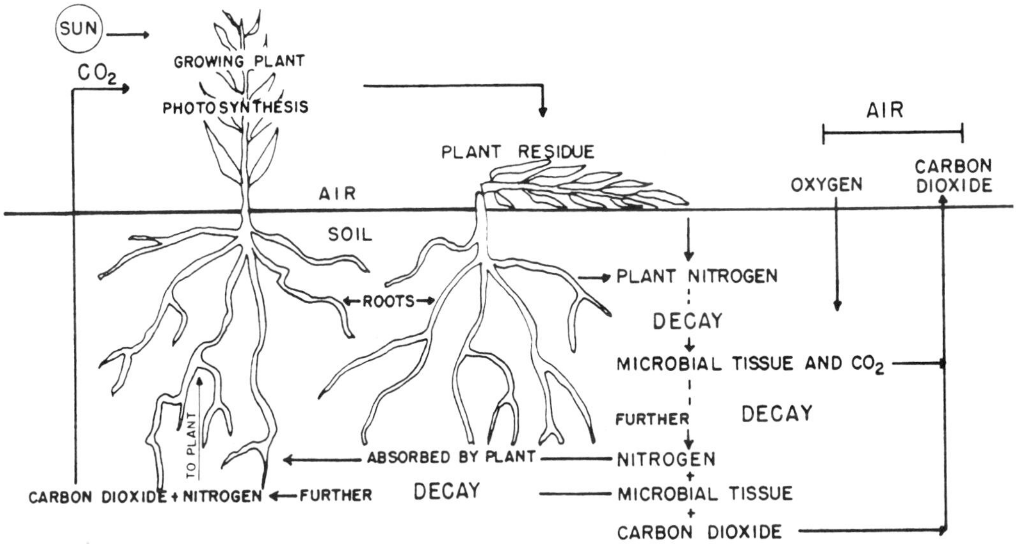

FIGURE 2 Production, decomposition, and recycling of plant residues in soil. (From Fuller, 1975a.)

Continued soil productivity depends upon the replenishment and maintenance of some organic matter. The maintenance of a given, fixed level is often not practical, nor is it advised. The practice of incorporating readily available plant residues, manures, straw, leaves, composts, and municipal wastes and sludges into the soil is more important than the establishment of a fixed level. In irrigated soils, including those of arid lands, the organic matter level may increase substantially above that under native conditions (Fig. 3), as a result of crop production and/or organic waste application. Only recently, municipal solid waste leachate in landfills has also been shown to slowly generate stable fulvic and humic compounds (Artiola and Fuller, 1982).

TABLE 2 A Comparison of the Plant Material Above and Below Ground Level Among Five Desert Soils and Four Humid Soils of Central Asia

Desert soil	Plant tops (t ha^{-1})	Plant roots (t ha^{-1})	Humid soil	Plant tops (t ha^{-1})	Plant roots (t ha^{-1})
Sandy sterozem	0.5	10.0	Light chestnut	0.8-1.5	—
Light gray sierozem	1.0	15.0	Dark chestnut	5.2	13.2
Light gray sierozem	0.5-1.5	21.4	Southern chernozem	7.0	25.0
Ordinary sierozem	2.1-4.5	—	Carbonate chernozem	—	25.8
Dark sierozem	4.5	30.0			

Source: From Rozanov (1961).

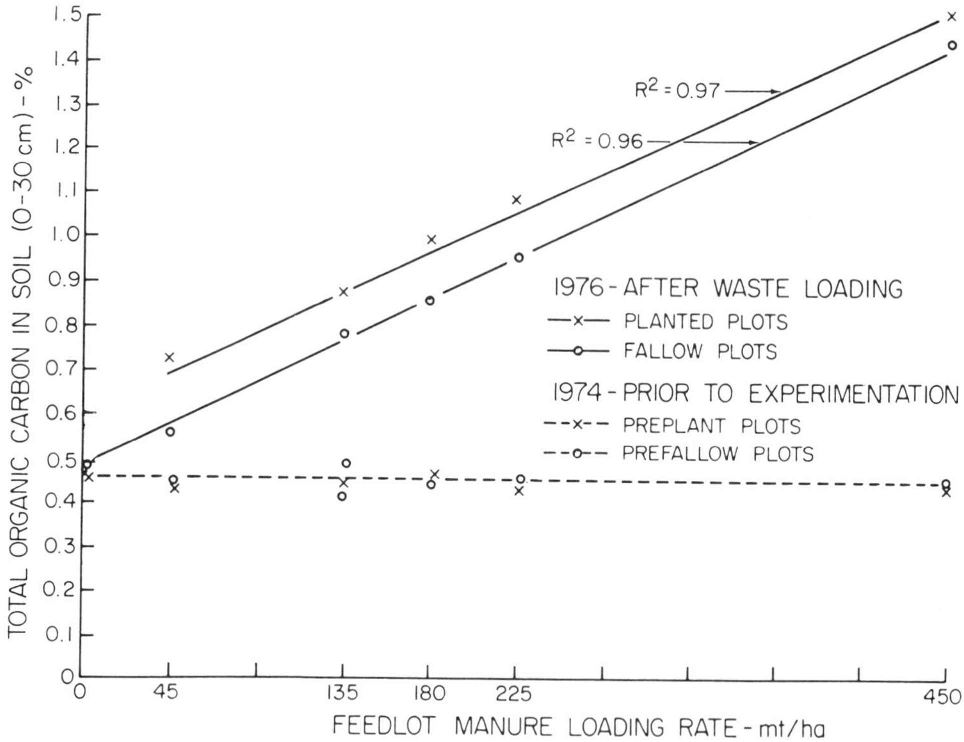

FIGURE 3 Influence on feedlot beef cattle manure loading rates and cropping practice on an irrigated desert soil organic matter content. (From Hathorn and Fuller, 1984.)

B. Functions of Organic Matter

Soil organic matter is known more by its functions than by its composition. Some of the more prominent functions are as follows:

- Direct contribution of nutrients for plants, such as nitrogen, phosphorus, and micronutrients
- Indirect contribution to plant nutrition by making soil elements more available
- Food and nutrient supply for microorganisms
- Soil structure development, and thereby water behavior and conservation, bulk density, infiltration, and reclamation of salt-affected soils
- Storehouse of elements attached to clay particles
- Biotic controls in soil through the presence of particular compounds, growth regulators, and antibiotics
- Effect on soil temperatures
- Increased water-holding capacity and increased hydrophobicity (alters drying)

Plant Nutrients

Organic matter contains such plant nutrients as nitrogen, phosphorus, potassium, and sulfur and trace elements of iron, manganese, copper, boron, and others. They need only to undergo decomposition to release elements available to plants. Plant nutrients in crop residues and soil organic matter are largely in organic combination and are mineralized or released slowly during the decay process (Fuller et al., 1950).

In the absence of chemical fertilizer, an element like N is almost wholly in the organic form. In unfertilized soils it represents the sole natural source for plant production. The importance of organic matter becomes more apparent when we realize that the nitrogen status is directly proportional to the amount of organic matter present. Thus, the higher the organic matter, the greater is the natural fertility status, and conversely, the lower the organic matter, the poorer is the fertility. Although seldom are there more than a few kilograms per hectare of mineral nitrogen present at one time, the total nitrogen in desert soils may range between 450 and 1800 kg ha^{-1} in the top furrow slice. Plants are dependent upon the conversion of organic nitrogen to inorganic nitrogen by soil microorganisms, since plants cannot directly assimilate the nitrogen as organic matter.

Food Source for Soil Organisms

Soil microorganisms, like other biological systems, require food to live. Organic matter supplies the necessary food for most of the soil microorganisms, except for a few that use carbon dioxide of the air and inorganic nutrients. The organisms obtain their necessary energy source from the degradation of the organic materials and residues that enter the soil (Fig. 2). Plants would starve for nitrogen and certain other elements if it were not for the mineralization action of the soil organisms in the organic matter cycle in soils (Huang and Schnitzer, 1986).

Organic matter and plant residues also act as food for the larger soil fauna. Earthworms are considered to benefit the soil by their burrowing, soil mixing, channel action, and castings. Collectively, these physical benefits offset the organic matter consumed. Earthworms create no fertility since they do not fix atmospheric nitrogen or any other unavailable elements and must live on the organic matter and plant residues found in the soil. Earthworms occur in desert soils whether irrigated or not. Seeding is not necessary. Wherever a food source and moisture are available, they appear in large numbers (Fuller, 1975a).

Benefit to Physical Properties

Individual soil particles are very small. If it were not for the tendency to combine into aggregates (soil structures), plants would find it difficult to grow in soil and water and air movement would all but cease except in coarse sands. Soil structure varies considerably, but not all structures provide optimal conditions for growing plants. Those soils most abundantly supplied with organic residues (including straw, manures, leaves, and composts) possess the most favorable structures. Soils of poor physical condition or tilth are hard and cloddy and bed with difficulty when worked with implements. Compaction in subsurface horizons occurs when organic matter declines to a critical level; aeration is poor and water penetrates

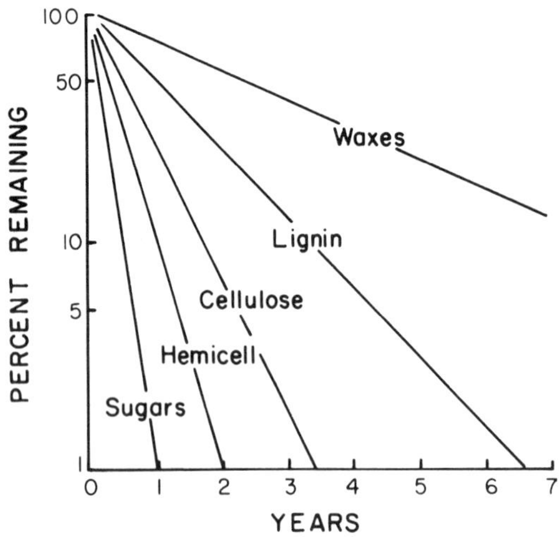

FIGURE 4 Rates of biodegradation of specific major organic plant constituents as affected by time. (From Minderman, 1968.)

slowly. Plants under these conditions grow with difficulty. Desert soils lose organic matter rapidly upon cultivation and irrigation, even with the return of cultivated plant residues to the soil.

The mere addition of organic residues to soil is not sufficient to produce good structure. It is the *continued* decomposition or organic matter and residues, for example, by organisms with the formation of *new* products that is important. Beneficial microorganisms live in soils in sufficient abundance, ready to work on organic materials. To increase their activity, they require mainly an abundant energy source, such as plant residues, other organic matter, favorable moisture relationships, and freedom from toxic substances (Fig. 3).

C. Decomposition and Biodegradation

Organic materials entering the soil are subject to microbial decay, often called biodegradation. Even the most resistant of substances, such as asphalt, phenol, petroleum oil, waxes, and resins, slowly decompose as a result of microbial attack. A host of different substances decay when plant residues mix with the soil. Some of these are cellulose, hemicellulose, lignin, proteins, waxes, resins, vitamins, growth substances, and water-soluble compounds (Fig. 4). The rate of decomposition is controlled primarily by the availability of the substance as food and energy. For example, water-soluble constituents decay very rapidly. In contrast, lignin is slowly decomposed. The more mature plant materials decompose less rapidly than younger materials, since lignin is more abundant in mature plants (Fig. 4). In addition, the C/N ratio must be favorable, preferably near 20, although ratios up to 50 may not adversely affect decomposition rates (Table 3). Older plant residues have wider C/N ratios than do young plants.

TABLE 3 Grouping of Common Crop Residues on the Basis of Their C/N Ratios

Residues	C%	N%	C/N ratio
High in both C and N	40	2.5	16
Legume residues			
Legume green manures			
Young cereal green manures			
Low C and high N	30	3.0	10
Animal feces			
High C and low N	40	0.5	80
Straw			
Stubble			
Strawy manure			
Mature cereal green manures			
Leaves, sawdust, or wood shavings			
Low C and low N	32	0.8	40
Composed straw			

D. Maintenance

Maintaining even a low content of organic matter is essential to a favorable plant economy. On the other hand, Fig. 5 shows a loss of organic matter and consequent loss of soil pore space and plant growth.

Maintenance may be illustrated by the example of a horse drinking from a trough of water. It is not the level of the water in the trough that concerns the horse as much as the rate of income of water in relation to its uptake. The trough may be 8 or 40 cm deep. If the water flows in at the rate the horse drinks, its thirst is satisfied at either depth. However, if there is *no* income of water, the 40 cm depth is the most desirable and satisfies the needs of the horse longer than the shallow depth. Furthermore, the exploitation of the first 30 cm is of little consequence, if replenishment at a rate equal to the outgo were begun and maintained at or above the minimum level for the horse's mouth to penetrate.

Desert soils being lower in natural organic matter, they need replenishment more frequently with organic carbon-containing material than those soils higher in organic matter under other climates less favorable to decomposition and unfavorable variability rates of mineralization. Organic matter in soils may be maintained by the addition of any biological material if it is decomposable and is in sufficient quantity. The problem is to find organic materials that are inexpensive, but suitable. Agriculturists use four main forms: (1) farm animal manures, (2) green manures, (3) mature crop residues, and (4) municipal wastes, either directly as sludge or as composts (Tables 2 and 3). Unlike the backyard gardener, in many farming circumstances it may not pay to use either 1 or 2 partly because of the high water

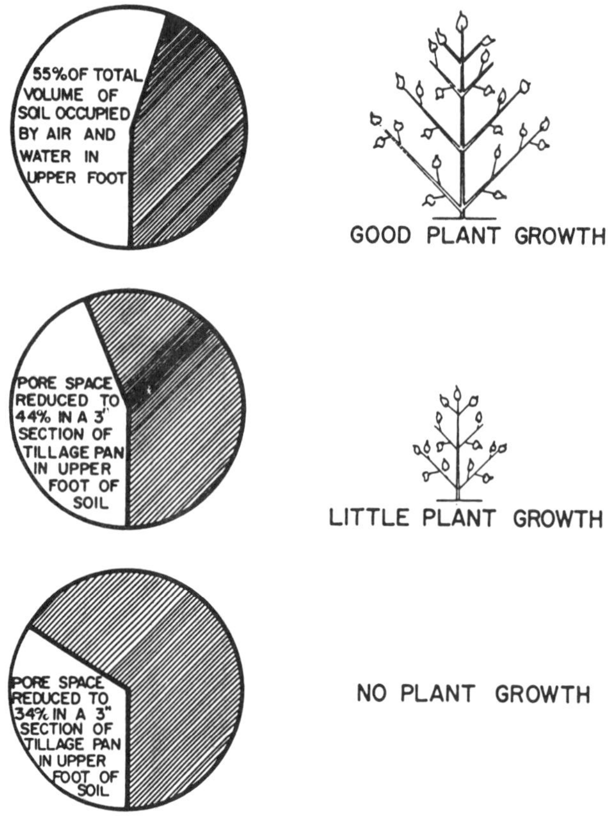

FIGURE 5 Plant growth on a Mohave Desert silty clay loam in Mesa, Arizona, as illustrated by loss of organic matter and resultant pore space. (From Fuller, 1975b.)

use, need for income, and so on, but it does pay to get as much of 3 and 4 into the soil as possible. Fertilization, which increases the *quantity* of top and root residues available to turn back into the soil, is definitely a practice that all growers cannot afford to overlook. The uses of municipal wastes (either directly or as composts and sludges) are developing as a result of overburdened waste disposal facilities and pressure by the public for more effective waste control. In the final analysis, the renewal of organic matter in desert soils is largely dependent upon the utilization of all suitable organic residues possible. This is often done by making a compost pile or incorporating plant residues into the soil at the termination of growth (Fuller, 1975a). In both instances water must be added to keep the material damp to permit decay to take place. Mulching can also be practiced where it is convenient and not too unsightly and where wind erosion control is possible.

III. FUEL USE VERSUS SOIL USE

A. Burning

The competition for the use of organic residues between soil improvement and fuel use has been largely responsible for permanent degradation of arid land soil for economic crop production and rehabilitation. The conflicts of this competition defy rational solution in many desert areas around the world. Not only has this conflict been responsible for historic desertification, but it is out of control today, continually expanding its degradation and devastation of land of crop production ability for food and fiber. Historical grazing lands have been denuded to bare soil to the extent that plant growth is not available for either soil or fuel use. In some areas, what scraps of straw and twigs exist are gleaned to heat water for cooking.

With fossil fuel so abundant near some of these world areas requiring rehabilitation, one might expect a reasonable trade-off between the agricultural and petroleum industries. Until some rational program of this problem is initiated, food, fiber, and shelter development will be at a critical standstill. Until the importance of soil organic matter enhancement as a key to soil water storage and conservation and plant nutrition is recognized, rehabilitation will remain stagnated near or at starvation levels.

With this warning, the use of organic materials from different sources can only apply to those areas where they are available or can be made available (U.S. Department of Energy, 1984; Sievers et al., 1975). Both great and small advantages can be expected depending on the sites and sources of organics, but continued organic matter enhancement of the soil always initiates positive results of the productivity of the land (Fuller, 1984; Unger and Fuller, 1984, 1985).

B. Biogas Generation

The use of organic residues, waste, and animal manures for biogas production as a fuel (heat, light, and electricity) is well known. Biogas in arid lands and desert areas is used primarily for heating purposes (Williams, 1980; Persson and Bartlett, 1981; Persson et al., 1979). Biogas burns similarly to natural gas but is dirtier and contains both moisture and carbon dioxide (CO_2). Although methane (CH_4) is the major gas generated, some carbon monoxide (CO) and hydrogen accumulate, usually in smaller amounts than methane, depending on the conditions of the digestion, the nature of the organic material, and the design of the gas generator.

A number of crude as well as sophisticated gas generators have been devised. In an attempt to satisfy both household fuel needs and farm needs for fertilizer and organic matter, many attempts of on-the-scene aid programmers have fashioned unsophisticated local scrap metals, stone, brick, and other inexpensive materials into anaerobic gas (methane, hydrogen, and carbon monoxide) generators for home appliances (heating and lighting units), conserving the gas-spent organic matter residue for applying as fertilizer to the land. The practical application of the crude units, although often poorly understood and operated by the user, have met with some success in the more stable rural desert areas.

The more sophisticated units manufactured with all the basic equipment and gas storage units have merit (U.S. Department of Energy, 1984;

Williams, 1980). Unfortunately, they represent a cost often well above the poverty level of nomadic agriculture seen in certain arid lands where stable farming has not been well maintained. Moreover, where such devices are needed, much of the land is barren of raw materials for the necessary continual recharging of the gas generators.

IV. ORGANIC RESOURCES FOR CULTIVATED LAND

Organic materials for land utilization programs may be placed into broad groups (Haug, 1980; Fuller and Warrick, 1985):

1. Agricultural
   a. Animal generated (manures, meat processing, and tanning)
   b. Crop produced (crop residues and organic materials)
   c. Forest residues (mill waste as bark, sawdust, and forest fuel)
   d. Food process discards (cannery waste, waste water, and whey)
2. Municipal
   a. Solid wastes of residential and commercial collections not associated with a product
   b. Sewage sludges and sewage waters
3. Industrial: Wastes from industrial manufacturing and production
4. Institutional: Schools, colleges, hospitals, and health centers
5. Microbial: Natural soil microorganisms and their products
6. Compost: Any biodegradable organic materials that has been biodegraded (Haug, 1980)

Certain constraints are required for waste material and wastewater use on crop land:

Plant and crop growth quality require that waste materials be free of toxic properties, applied at efficient loading rates, and free of excessive concentrations of salts of Na, K, Ca, Cl, Mg, and B that inhibit or limit growth.

Loading rates: Knowledge of optimum loading rates should be available to ensure maximal crop yields without leaving an excess of substances that migrate to underground waters.

Land application: The application and/or incorporation must not require unusual or additional farm practices or excessive energy needs over and beyond the expected return.

Amenability for biodegradation: There must be assurance that the waste will biodegrade within a reasonable period of time and leave no toxic or adverse residue.

C/N ratio: The waste material should have a favorable C/N ratio of 30 or less or, if low in N, add fertilizer.

Cost: The costs of hauling, application, and incorporation should not exceed the economy of return.

Off-site pollution: Susceptibility to wind, water and/or odor problems should be minimal.

Soil properties should not adversely affect soil physical properties of infiltration, structure, percolation, and aeration for economic crop production.

TABLE 4 Animal Manure Use on Cultivated Land

	Manure (dry weight basis, t ha^{-1})	
Region	Optimum	Maximum loading
Warm irrigated regions		
Livestock manure	34-56	80
Poultry manure	11-22	(to salt-tolerant crops)
Cold irrigated regions		
Livestock manure	34-45	50
Poultry manure	9-20	(to salt-tolerant crops)
Dryland regions		
Livestock manure	6.1-11	10
Poultry manure	Half or irrigated region	(to salt-tolerant crops)
Humid regions		
Livestock manure	34-45	34-45
Poultry manure	11-22	11-22
Slurry (5% cattle solids)	4 cm depth in 30 days	4 cm depth in 30 days

Source: Modified from Meeks et al. (1975).

Health factors: The process should provide no health hazard to the food chain, humans, or animals.

Successful organic matter utilization on cultivated land where farm practices have been established is (1) to avoid unnecessary development of extra work time or wasteful energy, and (2) to provide a better production than if the materials were not used. These constraints apply to animal manures, night soils, and crop residues as well as to municipal and industrial waste. A few examples of solid wastes commonly used on agricultural land are listed in Tables 1, 3, 4, and 5. Other examples, which compare nitrate production from application of different manures, municipal sludge, and commercial chemical mixes are illustrated in the study of a North American desert soil as related to time (Fig. 6). The natural organic materials released nitrogen as well as the commercial mixes of ammonium sulfate and gypsum.

Soil microorganisms have been shown to contribute substantial amounts of organic matter to arid land desert soils (Bolyshev, 1964; Sims and Dregne, 1962; Mayland et al., 1966; Durrell and Shields, 1961; Cameron and Fuller, 1960). Organic crusts (cryptogamic crusts), for example, develop quite rapidly in hot deserts, particularly after rainstorms (Fuller, 1974; Cameron and Blank, 1965; Cameron and Fuller, 1960; Fletcher and Martin, 1948; Reynaud and Lumpkin, 1988; Skujiņš and Klubek, 1978; Metting, 1986). Not only do soil cyanobacterial, algal, and algal-fungal

TABLE 5 Wastewater Management Relating Disposal Location, Type of Organic Waste, Mode, Rate, Time of Application, and Ground Cover

Location	Waste type	Application			Ground cover
		Mode[a]	Rate	Period	
Pennsylvania State College, State College, Pennsylvania	Domestic	SI	500 m^3 ha^{-1} week at 6.3 mm h^{-1}	32 week crops	Mixed crop land and forest land
Campbell Soup Co., Paris, Texas	Cannery	OR	635 m^3 ha^{-1} $week^{-1}$ summer, 320 m^3 ha^{-1} $week^{-1}$ winter; at 12.5 mm day^{-1} summer, 6.3 mm day^{-1} winter	52 weeks	Cropland (hay)
Muskegon, Michigan (under construction)	Municipal	SI	500 m^3 ha^{-1} $week^{-1}$, maximum 100 mm $week^{-1}$ rain plus spray	35 weeks	Cropland (corn)
Flushing Meadows Project, Phoenix, Arizona	Municipal	RI	1.2×10^6 m^3 ha^{-1} $year^{-1}$ at 2 weeks wet, followed by 10 days (summer) or 20 days (winter) dry-up	52 weeks	Open basin with Bermuda grass cover
Santee, California	Domestic	RI	3.4×10^5 m^3 ha^{-1} $year^{-1}$ intermittent spread and dry	52 weeks	1.2 ha spreading area
South Lake Tahoe, California	Domestic	OR	1100 m^3 ha^{-1} $week^{-1}$ at 2 days spray, 4-5 days rest	Study periods winter and summer 1963-1965	Forestland
North Lake Tahoe, California	Domestic	RI	Approximately 4700 m^3 day^{-1}, 6.9 ha site with 15,000 LF percolation trenches	52 weeks	Volcanic cinder cone
Celotex Corporation, L'Anse, Michigan	Industrial (insulation board)	SI	960 m^3 ha^{-1} $week^{-1}$ at 0.088 mm h^{-1}, 16 h day^{-1}	24 weeks	Cropland (Reed canary grass)
Seabrook Farms, Seabrook, New Jersey	Frozen vegetable	RI	2500 m^3 ha^{-1} $week^{-1}$ at 6.3 mm h^{-1}, 5 days $week^{-1}$	—	Original oak forest altered to marsh grasses

Westby, Wisconsin	Municipal	RF	5300 m^3 ha^{-1} $week^{-1}$ 2 weeks on, 2 weeks off	52 weeks	Reed canary grass (burned each spring)
Sunkist Growers, Inc., Corona, California	Lemon processing	RF	1500 m^3 ha^{-1}, load in 1 day, rest and dry 2 weeks	14 week canning season	Bare soil during spreading season; crops planted during 38 week dormant season
Campbell Soup Co., Napoleon, Ohio	Cannery (tomato)	OR	1800 m^3 ha^{-1} $week^{-1}$ at 2.1 mm h^{-1}, 6 h on, 6 h off	50-day canning season	Reed canary grass, seaside bent, and red top
Campbell Soup Co., Chestertown, Maryland	Cannery (poultry)	OR	Similar to Napoleon, Ohio	52 weeks	Similar to Napoleon, Ohio
Howard Paper Mills, Urbana, Ohio	Industrial (paper)	SI	330 m^3 ha^{-1} $week^{-1}$ at 5 mm h^{-1}	52 weeks	Alfalfa hay cover, underdrains for recapture
Sunapee State Park, Mt. Sunapee, New Hampshire	Domestic	SI	500 m^3 ha^{-1} $week^{-1}$ at 6.3 mm h^{-1}	20 weeks	Forested
Beardmore and Company, Toronto, Canada	Industrial	SI	1500 m^3 ha^{-1} $week^{-1}$ at 5 mm h^{-1}	30 weeks	Reed canary, bone, and timothy grass
Shoemaker's Dairies, Bridgeton, New Hampshire	Industrial (milk processing)	SI	50 m^3 ha^{-1} $week^{-1}$ at 0.5 mm h^{-1}	52 weeks	Hay and grasses
Commercial Solvents Corp., Terre Haute, Indiana	Industrial (fermentation)	SI	500 m^3 ha^{-1} in one 10 hr spray followed by 2 week rest	52 weeks	Bare ground
H. J. Heinz Co., Salem, New Jersey	Cannery (tomato)	SI	130 m^3 ha^{-1} $week^{-1}$ at 0.75 mm h^{-1}	Canning season	Reed canary grass

TABLE 5 (continued)

		Application			
Location	Waste type	Mode[a]	Rate	Period	Ground cover
Riegel Paper Co. Hughesville, New Jersey	Industrial	SI	800 m^3 ha^{-1} $week^{-1}$ at 5 mm 5 mm h^{-1} for 4 h and 20 h rest	52 weeks	Grass
Green Valley Farms, Avondale, Pennsylvania	Cattle wastes	SI	500 m^3 ha^{-1} $week^{-1}$ at 6.3 mm h^{-1}	52 weeks	Cropland (hay) in summer, forest in winter
Masonite Corp., Towanda, Pennsylvania	Industrial	SI	Approximately 130 m^3 ha^{-1} $week^{-1}$ at 1.3 mm h^{-1}		Reed canary grass

[a]Application mode: SI, spray irrigation; OR, overland runoff; RI, rapid infiltration; RF, ridge and furrow spreading.

Source: From Reed (1972), metricated.

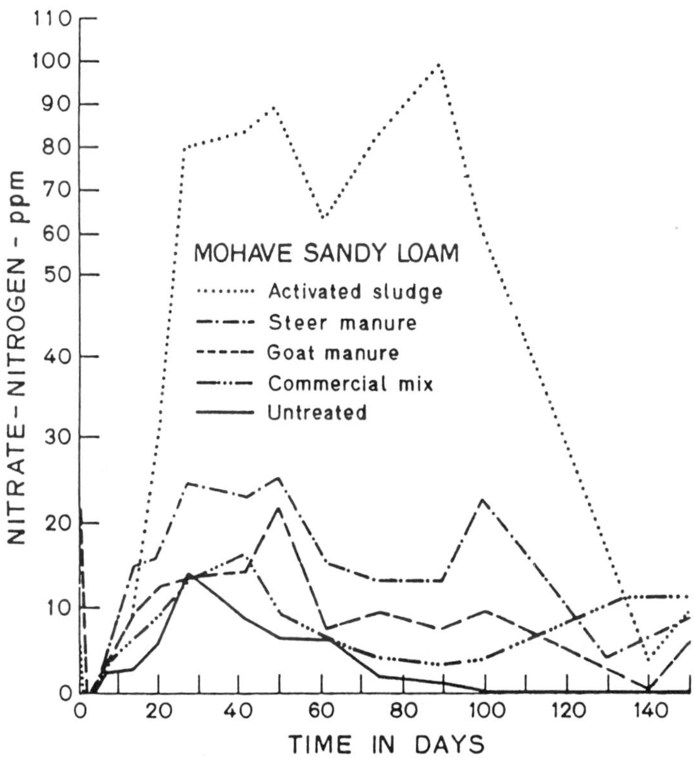

FIGURE 6 Change with time in nitrate nitrogen in Mohave soil (a desert soil) comparing various organic nitrogen fertilizers. (From Fuller and Hathorn, 1984.)

crusts high in organic matter form naturally from mineral matter on desert surface, but applications of liquid sulfuric and phosphoric acid to the surface of the Southwest U.S. deserts are found to initiate extensive algal bloom (growth) during short humid periods in the hot summers. Both carbon and nitrogen are fixed by cyanobacteria in substantial amounts in the cryptogamic crusts of all desert soils of the world (Durrell and Shields, 1960; Cameron and Fuller, 1960; Fletcher and Martin, 1948; Skujiņš and Klubek, 1978). In fact, cyanobacteria and algae become so dense in some arid lands that research has been performed in California by the USDA (personal communication from Dr. Hans Jenny, 1955) to scarify or break up surface crusts to ameliorate ephemeral storm drainage and streambed leaks for flood control and recharge of underground water storage.

The subject of organic matter contribution by the soil microflora needs extensive research before an exact contribution to organic matter accumulation can be assumed, as we think of manures and plant residues. Such possibilities cannot be overlooked under the desperate circumstances of some of our denuded and depressed deserts of the world.

V. LAND MANAGEMENT FOR SOLID ORGANIC MATERIALS APPLICATION

A. Procedures for Land Utilization of Organics

The recommended application rates of organic materials to croplands or arid areas depend largely on the decomposition rates as a function of climate and available equipment. Temperature and moisture are the most prominent factors in rate of biodegradation, except where irrigation is employed. Temperature, then, controls decomposition rates. Degree-days has been suggested (Meeks et al., 1975) as one of the better methods used to determine the *temperature-dependent decomposition potential*. The calculations are:

Fahrenheit degree-day = average month temperature in degrees Fahrenheit - 40°F

or

Fahrenheit degree-day in month = Fahrenheit degree-day (days in month)

Fahrenheit degree-day per year = sum of all Fahrenheit degree days for 12 months

"Cold" irrigated areas refer to those regions with less than 6000 annual Fahrenheit degree-days and "warm" irrigated areas are those with over 6000 annual Fahrenheit degree days.

Some simple examples of management practices effective on cropland, using feedlot manure, are (Hathorn and Fuller, 1984; Fuller, 1984) as follows:

Use high-quality manure that is low in moisture, salts, sand, and caliche.

Use aged manure on cropland in preference to fresh manure spread directly from feedlots; fresh manure may be placed into fallow or idle land.

Annually apply 25 t ha^{-1} of dry matter or equivalent, or 50 t ha^{-1} in alternate years.

Where water is abundant for irrigation, amounts up to 135 t ha^{-1} in alternate years.

Where water is abundant for irrigation, amounts up to 135 t ha^{-1} have been used successfully.

Apply before tilling, then work into the soil by plow, disk, or rototilling.

Irrigate thoroughly to leach excess salts below the root zone if manure is high in salts.

After irrigation, wait about 30 days before planting to allow biodegradation to occur and salts to redistribute.

Apply manure only where water penetration and soil permeability are good.

Consider that only about one-half of the N and P is available to the crop the first year.

Avoid applying commercial nitrogen and phosphorus with manure, unless it is needed for certain crops.

Management practices of the land, water, soil, fertilizers, and crop production in response to the kind of waste applied also determine the extent of biological degradation, metal attenuation, and environmental impact of the waste disposal system. For example, feedlot steer manure completely inhibited the growth of grain sorghum planted on a bed or ridge, but when planted on the flat area and irrigated to remove excess salts, yields were increased for applications up to about 180 kg ha^{-1} (Hathorn and Fuller, 1984).

B. Placement of Solid Organics

Land utilization requires that waste applications be on top of the soil surface or incorporated to a shallow depth. It does not include deep burial, as in landfills.

Surface Placement

Surface placement of solid organic materials includes (1) mulching, (2) minimum tillage, (3) stubble mulching, (4) "land farming" (as in oil and oil wastes), (5) strip mulching, and (6) vertical mulching. The loading rate from point sources is shallow and allows agricultural cropping. Deeper laying or stacking comes under the heading of composts (Haug, 1980). Temporary piling or stacking of organic residues for compost production is a means of ensuring controlled partial decomposition, with the resulting material well adapted to surface application. Surface placement of solid waste has certain advantages and disadvantages for crop production.

Advantages of surface placement
- Moisture conservation in the soil surfaces compared with bare surfaces
- Temperature modification of the soil surface
- Less nitrogen demand from the soil
- Lower cost of application
- Energy conservation because of nonincorporation power requirements can aid in preventing soil, water, and wind erosion to the land

Disadvantages of surface placement
- More susceptible to soil, water, and wind erosion
- Greater social nuisance in the forms of odors, volatilization of chemicals, dust and/or trash blowing, and microbial aerosol depending on the nature and particle size of the waste
- May be unsightly and attract vermin, flies, birds, and rodents
- Loss of certain plant nutrients to the atmosphere and erosion
- More available for surface runoff and pollution of surface waters
- Loading rate more critical to control
- Greater problems with disease spreading
- Potentially toxic or hazardous constituents more a problem when surface applied
- Loss of water by evaporation compared with subsurface

Some of the disadvantages can be reduced acceptably or eliminated by pretreatment of some organic materials, such as composting and/or pelletizing, as shown in Fig. 7.

Surface treatment is less expensive than subsurface incorporation because of the greater power demand for pulling equipment through the soil (Fuller, 1983). A great number of surface spreaders can be used, such as tank trucks, honey guns, rear-end truck spreaders, and blade and

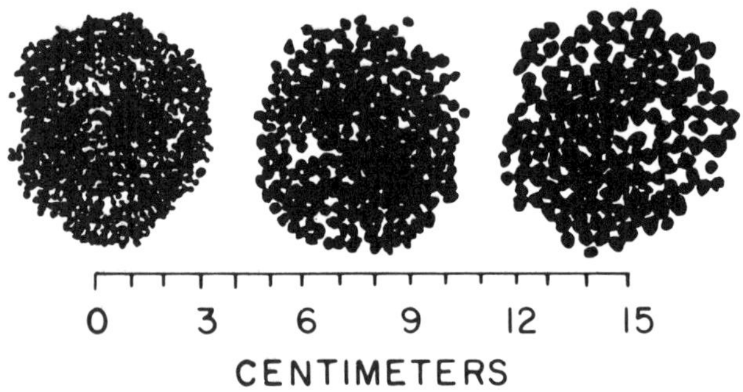

FIGURE 7 Pellets of biodigested and composted organic municipal solid waste used as fertilizer and soil conditioner. (From Fuller and Warrick, 1985.)

bulldozers, as well as farm manure spreaders. Following the Dust Bowl years in the United States (the 1930s), a waste utilization program developed out of a need for minimum tillage and mulch to protect the soil from from wind. With these systems, crops are seeded through the trash or grain stubble. Soils highly subject to wind and water erosion because of loose consistency and sloping topography benefit most from this type of surface waste utilization. This technique can be adapted to external wastes that are brought in, provided they are compatible with the crop being grown and remain in place against the eroding forces of wind and water. Similarly, wastes can be placed on pastures where the sod keeps them from moving. Also, vertical mulching, in which organic wastes and residues are placed in a trench, combines the advantages of both surface and subsurface application.

Other potential uses for organic matter application have not been fully researched. For example, asphalt and petroleum mulches have been shown to be effective for retaining moisture and elevating temperature. Earlier seedling emergence, better stands, and earlier maturity have been reported (Fig. 8) (Fuller, 1975b; Fuller and Warrick, 1985).

Municipal solid waste compost can be pelleted readily for surface application, giving it an added advantage over loose compost. The pellets are easier to apply, resist wind and water erosion, and lend advantage to fortification with fertilizer (Fig. 7).

Subsurface Placement

The physical condition of the soil, as a result of ripping the subsoil, applying organic matter, and deep injection of organic slurries, benefits crop production considerably. Soil incorporation (Fuller, 1983) serves in many ways to

Completely and neatly eliminate a waste disposal problem
Improve the physical condition of the soil for better crop production

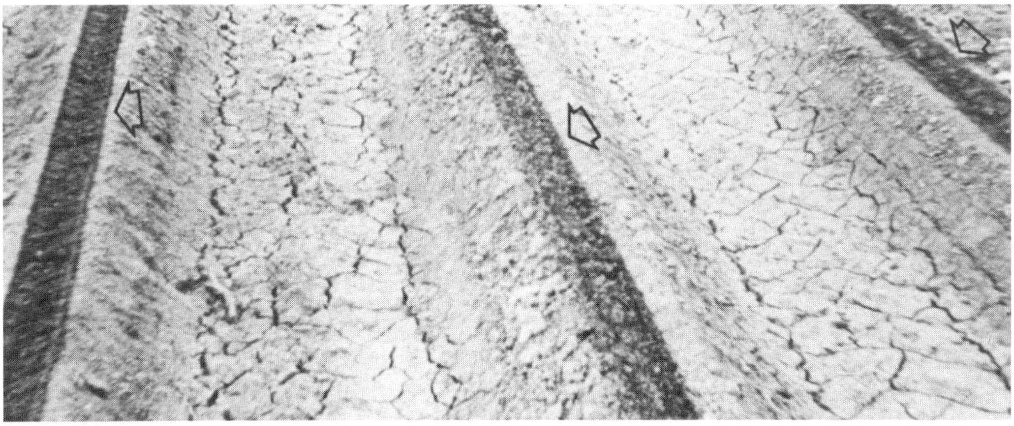

FIGURE 8 Asphalt mulch strips over the seed row. (From Fuller, 1975b.)

Conserve natural resources (organic matter, nitrogen, phosphorus, and water)
Conserve energy by reducing water pumping need
Provide a new soil injection technique for arid lands for better crop production and soil management
Ensure safe and sanitary disposal of human, animal, and plant pathogens below the soil surface
Place the organic matter directly into the active root zone
Enhance biodegradation and disease control
Eliminate noxious odors and aerosol problems
Eliminate the hazard of point source concentration of pollutants
Distribute the organics and waters in upland soils with favorable clay content to achieve maximum attenuation of heavy metals and retention of nitrates from reaching the groundwater

More slurries and solid wastes could be adapted to deep incorporation into the soil by pressure injection than are now used (Fig. 9).

The selection of a method and equipment for soil incorporation depends on the specific nature of the waste and crop to be grown. For example, municipal refuse should be subdivided and the large pieces of metal and glass removed for recycling. A large number of shredders and particle size-reducing machines are available to process almost any size object. The organic materials may then be placed on the land surface, incorporated, or composted. Composting is designed primarily for bulk reduction and pretreatment to reduce the volume of organic substances and disease organisms the waste may contain. Such wastes as those that originate from the pesticide, pharmaceutical, and other chemical industries may require pretreatment (Budzynski and Fuller, 1986). Calcium carbonate is inexpensive and readily available to use to neutralize acids, and it aids in precipitation and chemical binding of heavy metals. Synthetic organics (such as

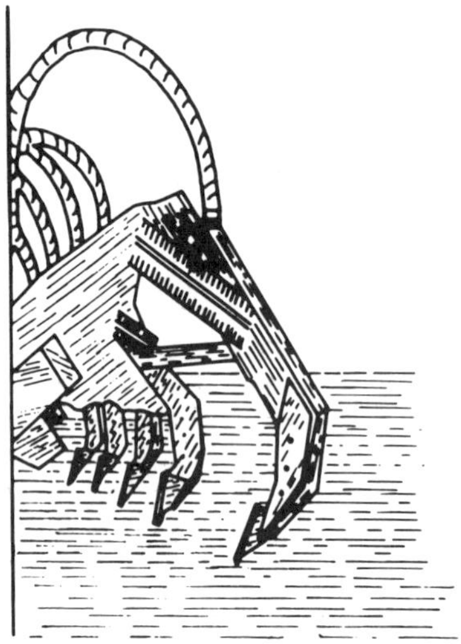

FIGURE 9 Soil injection technique for wastewaters, slurries, and semi-liquids. (From Fuller, 1983.)

pesticides, refinery oils, volatiles, and toxic wastes) may require considerable chemical pretreatment before they can safely be added to the land. In some instances, aqueous dilution may be used to mitigate the adverse effects of toxic pollutants on the soil microorganisms. Pretreatment for land-applied solid and hazardous waste materials is so highly specialized for each source that generalities here can have only limited value to the reader. Furthermore, the cost effectiveness of each waste varies considerably and must be determined as a case-by-case basis.

Most farm tillage equipment is suitable for subsurface application with a variety of wastes. On the other hand, incorporation may require more sophisticated equipment than conventional surface spreading. Highly engineered injection units appear in all sizes and power ranges. Vertical mulching, in which organic trash, sawdust, compost, and so on are placed into a narrow slit trench, must involve special equipment. The great versatility of the new rototillers makes them most popular for refuse incorporation. Wheel tractors, truck tractors, and refuse blades are also widely accepted for refuse soil management.

The power subsurface injector for sludges and wastewater is a good example of how rapidly waste management for soil improvement is progressing (Fuller, 1983). The attractiveness of the soil injector is due to the capability of precise placement of slurries and sludges of up to 10% solids and the convenience of wastewater placement. Other advantages relate to (1) conservation of water and other natural resource materials, (2) favorable placement of nutrients directly into the most active root zone, (3) improved soil physical condition due to subsurface organic matter incorporation, and (4) minimization of human and animal disease and vermin problems. Waste application can be designed to reclaim wasteland, such as eroded farmlands, overgrazed pasture, saline and alkaline soils, and other disturbed lands such as construction sites and mining waste disposals. The depth of injection can be adjusted by changing the shank depth, but power adjustments must be made.

A combination of surface and subsurface application must be obtained by use of disk or disk-plow procedures after the organic materials have been spread. Although disk incorporation is not preferred over rototilling, where rototillers are not available it may not be economical to buy special equipment. Indeed, arid land and desert organic matter application may often involve hand labor and simple hand tools or animal-drawn equipment for incorporation.

C. Organic Matter Loading Rate

The capacity to mix organic materials with the soil varies more with the bulkiness than any other single factor. Differences in bulkiness can be modified by adjusting the frequency of application to the rate of degradability. Application frequency depends on climatic conditions, land use, cropping pattern, and economics (Mays, 1977). The cost of application also varies with the nature of the material and method employed (Loehr et al., 1979). Loading rates are highly *site specific* and dependent on such conditions as

The microbiological reactivity of the waste and mineralization rates
The chemical properties of the waste
The physical properties of the waste
Site topographing features
Site climatological characteristics
Soil chemical, physical, and microbiological properties
Land use: irrigation methods and efficiency (for croplands)
Cropping plans
Crop nutrient requirements and sensitivities
Public ractions and sentiment

Animal Manures and Municipal Sludges

Organic matter loading rate information centers mainly around animal manures and municipal sludges. The loading rate recommendations for animal manures, as compiled in Table 4, depend largely on site climate (Meeks et al., 1975). Small amounts are applied over a longer period of time in cold (less than 6000 annual Fahrenheit degree-days) than in warm (more than 6000 annual Fahrenheit degree-days) climatic regions. In arid regions (less than 25 cm annual precipitation) less manure is recommended than in warm humid (more than 75 cm annual precipitation) regions, partly because of less leaching in native desert land and partly because less soil moisture is available under natural conditions. An irrigation water source could reverse this trend.

Municipal Solid Waste

Municipal solid wastes have been applied to arid land soils in the United States up to 900 t ha^{-1} (dry weight basis) (Hart et al., 1969; Volk, 1977). Levels above about 300 t ha^{-1} begin to demonstrate a risk to pasture yield improvement. Residues in the waste that are very slowly biodegradable require the municipal solid waste to be shredded into small particles if soil incorporation is to be effective. Pelleting has also been suggested for ease of application (Fuller, 1966). Incorporation of shredded materials can be effective soil builders, but costs are high (Rozanov, 1961). Dewatered sludges have been used effectively on all lands (Durlak, 1977; Unger and Fuller, 1985; Fuller and Larson, 1985; Fuller et al., 1985; Phung et al., 1978; Elliott and Stevenson, 1977).

Compost

Composting of raw organic residues from sewage sludge to shredded or chipped tree limbs and shrubs, leaves and sawdust, and other plant discards is a microbiological process for converting solid wastes into a more stable, humuslike material (Haug, 1980). Different constituents of the plant differ in rate of biodegradation. Slurries and manures have often been used to fortify the low nitrogen levels associated with tree and shrub composts to bring the C/N ratio closer to 25-30 for quality improvement. Almost any nonhazardous organic waste material can be made into compost, for example

Home garden trash, lawn clippings, leaves, shrub and tree prunings, vegetable trimmings, and kitchen wood ashes
Straw and waste hay
Animal manures, with and without straw and bedding
Wood: sawdust, bark, small chips
Paper and cardboard
Municipal solid waste
Food-processing waste: cannery and dairy

Compost originating from plant wastes represents a resource that benefits plant growth. Compost loading rates added to cropland are similar to those of sewage sludge and cattle manure (Unger and Fuller, 1984; Durlak, 1977).

Paper Pulp Mill Waste

Only a few examples of paper pulp mill waste are available in arid lands and desert areas. Paper pulp mill sludge is low in nitrogen with a C/N ratio of about 150. Nitrogen is a necessary additive to reduce the C/N ratio to about 20-30 before applying to croplands (Fuller et al., 1985; Fuller and Larson, 1985).

Wood and Forest Waste

Bark, sawdust, leaves, and forest litter are representative woody organic materials applied to lands. All are low in nitrogen and are used more extensively for mulching. Composted leaves and sawdust alone and with soil and/or sewage sludge have been used in the arid land and desert climates of North America and for container (can) plant production. Wood and wood products require considerable processing prior to application to cropland. Some wood materials contain toxic substances, necessitating composting prior to application to crops where seedings are made immediately after application.

Cannery Waste

Cannery waste occurs in only limited amounts as solids. Large aqueous disposals create many anaerobic and insect problems. Off-season applications are often successful. Cannery wastes are most suitable for pasture or forage irrigation (Elliott and Stevenson, 1977; Wallace, 1976; Reed et al., 1973; Spyridakis and Welch, 1973). The loading rate based on solids content and maintenance of aerobic conditions of the soil is similar to that of other plant residues. Soil injection is the most suitable method of incorporation (Fuller, 1983).

Soil Amendments

Soil amendments as used here and as distinguished from commercial fertilizers sold in the market refer to a heterogeneous host of materials, organic and inorganic, that may influence the biological, chemical, and physical nature of soil. Soil amendments are usually used to improve the physical properties of soil for soil reclamation. They are better known in arid than humid regions since their use for soil reclamation, improving water penetration, and excessive alkali and salt control extends over 75 years (1914-1989). Some examples include sulfur, gypsum, iron sulfate, compost, forest mulch, clay, bentonite, and rock mulch.

VI. LAND MANAGEMENT FOR LIQUID ORGANIC WASTEWATER APPLICATION

The basis for distinction between solid and liquid organic wastes (and other organic-containing land treatment materials) is not precise since the differentiation has been established primarily on the "handling" characteristics for convenience of application. For example, sludge solids content and handling nature may be classified as liquid, semisolid, or solid (dry).

Wastewaters must be used not only in a way that protects the environment but to realize actual improvement of local environmental conditions, even though the latter cannot always be achieved immediately (Table 5). Wastewater employment must therefore include (1) degree of renovation achieved, (2) best water use, (3) best land use, (4) potential for polluting groundwater, and (5) longtime effect on land productivity. Some additional detailed practical aspects influencing good wastewater irrigation and soil improvement use include

The rate of water application
Climatic patterns, particularly rainfall and temperature
Land topography as it relates to soil and water erosion
Method of soil incorporation
Permeability of the soil
Adequate drainage, both internally and externally
The establishment of new crop patterns
The enhancement of the soil productivity program
The mechanical land program of planting, tilling, and harvesting
Monitoring for pollution and environmental impact

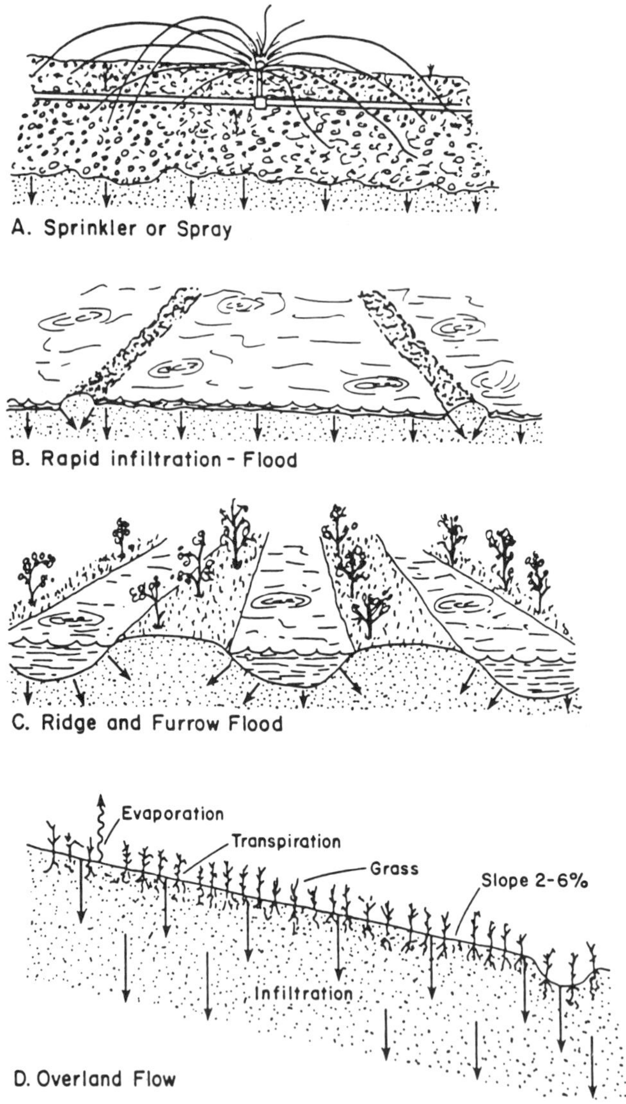

FIGURE 10 Illustration of wastewater application to the land for treatment and benefits of organic matter enhancement. (From Fuller and Warrick, 1985.)

Large quantities of suitable wastewater from industrial sources for irrigation use are generated in the United States.

A. Methods of Application

Applications of wastewater may be made to the land by a number of methods. Initially, they may be divided into two broad categories: (1) those made directly to the soil *surface* and incorporated into the upper few centimeters and (2) by *deeper incorporation* using injection (Fig. 9) for furrow placement to depths of 20-60 cm, depending on the rooting depth of the specific crop. Surface application methods of wastewaters containing organic matter to cultivated land have been suggested by a number of investigators (Fuller and Warrick, 1985; Elliott and Stevenson, 1977). Some of the most common application methods are diagrammed in Fig. 10.

Sprinkler or Spray Irrigation

Spray or sprinkler types use a large orifice on the nozzle. This requires a large power but lower labor requirement, with a wide selection of commercial equipment available. The sludge must be flushed from pipes when irrigation is completed (Table 5). An advantage of sloping land and pipes is that the system can be used yearround or drained for winter freeze. Some salt-intolerant crops are not adapted.

Ridge and Furrow

The land requires preparation, but this method has an advantage of lower power requirements than spray. The land cannot slope much. This method is adapted to 0.5-1.5% slopes depending on the solids content.

Overland Flow

This requires sloping land for flow purposes, but not excessive, since runoff must be controlled. The method is limited in use since it is difficult to obtain a truly uniform application.

Tank Truck

The capacities are variable (2000-7500 L tanks) depending on trucking tank size. At higher capacities flotation tires may be required. Pump discharge may be used to spray from the roadside onto the fields.

Farm Tank and Wagon and Tractor

This has a large capacity (1900-12,000 L) and thus it requires flotation-sized tires. The equipment is best used for temporary irrigation programs.

Flexible Irrigation Hose with Plow Furrow or Disk Cover

This system can be used with a pipeline or a tank truck with pressure or gravity discharge, or a hose connected to a manifold on a plow or disk for distribution into the soil. This system can also be useful for surface application.

Tank Truck with Plow Furrow Cover

This is a smaller unit of about 1000 L and may be used for sludge discharge in a furrow ahead of a plow mounted on a truck with four-wheel drive.

Farm Tank Wagon and Tractor Plow Furrow Cover

The sludge of this system is discharged into the furrow ahead of the plow mounted on a tank trailer for an application of 380-500 wet t ha^{-1}. Sludge may be spread in a narrow band on the ground surface and immediately plowed under at a rate of about 110-300 wet t ha^{-1}.

TABLE 6 Summary of Hydraulic and Organic Loading Rates Used in Existing Land Application Systems for Industrial Wastes

	Hydraulic load		Organic load (BOD)
Type of waste	$m^3\ ha^{-1}\ day^{-1}$	$cm\ week^{-1}$	$kg\ ha^{-1}\ day^{-1}$
Biological chemicals	14	0.99	330
Fermentation beers	12.6	0.89	151
Vegetable tanning			
Summer	505	35.33	320
Winter	75	5.30	48
Wood distillation	64	4.47	276
Nylon	16	1.12	257
Yeast water	141	9.88	—
Insulation board	138	9.68	123
Hardboard	56	3.94	76
Boardmill whitewater	141	9.88	34
Kraft mill effluent	130	9.17	23
RI[a]	3300	228.9	107
Semichemical effluent	670	47.09	180
Paperboard	71	4.90	27
Drinking	303	21.20	96
Poultry	370	26.16	89
Peas and corn			
57 day pack	460	32.11	212
35 day pack	322	22.50	1798
Dairy			
Low value	23	1.63	9
High value	290	19.60	890
Soup	63	4.42	43
Steam peel potato	180	12.40	71
Instant coffee and tea	54	3.78	82
Citrus	29	2.03	380
Cooling water: aluminum casting (RI)	890	62.13	31

[a]Rapid infiltration.

Source: From Wallace (1976), metricated.

The subject of wastewater application is a topic of broad and massive programs too complicated to completely include in this chapter. Here again, the method of application is highly site specific. Consequently, important details need further reference study before being programmed for a specific site. For example, a great variety of different methods and equipment are available, including (1) spray (sprinkle), (2) ridge and furrow spreading, (3) overland flow, and (4) rapid infiltration method. Application methods suitable for both humid and arid regions in the United States, suggesting wastewater management relating to disposal location, type of organic waste, mode, rate, time of application, and ground cover appear in Tables 5 and 6 and in a number of publications (Fuller and Warrick, 1985; Phung et al., 1978; Elliott and Stevenson, 1977; Wallace, 1976; Reed et al., 1973; Spyridakis and Welch, 1973). Not all methods of application reported can be recommended for use on other continents, where water is notoriously scarce and rainfall distribution is unpredictable. Site-specific innovations must be considered.

B. Organic Matter Content

With respect to biological oxygen demand (BOD), chemical oxygen demand (COD), and total organic carbon (TOC) of organic matter in wastewater application, much has been written relating to humid climates where wet soils often offer serious problems (Fuller and Warrick, 1985; Elliott and Stevenson, 1977). Less specialized attention, however, is required for wastewater containing organics when applied to dry desert soils. With this in mind, recent investigations make suggestions, again emphasizing the need for site-specific information before selecting a final method of application (Fuller and Warrick, 1985).

Preventing soluble organic matter (as measured by BOD, COD, and TOC) from reaching the groundwater is controlled almost wholly by biodegradation processes. Providing the soil with the most favorable environment and controlling the hydraulic loading rate in a manner compatible with the degradation rate is the best way to manage the soluble organic matter of wastes. The two most limiting elements in organic matter oxidation-reduction are nitrogen and oxygen. A continuous supply of soil moisture is also essential to provide a favorable microbial habitat for maximum biodegradation. Furthermore, organic matter decomposition stops when the soil freezes, a process linked to dehydration. Soil freezing is usually limited to the upper few centimeters except in extremely cold climates where the cold season remains prolonged.

Excessive loading rates of organic matter create problems. Soils become anaerobic, decomposition slows, odors form, and insects accumulate. A model of secondary effluent with a BOD concentration of 25 mg L^{-1} and an application rate of 5.1 cm $week^{-1}$ or 12.8 kg ha^{-1} $week^{-1}$ did not result in adverse accumulation of undecomposed organic matter in surface soil (Spryidakis and Welch, 1973). In addition, 1150 mg L^{-1} was spray irrigated on red canary grass sod on sandy soil at a BOD loading rate of 155 kg ha^{-1} day^{-1} with a resultant BOD removal of 99%. Yet at a BOD loading rate of 7.6 cm day^{-1}, the soil remained saturated and the grass died, a reminder that hydraulic loading rate and efficiency of organic matter degradation (or BOD loading) rate should be considered together in

TABLE 7 Comparison of Site Characteristics for Liquid Land Treatment Process

	Principal processes			Other processes	
Characteristics	Slow rate	Rapid infiltration	Overland flow	Wetlands	Subsurface
Slope	Less than 20% on cultivated land; less than 40% on noncultivated	Not critical; excessive slopes require much earthwork	Finished slopes 2–8%	Usually less than 5%	Not critical
Soil permeability	Moderately low to moderately rapid	Rapid (sands, loamy sands)	Slow (clays, silts, and soils with impermeable barriers	Slow to moderate	Slow to rapid
Depth to ground	0.6–1 m (minimum)	3 m (lesser depths are acceptable where underdrainage is provided)	Not critical	Not critical	Not critical
Climatic	Storage often needed for cold weather and precipitation	None (possible modify operation in cold weather)	Storage often needed for cold weather	Storage may be needed for cold weather	None

Source: From U.S. EPA (1977).

any long-term disposal program depending on climatic factors of rainfall and temperature.

VII. SITE SELECTION

The U.S. Environmental Protection Agency (U.S. EPA, 1977) offers some helpful suggestions of wastewater use in Table 7, which compares different site characteristics for land treatment processes (U.S. Department of Energy, 1984). Table 7 provides an aid in the selection of proper sites for the application of organic wastewater, slurries, and in some cases solid waste materials. The suggestions are quite detailed and, therefore, reasonably self-explanatory. A more recent interpretation of site selection fundamentals was published by Fuller with cooperation from the U.S. Environmental Protection Agency (Fuller, 1986).

VIII. CONCLUDING STATEMENT

Theories and practical techniques useful for the most effective means of applying organic materials to aid in the rehabilitation of degraded arid lands and desert soils have received less attention than for soils of other climates. However, some of the principles and methodologies can be carried over from humid to arid lands. Unfortunately, the necessary adaptations have not been widely implemented. Because certain arid lands have been left to the devastation of a harsh climate, deforestation, and overuse, it is important that techniques be applied to restore the soil's organic resources necessary for food and fiber productivity at a reasonably permanent level. These techniques are presented here to further encourage populations of arid regions to select the most desirable organic matter application methods available and to improve and adapt them to lands requiring renovation and rehabilitation, realizing they represent only one essential factor in maximizing food and fiber production. An example is the recent findings of the composition of cryptogamic crusts of central China (Reynaud and Lumpkin, 1988) and earlier findings (Metting, 1986; Skujiņš and Klubek, 1978) that suggest the real possibility of using the polysaccharide-producing *Chlamydomonas* spp. and cyanobacteria as inocula to improve soil productivity and moisture relationships.

REFERENCES

Artiola, J., and Fuller, W. H. (1982). Humic substances in landfill leachate. I. Humic acid extraction and identification. *J. Environ. Qual.* *11*: 633-669.

Bolyshev, N. N. (1964). Role of algae in soil formation. *Sov. Soil Sci.* *6*: 630-635.

Budzynski, J. W., and Fuller, W. H. (1986). Nitrogen and heavy metal uptake by field corn from processed spent acid. *Appl. Agr. Res.* *1*: 15-21.

Cameron, R. E., and Blank, G. W. (1965). A. Soil studies, microflora of desert regions. VIII. Distribution and Abundance of Desert Microflora, Technical Rep. California Institute of Technology, Pasadena, pp. 31-34, 193-202.

Cameron, R. E., and Fuller, W. H. (1960). Nitrogen fixation by some algae in Arizona soils. *Soil Sci. Soc. Am. Proc. 24*: 353-356.

Durlak, E. R. (1977). *Solid Waste Disposal by Landspreading Techniques.* TN No. N-149, Naval Construction Battalion Center, Port Hueneme, California.

Durrell, L. W., and Shields, Lora M. (1961). Characteristics of soil algae relating to crust formation. *Trans. Amer. Micros. Soc. 80*: 78-79.

Elliott, L. F., and Stevenson, F. J. (Eds.) (1977). *Soil Management of Organic Wastes and Wastewaters.* American Society Agronomy, Madison, Wisconsin.

Fletcher, J. E., and Martin, W. P. (1948). Some effects of algae and molds in the rain crust of desert soils. *Ecology 29*: 95-100.

Fuller, W. H. (1966). Pelleted compost now turns waste into useful mulch. *Prog. Agr. (Univ. Ariz. Exp. Sta.) 18*: 17.

Fuller, W. H. (1969). *Soil Organic Matter.* University of Arizona Exp. Sta. Tech. Bull. 240, Tucson.

Fuller, W. H. (1974). Desert soils. In *Desert Biology II,* G. W. Brown (Ed.). Academic Press, New York.

Fuller, W. H. (1975a). *Soils of the Desert Southwest.* University of Arizona Press, Tucson.

Fuller, W. H. (1975b). *Management of Southwestern Desert Soils.* University of Arizona Press, Tucson.

Fuller, W. H. (1983). *Soil Injection of Sewage Sludge for Crop Production.* University of Arizona Agr. Exp. Sta. Tech. Bull. 250, Tucson.

Fuller, W. H. (1984). *Use of Feedlot Manure and Sewage Sludge in Arid Lands.* University of Arizona, Coll. Agr. Tech. Bull. 255, Tucson.

Fuller, W. H. (1986). Site selection fundamentals for land treatment. In *Land Treatment: A Hazardous Waste Management Alternative*, R. C. Loehr and J. F. Molina (Eds.). Water Resources Symposium 13, Univsity of Texas, Austin, pp. 87-97.

Fuller, W. H., and Hathorn, S., III (1984). Beef cattle manure and municipal sludge compared. In *Use of Feedlot Manure and Municipal Sewage Sludge on Arizona Irrigated Land,* W. H. Fuller (Ed.). University of Arizona, Coll. Agric. Tech. Bull. No. 255, Tucson, pp. 28-29.

Fuller, W. H., and Larson, R. B. (1985). *Water Quality of Surface and Groundwater, and of Soil, and Vegetation.* University of Arizona Coll. Agr. Tech. Bull. 288, Tucson.

Fuller, W. H., and McGeorge, W. T. (1951). Phosphates in calcareous Arizona soils. II. Organic phosphorus contents. *Soil Sci. 71*: 45-49.

Fuller, W. H., and Tucker, T. C. (1977). Land utilization and disposal of organic wastes in arid regions. In *Soil for Management of Organic Wastes and Waste Waters,* L. F. Elliott and A. J. Stevenson (Eds.). American Society Agronomy, Madison, Wisconsin, pp. 473-489.

Fuller, W. H., and Warrick, A. W. (1985). *Soils in Waste Treatment and Utilization*, Vol. I. CRC Press, Boca Raton, Florida.

Fuller, W. H., Martin, W. P., and McGeorge, W. T. (1950). *Behavior of Nitrogenous Fertilizers in Alkaline Calcareous Soils. II. Field Experiments with Organic and Inorganic Nitrogenous Compounds.* University of Arizona Agr. Exp. Sta., Tech. Bull. 121, Tucson, pp. 471-500.

Fuller, W. H., Cameron, R. E., and Raica, N., Jr. (1961). Fixation of nitrogen in desert soils by algae. *Trans. 7th Int. Cong. Soil Sci.* Soil Sci. Soc. Am., Madison, Wisconsin, pp. 617-624.

Fuller, W. H., Bitterli, R., and Unger, M. (1985). *Attenuation of Paper-Pulp Mill Wastewater Constituents by Clay Soil.* University of Arizona Agr. Exp. Sta. Res. Rep. 287, Tucson.

Hart, S. A., Flocker, W. J., and York, G. K. (1969). Refuse stabilization on land. *Compost Sci. 11*: 82.

Hathorn, S., III, and Fuller, W. H. (1984). Feedlot beefcattle manure use on desert cropland. In *Use of Feedlot Manure and Municipal Sewage Sludge on Arizona Irrigated Land,* W. Fuller (Ed.). University of Arizona, Agr. Exp. Sta. Tech. Bull. No. 225, Tucson, p. 67.

Haug, R. T. (1980). *Compost Engineering Principal and Practice.* Ann Arbor Science Publishers, Ann Arbor, Michigan.

Huang, P. M., and Schnitzer, M. (Eds.) (1986). *Interactions of Soil Minerals with Natural Organics and Microbes.* Soil Sci. Soc. Am., Madison, Wisconsin.

Konovova, M. M. (1961). *Soil Organic Matter* (translated from Russian). Pergamon Press, Oxford.

Lobova, E. V. (1967). *Soils of the Desert Zone of the USSR* (translated from Russian by Israel Program of Scientific Translation, Jerusalem). Department of Commerce, Washington, D.C.

Loehr, R. C., Jewel, W. J., Novak, J. O., Clarkston, W. H., and Friedman, G. S. (1979). *Land Application of Wastes*, Vol. 7. Environmental Engineering Series, Van Nostrand-Reinhold, New York.

Mayland, H. F., McIntosh, T. H., and Fuller, W. H. (1966). Fixation of isotopic nitrogen on a semiarid soil by algal crust organisms. *Soil Sci. Soc. Am. Proc. 30*: 56-60.

Mays, D. A. (1977). Special problems and opportunities in use of waste heat for soil warming. In *Soils for Management of Organic Waste and Wastewaters*, L. F. Elliott and F. J. Stevenson (Eds.). American Society Agronomy, Madison, Wisconsin.

Meeks, B., Chesnin, L., Fuller, W. H., Miller, R., and Turner, D. (1975). *Guidelines for Manure Utilization in the Western Region USA.* Rept. W-124, Washington State University, Coll. Agr. Res. Center Bull. 814. Washington State University, Pullman.

Metting, B. (1986). Population dynamics of *Chlamydomonas cajao* and its influence on soil aggregate stabilization in the field. *Appl. Environ. Microbiol. 51*: 1161-1164.

Minderman, G. (1968). Addition, decomposition, and accumulation of organic matter in forests. *J. Ecol. 56*: 355.

Persson, S. P. E., and Bartlett, H. D. (1981). *Utilization of Farm-Produced Biogas.* Pennsylvania State University, College of Agr. and Eng., Paper 82-3620, University Park.

Persson, S. P. E., Bartlett, H. D., Branding, A. E., and Regan, R. W. (1979). *Agricultural Aerobic Digester—Design and Operations.* Pennsylvania State University, Penn. Agr. Exp. Sta. Bull. 827, University Park.

Pettygrove, G. S., and Asano, T. (1984). *Irrigation with Recycled Municipal Wastewater. A Guidance Manual.* California State Water Resources Control Board, Rep. No. 84-1, Sacramento.

Phung, T., Barker, L., Ross, D., and Bauer, D. (1978). *Land Cultivation of Industrial Wastes and Municipal Solid Wastes: State-of-the-Art Study 1*, EPA 600/12-78-140a, PB 278-080/As. U.S. Environmental Protection Agency, MERL, Cincinnati, Ohio.

Ponomareva, V. V. (1956). Gumus takyrov (humus of takyrs). In *Takyry Zapandnoi Turkmeii i ikh Selskohozyaistvennoe Osvoenie. Izd. Akad. Nauk, USSR*, 411-438.

Reed, A. D., Wildman, W. E., Seyman, W. S., Avers, R. S., Prato, J. D., and Rauschkolb, S. R. (1973). Soil recycling of cannery waste. *Calif. Agr. 27*: 6.

Reynaud, P. A., and Lumpkin, T. A. (1988). Microalgae of the Lanzhou (China) cryptogamic crust. *Arid Soil Res. Rehab. 2*: 145-155.

Rozanov, A. N. (1961). *Serozems of Central Asia* (translated from Russian by Israel Program for Scientific Translation, Jerusalem). Department of Commerce, Washington, D.C.

Shuvalov, S. A. (1966). In regards to the complexness of the Ust Urt soil vegetation cover (in Russian). *Pochv. Inst. Imeni V. V. Dokrechaeva, Trudi Yubileninoi Sessi 28*: 162-182.

Sievers, D. M., Fischer, J. R., Meador, N. F., Fulhage, C. D., and Shanklin, D. M. (1975). *Engineering and Economic Aspects of Farm Digester*. University of Missouri, Dep. Agr. Eng., Bull. 17, Columbia.

Sims, J. R., and Dregne, H. E. (1962). *Fertilizer Responses on a Sodium Soil*. New Mexico State University, Agr. Exp. Sta. Res. Rep. 63, Las Cruces.

Skujiņš, J., and Klubek, B. (1978). Nitrogen fixation and denitrification in arid soil and cryptogamic crust. In *Environmental Biogeochemistry and Geomicrobiology*, W. E. Krumbein (Ed.). Ann Arbor Science Publishers, Ann Arbor, Michigan, pp. 543-552.

Spyridakis, D. E., and Welch, F. D. (1973). Treatment process and waste effluent disposal on land. In *Land Treatment and Disposal of Municipal and Industrial Wastewater and Sludge Through Forest and Cropland*, W. E. Sopper and L. T. Kardos (Eds.). Pennsylvania State University, College Park, p. 232.

Tucker, T. C., and Fuller, W. H. (1971). Soil management, humid vs. arid areas. In *Food, Fiber, and the Arid Land*, W. G. McGinnies, B. J. Godman, and P. Paylore (Eds.). University of Arizona Press, Tucson, pp. 277-286.

Unger, M., and Fuller, W. H. (1984). Optimum utilization of sewage sludge of low and high metal content for grain production. In *Use of Feedlot Manure and Municipal Sewage Sludge on Irrigated Land*, W. Fuller (Ed.). University of Arizona, Coll. Agr. Tech. Bull. 225, Tucson, pp. 45-52.

Unger, M., and Fuller, W. H. (1985). Optimum utilization of sewage sludge of low and high metal content for grain production on arid land. *Plant Soil 88*: 321-332.

U.S. Department of Energy (1984). *An Introduction to Biogas Production on the Farm*. Report DOE/CE 25095-13. U.S. Government Printing Office, Washington, D.C.

U.S. EPA (1977). *Municipal Sludge Management: Environmental Factors*. EPA Report 430/9-77-004, MCO-28, U.S. Environmental Protection Agency, Washington, D.C.

Volk, V. V. (1977). Application of trash and garbage to land. In *Land Application of Waste Materials*, Soil Conserv. Soc. Am., Ankeny, Iowa, p. 15.

Wallace, A. T. (1976). Land disposal of industrial liquid wastes. In *Land Treatment and Disposal of Municipal and Industrial Wastewater*, A. T. Sank and T. Asavo (Eds.). Ann Arbor Science Publ., Ann Arbor, Michigan, p. 147.

Williams, D. W. (1980). *Operating Experiences with a Solar-Heated Anaerobic Digestor.* Report, Soil, Water, and Engineering Dept., University of Arizona, Tucson.

18

Inorganic Fertilizer Use and Its Effects on Semiarid and Arid Region Soils

G. A. PETERSON *Colorado State University, Fort Collins, Colorado*
S. E. WILLIAMS *University of Wyoming, Laramie, Wyoming*
L. E. MOSER *University of Nebraska, Lincoln, Nebraska*

I. INTRODUCTION

Soils are products of the interaction of climate and organisms on geological materials, in particular topographic settings. This interaction, integrated over time, has resulted in a wide variety of soils (Jenny, 1941). Any given soil is subject to disturbance—disturbance that may modify climate, organisms, and even topography. Disturbance may directly modify the milieu of solids, liquids, and gases, organics and inorganics, and the living and dead materials that compose the soil matrix. In this chapter we examine how additions of inorganic fertilizer materials modify soil, soil fertility, and soil processes. Such additions may be viewed as disturbance, but often this disturbance is vital to enhancing soil fertility and concomitant increases in plant and animal yields.

A. Cultivated Crop and Grassland Fertilization

Soils and climate regulate both the types of plants produced and the need for nutrient supplement via inorganic fertilizers. In turn, the use of inorganic fertilizer impacts the soil and surrounding environment. Fertilization can dramatically impact range species composition, soil organic matter, and the soil microbial community. The last two are affected indirectly through the amount of aboveground and belowground biomass returned to the soil.

Inasmuch as climate and soil environment buffer the impact of fertilization, the geographic boundaries of our discussion are established by soils and climates (Soil Survey Staff, 1975). Arid environments encompass a wide variety of naturally occurring soils. From the USDA Soil Conservation Service (1975) classification, soils can be categorized as arid or semiarid. The 1979 publication by UNESCO bases degree of aridity of land on the ratio of precipitation to evapotranspiration (P/Etp). Terrestrial environments in which this ratio is less than 0.03 are designated hyperarid environments. If a ratio falls between 0.03 and 0.20 the designation is

TABLE 1 Soils Classified as Semiarid, Arid, or Hyperarid According to the UNESCO (1979) Definition (P/Etp < 0.03)

Order	Suborder	Great group
Alfisols	Ustalfs	All
	Xeralfs	All
Andisols	Torrands	All
	Xerands	All
	Vitrands	Ustivitrands
	Ustands	All
Aridosols	All	All
Entisols	Psamments	Most Quartzipsamments
		Torripsamments
		Ustipsamments
		Xeropsamments
	Fluvents	Torrifluvents
		Ustifluvents
		Xerofluvents
		Probably some Tropofluvents
	Orthents	Torriorthents
		Ustorthents
		Xerorthents
Inceptisols	Andepts	Some Eutrandepts
		Some Durandepts
	Ochrepts	Ustochrepts
		Xerochrepts
		Some Eutrochrepts
	Tropepts	Ustropepts
	Umbrepts	Xerumbrepts
Mollisols	Ustolls	All
	Xerolls	All
	Borolls	Calciborolls
		Natriborolls
Vertisols	Torrerts	All
	Usterts	All
	Xererts	All

Soil taxonomy is according to the Soil Survey Staff (1975 and 1990).

arid, and a ratio between 0.20 and 0.50 denotes semiarid. Soils in areas with a P/Etp less than 0.50 are numerous (taxonomically) and extensive (physiographically). Portions of the world's so-called breadbaskets (e.g., the North American Great Plains, the Asian Steppe, and the South American Patagonia) fall into this classification. Certainly, the great deserts of the world are contained in this category, but so too are many areas described as tropical (e.g., east central Africa and the west coast of South America). Accordingly, using the Soil Survey Staff Soil Taxonomy (1975), all the subdivisions of the soil order Aridisols fall into the P/Etp less than 0.50 classification, as do varying portions of the subdivisions of the Inceptisol, Entisol, Mollisol, Vertisol, and Alfisol orders (Table 1). With irrigation, however, soils in even the most arid zones produce cultivated crops and forages and are often very productive.

Aridisols, having developed in climates where the P/Etp is between 0.03 and 0.20, remain high in bases and pH (Donahue et al., 1983). The pH levels between 7.5 and 8.2 keep P availability levels low. Low soil organic matter, which is related to the historically low level of plant production on these soils, is characteristic of all Aridisols. This translates into low N mineralization rates and generally low N fertility, especially under irrigation.

Moving from arid to semiarid regions greatly increases plant productivity potential. Most of the soils in semiarid regions (P/Etp = 0.20-0.50) belong to the ustic suborder of the Mollisoll order. They may receive up to 500 mm annual precipitation, but when associated with high temperatures, available water still limits plant production. Grazing, dryland cropping, and irrigated cropping all occur in these soil areas. Base saturation and pH tend to be high as with the Aridisols, but the Mollisols have higher organic matter contents compared to Aridisols because of the greater native plant productivity during their genesis. Fertility problems are usually related to N and P deficiencies, and under irrigation, the N deficiency is more pronounced. Both Aridisols and ustic Mollisols are N deficient at the elevated production levels associated with irrigation.

Interspersed with Aridisols and ustic Mollisols are soils of the Entisol order. They are found where sandy parent materials prevail and/or where steep, actively eroding slopes are present. Obviously they have the same nutrient deficiency problems associated with Aridisols and ustic Mollisols. They are more fragile, however, than either of the former and are easily degraded when improper grazing or cultivation practices are used. Erosion by wind and/or water is a problem when overgrazing is allowed or when clean tillage systems are used. Deterioration is worst under cultivated conditions, but Williams et al. (1968) report that degradation under rangeland conditions can be alarming.

The impact of inorganic fertilizers on these soils ranges from none, where no fertilizer is used because of very low precipitation, to potentially large where irrigation is practiced and large quantities of fertilizer are applied. The quantity of N and P fertilizer used is closely related to water availability, especially when cultivated agriculture is practiced. Where native rangeland grazing occurs, little fertilizer is used irrespective of rainfall level.

B. Soil Microorganism Response to Fertilization

Soil microbial communities control a variety of processes that are crucial for the survival of plants and maintenance of soil physical and chemical parameters important to retaining soil as a medium for plant growth. In arid terrestrial environments, these processes are no less important than they are in more humid environments. The rate of these processes is generally depressed in arid environments (Vitousek et al., 1982), but the steps in the cycles remain essentially unmodified.

The variation of microorganisms across the soils described as drier than semiarid is large. Very generally, one can say that with decreasing precipitation, microbial activity decreases in lands classified as semiarid or drier; hyperarid areas generally have very low relative microbial activity, arid areas have moderate activity, and semiarid environments have high relative activity. However, the diversity of soil microorganisms is great even in hyperarid environments.

A fundamental factor that often controls microbial activity is depth below the soil surface. Much work in soil microbiology has focused on the surface 15 cm of soil. Certainly, much microbial activity occurs in this zone, particularly in areas with udic and aquic soil moisture regimes. The activity of soil microorganisms is also high in this topsoil zone in environments with ustic, xeric, and torric moisture regimes, when the upper portions of such soils are moist. However, under dry conditions, microbial activity is higher deeper in the soil where moisture may be less depleted and temperature may be less variable. Many symbiotic associations between soil microorganisms and plant roots occur very deeply in arid soils (e.g., White et al., 1989; Virginia et al., 1986).

C. Objectives

The purpose of this chapter is to examine the impact of inorganic fertilizers on soils, plants, and soil microorganisms in environments with P/Ept of less than 0.50.

Cultivated lands in this chapter are defined as those areas where soils are tilled at least annually. Grasslands are defined as areas essentially managed exclusively for grass production. Cultivation may occur during grassland establishment or modification, but by and large these areas are seldom cultivated mechanically.

Rangelands are defined as land areas that by reason of physical limitations are not suited to cultivation. These areas may be sources of forage for free-ranging native and domestic animals, and they are often sources of water, wood, and wildlife (Stoddard et al., 1975). Under relatively semiarid circumstances, such areas may produce abundant forage at predictable intervals, are interspersed with patches of trees, and are moist enough that there is some runoff. Under hyperarid environments, rangelands may produce very little forage and only at long, unpredictable intervals, water is scarce, wood may be nonexistent, and wildlife sparse and specialized for survival under such conditions.

II. FERTILIZING CULTIVATED CROPS

A. Native Soil Condition

Fertilizer is usually not applied until a soil is incapable of supplying adequate nutrients for optimal plant growth. Therefore to understand the impact of inorganic fertilization on soils and on the overall environment, one must first establish the soil nutrient status in the native condition prior to cultivation. Vegetative type and amount of production in the native state were a function of climate and ranged from brushy plants to grasses in arid and semiarid regions, respectively. Total quantities of N and P, the two most limiting plant nutrients in dry areas, were usually in a state of equilibrium with the climatic environment when cultivation began (Figs. 1 and 2). Nitrogen was consumed and/or lost by the soil-plant system at the same rate it was added when the soils were in their native condition. Nitrogen mineralized from soil organic matter was absorbed by

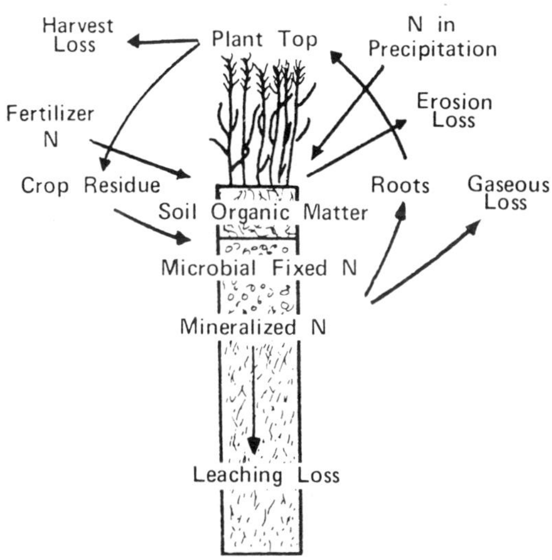

FIGURE 1 The nitrogen cycle.

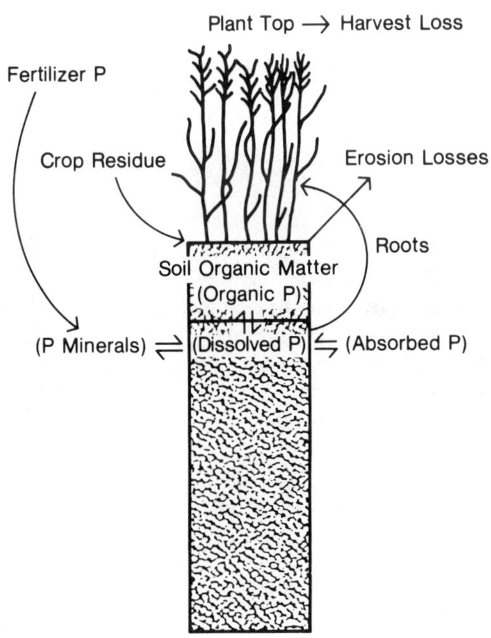

FIGURE 2 The phosphorus cycle.

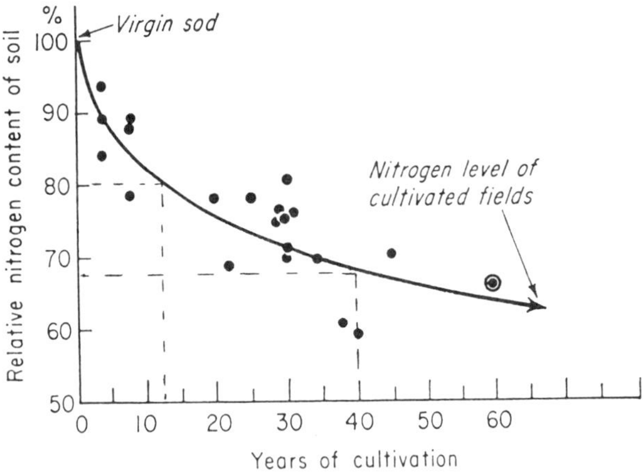

FIGURE 3 Decline of soil nitrogen with time of cultivation. (From Jenny, 1941.)

plants rapidly enough to minimize leaching potential in all but the wettest years. Phosphorus availability to plants was a function of P mineralized from organic matter and solubilization of phosphate minerals. Cycling of P through the soil organic matter pool reduced the opportunity for reactions with the soil mineral pool and thus kept the availability of P to plants at an optimal level for the plant community present on a given soil. Removal of N and P by animal grazing and nutrient losses due to erosion were minimal under native conditions (Figs. 1 and 2).

B. Cultivation and Soil Degradation

Cultivation of the soil radically changed the equilibrium conditions present in native soils. Perturbations of the soil surface by tillage exposed the soil surface to increased erosion potential, and simultaneously, the stirring decreased soil aggregate size. New organic surfaces were exposed to mineralization by stirring, and both N and P were released from organic pools at accelerated rates compared to the noncultivated systems. Removal of N and P from the system also became a larger factor compared to native systems grazed by undomesticated animals. Furthermore, cultivated crops, particularly under dryland conditions, did not return as much C to the soil as did the native plants. This stimulated N and P conversion to mineral forms because the steady-state C/N and C/P ratios were no longer possible. Jenny (1941) was one of the first to quantify this impact of cultivation in terms of N and soil organic matter status (Fig. 3). Note that it required 35-50 years for a new steady state to be established. A N budget reported by Peterson and Vetter (1971) shows the fate of the N in a typical cultivated system (Table 2). Note that erosion and crop removal account for most of the N losses, but that significant quantities of N were unaccounted for. Lamb et al. (1985), in a controlled study, conclusively demonstrated that some of the rapidly mineralized N in a newly cultivated soil can be leached below the root zone of winter wheat even in areas of

TABLE 2 Nitrogen Balance Sheet for Keith and Rosebud Soil Profiles (Values in kg ha^{-1})

	Keith		Rosebud	
Condition or Factor	Native	Cultivated	Native	Cultivated
Native N content	14,000		10,000	
N content, 1969		11,300		7,000
Crop removal		700		700
Erosion loss		1,500		1,500
N unaccounted for (leaching, gaseous loss)		500		800
Total	14,000	14,000	10,000	10,000

less than 400 mm annual precipitation. Olson et al. (1974) have also documented that dryland farming with the wheat-fallow system has resulted in significant quantities of N being leached below the 180 cm depth. Irrigation of these same soils accelerated leaching losses to soil zones below the rooting depths of most crop plants, except alfalfa, which can recover N from depths as great as 5 m in permeable soils.

Phosphorus is used by cultivated crops in smaller quantities than is N and is not subject to leaching. However, P is lost from soil-plant systems by erosional processes when soil cover is removed and aggregate size is reduced by tillage. Data reported by Tiessen et al. (1982) and depicted in Fig. 4 show that substantial losses of total P occurred that were attributable to erosion. Note that relatively large losses came from the organic P fraction. This is particularly serious in such high-pH, calcium carbonate-dominated systems as those found in arid and semiarid areas. Soluble P, whether released in organic matter mineralization or released in mineral dissolution, does not remain in solution but precipitates as low-solubility inorganic forms that are less plant available. The net result is that a soil capable of supplying P for the native plants, and even to cultivated crops in the early years immediately following conversion from the native to the cultivated condition, has a decreasing capacity to supply P needs as time under cultivation increases. This change is not due so much to P removal in harvested plant parts as it is to conversion of mineralized P to unavailable mineral forms. The upshot is that fertilizer supplements of both N and P begin to be needed as time under cultivation increases. For example, the transformation from a N- and P-rich condition to a deficient condition occurred in the North American Great Plains over a period of about 35-50 years (Sander and Peterson, 1968). Soils broken from the native condition in the 1890s had nutrient deficiencies by 1950. The time interval was shortened in irrigated areas where crop removal of nutrients was greater. Dryland areas of the Great Plains did not require inorganic fertilizers until well into the 1960s. Grain protein content declined more quickly than yield and was the first indicator of a N deficiency problem (Sander and Peterson, 1968).

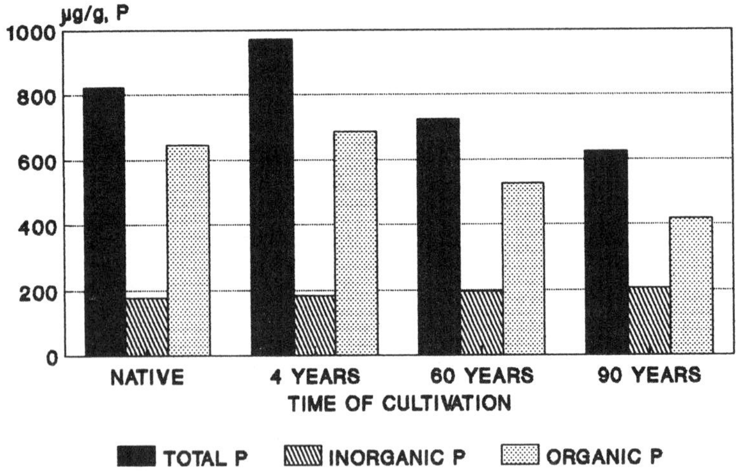

FIGURE 4 Changes in soil P fractions with time of cultivation (From Tiessen et al., 1982.)

C. Assessing Degree of Soil Degradation

High yields can be obtained through the use of inorganic fertilizers. However, does such usage maintain yield quality? Further, how does inorganic fertilizer usage impact soil?

Historically, farmers and scientists have judged soil quality and nutrient supply capacity by organic matter content. Organic matter logically has been associated with productivity because fertilizing with animal waste has been recognized as a means of restoring yield potential on soils whose nutrient supply capacity has been depleted. Manured soils have organic matter contents that are directly proportional to the quantity of manure added over time. Organic matter content is certainly a measure of a soil's capacity to supply nutrients, especially N and P.

When manure is added to soils there is a concomitant addition of N, P, and C. Increased crop residue left by the healthier crop plants created by the manuring also adds to the total C addition. Thus manured soils increase in organic matter content from both the organic materials in the manure and the increased crop residue, which results from the increased crop yield related to the manuring.

Legumes also improve the N status of soils. For millennia, rotating cereal crops with legumes was the standard method of N addition to the soil system and maintenance of acceptable cereal yields. The effect of the legume crops on soil productivity was also measurable in the soil organic matter levels. As with manure, the organic matter in soils managed with legume rotations was higher than in soils managed with no legumes. Increased organic matter, or at least maintenance of a level in a given soil, was the result of the additional N and C coming from the legume root system and/or the total plant production under these green manuring systems. Furthermore, inclusion of the legume plant in the rotation often increased the vegetative cover and decreased overall soil loss due to erosion. The combined effects were increased soil organic matter contents.

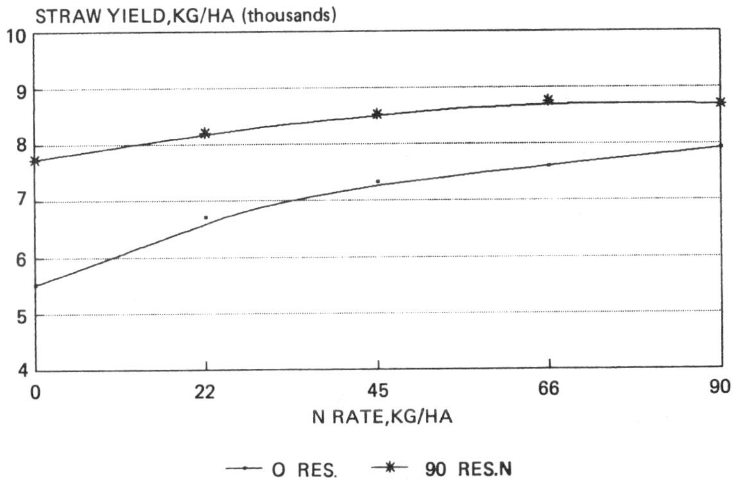

FIGURE 5 Straw yield response to nitrogen fertilizer at two soil test levels. (From Gogulwar, 1973.)

D. Inorganic Fertilizers and Soil Rejuvenation

Since soil organic matter levels can be used as an index of soil nutrient supply potential, it is possible to investigate the impact of inorganic fertilizer application on soil quality via soil organic matter changes. Obviously, additions of inorganic carriers of N and P do not have a C input and so their impact can only come through their effects on the production of crop residues. It is axiomatic that grain yields increase when N or P fertilizers are added to soils testing low in these nutrients. These fertilizers also increase the crop residue yields when used on nutrient-deficient soils. Gogulwar (1973) showed that N fertilizer increased wheat straw yields 2500 kg ha^{-1} on a soil testing low in residual N (Fig. 5). A response of 1000 kg ha^{-1} was still obtained at a medium residual N level. So, addition of a fertilizer nutrient to a soil deficient in that nutrient not only improved grain yields but also increased crop residue yields.

Obviously these residues can only affect soil organic matter if they are returned to the soil. If crop residues are burned or removed as silage, there are no positive effects on organic matter regardless of any improved productivity due to an inorganic fertilizer application. Hooker et al. (1982) reported declines in soil organic matter levels and decreases in N supply capacity after 11 years of burning and/or straw removal after harvest. This occurred on a soil that had been cultivated for 80 years. Declining organic matter was not related to erosion, but instead to reduced C and N additions to the soil.

Not only must the crop residue remain in the field, but its N content and/or the remaining N in the soil must satisfy the equilibrium C/N ratio of the soil. Under cultivated conditions the equilibrium C/N ratio of soil is about 10:1. So if a wide C/N ratio material like wheat straw at 80:1 is returned to the soil, the soil organic matter content is not readily increased unless additional N is available from soil carryover or fertilizer application.

Without the added N the C from the straw is oxidized and lost as CO_2. Anderson and Peterson (1973) reported that applications of N as ammonium nitrate for a period of 20 years up to rates of 180 kg ha^{-1} only slightly altered total soil N. Their highest rate of N maximized grain yields but left little extra N in the soil system. Even though there was a large input of C in the residues (Stover yield = 5 t ha^{-1}), it was not retained in the soil as organic matter because there was insufficient N to meet the C/N ratio requirement. Their land preparation techniques involved moldboard plowing annually plus numerous other cultivations. The latter practices enhance C and N loss. From their data one can conclude that use of inorganic N sources neither improved or detracted from the soil organic matter pool. Nitrogen fertilization only maintained the soil N status.

Lueking and Schepers (1985) reported that N fertilization of a coarse-textured soil actually increased soil organic matter levels. They reported increases in total weight of soil C and N of 7 and 5%, respectively, in the upper 30 cm of soil compared to the native soil condition prior to fertilization. These soils had received 220-280 kg ha^{-1} annually for 4-15 years depending on the field. This quantity of N probably exceeded crop need by 50-100 kg ha^{-1} annually. Thus soil organic matter was actually increased over native conditions. It is important to note that their soils were very low in organic matter in the native condition because of the environment and native vegetation. Other scientists have shown that soil organic matter levels usually cannot be increased by N fertilization in areas where organic matter levels are already very high (Hobbs and Thompson, 1971).

Interactions of tillage and fertilization are key to predicting how fertilization will affect soil organic matter and long-term fertility. Interestingly, Melsted in 1954 recognized the connection between the amount of soil tillage and N conservation. However, his insights appear to have been forgotten for 25 years. From 1954 to 1980 little attention was given to the fertilizer-tillage relationship. However, in the early 1980s the interrelationship has been rediscovered. Blevins et al. (1984) and Lamb et al. (1985) have recently documented the close link between amount of tillage and N conservation. In fact, a reduction in the degree of tillage is essential if inorganic N fertilization is to have a positive impact on soil organic matter content at the rates of N considered economically feasible in our current production systems.

Phosphorus fertilization with inorganic sources has been known to increase soil P levels in many different soil situations. Increases have not been related to improvements in soil organic matter in most cases but have been a function of amount applied and other soil characteristics. If soil pH is above 7.2 or below 6.0, P fertilizer tends to precipitate as insoluble compounds unavailable to plants. So, whether an application of an inorganic P fertilizer has altered a soil's plant-available P status depends on soil pH to a great extent. Obviously the rate of application is a large factor, too. The tendency has been for irrigated soils to receive P fertilizer at rates above the level needed by plants. This has resulted in very high available P levels in many irrigated soils. For example, Peterson et al. (1966) reported levels of bicarbonate extractable P $>$ 70 ppm in irrigated soils where native P levels had been $<$10 ppm. Levels judged adequate for crop production are near 12 ppm. Research in Kentucky by Thomas (1989) found that the percentage of soil samples testing high in P increased from 30 to 44% from 1975 to 1985. On high-cash-value crops the number of soils testing high had risen to 80%. Overapplication of P fertilizers is the apparent norm on irrigated soils and on soils used to produce high-cash-value crops. In contrast, under dryland farming conditions producers have been reluctant to apply even small amounts of P and many soils have become P deficient (Sander and Peterson, 1971) with time of cultivation.

E. Future Impacts of Fertilizers

Insertion of cultivated cropping into semiarid and arid environments has negatively impacted soils under the management systems used over the past 100 years. The use of inorganic fertilizers has slowed the degradation process but has not stopped it. There are a few examples in which the use of inorganic fertilizers has improved soil quality, but these are the exception. Overall the evidence points to a need for radical decreases in degree of cultivation to slow and eventually stop soil degradation. This must be done so that cultivated agriculture can be sustained in dry areas. The use of inorganic fertilizers in conjunction with reduced tillage can actually reverse the degradation process and improve soil quality as measured by soil organic matter content. Research efforts should focus on the development of total management packages that address the degree of cultivation, crop rotation, and the use of inorganic fertilizers.

III. FERTILIZATION OF GRASSLANDS

Nitrogen fertilization of grasslands in arid and semiarid regions increases forage production in many cases, but the magnitude of the response and the economics are dependent on many interacting factors. The amount and distribution of precipitation, soil water-holding capacity, natural soil fertility and pH, previous cropping history, species composition, age of stand, and potential for undesirable shifts in the vegetation or invasion by weeds all may affect the response. Target plants for fertilizers are usually grasses, but shrubs and forbs are also affected. Generally, grasslands respond little to phosphorus fertilization unless legumes are present. Other mineral elements are generally in adequate supply in most semiarid and arid grassland soils and do not limit growth. Grassland fertilization is often not considered as a management practice by itself but is considered part of a management system if it is to be profitable. In most situations where it is desirable to increase forage yield, livestock are necessary to harvest the forage, so livestock management and forage quality as affected by fertilization must be considered with any fertilization program. Fertilizer is sometimes considered a management tool to bring about a specific vegetation response and is not a practice that is used annually or even periodically. Stands that have been seeded either to introduced species or selected native species, especially cool-season grasses, are generally the most economically responsive to fertilization because they are on better land and have a higher potential yield than native range.

A. Nitrogen Cycling, Utilization, and Recovery

Nitrogen cycling in grassland ecosystems supplies N for new growth in arid and semiarid regions. Principal atmospheric additions under unfertilized conditions occur as dry fallout, such as NH_4^+, NO_2^-, and NO_3^-, and organic N (Reeder and Sabey, 1987). Atmospheric deposition is twice as much in the eastern compared to the western Great Plains, although deposition downwind of industrial regions can be very high in any climate. Symbiotic and nonsymbiotic N contributions are often small but can be significant. Losses of N from the ecosystem are from denitrification, volatilization of NH_3, leaching, erosion, volatilization during fires, and export of N in livestock tissues. Of these losses volatilization from plant residue, dung, and urine spots is the major source of loss on arid to semiarid grasslands. Large domestic ungulates can be a major vector in N loss (Woodmansee, 1978). The P cycle in most grasslands is closed with no major additions or losses (Clark et al., 1980). Mineralization of N is more rapid with warm soils, so more cycled soil N is available to warm-season grasses than to cool-season grasses. Therefore, most cool-season grasses are N deficient in the spring.

A large amount of fertilizer N is immobilized in grass roots, organic matter, and the microbial pool (Power, 1981, 1988). Fertilization alters the sink sizes and turnover rates, resulting in the eventual establishment of new equilibrium levels between rates of immobilization and mineralization in grassland ecosystems. Under semiarid to arid conditions in which there is no leaching below the root zone, high application rates of N applied infrequently have proven nearly as effective as smaller annual applications (Power and Alessi, 1971). As a result, the cost of application could be reduced and the residual effects of applied N would be important. With adequate rates of N under northern Great Plains conditions this immobilizing ability can be saturated in 3-4 years, which is the estimated rate of turnover of root-immobilized N. Fertilizer N in excess of the immobilizing capacity remained as inorganic N (Power, 1977; Smith and Power, 1985). With crested wheat grass (*Agropyron cristatum*) residual fertilizer N consisting of 30-50% of the ^{15}N that was applied, was incorporated into surface soil organic forms during the first season of application. Nitrogen immobilized in a shortgrass prairie ecosystem is not likely to be removed very quickly (Clark, 1977). Nitrogen use efficiency (kg forage produced kg^{-1} N applied) is greater at lower N rates. Nitrogen applied at 45 kg ha^{-1} to cool-season grasses was two to three times more efficient in producing yield than when applied at 225 kg ha^{-1} (Power, 1985).

Recovery of fertilizer N in the top growth of forages is low compared to cropland recovery. Nitrogen recovery in forage averaged 58% with calcium nitrate and 48% with urea, and N recovery from ammonium nitrate and ammonium sulfate was even less. Phosphorus recovery in plants ranged from 26.8% when it was applied with calcium nitrate to 18.2% when added with ammonium sulfate. Fertilizer N doubled P recovery (Power, 1983). In a 35 year N fertilization study on smooth bromegrass (*Bromus inermis*) in North Dakota, N recovery in plants ranged from 40 to 120% depending on rainfall. Nitrate N was undetectable below the root zone even though more than 1000 kg ha^{-1} was present in the root zone with the heavier rates (Meyer, 1989). Fertilizing crested wheat grass with 34 kg N ha^{-1}

or less in Oregon with 300 mm annual precipitation did not increase the nitrate concentration in the soil. With 56 kg N ha^{-1} or higher an accumulation of nitrate occurred just above the cemented caliche layer. With recommended rates of N for crested wheatgrass on semiarid soils, there should be little danger of nitrate accumulation on- or off-site (Sneva, 1977). In following ^{15}N recovery, Power and Legg (1984) found that, initially, N uptake was 20-25% greater from fertilized than from nonfertilized crested wheatgrass. The difference disappeared after the third year of fertilization, suggesting that at least 3 years was required for the N cycle to readjust to a new steady state. Of the ^{15}N applied, 12-52% was recovered in the year of application, which represented about 75% of the total recovered in the 5 years following application. About 15% of the N was in the roots the first season, and about two-thirds of the N appeared in the topgrowth within 3 years. Nitrogen recovery is also species dependent. Fertilizer N recovered in topgrowth was highest for smooth bromegrass (30%) and least for native range (15%). Crested wheatgrass was intermediate in percent N recovery (Power, 1980b).

There are limited studies showing the effects of long-term fertilization on grassland soil conditions. After 20 years of N fertilization of smooth bromegrass with N rates from 0 to 180 kg ha^{-1} soil pH dropped from 5.9 to 4.7 in the 0-15 cm layer but was unaffected at the 15-30 cm depth in fertilized plots (Owensby et al., 1969). Under several cool-season grasses, soil surface pH dropped 0.2-0.3 units after 9 years of fertilization with 225 kg N ha^{-1}, although surface pH under smooth bromegrass dropped about 0.7 pH units (Power, 1985).

B. Fertilization in the Establishment of Grasslands

Fertilization may be used to aid stand establishment where nutrient deficiency reduces seedling competitiveness. The use of fertilizer to aid establishment on reclaimed land is somewhat controversial. Some believe that if fertilizer is required to establish and maintain the plant community, such a plant community will not be adapted to the soil resource over the long run and will be replaced by weedy species (Brown et al., 1986). Phosphorus deficiencies are best corrected at seeding time when fertilizer may be incorporated, but often N fertilization is not used at seeding because undesirable species may be stimulated more than the seeded species. Nitrogen fertilization of seeded warm-season grasses 1 or 2 years after seeding increased the stand density and yields on marginal cropland in 12 locations across Nebraska for 9 years because the natural soil N on the reclaimed eroded cropland was low (Warnes et al., 1971).

In a field study in northwest Colorado neither N nor P increased stand establishment ratings of seeded grasses and legumes. Not until the third growing season did the amount and placement of N affect herbage yield. Alfalfa (*Medicago sativa*) yields were increased by 20% with P buried under 30 cm of topsoil and 44% with surface-applied P in the third growing season (McGinnies and Crofts, 1986). Much of the mine spoil in the northern Great Plains lacks P. Application of P before seeding is likely to be beneficial. Until a pool of mineralizable N is built up, annual application may be necessary to supply N to developing plant communities (Power et al., 1978). In annual grasslands in California where legumes can be used to supply significant N to the system, sulfur (S) and P are often needed to overcome deficiencies. Generally P and S are applied at seeding time and reapplied every 2 or 3 years (Murphy, 1986). The decision about whether to apply fertilizer elements at seeding time is dependent on climate, soil supplying capacity, potential for weed invasion, and the ability for a nutrient to move in the soil. Nitrogen fertilization at seeding may increase yield but does not generally improve stands in arid and semiarid areas.

C. Fertilization of Established Grasslands and Range

Introduced Cool-Season Grasses

Much of the fertilization of semiarid and arid grasslands has been in the northern Great Plains, targeted at the stimulation of productive introduced and native cool-season grasses. Generally, introduced cool-season grasses require N fertilization periodically to remain productive and to maintain stands. In many situations there is an economical response to annual applications of N at a moderate rate. Cool-season grasses make most of their growth in the spring when moisture from precipitation and stored soil moisture are adequate so N becomes limiting to growth.

Yields of introduced cool-season grasses, such as the wheatgrasses and smooth bromegrass, are increased markedly with N fertilization (Power, 1980a, 1983, 1985; Power and Alessi, 1970; Williams et al., 1979), particularly when application is made in spring. Growth and N uptake of smooth bromegrass tops increased until physiological maturity in July in North Dakota, decreased during a summer drought, and increased again after rains in August (Power, 1988). Cumulative 17 year dry matter (DM) production of crested wheatgrass was increased 15 Mg ha^{-1} with 90 kg N ha^{-1} and 20-25 Mg ha^{-1} with 180 kg N ha^{-1} applied annually for 5 years and the residual response followed for 12 more years. Most of the residual response came in the first 2 years after fertilization ceased. The cumulative 17 year yield increased 4-5 kg ha^{-1} with 19-39 kg P ha^{-1} for the first 5 years (Power, 1983). In a 35 year fertilization study with smooth bromegrass in North Dakota, yields were doubled with 74 kg N ha^{-1} and tripled with 298 kg N ha^{-1} (Meyer, 1989). Power and Alessi (1970) reported that DM production of crested wheatgrass increased with both N and N/P fertilization. At normally recommended rates there were no interactions with N source. They stated that the N response was 68-88 kg forage produced cm^{-1} water (with and without P) and the P response was 9 kg cm^{-1} water regardless of the water supply. Nitrogen efficiencies ranged from 5.9 to 45 kg forage kg^{-1} N depending on precipitation.

Water use efficiency (WUE), units of yield per unit of water used, is generally much greater for fertilized cool-season grasses compared to unfertilized grasses because of the increased DM production and more thorough water extraction from the soil. Power (1985) reported that WUE increased about three times with fertilized cool-season grass plots and that WUE increased 12 kg for each mm water used with N fertilization. The increase in WUE due to fertilization was greatest for high-yielding cool-season grasses and least for native range (Power, 1980a). Early in the season, N hastened the removal of stored water from fertilized crested wheatgrass plots. Later in the season the grass on fertilized plots grew at a lower soil water potential. Water use efficiency was much greater for fertilized than for the unfertilized grass (Williams et al., 1979). Nitrogen usually increases the yield of perennial cool-season grasses in arid and semiarid areas if there is adequate water. Introduced cool-season grass stands are generally maintained, and there is little danger of weed encroachment by timely N fertilization. However, stands of crested wheat grass were reduced when 90 kg N ha^{-1} was applied annually. The reduction of carbohydrate reserves due to N stimulation of topgrowth combined with heavy grazing was the cause of the stand decline (Rogler and Lorenz, 1969).

Crude protein (CP) content in cool-season grass forage increased due to N fertilization (Johnson and Nichols, 1969; Meyer, 1989). Effects on in vitro dry matter digestibility (IVDMD) are variable. Generally there is little effect on IVDMD due to fertilizer. Nitrogen fertilization of many cool-season grasses, especially crested wheatgrass, may be accompanied by an increased incidence of grass tetany in livestock. When 150 kg N ha^{-1} was applied to crested wheatgrass, forage Mg and Ca were increased more than the K concentration, thus reducing the K/Ca + Mg ratio. However, all forage K/Ca + Mg ratios were less than the 2.2 value, above which the incidence of grass tetany increased rapidly. Concentrations of K, N, higher fatty acids (HFA), aconitic acid, and the N-total water-soluble carbohydrate ratio were increased, all of which was associated with an increased incidence of grass tetany. Calculated serum Mg levels were lower for N-fertilized forage (Mayland et al., 1975). Similarly, Follett et al. (1977) reported that N fertilization of cool-season grass with either nitrate N or ammonium N increased the tetany hazard by raising the K/Ca + Mg ratio, total N, aconitate, and organic acid concentrations in the forage.

Livestock carrying capacity and beef production per hectare of cool-season grasses are generally increased significantly with N fertilization. Russian wild rye (*Elymus junceus*) pastures fertilized with 45 kg N ha^{-1} produced 100 kg additional beef per hectare compared with unfertilized pastures. Carrying capacity was increased 110 steer-days ha^{-1} (Rogler and Lorenz, 1970). Nitrogen fertilization causes earlier spring growth, and thus pasture is ready for grazing earlier than unfertilized pasture, reducing the hay-feeding season.

Native Cool-Season Range

Fertilization of native grassland ecosystems is more uncertain than with monocultures or simple mixtures because of the wide variation in soils and species composition across different areas. Soils are often less productive than those supporting introduced grasses, and there is a much greater risk of adversely affecting the complex plant community if an outside source of N is added to the ecosystem. Measurement of plant and animal response is also more complex with native range than with an introduced pasture consisting of one or a few species (Rogler and Lorenz, 1974).

Northern Great Plains range consists of a mixture of cool-season (*Agropyron* and *Stipa*) and warm-season (*Bouteloua*) species. With N fertilization in this plant community, it is possible to increase a high-yielding cool-season component (western wheatgrass, *Agropyron smithii*) at the expense of a low-yielding warm-season component (blue gramma, *Gouteloua gracilis*). This increases yield and shifts livestock production earlier in the growing season (Lorenz and Rogler, 1972; Goetz et al., 1978). The change can be rather rapid because readily available N stimulates early water extraction from the soil by the cool-season grass, placing severe moisture stress on the actively growing warm-season grass during the summer when the cool-season grass is dormant. Rogler and Lorenz (1974) reported that northern Great Plains range conditions improved from poor to excellent in 3 years with annual applications of 45 kg N ha^{-1}.

In a *Bouteloua-Carex* association in Montana, N- and P-fertilized plots had considerably more root biomass compared to unfertilized plots. The root system was a nutrient-deficient sink with the ability to immobilize large amounts of N and P fertilizer, which produced effects that may last for years. Nitrogen was the main limiting element, and a response to P occurred only when N was not limiting (Black and Wight, 1979). Power and Alessi (1971) showed a nearly equal response to equivalent units of N whether it was applied in one application or in several applications over a few years. When 270 or 540 kg N was applied to a northern Great Plains range in one application, plant growth responses were measured up to 9 years after fertilization ceased. Part of the residual response to N was due to a thicker stand produced by the fertilizer (Power, 1981) as well as stimulation of grass tiller growth. In Montana, Wight and Black (1979) also reported that single high-rate applications were nearly equal to annual applications in stimulating yield when compared on an annual rate basis. At rates of 336 kg N ha^{-1} or less, cool-season grasses increased in percentage composition by weight about the same as did forbs and shrubs. Unfertilized range produced an average of 2.60 kg ha^{-1} forage for each mm of precipitation compared to 5.81 on fertilized plots. However, Lorenz and Rogler (1972) indicated that lower levels of N were best because blue grama was kept in stand, which kept forage quality and summer livestock production higher than with the increased cool-season western wheatgrass growth on areas fertilized heavily with N. If an increase in forage yield is to be transferred to economic yield in grazing situations, there must be ample livestock available to utilize the increased forage production, so yield distribution is important. Fertilization of range tends to accent the natural forage yield fluctuations due to precipitation differences, so it is difficult to predict the N response in any given year to have additional livestock available to utilize the increased forage yield. For example, during a 10 year study in a Montana mixed prairie, the application of N and P increased herbage yields on the average of 114% ranging from 32% in a dry year to 218% in a wet year (Wight and Black, 1979). The protein content of forage was increased by N fertilization. Shrubs and forbs consistently had higher N and P contents regardless of the fertilization program compared to western wheatgrass and blue grama. Nitrogen and P contents of fertilized plants in a *Bouteloua-Carex* association in Montana were increased in proportion to the rates applied. Plant N and P contents were higher even after the effects of fertilization on yield had ceased (Black and Wight, 1979). Nitrate accumulation is a potential problem with fertilization of arid and semiarid grasslands if nitrate-accumulating forbs are increased. It is generally less of a problem with the perennial grasses.

Beef production per hectare was increased 68% with 45 kg N ha^{-1} and 75% with 90 kg N ha^{-1} in a 10 year study on northern Great Plains range. Average daily gain was not affected (Rogler and Lorenz, 1965). In another

11-year study in the northern Great Plains range, beef production was increased 95% with 45 kg N ha^{-1} and 150% with 90 kg N ha^{-1}. Gain per head was not affected (Rogler and Lorenz, 1974). On annual California range, triennial applications of S or S + N doubled and tripled the carrying capacity during wet seasons. Nitrogen alone increased beef production only during wet seasons. Average daily gain was not increased by fertilization (Wolters, 1986). Most of the increase in livestock production from fertilization of cool-season grass-dominated range comes from the increased carrying capacity, not from improved gains per head.

Native Warm-Season Range

Although most range in the central and southern Great Plains is dominated by warm-season grasses, there are also cool-season components. In Wyoming, blue grama populations varied from year to year due primarily to precipitation, not to fertilization. The frequency of cool-season western wheatgrass was higher on plots fertilized with 22 or 34 kg N ha^{-1} for only 2 of the 6 years. Annual species were more sensitive to long-term, low-rate N fertilization on semiarid rangeland than were the dominant perennial species. Blue grama yield was not increased with low levels of N (Samuel, 1986). With high rates of N on blue grama-dominated range in southeastern Wyoming, residual effects were evident 8 years after application. Western wheatgrass density increased and blue grama density decreased during the fertilization period and the first year of the residual period. Blue grama had increased to its original density 3 years after fertilization ceased. Although total forage over the 8 year period was increased 3052 and 3668 kg ha^{-1} for the 168 kg ha^{-1} and the 672 kg ha^{-1} N rates, it was not economical. Application of P did not increase the yields of western wheatgrass or blue grama (Rauzi, 1979). In Colorado, where water was applied to blue grama-dominated range and soil N level was kept 50 kg ha^{-1} above check plots, there were no major changes in botanical composition in 5 years. Cool-season annuals moved into the plots early in the study because of water, but they were replaced by perennial vegetation later. Although there would likely be major vegetational changes in the long term, the authors concluded that Colorado shortgrass range is very resistant to vegetation changes. Nitrogen treatments alone affected the vegetation composition only slightly and closely followed the climate-induced changes in the control plots (Lauenroth et al., 1978). Nitrogen at 36 kg ha^{-1} increased forage production of shortgrass range near Hays, Kansas, two to three times compared to unfertilized range (Launchbaugh and Owensby, 1978).

In southwestern U.S. ranges, precipitation is a more critical factor than fertilization. Fertilization of drought-stressed blue grama-dominated range did not improve herbage yield. The marked increase of annual plants with N fertilization under drought conditions may change species composition for an extended period of time (Fulgham et al., 1977). Although production was highly variable on New Mexico blue grama range owing to fluctuation in precipitation, herbage production was generally greater for fertilized range (Donart et al., 1978).

In the eastern part of the central and southern Great Plains the cool-season grasses that may increase with N fertilization are much less productive than those in the northern Great Plains. Hence, N fertilization has the potential to reduce range condition rather than improve it as in the northern Great Plains. Therefore, the timing of N application is critical. Nitrogen should be applied immediately before the greatest demand by the warm-season grass, which is after cool-season grass has made most of its growth. The rate of N must be closely regulated as well, so that carryover N to the fall or spring does not occur, stimulating cool-season invasion. Tall warm-season grasses were fertilized with 0-180 kg N ha^{-1} in northeastern Nebraska. Early spring application of N encouraged the encroachment of the cool-season, intermediate wheatgrass (*Agropyron intermedium*), which dominated the stand after just 2 years of N fertilization. At the lower N rates higher yields of warm-season grasses were produced when fertilizer was added in late spring. Nitrogen rates higher than 90 kg ha^{-1} caused cool-season encroachment due to carryover N to the fall and spring (Rehm et al., 1976). In another study the yield of seeded warm-season grass to 0-134 kg N ha^{-1} was linear where atrazine [6-chloro-*N*-ethyl-*N'*-(1-methylethyl)-1,3,5-triazine-2,4-diamine] was not used but curvilinear (increased more rapidly with lower N rates) where it was applied. Atrazine effectively controlled cool-season grass encroachment and annual species (Rehm, 1984). Where tall warm-season grasses were fertilized with N, P, K, S. and Zn, yields were increased by N and N/P combinations but K, S, and Zn had no effect (Rehm et al., 1972). After 4 years of fertilization there was no encroachment of weeds or cool-season species because the fertilizer was applied in late spring each year (May 15) at the time of N need by the warm-season grasses.

In the southern Great Plains the effects of atrazine, 2,4-D, and fertilizer were additive and produced a large yield increase in tall warm-season grasses, especially switchgrass (*Panicum virgatum*) (Baker and Powell, 1978). Taliaferro et al. (1975) suggested that in the southern Great Plains the selection of grasses for renovation be based on the future management program they will receive. Based on their N and P fertilization studies on low-fertility soils under low-fertility programs, weeping lovegrass (*Eragrostis curvula*) and Old World bluestems (*Bothriochloa ischaemum*) would produce more than Bermudagrass (*Cynodon dactylon*). The yield differences would likely diminish under high fertility and management.

Crude protein was increased by 3.5 percentage units in a seeded warm-season grass mixture with N fertilization. Atrazine or P did not increase CP (Rehm, 1984). However, in another study with seeded warm-season grasses CP or IVDMD was not changed with N, P, K, S, or Zn fertilization (Rehm et al., 1972). In a N and P fertilization study in the southern Great Plains all the grasses responded to N fertilizer with increased crude protein percentage. In vitro dry matter digestibility was increased in the three seeded grasses but not in the native range (Taliaferro et al., 1975).

Nitrogen fertilization of semiarid range in western Nebraska with 34 kg N ha^{-1} each year produced 51% more steer gains. However, fertilization shifted species composition to predominantly cool-season grasses (mostly needle-and-thread, *Stipa comata*) with an invasion of cool-season weeds. Fertilization of rangeland in western Nebraska is not recommended because of vegetation shifts from early N fertilizer applications, carryover N, and fertilizer cost (Stubbendieck, 1977). With low rates (22 and 34 kg ha^{-1}) of N applied to a blue grama-dominated mixed prairie in southeast Wyoming, forage yield, crude protein content, and utilization by cattle were increased. Spring applications of N were more effective than fall applications. During this 4-year study in a dry period, the 22 kg N ha^{-1} was more consistently effective. Cattle preferred to graze the fertilized range whenever fertilized and unfertilized range were contiguous (Samuel et al., 1980). Nitrogen fertilization might be used to attract cattle to underutilized range.

Livestock production from fertilized New Mexico range was two times greater with 45 kg N ha^{-1} (Donart et al., 1978). The diet composition of cattle on fertilized and unfertilized blue grama rangeland was examined during drought (Allison et al., 1977). Grass comprised an average of 62% of the cattle diet on fertilized compared to 73% on unfertilized range. Annual forbs made up 18% of the diet on fertilized but only 8% on unfertilized range. Cattle preferred fertilized blue grama to unfertilized.

When livestock were used to evaluate production in central Kansas shortgrass range, carrying capacity was increased by 50% and beef production ha^{-1} was increased over 50%. Japanese brome (*Bromus tectorum*) and western wheatgrass were stimulated in spring and early summer, and blue grama and buffalograss (*Buchloe dactyloides*) were productive during the summer (Launchbaugh and Owensby, 1978). They suggested taking advantage of the increased cool-season production and palatability by stocking the range more heavily early in the growing season and removing the animals in the summer after the range has been properly utilized. In the Flint Hills of Kansas, livestock performance was better on burned, N-fertilized than on unburned, unfertilized range (Launchbaugh and Owensby, 1978). In a 4 year trial in which weeping lovegrass was fertilized with 39 kg N ha^{-1}, gain per head was increased by 10%, carrying capacity by 25%, and beef production ha^{-1} was increased by 31%. Palatibility was increased markedly by fertilization (McIlvain and Shoup, 1970).

Generalizing about fertilization of arid and semiarid grasslands is difficult, especially in natural ecosystems, because of the diversity in soil and species composition and wide fluctuations in precipitation. Nitrogen is generally the only element that increases yield. In general, arid and semiarid rangelands are not fertilized with N owing to the lack of profit and potential damage to the natural ecosystem. Since N response is somewhat unpredictable, obtaining the right amount of livestock to utilize the additional forage is difficult. The right combination of a livestock management system, precipitation, and livestock availability and marketing with a particular grassland resource poses a complex economic problem. It is unlikely that extensive areas of semiarid grasslands will receive N in the future. Irrigated and subirrigated forages, introduced cool-season grasses, and some of the cool-season-dominated rangelands are most likely to respond economically to fertilizer in arid and semiarid climates.

IV. FERTILIZER EFFECTS ON SOIL MICROORGANISMS

The soil state equation (Jenny, 1941) can be used to obtain a general notion of how the modification of soil, parent materials, other organisms (e.g., plants), topography, climate, and time will also modify soil microbial communities. In this treatment it is accepted that a component of climate—precipitation—or, more accurately, plant-available soil moisture is low enough that the rates of microbial processes are depressed when compared to similar edaphic environments with higher available soil moisture. Obviously, when arid soils are irrigated, microbial processes are not as limited by water, although many irrigated arid soils are still very dry during the irrigation off-season. Other components of climate (e.g., temperature and length of growing season) also control microbial processes. Likewise, the other components of the soil state equation control microbial activity. For example, parent material may or may not provide basic nutrients for microbial processes; other organisms (usually plants) may provide substrates of varying levels of recalcitrancy for microbial growth or sites for microbial occupancy; topography frequently controls microclimate (and therefore macroorganisms, and thus, microorganisms); and the age of the site (time) also impacts on microbial processes as weathering reduces nutrients or, as so often occurs in arid climates, depositional events result in the accumulation of salts and clays. The salient points of this discussion are that the interaction of the components described in the soil state equation and the soil at a particular site predetermines the kinds of microorganisms at the site and the rate at which they can operate. Soil microbial activity must further be viewed in three dimensions: across landscapes and with depth, since soil depth, arrangement of horizons, texture, organic matter, and so on, also predetermine microbial activity.

A. Carbon Cycle

There are many reviews in print regarding the general nature of the C cycle. A general review of the involvement of microorganisms in this topic can be had from Paul and Clark (1989, Chap. 6). Other treatments can be found in Bolin et al. (1979), Ziekus (1982), Stevenson (1982), Zinke et al. (1984), and others.

A general notion of the rate of microbial C biocycling can be had from quantification of the organic C or organic matter present in a soil sample. Certainly, in most soils the organic constituents are concentrated near the surface of the soil, and this deposit is designated the *mollic epipedon*. The thickness of this organic layer, as well as the C content, generally correlates positively with the aboveground annual productivity of the soil (Munn et al., 1978) and can be used as an index of site stability when the organic matter content of the undisturbed soil is known.

In unfertilized rangelands and grasslands, the organic matter in the surface soil is relatively stable, and the rate of C fixed by autotrophs in the system is approximately equal to the rate of organic residue decomposition. When the system is at equilibrium (i.e., stable), the rate of humification of C substrates into the organic matter fraction of the soil is equal to the rate of decomposition of the organic matter fraction. This equilibrium is subject to perturbation by grazing animals, plowing, and fire, for example. The addition of N and/or P fertilizers and/or water is also a perturbation. It is relatively well recognized that soil organic matter stabilizes at a C/N ratio of approximately 10:1 (Alexander, 1977; Paul and Clark, 1989). When excess N is added to this system, other nutrients, including C, tend to become limiting factors of microbial growth. Carbon may be conserved along with P and other nutrients, whereas N, when in excess, is used for processes other than microbial anabolism. For example, ammonium is oxidized for energy acquisition and nitrate is used as a terminal electron acceptor. Both these processes allow general leakage of N from the system, either as nitrate movement through soil via leaching or loss of volatile N gases as N_2, N_2O, and NO. Additions of excess P to such a system are less dramatic. Compared to N, P is relatively insoluble in the soil solution and is not thought to be involved as a terminal electron acceptor or used as an energy source.

B. Nitrogen Cycle

As with the C cycle, the reader is referred to literature regarding the general nature of the N cycle. Good references include Paul and Clark (1989, Chaps. 8, 9, and 10), Focht and Verstraete (1977), Firestone (1982), Ahmadjian and Paracer (1986), and Becking (1982).

C. Nitrogen Fixation

In unfertilized, rainfed rangelands and grasslands, most of the N for plant growth can be traced initially to biological N fixation. Lightning has been implicated as a source of N in ecosystems, but the amount of N generated by lightening on a per hectare basis is likely negligible, particularly on an annual basis. Biological N fixation is generally subdivided into two categories: asymbiotic and symbiotic N fixation. Asymbiotic N fixation occurs via fixation by free-living cyanobacteria (blue-green algae), such as *Nostoc*, and bacteria, such as *Rhodopseudomonas*, *Chromatium*, *Clostridium*, and *Azotobacter*. Certain N-fixing organisms inhabit the rhizosphere. These include *Azospirilum*, *Azotobacter*, and *Bacillus* (Havelka et al., 1982). These organisms are generally thought to fix quite low amounts of N when figured on a per hectare-year basis, particularly in arid environments. However, there are exceptions. *Cyanobacteria* have been shown to form algal crust in some desert environments. When conditions are such that temperature and adequate moisture permit N fixation, these crusts may fix 10-100 kg N ha^{-1} annually (Rychert and Skujiņš, 1974). Perhaps more realistic estimates are those summarized by Skujiņš (1981) which range from 0.5 to 13 kg N ha^{-1} annually (mean of 6.45, n = 4). High annual rates of N fixation have also been reported by various researchers by *Azospirillum* in the rhizosphere of various grasses (Dobereiner and Day, 1976; Hubbell and Gaskins, 1984). However, it is generally considered that in arid environments N fixation by asymbiotic microorganisms is quite low when measured on an annual basis. In these low-productivity systems, however, this fixed N is probably crucial to ecosystem functioning.

In most environments, the majority of N fixation is accomplished by symbiotic systems. These include lichens, *Rhizobium*-nodulated plants, and *Frankia*-nodulated plants.

Lichens that fix N contain cyanobacteria as one of the symbiotic partners. In humid environments, lichens may contribute considerable N to the system, for example, 10 kg ha^{-1} (Pike et al., 1972) and 1.5-8 kg ha^{-1} (Forman, 1975), both annually. However, the quantity fixed in arid environments is very low. For example, Forman and Dowden (1977) show very low lichen density in sagebrush desert, pinyon-juniper woodland, ponderosa pine forest, and alpine tundra in New Mexico. Presumably,

TABLE 3 Actinorhizal Plants: Geographic Distribution and Nature of Sites Occupied

Family	Genera	No. sp. nod./total[a]	Geographic distribution	Sites
Betulaceae	*Alnus*[b]	33/35[b]	North America, Japan, the Andes	Europe, Siberia, poor soils, sand hills/dunes, glacial till, wet bogs, mine dumps, gravel, wasteland, volcanic ash
Casuarinaceae	*Casuarina*	24/45	Tropical Asia, Pacific Islands, widely introduced elsewhere	Australia, sand dunes, salt marshes, tropical forests, nondesert areas
Coriariaceae	*Coriaria*	13/15	Japan, New Zealand, Chile to Mexico	Mediterranean to lowlands and subalpine, sandy or gravelly soils or clay
Elseagnaceae	*Elaeagnus*	16/45	North America	Asia, Europe, disturbed areas, sand dunes, poor soils, riparian sites
	Hippophae	1/3	Asia, Europe, from Himalayas to Arctic Circle	Sand dunes, coastal areas
	Shepherdia	3/3	North America	Sand soils, disturbed areas, riparian, under *Pinus* forests
Myricaceae	*Myrica*	26/35	Many tropical, subtropical, and temperate regions extending nearly to Arctic Circle	Acidic bogs, sand dunes, mine wastes

Rhamnaceae	*Ceanothus*	31/55	North America, especially western United States	Dry forest and chaparral, subalpine zones
	Colletia	1/17	South America	
	Discaria	2/10	The Andes, Brazil, Oceania	Gravelly soils, arid zones
Rosaceae	*Cercocarpus*	4/20	Western United States to Mexico	Foothills, disturbed soils, rocky arid areas
	Chamaebatia	1/2	California	Chaparral (Heisey et al., 1980)
	Cowania	1/4	Utah, Arizona	Foothills, high arid plateaus
	Dryas	3/4-6	Alaska to Colorado in North America, Europe, circumpolar	High altitude and latitude, variable soils
	Purshia	2/2	Western North America	Disturbed soils on upland plains, foothills, and under *Pinus* forests, but also deserts
	Rubus	1/250	Indonesia	Sand, rock sites, bogs, North America

[a]Number of species nodulated over total number of species in genus.

[b]Total of 162 species in 16 genera in 7 families.

Source: From White and Williams (1985).

lichen N fixation in these systems is also very low. Even where lichen density was reported high by these authors (in New Mexico Douglas fir forest and spruce forest), N fixation attributed to lichens ranged from 0.04 to 3.3 kg N ha^{-1} annually.

Nitrogen fixation by actinomycete-nodulated plants can be important in rangelands, particularly shrubland environments. Nitrogen fixation by *Alnus* spp. has been reported to be as much as 300 kg ha^{-1} annually, averaged over 25 years (Newton et al., 1968); however, these high rates have been measured under the rather mesic conditions of the Pacific Northwest. *Ceanothus*, a genus peculiar to North America, inhabits a wide variety of environments from understory in coniferous forests to chaparral and open environments. Youngberg and Wollum (1976) established a N fixation for *Ceanothus velutinus* of 108 kg ha^{-1} annually, averaged over 10 years in the somewhat dry environment of the eastern Cascades. However, Kummerow et al. (1978) calculated a N fixation rate of 0.1 kg ha^{-1} annually for *Ceanothus greggii* in the very arid mountain ranges of Southern California. Other rates of N fixation by this plant range between these values (Conrad et al., 1985; Ellis and Kummerow, 1989). Other N-fixing plants (Table 3) that are actinomycete nodulated have been shown to fix variable amounts of N, but in arid environments the quantities are undoubtedly low.

Probably the greatest quantity of N fixed biologically in arid environments is that from the legume-*Rhizobium* symbiosis. Most of the plants that are nodulated by *Rhizobium* are in the Leguminosae (Fabaceae), although there is at least one known exception (Trinick, 1982a). Of the approximately 14,000 species in this family, only around 1200 have been examined for nodulation (Trinick, 1982b). Approximately 80% of the species examined were nodulated. Numerous leguminous species inhabit even very arid environments (Table 4).

The bacteria that are responsible for the nodulation of legumes are in the family Rhizobiaceae. Until recently, all bacteria that nodulated legumes were placed in the genus *Rhizobium*. In *Bergey's Manual* (Jordan and Allen, 1974) the genera *Rhizobium* and *Agrobacterium* are placed together in the Rhizobiaceae. More recently, in *Bergey's Manual* (Jordan, 1984) this placement was continued but divided *Rhizobium* into two genera, *Rhizobium* and *Bradyrhizobium* based on whether the infective agent is fast or slow growing (Greek adjective *bradus*, slow).

It is difficult to ascertain how much N is actually fixed by legumes in arid environments. Often, growing season length as well as precipitation and degree-days determines what legumes are active at a particular site and how much N they fix. Further, soil characteristics determine the quantity of N fixed. Certainly, it is well recognized that high levels of N in soils inhibit nitrogen fixation as well as nodule formation. In addition, some legumes root deeply and only nodulate deeply in soils. For example, Virginia et al. (1986) found that mesquite in the Jornada del Muerto of south central New Mexico nodulated very deeply in the alluvial soils of that environment. Beadle and Tchan (1955) also strongly suggest that conditions of temperature, moisture, or lack of endophyte often prevent nodulation in arid environments.

Fertilization of arid soils with N-containing fertilizers likely reduces the incidence of nodulation and probably the specific activity of the key

TABLE 4 Characteristics of the Family Leguminosae (Fabaceae)

Subfamily	No. of genera[a]	No. of species[a]	Percentage of species nodulated[b]
Caesalpinioideae[c]	150	2200	30
Mimosoideae[d]	40	2000+	90
Papilionodeae[e]	400	10000	90

[a]Conquist (1981) places the legumes in the order Fabales and separates this into the families Mimosaceae, Casalpiniaceae, and Fabaceae. Here, these families are condensed, as is traditionally done, into one family, Fabaceae, and three subfamilies.

[b]From Trinick (1982).

[c]Widespread in tropical and subtropical zones, especially in arid and semiarid environments. Major genera (number of species in parentheses): *Acacia* (700-800), *Mimosa* (450-500), *Albizia* (100-150), *Calliandra* (150), *Inga* (250), *Pithecellobium* (150-200).

[d]Widespread in tropical and subtropical zones; few species in arid environments. Major genera: *Baubinia, Chamaecrista, Senna* (250 each); *Caesalpinia, Swartzia* (125 each); *Cassia* (30).

[e]Widespread in cold, temperate, and tropical regions; many species in arid and semiarid environments. Major genera: *Astragalus* (2000), *Indigofera* (500), *Crotalaria* (500), *Trifolium* (300), *Dalea* (160+), *Phaseolus* (200), *Lupinus* (200), *Dalbergia* (200), *Vicia* (150), *Lathyrus* (150), *Onobrychis* (170), *Hedysarum* (160), *Psorealea* (150), *Lotus* (150), *Dolichos* (120), *Medicago* (110), *Lespedeza* (100).

enzyme nitrogenase (Paul and Clark, 1989; Chap. 10). Almost all the work done on the impact of N fertilizers on the biological nitrogen fixation process has been done on leguminous crop plants, most of which are found in mesic or humid environments (Allos and Bartholomew, 1959; Dart et al., 1977; Dart and Mercer, 1965; Dart and Wildon, 1970; Dazzo and Brill, 1978; Eaglesham et al., 1977, 1983; Mahon and Child, 1979; Munns, 1968).

Ammonium has a repressive impact on the production of nitrogenase by N-fixing organisms (Atkins and Rainbird, 1982). Further, nitrate has a negative impact on both nodulation and N fixation (Atkins and Rainbird, 1982). It seems reasonable to speculate that additions of mineral N to arid soils containing legumes would also impact on N, fixation by these plants. Direct measurements of N fixation repression or inhibition of nodulation by mineral N in arid soils are not present in the literature. However, several studies report that when rangelands were fertilized, the legumes disappeared. Houston and Hyder (1975) report that N fertilization over the course of 4 years eliminated *Astragalus shortianus*, *Astragalus missouriensis*, and *Lathyrus polymorphus* from a mixed-grass plain in southeastern Wyoming. Goetz et al. (1978) report a similar occurrence for *Astragalus caryocarpus* and *Lotus americanus* in western North Dakota.

TABLE 5 Means (± SD) of Surface Soil Properties (to 30 cm Depth) from Root Zones of Nodulated and Nonnodulated *Cercocarpus montanus* and *Purshia tridentata*

Soil component	Nodulated (n = 12)	Nonnodulated (n = 20)	F test[a]
Cercocarpus montanus			
Organic matter, %	0.61 (0.66)	2.94 (1.37)	4.25[b]
pH (1:1)	7.98 (0.41)	8.17 (0.25)	2.83[a]
Electrical conductivity, dS m^{-1}	0.44 (0.42)	0.38 (0.16)	6.00[b]
Extractable P, mg kg^{-1}	0.68 (0.37)	1.66 (0.87)	5.43[b]
Extractable NO_3, mg kg^{-1}	0.59 (0.50)	2.81 (1.73)	11.92[b]
Extractable K, mg kg^{-1}	51.62 (30.92)	111.00 (58.92)	3.63[a]
Purshia tridentata			
Organic matter, %	0.68 (0.59)	3.73 (2.24)	14.74[c]
pH (1:1)	7.23 (0.67)	7.15 (0.63)	1.13 NS[d]
Electrical conductivity dS m^{-1}	0.33 (0.13)	0.33 (0.14)	1.00 NS
Extractable P, mg kg^{-1}	2.29 (1.47)	3.50 (2.09)	2.03 NS
Extractable NO_3, mg kg^{-1}	0.92 (0.76)	3.50 (2.35)	9.53[b]
Extractable K, mg kg^{-1}	81.70 (60.48)	194.50 (86.0)	2.02 NS

[a]Significant at $p \leq 0.05$.
[b]Significant at $p \leq 0.01$
[c]Significant at $p \leq 0.001$.
[d]Nonsignificant.

Nonleguminous N-fixing plant systems, specifically the *Frankia* nodulated plants, have not been examined as to the impact of N-containing fertilizers on the N fixation process, although Rodriguez-Barrucco et al. (1970) demonstrated that, under laboratory conditions, nodulation of *Casuarina* and *Ceanothus* was inhibited by combined N. White and Williams (1985) report that the actinomycete-nodulated plants *Purshia tridentata* and *Cercocarpus montanus* do not nodulate in relatively high nitrate and organic matter soils (Table 5). It is tempting to extrapolate these findings to other actinomycete-nodulated plants; however, *Alnus* spp. very frequently are actinomycete nodulated in very high organic matter, riparian soils. It is, however, quite likely that such environments are quite low in mineral N.

D. Nitrification

Nitrifying processes are less studied than the process of nitrogen fixation. The basics of nitrification can be found in Paul and Clark (1989), Chap. 8. It seems likely that, in arid systems, most N that is mineralized (to NH_4) is further oxidized to nitrate. Skujiņš and Trujillo-y-Fulgham (1978) indicate that nitrification is a prominent process in N turnover in soils of the American Great Basin. These authors caution that climate regime has a substantial impact on nitrification and were able to demonstrate nitrate accumulation during in vitro studies. Wallace et al. (1978) observed the accumulation of nitrate in closed-basin soils of the Mohave Desert. This accumulation of nitrate likely occurred via nitrification of ammonium, followed by water transport of the resulting soluble nitrate salts into the closed basins. Evaporation of water resulted in elevated nitrate levels in surface soils. Undoubtedly, some approximation of this same scenario resulted in the very high accumulation in the Atacama Desert of northern Chile, which is one of the few places on earth where nitrates are of such concentration as to justify mining. In work done by Rauzi (1978), it was shown that N added as ammonium to range soils accumulated as nitrates.

One of the more unusual aspects of nitrification in arid environments is the partial inhibition that is a result of high pH. Martin et al. (1942) showed a buildup of nitrite in soils with an initial pH of 7.7 when the pH was raised. The implication of this work is that the *Nitrobacter* and related organisms that oxidize nitrite to nitrate are inhibited by high pH.

E. Denitrification

The fundamentals of denitrification can be found in a variety of reviews (e.g., Firestone, 1982) and other publications (McKenny et al., 1982; Robertson and Tiedje, 1984, 1987). Deserts are largely unstudied regarding their capacity to denitrify. This is not surprising since denitrification is normally an anaerobic process, often occurring in wet soils or sediments or in anaerobic microsites in mesic soils. Westerman and Tucker (1978) report that the denitrification potential for Sonoran Desert soils is high and also report that field studies in native desert ecosystems indicate high denitrification rates.

F. Phosphorus Cycle Organisms

Unlike N and C, P does not have major atmospheric sources, nor does it undergo oxidation and reduction. Again, major reviews on the biocycling of P can be accessed through the general text of Paul and Clark (1989). Although, as stated earlier, P does not undergo biologically mediated oxidation or reduction, there is some discrepancy on this point (Doetsch and Cook, 1973; Devai et al., 1988).

There really are no specific genera or groups of soil inhabitants noted solely for their involvement in the transport and biocycling of P. Many organisms are capable of producing phosphatases (acid and alkaline phosphomonesterases), the major enzyme(s) responsible for rendering organic phosphate to inorganic phosphate.

Some organisms are noted as capable of solubilizing inorganic phosphate. These are mostly noted as bacteria, such as *Pseudomonas* and *Bacillus* (Piccini and Azcon, 1987).

G. Phosphorus and V-A Mycorrhizae

Vesicular-arbuscular (V-A) mycorrhizal fungi infect most land plants (Gerdemann, 1975). These fungi are generally considered to benefit the plants they infect and therefore are considered mutualistic symbionts. The major benefit these fungi are considered to impart to plants is improved uptake of phosphate (Loree and Williams, 1984), although there is some discrepancy over this point. Further, it has been concluded by many authors that fertilization of soils with phosphate-containing fertilizers deters or inhibits the formation of vesicular-arbuscular mycorrhizae with plants inhabiting the fertilized soil (Holevas, 1966; Jasper et al., 1979; Johnson, 1976; Menge et al., 1978; Mosse, 1973; Pairunan et al., 1980; Pugh et al., 1981). This view of this particular plant-fungus interaction may be somewhat oversimplified, for it seems that most plants that are stressed for nutrients or for water develop V-A mycorrhizal associations, assuming that fungal propagules are present in the soil. Although the results are somewhat variable, plant uptake of other nutrients, such as K, Ca, Mg, S, Fe, Cu, Mn, Zn, Na, and B, have been shown to be mediated by mycorrhizal fungi (reviewed by Cooper, 1984). The most studied mycorrhizally mediated nutrient is P, and it is generally agreed that high soil P means low levels of root infection.

Much of the controversy regarding plant and V-A fungal response to phosphate fertilization is summarized by Menge (1984):

> . . . conclusive recommendations for specific soil P levels for mycorrhizal production (and presumably mycorrhizal infection) cannot be made. There are several reasons for this. First, it is recognized that it is not soil P, per se, that regulates mycorrhizal infection, but rather the amount of P absorbed by the host plant. Thus, it is the amount of available P, but not the form or source of P fertilization which influences mycorrhizal infection. Second, methods for evaluating available soil P often differ greatly, and plant tissue analysis is a far more reliable method for determining available soil P than most methods which analyze soil. Finally, since host plants vary in their ability to absorb P and mycorrhizal fungi vary in their response to P, each plant-soil-VAM symbiont system must be evaluated separately.

Although the specific nature of host-fungus interaction with P fertilizers is garbled, it is clear that under a wide variety of conditions, phosphate fertilization does inhibit the formation of V-A mycorrhizae. Presumably this also occurs in arid environments. A good review of the effect of phosphorus on mycorrhizal infection is that of Cooper (1984). A more general discussion of soil fertility on V-A fungi is that of Daniels-Hetrick (1984), who indicates that soil N may also depress root colonization by V-A fungi.

V. SUMMARY

Inorganic fertilization impacts plants, soil properties, and the microbial communities within the soil. We have attempted to show that all types of response are buffered by inherent soil properties as well as the climate in which the soil-plant system is located. Insertion of cultivated cropping into semiarid and arid environments has negatively impacted soils under the management systems used over the past 100 years. The use of inorganic fertilizers has slowed the degradation process, but it has not stopped it. Soil degradation can be slowed and eventually stopped by radical decreases in degree of cultivation and judicious use of inorganic fertilizers. Research efforts should focus on the development of total management packages that address the degree of cultivation, crop rotation, and the use of inorganic fertilizers.

Generalizing about the impact of inorganic fertilization on arid and semiarid grasslands is more difficult than for cultivated lands. Diversity in soil and species composition coupled with wide fluctuations in precipitation make response highly unpredictable. Generally, arid and semiarid rangelands are not fertilized because of lack of profit motive and potential damage to the natural ecosystem. A proper combination of livestock management system, precipitation, and livestock availability and marketing has proven to be such a complex problem that few hectares of rangeland are fertilized. It is unlikely that this will change in the near future. Irrigated and subirrigated forages, introduced cool-season grasses, and some cool-season grass-dominated rangelands are most likely to respond economically to fertilizer.

Soil microorganism responses to fertilization in arid and semiarid areas are equally complex. Additions of inorganic nitrogenous fertilizers generally inhibit biological N fixation, whether it is symbiotic or asymbiotic. Many organisms that fix N are probably competitive for resources within their respective niches, primarily because of their N_2-fixing abilities. Addition of inorganic N sources generally eliminates organisms or organism systems that acquire competitiveness from their N-fixing habit, although literature that addresses this in extensively managed environments is scanty.

Fertilization with inorganic P sources appears to depress infection of plant roots by V-A fungi. Presumably, under high P fertilization, infection is completely eliminated, although this is far from clear. V-A fungi are obligate symbionts, and propagules in soil decrease with time if infection of plants does not occur. Such a scenario may have considerable application where revegetation of devastated lands is accompanied by massive P fertilization. The hypothesis that high P eliminates or depresses V-A fungi infection has not been adequately tested in field settings.

The most obvious overall conclusion is that future management systems should be careful to address plant nutrient supply problems in arid and semiarid areas from the perspective of climatic conditions, soil properties, and the ultimate impact on the long-term sustainability of agriculture in these areas.

REFERENCES

Ahmadjian, V., and Paracer, S. (1986). *Symbioses: An Introduction to Biological Systems*. University Press of New England, Hanover.

Alexander, M. (1977). *Introduction to Soil Microbiology*, 2nd ed. John Wiley & Sons, New York.

Allison, C. D., Pieper, R. D., Donart, G. B., and Wallace, J. D. (1977). Fertilization influences cattle diets on blue grama range during drought. *J. Range Manag. 30*: 177-180.

Allos, H. F., and Bartholomew, W. V. (1959). Replacement of symbiotic function by available nitrogen. *Soil Sci. 87*: 61-66.

Anderson, F. N., and Peterson, G. A. (1973). Effects of continuous corn (*Zea mays* L.), manuring, and nitrogen fertilization on yield and protein content of the grain and on the soil nitrogen content. *Agron. J. 65*: 697-700.

Atkins, C. A., and Rainbird, R. M. (1982). Physiology and biochemistry in nitrogen fixation in legumes. In *Advances in Agricultural Microbiology*, N. S. Subba Rao (Ed.). Butterworth, London, pp. 26-52.

Baker, R. L., and Powell, J. (1978). Oklahoma tallgrass prairie responses to atrazine with 2,4-D and fertilizer. In *Proceedings of the First International Rangeland Congress*, D. E. Hyden (Ed.). Society for Range Management, Denver, Colorado, pp. 681-683.

Beadle, N. C. W., and Tchan (1955). Nitrogen economy in semi-arid plant communities, I. Environment and general considerations. *Proc. Linn. Soc. N.S.W. 80*: 62-70.

Becking, J. H. (1982). Nitrogen fixation in nodulated plants other than legumes. In *Advances in Agricultural Microbiology*, N. S. Subba Rao (Ed.). Butterworth, London, pp. 90-110.

Black, A. L., and Wight, J. R. (1979). Range fertilization, nitrogen and phosphorus fertilization uptake and recovery over time. *J. Range Manag. 32*: 349-353.

Blevins, R. L., Smith, M. S., and Thomas, G. W. (1984). Changes in soil properties under no-tillage. In *No-Tillage Agriculture*, R. E. Phillips and S. H. Phillips (Eds.). Van Nostrand-Reinhold, New York, pp. 190-230.

Bolin, B., Degens, E. T., Kempe, S., and Ketner, P. (Eds.) (1979). *The Global Carbon Cycle*. Wiley, New York.

Brown, D., Hallman, R. G., Lee, C. R., Skogerbee, J. G., Eskew, K., Price, R. A., Page, N. R., Clar, M., Kort, R., and Hopkins, H. (1986). *Reclamation and Vegetative Restoration of Problem Soils and Disturbed Lands*. Noyes Data Corp., Park Ridge, New Jersey.

Clark, F. E. (1977). Internal cycling of ^{15}nitrogen in shortgrass prairie. *Ecology 58*: 1322-1323.

Clark, F. E., Cole, C. V., and Bowman, R. A. (1980). Nutrient cycling. In *Grasslands*, A. J. Breymeyer and G. M. Van Dyne (Eds.). Cambridge University Press, Cambridge, pp. 659-712.

Conrad, S. G., Jaramillo, H. E., Cromack, K., Jr., and Rosa, S. (1985). The role of *Ceanothus* in western forest ecosystems. USDA-FS Gen. Tech. Report PNW-182, USDA, Washington, D.C.

Conquist, A. (1981). *An Integrated System of Classification of Flowering Plants*. Columbia University Press, New York.

Cooper, K. M. (1984). Physiology of VA mycorrhizal associations. In *VA Mycorrhiza*, C. L. Powell and D. J. Bagyaraj (Eds.). CRC Press, Boca Raton, Florida, pp. 155-186.

Daniels-Hetrick, B. A. (1984). Ecology of VA mycorrhizal fungi. In *VA Mycorrhiza*, C. L. Powell and D. J. Bagyaraj (Eds.). CRS Press, Boca Raton, Florida, Chap. 3, pp. 36-55.

Dart, P. J., and Mercer, F. V. (1965). The effect of growth temperature, level of ammonium nitrate, and light intensity on the growth and nodulation of cowpea (*Vigna sinensis* Endl. Ex Hassk.). *Aust. J. Agr. Res. 16*: 321-345.

Dart, P. J., and Wildon, D. C. (1970). Nodulation and nitrogen fixation by *Vigna sinensis* and *Vicia atropurpurea*: The influence of concentration, form and site of application of combined nitrogen. *Aust. J. Agr. Res. 16*: 321-345.

Dart, P. J., Huxley, P. A., Eaglesham, A. R. J., Minchin, F. R., Summerfield, R. J., and Day, J. M. (1977). Nitrogen nutrition of cowpea (*Vigna unguiculata*). II. Effects of short-term applications of inorganic nitrogen on growth and yield of nodulated and nonnodulated plants. *Exp. Agr. 13*: 241-252.

Dazzo, F. B., and Brill, W. J. (1978). Regulation by fixed nitrogen of host-symbiont recognition in the *Rhizobium*-clover symbiosis. *Plant Physiol. 62*: 18-21.

Devai, I., Felfoldy, L., Wittner, I., and Plosz, S. (1988). Detection of phosphine: New aspects of the phosphorus cycle in the hydrosphere. *Nature 333*: 343-345.

Dobereiner, J., and Day, J. M. (1976). Associative symbiosis and free-living systems. In *First International Symposium on Nitrogen Fixation*, Vol. 2, W. E. Newton and C. J. Nyman (Eds.). Washington State University Press, Pullman, pp. 518-538.

Doetsch, R. N., and Cook, T. M. (1973). *Introduction to Bacteria and Their Ecobiology*. University Park Press, Baltimore, pp. 310-311.

Donahue, R. L., Miller, R. W., and Schickluna, J. C. (1983). *Soils—An Introduction to Soils and Plant Growth*. Prentice Hall, Englewood Cliffs, New Jersey.

Donart, G. B., Parker, E. E., Pieper, R. D., and Wallace, J. D. (1978). Nitrogen fertilization and livestock grazing on blue grama rangeland. In *Proceedings of the First International Rangeland Congress*, D. E. Hyden (Ed.). Society for Range Management, Denver, Colorado, pp. 614-615.

Eaglesham, A. R. J., Minchin, F. R., Summerfield, R. J., Dart, P. J., Huxley, P. A., and Day, J. M. (1977). Nitrogen nutrition of cowpea (*Vigna unguiculata*). III. Distribution of nitrogen within effectively nodulated plants. *Exp. Agr. 13*: 369-380.

Eaglesham, A. R. J., Hassouna, S., and Seegers, R. (1983). Fertilizer-N effects on N_2 fixation by cowpea and soybean. *Agron. J. 75*: 61-66.

Ellis, B. A., and Kummerow, J. (1989). The importance of N_2-fixation in *Ceanothus* seedlings in early postfire chaparral. In *The California Chaparral, Paradigms Reexamined*, S. C. Keeley (Ed.). Natural History Museum of Los Angeles County, California, pp. 115-116.

Firestone, M. K. (1982). Biological denitrification. In *Nitrogen in Agricultural Soils*, F. J. Stevenson (Ed.). Am. Soc. Agron., Madison, Wisconsin, pp. 289-326.

Focht, D. D., and Verstraete, W. (1977). Biochemical ecology of nitrification and denitrification. *Adv. Microb. Ecol. 1*: 135-199.

Follett, R. F., Power, J. F., Grunes, D. L., and Klein, C. A. (1977). Effect of N, K, and P fertilization, N source, and clipping on potential tetany hazard of bromegrass. *Plant Soil 48*: 485-508.

Forman, R. T. T. (1975). Canopy lichens with blue-green algae: A nitrogen source in a Colombian rain forest. *Ecology 56*: 1176-1184.

Forman, R. T. T., and Dowden, D. L. (1977). Nitrogen fixing lichen roles from desert to alpine in the Sangre de Cristo Mountains, New Mexico. *Bryologist 80*: 561-570.

Fulgham, K. O., Donart, G. B., Pieper, R. D., Lentner, M., and Dwyer, D. D. (1977). Effects of types and rates of nitrogen fertilizer on range herbage production under drought conditions. New Mexico Agr. Exp. Sta. Res. Rep. 343, Las Cruces.

Gerdemann, J. W. (1975). Vesicular-arbuscular mycorrhizae. In *The Development and Function of Roots*, J. G. Torrey and D. T. Clarkson (Ed.). Academic Press, Longon, pp. 575-591.

Goetz, H., Nyren, P. E., and Williams, D. E. (1978). Implications of fertilizers in plant community dynamics of northern Great Plains rangelands. In *Proceedings of the First International Rangeland Congress*, D. N. Hyder (Ed.). Society for Range Management, Denver, Colorado, pp. 671-674.

Gogulwar, N. M. (1973). Effect of residual and applied N on the protein content and yield of winter wheat under three moisture regimes. Ph.D. dissertation, University of Nebraska, Lincoln.

Havelka, U. D., Boyle, M. G., and Hardy, R. W. F. (1982). Biological nitrogen fixation. *Agronomy 22*: 365-413.

Heisey, R. M., Delwiche, C. C., Virginia, R. A., Wrona, A. F., and Bryan, B. A. (1980). A new nitrogen-fixing nonlegume: *Chamaebatia foliolosa* (Rosaceae). *Am. J. Bot. 67*: 429-431.

Hobbs, J. A., and Thompson, C. A. (1971). Effect of cultivation on the nitrogen and carbon contents of a Kansas Argiustoll. *Agron. J. 63*: 66-68.

Holevas, C. D. (1966). The effect of a vesicular-arbuscular mycorrhiza on the uptake of soil phosphorus by strawberry (*Fragaria* sp., var. Cambridge Favourite). *J. Hort. Sci. 41*: 57.

Hooker, M. L., Herron, G. M., and Penas, P. (1982). Effects of burning, removal, and incorporation on irrigated cereal crop yields and soil chemical analysis. *Soil Sci. Soc. Am. J. 46*: 122-126.

Houston, W. R., and Hyder, D. N. (1975). Ecological effects and fate of N following massive N fertilization of mixed grass plains. *J. Range Manag. 28*: 56-60.

Hubbell, D. H., and Gaskins, M. H. (1984). Associative N_2 fixation with *Azospirillum*. In *Biological Nitrogen Fixation: Ecology, Technology and Physiology*, M. Alexander (Ed.). Plenum Press, New York, pp. 201-224.

Jasper, D. A., Robson, A. D., and Abbott, L. K. (1979). Phosphorus and the formation of vesicular-arbuscular mycorrhizas. *Soil Biol. Biochem. 11*: 501.

Jenny, H. (1933). Soil fertility losses under Missouri conditions. Missouri Agr. Exp. Sta. Bull. 324, Columbia.

Jenny, H. (1941). *Factors of Soil Formation.* McGraw-Hill, New York.
Johnson, J. R., and Nichols, J. T. (1969). Crude protein content of eleven grasses as affected by yearly variation, legume association, and fertilization. *Agron. J. 61*: 65-68.
Johnson, P. N. (1976). Effects of soil phosphate level and shade on plant growth and mycorrhizas. *N.Z. J. Bot. 14*: 333.
Jordan, D. C. (1984). Rhizobiacae. In *Bergey's Manual of Systematic Bacteriology*, Vol. I, N. R. Krieg and J. G. Holt (Eds.). Williams & Wilkins, Baltimore, pp. 345-356.
Jordan, D. C., and Allen, O. N. (1974). Rhizobiaceae. In *Bergey's Manual of Determinative Bacteriology*, R. E. Buchanan and N. E. Gibbons (Eds.). Williams & Wilkins, Baltimore, pp. 261-367.
Kummerow, J., Alexander, J. V., Neel, J. W., and Fishbeck, K. (1978). Symbiotic nitrogen fixation in *Ceanothus* roots. *Am. J. Bot. 65*: 63-69.
Lamb, J. A., Peterson, G. A., and Fenster, C. R. (1985). Wheat fallow tillage systems' effect on a newly cultivated grassland soils' nitrogen budget. *Soil Sci. Soc. Am. J. 49*: 352-356.
Lauenroth, W. K., Dodd, J. L., and Sims, P. L. (1978). The effects of water and nitrogen-induced stresses on plant community structure in semiarid grassland. *Oecologia (Berl.) 36*: 211-222.
Launchbaugh, J. L., and Owensby, C. E. (1978). Kansas rangelands: Their management based on a half century of research. Kansas Agr. Exp. Sta. Bull. 622.
Loree, M. A. J., and Williams, S. E. (1984). Vesicular-arbuscular mycorrhizae and severe land disturbance. In *VA Mycorrhizae and Reclamation of Arid and Semiarid Lands*, S. E. Williams and M. F. Allen (Eds.). Wyoming Agri. Exp. Sta. Rep. No. SA1261, Laramie, pp. 1-14.
Lorenz, R. J., and Rogler, G. A. (1972). Forage production and botanical composition of mixed prairie as influenced by nitrogen and phosphorus fertilization. *Agron. J. 64*: 244-249.
Lueking, M. A., and Schepers, J. S. (1985). Changes in soil carbon and nitrogen due to irrigation development in Nebraska's sandhill soils. *Soil Sci. Soc. Am. J. 49*: 626-630.
Mahon, J. D., and Child, J. J. (1979). Growth response of inoculated peas (*Pisum sativum*) to combined nitrogen. *Can. J. Bot. 57*: 1687-1693.
Martin, W. P., Buchrer, T. F., and Caster, A. B. (1942). Threshold pH value for the nitrification of ammonia in desert soils. *Soil Sci. Soc. Am. Proc. 7*: 223-228.
Mayland, H. F., Grunes, D. L., Waggoner, H., Florence, A., Hayes, D. A., and Joo, P. K. (1975). Nitrogen effects on crested wheatgrass as related to forage quality indices of grass tetany. *Agron. J. 67*: 411-414.
McGinnies, W. J., and Crofts, K. A. (1986). Effects of N and P fertilizer placement on establishment of seeded species on redistributed mine topsoil. *J. Range Manag. 39*: 118-121.
McIlvain, E. H., and Shoup, M. C. (1970). Fertilizing weeping lovegrass in eastern Oklahoma. In *Proceedings of the First Weeping Lovegrass Symposium*, Agr. Div., Roberts Nobel Foundation, Inc., Ardmore, Oklahoma.

McKenny, D. J., Shuttleworth, D. F., Vriesacker, J. R., and Findley, W. I. (1982). Production and loss of nitric oxide from denitrification in anaerobic Brookston clay. *Appl. Environ. Microbiol. 43*: 534-541.
Melsted, S. W. (1954). New concepts of management of Corn Belt soils. *Adv. Agron. 6*: 121-142.
Menge, J. A. (1984). Inoculum production. In *VA Mycorrhiza*, C. L. Powell and D. J. Bagyaraj (Eds.). CRC Press, Boca Raton, Florida, pp. 187-203.
Menge, J. A., Labanauskas, C. K., Johnson, E. L. V., and Platt, R. G. (1978). Partial substitution of mycorrhizal fungi for phosphorus fertilization in the greenhouse culture of citrus. *Soil Sci. Soc. Am. Proc. 42*: 926.
Meyer, D. W. (1989). Effects of nitrogen fertilization of bromegrass at six levels for 35 years. *Agron. Abstr.* 246.
Mosse, B. (1973). Plant growth responses to vesicular-arbuscular mycorrhiza. IV. In soil given additional phosphate. *New Phytol. 72*: 127.
Munn, L. C., Nielsen, S. A., and Mueggler, W. F. (1978). Relationships of soils to mountain and foothill range habitat types and production in western Montana. *Soil Sci. Soc. Am. J. 42*: 135-139.
Munns, D. N. (1968). Nodulation of *Medicago sativa* in solution culture. III. Effects of nitrate on root hairs and infection. *Plant Soil 28*: 33-47.
Murphy, A. H. (1986). Strategic use of legumes to improve annual grassland. In *Rangelands: A Resource Under Siege*, P. J. Joss, P. W. Lynch, and O. B. Williams (Eds.). *Proceedings of the Second International Rangeland Congress*, Adelaide, South Australia. Cambridge University Press, Cambridge, p. 302.
Newton, M., El-Hassan, B. A., and Zavitkovski (1968). Role of red alder in western Oregon forest succession. In *Biology of Alder*, J. M. Trappe, J. F. Franklin, R. F. Tarrant, and G. H. Hansen (Eds.). Pacific Northwest Forest Range Exp. Sta., Portland, Oregon, pp. 73-84.
Olson, R. A., Muir, J. H., Wesely, R. W., Peterson, G. A., and Boyce, J. S. (1974). Accumulation of inorganic nitrogen in soils and waters in relation to soil and crop management. In *Effects of Agricultural Production on Nitrates in Food and Water with Particular Reference to Isotope Studies*. IAEA-PL-539/3 Atomic Energy Agency, Vienna, pp. 19-38.
Owensby, C. E., Anderson, K. L., and Whitney, D. A. (1969). Some chemical properties of a silt loam soil after 20 years nitrogen and phosphorus fertilization of smooth bromegrass (*Bromus inermis* Leyss.). *Soil Sci. 108*: 24-29.
Pairunan, A. K., Robson, A. D., and Abbott, L. K. (1980). The effectiveness of vesicular-arbuscular mycorrhizas in increasing growth and phosphorus uptake of subterranean clover from phosphorus sources of different solubilities. *New Phytol. 84*: 327.
Paul, E. A., and Clark, F. E. (1989). *Soil Microbiology and Biochemistry*. Academic Press, New York.
Peterson, G. A., and Vetter, D. J. (1971). Soil nitrogen budget of the wheat-fallow area. *Farm Ranch Home Q. 12*(4): 23-25.

Peterson, G. A., Anderson, F. N., and Olson, R. A. (1966). A survey of the nutrient status of soils in the North Platte Valley of Nebraska for sugar beet production. *J. Am. Soc. Sugar Beet Technol. 14*: 48-60.

Piccini, D., and Azcon, R. (1987). Effect of phosphate-solubilizing bacteria and vesicular-arbuscular mycorrhizal fungi on the utilization of Bayovar rock phosphate by alfalfa plants using a sand-vermiculite medium. *Plant and Soil 101*: 45-50.

Pike, L. H., Tracy, D. M., Sherwood, M. A., and Nielsen, D. (1972). Estimates of biomass and fixed nitrogen of epiphytes from old-growth Douglas-fir. In *Proceedings of the Research on Coniferous Forest Ecosystems Symposium*, J. F. Franklin, L. J. Dempster, and R. H. Waring (Eds.). Pacific Northwest Forest and Range Exp. Sta., Portland, Oregon, pp. 117-187.

Power, J. F. (1977). Nitrogen transformation in the grassland ecosystem. In *Proceedings of the Symposium on Below-ground Ecosystems*, J. K. Marshall (Ed.). Colorado State University Range Sci. Serv. No. 26, Fort Collins, pp. 195-204.

Power, J. F. (1980a). Response of semi-arid grassland sites to nitrogen fertilization. I. Plant growth and water use. *Soil Sci. Soc. Am. J. 44*: 545-550.

Power, J. F. (1980b). Response of semi-arid grassland sites to nitrogen fertilization. II. Fertilizer recovery. *Soil Sci. Soc. Am. J. 44*: 550-555.

Power, J. F. (1981). Long-term recovery of fertilizer nitrogen applied to a native mixed prairie. *Soil Sci. Soc. Am. J. 45*: 782-786.

Power, J. F. (1983). Recovery of nitrogen and phosphorus after 17 years from various fertilizer materials applied to crested wheatgrass. *Agron. J. 75*: 249-254.

Power, J. F. (1985). Nitrogen- and water-use efficiency of several cool-season grasses receiving ammonium nitrate for 9 years. *Agron. J. 77*: 189-192.

Power, J. F. (1988). Seasonal changes in smooth bromegrass top and root growth and fate of fertilizer nitrogen. *Agron. J. 80*: 740-745.

Power, J. F., and Alessi, J. (1970). Effects of nitrogen source and phosphorus on crested wheatgrass growth and water use. *J. Range Manag. 23*: 175-178.

Power, J. F., and Alessi, J. (1971). Nitrogen fertilization of semi-arid grasslands: Plant growth and soil mineral N levels. *Agron. J. 63*: 277-280.

Power, J. F., and Legg, J. O. (1984). Nitrogen 15 recovery for five years after application of ammonium nitrate to crested wheatgrass. *Soil Sci. Soc. Am. J. 48*: 322-326.

Power, J. F., Ries, R. E., and Sandoval, F. M. (1978). Reclamation of coal-mined land in the Northern Great Plains. *J. Soil Water Conver. 33*: 69-74.

Pugh, L. M., Roncadori, R. W., and Hussey, R. S. (1981). Factors affecting vesicular-arbuscular mycorrhizal development and growth of cotton. *Mycologia 73*: 869.

Rauzi, F. (1978). High rates of nitrogen change composition of short grass rangeland in southeastern Wyoming. *J. Range Manag. 31*: 366-370.

Rauzi, F. (1979). Residual effects of phosphorus and high rates of nitrogen on shortgrass rangeland. *J. Range Manag. 32*: 470-474.

Reeder, J. D., and Sabey, B. (1987). Nitrogen. In *Reclaiming Mine Soils and Overburden in Western United States. Analytic Parameters and Procedures*, R. D. Williams and G. E. Schuman (Eds.). Soil Cons. Soc. Am., Ankeny, Iowa, pp. 155-184.

Rehm, G. W. (1984). Yield and quality of a warm-season grass mixture treated with N, P, and atrazine. *Agron. J. 76*: 731-734.

Rehm, G. W., Moline, W. J., and Schwartz, E. J. (1972). Response of a seeded mixture of warm-season prairie grasses to fertilization. *J. Range Manag. 25*: 452-456.

Rehm, G. W., Sorensen, R. C., and Moline, W. J. (1976). Time and rate of fertilizer application for seeded warm-season and bluegrass pastures. I. Yield and botanical composition. *Agron. J. 68*: 759-764.

Robertson, G. P., and Tiedje, J. M. (1984). Denitrification and nitrous oxide production in successional and old growth Michigan forests. *Soil Sci. Soc. Am. J. 48*: 383-389.

Robertson, G. P., and Tiedje, J. M. (1987). Nitrous oxide sources in aerobic soils: Nitrification, denitrification, and other biological processes. *Soil Biol. Biochem. 19:* 187-193.

Rodriguez-Barrueco, C., Mackintosh, A. H., and Bond, G. (1970). Some effects of combined nitrogen on the nodule symbiosis of *Casuarina* and *Ceanothus*. *Plant Soil 33*: 129-133.

Rogler, G. A., and Lorenz, R. J. (1965). Nitrogen fertilization of natural grasslands in the northern plains of the United States. In *Proceedings of the Ninth International Grassland Congress*, Z. Martins (Chm. Public. Comm.) Edicoes Alarico Limitada. Sao Paulo, Brazil, pp. 1327-1330.

Rogler, G. A., and Lorenz, R. J. (1969). Pasture productivity of crested wheatgrass as influenced by nitrogen fertilization and alfalfa. USDA Tech. Bull. No. 1402, USDA, Washington, D.C.

Rogler, G. A., and Lorenz, R. J. (1970). Beef production from Russian wildrye (*Elymus junceus* Fisch.) in the United States. In *Proceedings of the Eleventh International Grassland Congress*, M. J. T. Norman (Ed.), University of Queensland Press, St. Lucia, Queensland, Australia, pp. 835-838.

Rogler, G. A., and Lorenz, R. J. (1974). Fertilization of mid-continent range plants. In *Forage Fertilization*, D. A. Mays (Ed.). Am. Soc. Agron., Madison, Wisconsin, pp. 231-254.

Rychert, R. C., and Skujiņš, J. (1974). Nitrogen fixation by blue-green algae-lichen crusts in the Great Basin desert. *Soil Sci. Soc. Am. Proc. 38*: 766-771.

Samuel, M. J. (1986). Botanical composition changes from six years of low rate nitrogen fertilization. In *Rangelands: A Resource Under Siege*, P. J. Joss, P. W. Lynch, and O. B. Williams (Eds.). Proceedings of the Second International Rangeland Congress, Adelaide, South Australia. Cambridge University Press, Cambridge, p. 65.

Samuel, M. J., Rauzi, F., and Hart, R. H. (1980). Nitrogen fertilization of range: yield, protein content, and cattle behavior. *J. Range Manag. 33*: 119-121.

Sander, D. H., and Peterson, G. A. (1968). Western Nebraska wheat protein content dropping. *Farm Ranch Home Quart. 15*: 10-11.

Sander, D. H., and Peterson, G. A. (1971). Some wheat could have yielded 15 bu/A more. *Nebr. Farmer 113*: 20.

Skujiņš, J. (1981). Nitrogen cycling in arid ecosystems. *Ecol. Bull. (Stockh.) 33*: 477-491.

Skujiņš, J., and Trujillo-Fulgham, P. (1978). Nitrification in Great Basin desert soils. In *Nitrogen in Desert Ecosystems,* N. E. West and J. Skujiņš (Eds.). Dowden, Hutchinson and Ross, Stroudsburg, Pennsylvania, pp. 60-74.

Smith, S. J., and Power, J. F. (1985). Residual forms of fertilizer nitrogen on a grassland soil. *Soil Sci. 140*: 362-367.

Sneva, F. A. (1977). Soil nitrogen levels in a semi-arid climate following long-term nitrogen fertilization. *J. Range Manag. 30*: 328-330.

Soil Survey Staff (1975). Soil Taxonomy. USDA Soil Conservation Service Agriculture Handbook No. 436. U.S. Government Printing Office, Washington, D.C.

Soil Survey Staff (1990). *Keys to soil taxonomy,* 4th ed., Soil Services Technical Monograph No. 6. Blacksburg, Virginia.

Stevenson, F. J. (1982). *Humus Chemistry: Genesis, Composition, Reactions.* Wiley, New York.

Stoddard, L. A., Smith, A. D., and Box, T. W. (1975). *Range Management.* McGraw-Hill, New York.

Stubbendieck, J. (1977). Grazing systems and range fertilization. Nebraska Beef Cattle Report Ext. Circ. 77-218, pp. 18-20.

Taliaferro, C. M., Horn, F. P., Tucker, B. B., Totusek, R., and Morrison, R. D. (1975). Performance of three warm-season perennial grasses and native range mixture as influenced by N and P fertilization. *Agron. J. 67*: 289-292.

Thomas, G. W. (1989). The soil bank account and the farmer's bank account. *J. Prod. Agr. 2*: 122-124.

Tiessen, H., Stewart, J. W. B., and Bettany, J. R. (1982). Cultivation effects on the amounts and concentrations of carbon, nitrogen and phosphorus in grassland soil. *Agron. J. 74*: 831-835.

Trinick, M. J. (1982a). Host-rhizobium associations. In *Nitrogen Fixation in Legumes,* J. M. Vincent (Ed.). Academic Press, New York, pp. 111-122.

Trinick, M. J. (1982b). Competition between rhizobial strains for nodulation. In *Nitrogen Fixation in Legumes,* J. M. Vincent (Ed.). Academic Press, New York, pp. 229-238.

UNESCO (1979). Map of the world distribution of arid regions. Map Tech. Note 7. United Nations Educational, Scientific and Cultural Organization, New York.

Virginia, R. A., Jenkins, M. B., and Jarrell, W. M. (1986). Depth of root symbiont occurrence in soil. *Biol. Fertil. Soil 2*: 127-130.

Vitousek, P. M., Gosz, J. R., Grier, C. C., Melillo, J. M., and Reiners, W. A. (1982). A comparative analysis of potential nitrification and nitrate nobility in forest ecosystems. *Ecol. Monogr. 52*(2): 155-177.

Wallace, A., Romney, E. M., and Hunter, R. B. (1978). Nitrogen cycle in the northern Mohave desert: Implications and predictions. In *Nitrogen in Desert Ecosystems,* N. E. West and J. Skujiņš (Eds.). Dowden, Hutchinson and Ross, Stroudsburg, Pennsylvania, pp. 207-218.

Warnes, D. D., Newell, L. C., and Moline, W. J. (1971). Performance evaluation of some warm-season prairie grasses in Nebraska environments. Nebraska Agri. Exp. Sta. Bull. 241.

Westerman, R. L., and Tucker, T. C. (1978). Denitrification in desert soils. In *Nitrogen in Desert Ecosystems,* N. E. West and J. Skujiņs (Eds.). Dowden, Hutchinson and Ross, Stroudsburg, Pennsylvania, pp. 75-106.

White, J. A., and Williams, S. E. (1985). Symbiotic N_2-fixation by *Frankia:* Acyinorhizae establishment on disturbed lands. In *Proceedings, American Society of Surface Mining and Reclamation, National Conference,* Denver, Colorado, pp. 289-297.

White, J. A., Munn, L. C., and Williams, S. E. (1989). Edaphic and reclamation aspects of vesicular-arbuscular mycorrhizae in Wyoming Red Desert soils. *Soil Sci. Soc. Am. J. 53*: 86-90.

Wight, J. R., and Black, A. L. (1979). Range fertilization: Plant responses and water use. *J. Range Manag. 32*: 345-349.

Williams, R. E., Allred, B. W., Denio, R. M., and Paulsen, H. A., Jr. (1968). Conservation, development and use of the world's rangelands. *J. Range Manag. 21*: 355-360.

Williams, R. J., Broersma, K., and Van Ryswyk, A. L. (1979). The effects of nitrogen fertilization of water use by crested wheatgrass. *J. Range Manag. 32*: 98-100.

Wolters, G. L. (1986). Fertilization increases carrying capacity on green and dry season annual range. In *Rangelands: A Resource Under Siege,* P. J. Joss, P. W. Lynch, and O. B. Williams (Eds.). *Proceedings of the Second International Rangeland Congress,* Adelaide, South Australia. Cambridge University Press, Cambridge, pp. 312-313.

Woodmansee, R. G. (1978). Additions and losses of nitrogen in grassland ecosystems. *BioScience 28*: 448-453.

Youngberg, C. T., and Wollum, A. G. (1976). Nitrogen accretion in developing *Ceanothus velutinus* stands. *Soil Sci. Soc. Am. J. 40*: 109-112.

Ziekus, J. G. (1982). Lignin metabolism and the carbon cycle. *Adv. Microb. Ecol. 5*: 211-243.

Zinke, P. J., Stanenberger, A. G., Post, W. M., Emanuel, W. R., and Olson, J. S. (1984). Worldwide organic soil carbon and nitrogen data. Publ. 2217, Environ. Sci. Div., Oak Ridge Natl. Lab., Oak Ridge, Tennessee.

19

Weather-Predictive Models

GEORGE H. HARGREAVES *Utah State University, Logan, Utah*
ZOHRAB A. SAMANI *New Mexico State University, Las Cruces, New Mexico*

I. INTRODUCTION

Reclamation and use of semiarid lands and deserts can be significantly improved through a better understanding of the classifications of climate and use of weather predictive models. Models are useful in predicting future trends in climatic changes, the probable range in values of the various weather elements, relationships of one weather element to others, and the influence of weather and other environmental factors on crop yields, forage production, and the potential from agricultural technology transfer.

Models are more useful if they can be simple and easily understood and require a minimum of data. Weather variables required should, whenever possible, be those that can be measured with a minimum of observer and instrument error. The values of weather variables are influenced by aspect, slope, elevation, topography, and other site characteristics, such as whether the site is in an irrigated or nonirrigated area. Reporting of the weather should always include a site description.

Mean climatic values may have little meaning unless accompanied by a description of weather trends, local exposure factors, and the standard deviations of values or some indication of the range of probable values. If a probability distribution is used for extreme values, the distribution should be specified. If accurate measurements can be made available from well-standardized sites, a few measurements are frequently adequate for defining the climate for purposes of agricultural planning.

This chapter is an attempt to present useful equations, relationships, and models with a minimum of theoretical discussion. Some of the relationships presented have been developed by the authors; others, although potentially very useful, have remained relatively unused in literature that has been largely forgotten. The authors anticipate that this summary will provide incentive for further research, testing, and evaluation and for field trials designed to improve the utility of the weather and crop models described or mentioned here.

This chapter includes procedures for estimating dependable rainfall, reference crop evapotranspiration, a moisture availability index, solar radiation, and relative humidity. A classification of climate is proposed in an attempt to facilitate the use of terminology more consistent with agrotechnology transfer. The use of monthly mean values for generating synthetic daily weather values and of crop yield models as tools for agrotechnology transfer and for planning and management in agriculture is briefly described. Sections are included on predicting extreme rainfall and on the need for predicting future weather changes to facilitate long-range planning for future agricultural production.

II. DEPENDABLE RAINFALL

The climatic potential for dryland agricultural production is determined largely by the amount and distribution of rainfall and the vegetative water use requirements. The 75% probability of monthly assured rainfall has been used as dependable rainfall (PD). Values of PD are much superior to mean monthly rainfall (PM) as an index of the value of precipitation for agricultural production. The 75% probability for 5 or 10 day periods is significantly superior to monthly values. For the 75% probability it makes little difference which probability distribution is used. Experience indicates very similar values with ranking, log normal, and gamma distributions.

The 50% probability of assured rainfall (P_{50}) is usually significantly less than the mean monthly precipitation amounts. Linear regressions of P_{50} and PD with PM produce very satisfactory coefficients of determination (R^2). The equations can be written

$$P_{50} = a + b(PM) \tag{1}$$

and

$$PD = a + b(PM) \tag{2}$$

Calibration coefficients for Eq. (1) for several countries are given in Table 1 and those for Eq. (2) in Table 2. In Table 2 the standard deviation of values (S) is also given. Monthly rainfall amounts considered insufficient for economical or subsistence agricultural production were not included in the regression analyses. For Table 1 values of mean monthly precipitation of less than 50 mm were not used in the regression analyses. For Table 2 monthly rainfall amounts of less than one-third of estimated potential evapotranspiration were excluded. The coefficients for a in Tables 1 and 2 are for rainfall in mm month^{-1}.

Some methods for determining the length of growing season for dryland agriculture use monthly values of PM for comparisons with reference crop evapotranspiration (ETP). An inspection of Table 1 indicates that for a mean rainfall of 100 mm the 50% probability (P_{50}) may be as high as 90 mm or as low as 78 mm. Dependence upon mean rainfall implies accepting some degree of deficiency significantly more than 50% of the time. Use of a selected probability value of somewhat more than 50% is strongly recommended.

For agricultural development planning and for management, considerable improvement can be made provided rainfall records are kept by decades or

TABLE 1 Calibration Coefficients for Eq. (1)[a]

Country	a (mm)	b	Degrees of freedom (DOF)	Coefficient of determination (R^2, %)
Algeria	-5.0	0.92	34	96.2
Brazil	-8.9	0.96	460	99.0
Cameroon	-15.0	1.04	22	99.5
Central African Republic	-10.5	1.02	70	99.7
Chad	-11.1	1.01	56	99.4
Congo	-10.1	1.00	88	98.7
Ethiopia	-15.9	1.03	46	99.8
Guinea	-8.1	0.96	170	99.6
India	-20.0	0.98	277	99.6
Indonesia	-18.6	0.97	149	98.0
Mali and Niger	-10.0	1.00	56	99.5
Mauritania	-14.2	0.99	16	91.0
Nigeria	-9.6	0.98	63	99.3
Pakistan	-25.8	1.05	35	97.7
Senegal and Gambia	-16.0	0.99	52	99.4
Sudan	-12.5	1.02	52	99.2
South Africa	-13.5	1.03	46	99.8
Taiwan	-12.2	0.94	56	98.3
Zaire	-11.4	1.02	93	97.8

[a]Months with mean rainfall less than 50 mm were excluded from the regression analyses.

Source of data: From Hargreaves and Samani (1986).

by three periods for each month. For many soils and crops the potential moisture storage in the crop root zone can supply all or most of the crop water requirements for a 10 day period. Monthly probabilities of assured rainfall may appear to be fully adequate for crop growth, but the distribution within the month may be very undesirable. At the beginning of the rainy season, the first half of a month may have no rain and the last portion twice as much or more than required.

Table 2 indicates that for 77% of the countries evaluated the equation developed predicts 90% or more of the variance of the estimated values of PD or P_{75} for monthly values. When the values of R^2 are less than 90% it seems apparent that unlike rainfall distributions were evaluated together.

TABLE 2 Regression Coefficients for Eq. (2)[a]

Country	a (mm)	b	R^2	S	DOF
Bangladesh	-31	0.76	94	25	30
Brazil	-31	0.86	93	23	63
Burkina Faso	-15	0.83	99	8	7
Chad	-20	0.83	92	17	31
Congo Republic	-7	0.76	63	26	50
Dahomey	-40	0.93	92	14	25
Dominican Republic	-2	0.55	80	17	13
Ecuador	11	0.58	80	21	11
Ghana	-5	0.70	79	21	18
India	-19	0.71	98	47	109
Ivory Coast	-10	0.67	93	26	18
Mali	-18	0.83	96	24	35
Mauritius	-9	0.65	82	22	32
Niger	-10	0.77	95	10	8
Nigeria	-35	0.91	96	12	29
Philippines	-15	0.68	90	27	83
Senegal	-30	0.85	98	14	14
Sierra Leone	-6	0.79	96	36	15
Sri Lanka	-33	0.73	92	20	19
Sudan	-25	0.88	94	8	28
Taiwan	-10	0.58	90	25	32
Tanzania	-20	0.78	95	11	10
Thailand	-25	0.78	85	20	27
Togo	-25	0.83	93	11	26
Uganda	-22	0.84	97	10	14
Zambia	-38	0.90	91	17	17

[a]R^2, coefficient of determination; S, standard deviation; DOF, degrees of freedom.

Source of data: From Hargreaves and Samani (1986).

Samani and Hargreaves (1986) developed a procedure for estimating the full range of rainfall probabilities from average values for those regions or countries for which the relationship between PM and P_{50} has been defined as in Table 1. The values of PM and P_{50} are used to back-calculate the gamma probability distribution shape parameters. Once the value of the shape parameter is estimated, then rainfall at any other probability level can easily be estimated.

Adequacy of Length of Record

Rainfall is a variable parameter with respect to time and location. Any type of analysis based on the past record would not represent the climate accurately unless the length of record is adequate. Mockus (1960) presented an equation for evaluating the adequacy of the length of record for a given level of significance:

$$N = (4.3t \log R)^2 + 6 \tag{3}$$

where:

N = minimum acceptable years of record

t = Student's statistical parameter at 90% level of confidence with (N - 6) degrees of freedom

R = ratio of magnitude of the 100 year event to the 2 year event

The Mockus equation is a nonlinear equation with unknown N on both sides of the equal sign. The calculation of the minimum acceptable length of record is possible through an iterative search through the equation as described by Samani and Hargreaves (1988).

III. SOLAR RADIATION

The principal source of energy for crop growth and development is solar radiation at the earth's surface (RS). In many countries few if any measurements are made of RS, and where they are made the values reported may be questionable because of instrument drift and infrequent calibration. Hargreaves (1981b) proposed a procedure for estimating RS from the daily or monthly average temperature range of difference between maximum and minimum temperature (TD):

$$RS = KRA(TD)^{0.50} \tag{4}$$

where RS and RA are in the same units, RA is extraterrestrial radiation, and K is an empirical coefficient. K averages about 0.16 for temperature in degrees Celsius. Samani and Pessarakli (1986) made regression analyses to determine the value of K for 56 locations in the United States. The average value of K was 0.154, with a standard deviation of 0.019. With the exception of four locations, the coefficients of determination R^2 were in the range 90-100%. Equation (3) produces very good results during the agricultural growing season. Values of TD are less reliable as a climatic index in the presence of snow and/or ice. Average daily values of RA in equivalent millimeters of water evaporation by months and latitudes are available in numerous manuals and published papers and are therefore not repeated in this chapter. Samani and Pessarakli (1986) provided simplified equations for calculating RA.

The average value of extraterrestrial radiation (RA) for a given month can be estimated as

$$RA = 916.732 \left[\frac{CM \sin(Lat) \sin(Dec) + \cos(Lat) \cos(Dec) \sin(OM)}{ES \times 10/(595.9 - 0.55t)}\right] \quad (5)$$

where:

RA = extraterrestrial radiation in mm equivalent water evaporation day^{-1}
OM = arc cos[- tan (Lat) tan (Dec)]
Lat = latitude of the location in radians or degrees and decimal fractions divided by 57.3; positive for north Lat and negative for south Lat
Dec = $-0.00117 - 0.40117 \cos(\pi J/6) - 0.042185 \sin(\pi J/6) + 0.00163 \cos(\pi J/3) + 0.00208 \sin(\pi J/3)$; Dec is declination of the sun
J = number of the month, January = 1
ES = $1.00016 - 0.032126 \cos(\pi J/6) - 0.003354 \sin(\pi J/6)$; ES is relative distance of the earth from the sun compared to average distance
t = average temperature in degrees Celsius
π = 3.14

Values of RA for any particular day of the year can be calculated from Eq. (5), where:

Y = cos[0.0172142(D + 192)]
D = Julian day (January 1 = 1)
Dec = 0.40876Y
ES = 1.00028 + 0.03269Y

Powell et al. (1984) developed a model to estimate solar radiation from satellite imagery. In their method the required cloud cover information is visually extracted from nondigitized photographic forms of satellite imagery. This information is then used as input into the model to estimate the solar radiation. Although their method is useful in estimating solar radiation on a large scale, the accessing of quantitative and digitized forms of satellite imagery can become very expensive for many potential users.

IV. RELATIVE HUMIDITY

Relative humidity is used widely in procedures for estimating agricultural water requirements. Richardson (1988, and personal communication) made a study of the relationship between minimum relative humidity and the average daily range in temperature (TD) for 500 first-order stations in the United States. The resulting regression equation for average minimum monthly relative humidity (RH) as a percentage can be written

$$RH = 110.3 - 4.68TD \quad (6)$$

The limits of 100% and 5% are imposed to keep RH within appropriate bounds. The regression resulted in a coefficient of determination (R^2) of 88%.

V. REFERENCE EVAPOTRANSPIRATION

Many methods have been developed for estimating potential evapotranspiration or the amount of water use (evaporation plus transpiration) from a well-watered grass under specific management conditions. Doorenbos and Pruitt (1977) define reference crop evapotranspiration (ET_0) as the rate of evapotranspiration (ET) from an extensive surface of 8-15 cm tall green grass cover of uniform height, actively growing, completely shading the ground, and not short of water. Beard (1985) compared ET rates from various grass species and found significant differences in water use between species and within species. ET comparisons are presented for various Kentucky bluegrass cultivars. The relative ET rankings range from low to high. In comparisons among cultivars, higher nitrogen levels have a stronger effect on increasing the ET rate than higher cutting height. The evaporative demand of the region and the occurrence of environmental or pest stress can also cause substantial shifts in the ET rate and the resultant comparative rankings among species.

Beard (1985) describes the tall fescues as having relative high ET rates. In this chapter ETP is used as equivalent to or synonymous with ET_0 and refers to a reference potential calibrated using 8 years of Alta fescue grass lysimeter ET data from well-managed large weighing lysimeters at Davis, California.

Several methods for estimating the ETP provide satisfactory results for most conditions but may require some degree of local calibration due principally to differences in the interactions of various climatic variables. The method for estimating ETP presented in this chapter does not appear to require local calibration for the range of climatic conditions within which irrigation is required for good crop production.

The energy required to evaporate water varies with temperature. The latent heat of vaporization L (calories required to evaporate 1 g or 1 cm^3 water) is given by the equation

$$L = 595.9 - 0.55t \qquad (7)$$

Incoming radiation at the surface is usually expressed in Langleys (cal cm^{-2} day^{-1}) or in kJ m^{-2} (1 Langley = 41.84 kJ m^{-2}). At 20°C, 58.5 Langleys corresponds to 1 mm equivalent water evaporation.

Reference or potential evapotranspiration is determined by the air temperature and the energy available from solar radiation and from advection (sensible heat transfer). The energy available from advection is influenced by relative humidity or vapor pressure, temperature, radiation, and wind. Richardson (1988) developed regression equations for estimating class A pan evaporation in Utah for five different characteristic site conditions in Utah.

A different wind function was found for each month, and different interactions of the above-mentioned factors were determined for each of the site characteristics. Although advection can be either positive or negative, evaporation and evapotranspiration are increased by increasing wind run during hot dry conditions. However, generalized satisfactory equations for the increase have not been developed.

Relative humidity, vapor pressure, solar radiation, and mean temperature can all be determined or estimated from measured values of maximum and minimum temperatures. ETP can also be estimated using maximum and minimum temperatures. An equation for ETP proposed by Hargreaves et al. (1985) and Hargreaves and Samani (1985) is

$$ETP = 0.0023RA(t + 17.8)TD^{0.50} \tag{8}$$

where ETP and RA are in the same units of equivalent water evaporation, t is the average of maximum and minimum temperatures, and TD is in degrees Celsius. For temperature in degrees Fahrenheit the equation becomes

$$ETP = 0.00094RAtTD^{0.50} \tag{9}$$

Attempts to add a wind function and to find a need for local calibration for these equations have been inconclusive.

Allen et al. (1983) found an average increase in air temperature in nonirrigated sites from 0.9°C for May and October to 4.7°C for August using measured values from irrigated areas in southern Idaho. Temperature measurements should be from representative sites. Mean temperatures often decrease 5-6°C per 1000 m increase in elevation. TD may vary significantly with changes in elevation. Values are also influenced by topography and exposure.

For purposes of design, operation and management knowledge of how much monthly values of ETP may vary from the mean is desirable. Equations (8) and (9) were used with long records of monthly values of mean temperature and TD to calculate ETP and the standard deviation of ETP (SDET). These calculations resulted in the following conclusions:

1. SDET is equal to about 4.7% of mean ETP for latitudes N10° to S10°.
2. SDET increases with latitude departure in degrees (LD) from N10° or S10° and is approximately equal to (0.047 + 0.0012LD) times ETP.
3. The maximum SDET at extreme latitudes is about 10% of mean ETP.

Equations (8) and (9) are recommended for worldwide use for estimating ETP. However, temperature data should be collected from sites located in large irrigated areas or corrected for the differences in values between arid and irrigated areas. Some judgment is recommended for hot, dry, windy conditions. This applies particularly to relatively small irrigated areas surrounded by drylands. An increase in ETP of 15% in these conditions may be a reasonable assumption.

The amount of water required for maximum crop growth at 75% or more ground cover is usually in the range of 0.95-1.20 times ETP. For some crops and climates the maximum requirement may be as high as 1.4 times ETP. For a citrus crop with good weed control the maximum requirement may be only about 0.75 times ETP.

VI. MOISTURE AVAILABILITY INDEX

The moisture availability index (MAI) is a very useful concept for evaluating the potential for rainfed agriculture. MAI is the ratio of the 75% probability of assured precipitation (PD) divided by ETP:

$$MAI = \frac{PD}{ETP} \tag{10}$$

Hargreaves (1977) published values of MAI, PD, and ETP for 640 worldwide locations. Since 1977, various studies made at Utah State University have included values of MAI from the climatic data of specific countries and/or regions. Hargreaves and Samani (1986) give much more complete worldwide coverage of rainfall probabilities, MAI, and ETP.

VII. CLASSIFICATION OF CLIMATE

Classifying climates in a manner that clearly defines the climatic potential for agricultural production is difficult. Crop growth and development models can be used with monthly climatic tables and a model to synthesize daily weather data to model the influence of climate. Keller (1982, 1987) developed a WMAKER model to generate synthetic daily weather data from mean monthly values of ETP, temperature, number of days of rain, rainfall amounts, and the standard deviations of these values. The means and standard deviations should be from long records. Unfortunately, however, climatic classifications and the models do not give adequate consideration to the temperature range (TD). Two locations may have the same mean temperature, but one may have a daily temperature range twice that of the other. Similar mean temperatures do not necessarily indicate similar agricultural potentials.

Hargreaves (1977) reviewed various classifications of climate and definitions of drought and concluded that, although no simple classification can include all the variations, use of some rather simple generalizations would significantly improve the science of agrotechnology transfer. A more refined breakdown is required for most specific crops. The following climatic types based upon mean monthly temperatures were proposed:

Climatic type	Criteria
Polar	All months < 10°C
Boreal	1-3 months > 10°C
Subtemperate	4-5 months > 10°C
Temperate	6-9 months > 10°C
Subtropical	10-12 months > 10°C
Tropical	All months ≥ 17°C

Six climatic zones based on temperature are proposed. Each temperature zone is subdivided using up to seven criteria, relating to adequacy or excess of precipitation for agricultural production. These classifications are given in Table 3.

TABLE 3 Classification of Climate for Dryland Agricultural Productivity[a]

Criteria	Climate classification	Productivity classification
All months with MAI of 0.00-0.33	Very arid	Not suited for rainfed agriculture
1 or 2 months with MAI of 0.34 or above	Arid	Limited suitability for rainfed agriculture
3 or 4 months with MAI of 0.34 or above	Semiarid	Production possible for crops requiring a 3-4 month growing season
5 or more consecutive months with MAI of 0.34 or above	Wet-dry	Production possible for crops requiring a good supply of water during 5 or more months
1 or 2 months with MAI above 1.33	Somewhat wet	Natural or artificial drainage required for good production
3-5 months with MAI above 1.33	Moderately wet	Good drainage required for normal agricultural production
6 or more months with MAI above 1.33	Very wet	Very good drainage required for normal agricultural production

[a]See Eqs. (2), (8), (9), and (10) for definition of MAI.

Source: From Hargreaves (1977).

VIII. WEATHER GENERATING MODELS

Weather data are used in the design of hydraulic structures, erosion assessment, hydrological studies, alternative crop management, and many other aspects of water resources development. Methods that use a sequence of observed weather data alone may not be sufficient. One needs to explore other sequential alternatives that may occur. For some locations no weather data are available to make the desired assessment. For these reasons it is desirable to have the capability to generate synthetic weather data with the same statistical characteristics as the actual weather at the location.

Several weather generating models have been developed. Richardson and Wright (1984) developed a weather generating model (WGEN) to generate synthetic daily values of precipitation, maximum and minimum temperatures and solar radiation for any number of years. The WGEN model generates daily rainfall values independently of other variables and then proceeds to generate the other variables based on the wet and dry status of that day.

Richardson and Wright (1984) calibrated the WGEN model for 139 locations in the United States using 20 years of daily rainfall data. Four parameters are given for each month at each location. The model may be used at other locations by interpolation between parameters. However, use outside the United States of the WGEN model, modified versions, and similar weather generating models is seriously limited because of the unavailability of and/or the difficulty and expense in obtaining the daily weather records required for calibration.

Keller (1982, 1987) developed a weather generating model called WMAKER. The WMAKER model generates daily values of rainfall, radiation, and maximum and minimum temperatures from monthly climatic data tables. Because of the worldwide applicability of WMAKER and the availability of the required climatic data tables to be used by this model, the WMAKER model is described here in further detail.

IX. PREDICTING CROP YIELDS FROM MONTHLY WEATHER

Records of mean temperatures and rainfall are more easily obtained than those for maximum and minimum temperatures. For this reason one must often use long-term mean values of temperature to calculate long-term means of ETP and to use latitude to estimate the standard deviation of ETP from latitude as indicated in Sec. V.

The WMAKER model developed by Keller (1982, 1987) begins by generating the mean ETP for each day of the year. The ETP values are then randomized to simulate the variation already indicated. Daily temperature and RS are calculated from randomized ETP. Random rainfall events are generated to coincide with days of low solar radiation.

The procedure in more detail is as follows:

1. Long-term ETP values are generated for each day of the year. The mean daily ETP values for each month are converted to the expected value for the first of each month by integration. Mean ETP for each day is then estimated from a fourth-order Lagrangian approximation and the expected values for the first of each months. Random deviates are generated assuming a normal distribution.
2. Mean daily temperature is generated using a similar procedure, with the exception that a third-order Lagrangian function is used. Random average daily temperatures are generated based on the ratio of the randomly generated values of ETP to the mean daily ETP raised to the power of Z, where Z is given by the equation

$$Z = \frac{\ln(1 - CVT)}{\ln(1 - CVETP)} \tag{11}$$

where CVT = coefficient of variation of temperature and CVETP = 0.047 for latitude N10° to S10° and is 0.047 + 0.0012LD for latitudes further from the equator. LD is latitude departure from 10° north or south. Random daily solar radiation (RS) is calculated from temperature and ETP:

$$RS = \frac{ETP}{0.0135(t + 17.8)} \tag{12}$$

The daily extraterrestrial radiation (RA) is calculated using a computer routine [Eq. (5)], and the temperature difference (TD) is given by Eq. (4). Daily maximum and minimum temperatures are calculated from daily temperature values and TD. The simulated RS values are checked against the boundary limits of 25% of RA for completely cloudy days and 75% of RA for cloud-free days and adjusted if necessary.

3. The number of rainy days in each month is determined from a normal distribution as random deviates from the mean and standard deviation. The rainy days are predicted as those with the lowest ratios of RS to RA (cloudy days).
4. The depth of precipitation for each day with rain is generated using a log normal distribution from the monthly means and standard deviations and adjusted for the simulated number of rainy days.

The WMAKER model generates random weather events. This requires the processing of the complete set of synthetic daily values to calculate a given probability of crop yield. Efforts are now in progress to modify the WMAKER model to make possible the generation of any specified probability without using the full array of random values.

Samani et al. (1987) have provided a fairly detailed description of the WMAKER model. They compared relative yields of corn predicted by corn yield models using daily climiatic records from El Salvador and Brazil with those predicted from WMAKER-generated synthetic data. Comparisons were made for dryland yields and for one location with those predicted with one supplemental irrigation. It was concluded that yields predicted with actual data and generated data are roughly comparable. The range in differences in calculated relative yields for the various probabilities for the data evaluated was from zero to only 17%. Figure 1 compares the predicted yield values from synthetic weather data and measured weather data for a station in El Salvador.

X. CROP YIELD MODELS

Tsuji (1988) described the status of 11 crop growth and development models. Crops suitable for dryland agricultural production in semiarid climates include wheat, corn, sorghum, millet, peanuts, and dry beans. With irrigation, rice and potatoes also produce well in semiarid areas. Crop yield models for these crops are recommended for use with the WMAKER model and possible future modifications of the model.

Povoa and Karahaliloglu (1988) prepared monthly weather tables for use with the WMAKER model for 299 locations in Africa, 294 sites in Latin America, and 174 weather stations in Asia. An example of the monthly database is given for Marrakech, Morocco, in Table 4.

ETP was calculated from long-term mean maximum and minimum temperature values. A standard deviation of 8.5% of monthly ETP values is indicated by the procedure just given. The gamma probability distribution function of rainfall amounts (RPAR) is given for each location. This makes possible the calculation of any desired probability of rainfall amounts based on the historical records.

For ETP, mean temperature (T_{EMC}), and mean numbers of rainy days

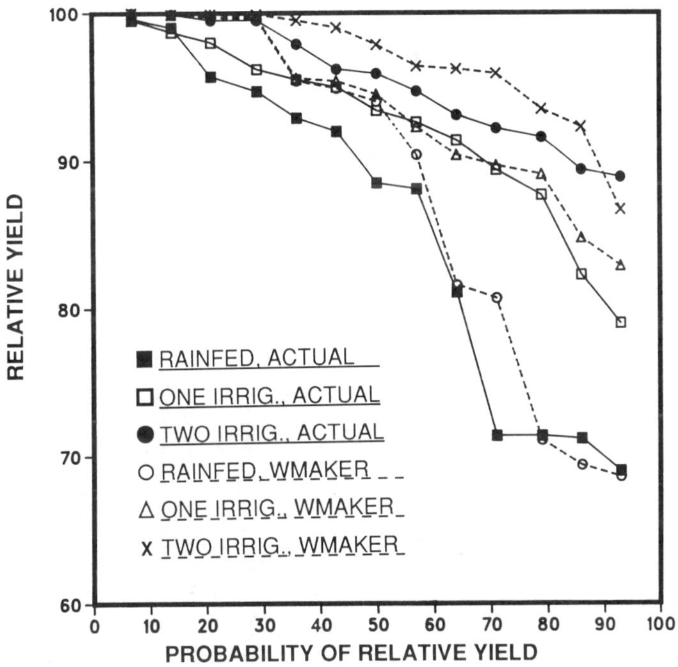

FIGURE 1 Probable relative yields of corn grain from rainfed agriculture compared with predicted yields from one or two irrigations, at La Union, El Salvador. The assumed planting date was May 1. The soil was a 60 cm deep medium-textured soil. Irrigations were at 75 and 85 days after planting. Yields estimated from historical daily climatic data are compared with those from WMAKER synthesized data.

(R_{day}), the mean values and standard deviations determine the shape of the probability distributions. For an increase or decrease in rainfall amounts there is a corresponding increase or decrease in ETP, T_{EMC}, and R_{day}. It is therefore feasible to modify the WMAKER model to predict any given probability of weather conditions related to probable crop yields. The correlations may not be as high as desired, but since rainfall amounts are often highly variable but ETP and T_{EMC} vary little from year to year for the same month, yields of dryland crops are largely determined by rainfall amounts compared with average or near-average water requirements.

The use of the WMAKER model and of crop growth and development models can be simplified providing the following assumptions produce satisfactory results:

1. The values of mean rainfall amounts correspond with mean numbers of rainy days and mean values of temperature and ETP.
2. There is 1 SD above or below the mean rainfall amount corresponding with 1 SD above or below the mean number of rainy days.
3. A similar relationship exists for temperature and ETP, but lower values of rainfall result in higher values of temperature and ETP.

TABLE 4 Typical Monthly Climatic Table for Use with WMAKER and Crop Yield Models[a]

	January	February	March	April	May	June	July	August	September	October	November	December
T_{max}	18.0	20.0	23.0	25.0	29.0	33.0	38.0	37.0	33.0	28.0	23.0	19.0
T_{min}	5.0	7.0	9.0	12.0	14.0	17.0	20.0	20.0	18.0	15.0	10.0	7.0
ETP	62.0	75.0	116.0	141.0	177.0	201.0	239.0	220.0	159.0	116.0	75.0	59.0
T_{EMC}	11.0	13.0	16.0	18.0	21.0	24.0	28.0	28.0	25.0	21.0	16.0	12.0
SDT	1.2	1.5	1.6	2.1	2.0	1.8	1.7	1.6	1.6	1.6	1.7	1.2
P_M	28.0	31.0	30.0	34.0	20.0	6.0	2.0	3.0	9.0	22.0	33.0	30.0
RPAR	0.7	0.7	0.8	0.9	0.5	0.2	0.2	0.2	0.3	0.6	0.6	0.8
SDPM	22.0	29.0	26.0	33.0	21.0	12.0	4.0	5.0	11.0	19.0	27.0	28.0
R_{day}	4.7	4.1	4.2	4.5	2.7	0.6	0.2	0.6	1.0	3.0	4.6	4.7
SDRD	3.7	2.9	3.6	3.3	1.9	0.7	0.5	0.7	1.0	2.4	3.2	3.6

[a]Station Marrakech: latitude 31°42' N; longitude 8°0' W; elevation 466.0 m; 54 years.

T_{max}, mean monthly maximum temperature, °C; T_{min}, mean monthly minimum temperature, °C; ETP, mean monthly potential evapotranspiration, mm; T_{EMC}, mean monthly temperature, °C; SDT, standard deviation of T_{EMC}, °C; P_M, mean monthly precipitation, mm; RPAR, gamma probability distribution factor (shape function); SDPM, standard deviation of P_M, mm; R_{day}, mean number of rainy days in the month; SDRD, standard deviation of R_{day} in days.

Source: From Povoa and Karahchiloglu (1988).

4. Daily synthetic climatic data generated for 3 years or more, consisting of the year with mean rainfall and those of 1 SD above and below normal, and, if desired, a few additional years corresponding to years of known probabilities of assured rainfall, can be used with crop models to define the shape of the influence of climate on probable relative yields over the full range of yield probabilities.

Relative probable yield distributions can be used to replace trial-and-error testing over many locations and years with varietal testing through crop simulation of multiyear performance. This approach would reduce the requirements for climatic data, lower costs, and accelerate the varietal testing process. Potential users could assess the suitability of each new cultivar in any location for any period, quickly and inexpensively.

It is proposed that models be used with crop genetic coefficients that summarize how a particular cultivar responds to any array of environmental factors, such as daylength, maximum and minimum temperatures, available soil water, and assimilated nitrogen. The use of models and genetic coefficients makes possible the prediction of the potential performance of diverse cultivars on a global scale, independent of location and management.

Caution should be used in generating synthetic data and in using weather data from short records. Wet or dry cycles of 10 years or more duration may be encountered in the records. The length of record required to provide a realistic probability distribution may be as short as 15 years or as long as 45 years. Data tables derived from short records have significant value, but long records are always much to be preferred.

The monthly climatic tables prepared by Povoa and Karahliloglu (1988) are from the historical records. The past provides an indication of the future. However, current predictions call for a doubling of atmospheric carbon dioxide as early as 2030. Significant resulting dislocations in the global distribution of climate are predicted. The largest effect on the water balance and on potential agricultural production is indicated for the central portions of North America and Eurasia. These regions comprise areas of major food grain and livestock production.

XI. PREDICTING EXTREME RAINFALL

In the design of irrigation, drainage, flood control, and related hydraulic structures, estimates of probable extreme rainfall amounts for various durations and return periods are desirable. The damage and destruction caused to irrigation projects and other hydraulic works by extreme rainfalls and the expenses incurred from overdesigning structures can be greatly reduced through improved estimates of depth-duration-frequency rainfall amounts.

Comparisons of nine different probability distributions with long records of daily rainfall amounts indicated that all distributions provided very similar predicted amounts for the 1 day-10 year return-period rainfalls. For other return periods the probability distributions produced significant differences in predicted amounts. In the range of return periods of 5-200 years, three probability distributions, the log-Pearson type III model of annual series, the censored log-normal of partial series, and an exponential method (Hargreaves, 1988), produced very similar predicted values.

There is no theoretically sound reason for selecting one rainfall intensity prediction model in preference to another. It is practical to evaluate models in terms of empirical fit of available data, computational ease, and consistency with various sample sizes. Hargreaves (1988) evaluated the Powell exponential model by comparisons with rainfall records and predicted censored log normal values from 22 locations. The average record length exceeded 100 years. The exponential equation is

$$D = KD(Tt)^{0.25} \tag{13}$$

where:

D = depth-duration-frequency amount
KD = a calibrated coefficient
T = frequency or return period in years
t = time or duration in hours

Values of D for the 1 day 10 year return-period rainfall maounts equal to those predicted from the censored log normal probability distribution were assumed. The extreme average maximum rainfall amounts in approximately 100 years are compared with those predicted by the censored log normal and the exponential distributions in Table 5.

Various authors have concluded that extreme rainfall amounts vary with $t^{0.25}$. This seems valid for periods of time of from 30 min to several days. The exponent of T varies somewhat with the probability distribution chosen to predict the extreme amounts. However, the comparisons presented and referenced by Hargreaves (1988) indicate that $T^{0.25}$ is a very practical choice.

Hershfield (1961) found that an average factor of 1.13 should be used to transform observation-day rainfall amounts to corresponding 1440 min return-period amounts (24 h amounts). An annual or partial series of observation-day rainfall extreme values can be multiplied by 1.13 and the value of KD calculated for $(Tt)^{0.25}$ or 3.94 (D = 3.94 KD) and D for the 10 year 24 h rainfall amount predicted from any frequently used probability distribution. The value 3.94 is the fourth root of 10 × 24. A ranking distribution is very convenient if computer facilities are not available. Depth-duration-frequency values can then be predicted from Eq. (13) for values of T from 5 to 200 years and of t from 0.5 to 100 h or more.

Samani and Hargreaves (1988) calculated extreme 1 day 30 year return-period rainfall amounts for four locations in the United States by using the WMAKER model. These values were divided by 2 to approximate 1 h extreme values and converted to 2, 50, and 100 year return periods by means of the equation

$$\frac{D_1}{D_2} = \left(\frac{t_1 T_1}{t_2 T_2}\right)^{0.25} \tag{14}$$

where D_1 and D_2 are rainfall amounts corresponding to duration t_1 and t_2 and return periods T_1 and T_2, respectively. These values generated from the WMAKER model were then compared with the extreme rainfall amounts given by Hershfield (1961) in the U.S. Weather Bureau atlas; the average

TABLE 5 Comparisons of Average of Two Probability Distributions with Maximum Recorded Rainfall Amounts (mm)[a]

Duration t (days)	Maximum recorded in duration t	Avg. censored log normal		$D = KD(Tt)^{0.25}$	
		T = 100	T = 200	T = 100	T = 200
1	137	135	158	135	161
2	162	184	216	178	212
3	186	205	240	199	237
7	216	244	285	241	286

[a]Maximum recorded rainfall in duration t, once in the length of record or approximately once in 100 years. In a censored probability distribution, values above a selected minimum are used. This may result in more than one extreme value in the same year.

difference was 12% (16% for T = 2, 11% for T = 50, and 8% for T = 100). In a field of much uncertainty, these results generated from the WMAKER model may be about as good as those from any other procedure or probability distribution. Therefore this procedure is recommended for further testing and evaluation.

XII. PREDICTING FUTURE WEATHER

Irrigation will have an increasing role in the reclamation of semiarid lands and deserts. However, for vast areas productivity will depend on the availability and use of water from precipitation. Planning the use of rainfall is made extremely difficult by the variations in distribution and amount. Precipitation during a wet year may be five to seven times that in a dry year. Planning and management can be much improved, provided more insight becomes available relative to the short-term and long-term future weather.

A. Response Farming

Stewart (1988) described a system of rainfall prediction as a guide to seasonal agricultural decisions. The amount of rainfall in the coming season as well as its duration and intensity are all correlated with the date of onset of the rains. The date of onset is defined as the date when a predetermined amount (say, 30 mm) of moisture is stored in the soil. Land preparation, rates of fertilizer applications, area planted, and related decisions can be influenced by the date of onset of rains and by the rainfall amount during the first few weeks or first 30 days of the rainy season.

Seasons with early onset in comparison with the historical records are usually of longer duration and produce rainfall amounts in the upper range of the historical record. For a given location historical records can be used to graphically show the interactions of onset date of rains, duration of rainy season, and seasonal rainfall amounts. These graphic representations provide very useful guidance relative to seasonal agricultural decisions.

B. Carbon Dioxide Buildup and Climate Change

Blasing (1985) presents the results from models indicating an increase in mean summer temperatures of 4-8°C for one model to an average of about 2°C for another model for the United States. Decker et al. (1985) state that a decrease in precipitation is indicated in latitudes 37-47° north. In various models, drier conditions are indicated for the central United States and Canada and for central Urasia. These changes are predicted due to the doubling of atmospheric carbon dioxide by as early as the year 2030. Although there is considerable uncertainty regarding the accuracy of the models used to predict these changes, the scientific community agrees that the buildup of carbon dioxide in the atmosphere and the consequent climatic changes comprise the most important environmental issue facing humankind.

California is the richest and most populous state in the United States. The principal source of wealth is agriculture. In winter the land surface is cooler than the ocean. Rain is produced as moist air is cooled as it moves inland. Continued cooling as the air is forced up over mountains in California and other western states and the influx of cooler air masses from the Gulf of Alaska produce more rain or snow. In summer the ocean surface is colder than the land, and the difference is accentuated by the cold California current. Air moving overland permits the retention of more water vapor, and rainfall is rare. It is anticipated that an increase in global temperatures will decrease precipitation in California and other western areas owing to a decreased cooling gradient between ocean and land during the winter and the increased warming gradient during the summer months. With higher global temperatures and with the larger increase in higher latitudes it seems probable that the Pacific high, the high-pressure zone over the Pacific Ocean, may extend farther north in summer and that the southward retreat of the Pacific high may be delayed, thereby delaying or thwarting the winter rainy season.

An increase in temperature will result in a significant increase in ETP. Decreasing rainfall will probably be associated with decreasing cloud cover, declining relative humidity, and increased solar radiation at the surface. Both mean temperatures and TD will be increased. Hargreaves (1981a) reviewed various predictive models and concluded that a doubling of CO_2 in the atmosphere would shift climatic zones for crop adaptation 400-480 km (250-300 miles) farther north, with a significant increase in the length of the growing season. Hargreaves (1981a) estimated that the predicted temperature changes combined with the longer frost-free growing season would increase crop water requirements significantly. The increase is predicted as 2.5% at the equator to 63% at 60 degrees north latitude. These changes are shown in Table 6.

A doubling of CO_2 will facilitate photosynthesis and increase potential crop yields by perhaps an average of 30-35%. Higher temperatures and possible lower rainfall will significantly increase the requirements for irrigation. Increased CO_2 also increases stomatal resistance and crop resistance to drought, but higher yields are associated with increased leaf area index and more efficient use of incoming radiation. All these interactions have not yet been well documented; however, it is certain that competition for available water supplies will be greatly increased.

TABLE 6 Predicted Increase in Crop Water Requirements Due to Doubling CO_2 in the Atmosphere

North latitude (degrees)	% Increase due to climate change		
	In ETP due to temperature	In growing season	Total seasonal increase in ETP
0	2.5	0	2.5
10	3	0	3
20	5	0	5
30	7	9	16
40	12	15	27
50	17	25	42
60	26	37	63

In the United States the world's most productive area for fruit and vegetable production and one of the world's principal food grain and livestock producing areas will be strongly affected by these projected weather changes. There is a potential for maintaining and increasing present levels of productivity through increased irrigation. Readily available surface water supplies are already under pressure. Almost all aquifers are declining, many seriously, so this will require expensive, large-scale surface water transfers. However, the planning, designing, and constructing of large projects for mass water transfer over long distances requires many years.

The first comprehensive plan for the Central Valley water project of California was completed in 1920. The major construction period was in the 1950s and 1960s. If the projected weather changes are to take place by the year 2030, planning as to how best to cope with the changes in climate should be initiated now.

Table 6 and the projections by Hargreaves (1981) are derived from general circulation models that do not include the annual insolation cycle. Schlesinger and Mitchell (1987) describe the results of including the annual insolation cycle. These simulations indicate an annual global mean warming of 3.5-4.2°C. An average increase of about 9% is indicated in global precipitation. There is, however, much uncertainty relative to the quantitative distribution of these changes. Probably temperature changes of this magnitude would shift the frost-free growing season more than 600 km farther north and increase the length of the growing season more than indicated by Hargreaves (1981a). However, adding the annual insolation cycle increases the uncertainty of the future quantitative distribution of rainfall in the middle latitudes.

Because of the large measure of uncertainty with respect to the future weather, contingency planning should be now initiated so that the possible benefits from change may be realized and the undesirable effects minimized.

The weather is also influenced by long-term trends or oscillations. Cycles in the earth's spin, tilt, and orbit produce glacial and interglacial periods. We are now at the end of an interglacial period. The earth's tilt is now decreasing so that the difference between the seasons will become less marked. Changes in radiation emitted by the sun and volcanic activity can significantly influence the weather. However, some computer models predict an average rise of up to 4.5°C in world temperature, making climates warmer than at any time in the past 100,000 years.

Trapped heat and temperature gradients are the sources of energy that create restlessness of the atmosphere, causing changes in weather conditions. With increasing trapped heat and a greater temperature difference between land and water, there will be more restlessness and greater violence of weather conditions. Large semiarid and desert areas may shift farther from the equator or become hotter and drier. In those areas of increased precipitation more variability should increase the frequency and intensity of excessive rainfall. In areas of bimodal precipitation the intermodal dry periods may become more pronounced. It therefore seems certain that within 50 years there will be a much greater need for water control and water management facilities and programs.

Increased restlessness of the atmosphere due to the effects of CO_2 buildup can be expected to increase the frequency of extreme events. A gradual change to a hotter and drier climate in much of the United States will be easier to adjust to than to an increase in the frequency of record high temperatures, frosts, floods, and droughts. The prediction of future variability and the planning necessary for the required future agricultural adjustments may be quite difficult.

XIII. WEATHER MODIFICATION

Ever since the dawn of history humans have dreamed of changing the weather to their own advantage. In recent years significant advances have been made in weather modification. Today most modern nations and even some developing nations have active weather modification programs. The weather modification programs are used to increase precipitation by 5-20% through cloud seeding, to reduce damage to drops, and to clear fog at airports. According to Brosten (1988), about 150×10^9 m^3 of water (in vapor form) passes over the United States each day from which only 10% precipitates and the rest remains airborne and ends up over oceans or Canada. Even though weather modification is in an early stage, a significant potential exists in this area. In a feasibility study for the state of California, Finlayson et al. (1987) concluded that the runoff can be increased by about 10^8 m^3 annually through cloud seeding. The modification of weather in the future will have dramatic legal, social, and economical impacts that will be difficult to predict.

XIV. CONCLUSION

This book provides much useful information on the use and reclamation of semiarid lands and deserts. Primary emphasis is on soils. This chapter describes various climatic interactions and useful relationships for enhancing rehabilitation efforts.

In comparing climates, too much emphasis is frequently given to mean values and not enough emphasis has been assigned to variability of weather, the probability distributions, or how the weather variables interact with each other. The temperature range is highly correlated with solar radiation, relative humidity, vapor pressure deficit, and crop water requirements, yet it is often omitted from climatic data summaries.

A classification of climate is proposed that consists of six types, determined by mean monthly temperatures, and up to seven divisions, depending on the adequacy of assured monthly rainfall for dryland agricultural productivity.

Use of monthly climatic data tables consisting of means and standard deviations of potential evapotranspiration, temperature, rainy days, and rainfall amounts is recommended for generating synthetic daily weather values. These synthetic values provide input for crop yield models, which together with crop genetic coefficients make possible the prediction of performance of diverse cultivars on a global scale, independent of location and management.

The buildup of carbon dioxide in the atmosphere and the consequent climate changes comprise the most important environmental issue facing humankind. Resulting higher temperatures and possibly lower rainfall for semiarid lands and deserts will greatly increase competition for available water supplies. Sound planning of large projects for mass water transfers requires many years. Planning for coping with probable future changes in climate should be initiated now.

REFERENCES

Allen, R. G., Brockway, C. E., and Wright, J. L. (1983). Weather station siting and consumptive use estimates. *J. Water Resources Planning Manag. ASCE 109*: 134-146.

Beard, J. B. (1985). An assessment of water use by turfgrass. In *Turfgrass Water Conservation.* Cooperative Extension, University of California Division of Agriculture and Natural Resources, Riverside.

Blasing, T. J. (1985). Background: Carbon cycle, climate, and vegetation responses. In *Characterization of Information Requirements for Studies of CO_2 Effects: Water Resources, Agriculture, Fisheries, Forests, and Human Health.* U.S. Department of Energy, Washington, D.C., pp. 9-22.

Brosten, D. (1988). Sow the seeds of rain (weather modification and cloud seeding). *Agrichem. Age, October.*

Dekker, W. L., Jones, V., and Achutuni, R. (1985). The impact of CO_2-induced climate change on U.S. agriculture. In *Characterization of Information Requirements for Studies of CO_2 Effects: Water Resources, Agriculture, Fisheries, Forests, and Human Health.*

Doorenbos, J., and Pruitt, W. O. (1977). *Crop Water Requirements.* Food and Agriculture Organization of the United Nations, FAO Irrigation and Drainage Papers 24, Rome.

Finlayson, D. J., Griffith, D. A., and Swart, H. R. (1987). Feasibility of weather modification for California's state water project irrigation systems for the 21st century. ASCE, New York, pp. 431-438.

Hargreaves, G. H. (1977). *World Water for Agriculture.* Agency for International Development, Washington, D.C.

Hargreaves, G. H. (1981a). Water requirements and man-induced climate change. *J. Irrigation Drainage Div. ASCE 107*: 247-255.

Hargreaves, G. H. (1981b). Responding to tropical climates—an approach. In *Food and Climate Review 1980-81.* Aspen Institute for Humanistic Studies, Boulder, Colorado.

Hargreaves, G. H. (1988). Extreme rainfall for Africa and other developing countries. *J. Irrigation Drainage Eng. ASCE 114*: 324-333.

Hargreaves, G. L., Hargreaves, G. H., and Riley, J. P. (1985). Irrigation water requirements for Senegal River Basin. *J. Irrigation Drainage Eng. ASCE 111*: 265-275.

Hargreaves, G. H., and Samani, Z. A. (1985). Reference crop evapotranspiration from temperature. *Appl. Eng. Agri. Trans. ASAE 1*: 96-99.

Hargreaves, G. H., and Samani, Z. A. (1986). *World Water for Agriculture—Precipitation Management.* International Irrigation Center, Utah State University, Logan.

Hershfield, D. M. (1961). Rainfall frequently atlas for the United States for durations for 30 minutes to 24 hours and return periods of 1 to 100 years. Technical Paper No. 40, U.S. Weather Bureau, Dept. of Commerce, Washington, D.C.

Keller, A. A. (1982). Development and analysis of an irrigation scheduling program with emphasis on forecasting consumptive use. M.S. thesis, Agricultural and Irrigation Engineering Department, Utah State University, Logan.

Keller, A. A. (1987). Modeling command area demand and response to water delivered by the main system. Ph.D. dissertation, Agricultural and Irrigation Engineering Department, Utah State University, Logan.

Mockus, V. (1960). Selecting a flood frequency method. *Trans. ASAE 3*: 48-51, 54.

Povoa, A. F. R., and Karahaliloglu, F. (1988). Appendixes. In *Manual for Dryland and Irrigated Crop Production.* International Irrigation Center, Department of Agricultural and Irrigation Engineering, Utah State University, Logan. Appendix I, Monthly climatic data tables for Africa; Appendix II, Monthly climatic data tables for Latin America; Appendix III, Monthly climatic data tables for selected Asian countries.

Powell, G., Brazel, A., and Pasquealetti, M. (1984). New approach to estimating solar radiation from satellite imagery. *Prof. Georgr. 36*(2): 227-233.

Richardson, C. W., and Wright, D. A. (1984). WGEN: A model for generating daily weather variables. USDA, ARS-8 (available from National Technical Information Service, Springfield, Virginia).

Richardson, E. A. (1988). Final Report to Bureau of Land Management on the Use and Development of the ASYMCUR Mathematical Model. Bureau of Land Management, Salt Lake City, Utah.

Samani, Z. A., and Hargreaves, G. H. (1986). Estimating rainfall probabilities from average values. *Appl. Eng. Agr. ASAE 2*: 141-147.

Samani, Z. A., and Hargreaves, G. H. (1988). Estimating rainfall intensities from a weather generating model. *J. Irrigation Drainage Eng. ASCE* (submitted). (Paper being revised for publication.)

Samani, Z. A., and Pessarakli, M. (1986). Estimating potential crop evapotranspiration with minimum data in Arizona. *Trans. ASAE 29*: 522-524.
Samani, Z. A., Hargreaves, G. H., Zuniga, E., and Keller, A. A. (1987). Estimating crop yields from simulated daily weather data. *Appl. Eng. Agr. ASAE 3*: 290-294.
Schlesinger, M. E., and Mitchell, J. F. B. (1987). Climate model simulations of the equilibrium climatic response to increased carbon dioxide. *Rev. Geophys. 25*: 760-798.
Stewart, J. I. (1988). *Response Farming in Rainfed Agriculture.* WHARF Foundation Press, Davis, California.
Tsuji, C. Y. (1988). *States Report of IBSNAT Crop Models.* IBSNAT Project, Department of Agronomy and Soil Science, University of Hawaii, Honolulu.

20

Soil and Water Management of Irrigated Lands

GAYLORD V. SKOGERBOE *Utah State University, Logan, Utah*

I. INTRODUCTION

The world has already entered an exciting era of irrigation development, not so much in developing new irrigated lands but toward an even more important goal—improving existing irrigated lands. Much of the emphasis in the future will be to increase crop yields on lands already under cultivation. There are several national goals that impinge on this subject. Besides continually increasing crop production, it is also necessary to reduce water pollution, to have cleaner water and to reduce energy consumption, at the same time sustaining the productivity of irrigated soils. Fortunately, these can be compatible goals! For irrigated croplands, this means moving away from the preoccupation with hardware development, such as dams and canals, to focusing on the management of soil and water resources on farms. This change in emphasis is very much needed everywhere.

Technological developments are now rapidly occurring regarding on-farm soil and water management, such as better methods of applying irrigation water, evaluating the surface and subsurface hydraulics or irrigated lands, modeling the plant-soil-water relationships, and pollutant transport. Technology can only be a partial solution, however. Consequently, the question of processes that can be employed to manage soil and water resources and to alleviate environmental degradation resulting from irrigated agriculture is more important.

II. PHYSICAL DESCRIPTION OF IRRIGATION SYSTEM

A. Subsystems

An irrigation system can be divided into four major subsystems, as shown in Table 1.

The farm subsystem is considered the "heart" of an irrigation system; it performs the system's primary function of food growth for humans and

TABLE 1 Irrigation Physical and Organizational Subsystems

Physical irrigation subsystems	Organizational irrigation subsystems
Water source(s)	
Watershed-catchment	
Collection and storage	
Surface flows	
Groundwater	
Water delivery system	
Main	Main water delivery subsystem
Tertiary	Unit command area(s) or tertiary subsystem(s)
Farm subsystem(s)	
Water application	
Precipitation	
Water applied	
Water infiltrated	
Root zone storage inflow	
Water use	
Evapotranspiration	
Root zone storage outflow	
Water outflow	
Tailwater runoff	
Deep percolation	
Water removal subsystem	
Surface	
Tertiary	
Main	Main water removal subsystem
Subsurface	

animals. This "biological action" takes place in the root zone and in the microclimate within the crop canopy immediately above the ground surface. In other words, the farm subsystem is where the physical system and the biological system come together. The water source(s) feed the water delivery subsystem, and the water delivery and water removal subsystems support the farm subsystem.

The farm subsystem begins at the point where water is delivered to the farm and continues to the point where surface water is removed from the farm (which does not occur with basin irrigation). Also, the farm subsystem is defined vertically as beginning at the ground surface and terminating at the bottom of the root zone.

The water removal subsystem consists of (1) the surface tailwater runoff from the lower end of the farm (which does not occur for basin irrigation unless there is severe flooding), and (2) water moving downward below the root zone as deep percolation. The underlying groundwater is a part of the water removal subsystem, but it often also serves as a water source. Thus, the groundwater aquifer(s) may be transporting deep percolation flows from croplands away from the irrigation system, may serve as an underground reservoir to be tapped by wells for pumping as either water supply or drainage, or may be providing all these functions.

The unit command area (UCA), or tertiary subsystem, is the cropland served by an outlet from the main water delivery subsystem. For most places in the world, the UCA is managed, operated, and maintained by the group of farmers receiving water from an outlet. The UCA consists of (1) a tertiary water delivery network; (2) a number of fields and farms; and (3) a tertiary water removal network. As a subsystem, it has been neglected by investigators. The effects of irrigation timing, frequency, duration, and discharge on optimal cropping mixes and higher levels of agricultural production are critically interdependent. Conjunctive use of groundwater and surface water resources is often significant within UCAs because of the irrigator's concern regarding the unreliability of irrigation project water. Other issues needing increased attention are (1) the priority of water uses within the UCA, including the equity in water distribution between the head and tail of the UCA; (2) formal and informal arrangements for operation and maintenance of the water delivery and water removal networks within the UCA, including the operating rules and communication linkages with the main water delivery and removal subsystems; and (3) changes in the nature of irrigation from season to season and year to year.

The primary conveyance and distribution metwork, or main water delivery subsystem, is the most deterministic of the irrigation subsystems (Merkley et al., 1988). As a result, a variety of operational strategies have emerged that can be broadly classified as continuous flow, rotation, and demand operation. Where farms are large and communications systems are advanced, as in the United States, canal operators can respond to requests for water from each individual user. In developing countries, farms are much smaller and communications with numerous individual farmers are difficult. Thus, important issues concern (1) the methods of improving water delivery flexibility at reasonable cost; (2) how to modify existing institutional arrangements to accommodate desirable changes; and (3) the preparation of trained personnel to implement the changes.

The nature of the catchment has an important bearing on irrigated systems. The size, shape, topography, and soil and vegetative characteristics of the drainage area dictate the runoff hydrographs at the head works of the main water delivery subsystem. These parameters also affect sediment and debris in the flows, which may increase the need for maintenance within the conveyance system. A few of the issues of concern with regard to the catchment area are (1) integrated management of storage and control facilities; (2) supplementing direct flow diversions; (3) forecasting and monitoring runoff; (4) treatments of the watershed to control sediment and debris; and (5) flow augmentation.

B. Hydraulic Performance

From an irrigation viewpoint, hydraulic performance is used as an indicator of the effectiveness of the irrigation system. More appropriate measures are crop yields, net income, and nutrition. When improving an existing irrigation system, however, hydraulic performance parameters are more appropriate indicators of the types of improvements needed for the various irrigation subsystems. The intent is to continually increase the hydraulic performance with the assumption that this will also result in increased agricultural production. No specific quantitative values of these parameters indicate whether a particular irrigation project is good or bad. Rather, this is a "site-specific" question that depends upon numerous factors, including the degree of maturity of the project and the degree of water scarcity compared with crop water requirements.

Field or Farm Level

When irrigating a crop, three efficiency terms and one infiltrated water distribution uniformity term are required to adequately describe the hydraulic performance of an irrigated field (Hart et al., 1979). Numerous (more than 20) efficiency terms are used, but the most common is irrigation application efficiency E_a, which is defined as

$$E_a = \frac{\text{volume of water added to root zone storage}}{\text{volume of water applied to field}}$$

Common values of E_a are 15-65% for surface irrigation methods (basins, borders, and furrows), 50-80% for sprinkle irrigation, and 80-90% for trickle (drip) irrigation.

Project Efficiency

The project efficiency E_p measures the amount of water diverted into the head works during a season that is consumed by plants as evapotranspiration during the season.

$$E_p = \frac{\text{volume of evapotranspiration by crops}}{\text{volume of water received at project head works}}$$

Common values of E_p are 20-60%. Some irrigation projects in mountainous regions may have values of E_p as low as 10%, but often there is no need for improvements because of plentiful water supplies and adequate natural drainage. For example, the Lam Nam Oon Irrigation Project in Northeast Thailand has E_p values of 15-20% but is considered a highly successful project. In contrast, some irrigation projects in California have values of E_p near 80%, yet because of environmental degradation, the project efficiencies need to be increased.

III. IRRIGATION TECHNOLOGY DEVELOPMENT

A. Pressurized Irrigation

The primary advantages of pressurized (sprinkle and trickle) irrigation systems, compared with surface irrigation methods, are (1) the design application rate is less than the soil infiltration rate so that the flow control remains in the closed conduits (pipelines) rather than in the soil; (2) small depths of application can be used by reducing the operating time; and (3) the closed conduit network can be easily automated.

During recent decades, considerable equipment has been developed for pressurized irrigation systems. Sprinkle irrigation equipment has quickly evolved from hand-move to wheel-move and, now, automated center-pivot and linear-move systems. A large assortment of highly reliable equipment is now available in the world market.

Trickle irrigation is a relatively recent innovation that is rapidly becoming popular for high-cash-value crops, such as vine plants and trees and, in some cases, vegetables. Trickle irrigation is becoming quite common for irrigating trees and shrubs in newly developed commercial landscapes. During the past decade, an assortment of reliable filters, valves, and emitters have been developed that will ensure the continued growth of this industry.

The vast majority of the literature on irrigation scheduling pertains to pressurized irrigation systems. Besides advising the farmer when to irrigate and how much water to apply, the private irrigation scheduling service also recommends to their clients the amount of fertilizer that should be applied, as well as weekly scouting each field for pests and making recommendations for chemical applications.

B. Surface Irrigation

Unfortunately, improvements in surface irrigation methods have lagged far behind pressurized irrigation system developments. The progress in surface irrigation technologies since 1975 is impressive, yet there is much research and development work remaining to be done.

In March 1975, the use of laser controls for land-leveling equipment was sufficiently perfected to be suitable for commercial use. Prior to this time, the only use for the basin irrigation method in the United States was for rice production. Now, laser land-leveling equipment is used to construct basins 4-16 ha in size. Large discharge rates are applied to a basin (e.g., 0.5 $m^3\ s^{-1}$), and very high irrigation application efficiencies are achieved (e.g., 80%). Since most of the irrigated land in the world uses the basin method and low irrigation application efficiencies usually occur (20-50%) because of large undulations (10-25 cm) in the land surface of only 0.1-0.5 ha as a result of animal cultivation, the development of small land-leveling equipment using laser controls is an extremely high priority. This would allow a farmer to apply small depths of water during each irrigation event, which in turn means that more basins can be irrigated during the time interval that the water is made available to the farmer.

Stringham and Keller (1979) introduced the concept of surge flow irrigation in 1979, which can be applied to furrow, border, or basin irrigated fields. This technology is readily understood by farmers because it corresponds with their field experiences while irrigating. This phenomenon varies considerably from one soil type to another and because of cultivation practices but essentially consists of pulsing the water onto a field in 10-60 min cycles in which the water flows into one field for half of the cycle time and is then diverted into an adjacent field for the other half of the cycle time. The result is that the water advances to the tail end of each field in roughly the same time as if all of the flow had run continuously onto only one field. So, the average depth of application during surge flow advance is only half of continuous flow advance, resulting in better irrigation application efficiencies and distribution uniformities. Surge flow irrigation can be easily automated with on-off valves. This technology has gained considerable favor in northwest Texas and will expand to other irrigated areas in the United States. Increased labor, rather than valves, can be used to implement this technology to other locations throughout the world.

There are two Saint-Venant equations derived from either momentum or energy and the continuity of mass, which describe fluid flow. A major breakthrough in describing the hydraulics of surface irrigation was reported by Strellkoff and Katopodes in 1977 when they solved the zero-inertia model, a simplification of the Saint-Venant equations, for border irrigation. By 1984, Haie had solved the complete Saint-Venant equations and developed a hydrodynamic model that was then used for comparing the hydraulic performance of continuous flow and surge flow in furrow irrigation. About the same time, the zero-inertia model was solved for basin irrigation, furrow irrigation, and basin-furrow irrigation. By 1985, Walker and Lee (1981) and others had applied the hydrodynamic model, zero-inertia model, and kinematic wave model (a further simplification of the zero-inertia model) to describing surge flow irrigation. These technological developments have completely changed the design procedures for surface irrigation methods.

The greatest difficulty in describing the hydraulics of surface irrigation is knowing the infiltration function. Field measurement techniques, such as cylinder infiltrometers, yield results that commonly have to be adjusted 20-50% based on volume balance measurements. Also, the infiltration function changes for each irrigation event during the season and from season to season. A concerted effort has been made in recent years (Elliott and Walker, 1982; Smerdon and Blair, 1989) to develop techniques for deriving the infiltration function from monitored field advance data.

Surface irrigation scheduling technology has recently been reported (Shafique and Skogerboe, 1987) that utilizes recent innovations in (1) deriving the infiltration function from monitored advance data, and (2) hydraulics of surface irrigation, particularly the hydrodynamic model and zero-inertia model. Knowing the advance function and the infiltration function for the forthcoming irrigation event is crucial to advise the farmer about how much time the water should be allowed to run onto the field knowing the inflow discharge rate.

Research efforts have been underway during the past two decades to develop equipment for automating the operation of surface-irrigated fields. At present, there are some gates (Haise et al., 1980) that can be placed in open channels and opened or closed using mechanical timers; there will be continual improvements made with these gates, including the use of electronic timers. The acceptance of surge flow irrigation has prompted the development of reliable on-off valves for automatic operation. However, there is still a tremendous need for research and development of more reliable equipment for automating surface irrigation. Each succeeding decade will see important strides made in automation equipment.

One of the recent applications (Merkley, 1988) of the hydrodynamic model is for computer operation of irrigation channel networks. This model accommodates unsteady flow phenomenon, such as the filling and emptying of channels, fluctuations with time of water surface levels as a result of changing gate settings, and time lags.

C. Application of Surface Irrigation Technology to India

India is a country with vast quantities of irrigated land with tremendous differences from one region of the country to another. Thus, the characteristics of irrigation in India cannot be generalized. However, there are numerous irrigation projects in which farmers receive their water supply on a fixed rotation schedule, called a *warabundi*. The common irrigation method is basins. Each farm is subdivided into numerous basins, say 10-30, with each basin perhaps 0.05-0.40 ha. Usually, the farmer cannot irrigate every basin during a rotation turn consisting of a few hours every week but can irrigate 40-70% of the basins. An ideal situation would be to irrigate every basin during each turn by applying small depths of application, say 40-75 mm.

The use of laser land-leveling equipment would allow the smallest depths of application to be applied, say 40-50 mm. Unfortunately, this equipment is at present quite expensive and more suited to basin sizes of 2-10 ha, but if smaller and lower cost laser leveling equipment could be developed, then this could be the most cost-effective solution.

Another technological improvement that overcomes many of the difficulties associated with ground surface undulations in basins is to cultivate beds and furrows within the basin, called basin-furrow irrigation. Any high spots in the furrow beds can be removed by the farmer using a shovel during the first irrigation event of the season. Because the irrigation water is only conveyed in furrows, rather than the entire ground surface, the advance to the end of the field is more rapid. The employment of surge flow irrigation in conjunction with basin-furrow irrigation should double, or more, the number of basins irrigated during the rotation turn. This combination of surface irrigation technologies should allow average depths of application to be 40-50 mm.

The use of surge flow irrigation on conventional basins cultivated with animals will have undulations that vary between 100 and 250 mm between low spots and high spots. In these cases, even though the advance to the end of the field will be more rapid using surge flow, the required depth of application is likely to be 75-200 mm to cover the high spots so they do not become salinized; thus, it would not be likely that all the basins could be irrigated during a single rotation turn.

IV. MODELING IRRIGATION SOIL AND WATER PROCESSES

The previous section presented some recent innovations in irrigation technology. This section presents the recent development of models that simulate soil and water processes in agriculture. The combination of irrigation technology and modeling capability greatly influences the processes that can be applied in implementing soil and water management programs that will increase the agricultural productivity of irrigated lands.

A. Plant Growth Simulation

In recent years, important developments have been achieved in simulating plant growth. An understanding of crop physiology is required for such simulations. These models predict on a daily basis the crop's response to its environment. Such predictive tools are highly important in evaluating management alternatives during the season upon crop yields and eventually net income.

Marani and Baker (1981) have developed comprehensive cotton growth simulation models. Similar plant growth simulation models can be expected for many of the grain crops, such as sorghum and wheat (Maas and Arkin, 1980; Jones and Kiniry, 1986; Jones et al., 1988). It is hoped that plant growth simulation models will be available for the major crops in the world during the next decade, but it could take longer. However, it will be necessary to locally calibrate any of these models if they are to be used for implementing field programs.

B. Salt Transport

An important development in characterizing salt transport in irrigated soils was the model reported by Dutt et al. (1972), which was subsequently modified by the U.S. Bureau of Reclamation (1976). Ayars et al. (1981) tested this model using field data from numerous large lysimeters (30 × 30 m ground area and 3-4 m deep).

Bresler et al. (1982) report recent models of salt transport that can be applied to large-scale agricultural problems. Certainly, a tremendous amount of knowledge has been acquired during the past two decades in applying soil physics and soil chemistry modeling to solving regional salinity problems.

C. Soil Erosion

Most of the research on soil erosion from agricultural lands in the United States has been done on rainfed areas. The most notable output from decades of soil erosion research is calibration of the universal soil loss equation (USLE) for nonirrigated agricultural waterbeds east of the Rocky Mountains (Wischmeier and Smith, 1978). The USLE is very useful for planning studies, but determination of actual sheet and rill erosion from specific agricultural lands requires site-specific field data for local calibration.

Most soil erosion on irrigated land results from furrow irrigation (Carter and Bondurant, 1976), which is the most common irrigation method in the western United States. Because the basin and border irrigation methods are practiced on either level or flat slopes, soil erosion is generally a minor problem.

Reported sediment concentrations in surface irrigation return flows vary widely. Carter (1976) has reported guidelines for sediment control in irrigated agriculture.

D. Nutrient Transport

The two nutrients most commonly investigated in irrigation systems are nitrogen and phosphorus. Phosphorus is tightly held by soil particles, and essentially all phosphorus encountered in surface runoff from irrigated lands is associated with sediment (Carter and Bondurant, 1976). Consequently, at best, the ability to predict phosphorus loads in surface irrigation return flows is commensurate with the ability to predict sediment loads in surface runoff, which is not very good at this time.

During the past decade, tremendous strides were made in developing models that describe nitrogen transformations, plant uptake, and movement through soils. Nielsen and MacDonald (1978a, b) have reported the results of an extensive research program by faculty in the University of California system in *Nitrogen in the Environment*. There are models available for predicting nitrogen concentrations and loads in subsurface irrigation return flows, provided the necessary field data has been collected for locally calibrating these models.

E. Transport of Biocides

A major portion of the U.S. research on the transport of biocides has been supported by the Environmental Research Laboratory at Athens, Georgia, which has focused upon rainfed agriculture. There is little published information on biocide concentrations in either surface irrigation return flows (Carter and Bondurant, 1976) or subsurface irrigation return flows. This is a major weakness in research programs to date. There is a definite need to assess research studies on biocide transport in rainfed agriculture as to its transferability to irrigated agriculture.

F. Surface Irrigation Hydraulics

Excessive irrigation water losses in most surface irrigation systems usually occur because of improper design and/or operation of the system. Traditional design and operation of surface irrigation methods have remained dependent on past experience. However, over the last decade, mathematical models have been developed to assist in designing irrigation systems, as described earlier. These models have been successfully applied in predicting the surface irrigation advance phase in border (Clemmens, 1979), basin (Clemmens et al., 1981), and graded furrow systems (Elliott et al., 1982). Zero-inertia modeling has also been applied to level basin-furrow fields to complete the application process of the model for all important surface irrigation methods. Walker (1989) has prepared a design manual with microcomputer software for all surface irrigation methods.

V. APPLICATION OF PROCESSES IN IMPLEMENTING SOIL AND WATER MANAGEMENT PROGRAMS

Recognizing the site-specific nature of irrigated agriculture, in which each project area is uniquely different, an effective approach for improving the performance of irrigated agriculture must be process oriented, rather than an emphasis on technology alone or a "prescriptive" approach that lists step-by-step procedures to be used on every irrigation project. Although a prescription is usually preferred by most irrigation officials, the disadvantages are (1) the procedures lead to less than optimal results for most projects; and (2) project field personnel do not "learn" how to accommodate the unique characteristics within their project area to improve the performance of the system. Instead, a process (or a series of processes) is required that is capable of being adapted to each site-specific situation to be transferable.

The maintenance and operations learning process provides one technological approach for effectively sustaining an irrigation network over a long time period (Skogerboe, 1988). The development process for improving irrigation water management on farms is directed toward increasing the productivity of existing irrigated lands, improving the equity of income distribution and resource conservation, thereby ensuring the long-term viability of the system (Skogerboe et al., 1980). A planning framework, or process, for waterlogging and salinity management in irrigated agriculture has been developed that delineates the combinations of technological and institutional solutions to be implemented in a salinity control program (Skogerboe et al., 1979).

A. Main Subsystem Management

Most countries in the world have already constructed the most economical irrigation projects. Consequently, the major focus for these countries in the future will be on improving management practices to increase crop production on existing lands. However, deteriorating irrigation channels and inadequate operating procedures often preclude any significant improvement. The irrigation maintenance and operations (M&O) learning process (Fig. 1) has been developed to provide guidelines that will (1) identify problems that commonly prevent irrigation systems from operating effectively; (2) develop solutions that treat the causes of these problems, rather than the symptoms alone, to more effectively sustain improved maintenance and operations; and (3) provide field experience and insights for further improvements in the irrigation system.

A common problem in irrigation systems is the continual cycle of irrigation system construction, followed by deterioration of the system because of inadequate maintenance, and then rehabilitation (which ends up being major reconstruction), followed again by deterioration. This significantly inhibits the development of irrigated agriculture. As the irrigation channel network deteriorates, there is a decreasing capability for equitably delivering water throughout the system. The adequacy and dependability of the irrigation water supply decreases for some farmers, depending upon their location; thus they are less likely to risk other expensive agricultural inputs, such as fertilizer. In some cases, a portion of the land must be abandoned for lack of water. The consequence is very slow progress in agricultural production.

The recurring cycle of construction-deterioration-rehabilitation (construction)-deterioration usually precludes the timely and equitable distribution of irrigation water supplies. Besides resulting in stagnant growth in agricultural production, the construction phase adds to the national debt burden, which has now become a painful problem for many countries. In

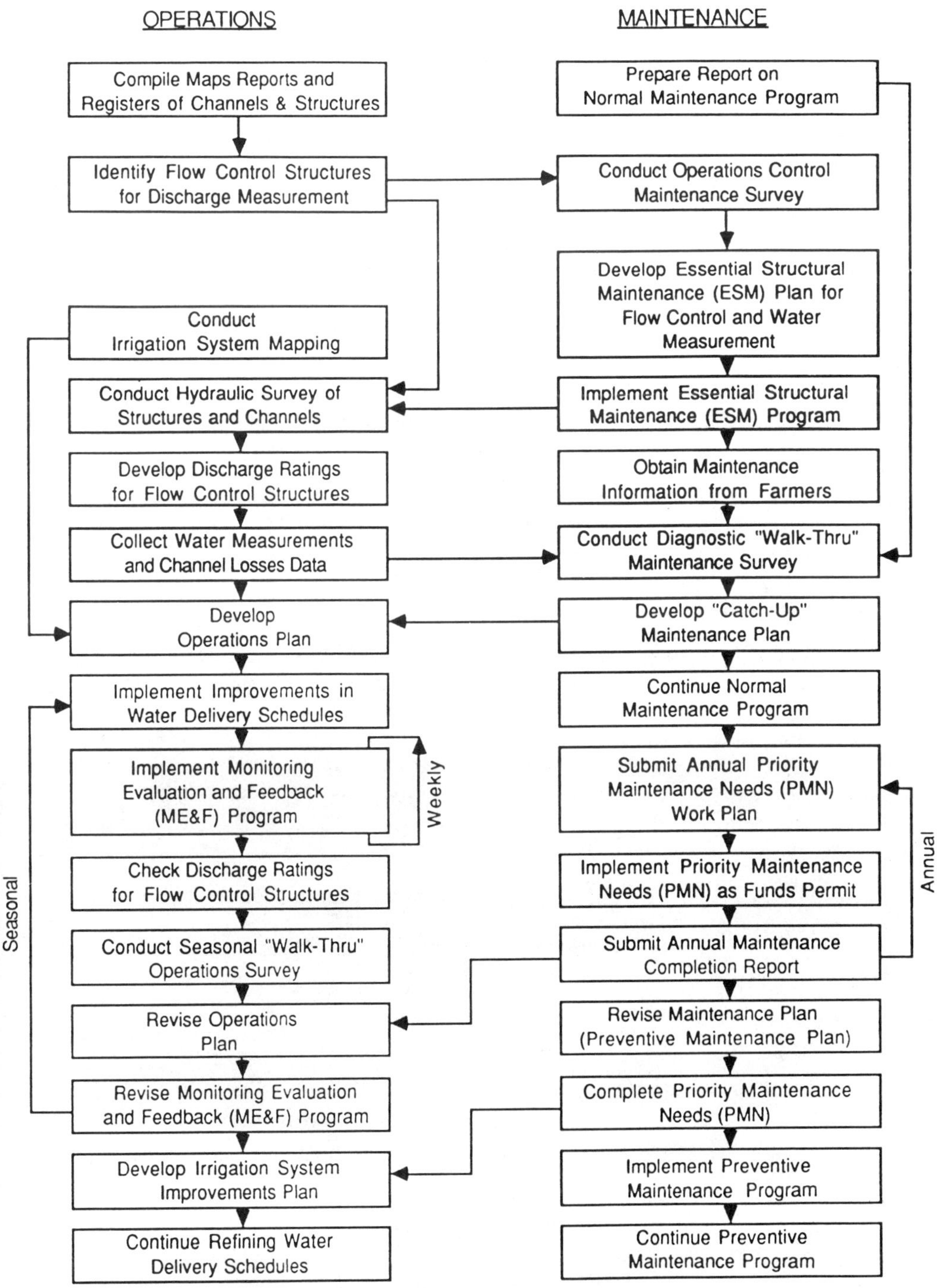

FIGURE 1 Irrigation maintenance and operations (M&O) learning process.

general, the economic benefits resulting from irrigation investments in recent decades have been dismal. Not only donors are questioning irrigation investments, but senior officials in a number of countries are raising the question: Isn't there a better way?

The maintenance and operations "learning process" provides one technological approach for effectively sustaining an irrigation network over a long time period. This process emphasizes (1) maintaining rather than rehabilitating; (2) documenting maintenance needs to improve financial management and accountability; (3) using existing flow control structures in irrigation channels for water measurement; (4) developing more detailed physical knowledge about what is occurring within the system; (5) increasing sensitivity about operating the system to meet the needs of farmers; and (6) documenting the needs and costs for irrigation system improvements.

B. On-Farm Soil and Water Management

The development process for improving irrigation water management on farms has two important themes: (1) an interdisciplinary approach, and (2) farmer-client involvement. Physical and social scientists work together with farmers in identifying major constraints to increasing agricultural productivity while conserving natural resources. Acceptable solutions are developed for priority problems in collaboration with farmers. Finally, a solution package is implemented that utilizes the resources of both the farmers and the government (Table 2).

Together, the three phases comprise an entire development process. In the problem identification phase (Fig. 2), the project staff seeks to understand and diagnose the agricultural system as it exists. In the development of solutions phase (Fig. 3), alternative designs to correct problems in the system are identified and evaluated. In the project implementation phase (figure not shown), the program staff attempts to change the present system to a better one by implementing a set of solutions.

TABLE 2 Phases and Subphases of the Development Process for Improving Irrigation Water Management on Farms

Phase	Subphase
Problem identification	Reconnaissance
	Problem diagnosis
Development of solutions	Identification of plausible solutions
	Testing and adaptation of solutions
	Assessment of solution packages
Project implementation	Project authorization
	Project organization
	Project operation

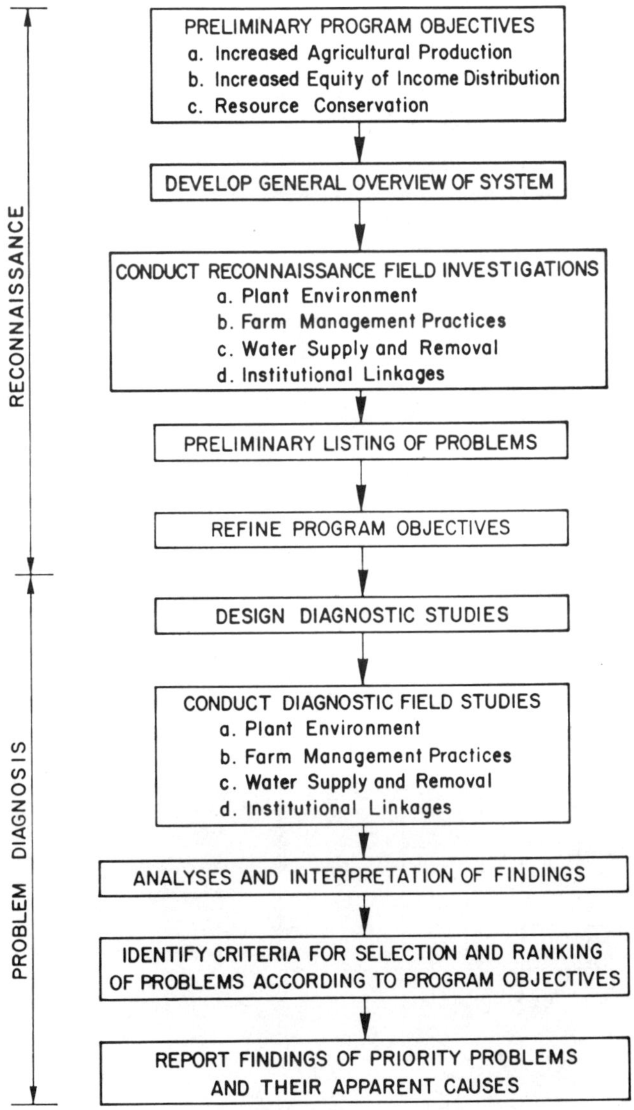

FIGURE 2 Problem identification phase of the development process for improving irrigation water management on farms.

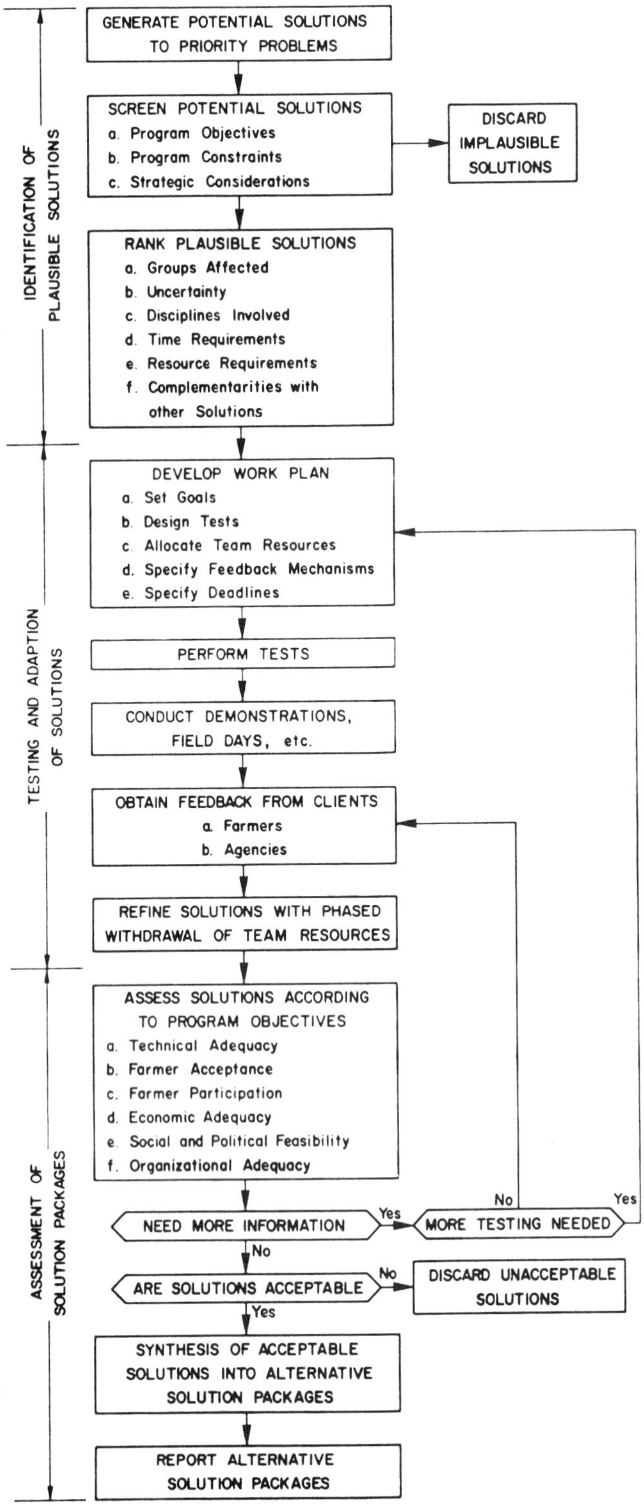

FIGURE 3 Development of solutions phase for the development process for improving irrigation water management on farms.

C. Waterlogging and Salinity Control

Planning Framework

An accurate assessment of waterlogging and salinity problems in an irrigated area requires detailed knowledge not only of the hydrology, but many other unique physical and institutional characteristics of the area being studied. Even though each area must be evaluated individually, because of the site-specific nature of irrigated agriculture, definite procedures are common to all evaluations. Thus, a framework for planning a waterlogging and salinity control program can be developed by modifying research reported by Skogerboe et al. (1979). This planning framework can be applied to any individual irrigated area (Fig. 4) or to an entire river basin. The waterlogging and salinity control planning framework consists of three phases: (1) identification of waterlogging and salinity problems; (2) development of best management practices; and (3) implementation of best management practices.

Waterlogging and Salinity Control in Pakistan

About 62% of the Indus Basin lies in Pakistan. The mean annual rainfall varies from 50 to 1000 mm. Approximately two-thirds of the total area receives less than 250 mm rainfall. The Indus River Basin plains start below the Salt Range about 800 miles from the sea. It is a vast land of about 21 million ha with very flat topography, having an average slope of 0.0002. The total irrigated area in the region is 12-14 million ha (Shafique and Skogerboe, 1984).

The Indus irrigation system is shown in Fig. 5. Its history began with agricultural practices carried out on the banks of rivers made moist by swollen and overflown streams, which evolved into the construction of "inundation canals." By the middle of the nineteenth century many such inundation systems existed. The British initiated a massive development program about the turn of the century that gave Pakistan the world's largest contiguous irrigation system. After independence in 1947, construction works continued with major dams, such as the Tarbela and Mangla, barrages, large canals linking rivers, irrigation canals, more than 200,000 tubewells, and 80,000 watercourses that deliver water to the farmers' fields. There are about 3 million farmers with a median farm size of 2.1 ha. An average watercourse serves about 40 farmers, and a normal village of 100-120 farms has approximately three watercourses.

The irrigation system is constructed so that water flows from minor canals through turnouts (*moghas*) supplying water to a village area. The farmers themselves operate and maintain their distribution system. There are no headgates at the moghas and if a particular canal has water in it, there is water in every watercourse (*khal*) along that canal. The farmers on the watercourse use the entire flow on turn (based on farm size), and there is a fixed time each week, 10 days or 2 weeks, to use this water. The system of rotation is called *warabundi*. The level basin method of irrigation is used extensively.

Natural drainage ways are not readily apparent. A few artificial drains have been constructed; therefore, surface drainage for irrigation water is virtually nonexistent from either artificial or natural drains. Irrigation water passing through each mogha, as well as surface runoff during the

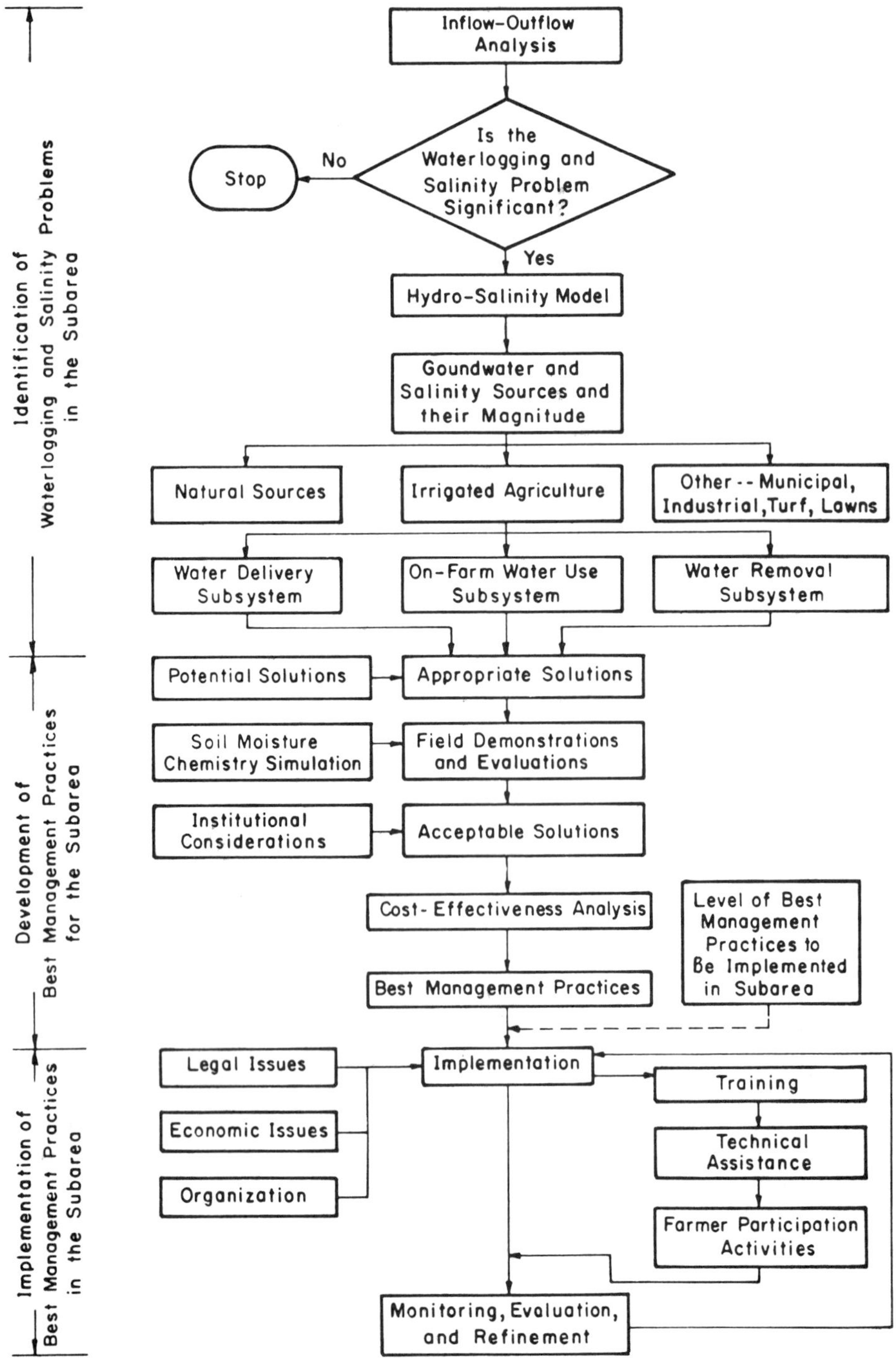

FIGURE 4 Planning framework for waterlogging and salinity control.

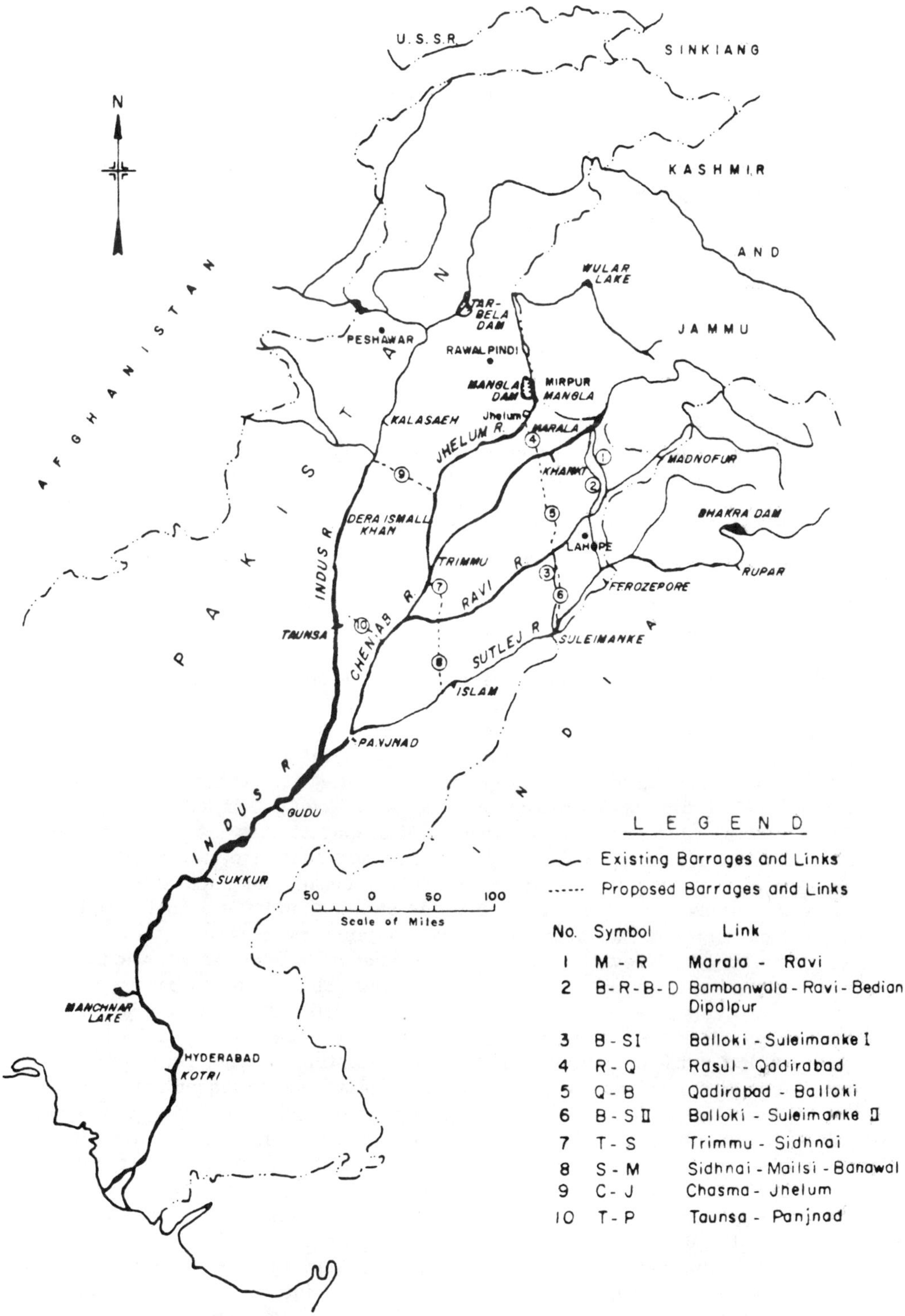

FIGURE 5 Indus River basin.

monsoon season, is therefore utilized on that particular watercourse area as there is no provision to let it pass to an area of lower elevation.

In this arid and semiarid region, the salts were deposited in varying degrees when the alluvium was laid down. Prior to the major conversion of inundation canals to weir control canals, the water table in the center of the *doabs* (land between two rivers) and in other portions of the Indus was 80 feet or more below the surface. The water table has gradually risen over a large area. A 1975 survey indicated that 6.9 million ha of canal-culturable commanded area (CCA) was underlaid by a water table within 3 m and nearly 80,000 ha had become unculturable because of the high water table. Because of the high water table and lack of drainage infrastructure, hundreds of thousands of hectares have become saline. Many thousands of hectares are going out of cultivation each year owing to waterlogging and increased salinity. The average rise of the groundwater table has been estimated between 15 and 60 cm $year^{-1}$.

The latest survey also reveals that of 14 million ha of CCA, nearly 5.2 million ha of land is severely saline and sodic. According to another survey supported by laboratory analysis, this badly salt-affected area is more than 5.8 million ha. Estimates prepared by FAO and UNESCO report 80% of the irrigated area of this region is affected by the problem to varying degrees.

The development of irrigated agriculture in the past has remained focused almost entirely upon the construction of water delivery subsystems. Moreover, in the past, even the increased water availability has not been matched by the expansion in acreage. As a matter of fact, increased water supplies have resulted in increased recharge to the groundwater aquifer because of increased system losses. More than 60% of the groundwater recharge takes place from watercourse seepage losses beyond the canal outlet and deep percolation from croplands. At present, recharge to the aquifer is at least two times greater than the water being pumped by tubewells. The on-farm water losses along with canal seepage losses are directly responsible for rising groundwater levels and degradation of return flows. This problem is steadily getting worse simply because of an inordinate focus on developing more and more water supplies and the almost complete neglect of improved soil-plant-water management practices.

Achieving high levels of water use efficiency to minimize canal deliveries and prevent or control waterlogging and salinity due to return flow is both difficult and expensive. Potential solutions and control measures involve physical changes in the system, which can be brought about by constructing sufficient improvements to the existing system, or placing new institutional influences on the system, or a combination of both.

The most significant improvements in controlling waterlogging and salinity will potentially come from watercourse improvement and improved on-farm water management practices. Pilot programs have been undertaken since 1977, primarily to increase crop production, which involve watercourse improvement, precision land leveling, and some emphasis on water management. By 1990, these programs had grown immensely.

In each subarea, the magnitude of the problem must be determined and the salinity sources delineated. This should include the magnitude of subsurface irrigation return flows resulting from seepage losses and deep percolation losses, along with changes in the chemical composition of these flows enroute to the groundwater reservoir and subsequently the Indus River. Appropriate solutions should be demonstrated and evaluated on farmers' fields to develop alternatives that are acceptable to both the farmers and the implementing agencies. In addition, cost-effectiveness analyses must be established.

The second step is to define the best management practices by optimizing the cost-effectiveness relationships from individual areas into a single strategy. This planning function delineates the level of salinity control required in each individual subbasin to achieve the overall goal for waterlogging and salinity control with the least cost. Relative levels of implementation among the areas are also determined.

Finally, the third step assigns the required level of salinity control that should be implemented in each subbasin or irrigated area. The basin planning process for salinity control involves integration of planning studies at the local level into the total basin framework to define the best management practices on a basinwide scale. By integrating the results outlined in the first step, the best management practices for salinity control in each area are identified.

D. Best Pollutant Management Practices

Agricultural technology is one of the real strengths of the United States. Although the population has increased steadily, food and fiber production has met the domestic needs and has also provided substantial amounts for export, which is so important to the U.S. trade balance. Fertilizers and pesticides have played a major role in this accomplishment because the acreage of cropland has changed little in the last 50 years: agricultural chemicals and other technological inputs have been substituted for land.

In today's concern for clean water, there are demands for preventing pollutants from entering ground and surface waters. Consequently, agricultural practices are being examined as to their contribution to water pollution. Also, there are several national goals that impinge on the subject of water quality management in agriculture. Besides reducing water pollution to have cleaner water, it is also necessary to continually increase crop production and to reduce energy consumption

Sources of pollutants are usually classed as "point" and "nonpoint." Point sources have received most of the attention and are covered by a permit system. The permit defines the requirements and the compliance schedule that must be followed. Nonpoint sources are diffuse in nature and discharge pollutants into waters by dispersed pathways. Examples are construction activities, mining areas, stormwater runoff from urban areas, and agricultural runoff.

Nonpoint sources are difficult to control if control is defined as the ability to establish and enforce effluent standards. However, in the United States, the emphasis for control of nonpoint source contributions is being placed upon "best management practices" (BMP) rather than on the collection, treatment, and effluent standards approach used for the control of point sources.

The nonpoint agricultural pollutants of primary concern are sediment, such nutrients as nitrogen and phosphorus compounds, and pesticides.

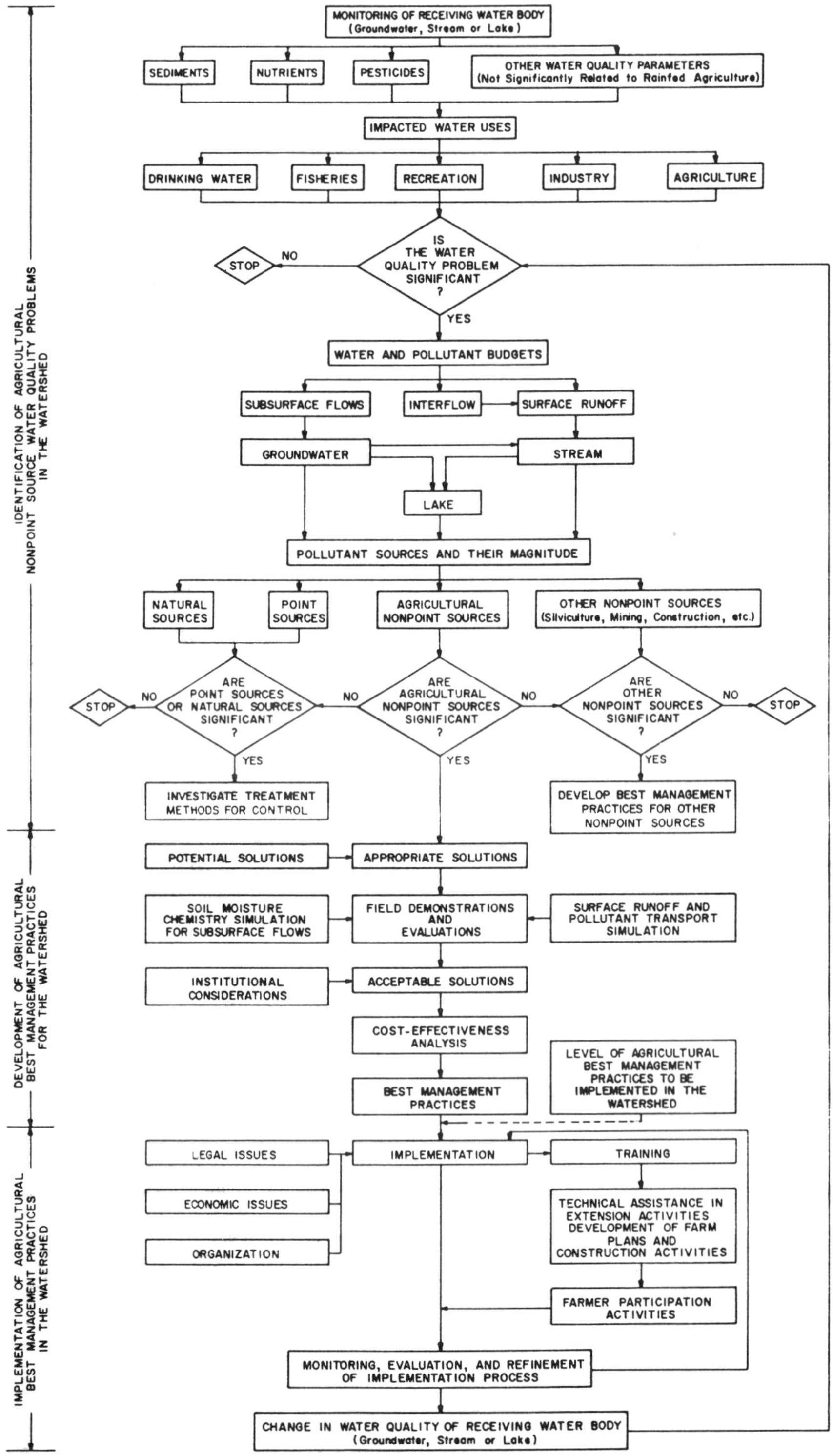

FIGURE 6 Planning framework for identifying, developing, and implementing solutions to water quality problems from agriculture. (From Skogerboe et al., 1979.)

Oxygen-consuming organic compounds, dissolved salts, and other pollutants are of less general concern. Pathogens and heavy metals may be of concern in some locations. It appears that sediment, nutrient, and pesticide losses can be controlled at an acceptable level by the selection of proper management systems.

Implementation of best management practices will not only reduce pollution from agriculture but may also increase crop production as a result of improved soil and water conservation practices, as well as improved management practices in the application of agricultural chemicals. The necessity for pollution control programs and the concomitant benefits to other national goals will continue to direct more and more attention to the control of agricultural nonpoint sources. Consequently, more agricultural areas will need to be evaluated and questions regarding the quantity and quality of agricultural nonpoint sources of pollution will need to be answered.

Because many of the nonpoint pollution parameters tend to be site specific, every area should be evaluated individually. The same conditions encountered in one area may result in an entirely different set of consequences in another area. The challenge, therefore, is to develop appropriate assessment techniques and institutional mechanisms so that proper controls are used where needed.

A rational framework (Skogerboe et al., 1983) is presented by which any agricultural area suspected of contributing significant amounts of nonpoint source pollution can be evaluated. The similarities between this process (Fig. 6) and waterlogging and salinity control (Fig. 4) are quite obvious. Both processes utilize the scientific approach in identifying constraints, developing solutions, and implementing solutions.

VI. STRATEGIES FOR IMPROVING THE PRODUCTIVITY OF IRRIGATED AGRICULTURE

A. River Basin Development

The purpose of irrigation is to increase crop production, but the real difficulty lies in sustaining increased crop production over a long time period. History is replete with hydraulic societies that flourished and then floundered, or became extinct. Why?

Maintaining or increasing agricultural productivity in an irrigated area requires, first, that a salt balance be achieved in the root zone; and second, that not too much water is applied such that the groundwater levels rise until the water table is near the ground surface, thereby resulting in waterlogging and increased salinity levels in the root zone. Thus, a balance must be reached in order that sufficient water is applied to the croplands to leach salts from the root zone, but not so much water that groundwater levels nearly reach the ground surface. The history of hydraulic societies has been primarily one of applying too much water. And over-irrigation continues today in most portions of the world.

In many regions and countries, increasing urbanization and industrialization require the reallocation of agricultural water uses to meet new water demands. To accomplish this water reallocation, improved water management practices must be instituted in the agricultural sector, thereby reducing agricultural water diversion requirements.

If the quantity of pollutants (e.g., salinity) in the return flow is large in relation to the river flow, then it is very likely that the water is not suitable for the next user unless the water is treated to remove objectionable constituents. Since water is diverted many times from the major rivers, the river flows show a continual degradation of quality in the downstream direction. As the water resources become more fully developed and utilized, without controls, the quality in the lower reaches of the river will likely be degraded to such a point that the remaining flows will be unsuitable for many uses, or previous uses of the waters arriving at the lower reaches of the river basin will no longer be possible.

As pressures on water resources become greater, because of higher degrees of water utilization in a river basin and due to increasing populations and the necessity to produce food in increased quantity and improved quality, there is a mounting concern for proper control of serious water quality deterioration and soil salinization. The need, then, for more precise information as a basis for wise policy action is a matter of critical importance.

B. Improving Existing Irrigation Systems

Much of the emphasis in the future will have to be on improving what we have. In other words, water management will have to be improved through better operations and maintenance practices and greater hydraulic performance from the farm subsystem, crop production increased by using better combinations of agronomic and irrigation practices, and waterlogging, salinity, and other pollutant problems alleviated by implementing the above-mentioned practices to an appropriate degree that also manages environmental problems.

The easiest program to implement and achieve dramatic results in a few years is the operations and maintenance learning process. The process for improving on-farm soil and water management practices is much more difficult and time consuming, mostly because of the complexity of the farm subsystem, which requires a highly client oriented and interdisciplinary approach. The on-farm program is enhanced if the O&M learning process is well underway. The program for waterlogging and salinity control is significantly enhanced if both the O&M learning process and on-farm programs are well underway or completed. Finally, implementation of a waterlogging and salinity control program will also benefit a program for controlling other pollutants, but more emphasis may be required on certain farm soil and water management practices.

C. Thailand Irrigation Development Strategy

The primary purpose of an irrigation system is to deliver water to individual farms on a timely basis that maximizes their welfare through increased crop production. In addition, the irrigation water supplies must be equitably distributed so that all farmers share equally in project benefits.

During recent decades, the emphasis has been on constructing new irrigation projects to increase the amount of cropland that can be served by irrigation networks. Now, increasing attention is being given to making existing irrigation projects more productive. This change in emphasis

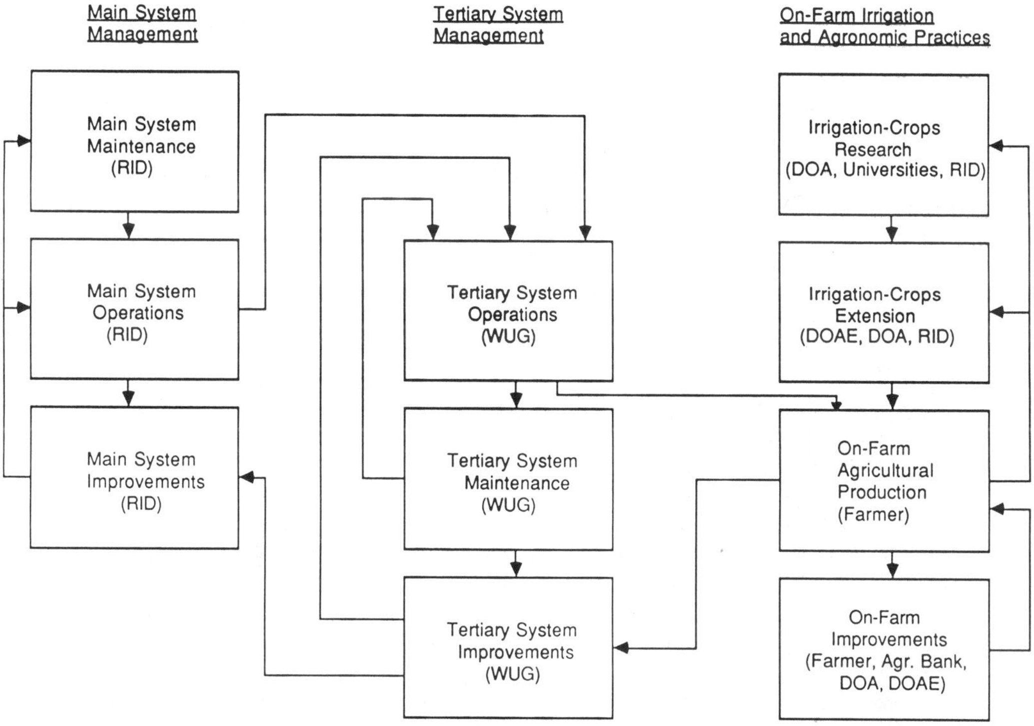

FIGURE 7 Improving the productivity of irrigation systems in Thailand.

is reflected in the major focus of the Royal Irrigation Department (RID) in Thailand now being "operation and maintenance" rather than "construction."

To improve the productivity of existing irrigation systems, the major thrusts shown in Fig. 7 will be: (1) main subsystem management; (2) tertiary subsystem management; and (3) on-farm irrigation and agronomic practices. Main subsystem management will be the primary responsibility of RID, and the water users groups (WUGs) will be responsible for tertiary subsystem management with technical assistance provided by RID. On-farm irrigation and agronomic practices will be targeted to individual farmer's fields with technical support from the Department of Agriculture (DOA), universities, Department of Agricultural Extension (DOAE), and RID. Both the DOA and the universities will provide agricultural research support (unpublished draft paper).

VII. FUTURE IRRIGATION DEVELOPMENT

Whether the goal is minimizing diversions to new croplands because of limited water supplies, reducing future diversions to irrigated agriculture to provide water supplies for new demands, maximizing the effectiveness of interbasin water transfers, minimizing water quality degradation in receiving streams resulting from irrigated agriculture, or maximizing agricultural production on existing croplands, the solutions are identical: improved soil and water management practices.

A. Environmental Management

The major problems resulting from irrigation are due to the basic fact that plants are large consumers of water resources. Growing plants extract water from the supply and leave salts behind, resulting in a concentration of the dissolved mineral salts present in all natural water resources. In addition to having a greater concentration of salts in the return flow resulting from evapotranspiration, irrigation also adds to the salt load by leaching natural salts arising from weathered minerals occurring in the soil profile or deposited below it. Irrigation return flows provide the vehicle for conveying the concentrated salts and other pollutants to a receiving stream or groundwater reservoir. It is then necessary to examine the waterlogging and salinity problems resulting from this process and to develop and implement measures to control or alleviate the detrimental effects.

Impaired crop production resulting from salinity is not limited to the western United States but is a major problem in many areas of the world. In fact, the more one observes problems in other countries, the more one is convinced that this is indeed an international problem. Unfortunately, the portions of the world now facing the greatest population pressures are the same areas that have the least amount of additional land available for agriculture and, in some cases, where waterlogging and salinization are a major problem. In such areas, increased food production must come from more intensive farming with consequent increased yields. On the one hand, there is a great need to increase the productivity of such lands, but on the other, agricultural production is continually being damaged due to rising groundwater tables (waterlogging) and increased salinity in the soils and groundwater supplies.

Whether irrigation return flows come from surface runoff or have returned to the system via the soil profile, the water can be expected to undergo a variety of quality changes due to varying exposure conditions. Drainage from surface sources consists mainly (there will be some precipitation runoff) of surface runoff from irrigated land. Because of its limited contact and exposure to the soil surface, the following changes in quality can be expected between application and runoff: (1) dissolved solids concentration only slightly increased; (2) addition of variable and fluctuating amounts of pesticides; (3) addition of variable amounts of fertilizer elements; (4) an increase in sediments and other colloidal materials; (5) crop residues and other debris floated from the soil surface; and (6) increased bacterial content.

Drainage water that has moved through the soil profile will experience different changes in quality from surface runoff. Because of its more intimate contact with the soil and the dynamic soil-plant-water regime, the following changes in quality are predictable: (1) considerable increase in dissolved solids concentration; (2) the distribution of various cations and anions may be quite different; (3) variation in the total salt load depending on whether there has been deposition or leaching; (4) little or no sediment or colloidal material; (5) generally, increased nitrate content unless the applied water is unusually high in nitrates; (6) little or no phosphorus content; (7) general reduction of oxidizable organic substances; and (8) reduction of pathogenic organisms and coliform bacterial. Thus, either type of return flow will affect the receiving water in proportion to respective discharges and the relative quality of the receiving water.

B. Software Development

The resource base for irrigated agriculture has not substantially changed since its inception thousands of years ago. Over the last century there has been little incentive for any major innovation to improve efficiency in the use of water, which now fortunately is considered a scarce resource because of more full utilization of the water supplies in many river basins scattered around the world. The provision of irrigation water since ancient times has been considered a governmental or collective responsibility, and the direct charges made for water have usually not been high enough to encourage proper budgeting of water or other necessary innovations. However, there are a few examples of extremely water-scarce areas in the world where considerable ingenuity has been applied in effectively utilizing water supplies.

Aggravating this situation is the fact that the development of irrigated agriculture in most places, even in the last few decades, has focused almost entirely upon the construction of water delivery subsystems. This preoccupation with the installation of "hardware" results from a naive single-discipline approach to water management (Wiener, 1972), and one discipline cannot begin to solve the complex physical, economic, and sociological problems involved. Probably the greatest deterrent to improved water management in most irrigation systems today is the inordinate focus on the water delivery subsystem and the almost complete neglect of other problems. In reality, though, especially in irrigated agriculture, we come face to face with a wide gap that frequently exists between "hardware development" and the development of all the other requisites for increased agricultural production, namely, "software development."

The approach that has been applied to irrigated agricultural development in the past is characterized by separating the development of water resources from the management aspects of water resource utilization. Therefore, the record shows development being emphasized greatly while management is most often neglected. This orthodox approach has been used almost exclusively in the western United States with reasonable success (but is no longer acceptable). However, as the water resources become more fully utilized, the necessity for meeting new water demands (along with physical, socioeconomic, and political problems of water quality degradation) require that we reject much of the conventional wisdom of the past. It should also be obvious that many other countries have neither the time nor resources many western irrigated regions had to utilize in their development. Pressures created by rapidly rising population rates alone will force them at some point to re-evaluate their approaches.

In contrast to the mere development of water resources approach, the "management" approach attempts to achieve water development objectives by applying a variety of measures after studying the entire system, thereby attempting to modify the total system to meet new and changing demands as well as estimated future demands. Therefore, instead of constructing new engineering works to meet new demands, the focus should be upon water resource management, with construction works considered only as a tool when necessary to meet water management objectives (Wiener, 1972). Unfortunately, in most cases, water management and the many disciplines required to produce efficient management are relegated to the postanalysis of engineering works, which aggravates not only the implementation of technology, but really constrains or makes extremely difficult the implementation of a host of services requiring strong institutional measures.

C. Interdisciplinary Approach

The importance of an interdisciplinary approach has been alluded to many times. This is particularly important for studying and designing programs and implementing field activities associated with the farm subsystem. Unfortunately, there are only a few success stories regarding irrigation interdisciplinary activities. As a consequence, numerous field programs have not been successful in achieving planned targets. When selecting personnel for interdisciplinary studies or field programs, it is important that such persons have a history of cooperating with others and are relatively unselfish, recognizing that it is human nature to be selfish.

D. Organizational Cooperation

A single organization may be responsible for the main subsystem (e.g., irrigation department) or the tertiary subsystem (e.g., farmers), but at the farm level, many organizations are involved, such as a farmer's WUA, irrigation department, agriculture department, agricultural extension, and credit. Thus, the farm subsystem becomes extremely complex. To continually improve farm soil and water management practices, cooperation between the various government agencies and private entrepreneurs becomes highly important. In fact, the degree of success in implementing field programs involving the farm subsystem is dictated by the degree of cooperation among the involved organizations. In the future, the success of irrigated agriculture in any particular country will be largely determined by the degree of cooperation among the government and private organizations impacting farmers, which will also establish the environment for interdisciplinary teamwork that can truly address the multitude of management issues required for sustaining irrigated agriculture and the population of a nation.

REFERENCES

Ayars, J. E., McWhorter, D. B., and Skogerboe, G. V. (1981). Modeling salt transport in irrigated soils. *Ecol. Model. 11*: 265-290.

Bresler, E., McNeal, B. L., and Carter, D. L. (1982). *Saline and Sodic Soils: Principles-Dynamics-Modeling*. Springer-Verlag, Berlin.

Carter, D. L. (1976). Guidelines for sediment control in irrigation return flows. *J. Environ. Qual. 5*: 119-124.

Carter, D. L., and Bondurant, J. A. (1976). Control of sediments, nutrients and adsorbed biocides in surface irrigation return flows. Report EPA-600/2-76-237, U.S. Environmental Protection Agency, Ada, Oklahoma.

Clemmens, A. J. (1979). Verification of the zero-inertia model for border irrigation. *Trans. ASAE 22*(6): 1306-1309.

Clemmens, A. J., Strelkoff, T., and Dedrick, A. R. (1981). Development of solutions for level-basin design. *J. Irrigation Drainage ASCE 107*(IR3): 265-279.

Dutt, G. R., Shaffer, M. J., and Moore, W. J. (1972). Computer simulation model of dynamic bio-physico-chemical processes in soils. University of Arizona, Soil Water Eng., Agr. Exp. Sta. Tech. Bull. No. 196, Tucson.

Elliott, R. L., and Walker, W. R. (1982). Field evaluation of furrow infiltration and advance functions. *Trans. ASAE 25*(2): 396-400.

Elliott, R. L., Walker, W. R., and Skogerboe, G. V. (1982). Zero-inertia modeling of furrow irrigation advance. *J. Irrigation Drainage ASCE 108*(IR3): 179-193.

Haie, N. (1984). Hydrodynamic simulation of continuous and surged surface flow. Ph.D. dissertation, Department of Agriculture and Irrigation Engineering, Utah State University, Logan.

Haise, H. R., Kruse, E. G., Payne, M. L., and Duke, H. R. (1980). Automation of surface irrigation: 15 years of USDA research and development at Fort Collins, Colorado. Prod. Res. Rep. 179, U.S. Department of Agriculture, Washington, D.C.

Hart, W. E., Peri, G., and Skogerboe, G. V. (1979). Irrigation performance: An evaluation. *J. Irrigation Drainage ASCE 105*(IRE): 275-288.

Jones, C. A., and Kiniry, J. R. (Eds.) (1986). *CERES-Maize: A Simulation Model of Maize Growth and Development*, Texas A&M University Press, College Station, Texas.

Jones, J. W., Boote, K. J., Jagtap, S. S., Hoogenboom, G., and Wilkerson, G. G. (1988). Soybean crop growth simulation model. Florida Agr. Exp. Sta. J. No. 8304, University of Florida, Gainesville.

Maas, S. J., and Arkin, G. F. (1980). TAMW: A wheat growth and development simulation model. Program and Model Documentation No. 80-3, Texas Agr. Exp. Sta., Blackland Research Center, Temple, Texas.

Marani, A., and Baker, D. N. (1981). Development of predictive dynamic simulation model of growth and yield in ACALA cotton. Res. Rep. Agr., Hebrew University of Jerusalem.

Merkley, G. P. (1988). Hydraulic modeling applications in main system management. Ph.D. dissertation, Department of Agriculture and Irrigation Engineering, Utah State University, Logan.

Merkley, G. P., Pojsoontoorn, C., Kawsard, K., Walker, W. R., and Skogerboe, G. V. (1988). USU irrigation main system hydraulic model: Replication of modeling capability in other countries. Water Management Synthesis Report 84, Utah State University, Logan.

Nielsen, D. R., and MacDonald, J. G. (Eds.) (1978a). *Nitrogen in the Environment*, Vol. 1, *Nitrogen Behavior in Field Soil*. Academic Press, New York.

Nielsen, D. R., and MacDonald, J. G. (Eds.) (1978b). *Nitrogen in the Environment*, Vol. 2, *Soil-Plant-Nitrogen Relationships*. Academic Press, New York.

Shafique, M. S., and Skogerboe, G. V. (1984). Planning and implementation framework for salinity control in the Indus River Basin. In *Salinity in Watercourses and Reservoirs*, R. H. French (Ed.). Butterworth, Boston, pp. 93-102.

Shafique, M. S., and Skogerboe, G. V. (1987). Surface irrigation scheduling. International Irrigation Center, Department of Agricultural and Irrigation Engineering, Utah State University, Logan (translated into Spanish: Programación del Riego Superficial).

Skogerboe, G. V. (1988). Implementating the irrigation maintenance and operations (M & O) learning process regionally or nationally. Water Management Synthesis Report 81, Utah State University, Logan.

Skogerboe, G. V., Walker, W. R., and Evans, R. G. (1979). Planning framework for alleviating waterlogging and salinity problems. *Water Supply Manag. 3*: 439-451.

Skogerboe, G. V., Lowdermilk, M. K., Sparling, E. W., and Hautaluoma, J. E. (1980). Development process for improving irrigation water management on farms: Executive summary. Water Management Technical Report No. 65A, Water Management Research Project, Colorado State University, Fort Collins.

Skogerboe, G. V., Walker, W. R., and Evans, R. G. (1983). Modelling water quality from irrigated and rainfed agricultural lands. In *Application of Ecological Modelling in Environmental Management, Developments in Environment Modelling*, Elsevier, Amsterdam, pp. 635-729.

Smerdon, E. T., and Blair, A. W. (1989). Infiltration from irrigation advance data. I. Theory. *J. Irrigation Drainage Eng. ASCE 114*(IR1): 4-17.

Strelkoff, T., and Katopodes, N. D. (1977). Border irrigation hydraulics with zero-inertia. *J. Irrigation Drainage ASCE 103*(IR3): 325-342.

Stringham, G. E., and Keller, J. (1979). Surge flow for automatic irrigation. In *Proceedings of the ASCE Irrigation Drainage Division Special Conference*, Albuquerque, New Mexico.

U.S. Bureau of Reclamation (1976). *Prediction of Mineral Quality of Irrigation Return Flow*, Vol. 5. U.S. Dept. of Interior, Washington, D.C.

Walker, W. R. (1989). Guidelines for designing and evaluating surface irrigation systems. Irrigation and Drainage Paper, Food and Agriculture Organization of the United Nations, Rome.

Walker, W. R., and Lee, T. S. (1981). Kinematic-wave approximations of surged furrow advance. Paper 81-2544. Abstracts, ASAE Winter Meeting, Chicago, Illinois, December 15-18.

Wiener, A. (1972). *The Role of Water in Development*. McGraw-Hill, New York.

Wischmeier, W. H., and Smith, D. D. (1978). *Predicting Rainfall Erosion Losses: A Guide to Conservation Planning*. Handbook No. 537, U.S. Department of Agriculture, Washington, D.C.

Author Index

A

B

C

D

G

H

K

M

P

S

Subject Index

A

S

XYZ